THE ELEMENTS

									VIIA	O

			IIIA	IVA	VA	VIA	1 H 1.00797 ± 0.00001	2 He 4.0026 ± 0.00005

	IB	IIB	5 B 10.811 ± 0.003	6 C 12.01115 ± 0.00005	7 N 14.0067 ± 0.00005	8 O 15.9994 ± 0.0001	9 F 18.9984 ± 0.00005	10 Ne 20.183 ± 0.0005

13 Al 26.9815 ± 0.00005	14 Si 28.086 ± 0.001	15 P 30.9738 ± 0.00005	16 S 32.064 ± 0.003	17 Cl 35.453 ± 0.001	18 Ar 39.948 ± 0.0005

28 Ni 58.71 ± 0.005	29 Cu 63.54 ± 0.005	30 Zn 65.37 ± 0.005	31 Ga 69.72 ± 0.005	32 Ge 72.59 ± 0.005	33 As 74.9216 ± 0.00005	34 Se 78.96 ± 0.005	35 Br 79.909 ± 0.002	36 Kr 83.80 ± 0.005

46 Pd 106.4 ± 0.05	47 Ag 107.870 ± 0.003	48 Cd 112.40 ± 0.005	49 In 114.82 ± 0.005	50 Sn 118.69 ± 0.005	51 Sb 121.75 ± 0.005	52 Te 127.60 ± 0.005	53 I 126.9044 ± 0.00005	54 Xe 131.30 ± 0.005

78 Pt 195.09 ± 0.005	79 Au 196.967 ± 0.0005	80 Hg 200.59 ± 0.005	81 Tl 204.37 ± 0.005	82 Pb 207.19 ± 0.005	83 Bi 208.980 ± 0.0005	84 Po (210)	85 At (210)	86 Rn (222)

63 Eu 151.96 ± 0.005	64 Gd 157.25 ± 0.005	65 Tb 158.924 ± 0.0005	66 Dy 162.50 ± 0.005	67 Ho 164.930 ± 0.0005	68 Er 167.26 ± 0.005	69 Tm 168.934 ± 0.0005	70 Yb 173.04 ± 0.005	71 Lu 174.97 ± 0.005

95 Am (243)	96 Cm (247)	97 Bk (247)	98 Cf (249)	99 Es (254)	100 Fm (253)	101 Md (256)	102 No (253)	103 Lw (257)

Atomic Weights are based on C^{12}—12.0000 and Conform to the 1961 Values

Printed in U.S.A.

INTRODUCTION TO ORGANIC LABORATORY TECHNIQUES

a contemporary approach

DONALD L. PAVIA

GARY M. LAMPMAN

GEORGE S. KRIZ, Jr.

Western Washington State College,
Bellingham, Washington

 SAUNDERS GOLDEN SUNBURST SERIES

W. B. SAUNDERS COMPANY
Philadelphia · London · Toronto

W. B. Saunders Company: West Washington Square
Philadelphia, PA 19105

1 St. Anne's Road
Eastbourne, East Sussex BN21 3UN, England

1 Goldthorne Avenue
Toronto, Ontario M8Z 5T9, Canada

Library of Congress Cataloging in Publication Data

Pavia, Donald L

Introduction to organic laboratory technique.

(Saunders golden sunburst series)
Includes index.

1. Chemistry, Organic—Laboratory manuals.
    I. Lampman, Gary M., joint author.   II. Kriz, George S.,
    joint author.   III. Title.

QD261.P38   547′.0028   75-22741

ISBN 0-7216-7121-7

Introduction to Organic Laboratory Techniques: A Contemporary Approach   0-7216-7121-7

Last digit is the print number: 18   17   16   15   14   13   12   11   10

To Neva-Jean, Marian, and Dian

PREFACE

At the time we first undertook the rather lengthy project of writing and adapting material for an organic laboratory textbook, we thought that a rather novel approach could have favorable results with students taking courses in organic chemistry. We speculated that a book less oriented toward purely chemical topics and more oriented toward a biological approach to chemistry would be successful. Our own experience suggests that this new approach can indeed be successful.

Enthusiasm for organic chemistry laboratory, on the part of the student, has been rather easy to generate. In the process of creating this enthusiasm, we have not varied significantly the content of a typical organic chemistry laboratory course. Our text presents all of the important laboratory techniques—all types of filtration, crystallization, extraction, distillation, and chromatography, as well as spectroscopy. Also included are many of the important reaction types—esterification, Grignard reaction, aromatic substitution, nucleophilic substitution, reduction, oxidation, and condensation. In short, we have not altered the content of a standard laboratory course in terms of the rigorous practical experience which it is expected to provide. What has been done differently is the effort to choose experiments which include as much material of topical interest as possible. Experiments which deal with the isolation of natural products occur early in the text. These experiments not only introduce the student to topics of an interesting nature, but they also permit the inclusion of a variety of important techniques early in the instructional program.

We have included topics which touch upon items of consumer, biological, and historical interest, as well as purely chemical interest. In choosing the topics for our text, we have attempted to recognize that many students taking an organic laboratory course are biology, pre-medical, pre-dental, and pre-pharmacy majors. Chemistry majors are often greatly outnumbered in most courses of this type. Much of the material in the text is designed to interest these students and to avoid "turning them off" to chemistry. Nevertheless, it is also designed to give the chemistry major all the practical experience needed to begin more advanced study.

Although this text is comparatively larger than other texts which seek to discuss the same area of chemistry, we feel that our text is actually three texts in one. First, there are the experiments. Essentially all of these experiments have been laboratory tested with our own students. Many of the ideas and experiments are totally new, while others are greatly reworked versions of experiments which have been used by many previous generations of students. In general, the experiments can be performed without a need for specialized equipment beyond the typical glassware kit. The experiments are each as fully self-contained as possible. Little is assumed about the student's prior knowledge of organic chemistry before attempting the experiment. Mechanisms of reactions, as well as their stoichiometry are discussed. The experiments have been designed to be used in almost any order, adding to the potential flexibility of the book. Many of the experiments have been selected to be consistent with our philosophy that they can be used as a vehicle for introducing the student to related areas of interest beyond those covered in a normal one-year lecture course.

Second, there are the techniques. We have sought to separate discussions of technique from the experiments themselves. This has been done to permit the substitution of new experiments into a laboratory course without the need for totally rewriting technique discussions. An instructor using this textbook has much greater flexibility in adding his own favorite experiments into the course. These chapters have been written carefully to include thorough discussions of each technique that might be encountered in a laboratory experience. We feel that the technique sections contain material of sufficient detail that the student is prepared even for independent research. Although it may appear that techniques are not included within the experiments, a careful examination of the text will show that not only are the techniques included, but that they are also quite extensively treated.

Third, there are the essays which precede many of the experiments. These essays might be useful as supplements to a lecture text, since they deal with subjects that most lecture texts ignore. The essays form a sort of "course within a course," and are particularly useful as a supplement for lecture courses in organic chemistry for science majors. With these essays, leading references have been included to permit students to pursue the subjects further if they so desire.

Our own experience with preliminary versions of this text has demonstrated that our approach is successful. We have observed that student interest improved greatly over what it had been during times when our course followed more traditional approaches. We believe that our approach can be successful in most institutions.

A very large number of people have helped us in the preparation of this book. While it may seem space-consuming, we feel that their contribution merits mentioning their names as we thank them for their efforts. We greatly appreciate the assistance in manuscript preparation provided by Gertrude Becker, Ann Drake, Dian Kriz, Mary Kriz, Neva-Jean Pavia, Dorothy Scott, and Judy Vinzant Widener.

Many students (and one faculty wife) were involved in the development and testing of experiments considered for inclusion in this book. We wish to thank Jennifer Andrews, Sandy Banks, Roger Blackman, Dave Bosell, Wayne Bratz, Brent Coleman, Sue Dufresne, Otto Hanssen, Larry Heimark, Connie Heimbuch, Tim Hoyt, Bruce Johnson, Ken Kelley, Linda Larson, Neva-Jean Pavia, Dana Perry, Kristi Pielstick, Joyce Rideout, Alan Schultz, Terry Smedley, Greg Smith, Carl Tuenge, Mark Watson, Rex Widener, Dan Wilson, and Sophia Zervas. These people all deserve a great deal of credit for their dedication and willingness to participate in this project with us. Special thanks must be given to the students who enrolled in our organic laboratory classes over the past three years and who patiently endured our on-the-spot modifications in experimental procedures.

A very special word of thanks and appreciation must be addressed to Mr. Robert LaRiviere, who illustrated the text. We feel that his work greatly enhances our manuscript. Special thanks should also go to Professors Arnold Krubsack and Thomas Cogdell, who reviewed an earlier version of our manuscript, and Professor John Miller, who class-tested many of the experiments. Their comments were very detailed and valuable.

Finally, we must thank our wives, Neva-Jean, Marian, and Dian, who patiently bore with us as we struggled to forge our preliminary ideas into a final manuscript. They suffered through endless hours of being ignored, having to deal with irritable spouses, and having to advise, type, and proofread.

DONALD L. PAVIA
GARY M. LAMPMAN
GEORGE S. KRIZ, JR.

CONTENTS

PART TWO   THE TECHNIQUES

APPENDICES

FOREWORD TO THE STUDENT AND WORDS OF ADVICE

Welcome to organic chemistry! Organic chemistry can be fun, and the authors of this textbook hope to prove it to you. That word "fun" may be misleading; it doesn't mean that you won't have to work hard. On the contrary, you should learn a lot from this laboratory course, and you will have to hustle to keep abreast of it all. For you to get the most out of the laboratory course there are several things you should strive to do. First, you need to understand the organization of this laboratory manual and how to use it effectively. It is your guide to learning. Second, you must try to understand both the purpose and the principles behind each of the experiments you will do. Third, you will need to attempt to organize your time effectively **in advance** of each laboratory period.

ORGANIZATION OF THE TEXTBOOK

Let us consider for a moment the manner in which this textbook is organized. There are four introductory sections of which this Foreword is the first; a section on laboratory safety is second; a section on advance preparation and laboratory records is third; and a section on laboratory glassware is fourth. Beyond these introductory sections, the text is divided into two major parts. Part I consists of about 60 experiments which may be assigned as part of your laboratory course. Interspersed within Part I are a variety of covering essays which provide background information related to the experiments. Part II consists of a series of rather detailed instructions and explanations dealing with the techniques of organic chemistry. While the experiments themselves may not appear to deal explicitly with instruction in the ordinary techniques of organic chemistry, in fact, those techniques **are** extensively developed and utilized. They are developed and used **in the context of the experiments.** Within each experiment, you will find a section, titled "Special Instructions," which indicates which techniques should be studied in order to perform that experiment. That section will also list special safety precautions and specific instructions to the student. An appendix to this text consists of sections dealing with infrared spectroscopy and nuclear magnetic resonance spectroscopy. While only a few experiments specifically using these spectroscopic techniques are included in Part I, they may be incorporated into the experiments with little difficulty by your instructor.

ADVANCE PREPARATION

Earlier you were told that you would have to hustle to keep abreast of the material which you might expect to learn in your organic chemistry laboratory course. This means that you should not treat these experiments as a novice cook would treat a Fanny Farmer Cookbook. You should come to the laboratory with a plan for the use of your time and some understanding of what you are about to do. A **really good** cook does not follow the recipe line-by-line with a finger, nor does a good mechanic fix your car with the instruction manual in one hand and a wrench in the other. In addition, it is unlikely that you will learn much if you blindly try to follow the instructions without understanding them. It can't be emphasized strongly enough that you should come to the lab **prepared.**

If there are items or techniques you don't understand, you should not be hesitant to ask questions. Since other students are not always reliable sources, you should ask an instructor or a laboratory assistant. You will also learn more if you figure things out on your own. Don't rely on others to do your thinking for you.

You should read the section entitled "Advance Preparation and Laboratory Records" right away. Although your instructor will undoubtedly have his own preferred format, much of the material here will be of help to you in learning to think constructively about laboratory experiments in advance. You will also have to learn to keep complete records **for your own use,** no matter what the results may be. It would also be a time saver if, as soon as possible, you read at least the first four technique sections in Part II. These techniques are basic to all the experiments in this text. The laboratory class will begin with experiments almost immediately, and a thorough familiarity with this particular material will save you much valuable laboratory time. You should also read the section on safety. Knowledge of what to do and what not to do in the laboratory is of paramount importance, since the laboratory does have a number of potential hazards associated with it.

BUDGETING TIME

As was mentioned in the Advance Preparation section of this Foreword, you should have read a number of chapters of this book even before your first laboratory class meeting. In addition, you should read the assigned experiment carefully before every class meeting. Having read the experiment, you should schedule your time wisely.

Often you will be doing more than one experiment at a time. Experiments like the fermentation of sugar require a few minutes of advance preparation **one week** ahead of the actual experiment. At other times you will have to catch up on some unfinished details of a previous experiment. For instance, usually it is not possible to determine accurately a yield or a melting point of a product immediately after you first obtain it. Such products must be free of solvent to give an accurate weight or melting point range; they have to be "dried." This drying is done usually by leaving the product in an open container in your desk. Then when you have a pause in your schedule during the subsequent experiment, you may determine these missing data on a dry sample.

THE PURPOSE

The main purpose of an organic chemistry laboratory is to teach you the basic techniques necessary to a person dealing with organic chemicals. You will learn how to handle equipment that is basic to every research laboratory. You will also learn the techniques necessary to separate and purify organic compounds. If the appropriate experiments are included in your course, you may also learn how to identify unknown compounds. The experiments themselves are only the **vehicle** for learning these techniques. The technique chapters in Part II are the heart of this text, and you should learn them thoroughly. Your instructor may provide laboratory lectures and demonstrations explaining these techniques, but the burden is on you to master them fully by familiarizing yourself with the technique chapters. Your instructor simply will not be able to explain or demonstrate everything.

In choosing experiments we have, whenever possible, tried to make them relevant and, more importantly, interesting. To that end, we have tried to make them a learning experience of a different kind. Most experiments are prefaced by a background essay which we hope will place things in context and provide you with some new information. We hope to show you that organic chemistry is ubiquitous (drugs, foods, plastics, perfumes, etc.). We have probably not succeeded totally in this text, but we have made a good beginning. We hope that organic chemistry itself is the dominant aspect of this text and that the level of expected performance has not been downgraded one iota. You should emerge from your course well-trained in organic laboratory techniques. We are enthused about our text and hope you will enter into it with the same spirit. Fun may be had by all. Welcome aboard!

DONALD L. PAVIA
GARY M. LAMPMAN
GEORGE S. KRIZ, JR.

LABORATORY SAFETY

In any laboratory course, the most important piece of advance preparation is to have some familiarity with the fundamentals of laboratory safety. Any chemistry laboratory, particularly an organic chemistry laboratory, can be a dangerous place in which to work. Understanding what to do and what not to do will serve well in minimizing that danger for you. Avoiding accidents is always a good policy. Remember that if you have a serious accident it will not be reversible. You won't get a second chance!

EYE SAFETY

First and foremost, ALWAYS WEAR APPROVED SAFETY GLASSES OR GOGGLES. This sort of eye protection must be worn whenever you are in the laboratory. Even if you are not actually carrying out an experiment, the person near you might have an accident which could endanger your eyes, so eye protection is essential. Even dishwashing may be hazardous. Cases are known where a person had been cleaning glassware only to have an undetected piece of reactive material explode, sending fragments into the person's eyes. To avoid this sort of accident it is wise to wear your safety glasses at all times.

If there are eyewash fountains in your laboratory, you should learn the location of the one nearest to you. In case any chemical enters your eyes, proceed immediately to the eyewash fountain and flush your eyes and face with large amounts of water. If an eyewash fountain is not available, the laboratory will usually have at least one sink fitted with a piece of flexible hose. When the water is turned on, this hose may be aimed upward and directly into the face, thus working much like an eyewash fountain.

FIRES

Next in importance is the need to stress caution with respect to fire. Because an organic chemistry laboratory course deals with flammable organic solvents at all times, the danger of fire is always present. Because of this danger, DO NOT SMOKE IN THE LABORATORY. Furthermore, exercise a great deal of caution when you light matches or use any open flame. Always check to see if your neighbors on either side, across the bench, and behind you are using flammable solvents—if so, either delay your use of a flame or move to a safe location, such as a fume hood, to use your open flame. Many flammable organic substances have rather dense vapors which may travel for some distance down a bench. These vapors present a fire danger, and you should be careful, since the source of those vapors may be located at a great distance from you. Do not use the bench sinks to dispose of flammable solvents. If your bench has a trough running along it, only **water** (no flammable solvents!) should be poured into it. The trough is designed to carry the water from condenser hoses and aspirators, not to carry flammable materials.

For your own protection in case of a fire, you should learn immediately the location of the nearest fire extinguisher, fire shower, and fire blanket. You should learn how these safety devices are operated, particularly the fire extinguisher. Your instructor can demonstrate the operation of the extinguisher.

If you have a fire, the best advice is to **get away from it** and let the instructor or laboratory assistant take care of it. Don't panic! Time spent in thought before action is never wasted. If it is a small fire in a flask it usually can be extinguished quickly by placing an asbestos pad or a watch glass over the mouth of the flask. It is a good practice to have an asbestos pad or a watch glass handy whenever you are using a flame. If this method does not take care of the fire and if help from an experienced person is not readily available, then extinguish the fire yourself with a fire extinguisher.

Should your clothing catch on fire, do not run, but rather walk **purposefully** toward the nearest fire blanket or fire shower station. Running will fan the flames and intensify them. Wrapping yourself in the fire blanket will smother the flames quickly.

INADVERTENTLY MIXED CHEMICALS

To avoid unnecessary hazards of fire and explosion, never pour any reagent back into a stock bottle. There is always the chance that you may accidentally pour back some foreign substance which may react with the chemical in the stock bottle in an explosive manner. Of course, by pouring reagents back into stock bottles you may introduce impurities which could spoil the experiment for the person using the stock reagent after you, so pouring things back into bottles is not only a dangerous practice, it is also inconsiderate.

UNAUTHORIZED EXPERIMENTS

You should never undertake any unauthorized experiments. Again, the chances of an accident are high, particularly with an experiment which has not been completely checked to reduce the hazard.

FOOD IN THE LABORATORY

Because all chemicals are toxic to some extent, you should avoid accidental ingestion of toxic substances by never eating or drinking anything in the laboratory. There is always the possibility that whatever you are eating or drinking may become contaminated with a potentially hazardous material.

SHOES

You should always wear shoes in the laboratory. For that matter, even open-toed shoes or sandals offer inadequate protection against spilled chemicals or broken glass.

ORGANIC SOLVENTS: THEIR HAZARDS

The first major thing to remember about organic solvents is that many are **flammable** and will burn if they are exposed to an open flame or a match. The second major thing to remember is that many are **toxic.** For example, many chlorocarbon solvents, when accumulated in the body, cause liver deterioration similar to the cirrhosis caused by the excessive use of ethanol. The body does not rid itself easily of chlorocarbons nor does it **detoxify** them, thus they build up over a period of time and cause trouble. MINIMIZE YOUR EXPOSURE. Constant and excessive exposure to benzene may cause a form of leukemia. Don't sniff it because it has a nice odor, and don't spill it on yourself. Many other solvents, such as chloroform and ether, are good anesthetics and will put you to sleep if you breathe too much of them. Later they cause nausea. Many of these solvents exhibit a **synergistic** effect with ethanol. They enhance its effect. Pyridine causes **temporary** impotence. In other words, organic solvents are just as dangerous as corrosive chemicals, such as sulfuric acid, but manifest their hazardous nature in other, more subtle, ways. Minimize your direct exposure to solvents, and treat them with respect. The laboratory room should be well-ventilated, and normal cautious handling of solvents should not cause any health problem. However, if you are trying to evaporate a solution in an open flask you should perform the evaporation in the hood. Excess solvents should be discarded in a hood sink or in a container specifically intended for waste solvents, rather than down the drain at the laboratory bench.

If you should wish to check the odor of a substance, you should be careful not to inhale very much of the material. The technique for smelling flowers is **not** advisable here; you could inhale dangerous amounts of the compound. Rather, a technique for smelling minute amounts of a substance is used. You should pass a stopper moistened with the substance (if it is a liquid) under your nose. Alternatively, you may hold the substance away from you and waft the vapors toward you with your hand. But you should never hold your nose over the container and inhale deeply!

USE OF FLAMES

Even though organic solvents are frequently flammable (e.g., hexane, ether, dioxane, methanol, acetone, petroleum ether), there are certain laboratory procedures for which a flame may be used. Most often these procedures will involve dealing with an aqueous solution. In fact, as a general rule a flame should be used only to heat aqueous solutions. Most organic solvents boil well below the boiling point of water (100°C), and a steam bath may be used to heat these solvents. A table of commonly used organic solvents is given in Table 1.1 of Technique 1. Solvents marked in that table with bold face type will burn. Ether, pentane, and hexane are especially dangerous since, in combination with the correct amount of air, they will explode.

Some common sense rules apply to the use of a flame in the presence of flammable solvents. Again, it should be stressed that you should check to see if anyone in your vicinity is using flammable solvents before you ignite any open flame. If someone is using such a solvent, move to a safer location before you light your flame. Remember, IF THE SOLVENT BOILS BELOW 80–85°C, USE A STEAM BATH FOR HEATING. The drainage troughs or sinks should never be used to dispose of flammable organic solvents. They will vaporize if they are low-boiling and may encounter a flame further down the bench on their way to the sink. If it is not prudent to use a flame at your bench, move to a safer location for your operations. Methods for heating solvents safely will be discussed in detail in Technique 1.

FIRST AID: CUTS, MINOR BURNS, AND ACID/BASE BURNS

If any chemical enters your eyes, immediately irrigate the eyes with copious quantities of water. Slightly warm water, if it is available, is preferable. Be sure that the eyelids are kept open. Continue flushing the eyes in this manner for 15 minutes.

In case of a cut, wash the wound well with water, unless you are specifically instructed to do otherwise. If necessary, apply pressure to the wound to stop the flow of blood.

Minor burns caused by flames or contact with hot objects may be soothed by immediately immersing the burned area into cold water or ice for about five minutes. The application of salves to burns is to be discouraged. Severe burns must be examined and treated by a physician.

In the case of chemical burns, the chemical should be neutralized. For acid burns, the application of a dilute solution of sodium bicarbonate is recommended. For burns by alkali, the application of a **dilute** solution of a weak acid, such as a 2% acetic acid solution, a 25% vinegar solution, or a 1% boric acid solution, will neutralize the alkali. Most well-equipped laboratories will always have these safety solutions available in containers which are clearly marked FOR ACID BURNS and FOR BASE BURNS. You should learn the location of these safety solutions. After application of either sodium bicarbonate or dilute acid solution, flush the affected area with water for 10 to 15 minutes. Do not use the safety solutions for chemical purposes. If you do, they may not be available when they are really needed!

ADVANCE PREPARATION AND LABORATORY RECORDS

In the Foreword to the Student, some mention was made of the importance of advance preparation in performing laboratory experiments. In this chapter you will find some suggestions about what specific information you should try to obtain in the course of your advance studying. Much of this information must be obtained while preparing your notebook, and so the two subjects, advance study and notebook preparation, will be developed simultaneously here.

An important part of any laboratory experience is learning to maintain very complete records of every experiment undertaken and every datum obtained. Far too often, careless recording of data has resulted in mistakes, frustration, and lost time due to needless repetition of experiments. In cases where reports are required, proper collection and recording of data can facilitate report-writing greatly. In addition to learning good laboratory technique and the methods of carrying out basic laboratory procedures, among the many things you should learn from this laboratory course are:

1. How to take data carefully
2. How to record relevant observations
3. How to use your time effectively
4. How to assess the efficiency of your experimental method
5. How to plan for the isolation and purification of the substance you prepare.

Because organic reactions are seldom quantitative in nature, special problems result. Frequently reagents have to be used in large excess in order to increase the amount of product obtained. In addition some of the reagents are rather expensive, which necessitates care in the amounts of these substances used. Very often many more reactions may take place than you may desire. These extra reactions, or **side reactions,** may form products in addition to the desired product. These are called **side products.** For these reasons it will be necessary to plan your experimental procedure carefully before undertaking the actual experiment.

THE NOTEBOOK

For recording data obtained during experiments, A BOUND NOTEBOOK SHOULD BE USED. The notebook should have consecutively numbered pages. If it does not, you should number the pages immediately. A spiral-bound notebook or any other type from which the pages can be removed easily is not acceptable, since the possibility of losing the pages is great.

The notebook is the place where all primary data must be recorded. Paper towels, napkins, toilet tissue, or scratch paper have a tendency to become lost or destroyed. It is bad laboratory practice to record primary data on such random and perishable pieces of paper. Data must be recorded in **permanent** ink. It can be frustrating to have important data disappear from the notebook because they were recorded in washable ink and could not survive a flood caused by the student at the next position on the bench. Because you will be using the notebook in the laboratory during the course of your experiments, it is quite likely that the book will become soiled or stained by chemicals, filled with scratched out entries, or even slightly burned. This damage is to be expected.

Your instructor may wish to check your notebook at any time, so you should always have it up to date. The preparation of reports, if these are required by your instructor, is then accomplished using the material recorded in the laboratory notebook.

NOTEBOOK FORMAT: ADVANCE PREPARATION

Individual instructors will vary greatly in the type of format which they prefer. Such variation stems from differences in philosophies and experience. You will have to get specific directions in preparing a notebook from your own instructor. However, there are certain features which are common to most notebook formats. What follows is a discussion of what **might** be included in a typical notebook.

You can save a great deal of time in the laboratory if you understand fully the procedure of the experiment and the theory underlying it. Knowledge of the major reactions, the potential side reactions, the mechanism, the stoichiometry, and the procedure before you come to the laboratory is very helpful. Understanding the procedure by which the desired product is to be separated from undesired materials is also very important. By examining each of these topics before coming to class, you will be prepared to perform the experiment efficiently. You will have your equipment and reagents prepared in advance of the time at which they are to be used. Your reference material will be at hand when needed. Finally, with your time efficiently organized, you will be able to take advantage of long reaction or reflux periods to perform other tasks, such as doing shorter experiments or finishing previous experiments.

For those experiments where a compound is being synthesized from other reagents, that is, a **preparative** experiment, a knowledge of the main reaction is essential. To facilitate the stoichiometric calculations which must be performed later, the equation for the main reaction should be balanced. Therefore, before you begin the experiment, your notebook should contain the balanced equation for the pertinent reaction. Using the preparation of aspirin (Experiment 1) as an example, you would write:

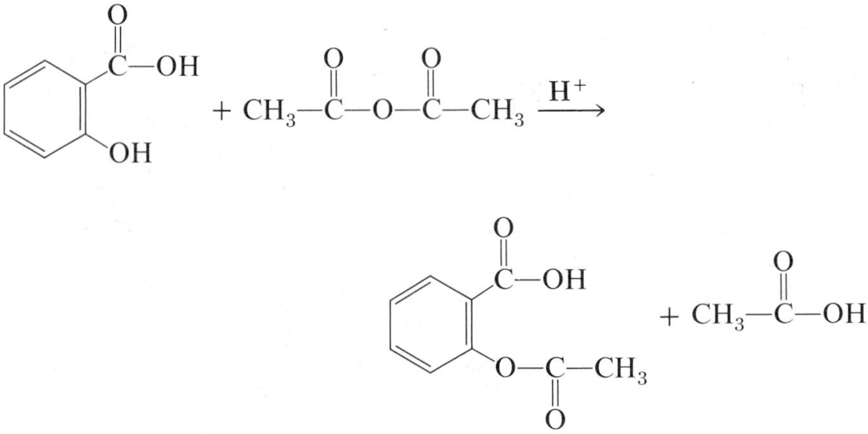

When known, the possible side reactions which divert reagents into contaminants (side products) should also be entered in the notebook before the experiment is begun. These side products will have to be separated from the major product during the purification procedures. In the case of the preparation of aspirin, the principal side reaction is the reaction of salicylic acid with itself to form a polymer:

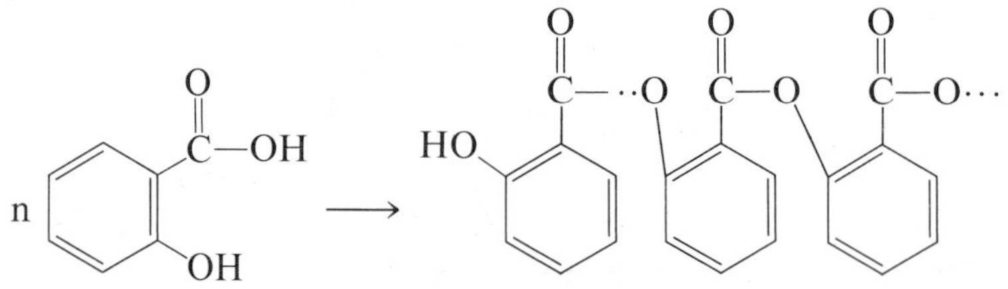

Other forms of advance preparation should also be included in the notebook. The stoichiometry of the reaction and the actual experimental procedure should be examined to determine the **limiting reagent.** The limiting reagent is that reagent which is not present in excess and upon which the overall yield of product depends. You will need this information later as a basis for your calculations of yield. In the aspirin experiment, the limiting reagent is **salicylic acid.** Such useful data as melting points, boiling points, molecular weights, and densities might also be recorded. These data are located in such sources as **Handbook of Chemistry and Physics, The Merck Index,** or Lange's **Handbook of Chemistry.**

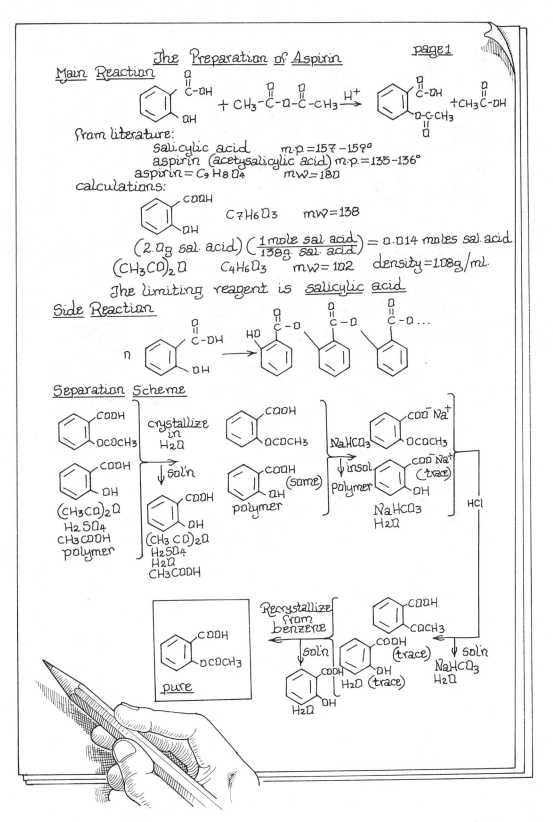

The Preparation of Aspirin — page 1

Main Reaction

From literature:
- salicylic acid   m.p. = 157–159°
- aspirin (acetysalicylic acid) m.p. = 135–136°
- aspirin = $C_9H_8O_4$ mw = 180

calculations:

$C_7H_6O_3$ mw = 138

$(2.0g \text{ sal. acid}) \left(\dfrac{1 \text{ mole sal. acid}}{138g. \text{ sal. acid}}\right) = 0.014 \text{ moles sal. acid}$

$(CH_3CO)_2O$ $C_4H_6O_3$ mw = 102   density = 1.08g/ml.

The limiting reagent is <u>salicylic acid</u>

Side Reaction

Separation Scheme

A sample notebook page 1

page 2

Procedure:

Salicylic acid: sample + paper 3.12g.
paper 1.12g.
sample 2.00g. salicylic acid

2.00g.   = 0.014 moles
1.38g./mole

The salicylic acid was placed in a 125ml. Erlenmeyer flask. Acetic anhydride (5ml) was added, along with 5 drops of conc. H_2SO_4. The flask was swirled until the salicylic acid was dissolved. The solution was heated on the steambath for 10 minutes. The flask was allowed to cool to room temp., and some crystals appeared. Water (50 ml.) was added and the mixture was cooled in an ice bath. The crystals were collected by suction filtration, rinsed three times with cold H_2O, and dried. Crude yield:

product + paper 3.07g.
paper 1.15g.
1.92g. aspirin

Theoretical yield = (0.014 moles)(180g. aspirin/mole) = 2.60g.
Actual yield   = 1.92g.
Percentage yield = 74%

The crude product gave a faint color with $FeCl_3$. Phenol and salicylic acid gave strong positive tests.

The crude solid was placed in a 150 ml. beaker, and 25 ml. of saturated $NaHCO_3$ was added. When the reaction had ceased, the solution was filtered by suction. The beaker and funnel were washed with ca. 2ml. of H_2O. Dilute HCl was prepared by mixing 3.5ml. conc. HCl, and 10 ml. H_2O in a 150 ml. beaker. The filtrate was poured into the dilute acid, and a precipitate formed immediatly. The mixture was cooled in an ice bath. The solid was collected by suction filtration, washed three times with cold H_2O, and placed on a watch glass to dry overnight. Yield:

paper and product   2.78g.
paper   1.08g.
product   1.70g.

Theoretical yield = 2.60g.
Actual yield   = 1.70g.
percentage yield = 65%   m.p. 133–135°C.

The solid did not give a positive $FeCl_3$ test.

The final product (ca. 0.75g.) was dissolved in a minimum amount of hot benzene. The hot solution was filtered by gravity and allowed to cool. Crystals appeared. The crystals were collected by suction and dried. m.p. 135–136° C.

Advance preparation might also include examination of some subjects which would not necessarily be recorded in the notebook, but which would nevertheless prove useful in understanding the experiment. Included in a list of such subjects would be an understanding of the mechanism of the reaction, an examination of other methods by which the same compound might be prepared, and a detailed study of the experimental procedure. Many students find that an outline of the procedure, prepared **before** they come to class, helps them use their time more efficiently once they begin the experiment. Such an outline could very well be prepared on some loose sheet of paper, rather than in the notebook itself.

Once the reaction has been completed, the desired product does not magically appear as purified material; rather, it must be isolated from a frequently complex mixture of side products, unreacted starting materials, solvents, and catalysts. You should try to outline a separation scheme in your notebook for the isolation of the product from its contaminants. At each stage you should try to understand the reason for the particular instruction that is given in the experimental procedure. This will not only familiarize you with the basic separation and purification techniques used in organic chemistry, but it will also help you to develop an understanding of when to use these techniques. Such an outline might take the form of a flow chart, an example of which, again taken from the aspirin experiment, is:

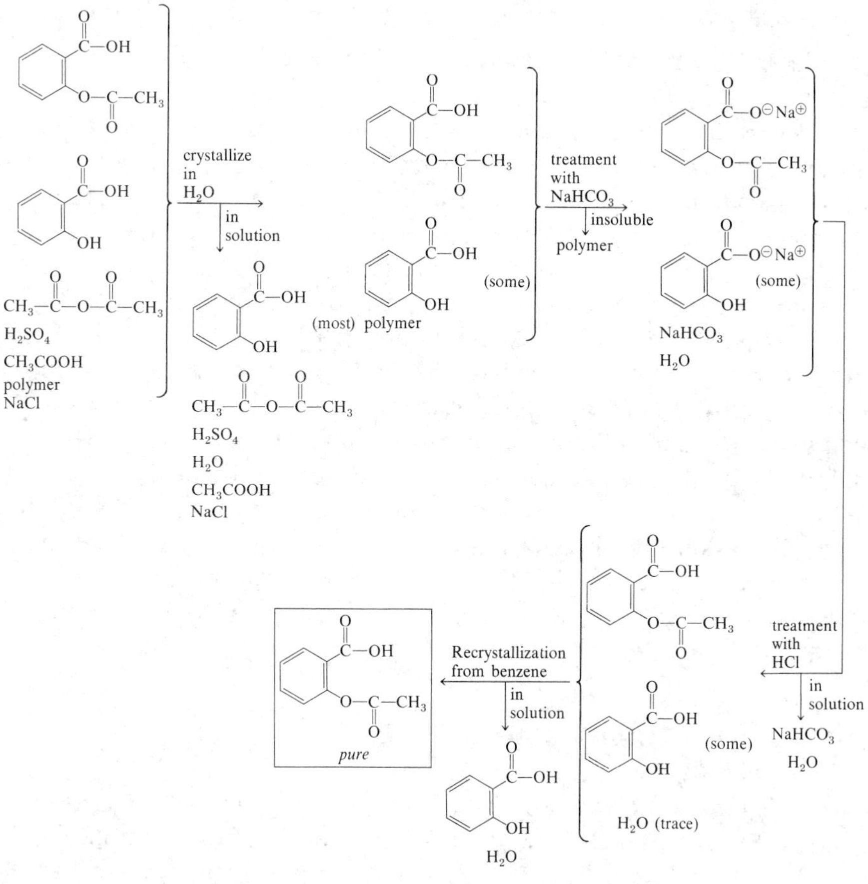

Besides serving the objective of familiarizing you with the procedure by which the desired product is separated from impurities in your particular experiment, careful attention to understanding the separation scheme may also prepare you for original research, where no experimental procedure exists.

In designing the separation scheme it is important to note that the separation scheme outlines those steps which are undertaken once the reaction period has been concluded. For this reason the above scheme did not include such steps as the treatment of salicylic acid with acetic anhydride, the addition of sulfuric acid, or the heating of the reaction mixture.

For experiments where a compound is being isolated from a particular source, but is not being prepared from other reagents, a great deal of the information described in this section will not be appropriate. These experiments are called **isolation experiments.** A typical isolation experiment might involve the isolation of a pure compound from a natural source. Some examples of this type of experiment might include the isolation of caffein from coffee or the isolation of nicotine from tobacco. Isolation experiments require somewhat different advance preparation, but this advance study might include a determination of physical constants for the compound isolated and an outlining of the isolation procedure. A detailed examination of the separation scheme is most important here, since this is the heart of such an experiment.

NOTEBOOK FORMAT: LABORATORY RECORDS

When you begin the actual experiment, your notebook should always be kept nearby in order that you will be able to record in it those operations which you perform. When working in the laboratory, the notebook serves as a place where a rough transcript of your experimental method is recorded. In it data from actual weighings, volume measurements, and determinations of physical constants, etc., are noted. This section of your notebook should **not** be prepared in advance. The purpose here is not to write a recipe; rather it is to provide a record of what you did and what you **observed.** These observations will help you to write reports without having to resort to memory. They will also help you or other workers to repeat the experiment **exactly.**

When your product has been prepared and purified or, as in the case of isolation experiments, has been isolated, you should record such pertinent data as the melting point or boiling point of the substance, its density, its index of refraction, the conditions under which spectra were determined, etc.

NOTEBOOK FORMAT: CALCULATIONS

A chemical equation for the overall conversion of the starting materials into products is written on the assumption of simple ideal stoichiometry. Actually the ideal assumption is seldom true. Side reactions or competing reactions will also occur to give other products. For some synthetic reactions, an equilibrium state will be reached in which an appreciable amount of starting material still is present and can be recovered. Some of the reactant may also remain if it is present in excess or if the reaction was incomplete. A reaction involving an expensive reagent illustrates another need for knowledge about the extent to which a particular type of reaction converts reactants to products. In such a case it is preferable to use the method which is the most efficient in this conversion. Thus, information about the efficiency of conversion for various reactions is of interest to the person contemplating using these reactions.

The quantitative expression for the efficiency of a reaction is given by a calculation of the **yield** for the reaction. The **theoretical yield** is the number of grams of product expected from the reaction on the basis of ideal stoichiometry, ignoring side reactions, reversibility, losses, etc. The theoretical yield is calculated from the expression:

Theoretical yield =
(moles of limiting reagent) (ratio) (molecular weight of product)

The ratio, here, is the stoichiometric ratio of product to limiting reagent. In the case of the aspirin preparation that ratio is **one.** One mole of salicylic acid, under ideal circumstances, would yield one mole of aspirin.

The **actual yield** is simply the number of grams of desired product obtained. The **percentage yield** describes the efficiency of the reaction and is determined by:

$$\text{Percentage yield} = \frac{\text{Actual yield}}{\text{Theoretical yield}} \times 100$$

A sample calculation, using a hypothetical value for the actual yield, can be provided, again using the case of the aspirin preparation as an example:

Theoretical yield =

$$(0.015 \text{ mole salicylic acid})\left(\frac{1 \text{ mole aspirin}}{1 \text{ mole salicylic acid}}\right)\left(\frac{180 \text{ g aspirin}}{1 \text{ mole aspirin}}\right)$$

$$= 2.7 \text{ g aspirin}$$

Actual yield = 1.7 g aspirin

$$\text{Percentage yield} = \frac{1.7 \text{ g}}{2.7 \text{ g}} \times 100 = 63\%$$

For experiments where the principal objective is the isolation of a substance such as a natural product, rather than the preparation and purification of some reaction product, the **weight percent recovery,** rather than the percentage yield, is of interest. This value is determined by:

$$\text{Weight percent recovery} = \frac{\text{weight of substance isolated}}{\text{weight of original material}} \times 100$$

Thus, for instance, if 0.08 g of caffein were obtained from 35.0 g of coffee, the weight percent recovery of caffein would be:

$$\text{Weight percent recovery} = \frac{0.08 \text{ g caffein}}{35.0 \text{ g coffee}} \times 100 = 0.23\%$$

LABORATORY REPORTS

Various formats for reporting the results of laboratory experiments may be used. In cases where original research is being performed, these reports should include a detailed description of all of the experimental steps undertaken. Frequently the style used in scientific periodicals, such as the **Journal of the American Chemical Society,** is applied to the writing of laboratory reports. Your instructor is likely to have his own requirements for laboratory reports, and he should detail his requirements to you.

SUBMISSION OF SAMPLES

In all preparative experiments, and in some of the nonpreparative ones, you will be required to submit the sample of the substance which you prepared to your instructor. How this sample is labeled is very important. Again, learning a correct method of labeling bottles and flasks can save a great deal of time in the laboratory because fewer mistakes will be made. More importantly, learning to label properly can decrease greatly the danger inherent in any chemical laboratory.

Solid materials should be stored and submitted in containers which permit the easy removal of the substance. For this reason, narrow-mouthed bottles are not used for solid substances. Liquids should be stored in containers which will not permit their escape through leakage. You should be careful not to store volatile liquids in containers which have plastic caps. The vapors from the liquid are likely to come in contact with the plastic and dissolve some of it, thus contaminating the substance being stored.

On the label are printed the name of the substance, its melting or boiling point, the actual and percentage yields, and the student's name. An illustration of a properly prepared label follows:

```
Aspirin
M.P. 135–136°C
Yield 1.7 g (63%)
Joe Schmedlock
```

LABORATORY GLASSWARE

Since your glassware is expensive and since you are responsible for it, you will want to give it proper care and respect. Needless maltreatment of your equipment may cost you money, so if you read this section carefully you may be able to avoid some unnecessary expense. Mistreatment of equipment can also cause lost time in the laboratory. Cleaning problems and/or the replacement of destroyed glassware are time-consuming.

For those of you who are unfamiliar with the equipment typically found in an organic laboratory, or who are uncertain about how such equipment should be treated, this section should provide useful information. Included within this section will be such topics as the cleaning of glassware, the greasing of ground-glass joints, and the care of glassware when using corrosive or caustic reagents. Included at the end of this section are illustrations of most of the pieces of apparatus which you are likely to find in your drawer or locker. Names are given to help you identify them.

CLEANING OF GLASSWARE

Glassware can be cleaned more easily if it is cleaned immediately after use. It is good practice to do your "dishwashing" right away. With time, the organic tarry materials left in a flask begin to attack the surface of the glass. The longer you wait to clean glassware, the more extensively this interaction of the residue with the glass will have progressed. The glassware then becomes more difficult to wash because water will no longer wet the surface of the glass as effectively as it did when the glass was clean.

Various washing powders and cleansers may be used as aids in the dishwashing process. Synthetic detergents may also be used, although the fine abrasives contained in cleansing powders aid in the cleaning process. Organic solvents may also be used in cleaning, since the residue is likely to be soluble in some organic solvent. After the solvent has been used, the flask probably will have to be washed with detergent (or cleanser) and water to remove the residual solvent. The need to wash the glassware after using an organic solvent is particularly great with the chlorocarbon solvents, such as carbon tetrachloride or chloroform. When solvents are used in cleaning glassware, caution should be exercised since the solvents are hazardous (see the section entitled "Laboratory Safety"). You should try to use fairly small amounts of a solvent for cleaning purposes. Acetone is a commonly used solvent for cleaning glassware, but it is expensive. For this operation either **waste acetone** (recycled) or **wet acetone** should be used. Do not use reagent grade acetone for cleaning purposes. Your **wash acetone** can be used effectively several times before it is "spent."

For troublesome stains and residues, which insist upon adhering to the glass in spite of your best efforts, a cleaning solution may be used. The most common cleaning solution consists of 35 ml of saturated aqueous sodium or potassium dichromate solution dissolved in one liter of concentrated sulfuric acid. If you prepare the cleaning solution yourself, remember to ADD THE SULFURIC ACID TO THE SATURATED DICHROMATE SOLUTION! It is much more hazardous to add the aqueous solution to the concentrated acid, since spattering may result. The preparation of cleaning solution is quite hazardous; it should not be undertaken without careful preparation. SAFETY GLASSES MUST BE WORN WHEN YOU ARE PREPARING OR USING CLEANING SOLUTION. Do not allow the cleaning solution to come into contact with your flesh or your clothing. When you are using the cleaning solution, use only small quantities. Swirl the cleaning solution in the glassware for a few minutes, and pour the remaining cleaning solution into the sink. Be certain that both the flask **and the sink** are rinsed with copious amounts of water. For most common organic chemistry applications, any stain which survives this treatment is not likely to cause any difficulties in subsequent laboratory procedures.

The easiest way to dry glassware is to allow it to stand overnight. Flasks and beakers should be stored upside down to permit the water to drain from them. Drying ovens may be used to dry glassware, if they are available and if they are not being used for other purposes. Rapid drying may be achieved by rinsing the glassware with acetone and air-drying it. For this latter method, the glassware is thoroughly drained of water. The glassware is rinsed with one or two **small** portions (about 10 ml) of "wash acetone," which is a cheaper grade of acetone suitable for the cleaning of glassware. Reagent grade acetone should not be used. You should not use any more acetone than is suggested here. Using great quantities of acetone does not improve the drying procedure, but it does waste the increasingly expensive acetone. The used acetone is either poured down the drain or returned to a waste acetone container. Waste acetone can be used for cleaning as has been mentioned previously. It can also be recycled for further use in the cleaning of glassware.

After rinsing with acetone, the flask is then dried by blowing a **gentle** stream of compressed air into the flask. Before drying the flask with air, you should make sure that the air line is not filled with oil. Otherwise the oil will be blown into the flask, and you will have to clean it all over again. It is not necessary to blast the acetone out of the flask with a wide-open stream of air; a gentle stream of air is just as effective and will not offend the ears of other people in the room. The acetone may also be removed by aspirator suction. With either procedure, as the acetone evaporates, it carries the residual water away with it.

LUBRICATING GROUND-GLASS JOINTS

Standard-taper glassware has many advantages. It can be assembled easily into a completed apparatus without the need for laborious cork-boring, and it can be made to fit together tightly for use under reduced pressure or with inert atmospheres. It can be disassembled and cleaned easily. However, one disadvantage of apparatus with ground-glass joints, besides the expense you may incur if you break a piece, is that the joints have some tendency to stick together.

To prevent this sticking, a lubricant may be used. Typical lubricants include a variety of hydrocarbon-based stopcock greases such as Lubriseal and Apiezon. Silicone greases are also used. To apply the grease, the inner (male) joint is coated with a **very thin** film of lubricant. You should be careful not to apply too much grease, since the excess grease may contaminate the materials contained within the apparatus. A properly lubricated joint appears clear with no striations visible. Contamination due to excess grease is a particular problem with silicone grease, which is much harder to remove than hydrocarbon-based greases. Because of this possible contamination, many chemists prefer to use grease only when necessary. The joints usually seal well without the use of grease. Necessary situations include distillations under reduced pressure, where the joints must be air-tight, and reactions involving strong caustic solutions (see below). Whenever there is doubt, however, it never hurts to use grease. Before lubricating the joints, always make sure that they are free of any adhering liquid or solid.

ETCHING OF GLASSWARE

Glassware which has been used for reactions involving the use of strong bases such as sodium hydroxide or sodium alkoxides must be cleaned thoroughly **immediately** after being used. If these caustic materials are allowed to remain in contact with the glass, they will etch the glass permanently. This etching makes later cleaning attempts more difficult, since dirt particles may become trapped within the microscopic surface irregularities of the etched glass. Furthermore, the etched glass is weakened, so the lifetime of the glassware is decreased. If caustic materials are allowed to come into contact with ground-glass joints without being removed promptly, the joints will become fused. Fused joints are extremely difficult to separate without breaking them.

FROZEN JOINTS

Occasionally, ground-glass joints may become "frozen," or stuck together so strongly that they are difficult to separate. The same is true of glass stoppers in bottles. If a joint becomes frozen, it may sometimes

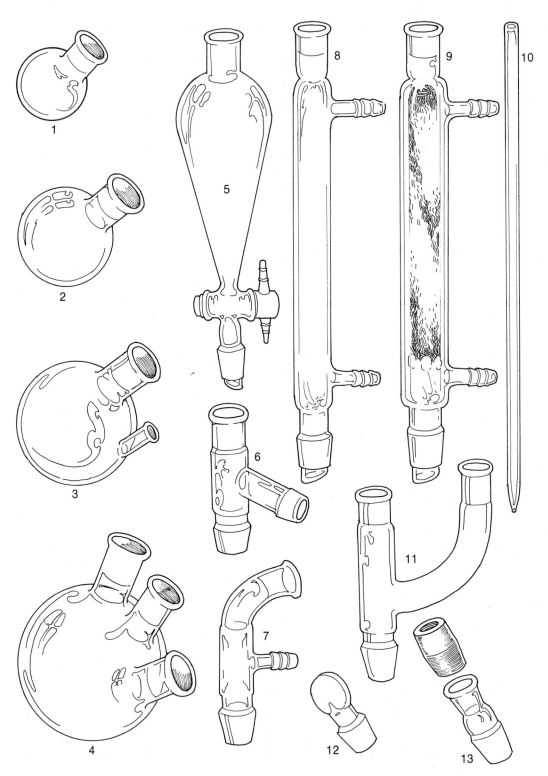

Components of the organic kit: 1) 50 ml round bottom boiling flask; 2) 100 ml round bottom boiling flask; 3) 250 ml round bottom boiling flask; 4) three-neck flask; 5) 125 ml separatory funnel; 6) distillation head; 7) vacuum adapter; 8) condenser (West); 9) fractionating column; 10) ebulliator tube; 11) Claisen head; 12) stopper; 13) thermometer adapter.

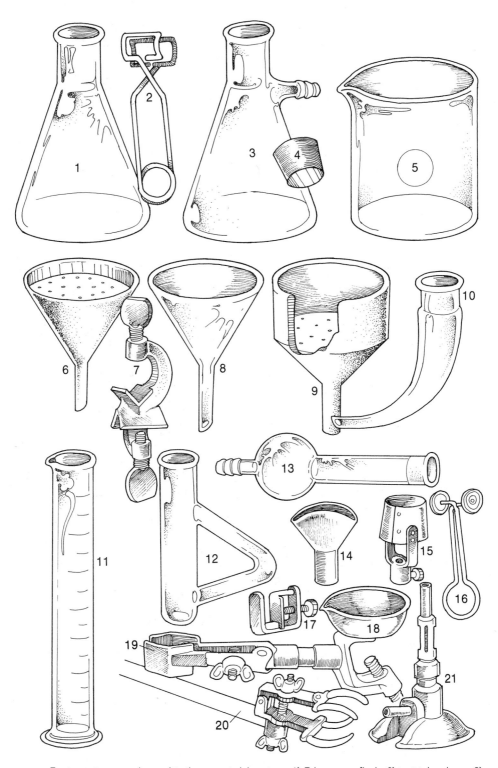

Equipment commonly used in the organic laboratory: 1) Erlenmeyer flask; 2) test tube clamp; 3) filter flask; 4) neoprene adapter; 5) beaker; 6) Hirsch funnel; 7) clamp holder; 8) conical funnel; 9) Büchner funnel; 10) bent adapter; 11) graduated cylinder; 12) melting point tube (Thiele); 13) drying tube; 14) wing top; 15) burner chimney; 16) pinch clamp; 17) screw clamp; 18) crystallizing dish (evaporating dish); 19) utility clamp; 20) condenser clamp; 21) micro burner.

be loosened by tapping it **gently** with the wooden handle of a spatula or by tapping it **gently** on the edge of the bench-top. If this procedure fails, you may try heating the joint in hot water or a steam bath. A heated joint will often free more easily. If this heating fails, the instructor may be able to give some advice. As a last resort, you may try heating the joint in a flame. You should not attempt this latter heating procedure unless the apparatus is hopelessly stuck. Heating by flame often causes the joint to expand rapidly and to crack or break. The joint should be clean and dry. The outer (female) part of the joint is heated slowly, using the yellow portion of a low flame, until it expands and breaks away from the inner section. The joint should be heated very slowly and carefully, or it may break due to rapid expansion.

APPARATUS ASSEMBLY

One final note of advice concerns the assembly of glass components into the desired apparatus. You should always remember that Newtonian physics applies to assemblies of chemical apparatus and that unsecured pieces of glassware are certain to respond to the forces of gravity. You should be careful to clamp the glassware securely to a ring stand. Do not rely on friction to hold the apparatus together. This is particularly true of the vacuum adapter, which has a suicidal tendency to fall if it is not clamped. Throughout this text, the illustrations of the various glassware arrangements include the clamps which attach that apparatus to a ring stand. You should assemble your apparatus using the clamps in the manner shown in the illustrations.

When assembling ground-glass equipment, you should make sure that no solid or liquid material is on the joint surfaces. Such materials will lessen the efficiency of the seal, and the joints may leak. Also, if the apparatus is to be heated, material caught between the joint surfaces will increase the tendency for the joints to stick. If the joint surfaces are coated with liquid or adhering solid, they should be wiped with a paper towel or cloth before assembly.

THE EXPERIMENTS

ESSAY

Aspirin

Aspirin is one of the most popular panaceas (cure-alls) of modern life. And, even though its curious history began over 200 years ago, we still have much to learn about this enigmatic remedy. No one yet knows exactly how or why it works, yet more than 20 million pounds of aspirin are consumed each year in the United States.

The history of aspirin began on June 2, 1763 when Edward Stone, a clergyman, read a paper to the Royal Society of London entitled "An Account of the Success of the Bark of the Willow in the Cure of Agues." By ague, Stone was referring to what we now call malaria, but his use of the word "cure" was optimistic; what his extract of willow bark actually did was to reduce the feverish symptoms of the disease. Almost a century later a Scottish physician was to find that extracts of willow bark would also alleviate the symptoms of acute rheumatism. This extract was ultimately found to be a powerful **analgesic** (pain reliever), **antipyretic** (fever reducer), and **anti-inflammatory** (reduces swelling) drug.

Soon thereafter, organic chemists working with willow bark extract and flowers of the meadowsweet plant (which gave a similar principle) isolated and identified the active ingredient as salicylic acid; from **salix,** the Latin botanical name for the willow tree. The substance could then be chemically produced in large quantity for medical use. It soon became apparent that the use of salicylic acid as a remedy was severely limited by its acidic properties. The substance caused severe irritation of the mucous membranes lining the mouth, gullet, and stomach. The first attempts at circumventing this problem by using the less acidic

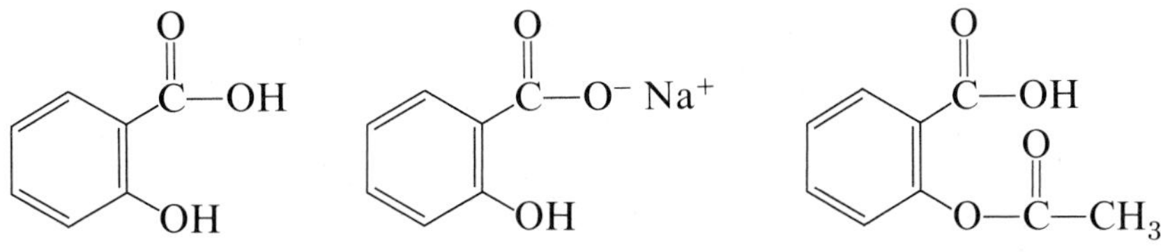

Salicylic Acid   **Sodium Salicylate**   **Acetylsalicylic Acid (Aspirin)**

sodium salt (sodium salicylate) were only partially successful. This substance gave less irritation but had such an objectionable sweetish taste that most people could not be induced to take it. The breakthrough came around the turn of the century (1893) when Felix Hofmann, a chemist for the German firm of Bayer, devised a practical route for the synthesis of acetylsalicylic acid, which was found to exhibit all the same medicinal properties without the highly objectionable taste or the high degree of mucosal membrane irritation. The firm of Bayer called its new product ASPIRIN, a name derived from **a** for acetyl, and the root **spir,** from the Latin name, **spirea,** for the meadowsweet plant.

The history of aspirin is typical of many of the medicinal substances in current use. Many began as a crude plant extract or folk remedy whose active ingredient was isolated and its structure determined by chemists who then improved upon the original.

In the last few years the mode of action of aspirin has just begun to unfold. A whole new class of compounds called **prostaglandins** has been found to be involved in the body's immune responses. Their synthesis is provoked by interference with the body's normal functioning due to foreign substances or by unaccustomed stimuli. These sub-

Prostaglandin E_2 Prostaglandin $F_{2\alpha}$

stances are involved in a wide variety of physiological processes and are thought to be responsible for evoking pain, fever, and local inflammation. Aspirin has been recently shown to prevent bodily synthesis of prostaglandins and thus to alleviate the symptomatic portion (fever, pain, inflammation) of the body's immune responses (i.e., the ones that let you know something is wrong). A rather startling discovery is that the prostaglandin $F_{2\alpha}$ can induce smooth muscle contraction in the uterus, thus causing abortion. In fact, according to one hypothesis, the IUD (intrauterine device for birth control) is thought to act by provoking a continuous local synthesis of prostaglandins due to the slight irritation of the uterine membrane caused by the device. The prostaglandin production prevents implantation of the ovum and, hence, pregnancy is averted. The link between aspirin and prostaglandins mentioned here leads one to the suspicion that perhaps women who are chronic users of aspirin should not put their confidence in the IUD as a means of contraception. However, to date, no definite link has been shown between IUD failure and the use of aspirin.

Aspirin tablets (5 grain) are usually compounded of about 0.32 g of acetylsalicylic acid pressed together with a small amount of starch, which serves to bind the tablet together. Buffered aspirin usually contains a basic buffering agent to reduce the acidic irritation of mucous membranes in the stomach, since the acetylated product is not totally free of this irritating effect. Bufferin contains 5 grains of aspirin

(1 grain = 0.0648 g), 0.75 grains of aluminum dihydroxyaminoacetate, and 1.5 grains of magnesium carbonate. Combination pain relievers usually contain aspirin, phenacetin (or a related compound), and caffein. Empirin, a typical APC tablet, for instance, contains 0.233 g aspirin, 0.166 g phenacetin, and 0.030 g caffein.

REFERENCES

"Aspirin Action Linked to Prostaglandins." *Chemical and Engineering News* (August 7, 1972), 4.
"Aspirin—Good News, Bad News." *Saturday Review* (November 25, 1972), 60.
Chaudhuri, G. "Intrauterine Device: Possible Role of Prostaglandins." *The Lancet* (March 6, 1971), 480.
Collier, H. O. J. "Aspirin." *Scientific American, 209* (November, 1963), 96.
Collier, H. O. J. "Prostaglandins and Aspirin." *Nature, 232* (July 2, 1971), 17.
Pike, J. E. "Prostaglandins." *Scientific American, 225* (November, 1971), 84.
"The Uses of Aspirin." *The Medicine Show*. Mt. Vernon, New York: Consumers Union, 1971.
Vane, J. R. "Inhibition of Prostaglandin Synthesis as a Mechanism of Action for Aspirin-like Drugs." *Nature-New Biology, 231* (June 23, 1971), 232.

Experiment **1**

ACETYLSALICYLIC ACID

Crystallization
Vacuum Filtration
Esterification
    (Acetylation)

Salicylic acid (**o**-hydroxybenzoic acid) is a bifunctional compound. It is a phenol (hydroxybenzene) and a carboxylic acid. Hence, it can undergo two different types of esterification reactions, acting as **either** the alcohol **or** the acid partner in the reaction. In the presence of acetic anhydride, acetylsalicylic acid (aspirin) is formed; whereas, in the presence of excess methanol, the product is methyl salicylate (oil of wintergreen). In this experiment we will use the former reaction to

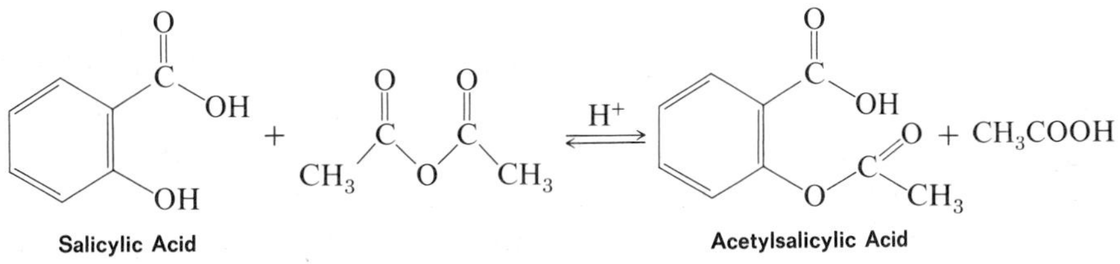

Salicylic Acid   Acetylsalicylic Acid

Salicylic Acid   Methyl Salicylate

prepare aspirin. (The synthesis of oil of wintergreen is described in another experiment.) The reaction is complicated by the fact that salicylic acid has a carboxyl as well as a phenolic hydroxyl group, and a small amount of polymeric by-product is also formed. Acetylsalicylic acid will react with sodium bicarbonate to form a water-soluble sodium salt while the polymeric by-product is insoluble in bicarbonate. This difference in behavior will be utilized for the purification of the product aspirin.

The most likely impurity in the final product is salicylic acid itself, which can arise from incomplete acetylation or from hydrolysis of the product during the isolation steps. This material is removed during the various stages of the purification and in the final crystallization of the product. Salicylic acid, like most phenols, forms a highly-colored complex with ferric chloride (Fe^{+3} ion). Aspirin, which has this group acetylated, will not give the color reaction. Thus, the presence of this impurity in the final product is easily detected.

SPECIAL INSTRUCTIONS

Before beginning this experiment, one should read Techniques 1, 2, 3, and 4. There are many convenient stopping places in this experiment, but one should not leave the product in contact with sodium bicarbonate for prolonged periods.

PROCEDURE

Weigh 2.0 g (0.015 mole) of salicylic acid crystals and place them in a 125 ml Erlenmeyer flask. Add 5 ml (0.05 mole) of acetic anhydride, followed by 5 drops of concentrated sulfuric acid from a dropper, and swirl gently until the salicylic acid dissolves. Heat gently on the steam bath for 5 to 10 minutes. Allow the flask to cool to room temperature, during which time the acetylsalicylic acid should begin to crystallize from the reaction mixture. If it does not, scratch the walls of the flask with a glass rod and cool the mixture slightly in an ice bath until crystallization has occurred. Add 50 ml of water and cool the mixture in an ice bath to complete the crystallization. Collect the product by vacuum filtration on a Büchner funnel (see Technique 2, Section 2.3 and Figure 2.4). The filtrate may be used to rinse the Erlenmeyer flask repeatedly until all crystals have been collected. Rinse the crystals several times with small portions of **cold** water. Continue drawing air through the crystals on the Büchner funnel by suction until the crystals are free of solvent. Remove the crystals for air drying. Weigh the crude product, which may contain some unreacted acid, and calculate the yield of crude product.

PURIFICATION

Into each of three separate test tubes containing 5 ml of water, dissolve a few crystals of either phenol, salicylic acid, or your crude product. Add a drop or two of 1% ferric chloride solution to each tube and note the color. Formation of an iron-phenol complex with Fe(III) gives a definite color ranging from red to violet, depending on the particular phenol present. At the end of the purification procedure you will be asked to repeat this test and note the difference.

Transfer the crude solid to a 150 ml beaker and add 25 ml of a saturated aqueous sodium bicarbonate solution. Stir until all signs (listen) of reaction have ceased. Filter the solution by suction through a Büchner funnel. Any polymer should be left behind at this point. Wash the beaker and funnel with 5 to 10 ml of water. Prepare a mixture of 3.5 ml of concentrated hydrochloric acid and 10 ml of water in a 150 ml beaker. Carefully pour the filtrate into this mixture while stirring. The aspirin should precipitate. Cool the mixture in an ice bath, filter the solid by suction using a Büchner funnel (see Technique 2, Section 2.3), press the liquid from the crystals with a clean stopper or cork, and wash the crystals well with **cold** water. It is essential that the water used for this step be ice cold. Place the crystals on a watch glass to dry. Weigh the product, determine its melting point (135–136°C), and calculate the percentage yield. Test for the presence of unreacted salicylic acid using the ferric chloride solution, as described above.

RECRYSTALLIZATION

The product which was prepared according to the above instructions was isolated by **precipitation;** it should now be prepared as a pure **crystalline** substance. Water is not a suitable solvent for crystallization because aspirin will undergo partial hydrolysis when heated in water. Following the general instructions given in Technique 3, dissolve a small sample of the final product in a **minimum** amount of hot benzene, while gently and continuously heating the mixture on the steam bath. If any solid material remains, filter the hot solution by gravity through a fluted filter placed in a short-stemmed funnel which has previously been preheated by pouring hot benzene through it (see Technique 2, Section 2.1 and Technique 3, Section 3.4). Let the filtered solution stand. On cooling to room temperature, the aspirin should recrystallize. If it does not, add a little petroleum ether and cool the solution **slightly** (benzene freezes at 5°C) in ice water, while scratching the inside of the glass with a glass rod (un-fire-polished). Collect the product by vacuum filtration using a Hirsch funnel. Submit the crystalline sample, along with a record of its melting point, in a small vial to your instructor. Do not forget to test the crystals with $FeCl_3$.

ASPIRIN TABLETS

Aspirin tablets are acetylsalicylic acid pressed together with a small amount of inert binding material (usually starch). One may test for the presence of starch by boiling a 10 mg piece of aspirin tablet with 2 ml of water. The liquid is cooled, and a drop of iodine solution is added. If starch is present, it will form a complex with the iodine. The starch-iodine complex is deep blue-violet. Repeat this test with a commercial aspirin tablet and with the aspirin prepared in this experiment.

QUESTIONS

1. What is the purpose of the concentrated sulfuric acid used in the acetylation reaction?

2. Give a possible structure for the polymeric by-product obtained in the reaction.

3. Why is the polymeric by-product not soluble in sodium bicarbonate solution, while salicylic acid itself is soluble?

4. If one were to use 5.0 g of salicylic acid and an excess of acetic anhydride in the above synthesis of aspirin, what would be the theoretical yield of acetylsalicylic acid in moles? In grams?

5. When heated in boiling water, aspirin (acetylsalicylic acid) decomposes to give a solution which gives a positive $FeCl_3$ test. Why? What is the reaction? Give an equation for the reaction.

ESSAY

Analgesics

Acylated aromatic amines (those having an acyl group, $R-\overset{\overset{O}{\|}}{C}-$, substituted on nitrogen) are important in over-the-counter headache remedies. Over-the-counter drugs are those which you may buy without a prescription. Acetanilide, phenacetin, and acetaminophen are mild analgesics (relieve pain) and antipyretics (reduce fever) and have had importance, along with aspirin, in many non-prescription drugs.

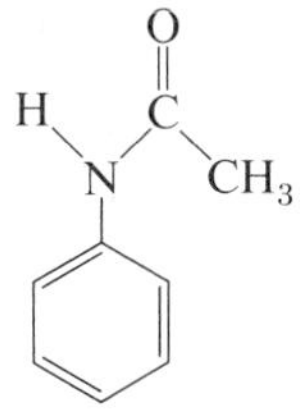

Acetanilide 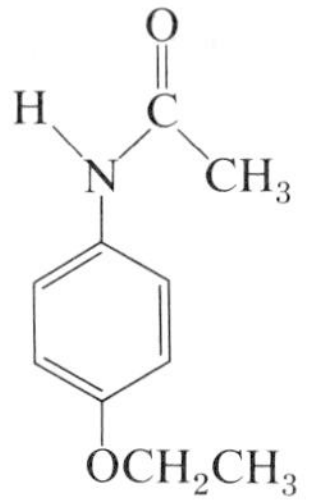   **Phenacetin** 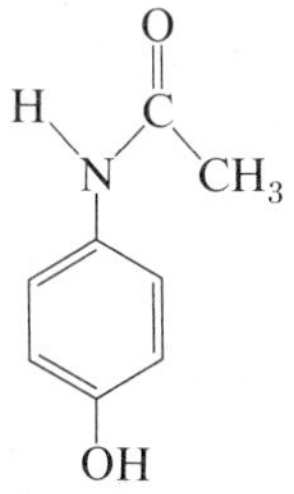   **Acetaminophen**

The discovery that acetanilide was an effective antipyretic came about by accident in the year 1886. Two doctors, Cahn and Hepp, had been testing naphthalene as a possible **vermifuge** (an agent that expels worms). Their early results on simple worm cases were very discouraging, so Dr. Hepp decided to test the compound on a patient with a larger variety of complaints, including worms—a sort of "shotgun" approach. A short time later, Dr. Hepp excitedly reported to his colleague Dr. Cahn that naphthalene had miraculous fever-reducing properties.

In trying to verify this observation, the doctors discovered that the bottle they thought contained naphthalene had apparently been mislabeled. In fact, the bottle brought to them by their assistant had a label so faint as to be illegible. They were sure that the sample was not naphthalene since it had no odor. Naphthalene has a strong odor reminiscent of moth balls. Being close to an important discovery, the doctors were nevertheless stymied, and they appealed to a cousin of Hepp, who was a chemist in a nearby dye factory, to help them identify the unknown compound. This compound turned out to be acetanilide, a compound with a structure not at all like that of naphthalene. Cer-

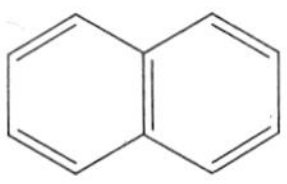

Naphthalene

tainly, Hepp's unscientific and risky approach would be frowned upon

31

by doctors today, and, to be sure, the Food and Drug Administration (FDA) would never allow human testing prior to extensive animal testing (consumer protection **has** progressed); nevertheless, Cahn and Hepp made an important discovery.

In another instance of serendipity, the publication of Cahn and Hepp, describing their experiments with acetanilide, caught the attention of Carl Duisberg, director of research at the Bayer Company in Germany. Duisberg was confronted with the problem of profitably getting rid of nearly 50 tons of p-aminophenol, a by-product from the synthesis of one of Bayer's other commercial products. He immediately saw the possibility of converting p-aminophenol to a compound similar in structure to acetanilide, by putting an acyl group on the nitrogen. It was, however, believed at that time that all compounds having a hydroxyl group on a benzene ring (i.e., phenols) were toxic. Duisberg devised the following scheme of structural modifications of p-amino-phenol to get the compound phenacetin. Phenacetin turned out to be

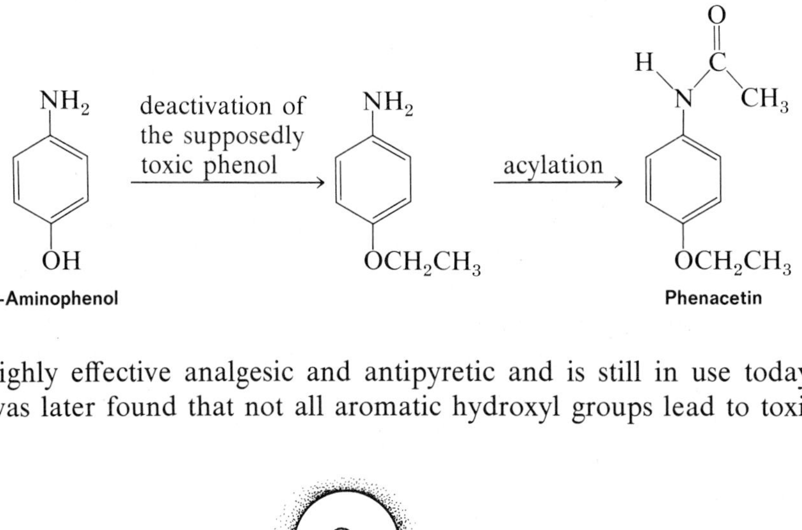

a highly effective analgesic and antipyretic and is still in use today. It was later found that not all aromatic hydroxyl groups lead to toxic

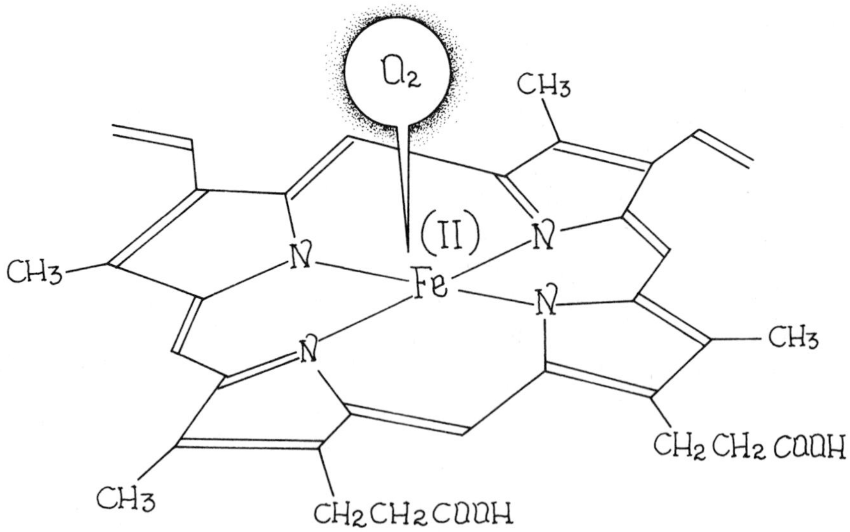

Heme portion of blood-oxygen carrier, hemoglobin

USE OF THE MAJOR ANALGESICS AND CAFFEIN IN SOME COMMON PREPARATIONS

(Non-analgesic ingredients, e.g., buffers, are not listed.)

	ASPIRIN	PHENACETIN	CAFFEIN	ACETAMINOPHEN
ASPIRIN[1]	0.324 g	—	—	—
ANACIN	0.400 g	—	0.0325 g	—
BUFFERIN	0.324 g	—	—	—
COPE	0.420 g	—	0.032 g	—
EXCEDRIN[2]	0.190 g	—	0.060 g	0.090 g
EMPIRIN	0.233 g	0.166 g	0.030 g	—
TYLENOL	—	—	—	0.325 g

[1] 5 grain tablet (1 grain = 0.0648 g)

[2] Also contains 0.13 g salicylamide, the amide of salicylic acid, which is also an analgesic.

compounds, and today the compound acetaminophen is also a widely-used analgesic.

On continued or excessive use, acetanilide can cause a serious blood disorder called **methemoglobinemia.** In this disorder, the central iron atom in hemoglobin is converted from Fe(II) to Fe(III) to give methemoglobin. Methemoglobin will not function as an oxygen carrier in the blood stream. The result is a type of anemia (deficiency of hemoglobin or lack of red blood cells). Phenacetin and acetaminophen cause the same disorder, but to a much lesser degree. Since they are also somewhat more effective as antipyretic and analgesic drugs than acetanilide, their use is preferred. Acetaminophen is marketed under the trade name Tylenol and is often successfully used by persons who are allergic to aspirin.

Many over-the-counter preparations, as can be seen from the accompanying table, employ a combination of one or more analgesic drugs and caffein, which is thought to "perk-up" the user. Many drug houses sell a combination pain reliever called an "APC tablet." Empirin is an example of these types of tablets, which usually contain **A**spirin, **P**henacetin, and **C**affein (hence, **APC**).

REFERENCES

Barr, W. H., and Penna, R. P. "O-T-C Internal Analgesics." G. B. Griffenhagen (Ed.), *Handbook of Non-Prescription Drugs.* American Pharmaceutical Association, 1969, 26–36.

Hansch, C. "Drug Research or the Luck of the Draw." *Journal of Chemical Education, 51,* (1974), 360.

Ray, O. S. "Internal Analgesics." *Drugs, Society, and Human Behavior,* St. Louis: Mosby, 1972, 122–132.

"The Uses of Aspirin." *The Medicine Show.* Consumers Union, Mt. Vernon, New York: Consumers Union, 1971.

Woodbury, D. M. "Analgesic-Antipyretics, Anti-inflammatory Agents, and Inhibitors of Uric Acid Synthesis." L. S. Goodman and A. Gilman, *The Pharmacological Basis of Therapeutics* (4th edition). New York: Macmillan, 1970.

ACETANILIDE

Crystallization
Vacuum Filtration
Acylation
 (Preparation Of An Amide)

  An amine may be treated with an acid anhydride to form an amide. In this particular example, the amine, aniline, is treated with acetic anhydride to form the amide, acetanilide. The impure liquid amine, aniline, is converted to a solid product, acetanilide, which is readily purified by crystallization. The acylated amine precipitates as a crude

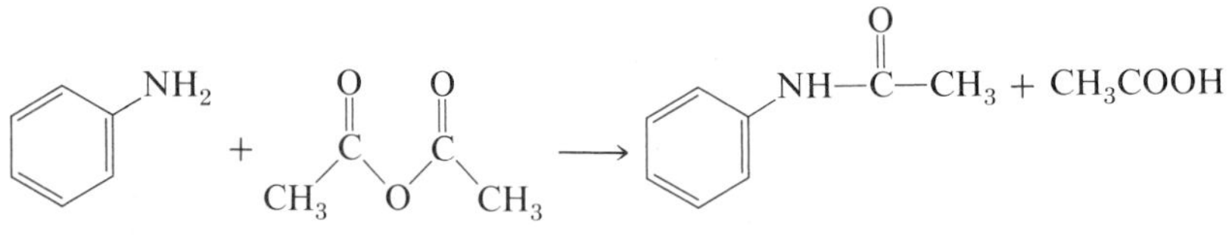

Aniline   Acetanilide

solid which is dissolved in hot water, decolorized with charcoal, filtered, and recrystallized.

THE REACTION

  Amines can be acetylated by several means. Among these are the use of acetic anhydride, acetyl chloride, or glacial acetic acid (with removal of the water formed in the reaction). This last procedure is of commercial interest since it is economical. However, it requires a long period of heating. Acetyl chloride is unsatisfactory for several reasons. Principally, it reacts vigorously liberating HCl, which converts half of the amine to its hydrochloride salt, thus rendering it incapable of participating in the reaction.

  Acetic anhydride is preferred for a laboratory synthesis. Its rate of hydrolysis is slow enough to allow the acetylation of amines to be carried out in aqueous solutions. The procedure gives a product of high purity and in good yield, but it is not suitable for use with deactivated amines (weak bases) such as ortho- and para-nitroanilines.

  Acetylation is often used to "protect" a primary or secondary amine functional group. Acylated amines are less susceptible to oxidation, less reactive in aromatic substitution reactions, and less prone to participate in many of the typical reactions of free amines, since they are less basic. The amino group can be regenerated readily by hydrolysis in acid or base.

SPECIAL INSTRUCTIONS

Before beginning this experiment, one should read Techniques 1, 2, 3, and 4. One should be careful not to stop the experiment at any place where solid has not been filtered from solution.

PROCEDURE

Measure 4 ml of technical grade aniline in a 10 ml graduated cylinder. Aniline is a toxic substance, and care should be taken to avoid contact with the skin. Weigh the cylinder to within the nearest 0.1 g, pour the aniline into a 250 ml Erlenmeyer flask, and reweigh the cylinder to determine accurately the amount of aniline used in the reaction. Add 30 ml of water to the flask, and then, while swirling the flask, add 5 ml of acetic anhydride (density = 1.08 g/ml) in several small portions. Record any changes which might be observed.

The crude acetanilide, which precipitates in the reaction, will now be recrystallized in the same flask. Add 100 ml of water and a boiling stone, and heat the mixture with a burner, using a wire gauze under the flask, until all of the solid and oily materials have dissolved. Remove the flame, pour about 1 ml of the hot solution into a small beaker, and set this material aside to cool. The hot flask should be handled with a towel. Before the mixture in the Erlenmeyer flask is boiled again, about 1 g (approximately 1/2 teaspoon) of activated charcoal, or Norite, is added. The charcoal must not be added to a vigorously boiling solution, or frothing will occur. Swirl the mixture and boil gently for a few minutes. Meanwhile, an apparatus for gravity filtration, using a stemless funnel and fluted filter paper placed into a 250 ml or 500 ml Erlenmeyer flask, should be assembled (see Technique 2, Section 2.1). Also, about 50 ml of boiling water, to be used for washing, will be needed for the steps which follow.

Warm the funnel by pouring about 20 ml of hot water through it. Discard this warming solution and then filter the acetanilide solution a little at a time. Keep the solution warm on a steam bath until it is poured into the funnel. Also, keep the collection flask warmed on a steam bath. Solvent vapors will keep the funnel stem warm and prevent premature crystallization. If crystallization begins in the funnel, add some hot water to the funnel to dissolve the crystals. Rinse the flask and the solids in the funnel with a little hot water, and set the filtrate aside to cool. To complete the crystallization, chill the flask in an ice bath for 10 to 15 minutes. Prepare a Büchner funnel with filter paper according to the instructions given in Technique 2, Section 2.3. Collect the crystals by vacuum filtration, dry them as thoroughly as possible by drawing air through them on the Büchner funnel, and complete the drying by spreading them on paper and allowing them to stand overnight.

In a Hirsch funnel collect the crystals which were not treated with charcoal and compare the color of these crystals with that of those crystals which had been treated with charcoal.

Record the weight, percentage yield, and melting point of the pure and impure dried acetanilide. Place the crystals in a labeled vial, and submit them to the instructor.

QUESTIONS

1. Aniline is basic while acetanilide is not basic. Explain this difference.

2. Calculate the theoretical yield of acetanilide which would be expected if 10 g of aniline were allowed to react with an excess of acetic anhydride.

3. Give equations for the reactions of aniline with acetyl chloride and with acetic acid, to give acetanilide.

4. In the introduction to this experiment, the **hydrolysis** of acetic anhydride is mentioned as a competing reaction. Write an equation for this reaction.

Experiment 3
PHENACETIN

Crystallization
Vacuum Filtration
Acylation
    (Preparation Of An Amide)

The preparation of phenacetin involves the treatment of an amine with an acid anhydride to form an amide. In this case, the amine, **p**-phenetidine (**p**-ethoxyaniline), is treated with acetic anhydride to form the amide, phenacetin. Much of the material presented in the introduction to Experiment 2, Aniline, also applies to the preparation of phenacetin, so that material should be read. The acylated amine precipitates as a crude solid which is crystallized from hot water.

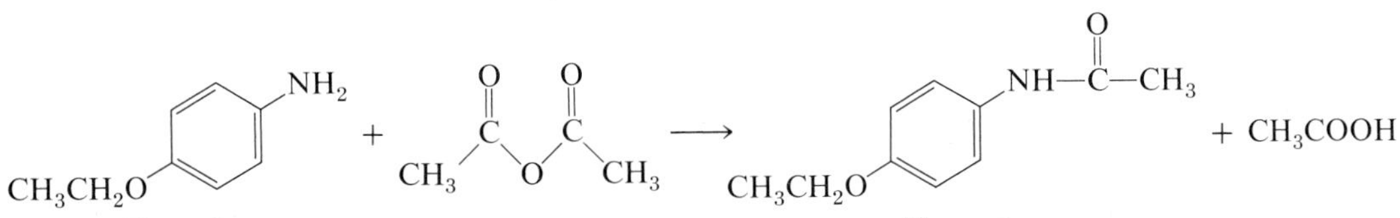

SPECIAL INSTRUCTIONS

Before beginning this experiment, one should read Techniques 1, 2, 3, and 4. One should be careful not to stop the experiment at any place where solid has not been filtered from solution.

PROCEDURE

Place 4 g of **p**-phenetidine into 75 ml of water. Add 2.5 ml of concentrated hydrochloric acid, which should dissolve the amine completely. If some amine remains undissolved, add a few more drops of concentrated hydrochloric acid until the amine dissolves. Add 1 g of decolorizing charcoal (Norite) to the solution, swirl it for a few minutes, and filter the charcoal from the solution by gravity filtration using a fluted filter (see Technique 2, Section 2.1).

Prepare a solution for use as a buffer by dissolving 4.5 g of sodium acetate ($CH_3COONa \cdot 3H_2O$) in 15 ml of water, and clarify the solution by gravity filtration. Set this solution aside.

Transfer the **p**-phenetidine hydrochloride solution to a 250 ml Erlenmeyer flask, and warm it on a steam cone. Add 3.5 ml of acetic anhydride while swirling the solution. Add the sodium acetate buffering solution **at once,** and swirl the solution vigorously to assure mixing.

Cool the reaction mixture by immersing the flask in an ice-water bath, and stir the mixture vigorously until the crude phenacetin crystallizes. Collect the crystals in a Büchner funnel by vacuum filtration (see Technique 2, Section 2.3). Wash the crystals with a portion of cold water. Weigh the crude crystals.

Using the crystallization procedures described in Technique 3, Section 3.4, dissolve about 2 g of the crude phenacetin in the minimum amount of **boiling** water. Allow the solution to cool slowly. When the first crystals appear, immerse the flask in an ice bath for 15 to 20 minutes. Collect the crystals by vacuum filtration using a Büchner funnel (see Technique 2, Section 2.3), and dry them. Record the weight of the phenacetin obtained before the crystallization procedure, then the percentage yield of phenacetin obtained in the experiment, and finally, the melting point of the recrystallized sample. Submit the sample to the instructor in a labeled vial.

QUESTIONS

1. **p**-Phenetidine is purified by adding concentrated hydrochloric acid and decolorizing the resulting solution with Norite. Write the chemical equation for the reaction of **p**-phenetidine with hydrochloric acid. Explain why the **p**-phenetidine dissolves.

2. **p**-Phenetidine is basic while phenacetin is not basic. Explain this difference.

3. Acetaminophen has the structure shown. How might it be prepared?

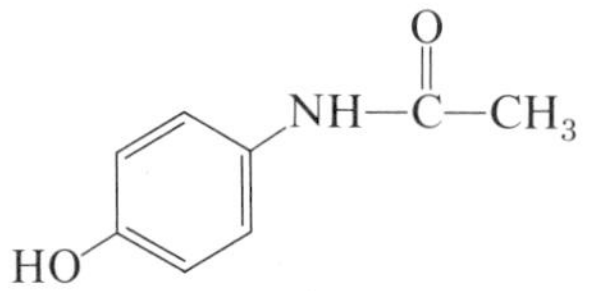

4. Calculate the theoretical yield of phenacetin which would be expected if 10 g of **p**-phenetidine were allowed to react with an excess of acetic anhydride.

ESSAY

Identification of Drugs

Frequently a chemist may be called upon to **identify** a particular unknown substance. If there is no prior information to work from, this can be a formidable task. There are several million known compounds, both inorganic and organic. For a completely unknown substance, the chemist must often use every method at his disposal. If the unknown substance is a mixture, then the mixture must be separated into its components and each component identified separately. A pure compound can often be identified from its physical properties (mp, bp, density, refractive index, etc.) and a knowledge of its functional groups. These may be identified by the reactions that the compound is observed to undergo, or by spectroscopy (infrared, ultraviolet, nmr, and mass spectroscopy). The techniques necessary for this type of identification are introduced in a later section of this text.

A somewhat simpler situation often arises in drug identification. The scope of the problem is more limited here, and the chemist working in a hospital trying to identify the source of a drug overdose, or the law enforcement officer trying to identify a suspected illicit drug or a poison, usually has some prior clues to work from. So does the medicinal chemist working for a pharmaceutical manufacturer who might be trying to discover why a competitor's product is better than his.

As an example, consider a drug overdose case. The patient is brought into the emergency ward of a hospital. This person may either be in a coma, in a hyperexcited state, have an allergic rash, or clearly be hallucinating. These physiological symptoms are themselves a clue to the nature of the drug. Often samples of the drug may be found in the patient's possession. Correct medical treatment may require a rapid and accurate identification of a drug powder or capsule. If the patient is conscious, the necessary information may be elicited orally; if not, an examination of the drug is required. If the drug is a tablet or a capsule, the process is often simple, since many drugs are coded by a manufacturer's trademark or logo, by shape (round, oval, bullet shape, etc.), by formulation (tablet, gelatin capsule, time release micro-capsules, etc.), and by color.

A powder is more difficult to identify, but under some circumstances it may be easy. Plant drugs are often easy to identify since they contain microscopic bits and pieces of the plant from which they are obtained. This cellular debris is often quite characteristic for certain types of drugs, and they can be identified on this basis alone. A microscope is all that is required. Sometimes chemical color tests may be used as a confirmation. Certain drugs give rise to characteristic colors when treated with special reagents. Other drugs form crystalline precipitates of characteristic color and crystal structure when treated with appropriate reagents.

If the drug itself is not available and the patient is unconscious (or dead), the identification may be more difficult. It may be necessary to pump the stomach or bladder contents of the patient (or corpse), or to obtain a blood sample, and to work on these. These samples of stomach fluid, urine, or blood would be extracted with an appropriate organic solvent, and the extract would be analyzed.

Often the final identification of a drug, either as a urine, serum, or stomach fluid extract, hinges on some type of **chromatography.** Thin layer chromatography is often used. Under specified conditions, many drug substances can be identified by their R_f values and by the colors that their tlc spots turn when treated with various reagents, or when they are observed under certain visualization methods. In the experiment which follows, thin layer chromatography is used in a simple example of these techniques as applied to the analysis of an unknown analgesic drug.

REFERENCES

Keller, E. "Forensic Toxicology: Poison Detection and Homicide." *Chemistry, 43* (1970), 14.
Keller, E. "Origin of Modern Criminology." *Chemistry, 42* (1969), 8.
Lieu, V. T. "Analysis of APC Tablets." *Journal of Chemical Education, 48* (1971), 478.
Neman, R. L. "Thin Layer Chromatography of Drugs." *Journal of Chemical Education, 49* (1972), 834.
Rodgers, S. S. "Some Analytical Methods Used in Crime Laboratories." *Chemistry, 42* (1969), 29.
Tietz, N. W. *Fundamentals of Clinical Chemistry.* Philadelphia: Saunders, 1970.
Walls, H. J. *Forensic Science.* New York: Praeger, 1968.

TLC ANALYSIS OF ANALGESIC DRUGS

Thin Layer Chromatography

In this experiment, thin layer chromatography (tlc) will be used to determine the composition of various over-the-counter analgesics. If the instructor chooses, you may also be required to identify the components and actual identity (trade name) of an unknown analgesic. You will be given two commercially pre-prepared tlc plates with a flexible backing and a silica gel coating with a fluorescent indicator. On one tlc plate you will spot five standard compounds often used in analgesic formulations. In addition, a standard reference mixture, containing these same compounds, will also be spotted. The reference substances are:

> Acetaminophen (Ac)
> Aspirin   (Asp)
> Caffein   (Cf)
> Phenacetin   (Ph)
> Salicylamide   (Sal)

They will all be available as solutions of 1 g of each dissolved in 20 ml of a 50/50 mixture of chloroform and ethanol. The purpose of the first plate is to determine the order of elution (R_f values) of the known substances and to index the standard reference mixture. On the second plate the standard reference mixture will be spotted along with several solutions prepared from commercial analgesic tablets. The crushed tablets will also be dissolved in a 50/50 chloroform–ethanol mixture. At your instructor's option, one of the analgesics to be spotted on the second plate may be an unknown.

Two methods of visualization will be used to observe the positions of the spots on the developed tlc plates. First, the plates will be observed while under illumination from a short wavelength ultraviolet lamp. This is done best in a darkened room or in a fume hood which has been darkened by taping butcher paper or aluminum foil over the lowered glass cover. Under these conditions, some of the spots will appear as dark areas on the plate, while others will fluoresce brightly. This difference in appearance under UV illumination will help to distinguish the substances from one another. You will find it convenient to outline very lightly in pencil the spots observed and to place a small x inside those spots which fluoresce. For a second means of visualization, iodine vapor will be used. Not all of the spots will become visible on treatment with iodine, but at least two will develop a deep brown color. The differences in the behaviors of the various spots with iodine can be used to further differentiate between them.

It is possible to use several developing solvents for this experiment. The use of ethyl acetate is preferred. Acetone and cyclohexanone also give fairly good results. Cyclohexanone, however, is a fairly viscous liquid and it takes a good while longer than the other solvents to diffuse up the plate ($>$1 hour). It is also more difficult to evaporate from the slide. Ethyl acetate requires only about 20 minutes to ascend the plate and evaporates quickly. A more complicated developing solvent of 120:60:18:1—benzene, ether, glacial acetic acid, and methanol—has been recommended in the pharmaceutical literature.

In some analgesics you may encounter ingredients in addition to the five mentioned above. Some include an antihistamine and others include a mild sedative. For instance, Midol contains N-cinnamyleph-edrine (cinnamedrine), an antihistamine, while Excedrin PM contains the sedative methapyrilene hydrochloride. Cope contains the related sedative methapyrilene fumarate.

SPECIAL INSTRUCTIONS

To fully understand this experiment, the two essays "Aspirin" and "Analgesics" should be read. These precede Experiments 1 and 2 in this text. In addition, it is necessary to have read Technique 11 which explains the principles and methods of thin layer chromatography. Note that the UV examination of the developed tlc plates must be done **before** the iodine visualization method. The iodine permanently affects some of the spots. Aspirin presents some special problems since it is present in large amount in many of the analgesics and since it hydro-lyzes easily. For these reasons, the aspirin spots often show excessive tailing.

PROCEDURE

To begin, at least 12 capillary micropipets will be needed to spot the plates. The preparation of these pipets is described and illustrated in Technique 11, Section 11.3.

After preparing the micropipets, obtain two 10 cm $\times$ 6.6 cm tlc plates (Eastman Chromagram Sheet, No. 13181) from your instructor. These plates have a flexible backing, but should not be bent excessively. They should be handled carefully or the adsorbent may flake off of them. In addition, they should only be handled by the edges. The surface should not be touched. Using a lead pencil (not a pen) **lightly** draw a line across the plates (short dimension) about 1 cm from the bottom. Using a centimeter ruler, move its index about 0.6 cm in from the edge of the plate and lightly mark off six 1 cm intervals on the line (see figure). These are the points at which the samples will be spotted. On the first plate, starting from left to right, spot first

acetaminophen, then aspirin, caffein, phenacetin, and salicylamide. This order is alphabetic and will avoid any further memory problems or confusion. Solutions of these compounds will be found in small bottles on the side shelf. The standard reference mixture, also found on the side shelf, is spotted in the last position. The correct method of spotting a tlc slide is described in Technique 11, Section 11.3. It is important that the spots be made as small as possible and that the plates not be overloaded. If either of these errors is made, the spots will tail and will overlap one another after development. The applied spot should be about 1 or 2 mm (1/16 in) in diameter. If scrap pieces of the tlc plates are available, it would be a good idea to practice spotting on these prior to preparing the actual sample plates.

When the first plate has been spotted, obtain a 16 oz wide mouth screw cap jar for use as a development chamber. The preparation of a development chamber is described in Technique 11, Section 11.4. For this experiment, a slight modification of the development chamber must be made. Since the backing on the tlc slides is very thin, if they touch the filter paper liner of the development chamber **at any point,** solvent will begin to diffuse onto the absorbent surface at that point. To avoid this, a very narrow liner strip of filter paper (approx. 5 cm wide) should be used. It should be folded into an L-shape and should be long enough to traverse the bottom of the jar and extend up the side to the top of the jar. When placing slides in the jar for development, they should straddle this liner strip, but not touch it.

When the development chamber has been prepared, saturate the filter paper liner with ethyl acetate (or other solvent) and fill the jar to a depth of about 0.5 cm with the same solvent. Recall that the solvent level must not be above the spots on the plate. Place the spotted plate in the development chamber (straddling the liner) and allow the plate to develop. When the solvent has risen on the plate to a level about 0.5 cm from the top, remove the plate from the chamber and, using a lead pencil, mark the position of the solvent front. Allow the plate to dry and then observe it under a short wavelength UV lamp. Lightly mark the observed spots with a pencil. Next, place the plate in a jar containing a few iodine crystals, cap the jar, and warm it **gently** on a steam bath until spots begin to appear. Notice which spots become visible and record that information in your notebook. Also, note which spot(s) fluoresced under the UV lamp. Remove the plate from the iodine jar and, using a ruler marked in millimeters, measure the distance that each spot has traveled relative to the solvent front and calculate its R_f value (see Technique 11, Section 11.8).

Next, obtain one-half of a tablet of each of the analgesics to be analyzed. If you were issued an unknown, you may analyze four other analgesics; if not, you may analyze five. Five choices that lead to a particularly wide spectrum of results would be Anacin, Bufferin, Empirin, Excedrin, and Tylenol. However, if you have a favorite analgesic, you may prefer to analyze it. Take each analgesic half tablet, place it on a smooth piece of notebook paper, and crush it well with a spatula. Transfer each crushed half tablet to a small labeled test tube or Erlenmeyer flask. Using a graduated cylinder, mix together 15 ml of absolute ethanol and 15 ml of chloroform. Mix the solution well. Add about 5 ml of this solvent mixture to each of the crushed half tablets, and then heat each of them **gently** for a few minutes on a steam bath. Not all the tablet will dissolve since the analgesics usually contain starch as a binder for the tablet. In addition, many of them contain inorganic buffering agents which are insoluble in this solvent mixture. After heating, allow the samples to settle and then spot the clear liquid extracts at five of the positions on the second plate. At the sixth position, spot the standard reference solution. Develop the plate in ethyl acetate (or other solvent) as before. Observe the plate under UV illumination and mark the visible spots. Repeat the visualization using iodine. Record your conclusions as to the content of the various analgesics. If you were issued an unknown, attempt to determine its actual identity (trade name).

ESSAY

Natural Products

From the earliest times, man has always found and invented uses for substances from his natural environment. The caveman, perhaps, sought merely food and shelter, but subsequent generations of man sought tools, things of beauty or art, and objects of more sophisticated utility. They learned how to grow food and domesticate animals rather than hunt, and to use the fibers of plants rather than the skins of animals to make clothing. Man learned to **manipulate** his environment. He learned to utilize minerals from the earth and to convert them to useful metals and chemicals. Likewise, he learned to utilize the largess of the plant and animal world. From these crude beginnings, and from an innate curiosity in man to understand the relationships between substances found in his natural environment, chemistry was born.

Originally, and as late as 1850, chemistry was divided into two large branches—organic and inorganic. Until 1828 when Wöhler synthesized urea starting from inorganic materials, it was thought that only living organisms could produce organic substances. The earliest "organic" chemists were, therefore, naturally interested in substances which could be derived from living sources, that is, from plants and animals. Many of these substances gave them good reason to be curious. Many plants were known to yield drug substances which displayed spectacular physiological effects. Others yielded valuable spices, perfumes, and dyes. Some were important foodstuffs. Several could be made into important fibers for manufacturing clothing. The initial curiosity in these substances excited even more complex questions. Why did some plants produce physiologically active substances? What were these substances? Why did some plants yield dyes? What was a dye that it should have color? Why do some objects (spices, flowers, pine trees) have a pleasant odor? Why is there a difference in the behavior of fibers like wool and cotton? Why do some substances taste sweet or have a characteristic flavor? Questions of this sort spurred chemists to isolate the distinctive chemical substances from their environment and study them.

Substances or chemicals derived from natural plant or animal sources are called **natural products.** Although we are able to synthesize many of these substances today, they were first encountered in nature, and many were utilized by primitive peoples even before modern chemists understood their nature.

Since a wide variety of substances occurs in nature, there are many types or classes of natural products. Generally, however, the chemist recognizes several classes of natural products based on either the structural similarities of the substances to one another, their related chemical behavior, or their similar (in the broadest sense) biological behavior and origin. All of these substances either represent necessary cellular material of plants and animals, or are the synthetic end products of the metabolism of these organisms. Although classifications are not always easy, most chemists recognize the following broad categories of natural products:

> Terpenes and Carotenoids
> Steroids
> Lipids (Fats and Oils)
> Alkaloids
> Carbohydrates and Sugars
> Proteins and Amino Acids
> Vitamins

The list is not comprehensive, as many chemists might wish to add other categories, and some to even delete categories.

To define each category and to give a description of the origin and similar properties of the members of each group listed would be very difficult in a short space. Nevertheless, if you read many of the essays in this text, you will quickly find that almost all of the categories are covered. Wherever possible, experiments have been based on these natural substances. Some of the experiments have as their focus the **isolation** of a natural product from its plant or animal source. Experiments 5 through 9 are of this type. Many of the other experiments are isolation experiments as well, but have a different point of emphasis, such as the chemistry or the physical properties (stereochemistry) of the substance.

ESSAY

Nicotine

Nicotine

Tobacco that outlandish weede
It spends the braine and spoiles the seede
It dulls the spirite, it dims the sight
It robs a woman of her right

William Vaughn (1617)

Tobacco is one of the few major contributions to civilization which the New World can claim. Prior to the discovery of this continent tobacco was unknown to Europeans. In fact, the plants **Nicotiana tobacum** and **Nicotiana rustica** are indigenous to America, the former to South and Central America and the latter to eastern North America and the West Indies.

Nicotiana was used by Indians in both the northern and southern hemispheres in religious, curing, and intoxicating practices. The literature from 1492 to the present leaves little doubt that the native peoples of nearly all parts of the New World accorded supreme ritualistic and mythological status to these plants. There is also evidence that in many areas tobacco has been employed to trigger ecstatic states very similar to those induced by the true hallucinogens. Even today the shamans ("witch doctors") of the Warao Indians of Venezuela employ a psychotomimetic snuff, composed at least in major part of tobacco, to achieve the trance states which are part of the purification, initiation, and supernatural curing rituals. The American Indian Peace Pipe ritual is also well known.

Tobacco is not generally considered to be a hallucinogen. In fact, it is not clear that tobacco is the only component that the Indian shamans use in their secret preparations. In those cases where the preparation is smoked, oxygen starvation in the lungs, at least in part, may well be the cause of the trance-like state.

It is interesting to note that the Indians held the tobacco plant in high esteem and its use was limited to ritualistic and religious ceremony. Not so with civilized culture! Its acceptance into civilized society was both rapid and thorough. Tobacco smoking was introduced into Europe by Rodrigo de Jerez. He was the first European to land in Cuba, where he picked up the smoking habit. When he returned to Portugal and continued the practice, people who saw the smoke coming out of his nose and mouth thought he was possessed by the devil. Rodrigo spent the next several years in jail. During this time the spread of tobacco use was so rapid that when Rodrigo was released from jail he was amazed to see everyone indulging in the very offense for which he had been confined.

46

The active ingredient in tobacco, (−)-nicotine, was first isolated in 1828. Nicotine is a colorless liquid **alkaloid,** and the natural material is levorotatory due to a single chiral (asymmetric) center. Alkaloids are a class of naturally occurring compounds containing nitrogen and having the properties of an amine base—alkaline, hence, "alkaloid." As with almost all alkaloids, nicotine has dramatic effects on the human system. One quickly develops tolerance to the effects of small amounts of nicotine (tobacco), along with a dependency. Mark Twain remarked how easy it is to stop smoking—"I've done it a thousand times!" Even in the face of a highly condemning report by the Surgeon General of the United States, millions of Americans continue to smoke.

Nicotine itself is one of the most toxic drugs known to man. A dose of 60 mg is lethal with death following intake by only a few minutes. A cigar contains enough nicotine for two lethal doses. However, it is estimated that only 10% of that amount is absorbed on inhalation of the smoke and that this dose is absorbed over a relatively long time period. The typical filter cigarette contains 20–30 mg of nicotine. That smoking can be practiced at all is explained by the body's ability to degrade or metabolize nicotine rapidly and eliminate it, thus preventing its accumulation. In acute poisoning nicotine causes tremors which turn into convulsions and frequently causes death. Death comes about by paralysis of the muscles used in respiration. This results from a blocking effect on the motor nerve system that normally activates these muscles. Curiously enough, with lower doses there is actually an increase in respiratory rate because the body attempts to counteract the effects of the nicotine. The oxygen need of the carotid artery is stimulated, and an increased heart rate, blood flow, and constriction of the arteries is observed. In smoking, this effect is enhanced by the fact that some of the combustion products, namely carbon monoxide and hydrogen cyanide, irreversibly bind to hemoglobin in the blood so that it can no longer carry oxygen. In a regular smoker, up to 10% of all hemoglobin can be inactivated in this way. This is the basis of the "shortness of breath" phenomenon familiar to smokers. At higher doses, nicotine interferes with the transmission of nerve impulses across the nerve junctions (synapses) of the cholinergic system. That is, nicotine competes with the body's natural **motor nerve** transmission agent, acetylcholine, for the stimulation of voluntary muscles (other types of nerves may use other chemical transmitter agents).

Nicotine
(which is 90%
protonated at pH 7)

Acetylcholine

Nerve fibers are connected to muscle by means of a special kind of junction called a synapse. Conduction of messages along nerve fibers or axons is primarily electrical. However, transmission across the synapse is chemical in nature. The chemical agent released from the nerve ending by the electrical pulse is acetylcholine. This molecule travels through the membrane, which is somehow opened to it, and across the synaptic gap to a **receptor site** where it binds. When it binds, muscle contraction is initiated. Soon afterward an enzyme, acetylcholinesterase (AChE), destroys the acetylcholine by removing the acetyl group (hydrolysis) and producing choline:

$$H_2O + CH_3 \overset{\overset{\displaystyle O}{\underset{\displaystyle\|}{C}}}{} O-CH_2CH_2-\overset{+}{N}(CH_3)_3 \xrightarrow{\text{AChE}}$$

Acetylcholine

$$HO-CH_2CH_2-\overset{+}{N}(CH_3)_3 + CH_3COOH$$

Choline

This enzyme clears the receptor sites, opening them up for another pulse. The choline finds its way back into the nerve fiber where other enzymes resynthesize acetylcholine which is stored for future use. (See Diagram.)

Nicotine has the same ability as acetylcholine to bind to acetylcholine receptor sites and generate muscle stimulation. However, the enzyme AChE cannot destroy the nicotine. Hence, in large enough quantity, nicotine can block all the receptor sites and cause paralysis. Thus, one may observe, for instance, an initial stimulation of heart muscle, but ultimately a paralysis of the heart leading to death. Other cholinergic centers, such as those in the brain, are affected similarly.

The ability of nicotine to block cholinergic receptors is the basis of its use as a natural insecticide, the insect dying of nerve poisoning and respiratory failure. Extracts of tobacco plant were long used as a potent insecticide against certain types of insects. Today this extract has largely been replaced by synthetic insecticides. Many of these synthetics are also nerve poisons and block nerve transmissions in one way or another. For instance, Malathion, an organophosphate, inhibits or stops the action of AChE. The mode of action of DDT is, unfortunately, not currently well understood. However, it is also known to be a potent nerve poison.

REFERENCES

Axelrod, J. "Neurotransmitters." *Scientific American, 230* (June, 1974), 59.
Emboden, W. "Tobaccos and Snuffs." *Narcotic Plants.* New York: Macmillan, 1972.
Fried, R. "Introduction to Neurochemistry." *Journal of Chemical Education, 45* (1968), 322.
Hammond, C. "The Effects of Smoking." *Scientific American, 207* (July, 1962), 39.
Ray, O. S. "Nicotine." *Drugs, Society, and Human Behavior.* St. Louis: Mosby, 1972.

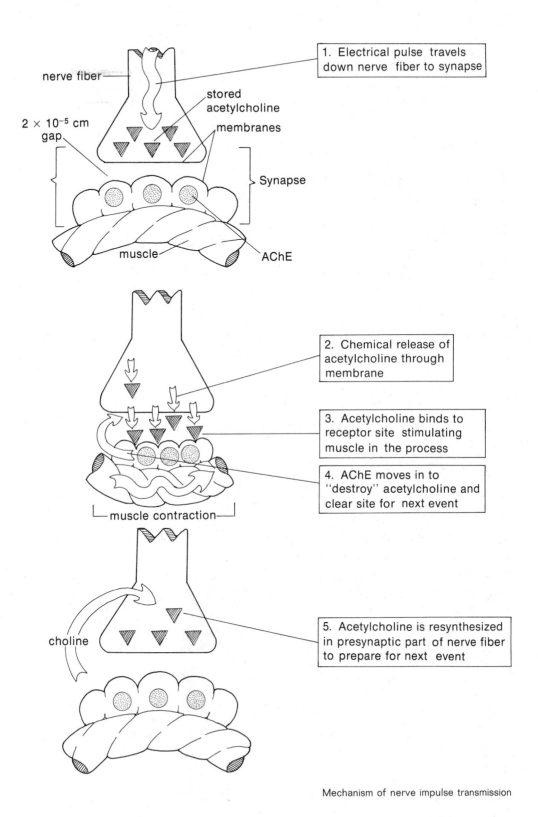

1. Electrical pulse travels down nerve fiber to synapse

2. Chemical release of acetylcholine through membrane

3. Acetylcholine binds to receptor site stimulating muscle in the process

4. AChE moves in to "destroy" acetylcholine and clear site for next event

5. Acetylcholine is resynthesized in presynaptic part of nerve fiber to prepare for next event

Mechanism of nerve impulse transmission

Robinson, T. "Alkaloids." *Scientific American, 201* (July, 1959), 113.

Taylor, N. "The Lively Image and Pattern of Hell—The Story of Tobacco." *Narcotics—Nature's Dangerous Gifts.* New York: Dell, 1970. (The paper bound revision of *Flight From Reality.*)

Volle, R. L., and Koelle, G. B. "Ganglionic Stimulating and Blocking Agents." L. S. Goodman and A. Gilman, *The Pharmacological Basis of Therapeutics* (4th edition). New York: Macmillan, 1970.

Wilbert, J. "Tobacco and Shamanistic Ecstasy among the Warao Indians." P. Furst, (ed.), *Flesh of the Gods.* New York: Praeger, 1972.

Experiment **5**

NICOTINE FROM TOBACCO

Isolation of A Natural Product
Filtration
Extraction
Crystallization

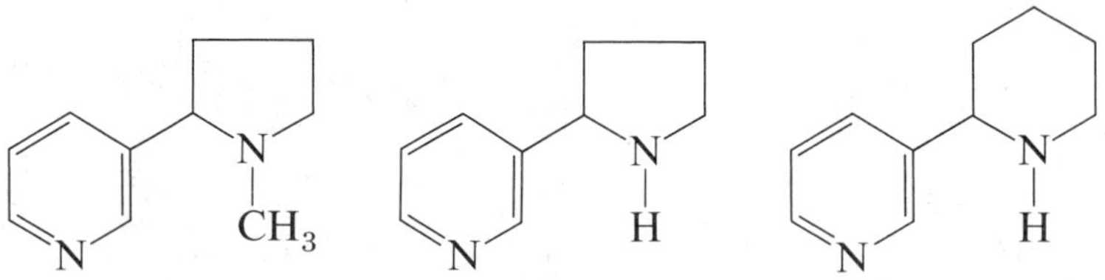

| Nicotine | Nornicotine | Anabasine |

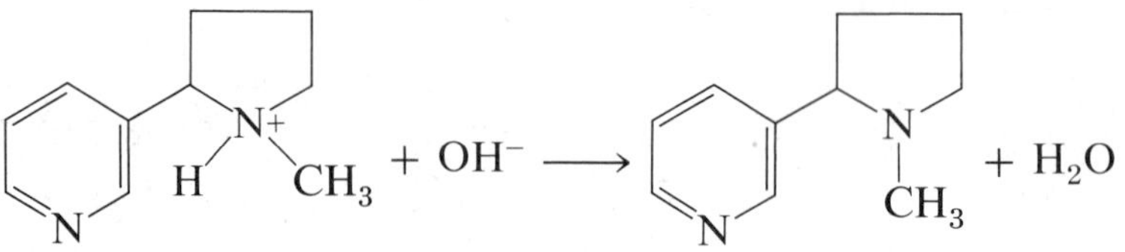

The major alkaloidal component of the tobacco leaf is nicotine, along with smaller amounts of nornicotine, anabasine, and at least seven other minor alkaloids. In this experiment we shall isolate and characterize the pure chemical nicotine from its natural source. It is a nitrogen-containing base built from two heterocyclic rings (pyridine and pyrrolidine). Both amine functions are tertiary and can be protonated to form salts. In fact, the pyrrolidine ring has a pK_a of about 8 (pyridine has $pK_a = 3$) which means that at pH 7 this ring would be expected to be about 90% protonated—one of the factors making nicotine a good mimic for the quaternary acetylcholine. This is also probably the major reason for its ready solubility in water (ionic substance). We will extract tobacco leaf with a strong base (5% NaOH) to generate the free base which can then be extracted into the organic solvent, ether. Evaporation of the ether leaves nicotine as an oil (bp 246°) which in small amounts is not easily purified or easy to handle. Hence, we shall convert it to its salt with picric acid (a phenol). This will give a crystalline material which can be purified and characterized by its melting point. Nicotine forms a dipicrate salt by reaction at both basic centers, and hence, the molecular weight of the product is more than three times as great as that of nicotine. This has the advantage of providing more material to work with.

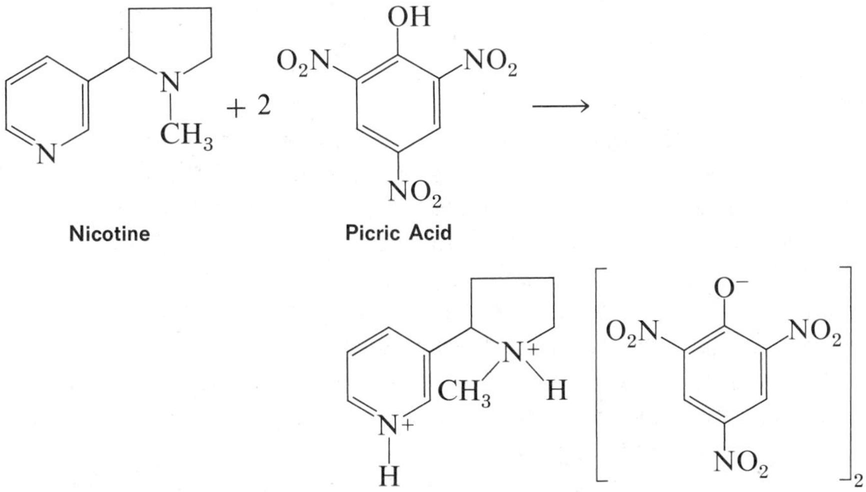

Nicotine   Picric Acid

Nicotine Dipicrate

Although cigarette tobacco could be used, most manufacturers attempt to remove as much "nicotine and tar" as possible, hence, cigars or snuff provide a better source. Dried and crushed cigar tobacco is the more convenient source of the two. Other components of the tobacco leaf are cellulose and tannic acid, which will not extract into ether from a basic solution because it forms a salt, along with oxidation products of the once green chlorophyll content of the leaf. These latter two components are primarily responsible for the brown color of the extract. A structure for tannic acids can be found in Experiment 6, Caffein from Tea.

SPECIAL INSTRUCTIONS

Read the essay which precedes this experiment. In addition, it is necessary to have read Techniques 1 through 5. Old cigars work well for a source of nicotine. They can be further dried by crushing them and allowing them to be exposed to air for at least one week.

CAUTION: Nicotine is extremely toxic. A dose of 60 mg is lethal. Handle it with extreme care.

PROCEDURE

ISOLATION OF NICOTINE DIPICRATE

Place 8.5 g of crushed cigar tobacco and 100 ml of 5% sodium hydroxide solution into a 400 ml beaker. Stir the mixture for 15 minutes. Assemble an apparatus for vacuum filtration (Technique 2, Section 2.3) using a Büchner funnel and a 250 ml filter flask. Do not use filter paper because it will swell and cease to function properly when it comes in contact with strong base. Turn on the aspirator, and filter the mixture. Press the tobacco with the bottom of a beaker or a large clean cork to remove the remaining base. Transfer the tobacco back to the original beaker and add 30 ml of water. Stir the mixture and refilter it with suction (no paper). Discard the tobacco into a waste container (not a sink!). Remove any small particles which passed through the Büchner funnel by passing the basic extract through a thick layer of loosely compacted glass wool in a short stem funnel. Wash the glass wool with a little water and add the washings to the aqueous extract. If solid particles remain in the filtrate (extract), repeat the filtration process using a new layer of glass wool.

Transfer the filtrate to a large (250 ml) separatory funnel and extract the aqueous phase with a 25 ml portion of ether (Technique 5, Section 5.4). When extracting, separate the layers and save them both. Collect and save the lower aqueous phase in a beaker. When the upper, ether layer begins to approach the stopcock, some dark oily material (emulsion) may collect at the apex. Carefully decant the ether from the emulsion through the top of the separatory funnel into a dry beaker. Transfer the saved aqueous layer back into the funnel and add another 25 ml portion of ether. Shake and separate the layers. Again, decant the ether layer away from any remaining emulsion into the beaker containing the first ether extract. Return the aqueous layer to the separatory funnel and again extract it with a third 25 ml portion of ether. Separate the phases and decant the ether layer away from the emulsion into the storage vessel. Discard the aqueous layer. Carefully decant the combined ether extracts away from any remaining emulsion or water into a dry container. Repeat if necessary until no emulsion or water remains.

Remove the ether solvent under vacuum to prevent decomposition of the nicotine; this must be done in portions. Do not evaporate the solvent using heat. Using a 125 ml filter flask, assemble the apparatus shown in Technique 1, Figure 1-7. Add about 25 ml of the clear ether extract to the 125 ml filter flask and place a wooden applicator stick in it, then cork the flask securely. Attach the flask to a trap and then to the aspirator. The trap should be cooled in an ice bath. Start the aspirator and remove the ether under vacuum. Gently warm the filter flask on the steam bath (with constant swirling) to keep the flask from becoming too cold. When the volume is reduced to about 5 ml, add another 25 ml portion of the ether extract from the storage vessel. Remove the old applicator stick and add a new one. Again reduce the volume under vacuum with the aspirator to about 5 ml. Finally, transfer the remaining ether extract to the filter flask, and once again replace the applicator stick with a new one. Remove all of the remaining ether.

A small residue of oil and solid will remain after the ether is removed. Add 1 ml of water and swirl gently to dissolve the residue. Then add 4 ml of methanol and filter the solution into a small beaker through a **small** plug of glass wool placed loosely in the apex of a short stem funnel. Rinse the funnel and glass wool with an additional 5 ml of methanol and combine the two methanol solutions. The solution must be **clear** (no suspended particles) at this point; if not, refiltering is necessary. Add 10 ml of saturated methanolic picric acid. Nicotine dipicrate will immediately appear as a pale yellow precipitate. Without delay, collect the precipitate by vacuum filtration using a Hirsch funnel (use filter paper). Dry the precipitate in the funnel by drawing air through the filter. The melting point of the air-dried nicotine dipicrate is 216 to 217°. If allowed to dry overnight, the melting point is raised to 217 to 220°. Obtain the melting point of your solid. Weigh the dry solid and calculate the percentage yield of actual nicotine from tobacco. The yield of nicotine dipicrate may be as high as 0.42 g. The precipitate which appears upon evaporation of the methanol in the filter flask is mostly excess picric acid and is discarded.

RECRYSTALLIZATION OF NICOTINE DIPICRATE

Transfer the crude nicotine dipicrate to a 50 ml Erlenmeyer flask. Add hot 50% (by volume) ethanol-water until the solid just dissolves. Keep both containers hot during this process. Allow the contents of the flask to cool slowly. Since crystallization is slow, it is best to stopper the flask lightly and to allow it to stand until the next laboratory period. Collect the product by vacuum filtration using a Hirsch funnel and allow it to dry, preferably overnight. Obtain the melting point. The melting point of pure nicotine dipicrate is 222 to 223°. Submit the sample to the instructor in a labeled vial.

QUESTIONS

1. Outline a separation scheme for the isolation of nicotine from tobacco. Use a flow chart similar in format to that shown in the **Advance Preparation and Laboratory Records** section at the beginning of the book.

2. Explain why the pyrrolidine part of the nicotine molecule is more basic than the pyridine part.

3. A structure for tannic acid (pentadigalloyl derivative of glucose) is shown in

Experiment 6. Which one(s) of the hydroxyl groups would you expect to be the more acidic, and why?

4. Indicate how you might isolate gallic acid from tannic acid.

5. Identify the chiral (asymmetric) center in nicotine.

ESSAY
Caffein

The origins of coffee and tea as beverages are so old that they are lost in legend. Coffee is said to have been discovered by an Abyssinian goatherd who noticed an unusual friskiness in his goats when they consumed a certain little plant with red berries. He decided to try the berries himself and discovered coffee. The Arabs soon cultivated the coffee plant and one of the earliest descriptions of its use is found in an Arabian medical book circa 900 A.D. The great systematic botanist, Linnaeus, named the tree **Coffea arabica.**

One legend of the discovery of tea—from the Orient, as one might expect—attributes the discovery to Daruma, the founder of Zen-Buddhism. Legend has it that he inadvertently fell asleep one day during his customary meditations. To be assured that this indiscretion would not reoccur, he cut off both eyelids. Where they fell to the ground, a new plant took root that had the power to keep a person awake. Although some experts assert that the medical use of tea was reported as early as 2737 B.C. in the pharmacopeia of Shen Nung, an emperor of China, the first indisputable reference is from the Chinese dictionary of Kuo P'o which appeared in 350 A.D. The non-medical or popular use of tea appears to have spread slowly. It was not until around 700 A.D. that tea was widely cultivated in China. Since tea is native to upper Indochina and upper India, it must have been cultivated in these places prior to its introduction to China. Linnaeus named the tea shrub **Thea sinensis;** however, tea is more properly a relative of the camellia, and botanists have renamed it **Camellia thea.**

The active ingredient that makes tea and coffee valuable to man is **caffein.** Caffein is an **alkaloid,** a class of naturally occurring compounds containing nitrogen and having the properties of an organic amine base (alkaline, hence, **alkaloid**). Tea and coffee are not the only plant sources of caffein. Others include: kola nuts, maté leaves, guarana seeds, and, in small amount, cocoa beans. The pure alkaloid was first isolated from coffee in 1821 by the French chemist Pierre Jean Robiquet.

Caffein belongs to a family of naturally occurring compounds

XANTHINES
Xanthine R=R′=R″=H
Caffein   R=R′=R″=CH$_3$
Theophylline R=R″=CH$_3$, R′=H
Theobromine R=H, R′=R″=CH$_3$

called **xanthines.** In the form of their plant progenitors, the xanthines are possibly the oldest stimulants known to man. They all, to varying extents, cause stimulation of the central nervous system and the skeletal muscles. This stimulation results in an increased alertness, the ability to put off sleep, and an increased capacity for thinking. Caffein is the most powerful xanthine in this respect. It is the main ingredient of the popular No-Doz keep alert tablets. While caffein has a powerful effect on the central nervous system, not all xanthines are as effective. Thus, theobromine, a xanthine found in cocoa, has fewer central nervous system effects. It is, however, a strong **diuretic** (induces urination) and is useful to doctors in treating patients with severe water retention problems. Theophylline, a second xanthine found along with caffein in tea, also has few central nervous system effects, but it is a strong myocardial (heart muscle) stimulant and it dilates (relaxes) the coronary artery that supplies blood to the heart. Theophylline, also called amino-phylline, is often used in the treatment of congestive heart failure. It is also used to alleviate and to reduce the frequency of attacks of angina pectoris (severe chest pains). In addition, it is a more powerful diuretic than theobromine. Since it is also a vasodilator (relaxes blood vessels) it is often used in the treatment of hypertensive headaches and bronchial asthmas.

One can develop both a tolerance and a dependence on the xanthines, particularly caffein. The dependence is real, and a heavy (>5 cups of coffee per day) user will experience lethargy, headache, and perhaps nausea after about 18 hours abstinence. An excessive intake of caffein may lead to restlessness, irritability, insomnia, and muscular tremor. Caffein can be toxic, but it has been estimated that to achieve a lethal dose of caffein one would have to drink about 100 cups of coffee over a relatively short period of time!

Caffein occurs naturally in coffee, tea, and kola nuts (**Kola nitida**). Theophylline is found as a minor constituent of tea. The major constituent of cocoa is theobromine. The amount of caffein in tea varies from 2 to 5%. In one analysis of black tea, the following compounds were found: caffein, 2.5%; theobromine, 0.17%; theophylline, 0.013%; adeneine, 0.014%; with traces of guanine and xanthine. Coffee beans can contain up to 5% by weight of caffein, and cocoa contains around 5% theobromine. Commercial cola is a beverage based on a kola nut extract. We cannot easily obtain kola nuts in this country, but we can obtain the ubiquitous commercial extract as a syrup. The syrup can be converted into "cola." The syrup contains caffein, tannins, pigments, and sugar. Phosphoric acid is added and caramel is also added to give the syrup a deep color. The final drink is prepared by the addition of water and carbon dioxide under pressure to give the bubbly mixture. The Food and Drug Administration currently requires that a "cola" contain **some** caffein, but limits this amount to a maximum of 5 milligrams per ounce. To achieve a regulated level of caffein, most manufacturers remove all caffein from the kola extract and then re-add the correct amount to the syrup. The caffein content of various beverages is listed in the table which follows.

THE AMOUNT OF CAFFEIN (mg/oz) FOUND IN BEVERAGES*

Brewed Coffee	18–25	Tea	5–15
Instant Coffee	12–16	Cocoa	1
		(but 20 mg/oz	
Decaffeinated	5–10	theobromine)	
Coffee		Coca-Cola	3.5

*The average cup of coffee or tea contains about 5 oz of liquid. The average bottle of cola contains about 12 oz of liquid.

Because of the central nervous system effects that caffein causes, many persons prefer to use **decaffeinated** coffee. The caffein is removed from coffee by extracting the whole beans with trichloroethylene at 71°. Following this, the solvent is drained off, and the beans are steamed to remove any residual solvent. Then, the beans are dried and roasted to bring out the flavor. Decaffeination reduces the caffein content of coffee to the range of 0.03% to 1.2% caffein. The extracted caffein is used in various pharmaceutical products such as APC tablets.

Caffein has always been a controversial compound. Some religions forbid the use of beverages containing caffein, which they consider to be an addictive drug. In fact, many people consider caffein to be an addictive drug. Recently there has been concern because caffein is structurally similar to the purine bases adenine and guanine, which are two of the five major bases used by organisms to form the nucleic acids DNA and RNA. It is feared that the substitution of caffein for adenine or guanine in either of these genetically important substances could lead to chromosome defects.

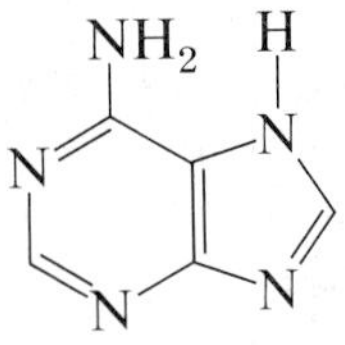

Adenine 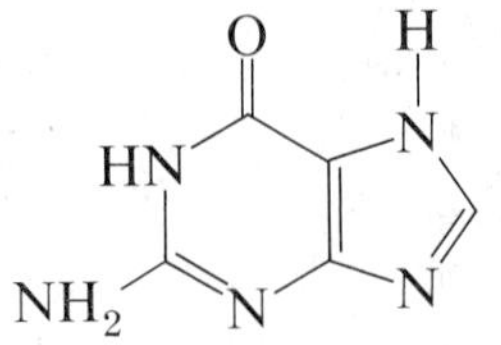   Guanine 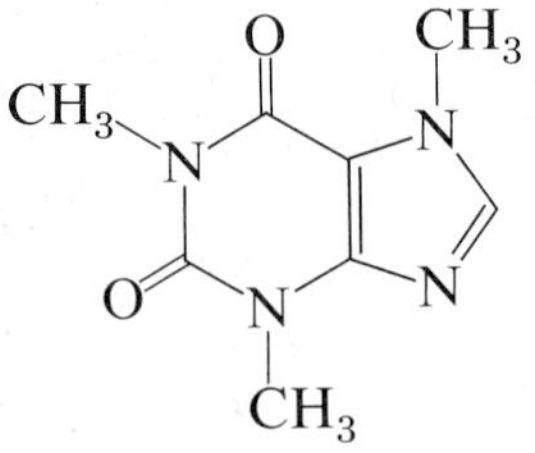   Caffein

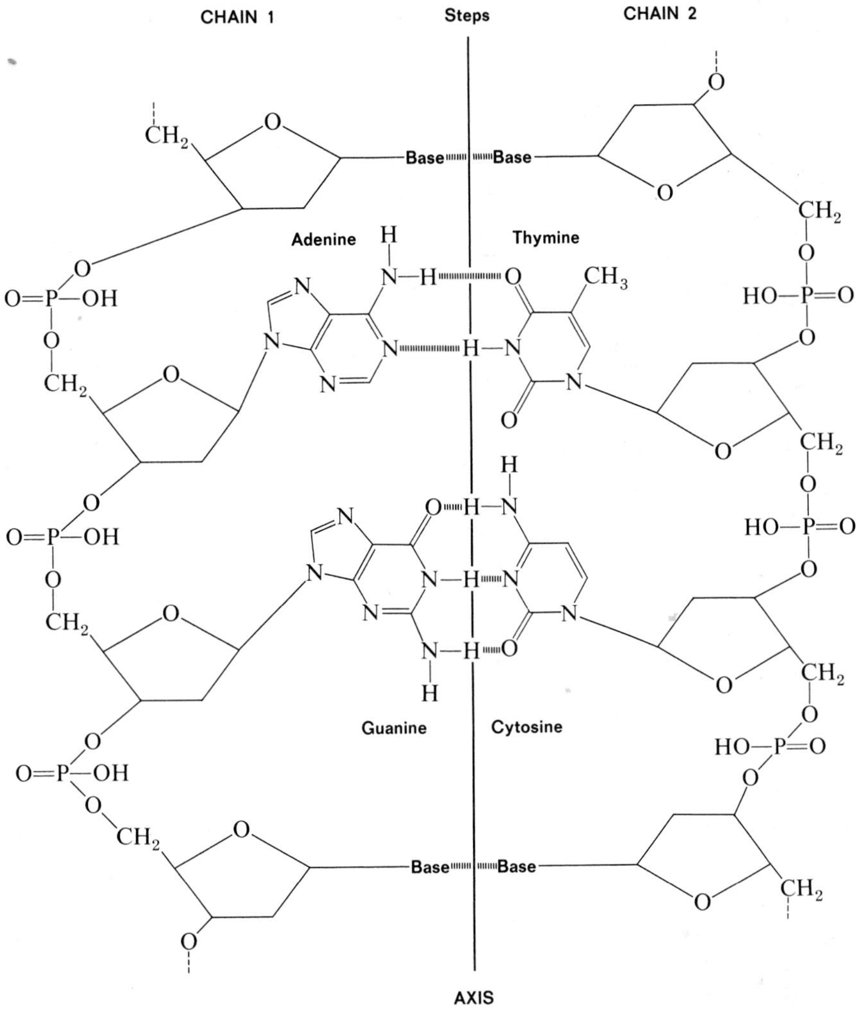

CHAIN 1   Steps   CHAIN 2

Base ⅲⅲ ⅲⅲ Base

CH₂   O   O

Adenine   H   Thymine

O=P—OH   HO—P=O

CH₂   CH₂

O=P—OH   HO—P=O

CH₂   CH₂

Guanine   Cytosine

O=P—OH   HO—P=O

CH₂   O   CH₂

O   O

Base ⅲⅲ ⅲⅲ Base

AXIS

A portion of a DNA molecule

A portion of the structure of a DNA molecule is shown in the figure. The typical mode of incorporation of both adenine and guanine is specifically shown. If caffein were substituted for either of these, the hydrogen bonding necessary to link the two chains together would be disturbed. Although caffein is most similar to guanine, it could not form the central hydrogen bond as it has a methyl group rather than a hydrogen in the necessary position. Hence, the genetic information would be garbled, and there would be a **break** in the chain. Fortunately, there exists little evidence of chromosome breaks due to the use of caffein. Many cultures have been using tea and coffee for centuries without any apparent genetic problems.

REFERENCES

"Caffein May Damage Genes by Inhibiting DNA Polymerase." *Chemical and Engineering News, 45* (November 20, 1967), 19.

Emboden, W. "The Stimulants." *Narcotic Plants.* New York: Macmillan, 1972.

Ray, O. S. "Caffein." *Drugs, Society and Human Behavior.* St. Louis: Mosby, 1972.

Ritchie, J. M. "Central Nervous System Stimulants, II: The Xanthines." L. S. Goodman and A. Gilman, *The Pharmacological Basis of Therapeutics* (4th edition). New York: Macmillan, 1970.

Taylor, N. *Plant Drugs that Changed the World.* New York: Dodd, Mead, and Company, 1965. Pp. 54–56.

Taylor, N. "Three Habit-Forming Nondangerous Beverages." *Narcotics—Nature's Dangerous Gifts.* New York: Dell, 1970. (The paperbound revision of *Flight from Reality.*)

Experiment 6

ISOLATION OF CAFFEIN FROM TEA

Isolation Of A Natural Product
Heating Under Reflux
Filtration, Extraction
Simple Distillation

In this experiment, caffein will be isolated from tea leaves. The major problem of the isolation is that caffein does not occur alone in tea leaves, but is accompanied by other natural substances from which it must be separated. The major component of tea leaves is cellulose, which is the major structural material of all plant cells. Cellulose is a polymer of glucose. Since cellulose is virtually insoluble in water, it presents no problems in the isolation procedure. Caffein, on the other hand, is water soluble and is one of the major substances extracted into the solution called "tea." Caffein comprises as much as 5 percent by weight of the leaf material in tea plants. Tannins also dissolve in the hot water used to extract tea leaves. The term **tannin** does not refer to a single homogeneous compound, or even to substances which have similar chemical structure. It refers to a class of compounds which have certain properties in common. Tannins are phenolic compounds having molecular weights between 500 and 3000. They are widely used to "tan" leather. They precipitate alkaloids and proteins from aqueous solutions. Tannins are usually divided into two classes: those which can be hydrolyzed and those which cannot. Tannins of the first type which are found in tea generally yield glucose and gallic acid when they are hydrolyzed. These tannins are esters of gallic acid and glucose. They represent structures in which some of the hydroxyl groups in glucose have been esterified by digalloyl groups. The non-hydrolyzable tannins found in tea are condensation polymers of catechin. These polymers are not uniform in structure, but catechin molecules are usually linked together at ring positions 4 and 8.

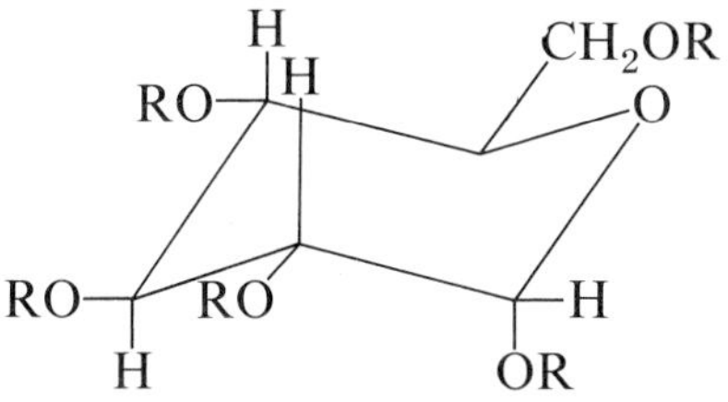

Glucose if R = H
A Tannin if some R = Digalloyl

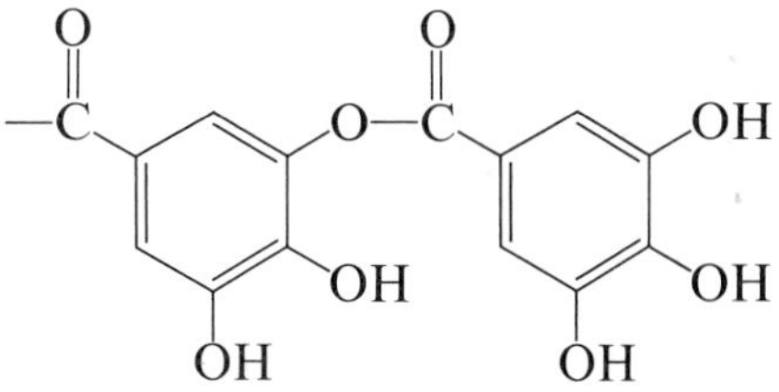

A Digalloyl Group

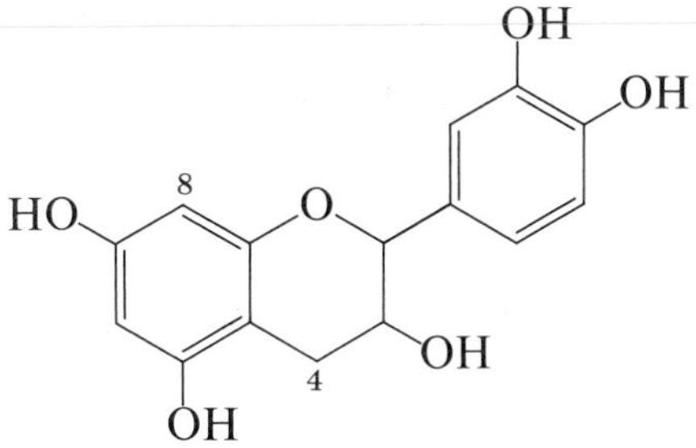

Catechin

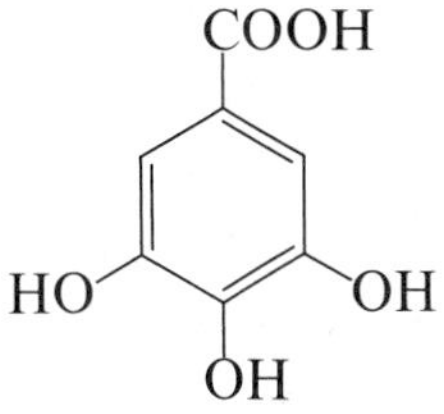

Gallic Acid

When tannins are extracted into hot water, the hydrolyzable ones are partially hydrolyzed, meaning that free gallic acid is also found in tea. The tannins, by virtue of their phenolic groups, and gallic acid by virtue of its carboxyl groups, are both acidic. If calcium carbonate, a base, is added to tea water, the calcium salts of these acids are formed. Caffein can be extracted from the basic tea solution with chloroform, but the calcium salts of gallic acid and the tannins are not chloroform soluble and remain behind in the aqueous solution.

The brown color of a tea solution is due to flavonoid pigments and chlorophylls, as well as their respective oxidation products. Although chlorophylls are somewhat chloroform soluble, most of the other substances in tea are not. Thus, the chloroform extraction of the basic tea solution removes nearly pure caffein. The chloroform is easily removed by distillation (bp 61°) to leave the crude caffein. The caffein may be purified by recrystallization or by sublimation.

In a second part of this experiment, caffein will be converted to a **derivative.** A derivative of a compound is a second compound, of known melting point, formed from the original compound by a simple chemical reaction. In trying to make a positive identification of an organic compound, it is often customary to convert it into a derivative. If the first compound, caffein in this case, and its derivative both have melting points which match those reported in the chemical literature (e.g., a handbook), it is assumed that there is no coincidence and that the identity of the first compound, caffein, has been definitely established.

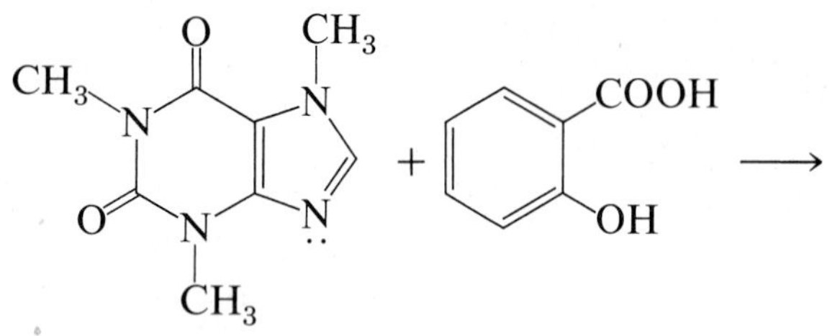

Caffein   Salicylic Acid

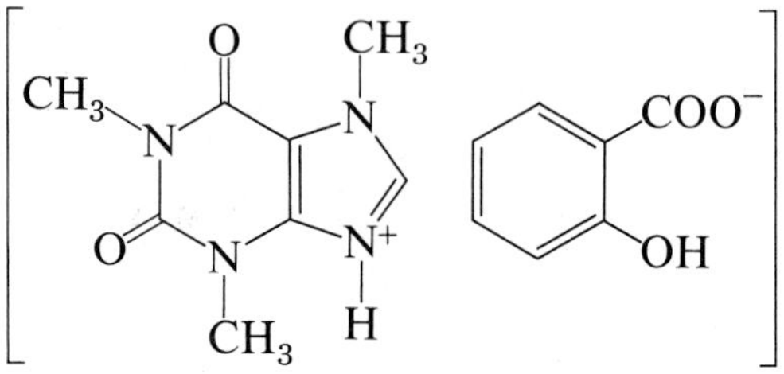

Caffein Salicylate

Caffein is a base and will react with an acid to give a salt. Using salicylic acid, a derivative **salt** of caffein, caffein salicylate, will be made in order to establish the identity of the caffein isolated from tea leaves.

SPECIAL INSTRUCTIONS

As an introduction, you should have read the essay "Caffein" which precedes this experiment. In order to perform this experiment, you should have read Techniques 1 through 5. In addition, Sections 6.3 and 6.4 of Technique 6 are also necessary. Be careful when handling chloroform. It is a toxic solvent, and you should not breathe it excessively or spill it on yourself. When discarding spent tea leaves, do not put them in the sink because they will clog the drain. Dispose of them in a waste container.

PROCEDURE

Place 25 g of dry tea leaves, 25 g of calcium carbonate powder, and 250 ml of water in a 500 ml three neck round bottom flask equipped with a condenser for reflux (Technique 1, Figure 1–4). Stopper the unused openings in the flask and heat the mixture under reflux (Technique 1, Section 1.7) for about 20 minutes. Use a Bunsen burner to heat. While the solution is still hot, filter it by gravity through a fluted filter (Technique 2, Figure 2–3) using a fast filter paper such as E&D No. 617 or S&S No. 595. You may need to change the filter paper if it clogs.

Cool the filtrate (filtered liquid) to room temperature and, using a separatory funnel, extract it twice (Technique 5, Section 5.4) with 25 ml portions of chloroform. Combine the two portions of chloroform in a 100 ml round bottom flask. Assemble an apparatus for simple distillation (Technique 6, Figure 6–6) and remove the chloroform by distillation. Use a steam bath to heat. The residue in the distillation flask contains the caffein and is purified as described below (crystallization). Save the chloroform that was distilled. You will use some of it in the next step. The remainder should be placed in a collection container.

CRYSTALLIZATION (PURIFICATION)

Dissolve the residue obtained from the chloroform extraction of the tea solution in about 10 ml of the chloroform that you saved from the distillation. It may be necessary to heat the mixture on a steam bath. Transfer the solution to a 50 ml beaker. Rinse the flask with an additional 5 ml of chloroform and combine this in the beaker. Evaporate the now light-green solution to dryness by heating it on a steam bath **in the hood.**

The residue obtained on evaporation of the chloroform is next crystallized by the mixed solvent method (Technique 3, Section 3.7). Dissolve it in a small quantity (about 2 to 4 ml) of hot benzene and add just enough high-boiling (60 to 90°) petroleum ether (or ligroin) to turn the solution faintly cloudy. Alternatively, acetone may be used for simple crystallization without a second solvent. Cool the solution and collect the crystalline product by vacuum filtration using a Hirsch funnel. Crystallize the product the same way a second time if necessary, and allow the product to dry by allowing it to stand in the suction funnel for a while. Weigh the product. Calculate the weight percentage yield based on tea and determine the melting point. If desired, the product may be further purified by sublimation as described in the next experiment.

THE DERIVATIVE

Dissolve 0.20 g of caffein and 0.15 g of salicylic acid in 15 ml of benzene in a small beaker by warming the mixture on a steam bath. Add about 5 ml of high boiling (60 to 90°) petroleum ether and allow the mixture to cool and crystallize. It may be necessary to cool the beaker in an ice water bath or to add a small amount of extra petroleum ether to induce crystallization. Collect the crystalline product by vacuum filtration using a Hirsch funnel. Dry the product by allowing it to stand in the air, and determine its melting point. Check the value against that in the literature. Submit the sample to the instructor in a labeled vial.

QUESTIONS

1. Outline a separation scheme for the isolation of caffein from tea. Use a flow chart similar in format to that shown in the **Advance Preparation and Laboratory Records** section at the beginning of the book.

2. Carefully inspect the structure of caffein and caffein salicylate. Why is the particular nitrogen which became protonated with salicylic acid more basic than the other three nitrogens?

3. Inspect the structures given for tannins. Which one(s) of the hydroxyl groups would you expect to be the more acidic, and why?

4. Indicate how you might isolate gallic acid from tannic acid.

5. In the crystallization of crude caffein, petroleum ether (ligroin) is added to a benzene solution of caffein. Why was the petroleum ether added? How does it work?

EXPERIMENT **7**

ISOLATION OF CAFFEIN FROM COFFEE

Isolation Of A Natural Product
Heating Under Reflux
Filtration, Extraction
Crystallization
Sublimation

In this experiment, we shall isolate caffein from coffee. The major problem of the isolation is that caffein does not occur alone in coffee, but it is accompanied by other natural substances from which it must be separated. The green coffee bean (seed) contains caffein (1 to 2%), tannins, glucose, fats, proteins and cellulose. In the roasting process, the coffee beans swell and change their color to a dark brown. The roasting process also develops the characteristic odor and flavor. The aroma is due to an oil called **caffeol.** The volatile oil contains mainly furfural with traces of several hundred other compounds. The roasting process also liberates the caffein from a combination with chlorogenic acid in which it exists in the unroasted bean. Some caffein sublimes from the beans during the roasting process.

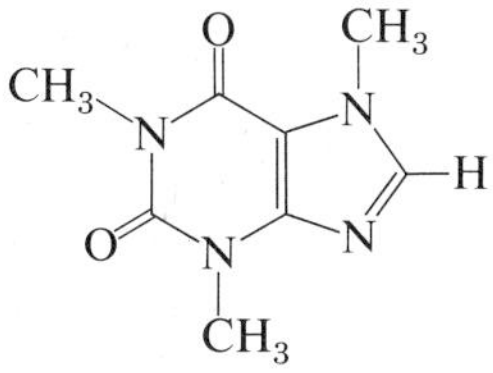

Caffein 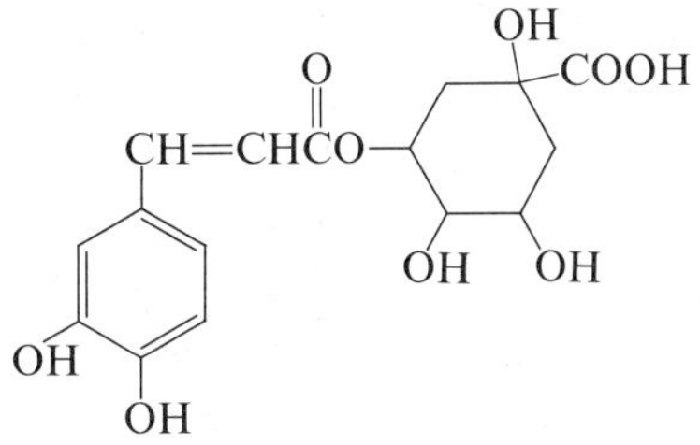   Chlorogenic Acid 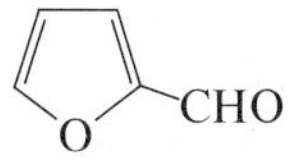   Furfural

The separation depends upon differences in the solubilities of the various constituents in coffee. The polymeric substances, protein and cellulose, are insoluble in water. Proteins are the polymers formed from amino acids, while cellulose is the polymer formed from glucose. Fats, which are esters formed from glycerol and three long chain carboxylic acids, are also insoluble in water. On the other hand, caffein, tannins, glucose, and chlorogenic acid are quite soluble in hot water. Thus, the first crude separation may be accomplished by filtration of the hot extract of ground coffee.

Both the tannins (structures shown in Experiment 6) and chlorogenic acid are acidic. The addition of a lead acetate solution will precipitate these acidic substances as their lead salts and will remove them from the aqueous solution. The lead salts can then be removed by filtration. Caffein and glucose will remain behind in the aqueous solution.

$$2\,RCOOH + Pb(OAc)_2 \longrightarrow Pb^{+2}\,(RCOO^-)_2 + 2\,HOAc$$

Finally, the caffein is removed from the aqueous solution by extraction with chloroform. Glucose is not soluble in chloroform. The chloroform is easily removed by evaporation to yield crude caffein. The caffein is purified by recrystallization followed by sublimation.

SPECIAL INSTRUCTIONS

Read the essay which precedes Experiment 6. In order to perform this experiment, you should have read Techniques 1 through 5, and 13. Be careful when handling chloroform. It is a toxic solvent, and you should not breathe it or spill it on yourself. When discarding spent coffee grounds, do not put them in a sink because they will clog the drain. Dispose of them in a waste container.

PROCEDURE

ISOLATION OF CAFFEIN

Place 35 g of ground coffee (regular grind), a boiling stone, and 125 ml of water into a 500 ml three-neck round bottom flask equipped with a condenser for reflux (Technique 1, Figure 1–4). Stopper the unused openings in the flask and heat the mixture under reflux (Technique 1, Section 1.7) for about 20 minutes. Use a Bunsen burner to heat. During the heating period, assemble a vacuum filtration apparatus (Technique 2, Figure 2–4). When boiling action has stopped and the coffee grounds have settled somewhat, but while the solution is still hot, filter the solution through a Büchner funnel by vacuum filtration (Technique 2, Section 2.3). Use a fast filter paper such as E&D No. 617 or S&S No. 595.

Transfer the filtrate (filtered liquid) to a 250 ml Erlenmeyer flask or beaker and add 20 to 25 ml of a 10% lead acetate solution. Using a Bunsen burner, bring the solution to a boil and then place it on a hot steam bath to keep it warm for about 10 minutes. During this period, swirl or stir the solution frequently to coagulate the precipitate. Filter the hot solution through a Büchner funnel by vacuum filtration. Allow the hot filtrate to cool to room temperature and transfer it to a separatory funnel. Extract the solution with a 25 ml portion of chloroform (Technique 5, Section 5.4). In performing the extraction, if the separatory funnel is shaken too vigorously, an emulsion may form. To avoid this problem, **gently** swirl the mixture to mix the layers and **carefully** invert the funnel several times. Do this for a period of about 5 minutes.

Allow the layers to separate and drain the lower layer, which is chloroform. Save this chloroform layer. Add another 25 ml portion of chloroform to the remaining aqueous layer. Again **gently** swirl and invert the mixture for 5 minutes. Allow the layers to separate and drain the lower chloroform layer. Combine the two chloroform extracts. Pour the combined extracts back into the separatory funnel and wash (extract) them, first with 10 ml of 5 or 10% sodium hydroxide solution, and then with 10 ml of water.

Pour the washed chloroform layer into a dry 125 ml Erlenmeyer flask and add about 1 g of anhydrous sodium sulfate to dry it (Technique 5, Section 5.6). Swirl the mixture several times over a period of about 10 minutes, or until the solution is dry. When the solution is dry it will be clear. Remove the drying agent by gravity filtration of the solution through a fluted filter into a tared (pre-weighed) 100 ml beaker (Technique 2, Figure 2–3). Rinse the sodium sulfate crystals with a small (about 10 ml) portion of chloroform and add this chloroform to the filter. Add a boiling stone and evaporate the combined filtrates to dryness on a steam bath **in the hood.** The light tan residue that results is crude caffein. Dry the outside of the beaker and remove the boiling stone. Weigh the beaker and calculate the weight of crude product and the weight percent of caffein which was recovered from the coffee. Determine the melting point of this crude material. Since the product sublimes, you may have to seal both ends of the melting point capillary tube in order to determine the melting point successfully. Purify the product by crystallization, followed by sublimation.

CRYSTALLIZATION OF CAFFEIN

The crude caffein is crystallized by the mixed solvent method (Technique 3, Section 3.7). Dissolve the solid in a minimum amount of hot benzene and add just enough high-boiling (60 to 90°) petroleum ether or ligroin to turn the solution faintly cloudy. Cool the solution and collect the crystals by vacuum filtration using a Hirsch funnel. Allow the product to stand in the suction funnel until dry. Determine the melting point.

SUBLIMATION OF CAFFEIN

The final purification of caffein is accomplished by sublimation (Technique 13). Assemble a sublimation apparatus such as that shown in Technique 13, Figure 13–2a or 13–2b. Place all or part of your caffein in the bottom of the outer tube and reassemble the sublimation apparatus so that the vertical gap between the two tubes is no more than 2 cm. Using an aspirator vacuum to evacuate the apparatus, heat the sample gently and carefully with a Bunsen burner to sublime the caffein. Hold the burner in your hand (hold it at its base, **not** by the hot barrel) and apply heat by moving the flame back and forth under the outer tube. If the sample begins to melt, remove the flame for a few seconds before resuming the heating operation. When the sublimation is complete, remove the burner and allow the apparatus to cool.

When the apparatus has cooled, stop the aspirator and **carefully** remove the inner tube of the sublimation apparatus. If this operation is performed carelessly, the sublimed crystals may be dislodged from the condenser tube back into the residue. Place the condenser tube (cold finger) with the sublimed material on a weighing paper and scrape the sublimed caffein onto the weighing paper using a small spatula. Determine the melting point of this purified caffein and compare both its melting point and color to that of the caffein obtained at each stage above. The melting point of pure caffein is 236°. Submit the sample to the instructor in a labeled vial.

QUESTIONS

1. Outline a separation scheme for the isolation of caffein from coffee. Use the flow chart format shown in the **Advance Preparation and Laboratory Records** section at the beginning of the book.

2. In the essay preceding Experiment 6, the decaffeination method is explained. Explain why the decaffeination procedure is conducted **before** roasting and not after. In answering this question, consider the differences between the chemical content of the green bean and the roasted bean. Would something desirable be extracted?

3. A structure of tannic acid (pentadigalloyl derivative of glucose) is shown in Experiment 6. Which one(s) of the hydroxyl groups would you expect to be the more acidic and why?

4. Indicate how you might isolate gallic acid from tannic acid.

5. Experimentally, how might the isolation procedure be modified if instant coffee were used rather than regular grind coffee?

ESSAY
Steroids

Steroids are an important class of natural products whose structures are related in that they all have the same basic four-ring carbon skeleton of perhydrocyclopentanophenanthrene. Steroids occur widely in both

THE BASIC STEROID RING SKELETON

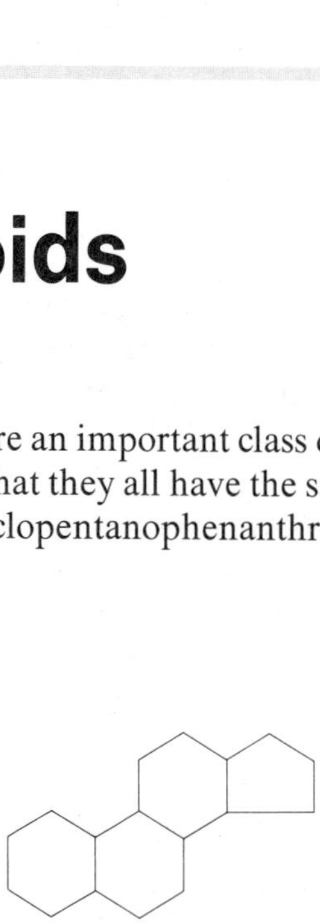

Perhydrocyclopentanophenanthrene

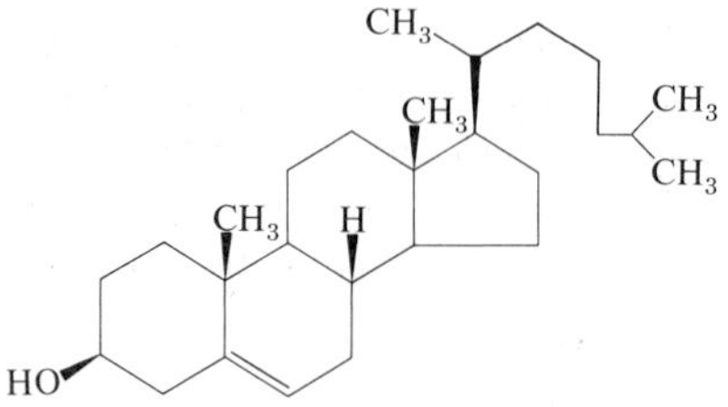

Cholesterol

plants and animals, however, the most important steroids are found in animals where they perform various essential biological functions. The male and female sex hormones in mammals are steroids, as are the bile acids and the adrenal hormones. Each of these types of steroids is illustrated and discussed below.

The most abundant steroid is cholesterol. Cholesterol occurs in all tissues of the mammalian body. It is particularly abundant in the spinal chord, the brain, and, when they occur, in gallstones. The total cholesterol content of a 165 pound man averages about 250 grams. Because of its wide availability, cholesterol was the first steroid to be isolated and identified. Because of its complicated structure, however, more than 160 years intervened between its discovery in 1770 and the realization of its structure in 1932. Since cholesterol has eight asymmetric centers, and the possibility of 2^8 or 256 possible stereoisomers, it required another 23 years (1955) before its total three dimensional structure was fully elucidated.

Mammals have the ability to absorb cholesterol from dietary sources, but early studies of cholesterol showed that an animal will often excrete more cholesterol than was consumed in its diet. Therefore, animals must be able to synthesize cholesterol. It actually turns out that all tissues in the body have, to some degree, the ability to synthesize cholesterol.

Biochemical studies have shown that cholesterol is important because it is one of the first intermediates which the body synthesizes along the routes to the production of all the other steroid hormones. Cholesterol, for example, is the precursor of all the **bile acids.** Cholic acid and deoxycholic acid are the most important bile acids, but there are many different, though related, compounds found among the various species of animals. The bile acids are synthesized from cholesterol in the liver and converted to conjugates with simple peptides and amino acids. Glycocholic acid and deoxyglycocholic acid, formed by the conjugation (amide formation) of the above-mentioned bile acids with glycine, are representative of these conjugates. The bile acid conjugates are secreted as "liver bile" into the intestines in the form of their sodium salts. They are natural detergent-like substances, and their main purpose is to emulsify (carry into solution) fats and oils so that they may be more easily digested and transported across the intestinal cell wall into the blood stream. Because of this, cholesterol (the precursor of the bile acids) is inextricably involved in the metabolism of fats and oils.

Cholic Acid   R = OH
Deoxycholic Acid   R = H

Glycocholic Acid   R = OH
Deoxyglycocholic Acid   R = H

Two diseases are commonly associated with an improper balance between cholesterol and the bile acids, on the one hand, and fats and oils in the diet, on the other. Gallstones represent a condition where the liver is induced to secrete a bile with a high composition of cholesterol into the gall bladder. In the gall bladder, cholesterol crystallizes from the bile producing large crystalline aggregates of cholesterol.

The second disease results from an elevated level of cholesterol in the blood. Many persons have difficulty in maintaining a proper cholesterol balance within their systems. Excesses of cholesterol in the blood lead to **atherosclerosis,** a hardening of the arteries due to cholesterol deposits. Atherosclerosis (also called arteriosclerosis) is a dangerous condition. The loss of elasticity and the narrowing of the channel in the arteries leads to a strain on the heart and a likelihood of heart disease. Should some of the cholesterol deposits break loose, they could either block the flow of blood to the heart, causing a heart attack, or block the flow of blood to the brain, causing a stroke. For these reasons, the determination of the level of cholesterol in the blood is a relatively routine medical test.

The cholesterol level in the blood is often related to diet and physiology. Fat persons have generally higher serum cholesterol levels than lean ones. Populations whose diets include a high proportion of animal fats (saturated fats) or foods containing cholesterol tend also to have higher levels. The role of cholesterol itself and of diet have both been studied with regard to atherosclerosis, but the role of neither is clearly defined. It is clear that those persons having the disease have an excess of cholesterol and do not metabolize it correctly. The cholesterol, however, probably did not **cause** the condition, but it is more likely a **result** of an abnormality in lipid (fat) metabolism. Persons who have this abnormality must avoid both cholesterol and saturated fats in their diets.

A recent finding has shown that vitamin C will reduce the level of cholesterol in the blood and that it is also a required cofactor for the production of bile acids from cholesterol. Persons deficient in vitamin C usually have high cholesterol levels. This may result from the inability to produce bile acids, which would build up cholesterol, and would lead to an incorrect metabolism of fats.

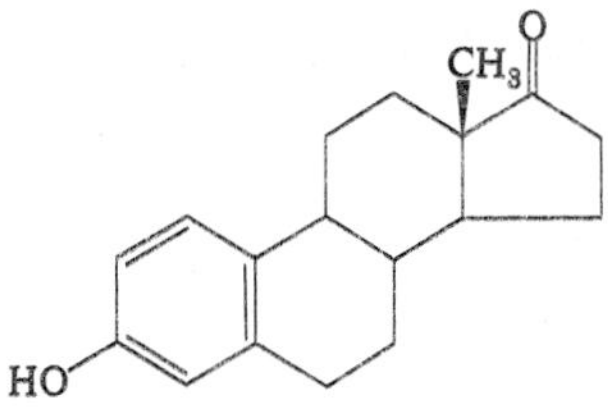

Estrone 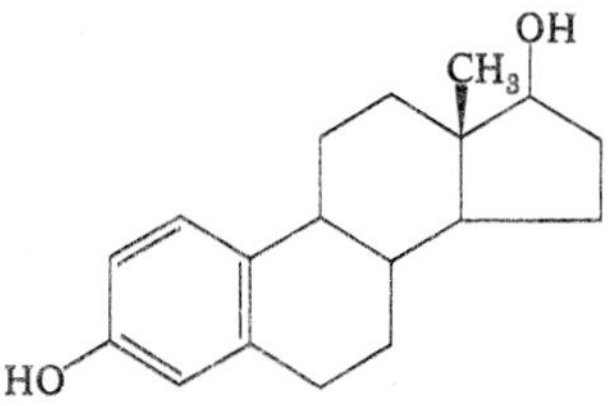   Estradiol 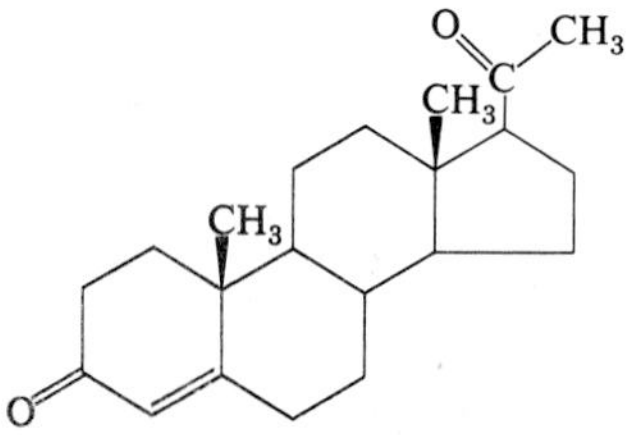   Progesterone

A second class of important steroids which the body produces from cholesterol includes the male and female sex hormones. The principal female hormones (estrogens) are estrone and estradiol, along with a third compound, progesterone. These compounds are responsible for the secondary female sex characteristics and the maintenance of the female menstrual cycle. Progesterone, the pregnancy hormone, suppresses ovulation in the female. Synthetic compounds of similar structure are now used as birth control agents (e.g., Norlutin and Norethynodrel).

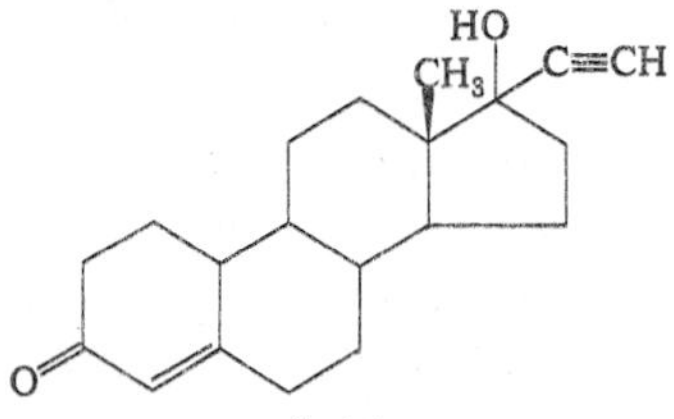

Norlutin 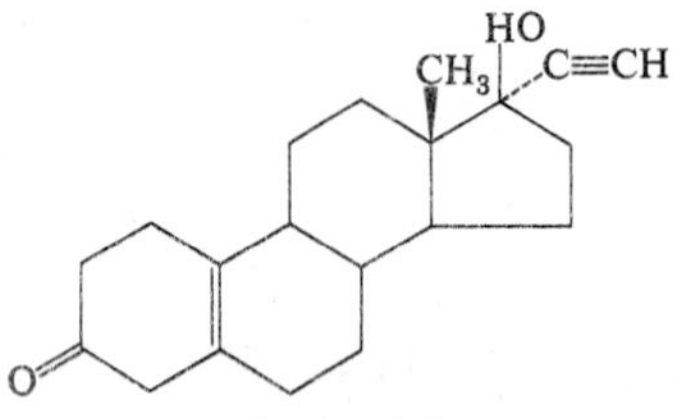   Norethynodrel

These two compounds mimic progesterone in their action and are more effective orally, since they are better absorbed from the stomach and intestines than is progesterone itself. The two principal male hormones are testosterone and androsterone.

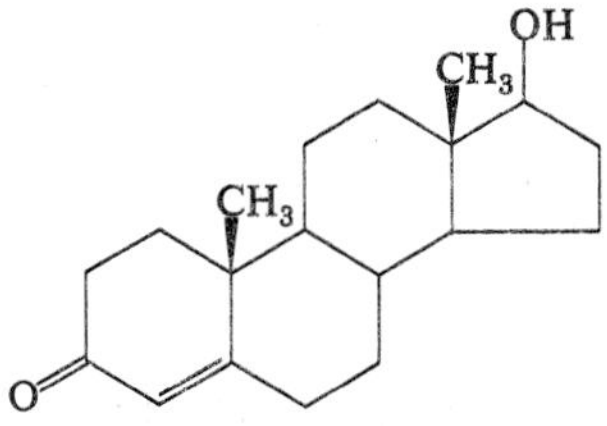

Testosterone

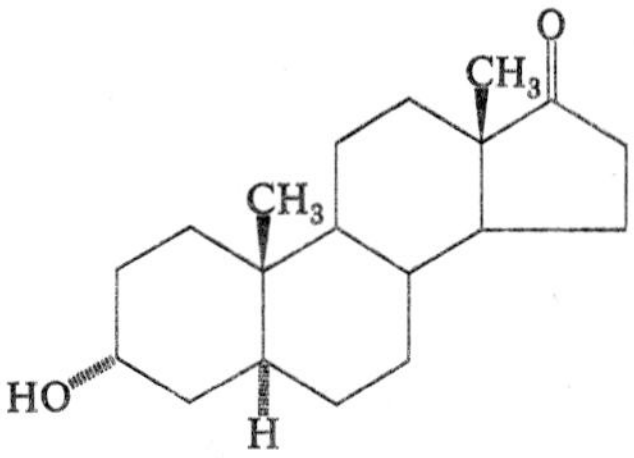

Androsterone

A third class of steroids biochemically synthesized from cholesterol is that of the adrenal hormones. Two representative members of this class are corticosterone and cortisone. The adrenal hormones govern a wide variety of metabolic processes which are too extensive to detail here. Cortisone is medically used for the treatment of rheumatoid arthritis. It reduces swelling in inflamed joints.

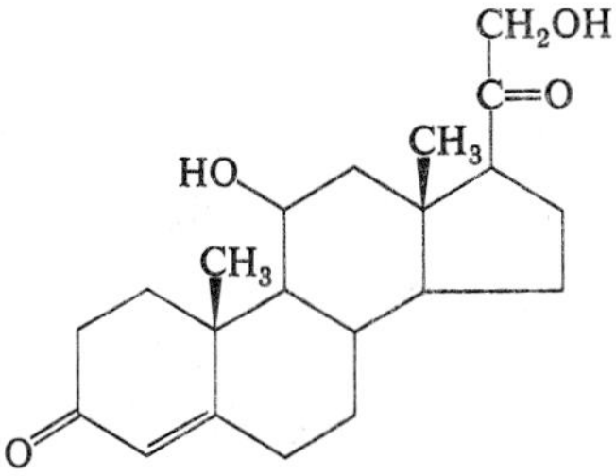

Corticosterone

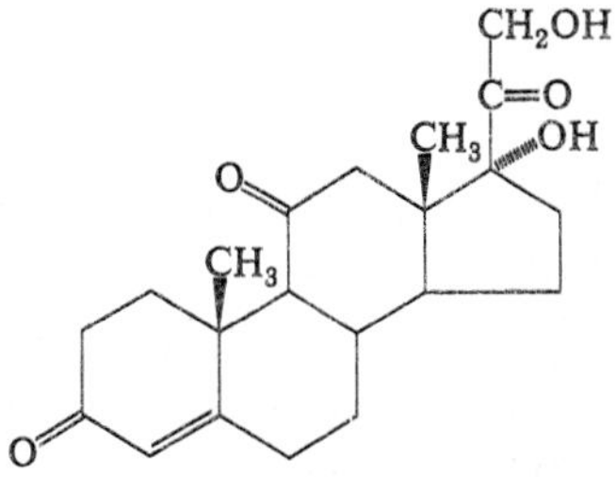

Cortisone

A variety of other important steroids are found in the plant world where they serve many purposes.

REFERENCES

"Biochemistry of the Pill Largely Unknown." *Chemical and Engineering News, 45* (March 27, 1967), 44.

Bloch, K. "The Biological Synthesis of Cholesterol." *Science, 150,* (1965), 19.

Csapo, A. "Progesterone." *Scientific American, 198,* (April, 1958), 48.

Fieser, L. F. "Steroids." *Scientific American, 192,* (January, 1955), 52.

Ginter, E. "Cholesterol: Vitamin C Controls its Transformation to Bile Acids." *Science, 179* (1973), 704.

Petrow, V. "Current Aspects of Fertility Control." *Chemistry in Britain, 6* (1970), 167.

Pincus, G. "Control of Conception by Hormonal Steroids." *Science, 153,* (1966), 493.

Zuckerman, S. "Hormones," *Scientific American, 196,* (March 1957), 76.

Experiment 8

CHOLESTEROL FROM GALLSTONES

Isolation Of A Natural Product
Bromination–Debromination
Filtration, Extraction

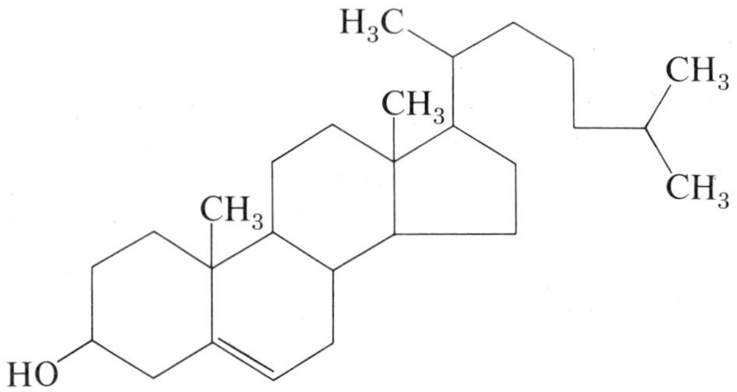

Cholesterol

In this experiment we shall isolate and purify cholesterol from human gallstones. The gallstones are crushed in a mortar and pestle, and the cholesterol is extracted with hot dioxane. Recrystallization is, unfortunately, not sufficient to purify cholesterol. This is because three other steroids, albeit in small amounts (0.1 to 3%), are found along with the cholesterol. These compounds are so similar to cholesterol in their solubility characteristics that they are not separable by crystallization. The structure of one of the impurities, cholestanol, is shown below. The other impurities have exactly the same structure as cholestanol, except that they contain double bonds. 7-Dehydrocholesterol has double bonds between carbons 5 and 6 and between carbons 7 and 8, while Δ^7-cholesten-3β-ol has one double bond between carbons 7 and 8.

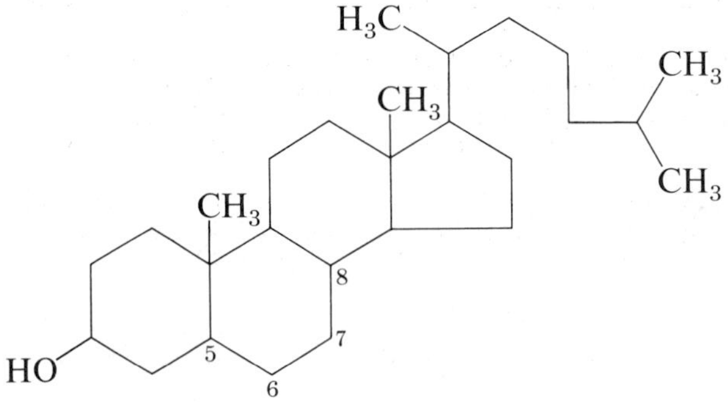

Cholestanol

The separation is accomplished chemically. Cholesterol is converted to cholesterol dibromide by the addition of bromine across the double bond. The impurities have either no double bond (cholestanol) or less active ones, probably for steric reasons. If 7-dehydrocholesterol or Δ^7-cholesten-3β-ol should react with bromine, they would be dehydrobrominated to dienes and trienes which would stay, along with cholestanol, in the mother liquors when cholesterol dibromide was recrystallized. The dibromide has solubility properties different from any of the impurities.

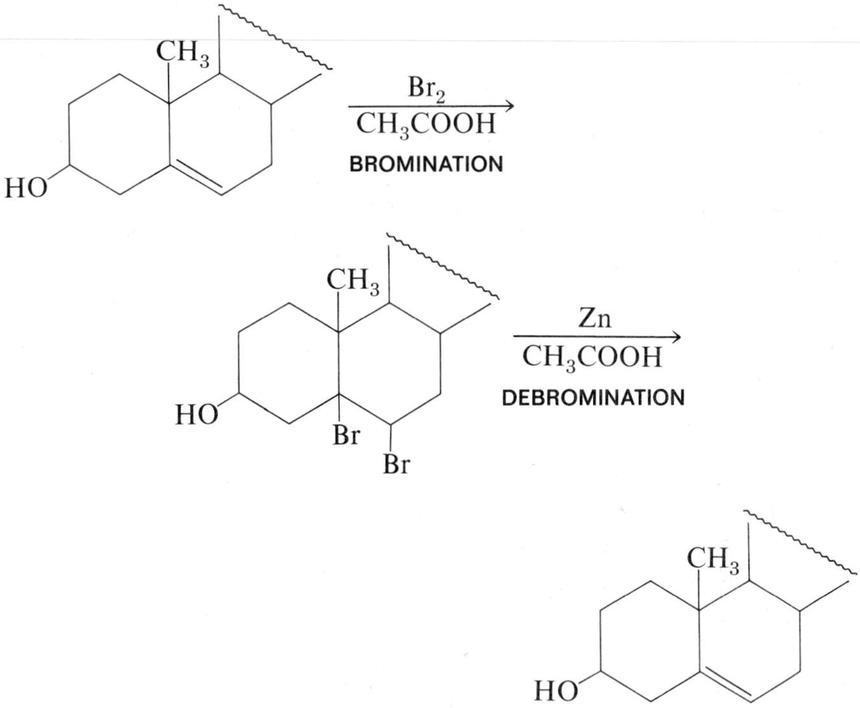

The cholesterol dibromide, now pure, can be converted back to cholesterol by debromination with zinc metal in acetic acid. Recrystallization yields pure cholesterol.

SPECIAL INSTRUCTIONS

Before beginning this experiment you should read the essay which preceeds this experiment and Techniques 1, 2, 3, 4, and 5. The brominating solution can cause unpleasant burns if it comes in contact with skin, so care should be exercised when handling it.

PROCEDURE

ISOLATION

Weigh 2 g of crushed gallstones and place them in a 25 ml Erlenmeyer flask. Add 10 ml of dioxane and heat the mixture, with swirling, on a steam bath until all the solid has disintegrated, and the cholesterol has dissolved (Technique 1, Section 1.7 and Figure 1–5). Filter the brown-yellow solution through a fluted filter while it is still hot (Technique 2, Section 2.1). It may be advisable to preheat the funnel. The brown residue which collects is a metabolic oxidation product of hemoglobin—a bile pigment called **bilirubin.** Dilute the filtrate with 10 ml of methanol, add a little Norite to decolorize the solution, and heat the mixture on the steam bath. Preheat a funnel and filter the hot solution through a fluted filter paper. Reheat the greenish-yellow filtrate and add just enough water (dropwise) to make the solution cloudy (Technique 3, Section 3.7). The solution is now saturated at the boiling point, and cholesterol will crystallize on cooling. Collect the crystals by vacuum filtration using a small Büchner funnel (Technique 2, Section 2.3). Wash them once with cold methanol and let them stand for a time in the open Büchner funnel to allow all the solvent to dry. Weigh the product and determine the melting point. Approximately 1.5 g of impure cholesterol (mp ca. 146 to 148°C) should have been collected.

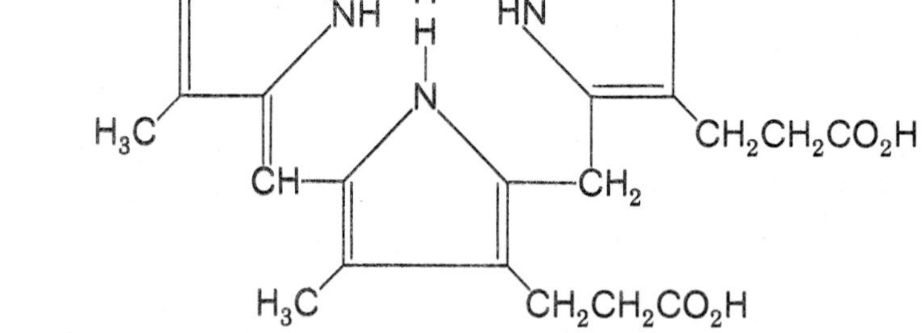

Bilirubin

BROMINATION

Dissolve 1.0 g of the crude cholesterol obtained from the gallstones in 10 ml of **anhydrous** ether in a 25 ml Erlenmeyer flask. Gentle warming on the steam bath will probably be necessary. Using an eyedropper, slowly add 5 ml of a solution of bromine and sodium acetate in glacial acetic acid (CAUTION: This solution causes bad burns if it comes in contact with the skin). The solution should be added until the yellowish color persists. The cholesterol dibromide should begin to crystallize in a minute or so. Cool the flask in an ice bath and stir the solution to ensure complete crystallization. Also cool a mixture of 3 ml ether and 7 ml glacial acetic acid in a separate flask in the ice bath. Collect the crystals by vacuum filtration using a small Büchner or Hirsch funnel and wash them with the cold ether–acetic acid solution. Wash the crystals a second time with 5 to 10 ml cold methanol and leave them in the funnel long enough for the passage of air to dry them.

DEBROMINATION

Transfer the dibromide crystals to a 50 ml Erlenmeyer flask and add 20 ml ether, 5 ml glacial acetic acid, and 0.2 g zinc dust. If the reaction is slow, add a little more zinc dust. Zinc deteriorates on exposure to air, so it is best to use a freshly opened bottle. Swirl the mixture. Gentle heating may be required. The dibromide should dissolve, and zinc acetate should separate as a white paste (5 to 10 min). Stir for an extra 5 min and then add enough water (dropwise) to dissolve any solid and give a clear solution. Decant the solution from the unreacted zinc into a separatory funnel (stopcock closed!) and extract the ether solution twice with water to remove the zinc acetate and zinc dibromide (Technique 5, Section 5.4). Then wash the ether solution with 10% sodium hydroxide to remove acetic acid. Repeat until a drop of the ether solution no longer turns blue litmus red. Shake the solution with an equal volume of a saturated aqueous NaCl solution to reduce its water content, separate the ether layer and filter it through a filter paper containing anhydrous sodium sulfate to complete the drying.

RECRYSTALLIZATION

Add 10 ml of methanol to the dry ether solution and evaporate the solution on a steam bath until most of the ether is removed and the purified cholesterol begins to crystallize. Let the solution stand at room temperature for a while, then cool it in an ice bath to complete the crystallization. Collect the crystals by vacuum filtration, wash them with a little cold methanol, and leave them to dry. Weigh the product and determine the melting point. The melting point should be determined relatively quickly. Otherwise air oxidation may produce impurities which lower the melting point (Pure cholesterol: mp 149 to 150°C). Calculate the yield from gallstones, and submit the cholesterol to the instructor in a labeled vial.

QUESTIONS

1. Outline a separation scheme for the isolation of cholesterol from gallstones. Use a flow chart similar in format to that shown in the **Advance Preparation and Laboratory Records** section at the beginning of the book.

2. Bromine adds to the double bond in cholesterol with **trans** stereochemistry. Draw a three-dimensional representation of the mechanism of this process, showing the two rings involved in the reaction in their chair conformations.

3. Write a balanced equation for the formation of cholesterol from cholesterol dibromide and zinc metal. How is zinc acetate formed in the reaction?

4. Write a mechanism for the debromination of a dibromide by zinc metal. What stereochemistry is required for this reaction? Why?

ESSAY
The Chemistry of Vision

An interesting and challenging topic for investigation by chemists is the determination of how the eye functions. What chemistry is involved in the process of detection of light and transmission of that information to the brain?

The first definitive studies on the functioning of the eye were begun in 1877 by Franz Boll. Boll demonstrated that the red color of the retina of a frog's eye could be bleached yellow by strong light. If the frog was then kept in the dark, the red color of the retina slowly returned. Boll recognized that a bleachable substance had to be somehow connected with the ability of the frog to perceive light.

Most of what is presently known about the chemistry of vision is the result of the elegant work of George Wald, of Harvard University, whose studies, which began in 1933, ultimately resulted in his being awarded a Nobel Prize in biology. Wald identified the means by which light is converted into some form of chemical information which can be transmitted to the brain. What follows is a brief outline of that process.

The retina of the eye is made up of two types of photoreceptor cells, **rods** and **cones.** The rods are responsible for vision in dim light, while the cones are responsible for color vision in bright light. The same principles apply to the chemical functioning of the rods and cones, but the details of that functioning are less well understood for the cones than for the rods.

Each rod contains several million molecules of **rhodopsin.** Rhodopsin is a complex of a protein, **opsin,** and a molecule derived from vitamin A, 11-**cis**-retinal (sometimes called **retinene**). Very little is known about the structure of opsin. The structure of 11-**cis**-retinal is shown.

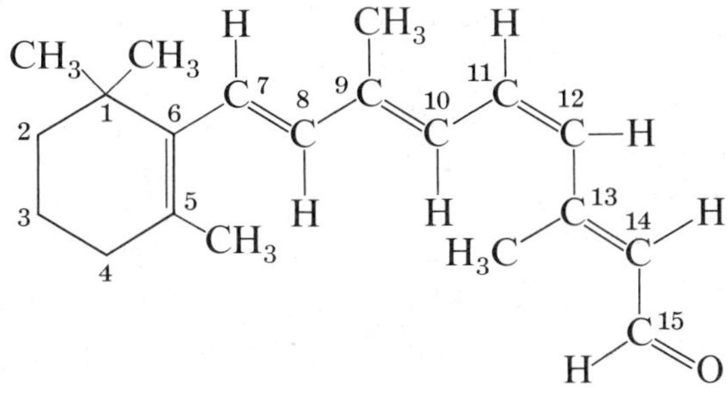

11-*cis*-**retinal**

The detection of light involves the initial conversion of 11-**cis**-retinal to its all-**trans**-isomer. This is the only role played by light in this process. The high energy of a quantum of visible light promotes the fission of the π-bond between carbons 11 and 12. When the π-bond breaks, free rotation about the σ-bond in the resulting radical is possible. When the π-bond reforms after such rotation, all-**trans**-retinal results. All-**trans**-retinal is more stable than is 11-**cis**-retinal, which is why the isomerization proceeds spontaneously in the direction shown.

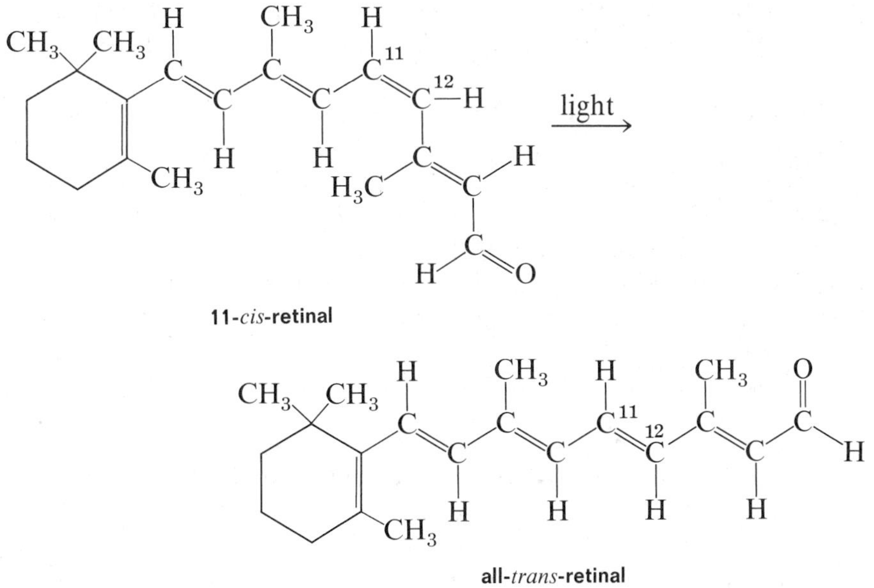

11-*cis*-**retinal**

all-*trans*-**retinal**

The two molecules possess different shapes due to their different structures. 11-**cis**-Retinal has a fairly curved shape, and the parts of the molecule on either side of the **cis**-double bond tend to lie in different planes. Because proteins have very complex and specific three-dimensional shapes (tertiary structures), 11-**cis**-retinal will associate with the protein opsin in a particular manner. All-**trans**-retinal has an elongated shape, and the entire molecule tends to lie in a single plane. This rather

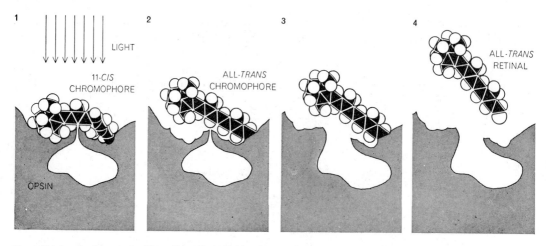

different shape from the 11-**cis** isomer means that all-**trans**-retinal will have a quite different association with the protein opsin.

In fact, all-**trans**-retinal associates very weakly with opsin because its shape does not fit the protein. Consequently, the next step after the isomerization of retinal is the dissociation of all-**trans**-retinal from opsin. The protein undergoes a change of conformation during this dissociation, and this change in conformation triggers some sort of message which is transmitted by the optic nerve to the brain, where the detection of a quantum of light is registered. The means by which the conformation change is translated into nerve impulses is not understood at this time. The figure illustrates this process in a more easily visualized manner.

All-**trans**-retinal is enzymatically converted back to 11-**cis**-retinal. The regeneration of 11-**cis**-retinal involves the formation of all-**trans**-vitamin A, which isomerizes to the 11-**cis** isomer. The vitamin A is oxidized to 11-**cis**-retinal, which recombines with opsin to form rhodopsin. This process is illustrated below.

$$\nearrow \text{RHODOPSIN} \searrow$$

11-*cis*-retinal + opsin ⇌ all-*trans*-retinal + opsin

⇅ ⇅

11-*cis*-Vitamin A + opsin ⇌ all-*trans*-Vitamin A + opsin

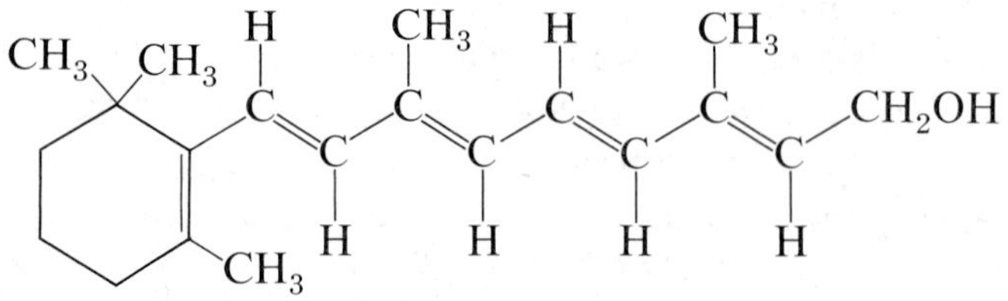

All-*trans*-Vitamin A

By means of this process, as little light as 10^{-14} of the number of photons emitted from a typical flashlight bulb can be detected. The conversion of light into isomerized retinal exhibits an extraordinarily high quantum efficiency. Virtually every quantum of light absorbed by a molecule of rhodopsin causes the isomerization of 11-**cis**-retinal to all-**trans**-retinal.

As may be observed in the above reaction scheme, the retinal derives from vitamin A, which merely requires the oxidation of a —CH_2OH group to a —CHO group to be converted to retinal. The precursor in the diet which is transformed into vitamin A is β-carotene. β-carotene is the yellow pigment of carrots, and is an example of a family of long-chain polyenes called **carotenoids.**

In 1907 Wiltstätter established the structure of carotene, but it was not known until 1931–1933 that there were actually three isomers of carotene. α-Carotene differs from β-carotene in that the α-isomer has a double bond between C_4 and C_5 rather than between C_5 and C_6, as in the β-isomer. The γ-isomer has only one ring, identical to the ring in the β-isomer, while the other ring is opened in the γ-form between $C_{1'}$ and $C_{6'}$. The β-isomer is by far the most common of the three.

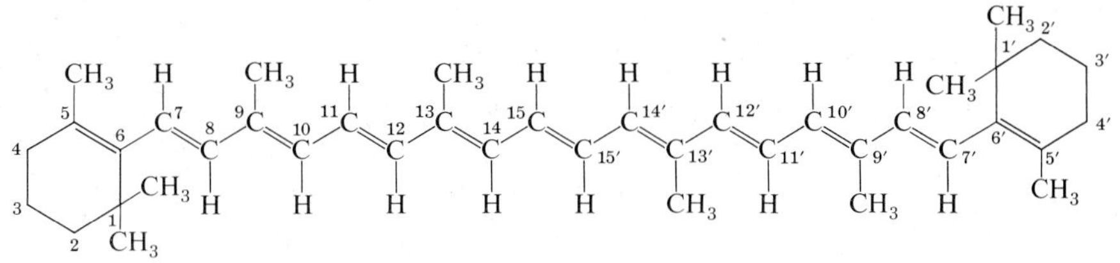

β-**Carotene**

β-Carotene is converted to vitamin A in the liver. Theoretically, one molecule of β-carotene should give rise to two molecules of the vitamin by cleavage of the C_{15}-$C_{15'}$ double bond, but actually only one molecule of Vitamin A is produced from each molecule of the carotene. The vitamin A thus produced is converted to 11-**cis**-retinal within the eye.

An interesting problem remaining to be solved is the manner in which the change in conformation of opsin results in a message to the brain. Some knowledge of the complete structure of opsin would be required. The nature of the color vision characteristic of the cones of the retina is also an interesting research problem for the future. The study of other materials which can absorb light with the same high quantum efficiency characteristic of 11-**cis**-retinal can have useful applications in such areas as photography.

REFERENCES

Clayton, R. K. *Light and Living Matter. Volume 2: The Biological Part.* New York: McGraw-Hill, 1971.
Hubbard, R., and Kropf, A. "Molecular Isomers in Vision." *Scientific American, 216* (June, 1967), 64.

Hubbard, R., and Wald, G. "Pauling and Carotenoid Stereochemistry." A. Rich and N. Davidson (eds.), *Structural Chemistry and Molecular Biology.* San Francisco: Freeman, 1968.

MacNichol, E. F., Jr. "Three Pigment Color Vision." *Scientific American, 211* (December, 1964), 48.

Rushton, W. A. H. "Visual Pigments in Man." *Scientific American, 207* (November, 1962), 120.

Wald, G. "Life and Light." *Scientific American, 201* (October, 1959), 92.

Wolken, J. J. *Photobiology.* New York: Reinhold Book Co., 1968.

Experiment **9**

LYCOPENE AND β-CAROTENE

Isolation Of A Natural Product
Column Chromatography
Thin Layer Chromatography

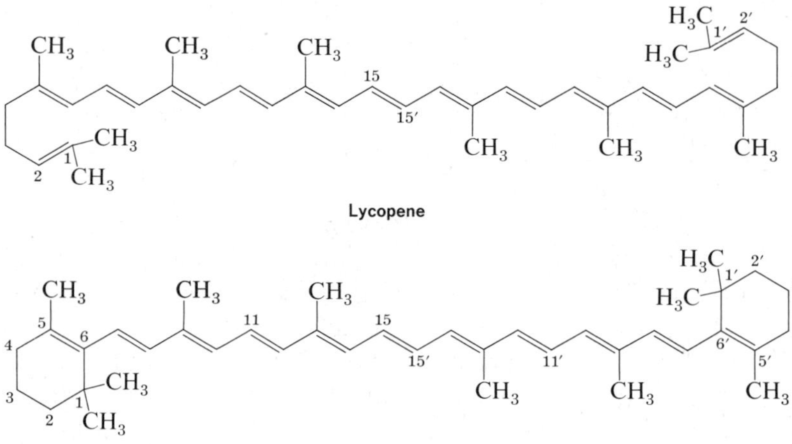

Lycopene

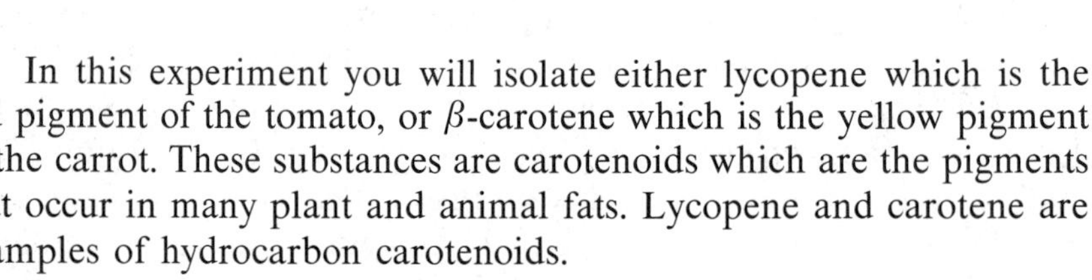

β-Carotene

In this experiment you will isolate either lycopene which is the red pigment of the tomato, or β-carotene which is the yellow pigment of the carrot. These substances are carotenoids which are the pigments that occur in many plant and animal fats. Lycopene and carotene are examples of hydrocarbon carotenoids.

THE EXPERIMENT

Fresh tomatoes contain about 96% water, and in 1907 Willstätter isolated 20 mg of lycopene per kilogram of fruit from this source. Today a more convenient source is found in commercial tomato paste, from which seeds and skin have been eliminated. In tomato paste, the water content of the tomatoes has been reduced by evaporation in vacuum to the point where the solids represent 26% of the total weight. As much as 150 mg of lycopene per kilogram of paste can be isolated. Tomato paste also contains some β-carotene. The expected yield in the present experiment is about 0.75 mg of the mixed carotenoids.

A jar of strained carrots, sold as baby food, serves as a convenient source of β-carotene. Originally, Willstätter isolated 1 g of β-carotene per kilogram of "dried" shredded carrots of unstated water content.

The procedure in this experiment calls for dehydration of tomato or carrot paste with ethanol, followed by extraction with methylene chloride which is an efficient solvent for lipids. The initial dehydration is performed in order to remove water from the tomato or carrot tissue so that the extraction with methylene chloride will be more efficient. Methylene chloride, being immiscible with water, will not effectively extract the carotenoids from the tissue until the water is removed. The final purification by column chromatography illustrates an extremely important technique. The α and γ isomers are not detected in this experiment. This isolation procedure affords amounts of pigments that are unweighable except on a microbalance but are more than adequate for the thin layer chromatography experiment which you will conduct after the isolation procedure has been accomplished.

SPECIAL INSTRUCTIONS

Before beginning this experiment, the preceding essay describing the chemistry of vision should be read. Also read Techniques 1, 2, 10, and 11. Care should be exercised when handling such toxic solvents as methylene chloride, ligroin, benzene, and cyclohexane. Time should be budgeted so that the entire column chromatography, including the packing of the column, can be completed without interruption. Allow 1 to $1\frac{1}{2}$ hours for the chromatography procedure.

PROCEDURE

Place a 5 g sample of tomato or carrot paste (your choice) in an Erlenmeyer flask with the provided spoon. Add 10 ml of 95% ethanol and heat the mixture to boiling for 5 minutes. The mixture should be boiled gently so that the amount of alcohol is not reduced appreciably. After boiling, filter the hot mixture through a small Hirsch funnel (no vacuum). Be sure to scrape out the flask with a spatula and to let the flask drain thoroughly. Press the liquid out of the semi-solid residue collected in the funnel by pressing it gently with the flat side of a spatula. Pour the yellow filtrate into a 125 ml Erlenmeyer flask and save it. Return the solid residue, with or without the adhering filter paper, to the original flask, add 10 ml of methylene chloride, and heat the mixture under reflux for 3 to 4 minutes (Technique 1, Figure 1–5). Methylene chloride has a low boiling point (41°), so reflux the mixture gently or all of the solvent will be lost. Methylene chloride is also toxic; be careful not to fill the room with its vapors. Once again, filter the yellow extract as above and add the filtrate to that previously saved in the Erlenmeyer flask. In this operation the **liquid** may simply be decanted from the solids and passed through the filter. It is not necessary to filter the solid residues which can by this process be left in the flask. Repeat the extraction of the residues left in the flask with two or three additional portions of methylene chloride. Pour the combined methylene chloride extracts into a separatory funnel, add a few milliliters of saturated sodium chloride solution (to aid in layer separation), and shake. Allow the layers to separate and then slowly drain the colored lower layer (methylene chloride is more dense than water) through a funnel which has a loose plug of cotton in the neck and a small layer (1 cm) of anhydrous sodium sulfate on top of it. This procedure removes water from the solution. Store the solution in a dry, stoppered flask until it is to be chromatographed. Prior to chromatography, the solution should be evaporated to dryness on a steam bath in the hood.

COLUMN CHROMATOGRAPHY

The crude carotenoid is chromatographed on a 15 cm column of acid-washed alumina prepared with ligroin (60 to 75°) as the solvent (see Technique 10, especially Sections 10.5 and 10.6). Place a short (3 cm) piece of rubber tubing on the bottom of the dry column and close it with a screw clamp. Place a piece of cotton in the bottom of the column and add a small amount of sand to form a layer about 3 mm thick at the bottom (Technique 10, Section 10.6A). Weigh 10 g of alumina and place it in an Erlenmeyer flask. Add 15 ml of ligroin. Swirl the mixture vigorously to form a slurry and then add it quickly to the top of the column (Technique 10, Section 10.6B). Open the stopcock and allow the liquid to drain into the Erlenmeyer flask. Again swirl the flask and then add more of the slurry to the column until a 15 cm column of alumina is obtained. The column should be tapped gently during this process. Do not allow the column to run dry. Try to fill the column at a steady rate. This slurry method eliminates air bubbles from the surface of the alumina. Drain the excess solvent until it just reaches the top level of the alumina. Close the screw clamp. Dissolve the crude carotenoid in a maximum of 1 to 2 ml of benzene and transfer **most** of the solution onto the column with a small disposable pipet (Technique 10, Section 10.7). The solution should be transferred all at once. Save a few drops for later analysis by thin layer chromatography. Open the screw clamp and allow the colored material to pass onto the column. When the top of the column just becomes dry, close the screw clamp. Using a capillary pipet, add a few milliliters of ligroin to wash down the sides of the column and allow this to pass onto the column by draining more of the solvent from the column. When the surface of the alumina is exposed above the level of the solvent, add a large amount of ligroin to elute the column (Technique 10, Section 10.8). The yellow β-carotene should move rapidly through the column, while the red lycopene moves more slowly. Tomatoes contain a little β-carotene in addition to lycopene, and a separation should be observed if the column is run carefully. Carrots, on the other hand, contain mostly β-carotene, with a small amount of lycopene. Only with very careful work will the lycopene be observed. Discard the colorless eluants and collect the yellow and/or red eluant(s) in separate flasks. It may be necessary to use benzene, the more polar solvent, after all the β-carotene is removed from the column in order to remove the more polar lycopene. This is a common practice with column chromatography where a more polar solvent is used when removing more polar substances following removal of less polar substances (Technique 10, Section 10.8). Evaporate the various fractions which were collected on a steam bath in the hood.

Dissolve the samples in the smallest possible amount of methylene chloride, hold the flasks in a slanting position for drainage into a small area, and transfer the solution with a disposable pipet into a small test tube. The few drops of material which were not column chromatographed should also be transferred to a small test tube. These materials should be analyzed by thin layer chromatography as soon as possible. Since the chromatography usually requires an hour or more, you will probably have to store the samples until the next class period. If so, cork the test tubes tightly, label them, and store them. Carotenoids are light and air sensitive, and they may decompose somewhat upon standing. They should be analyzed in the very next laboratory period.

THIN LAYER CHROMATOGRAPHY

Technique 11 describes the procedures needed for thin layer chromatography (tlc). Silica gel G will be used. Use the dipping method to prepare the plates (Technique 11, Section 11.2A). Analyze the materials which were extracted from tomatoes and carrots as described below.

If the solvent has evaporated from the carotenoid samples obtained above, add a **few** drops of methylene chloride and shake the mixture to dissolve the material. The sample may not completely dissolve because of partial air oxidation of the carotenoids to form insoluble material, but there should be enough colored supernatant liquid for numerous experiments. Keep the solutions out of the light when they are not in use. The highly unsaturated hydrocarbons are subject to rapid photochemical autoxidation, but during chromatography they are protected by solvent vapor. However, on removal of a plate from the chamber a spot may disappear rapidly, and it should be outlined in pencil immediately. Save a plate from which the outlined spots of the carotenoids have disappeared and develop it in an iodine chamber as described in Technique 11, Section 11.6.

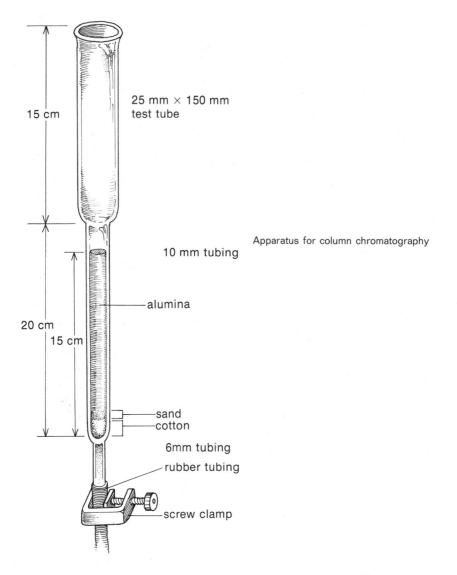

15 cm

25 mm × 150 mm
test tube

Apparatus for column chromatography

10 mm tubing

alumina

20 cm

15 cm

sand
cotton

6mm tubing

rubber tubing

screw clamp

The materials can be spotted on a single plate (Technique 11, Section 11.3). There should be three samples (two purified by column chromatography and one unpurified) from either the tomato or the carrot experiment, although sometimes it may be difficult to obtain a lycopene fraction from carrots. Develop the tlc plate in cyclohexane. In addition try a more polar solvent system such as benzene-cyclohexane (10:90). Remove the plates and immediately outline the spots with a pencil point. Calculate the R_f values for each spot. Since the tomato paste is reported to contain both β-carotene and lycopene, it should be possible to detect their presence in the tomato sample by analyzing the sample before column purification (at least two spots) and comparing this to the two fractions obtained by column chromatography (one spot for each fraction). By this technique one can tell if he has obtained a good separation of the two materials by the column technique. A similar result should be obtained with the carrot sample, although lycopene may be more difficult to detect in carrots than in tomatoes. It may be of interest to compare the R_f values of each of the components with those of a neighbor who has used the other vegetable source in his experiment. Are they the same? A better comparison would be to spot both samples (from carrots and from tomatoes) being compared on the same plate and to develop them under identical conditions. Are the R_f values the same under these conditions? The R_f values obtained in these experiments should be reported to the instructor.

QUESTIONS

1. Why is lycopene less mobile on column chromatography and why does it have a lower R_f value than β-carotene?

2. Why is lycopene red, while β-carotene is yellow?

3. Explain the differences in R_f values observed for the carotenoids when one changes solvents from cyclohexane to benzene-cyclohexane.

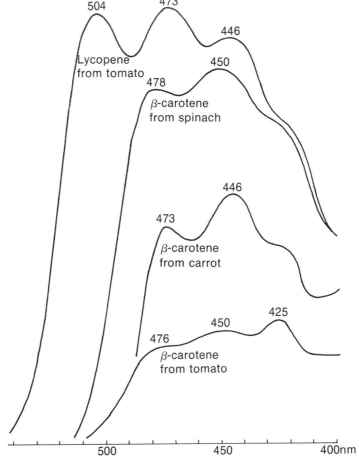

Visible light absorption spectra of isolated lycopene and β-carotene in petroleum ether. An unidentified absorption appears at 425 nm in the β-carotene from tomato sample.

Ethanol and Fermentation Chemistry

The fermentation processes involved in bread making, wine making, and brewing are among the oldest chemical arts. Even though the process of fermentation had been known as an art for centuries, it was not until the nineteenth century that chemists began to understand this process from the point of view of science. In 1810 Gay-Lussac discovered the general chemical equation for the breakdown of sugar into ethanol and carbon dioxide. The manner in which the process took place was the subject of much conjecture until Louis Pasteur began his thorough examination of fermentation. Pasteur demonstrated that yeast was required in the fermentation. He was also able to identify other factors which controlled the action of the yeast cells. His results were published in 1857 and 1866.

For many years it was believed that the transformation of sugar into ethanol and carbon dioxide by yeasts was inseparably connected with the life process of the yeast cell. This view was abandoned in 1897 when Büchner demonstrated that yeast extract will bring about alcoholic fermentation in the absence of any yeast cells. The fermenting activity of yeast is due to a remarkably active catalyst of biochemical origin, the enzyme zymase. It is now recognized that most of the chemical transformations that go on in living cells of plants and animals are brought about by enzymes. The enzymes are organic compounds, generally proteins, and establishment of structures and reaction mechanisms of these compounds is an active field of present-day research. Zymase is now known to be a complex of at least twenty-two separate enzymes, each of which catalyzes a specific step in the fermentation reaction sequence.

Enzymes show an extraordinary specificity—a given enzyme acts on a specific compound, or closely related group of compounds. Thus, zymase acts only on a few select sugars and not on all carbohydrates; the digestive enzymes of the alimentary tract are equally specific in their activity.

The chief sources of sugars for fermentation are the various starches and the molasses residue obtained from sugar refining. Corn **(maize)** is the chief source of starch in the United States, and ethyl alcohol made from corn is known commonly as **grain alcohol.** In preparing alcohol from corn, the grain, with or without the germ, is ground and cooked to give the **mash.** The enzyme diastase is added in the form of **malt** (sprouted barley that has been dried in air at 40° and ground to a powder) or of a mold such as **Aspergillus oryzae.** The mixture is kept at 40° until all of the starch has been converted into the sugar **maltose** by hydrolysis of ether and acetal bonds. This solution is known as the **wort.**

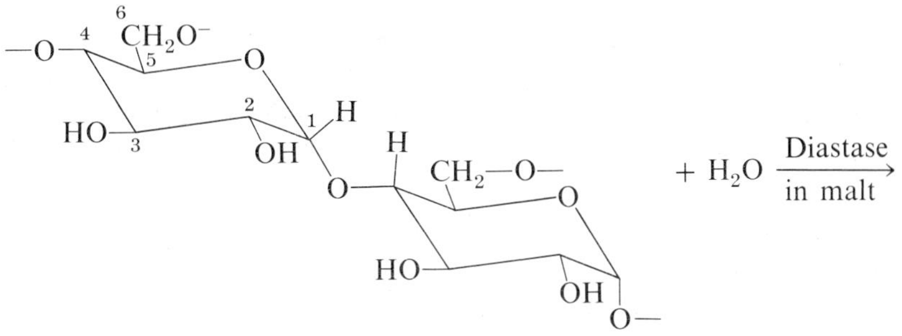

Starch

This is a glucose polymer with 1,4- and 1,6-glycosidic linkages. The linkages at C_1 are α.

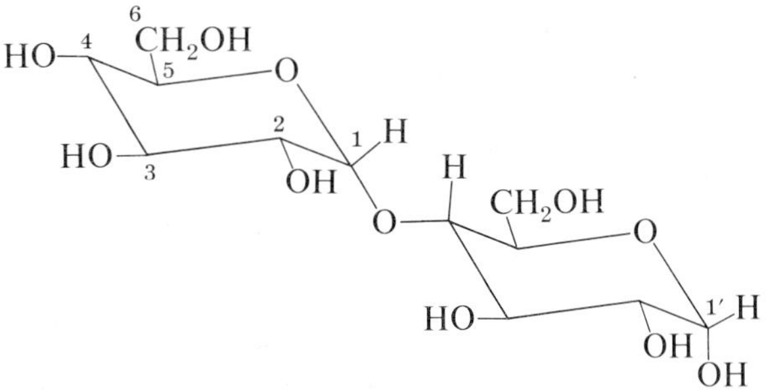

Maltose ($C_{12}H_{22}O_{11}$)

The α linkage still exists at C_1.
The —OH is shown α at the 1' position (axial), but it can also be β (equatorial).

The wort is cooled to 20°, diluted with water to 10% maltose, and a pure yeast culture is added. The yeast culture is usually a strain of **Saccharomyces cerevisiae** (or **ellipsoldus**). The yeast cells secrete two enzyme systems, maltase, which converts the maltose into glucose, and zymase, which converts the glucose into carbon dioxide and alcohol.

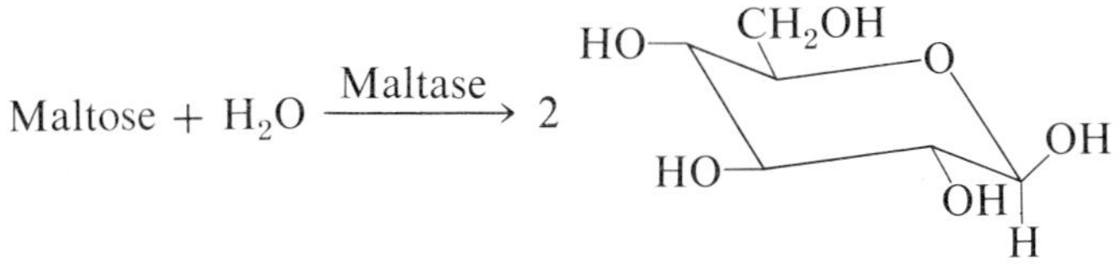

$$\text{Maltose} + H_2O \xrightarrow{\text{Maltase}} 2$$

β-D-(+)-Glucose

(α-D-(+)-Glucose, with an axial —OH, is also produced.)

$$\text{glucose} \xrightarrow{\text{Zymase}} 2CO_2 + 2CH_3CH_2OH + 26 \text{ kcal}$$
$$C_6H_{12}O_6$$

Heat is liberated, and the temperature must be kept below 32° by cooling to prevent destruction of the enzymes. Oxygen in large amounts is initially necessary for the optimum reproduction of yeast cells, but the production of alcohol is anaerobic. During fermentation the evolution of carbon dioxide soon establishes anaerobic conditions. If oxygen were freely available, only carbon dioxide and water would be produced.

After 40 to 60 hours, fermentation is complete, and the product is distilled to remove the alcohol from solid matter. The distillate is fractionated by means of an efficient column. A small amount of acetaldehyde, bp 21°, distills first and is followed by 95 per cent alcohol. Fusel oil is contained in the higher boiling fractions. The fusel oil consists of a mixture of higher alcohols, chiefly 1-propanol, 2-methyl-1-propanol, 3-methyl-1-butanol, and 2-methyl-1-butanol. The exact composition of fusel oil varies considerably, being particularly dependent on the type of raw material that is fermented. These higher alcohols are not formed by fermentation of glucose. They arise from certain amino acids derived from the proteins present in the raw material and in the yeast. These fusel oils cause the headaches familiarly associated with the drinking of alcoholic beverages.

Industrial alcohol is ethyl alcohol used for nonbeverage purposes. Most of the commercial alcohol is denatured in order to avoid payment of taxes, the major cost in the price of liquor. The denaturants render the alcohol unfit for drinking. Methanol, aviation fuel and other substances are used for this purpose. The difference in price between taxed and nontaxed alcohol is more than $20 per gallon. Before the development of efficient synthetic processes, the chief source of industrial alcohol was fermented blackstrap molasses, the noncrystallizable residue from the refining of cane sugar (sucrose). Most industrial ethanol in the United States is presently manufactured from ethylene, a product of the "cracking" of petroleum hydrocarbons. By reaction with concentrated sulfuric acid, ethylene is converted to ethyl hydrogen sulfate, which is hydrolyzed to ethanol by dilution with water. 2-Propanol, 2-butanol, 2-methyl-2-propanol and higher secondary and tertiary alcohols also are produced on a large scale from alkenes derived from the "cracking" process.

Yeasts, molds, and bacteria are used commercially for the large-scale production of various organic compounds. An important example, in addition to ethanol, is the anaerobic fermentation of starch by certain bacteria to yield 1-butanol, acetone, ethanol, carbon dioxide, and hydrogen.

REFERENCES

Amerine, M. A. "Wine." *Scientific American, 211* (August, 1964), 46.
"Chemical Technology: Key to Better Wines." *Chemical and Engineering News, 51* (July 2, 1973), 14.
"Chemistry Concentrates on the Grape." *Chemical and Engineering News, 51* (June 25, 1973), 16.

Church, L. B. "The Chemistry of Winemaking." *Journal of Chemical Education, 49* (1972), 174.
Ray, O. S. "Alcohol." *Drugs, Society, and Human Behavior.* St. Louis: Mosby, 1972.

Students wishing to investigate the subject of alcoholism and possible chemical explanations for alcohol addiction may wish to consult the following references:

Cohen, G., and Collins, M. "Alkaloids from Catecholamines in Adrenal Tissue: Possible Role in Alcoholism." *Science, 167* (1970), 1749.
Davis, V. E., and Walsh, M. J. "Alcohol Addiction and Tetrahydropapaveroline." *Science, 169* (1970), 1105.
Davis, V. E., and Walsh, M. J "Alcohols, Amines, and Alkaloids: A Possible Biochemical Basis for Alcohol Addiction." *Science, 167* (1970), 1005.
Seevers, M. H., Davis, V. E., and Walsh, M. J. "Morphine and Ethanol Physical Dependence: A Critique of a Hypothesis." *Science, 170* (1970), 1113.
Yamanaka, Y., Walsh, M. J., and Davis, V. E. "Salsolinol, An Alkaloid Derivative of Dopamine Formed in vitro during Alcohol Metabolism." *Nature, 227* (1970), 1143.

Experiment **10**

ETHANOL FROM SUCROSE

Fermentation
Simple Distillation
Fractional Distillation
Azeotropes

In this experiment you will use sucrose, rather than maltose, as a starting material. It is a disaccharide with the formula $C_{12}H_{22}O_{11}$. Sucrose has one glucose molecule combined with fructose, rather than two glucose molecules as found in maltose. The enzyme **invertase** catalyzes the hydrolysis of sucrose, rather than maltase, which hydrolyzes maltose. The hydrolysis of maltose is discussed in the previous essay. Zymase is used to convert the sugars into alcohol and carbon dioxide. Pasteur observed that growth and fermentation were promoted by the addition of small amounts of mineral salts to the nutrient medium. Later it was found that, prior to fermentation, the hexose sugars combine with phosphoric acid, and the resulting hexose–phosphoric acid combination is then degraded into carbon dioxide and ethanol. The carbon dioxide is not wasted in the commercial process but is converted to Dry Ice.

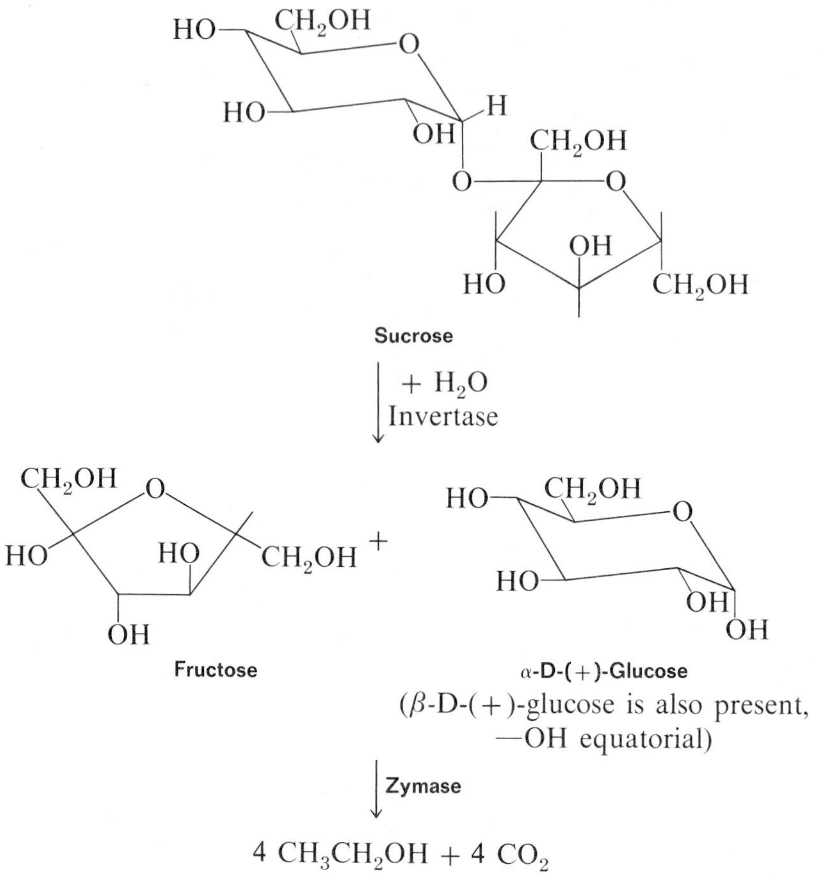

Sucrose

$+ H_2O$
Invertase

Fructose

$+$

α-**D**-(**+**)-**Glucose**
(β-D-($+$)-glucose is also present,
—OH equatorial)

Zymase

$$4 \ CH_3CH_2OH + 4 \ CO_2$$

The fermentation process is inhibited by ethanol, and it is not possible to prepare solutions containing more than 10 to 15 percent ethanol by this method. Isolation of more concentrated ethanol is accomplished by fractional distillation. Ethanol and water form an azeotropic mixture consisting of 95 percent ethanol and 5 percent water by weight, which is the most concentrated ethanol that can be obtained by fractionation of dilute ethanol-water mixtures.

SPECIAL INSTRUCTIONS

Before beginning this experiment, the preceding essay and Techniques 1, 2, 6, and 7 should be read. The fermentation must be begun at least one week before the actual isolation of ethanol is to be undertaken. When the aqueous ethanol solution is to be separated from the yeast cells, it is important to carefully siphon as much of the clear supernatant liquid as possible, without agitating the mixture. The crude ethanol obtained by two or more students may be combined for the final fractional distillation, at the instructor's option, if shortages of glassware prevent each student from performing his own fractional distillation.

PROCEDURE

Place 80 g of sucrose (common granulated sugar) in a 1-liter flask or bottle. Add 700 ml of water, warmed to room temperature, 70 ml of Pasteur's salts[1] and one-half package of dried baker's yeast or 15 g of cake yeast. Shake vigorously and fit the flask with a one-hole rubber stopper with a glass tube leading to a beaker or test tube containing a solution of barium hydroxide. Protect the barium hydroxide from air by adding some kerosene or xylene to form a layer above the barium hydroxide. A precipitate of barium carbonate will form indicating that CO_2 is being evolved. Alternatively a balloon may be substituted for the barium hydroxide trap assembly. The gas will cause the balloon to expand as the fermentation proceeds. Oxygen from the atmosphere is excluded by these techniques. If oxygen were allowed to continue to be in contact with the fermenting solution, the ethanol could be further oxidized to acetic acid or even all the way to carbon dioxide and water. As long as carbon dioxide continues to be liberated, alcohol is being formed.

---

[1] A solution of Pasteur's salts consists of: potassium phosphate, 2.0 g; calcium phosphate, 0.20 g; magnesium sulfate, 0.20 g; and ammonium tartrate, 10.0 g, dissolved in 860 ml of water.

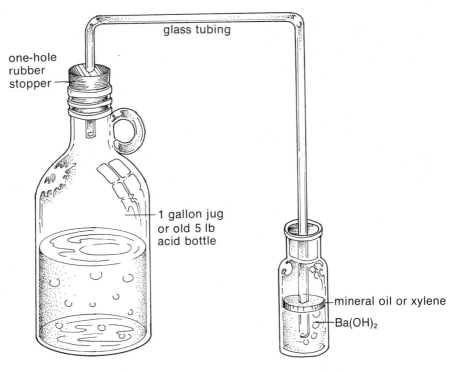

Apparatus for fermentation experiment

Allow the mixture to stand at a temperature of about 25° until fermentation is complete, as indicated by the cessation of gas evolution. Usually about one week is required. After this time, carefully move the flask to a desk and remove the stopper. Siphon the liquid out of the flask so that none of the sediment is removed. A siphon is easily started by filling a short section of rubber tubing with water, pinching one end closed, and placing the other end into the liquid in the flask. Release the end of the tubing which is not in the flask and hold it over the edge of the desk top. Allow the ethanol–water solution to run into a large beaker or flask. When the level of the liquid approaches the sediment, slow the rate of siphoning by pinching the rubber tubing slightly. It is better to leave some liquid behind than to draw some sediment out of the flask. If the siphoned liquid is not clear, clarify it by the following procedure. Place about 2 tablespoons of Filter Aid (Johns-Manville Celite) in a beaker with about 200 ml of water. Stir the mixture vigorously and then pour the contents into a Büchner funnel (with paper) while applying a vacuum, as in a vacuum filtration (Technique 2, Section 2.3). This procedure will result in a thin layer of Filter Aid being deposited on the filter paper (Technique 2, Section 2.4). The water which passes through the filter is discarded. The siphoned liquid is then passed through the filter under gentle suction. The extremely tiny yeast particles are trapped in the pores of the Filter Aid (Technique 2, Section 2.4). The liquid contains ethanol in water plus smaller amounts of dissolved metabolites (fusel oils) from the yeast.

The liquid is transferred to a simple distillation apparatus (Technique 6, Section 6.3). About 200 to 250 ml of distillate are collected, or the distillation is continued until the temperature reaches 100°. Discard the residue remaining in the flask and redistill the distillate through a fractionating column packed with a metal sponge (Technique 7, Section 7.5). Distill slowly and carefully in order to obtain the best possible separation. Collect the fraction boiling between 78° and 88° and discard the residue in the distilling flask. The extent of purification of the ethanol is limited by the fact that ethanol and water form a constant boiling mixture, or azeotrope, with a composition of 95% ethanol and 5% water (bp 78.1). No amount of distillation will remove the last 5% of water.

Calculate the percentage yield of alcohol, assuming that the product is 80% alcohol/20% water,[2] and submit the ethanol to the instructor in a labeled vial.

---

[2] A careful analysis by flame ionization gas chromatography and a water analysis on a typical student-prepared ethanol sample provided the following results:

acetaldehyde	.03%
methyl ethyl acetal of acetaldehyde	.23
diethyl acetal of acetaldehyde	.12
ethanol	77.52
1-propanol	.33
2-propanol	.51
5-carbon and greater alcohols	.56
water	20.7

QUESTIONS

1. By doing a little library research, see if you can find out the method(s) which is(are) commercially used to produce **absolute** ethanol.

2. Why is the air trap necessary in the later stages of the fermentation?

3. The percentage of water in the alcohol can be estimated by measuring the density of the solution. Explain how this procedure works.

4. How does the acetaldehyde impurity arise in the fermentation process?

ESSAY
Esters—Flavors and Fragrances

Esters are a class of compounds widely distributed in nature. They have the general formula:

$$R-\overset{\overset{\textstyle O}{\|}}{C}-OR'$$

The simpler esters tend to have pleasant odors. In many cases, although not exclusively so, the characteristic flavors and fragrances of flowers and fruits are due to compounds with the ester functional group. An exception is the case of the essential oils (see the essay which precedes Experiment 19). The organoleptic qualities (odors and flavors) of fruits and flowers may often be due to a single ester, but, more often, the flavor or the aroma is due to a complex mixture in which a single ester predominates. Some common flavor principles are listed in Table 1. Food and beverage manufacturers are thoroughly familiar with these esters and often use them as additives to spruce up the flavor or odor of a dessert or beverage. Many times such flavors or odors are not even naturally occurring as is the case with the "juicy fruit" principle, isopentenyl acetate. An instant pudding which has "rum" flavor may never have seen its alcoholic namesake—this flavor can be duplicated by the proper admixture, along with other minor components, of ethyl formate and isobutyl propionate. The natural flavor and odor are not **exactly** duplicated, but most people can be fooled. Often only a trained person with a high degree of gustatory perception, a professional taster, can tell the difference.

TABLE 1. ESTER FLAVORS AND FRAGRANCES

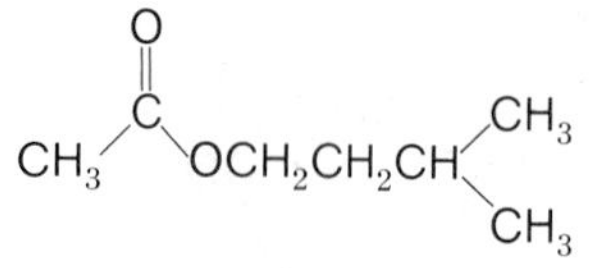

Isoamyl acetate

banana

(alarm pheromone of honeybee)

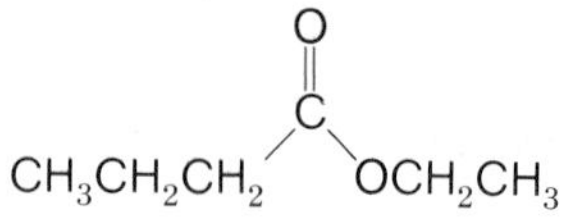

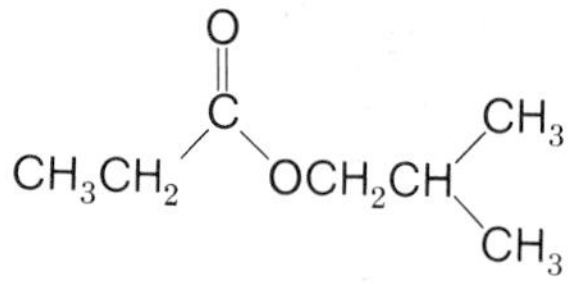

Isobutyl propionate

rum

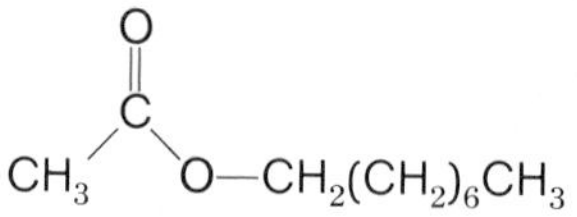

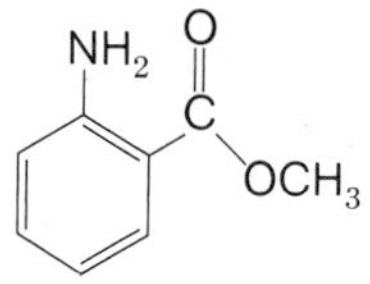

Methyl anthranilate

grape

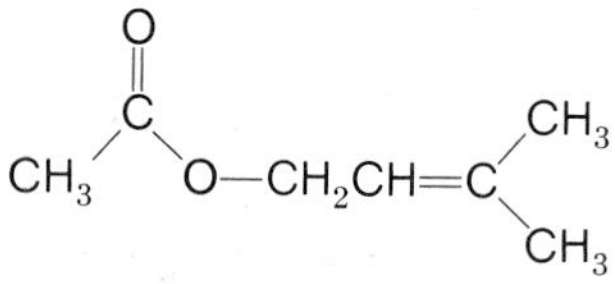

Benzyl acetate

peach

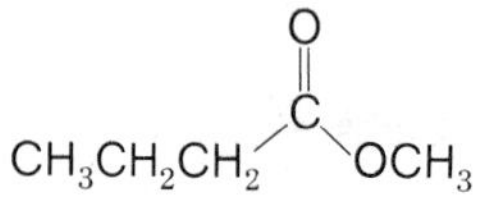

Methyl butyrate

apple

Ethyl butyrate

pineapple

Octyl acetate

oranges

Isopentenyl acetate

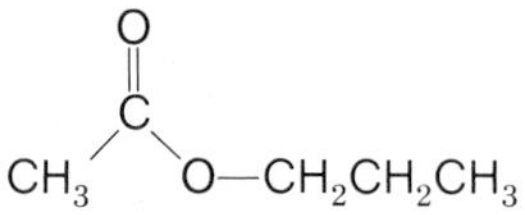

"juicy fruit"

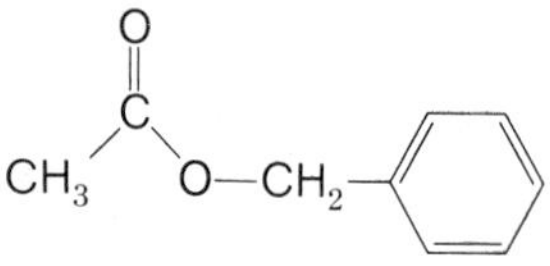

n-Propyl acetate

pear

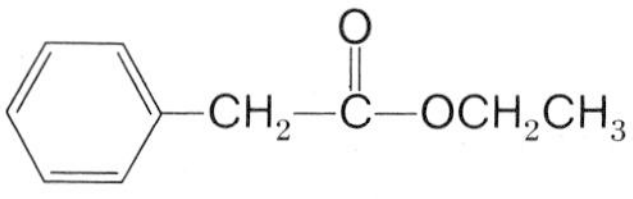

Ethyl phenylacetate

honey

A single compound is rarely used in good quality imitation flavoring agents. A formula for an imitation pineapple flavor, which might fool an expert, is listed in Table 2. The formula includes ten esters and carboxylic acids which may easily be synthesized in the laboratory. The remaining seven oils are isolated from natural sources.

Flavor is a combination of taste, sensation, and odor transmitted by receptors in the mouth (taste buds) and nose (olfactory receptors). The stereochemical theory of odor is discussed in the essay which precedes Experiment 21. The four basic tastes, sweet, sour, salty, and bitter are perceived in specific areas of the tongue. The sides of the tongue perceive sour and salty tastes, the tip is most sensitive to sweet tastes and the back of the tongue detects bitter tastes. The perception of flavor, however, is not that simple. If it were, it would only require

TABLE 2. ARTIFICIAL PINEAPPLE FLAVOR

PURE COMPOUNDS	%	ESSENTIAL OILS	%
Allyl caproate	5	Oil of sweet birch	1
Isoamyl acetate	3	Oil of spruce	2
Isoamyl isovalerate	3	Balsam Peru	4
Ethyl acetate	15	Volatile mustard oil	1
Ethyl butyrate	22	Oil cognac	5
Terpinyl propionate	3	Concentrated orange oil	4
Ethyl crotonate	5	Distilled oil of lime	2
Caproic acid	8		19
Butyric acid	12		
Acetic acid	5		
	81		

the formulation of various combinations of four basic substances: a bitter substance (a base), a sour substance (an acid), a salty substance (sodium chloride), and a sweet substance (sugar), to duplicate **any** flavor! In fact, we cannot duplicate flavors in this way. The human actually possesses 9000 taste buds. It is a combined response of these taste buds which allows us to perceive a particular flavor.

Although the "fruity" tastes and odors of esters are pleasant, they are seldom used in perfumes or scents that are applied to the body. The reason for this is a chemical one. The ester group is not as stable to perspiration as the ingredients of the more expensive essential oil perfumes. The latter are usually hydrocarbons (terpenes), ketones, and ethers extracted from natural sources. Esters, however, are only used for the cheapest toilet waters, since on contact with sweat, they undergo a hydrolysis reaction to give organic acids. These acids, unlike their

$$
\underset{\substack{|\\ OR'}}{R-\overset{\displaystyle O}{\overset{\|}{C}}} + H_2O \longrightarrow \underset{\substack{|\\ OH}}{R-\overset{\displaystyle O}{\overset{\|}{C}}} + R'OH
$$

precursor esters, generally do not have a pleasant odor. Butyric acid, for instance, has a strong odor similar to that of rancid butter (of which it is an ingredient) and is, in fact, a component of what we normally call "body odor." It is this substance that makes foul-smelling humans so easy for an animal to detect when he is downwind of them. It is also of great help to the bloodhound which is trained to follow small traces of this odor. The **esters** of butyric acid, ethyl butyrate and methyl butyrate, however, smell like pineapple and apple, respectively.

A sweet fruity odor also has the disadvantage that it may attract fruit flies and other insects in search of food. The case of isoamyl acetate, the familiar solvent called banana oil, is particularly interesting. It is identical to the alarm **pheromone** of the honeybee. Pheromone is the name applied to a chemical secreted by an organism which evokes a specific response in another member of the same species. This kind of communication is common between insects who otherwise lack means of intercourse. When a honeybee worker stings an intruder, an alarm pheromone, composed in part of isoamyl acetate, is secreted along with the sting venom. This chemical causes aggressive attack on the intruder by other bees, who swarm after the intruder. Obviously it wouldn't be wise to wear a perfume compounded of isoamyl acetate near a beehive. Pheromones are discussed in more detail in the essay preceding Experiment 17.

REFERENCES

Benarde, M. A. *The Chemicals We Eat.* New York: American Heritage Press, 1971. Pp. 68–75.
The Givaudan Index. New York: Givaudan-Delawanna, Inc., 1949. (Gives specifications of synthetics and isolates for perfumery.)
Gould, R. F., Ed. *Flavor Chemistry, Advances in Chemistry No. 56.* Washington: American Chemical Society, 1966.
Pyke, M. *Synthetic Food.* London: John Murray, 1970.
Shreve, R. N. *Chemical Process Industries* (3rd edition). New York: McGraw-Hill, 1967.

Experiment **11**
ISOAMYL ACETATE (BANANA OIL)

Esterification
Heating Under Reflux
Extraction
Simple Distillation

In this experiment, we will prepare an ester, isoamyl acetate. This ester is often referred to as banana oil since it has the familiar odor of this fruit.

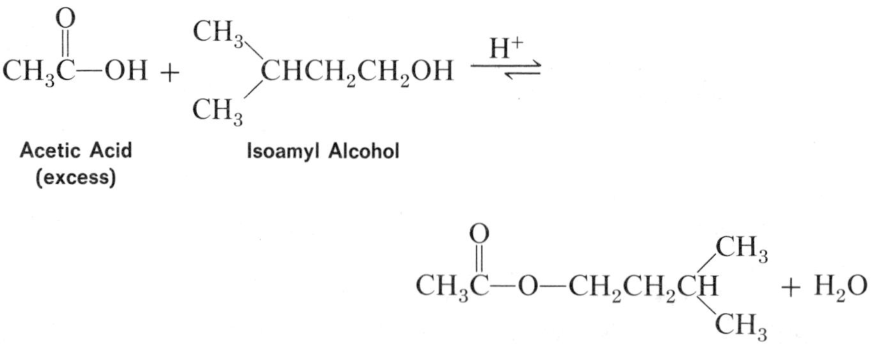

$$CH_3\overset{O}{\overset{\|}{C}}{-}OH + \underset{CH_3}{\overset{CH_3}{>}}CHCH_2CH_2OH \overset{H^+}{\rightleftharpoons}$$

Acetic Acid   **Isoamyl Alcohol**
(excess)

$$CH_3\overset{O}{\overset{\|}{C}}{-}O{-}CH_2CH_2\underset{CH_3}{\overset{CH_3}{CH}} + H_2O$$

Isoamyl Acetate

Isoamyl acetate is prepared by the direct esterification of acetic acid with isoamyl alcohol. Since the equilibrium does not favor the formation of the ester, it must be shifted to the right, in favor of the product, by the use of an excess of one of the starting materials. Acetic acid is used in excess because it is less expensive than isoamyl alcohol and is easier to remove from the reaction mixture.

In the isolation procedure, much of the excess acetic acid and the remaining isoamyl alcohol are removed by extraction with water. Any remaining acid is removed by extraction with aqueous sodium bicarbonate. The ester is purified by distillation.

SPECIAL INSTRUCTIONS

Read the essay which precedes this experiment. In order to perform this experiment, you should have read Techniques 1, 5 and 6. Since there is a one hour reflux in this experiment, the experiment should be started at the very beginning of the laboratory period. During the reflux period, other experimental work may be performed. Be careful when handling concentrated sulfuric acid. It will cause extreme burns if it is spilled on the skin.

PROCEDURE

Pour 18 ml (14.6 g, 0.166 mole) of isoamyl alcohol (also called isopentyl alcohol or 3-methyl-1-butanol) and 24 ml (24 g, 0.4 mole) of glacial acetic acid into a 100 ml round bottom flask. Carefully add 4 ml of concentrated sulfuric acid to the contents of the flask, with swirling. Add several boiling stones to the mixture.

Extreme care must be exercised in order to avoid contact with concentrated sulfuric acid. It will cause serious burns if it is spilled on the skin. If it comes in contact with the skin or clothes, it must be washed off immediately with excess water. In addition, sodium bicarbonate may be used to neutralize the acid. Clean up all spills, immediately.

Assemble a reflux apparatus as shown in Technique 1, Figure 1–4. Bring the mixture to a boil with a suitable heating source, such as a heating mantle or oil bath (Technique 1, Sections 1.4 and 1.5). Allow the mixture to reflux for one hour (Technique 1, Section 1.7). Remove the heating source and allow the mixture to cool to room temperature. Pour the cooled mixture into a separatory funnel and carefully add 55 ml of cold water. Rinse the reaction flask with 10 ml of cold water and pour the rinsings into the separatory funnel. Use a stirring rod to mix the materials somewhat. Stopper the separatory funnel and shake it several times (Technique 5, Section 5.4). Separate the lower aqueous layer from the upper organic layer (density = 0.87 g/ml). Discard the aqueous layer after making certain that the correct layer has been saved.

The crude ester in the organic layer contains some acetic acid which can be removed by extraction with 5% aqueous sodium bicarbonate solution. Carefully add 30 ml of the 5% base to the organic layer contained in the separatory funnel. Swirl the separatory funnel gently until carbon dioxide gas is no longer evolved. Stopper and gently shake the funnel once or twice, and then vent the vapors. Shake the funnel until no vapors are evolved when the separatory funnel is vented. Remove the lower layer, and again repeat the above extraction with 30 ml of 5% sodium bicarbonate solution. Remove the lower layer and check to see if it is basic to litmus. If it is not basic, repeat the procedure with additional 30 ml portions of 5% base until the aqueous layer is basic. Discard the basic washings and extract the organic layer with one 25 ml portion of water. Add 5 ml of saturated aqueous sodium chloride to aid in layer separation. Stir the mixture gently; do not shake it. Carefully separate the lower aqueous layer and discard it. When the water has been removed, pour the ester from the top of the separatory funnel into a flask. Add about 2 g of anhydrous magnesium sulfate to dry the ester (Technique 5, Section 5.6). Stopper the flask and swirl it gently. Allow the crude ester to stand until the liquid is clear. About 15 minutes will be required in order for complete drying to occur. If the solution is still cloudy after this period, decant the solution, and add to it a fresh 1 g quantity of drying agent.

Assemble a simple distillation apparatus as shown in Technique 6, Figure 6–6. Dry all glassware thoroughly before use. Carefully decant the ester into the distilling flask so that the drying agent is excluded. Add several boiling stones and distill the ester (Technique 6, Section 6.4). The receiver must be cooled in an ice bath. Collect the fraction boiling between 134° and 141° in a dry flask. Weigh the product and calculate the percentage yield.

At your instructor's option, obtain an infrared spectrum (see Technique 17). Compare the spectrum to the one reproduced in this experiment. Include it with your report to the instructor. Submit the prepared sample to the instructor in a labeled vial.

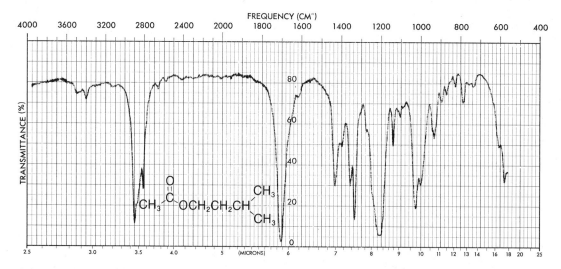

Infrared spectrum of isoamyl acetate, neat

QUESTIONS

1. Give a mechanism for the formation of isoamyl acetate.

2. One method for favoring the formation of the ester is to add an excess of acetic acid. Suggest another method, involving the right-hand side of the equation, which will favor the formation of the ester.

3. Outline a separation scheme for the isolation of pure isoamyl acetate from the reaction mixture.

4. Why is excess acetic acid easier to remove from the products than excess isoamyl alcohol would be?

5. Interpret the major absorption bands in the infrared spectrum of isoamyl acetate.

Experiment **12**
METHYL SALICYLATE (OIL OF WINTERGREEN)

Synthesis Of An Ester
Extraction
Vacuum Distillation

In this experiment you will prepare a familiar-smelling organic ester—oil of wintergreen. Methyl salicylate was first isolated in 1843 by extraction from the Wintergreen plant (**Gaultheria**). It was soon found that this compound had analgesic and antipyretic character almost identical to that of salicylic acid (see the essay, **Aspirin**) when taken internally. This medicinal character is probably due to the fact that methyl salicylate is easily hydrolyzed to salicylic acid under the alkaline conditions found in the intestinal tract. Salicylic acid is known to have such analgesic and antipyretic properties. Methyl salicylate can be taken internally or absorbed through the skin. Thus it finds much use in liniment preparations. When applied to the skin it produces a mild burning or soothing sensation, which is probably due to the action of its phenolic hydroxyl group. This ester also has a pleasant odor, and it is used to a minor extent as a flavoring principle.

Salicylic Acid

Methyl Salicylate
(Oil of Wintergreen)

Methyl salicylate will be prepared from salicylic acid, which is esterified at the carboxyl group with methanol. You should recall from your organic chemistry lecture course that esterification may be an acid-catalyzed equilibrium reaction. The position of the equilibrium does not lie sufficiently far to the right to favor the formation of the ester in high yield. More product may be formed by increasing the concentrations of one of the reactants. In this experiment the use of a large excess of methanol will shift the equilibrium to favor more completely the formation of the ester.

This experiment also illustrates the use of distillation under reduced pressure for the purification of high-boiling liquids. Distillation of high-boiling liquids at atmospheric pressure is often unsatisfactory. At the high temperatures required, the material being distilled (the ester, in this case) may partially, or even totally, decompose and result in a loss of product and contamination of the distillate. However, when the total pressure inside the distillation apparatus is reduced, the boiling point of the substance is lowered. In this way the substance may be distilled without decomposing it.

SPECIAL INSTRUCTIONS

Before beginning this experiment you should read the essay which precedes Experiment 11 as well as Techniques 1, 5, 6, and 9. Handle concentrated sulfuric acid with caution, since it can cause severe burns. The experiment must be started at the beginning of the laboratory period, since a long reflux time is required to esterify salicylic acid and obtain a respectable yield. Enough time should remain at the end of the reflux to perform the extractions and to place the product over the drying agent. A supplementary experiment may be performed during this reaction period, or work from previous experiments which may still be pending may be completed. When a distillation is conducted under reduced pressure, it is important to guard against the dangers of an implosion. Inspect the glassware for flaws or cracks, and replace any glassware which is found to be defective. WEAR YOUR SAFETY GLASSES. Because the amount of crude methyl salicylate obtained in this reaction is rather small, your instructor may wish to have four students combine their products for the final vacuum distillation.

PROCEDURE

A reflux apparatus using a 100 ml round bottom flask will be needed (Technique 1, Section 1.7). Place 6.9 g (0.05 moles) of salicylic acid and 24 g (30 ml, 0.75 moles) of methanol in the flask. **Carefully** add 8 ml of concentrated sulfuric acid to the mixture. Swirl the flask gently to thoroughly mix the reactants, add a boiling stone to the flask, and assemble the apparatus by attaching the condenser. Heat the mixture to boiling using a heating mantle or an oil bath. Allow the mixture to reflux for 2 to 2-1/2 hours. Cool the solution in the reaction flask by immersing the flask in a cold water bath; then add 50 ml of water. Pour the reaction mixture into a 125 ml separatory funnel and separate the layers (Technique 5, Section 5.4). Be careful to save the liquid layer which contains the ester. Wash the crude ester with 50 ml of 5% $NaHCO_3$ by transferring the ester and the $NaHCO_3$ solution to the separatory funnel and shaking the mixture several times. Separate and discard the aqueous layer. Wash the ester a third time with 30 ml of water. Separate the layers and transfer the ester to a 25 ml Erlenmeyer flask. Dry the product by allowing it to stand over 0.5 g of anhydrous calcium chloride until the next laboratory period. A typical yield of crude ester is approximately 7.0 g.

Assemble an apparatus for distillation under reduced pressure (Technique 6, Figure 6–7). Install a manometer in the system if one is available (Technique 9, Section 9.4 and Figure 9–6). Use the smallest round bottom flask as the distilling flask. Fit the distilling flask with a capillary ebulliator tube which reaches to the bottom of the flask (Technique 6, Figure 6–7). The bleeder tube supplied with most organic glassware kits will serve as an ebulliator if it is equipped with a piece of rubber tubing. The tubing is closed by means of a screw clamp until a gentle stream of bubbles is passed into the solution when the vacuum is applied to the system. A trap must be used between the aspirator and the distillation apparatus (Technique 6, Figure 6–7 or Technique 9, Figure 9–6).

The combined yields of crude methyl salicylate obtained by four students are placed in the distilling flask. It is important that all tubing connections and ground glass joints fit tightly. Apply a vacuum to the apparatus by turning on the water aspirator to the maximum extent. Adjust the screw clamp on the ebulliator until a small stream of bubbles issues from the tube. Apply heat from a heating mantle or an oil bath to the distilling flask. Collect all of the liquid which distills at a temperature of 100° or higher[1]. If a manometer was used, be certain to record the pressure at which the liquid distills. When almost all the liquid has

---

[1]The boiling point of methyl salicylate at atmospheric pressure is 222°, but the ester will distill at 105° at a pressure of 14 mm. A pressure of 14 mm is the lowest that may be obtained with a good aspirator and with a water temperature of 16.5°. A manometer should be connected to the system in order to measure the pressure precisely (Technique 9, Section 9.4).

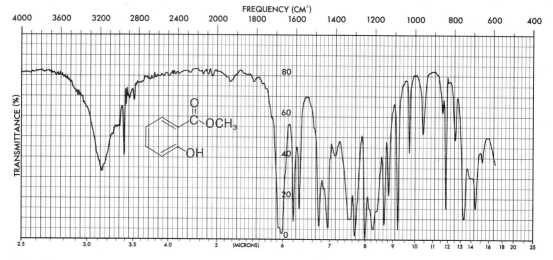

Infrared spectrum of methyl salicylate, neat

been distilled, stop the heating, remove the hose from the aspirator or open the trap valve, and turn off the water. Always disconnect the vacuum before turning off the aspirator. Weigh the product and submit it to the instructor in a labeled vial. The report should include the names of the persons who combined their samples of crude ester for the vacuum distillation.

At your instructor's option, record the infrared spectrum of methyl salicylate (Technique 17). Compare the spectrum to the one reproduced in this experiment.

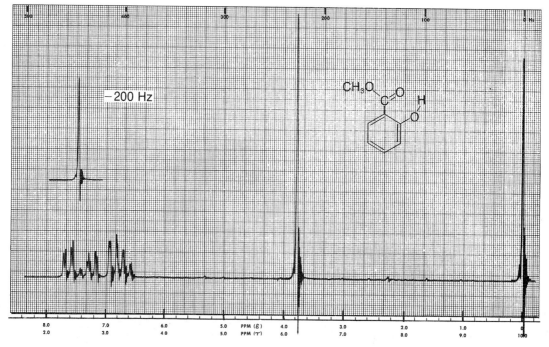

NMR spectrum of methyl salicylate

QUESTIONS

1. Write a mechanism for the acid-catalyzed esterification of salicylic acid with methanol.

2. What is the function of the sulfuric acid in this reaction? Is it consumed in the reaction?

3. In this experiment an excess of methanol was used to shift the equilibrium toward the formation of more ester. Describe other methods for achieving the same result.

4. How are sulfuric acid and the excess methanol removed from the crude ester after the reaction has been completed?

5. Why was 5% $NaHCO_3$ used in the extraction? What would have happened if 5% NaOH had been used?

6. Interpret the major absorption bands in the infrared and NMR spectra of methyl salicylate.

Experiment **13**

THE HYDROLYSIS OF METHYL SALICYLATE

Hydrolysis Of An Ester
Heating Under Reflux
Filtration, Crystallization
Melting Point Determination

Esters may be hydrolyzed into their constituent carboxylic acid and alcohol parts under either acidic or basic conditions. In this experiment, **methyl salicylate,** an ester which is known as **oil of wintergreen** because of its natural source, will be treated with aqueous base. The immediate product of this hydrolysis, besides methanol and water, is the sodium salt of salicylic acid. The reaction mixture is acidified with sulfuric acid, which converts the sodium salt to the free acid. The overall organic products of the reaction, therefore, are salicylic acid and methanol. The salicylic acid is a solid, and it can be isolated and purified by crystallization. The chemical equations which describe this experiment are:

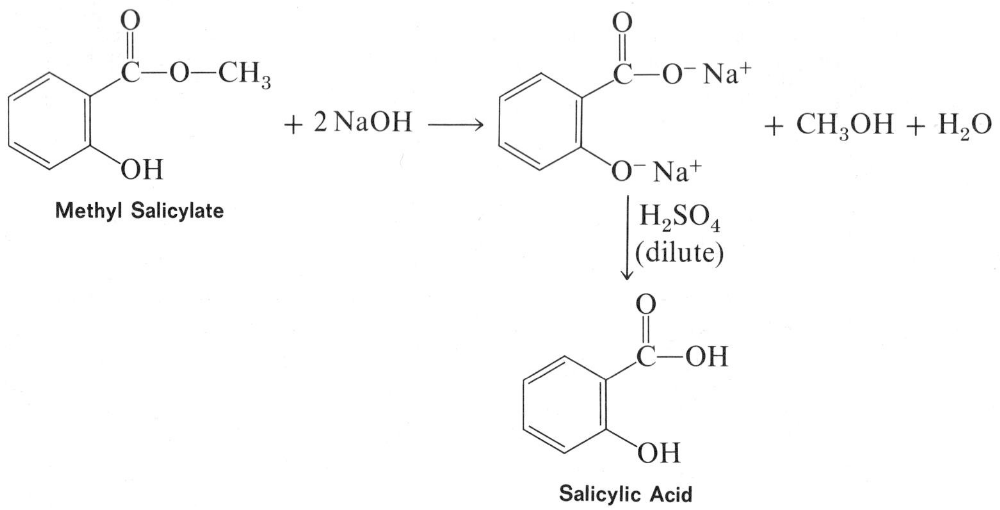

Methyl Salicylate

Salicylic Acid

Because the phenolic hydroxyl group is acidic, it is also converted to the corresponding sodium salt in the course of the basic hydrolysis. In the subsequent acidification, this group becomes reprotonated.

SPECIAL INSTRUCTIONS

Before proceeding with this experiment, you should have read Techniques 1, 2, 3, and 4. This procedure may be stopped at nearly any place.

PROCEDURE

Dissolve 10 g of sodium hydroxide in 50 ml of water. When it has cooled, place this solution, along with 5.0 g (0.033 mole) of methyl salicylate, in a 250 ml round bottom flask. A white solid may form at this point, but it will dissolve on heating. Attach a reflux condenser to the flask, according to the instructions given in Technique 1, Section 1.7. The standard taper joints should be greased lightly.

Add one or two boiling stones to the reaction mixture to prevent bumping when the solution is heated. Heat the solution at its boiling point for about 20 minutes with a microburner. Use a wire gauze beneath the round bottom flask to disperse the flame. After heating, allow the mixture to cool to room temperature. When the solution is cool, transfer it to a 250 ml beaker, and carefully add enough 1 M sulfuric acid to make the solution acidic to litmus paper (blue litmus turns pink). As much as 150 ml of 1 M sulfuric acid may have to be added at this stage. When the litmus turns pink, add an extra 15 ml of the sulfuric acid, thus causing the salicylic acid to precipitate from the solution. Cool the mixture in an ice water bath to a temperature of about 0°C. Allow this cold mixture to settle. Collect the product by vacuum filtration using a Büchner funnel with filter paper. See Technique 2, Section 2.3 for details of this method. The filtration may be conducted more easily by decanting most of the supernatant liquid through the Büchner funnel before adding the mass of crystals.

Recrystallize the crude salicylic acid from water in a 125 ml Erlenmeyer flask. Add 100 ml of hot water and a boiling stone, and heat the mixture to boiling to dissolve the solid.

If the solid does not dissolve on boiling, add enough extra water to dissolve the solid. Filter the hot solution by gravity filtration through a fluted filter paper (Technique 2, Section 2.1 and Figure 2–3) using a fast filter paper, and set the solution aside to cool.

This gravity filtration must be carried out carefully. Filter the hot solution using only a small quantity at a time. Use a short stem funnel for the filtration, in order to reduce the probability that crystals might form in the stem and clog the funnel. The filtration assembly should be placed on a steam bath. If salicylic acid begins to crystallize in the funnel, add the **minimum** amount of boiling water to the filter which is required to redissolve the crystals.

After the filtered solution has cooled, place the flask in an ice water bath to aid crystallization. When the crystals of salicylic acid have formed, collect them by means of vacuum filtration (Technique 2, Section 2.3). Allow the crystals to dry overnight on a watchglass. When the crystals are thoroughly dry, weigh them and determine the percentage yield. Determine the melting point of the pure material (Technique 4, Sections 4.5 and 4.6). The melting point of pure salicylic acid is 159 to 160°C. Place the sample of product in a labeled vial and submit it to the instructor.

QUESTIONS

1. What is the white solid which forms when methyl salicylate is added to an aqueous solution of sodium hydroxide?

2. Give a mechanism for the hydrolysis of methyl salicylate.

3. Is sodium hydroxide consumed in the reaction or is it a catalyst? Explain.

ESSAY
Fats and Oils

In a normal diet, about 25 to 50 percent of the caloric intake of man consists of fats and oils. These substances are the most concentrated form of food energy in our diet. When metabolized, fats produce about 9.5 kcal of energy per gram. Carbohydrates and proteins produce less than half this amount. For this reason, animals tend to build up fat deposits as a reserve source of energy. They do this, of course, only when their food intake exceeds their energy requirements. In times of starvation, the body will metabolize these stored fats. Even so, some fats are required by animals for bodily insulation and as a protective sheath around some of the more vital organs.

The constitution of fats and oils was first investigated by the French chemist Chevreul during the years 1810–1820. He found that, when hydrolyzed, fats and oils give rise to several "fatty acids" and the trihydroxylic alcohol glycerol. Thus, fats and oils are **esters** of glycerol, called **glycerides** or **acylglycerols.** Since glycerol has three hydroxyl groups, it is possible to have mono-, di-, and triglycerides. Fats and oils are predominantly triglycerides (triacylglycerols), constituted as follows:

$$
\begin{array}{c}
\text{O} \\
\parallel \\
R_1\text{—C}\text{(OH} \quad \text{H)—O—CH}_2
\end{array}
\qquad
\begin{array}{c}
\text{O} \\
\parallel \\
R_1\text{—C—O—CH}_2
\end{array}
$$

$$
\begin{array}{c}
\text{O} \\
\parallel \\
R_2\text{—C}\text{(OH} \quad \text{H)—O—CH}
\end{array}
\;\longrightarrow\;
\begin{array}{c}
\text{O} \\
\parallel \\
R_2\text{—C—O—CH}
\end{array}
$$

$$
\begin{array}{c}
\text{O} \\
\parallel \\
R_3\text{—C}\text{(OH} \quad \text{H)—O—CH}_2
\end{array}
\qquad
\begin{array}{c}
\text{O} \\
\parallel \\
R_3\text{—C—O—CH}_2
\end{array}
$$

3 FATTY ACIDS + GLYCEROL   =   A TRIGLYCERIDE

Thus, most fats and oils are esters of glycerol, and their differences are due to the differing fatty acids with which glycerol may be combined. The most commonly found fatty acids have either 12, 14, 16, or 18 carbons, although acids with both lesser and greater numbers of carbons are found in several fats and oils. These common fatty acids are listed in the table below, along with their structures. As is seen, these acids are both saturated and unsaturated. The saturated acids tend to be solids, while the unsaturated acids are usually liquids. This circumstance also extends to fats and oils. Fats are made up of fatty acids which are for the most part saturated, while oils are primarily composed of fatty acid portions which have greater numbers of double bonds. In other words, unsaturation lowers the melting point. Fats (solids) are usually obtained from animal sources, while oils (liquids) are commonly

104

obtained from vegetable sources. Vegetable oils, therefore, usually have a higher degree of unsaturation.

THE COMMON FATTY ACIDS

C_{12} Acids	Lauric	$CH_3(CH_2)_{10}COOH$
C_{14} Acids	Myristic	$CH_3(CH_2)_{12}COOH$
C_{16} Acids	Palmitic	$CH_3(CH_2)_{14}COOH$
	Palmitoleic	$CH_3(CH_2)_5CH{=}CH{-}CH_2(CH_2)_6COOH$
C_{18} Acids	Stearic	$CH_3(CH_2)_{16}COOH$
	Oleic	$CH_3(CH_2)_7CH{=}CH{-}CH_2(CH_2)_6COOH$
	Linoleic	$CH_3(CH_2)_4(CH{=}CH{-}CH_2)_2(CH_2)_6COOH$
	Linolenic	$CH_3CH_2(CH{=}CH{-}CH_2)_3(CH_2)_6COOH$
	Ricinoleic	$CH_3(CH_2)_5CH(OH)CH_2CH{=}CH(CH_2)_7COOH$

There are about 20 to 30 different fatty acids which are found in fats and oils, and it is not uncommon for a given fat or oil to be composed of as many as 10 to 12 (or more) different fatty acids. Typically, these fatty acids are randomly distributed among the triglyceride molecules, and one cannot identify anything more than an average composition for a given fat or oil. The average fatty acid composition of some selected fats and oils is given in the second table. As indicated, all of the values in the table may vary in percent depending, for instance, on the locale in which the plant was grown, or on the particular diet on which the animal subsisted. Thus, perhaps there is a basis for the claims that corn-fed hogs or cattle taste better than those maintained on other diets.

Vegetable fats and oils are usually found in fruits and seeds, and are recovered by three principal methods. In the first method, **cold pressing,** the appropriate part of the dried plant is pressed in a hydraulic press to literally squeeze out the oil. The second method is **hot pressing,** which is the same as the first method, but performed at a higher temperature. Of the two methods, cold pressing usually gives a better grade of product (more bland), while the hot pressing method gives a higher yield, but with more undesirable constituents (stronger odor and flavor). The third method is **solvent extraction.** Solvent extraction gives the highest recovery of all and can now be regulated to give bland, high grade food oils.

Animal fats are usually recovered by **rendering,** which involves cooking the fat out of the tissue by heating it to a high temperature. An alternate method is carried out by placing the fatty tissue in boiling water. The fat floats to the surface and is easily recovered. The most common animal fats, lard (from hogs) and tallow (from cattle), can be prepared in either way.

Of course, many of the triglyceride fats and oils are used for cooking purposes. We use them to fry meats and other foods and to make sandwich spreads. Almost all commercial cooking fats and oils, other than lard, are prepared from vegetable sources. Vegetable oils

AVERAGE FATTY ACID COMPOSITION (BY PERCENTAGE) OF SELECTED FATS AND OILS

	C_{10} C_8 C_6 C_4	LAURIC C_{12}	MYRISTIC* C_{14}	PALMITIC* C_{16}	STEARIC* C_{18}	C_{20} C_{22} C_{24}	PALMITOLEIC C_{16}	OLEIC* C_{18}	RICINOLEIC C_{18}	LINOLEIC* C_{18} (2)	LINOLENIC C_{18} (3)	ELEOSTEARIC C_{18} (3)	UNSATURATED C_{20} C_{22} C_{24}
				SATURATED FATTY ACIDS (NO DOUBLE BONDS)			UNSATURATED (1 DOUBLE BOND)			UNSATURATED (>1 DOUBLE BOND)			UNSAT-URATED
ANIMAL FATS													
Tallow			2–3	24–32	14–32		1–3	35–48		2–4			
Butter	7–10	2–3	7–9	23–26	10–13		5	30–40		4–5			2
Lard			1–2	28–30	12–18		1–3	41–48		6–7			2
ANIMAL OILS													
Neat's Foot				17–18	2–3	3–10	13–18	74–77					
Whale			4–5	11–18	2–4		6–15	33–38					17–31
Sardine			6–8	10–16	1–2					24–30			12–19
VEGETABLE OILS													
Corn			0–2	7–11	3–4		0–2	43–49		34–42			
Olive			0–1	5–15	1–4		0–1	69–84		4–12			
Peanut				6–9	2–6		0–1	50–70		13–26			
Soybean			0–1	6–10	2–6			21–29		50–59	4–8		
Safflower				6–10	1–4			8–18		70–80	2–4		
Castor Bean				0–1				0–9	80–92	3–7			
Cottonseed			0–2	19–24	1–2		0–2	23–33		40–48			
Linseed				4–7	2–5			9–38		3–43	25–58		
Coconut	10–22	45–51	17–20	4–10	1–5			2–10		0–2			
Palm			1–3	34–43	3–6			38–40		5–11			
Tung					2–6			4–16		0–1		74–91	

* Acids used to prepare the standards used in Experiment 14.

are liquids at room temperature. If the double bonds in a vegetable oil are hydrogenated, the resultant product becomes solid. In making commercial cooking fats (Crisco, Spry, Fluffo, etc.) the manufacturers hydrogenate a liquid vegetable oil until the desired degree of consistency is achieved. This results in a product that still has a high degree of unsaturation (double bonds) left. The same technique is used for margarine. "Polyunsaturated" oleomargarine is produced by the partial hydrogenation of oils from corn, cottonseed, peanut, and soybean sources. The final product has a yellow dye (β-carotene) added to make it look like butter, and milk, about 15 percent by volume, is mixed into it to form the final emulsion. Vitamins A and D are also commonly added. Since the final product is tasteless (try Crisco), salt, acetoin, and biacetyl are often added. The latter two additives mimic the characteristic flavor of butter.

$$CH_3-\underset{\underset{H}{|}}{\overset{\overset{HO}{|}}{C}}-\overset{\overset{O}{||}}{C}-CH_3 \qquad CH_3-\overset{\overset{O}{||}}{C}-\overset{\overset{O}{||}}{C}-CH_3$$

Acetoin   **Biacetyl**

Many of the producers of margarine claim it to be more beneficial to health because it is "high in polyunsaturates." Animal fats are low in unsaturated fatty acid content and are generally excluded from the diets of persons who have a high cholesterol level in the blood (see the essay preceding Experiment 8). These people have a difficulty in metabolizing saturated fats correctly and should avoid them since they encourage cholesterol deposits to form in the arteries. This ultimately leads to high blood pressure and heart trouble. Persons with normal metabolism, however, have no real need to avoid saturated fats.

Butter, when unrefrigerated and left exposed to air, will turn rancid, giving an unpleasant odor and taste. This is due to the hydrolysis of the triglycerides by the moisture in air and to oxidation occurring at the double bonds in the fatty acid components. Oxygen will add to allylic positions by an abstraction-addition reaction to give hydroperoxides. Rancid butter smells bad compared to a partially hydrolyzed

$$R-CH=CH-CH_2-CH=CH-(CH_2)_7COOH + O_2 \longrightarrow$$

$$R-CH=CH-\overset{\overset{OOH}{|}}{CH}-CH=CH-(CH_2)_7COOH$$

margarine or cooking fat, because it contains, along with the fatty acids listed in the table, triglycerides which are composed of butyric (C_4), caproic (C_6), caprylic (C_8), and capric (C_{10}) acid moieties as well. These low molecular weight carboxylic acids are the source of the well-known objectionable odor. The hydroperoxides also decompose to low molecular weight aldehydes which also have objectionable odors and tastes. The same reaction takes place when fats are burned, as in an oven fire. On combustion, an unsaturated fat produces large amounts of acrolein, a potent **lachrymator** (tear inducer), and other aldehydes which also irritate the eyes.

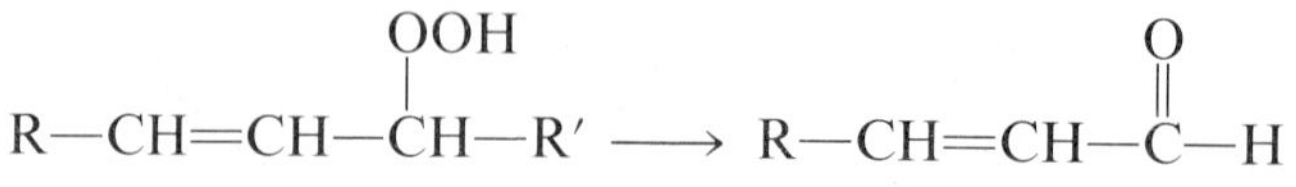

$$R-CH=CH-\overset{\overset{\displaystyle OOH}{|}}{CH}-R' \longrightarrow R-CH=CH-\overset{\overset{\displaystyle O}{\|}}{C}-H$$

If R=H, Acrolein

Some oils, mainly those which are highly unsaturated (e.g., linseed oil), thicken on exposure to air and eventually harden to give a smooth, clear resin. Such oils are called **drying oils,** and they are widely used in the manufacture of shellacs, varnishes, and paints. Apparently the double bonds in these compounds undergo both partial oxidation and polymerization on exposure to light and air.

REFERENCES

Dawkins, M. J. R., and Hull, D. "The Production of Heat by Fat." *Scientific American, 213* (August, 1965), 62.
Eckey, E. W., and Miller, L. P. *Vegetable Fats and Oils.* ACS Monograph No. 123. New York: Reinhold, 1954.
Green, D. E. "The Metabolism of Fats." *Scientific American, 190* (January, 1954), 32.
Green, D. E. "The Synthesis of Fats." *Scientific American, 202* (February, 1960), 46.
Shreve, R. N. "Oils, Fats, and Waxes." *The Chemical Process Industries* (3rd edition). New York: McGraw Hill, 1967.

Experiment **14**

GAS CHROMATOGRAPHIC ANALYSIS OF THE FATTY ACID COMPOSITION OF FATS AND OILS

Transesterification
Sealed Tube Reaction
Gas Chromatography

NOTE TO THE STUDENT: Although a number of fats and oils will be available for this experiment, the number of these will be limited. If you could bring a small (10 ml) sample of your own favorite cooking oil, shortening (Crisco, peanut oil, Saffola, lard, Wesson Oil, Mazola, soybean oil, butter) or other oil (castor bean, cottonseed, linseed), the range of the experiment can be extended. If you work in a place that uses a commercial cooking oil, this oil would also be interesting. If a wider variety of fats and oils is available, more can be learned from the experiment.

The purpose of this experiment is to determine the **average** fatty acid composition of various commonly used fats and oils. Since the typical fat or oil (triglyceride) is a triester of glycerol, each molecule of the fat or oil is composed of three fatty acids combined with glycerol. In any given molecule, these three fatty acids may either be the same, or they may all be different. Additionally, there may be as many as 10 different fatty acids involved in the formation of the triglyceride mixture that represents a fat or oil. To solve the problems inherent in the analysis of a great many different triglycerides, it will be necessary to simplify the problem by "breaking apart" the triglyceride molecules to liberate the individual fatty acids.

The classical procedure for obtaining the fatty acids from a fat is **saponification** (soap making), or hydrolysis with alkali to give the sodium salts of the fatty carboxylic acids mixed with glycerol. This mixture constitutes a soap (see the essay preceding Experiment 15). Acidification of the soap mixture would give the fatty acids:

$$RCOO^-Na^+ + HCl \longrightarrow RCOOH + NaCl$$

Unfortunately, saponification is rather slow at room temperature, and at higher temperatures the unsaturated acids show a tendency to isomerize the positions of their double bonds. This would clearly interfere with the analysis. An additional complication would be that, while some of the acids are solids, others are liquids, and all are high boiling.

Instead of saponification, we will make use of the **transesterification** reaction, where one alcohol group of an ester is exchanged for another. If we use a monohydroxylic alcohol like methanol, **three** different methyl esters and one molecule of glycerol will be produced.

Triglyceride (a fat or an oil)   **Three Methyl Esters**   **Glycerol**

The methyl esters have lower boiling points than the acids, and can easily be analyzed by gas chromatography. By using a large excess of methanol, the equilibrium is shifted completely to the right. As a catalyst for the reaction, the Lewis acid boron trifluoride will be used rather than a protonic acid. The reaction will be carried out in a **sealed tube** so that the mixture may be heated without loss of either methanol or boron trifluoride, both of which are low boiling. At the end of the transesterification, the sealed tube will be opened and a small amount of water added. A layer containing the water insoluble methyl esters will float on top of the water. The composition of this layer will be determined by gas chromatographic analysis.

Three representative gas chromatograms are shown below. The first of these is a gas chromatogram of a synthetic reference mixture, made up from the methyl esters of myristic, palmitic, stearic, oleic, and linoleic acids. The peak corresponding to each of these substances is labeled with either M, P, S, O, or L to designate its origin. The gas chromatograms of the mixtures of methyl esters prepared from samples of safflower oil and of soybean oil are also shown. You will prepare a similar gas chromatogram, starting with a fat or oil of your choice, and from the chromatogram you will calculate the percentage of each of the five fatty acids M, P, S, O, and L. These can quickly be identified because their retention times will have been established using the chromatogram obtained from the reference mixture. From the table in the essay which precedes this experiment, it can be seen that these five acids are the major constituents of most of the fats and oils. We will ignore any minor constituents. Some of the triglycerides have a major component which does not correspond to these five reference compounds. You will find it easy to identify any major components of this type simply by referring to the table in the essay and by using a process of elimination. Any major component of this type should also be included in the calculations. Analysis by this method requires only a small amount of material—about 8 drops will suffice. This is a major advantage of the gas chromatographic method.

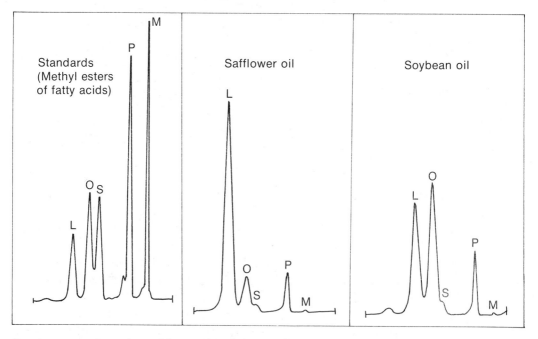

Gas chromatographic analysis of fatty acid methyl esters (210°, Varian-aerograph A90P gas chromatograph, 6.5 ft × 1/4 in D column, packed with 8% diethyleneglycol succinate (DEGS) on chromosorb W).

SPECIAL INSTRUCTIONS

Read the essay immediately preceding the experiment. In addition, it will be necessary to read Technique 12 ("Gas Chromatography"). If you bring a sample of cooking fat or oil from home, you should carefully inspect the label to see if it reveals the source. For instance, Mazola is corn oil and Wesson Oil is derived partly from soybean oil. You will need help from the instructor or a laboratory assistant when sealing the reaction tube. It must be sealed tightly so that it does not leak or break when pressure builds up inside. Once the tube has been sealed, there is always the possible (but small) danger that an incorrectly sealed tube may explode when pressure builds inside. Wear safety goggles and be sure that the tube is behind a shield when heated. Treat the sealed tube with respect. The instructor or an assistant will either perform the gas chromatographic analysis for you or show you how to do it.

> NOTE TO THE INSTRUCTOR: Since the column must be heated to a relatively high temperature (210°), it will be necessary to begin heating the gas chromatograph several hours **before** the laboratory begins in order to allow the columns time to equilibrate.

PROCEDURE

TRANSESTERIFICATION OF THE TRIGLYCERIDES

Prepare a container for the reaction by taking a 25 cm length of 8 mm glass tubing and sealing one end. Using a gas–oxygen torch, heat the tube near the end and, with tongs, pull a short section of the tubing off so as to seal the end. Return the sealed end of the tube to the flame briefly and rotate it in the flame until it is clear that a good seal has been made. This method of sealing the tube is much more reliable than simply rotating the very end of the tube in the flame until it closes. Be sure that the glass has not been excessively thickened at the sealed end.

Add about 8 drops of the oil to the tube with an eye dropper. Add 2 ml of benzene and 1 ml of a solution of boron trifluoride in methanol. Cool the mixture in a dry ice/methylene chloride bath until the sample solidifies. While the sample is still frozen, quickly heat the unsealed end of the tube **near the end** and pull off a short section so as to seal the end. Once again, tongs should be used to pull off the heated section.

When the tubing is cool, shake the sample well and place it in a bath of boiling water behind a shield **in the hood.** After placing the sample in the hood, the hood window should be closed and left closed until it is time to remove the sample. Allow the sample to heat for about

15–20 minutes. While you are waiting, you will have time to do the unsaturation tests described in the third part of the experimental procedures (below).

After the heating period, remove the sealed tube from the heating bath and cool it in a bath of ice water. Wrap the tube in a piece of cloth towel so that only about 2 or 3 cm of the empty end protrudes. Using a file, score a line on the tubing about 2 cm from the end. Then, in an empty hood, place the score mark on the edge of a wooden block with the tube tilted so that the liquid is at the opposite end, and strike the tube with either a hammer or another piece of wood so as to break off the scored section and open the tube. This entire operation should be done quickly enough so that the contents of the tube are kept cold. If the tube has become warm during the scoring operation, cool it again before it is broken open. If the tubing did not break cleanly, score it again and break off another section by flexing the tubing between your thumbs (the usual manner of breaking tubing).

Add 1 ml of water to the tube, stopper it with a small cork, and shake it vigorously. Allow the layers to separate. The upper layer (benzene) contains the methyl esters of the fatty acid. Submit it for gas chromatographic analysis.

GAS CHROMATOGRAPHIC ANALYSIS OF THE METHYL ESTERS

With the help of the instructor or an assistant, inject a small sample (about 2 μl) from the benzene layer into the gas chromatograph. From the gas chromatogram obtained, identify the various peaks by comparison of their retention times to those of the known substances in the reference mixture. The reference chromatogram will either be posted or the numerical retention time values will be made available. Using the method explained in Technique 12, Section 10, calculate the approximate percentage composition of the mixture with respect to its major components. Areas of each peak may be estimated by multiplying the peak height by the width of the peak at half height. Sum all the areas estimated in this way and divide each individual value by this sum. Multiply each fraction obtained in this way by 100 to get the percentage that component represents.

UNSATURATION TESTS

Dissolve 0.5 ml (approximately 10 drops) of cottonseed oil in 3 ml of carbon tetrachloride. Add a 5% bromine–carbon tetrachloride solution **drop by drop with shaking,** until the bromine color is no longer rapidly discharged. Record the number of drops required. Repeat this procedure using 0.5 ml (10 drops) of a melted cooking fat or shortening

(Crisco or margarine). Compare the result and explain your observations.

QUESTIONS

1. Why was boron trifluoride used rather than a protonic acid?

2. Draw a mechanism that shows the function of boron trifluoride.

3. Why does neither methanol nor glycerol give rise to a peak in the gas chromatogram?

4. If an oil were composed of a mixture of triglycerides formed from only *two* different fatty acids, how many different possible triglycerides might be found in the oil? Draw them.

5. Repeat problem 4 assuming that *three* different fatty acids compose the mixture of triglycerides.

ESSAY
Soaps and Detergents

Soaps as we know them today were virtually unknown prior to the first century A.D. Before this time, clothes were cleaned primarily by the abrasive action of rubbing them on rocks in water. Somewhat later, it was discovered that certain types of leaves, roots, nuts, berries and barks form soapy lathers which solubilize and remove dirt from clothes. We now know these natural materials which lather as **saponins.** Many of these saponins contain pentacyclic triterpene carboxylic acids, such as oleanolic acid or ursolic acid, chemically combined with a sugar molecule. In addition, these acids also appear in the uncombined state. Saponins were probably the first known "soaps." They may have also been an early source of pollution in that they are known to be toxic to fish. The pollution problem associated with the development of soap and detergents has been a long and controversial one.

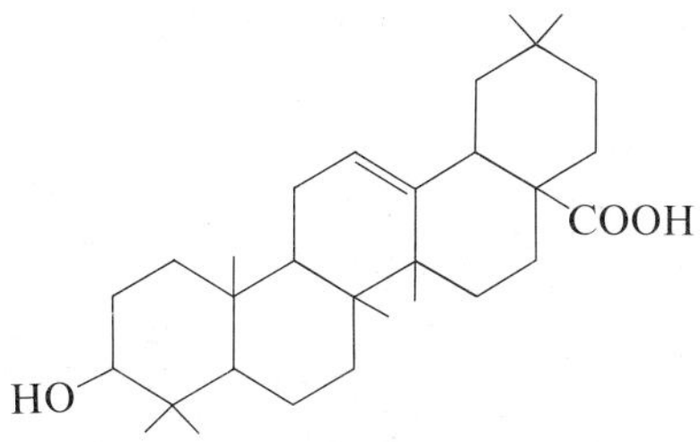

Oleanolic Acid 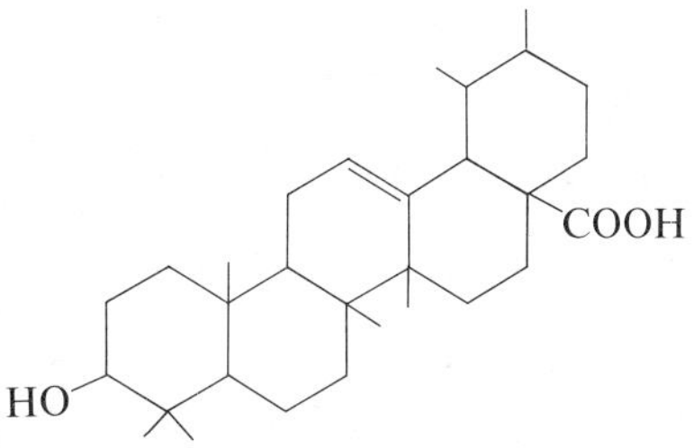   Ursolic Acid

The discovery of soap as we know it today probably evolved over a number of centuries from experimentation with crude mixtures of alkaline and fatty materials. Pliny the Elder described the manufacture of soap during the first century A.D. A modest soap factory was even built in Pompeii. During the Middle Ages, cleanliness of the body or clothing was not considered to be important. Those who could afford it used perfumes to hide their body odor. Perfumes, like fancy clothes, were a status symbol for the rich. An interest in cleanliness again re-emerged during the eighteenth century when disease-causing micro-organisms were discovered.

SOAPS

The process of soapmaking has remained practically unchanged for 2000 years. The procedure involves the basic hydrolysis (saponification) of a fat. Chemically, fats are usually referred to as **triglycerides.** They contain ester functional groups. The saponification process involves the heating of the fat with an alkaline solution. This alkaline solution was originally obtained by leaching wood ashes or from the evaporation of naturally occurring alkaline waters. Today, lye (sodium hydroxide) is used as the source of the alkali. The alkaline solution hydrolyzes the fat into its component parts, the salt of a long chain carboxylic acid (soap) and an alcohol (glycerol). When common salt is added, the soap precipitates. The soap is washed free of unreacted sodium hydroxide, and molded into bars. The carboxylic acid salts of soap usually contain from 12 to 18 carbons arranged in a straight chain. The carboxylic acids containing even numbers of carbon atoms predominate, and the chains may contain unsaturation. The chemical structures of fats, and the related oils, are shown in the essay which precedes Experiment 14. The equation given below shows how soap is produced from a fat. It is an idealized reaction, where the R groups are the same. Usually, the R groups are **not** the same. Therefore, soap is actually a mixture of salts of carboxylic acids.

$$
\begin{array}{l}
\underset{\text{Triglyceride}}{\underset{\text{(Fat)}}{
\begin{array}{l}
R-\overset{\overset{\textstyle O}{\|}}{C}-O-CH_2 \\[2mm]
R-\overset{\overset{\textstyle O}{\|}}{C}-O-CH \\[2mm]
R-\overset{\overset{\textstyle O}{\|}}{C}-O-CH_2
\end{array}}}
\;+\; \underset{\text{Lye}}{3\,NaOH} \;\longrightarrow\;
\underset{\substack{\text{Sodium Salt}\\\text{of an Acid (Soap)}}}{3\,R-\overset{\overset{\textstyle O}{\|}}{C}-O^-Na^+}
\;+\;
\underset{\text{Glycerol}}{
\begin{array}{l}
HO-CH_2 \\[2mm]
HO-CH \\[2mm]
HO-CH_2
\end{array}}
$$

A disadvantage of soap is that it is an ineffective cleanser in hard water. Hard water contains salts of magnesium, calcium and iron in solution. When soap is used in hard water, "calcium soap," the insoluble calcium salts of the fatty acids, and other precipitates are deposited as **curds.** This precipitate or curd is referred to as "bathtub ring." Although it is a poor cleanser in hard water, soap is, nevertheless, an excellent cleanser in soft water.

$$2 \underset{\text{Soap}}{R-\overset{\overset{\textstyle O}{\|}}{C}-O^- Na^+} + Ca^{++} \longrightarrow \left(\underset{\text{Curd}}{R-\overset{\overset{\textstyle O}{\|}}{C}-\overset{-}{O}} \right)_2 Ca^{++}$$

Water softeners are added to soaps to help remove the troublesome hard water ions so that the soap will remain effective in hard water. Sodium carbonate or trisodium phosphate will precipitate the ions as the carbonate or phosphate. Unfortunately, the precipitate may become lodged in the fabric of the clothing, causing an off-white color.

$$Ca^{++} + CO_3^= \longrightarrow CaCO_3 \downarrow$$

$$3 Ca^{++} + 2PO_4^{-3} \longrightarrow Ca_3(PO_4)_2 \downarrow$$

An important advantage of soap is that it is **biodegradable.** Microorganisms can consume the linear soap molecules and convert them to carbon dioxide and water. The soap is thus eliminated from the environment.

ACTION OF SOAP IN CLEANING

Dirty clothes, skin, or other surfaces have particles of dirt suspended in a layer of oil or grease. Polar water molecules cannot remove the dirt embedded in nonpolar oil or grease. However, one may remove the dirt with soap because of its dual nature. The soap molecule has a polar, **water soluble** head (carboxylate salt) and a long, **oil soluble** tail, the hydrocarbon chain. The hydrocarbon tail of soap dissolves itself into the oily substance but leaves the ionic end outside the oily surface. When sufficient numbers of soap molecules have oriented themselves around an oil droplet with their hydrocarbon ends dissolved in the oil, the oil droplet, together with the suspended dirt particles, will be removed from the surface of the cloth or skin. The oil droplet is removed because the heavily negatively charged oil droplet is now strongly attracted to and solvated by water. The resulting solvated oil droplet is called a **micelle,** and is shown in the figure.

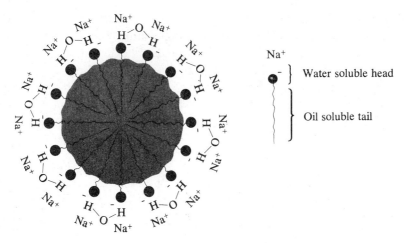

A soap micelle solvating a droplet of oil (from Linstromberg, Walter W.; *Organic Chemistry; a brief course,* Heath, 1974.)

DETERGENTS

Detergents are synthetic cleaning compounds, often referred to as "syndets." They were developed as an alternative to soaps because they are effective in **both** soft and hard water. No precipitates form when calcium, magnesium, or iron ions are present in a detergent solution. One of the earliest detergents which was developed was sodium lauryl sulfate. It is prepared by the action of sulfuric acid or chlorosulfonic acid on lauryl alcohol (1-dodecanol). However, this detergent is relatively expensive. One industrial method of preparation is shown in the reactions which follow:

$$CH_3(CH_2)_{10}CH_2OH + H_2SO_4 \longrightarrow CH_3(CH_2)_{10}CH_2OSO_3H + H_2O$$
Lauryl Alcohol

$$CH_3(CH_2)_{10}CH_2OSO_3H + Na_2CO_3 \longrightarrow$$
$$CH_3(CH_2)_{10}CH_2OSO_3{}^-Na^+ + NaHCO_3$$
Sodium Lauryl Sulfate

The first of the inexpensive detergents appeared around 1950. These detergents, called alkylbenzenesulfonates (ABS), can be prepared from inexpensive petroleum sources by the following set of reactions.

$$4\ CH_3CH{=}CH_2 \xrightarrow{\ H^+\ }$$

$$CH_3{-}CH{-}\left(CH_2{-}CH\right)_2 CH{=}CH$$
with CH$_3$ groups

ALKENE POLYMERIZATION

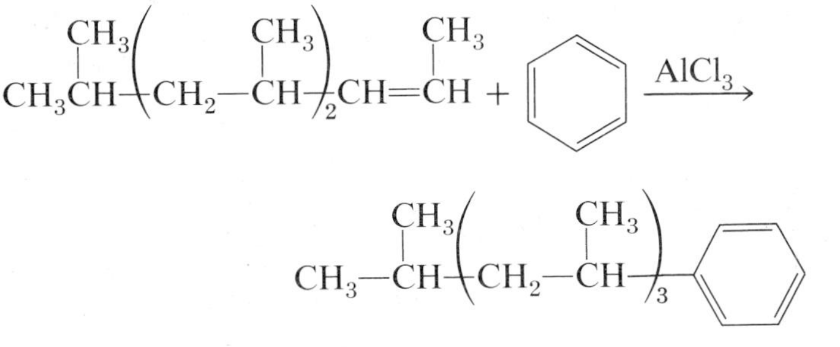

$$CH_3CH-(CH_2-CH)_2CH=CH \begin{matrix} CH_3 \\ | \end{matrix} \begin{matrix} CH_3 \\ | \end{matrix} \begin{matrix} CH_3 \\ | \end{matrix} + \bigcirc \xrightarrow{AlCl_3}$$

$$CH_3-CH-(CH_2-CH)_3 \bigcirc \qquad \begin{matrix} \text{FRIEDEL-} \\ \text{CRAFTS} \\ \text{ALKYLATION} \end{matrix}$$

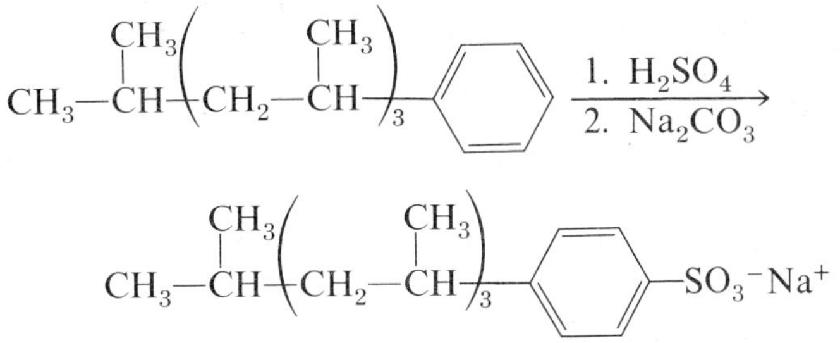

$$CH_3-CH-(CH_2-CH)_3 \bigcirc \xrightarrow[\text{2. } Na_2CO_3]{\text{1. } H_2SO_4}$$

$$CH_3-CH-(CH_2-CH)_3 \bigcirc -SO_3^-Na^+ \qquad \text{SULFONATION}$$

An Alkylbenzenesulfonate (ABS)

These detergents became very popular because they could be used effectively in all types of water and were cheap. They rapidly displaced soap as the most popular cleaning agent. A problem with these detergents was that they passed through sewage treatment plants without being degraded by the treatment process microorganisms. Rivers and streams in many sections of the country became polluted with soap foam. The detergents even found their way into the drinking water supplies of numerous cities. The reason for the persistence of the detergent was that bacterial enzymes, which could degrade straight chain soaps and sodium lauryl sulfate, could not destroy the highly branched detergents such as the alkylbenzenesulfonates (ABS).

It was soon found that the bacterial enzymes could only degrade a chain of carbons which contained, at the most, **one** branch. As numerous cities and states banned the sale of the non-biodegradable detergents, they were replaced by 1966 with the new biodegradable detergents called linear alkylsulfonates (LAS). One example of a LAS detergent is shown below. Notice that there is one branch next to the aromatic ring.

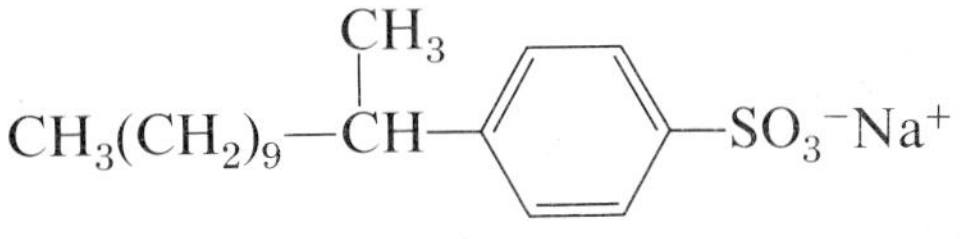

$$CH_3(CH_2)_9-CH \begin{matrix} CH_3 \\ | \end{matrix} \bigcirc -SO_3^-Na^+$$

A Linear Alkylsulfonate Detergent (LAS)

NEW PROBLEMS WITH DETERGENTS

Detergents (also soaps) are not sold as pure compounds. A typical heavy duty, controlled "sudser" may contain only 8 to 20% of the linear alkylsulfonate. A large quantity (30 to 50%) of a "builder" such as sodium tripolyphosphate ($Na_5P_3O_{10}$) may be present. Other additives include corrosion inhibitors, antideposition agents, and perfumes. Optical brighteners are also added. Brighteners absorb invisible ultraviolet light and re-emit it as visible light so that laundry appears white, and thus "clean." The phosphate builder is added to complex the hard water ions, calcium and magnesium, and keep them in solution. Builders seem to enhance the washing ability of the LAS and also act as a cheap filler.

Unfortunately, phosphates speed the **eutrophication** of lakes and other bodies of water. They, along with other substances, serve as nutrients for the growth of algae. When the algae begin to die and decompose, they consume so much dissolved oxygen from the water that no other life can exist in that water. The lake rapidly "dies."

Because of the undesirable effect of phosphates, a search was initiated for a replacement of the phosphate builders. Some replacements have been made, but most of these also have problems associated with them. Two replacement builders, sodium metasilicate and sodium perborate, are highly basic substances, and they have caused injuries to children. In addition, they appear to destroy bacteria in sewage treatment plants and may have other unknown environmental effects.

Many have suggested a return to soap. The main problem is that we probably cannot produce enough soap to meet the demand because of the limited amount of animal fat available. Where do we go from here?

REFERENCES

Davidsohn, J. et al., *Soap Manufacture*. New York: Interscience, 1953.
Garner, W. *Detergents*. Amsterdam, Elsevier, 1966.
Kushner, L. M., and Hoffman, J. I. "Synthetic Detergents." *Scientific American, 185* (October, 1951), 26.
"LAS Detergents End Stream Foam." *Chemical and Engineering News, 45* (1967), 29.
Levey, M. "The Early History of Detergent Substances." *Journal of Chemical Education, 31* (1954), 521.
"Makers Plan New Detergent Formulas." *Chemical and Engineering News, 48* (1970), 18.
Puplett, P. A. R. *Synthetic Detergents*. London: Sidgwick and Jackson, 1957.
Snell, F. D. "Soap and Glycerol." *Journal of Chemical Education, 19* (1942), 172.
Snell, F. D., and Snell, C. T. "Syndets and Surfactants." *Journal of Chemical Education, 35* (1958), 271.

Experiment **15**
PREPARATION OF SOAP

Hydrolysis Of A Fat (Ester)
Filtration

In this experiment we will prepare soap from animal fat (lard). Animal fats and vegetable oils are esters of carboxylic acids with high molecular weight and the alcohol glycerol. Chemically, these fats and oils are called **triglycerides.** The principal acids in animal fats and vegetable oils can be prepared from the naturally occurring triglycerides by alkaline hydrolysis (saponification). The naturally occurring acids are

$$
\begin{array}{c}
\underset{\text{O}}{\overset{\text{O}}{\underset{\parallel}{\text{R}_1\text{C}}}}{-}\text{O}{-}\text{CH}_2 \\
\underset{\text{O}}{\overset{\text{O}}{\underset{\parallel}{\text{R}_2\text{C}}}}{-}\text{O}{-}\text{CH} \\
\underset{}{\overset{\text{O}}{\underset{\parallel}{\text{R}_3\text{C}}}}{-}\text{O}{-}\text{CH}_2
\end{array}
\xrightarrow[\substack{\text{saponification} \\ \text{or} \\ \text{hydrolysis}}]{\text{NaOH}}
\begin{array}{l}
\text{R}_1\text{COO}^-\text{Na}^+ \\
{}+ \text{R}_2\text{COO}^-\text{Na}^+ + \\
\text{R}_3\text{COO}^-\text{Na}^+
\end{array}
\begin{array}{l}
\text{HO}{-}\text{CH}_2 \\
\text{HO}{-}\text{CH} \\
\text{HO}{-}\text{CH}_2
\end{array}
$$

Triglycerides (fat or oil)	**Carboxylic Acid Salts** (Soap)	**Glycerol**

rarely of a single type in any given fat or oil. In fact, a single triglyceride molecule in a fat may contain three different acid residues (R_1COOH, R_2COOH, R_3COOH), and not every triglyceride in the substance will be identical. Each fat or oil, however, has a characteristic **statistical distribution** of the various types of acids possible. The composition of the common fats and oils are given in the essay which precedes Experiment 14.

The fats and oils that are most commonly used in soap preparations are lard and tallow from animal sources, and coconut, palm, and olive oils from vegetable sources. The length of the hydrocarbon chain and the number of double bonds in the carboxylic acid portion of the fat or oil determine the properties of the resulting soap. For example, a salt of a saturated long chain acid makes a harder, more insoluble soap. Chain length also affects solubility.

Tallow is the principal fatty material used in soapmaking. The solid fats of cattle are melted with steam, and the tallow layer formed at the top is removed. Soapmakers usually blend tallow with coconut oil and saponify this mixture. The resulting soap mainly contains the salts of palmitic, stearic, and oleic acids from the tallow, and the salts of lauric and myristic acids from the coconut oil. The coconut oil is added in order to produce a softer, more soluble soap. Lard (from hogs) differs from tallow (from cattle or sheep) in that lard contains somewhat more oleic acid.

TALLOW $CH_3(CH_2)_{14}COOH$ $CH_3(CH_2)_{16}COOH$
                Palmitic Acid   Stearic Acid

$$CH_3(CH_2)_7CH=CH(CH_2)_7COOH$$
                            Oleic Acid

COCONUT OIL $CH_3(CH_2)_{10}COOH$ $CH_3(CH_2)_{12}COOH$
                          Lauric Acid   Myristic Acid

Pure coconut oil yields a soap which is very soluble in water. The soap contains essentially the salt of lauric acid, with some myristic acid. It is so soft (soluble) that it will lather even in sea water. Palm oil contains mainly two acids, palmitic acid and oleic acid, in about equal amounts. Saponification of this oil yields a soap which is an important constituent in toilet soaps. Olive oil contains mainly oleic acid. It is used to prepare Castile soap, named after the region in Spain where it was first made.

Toilet soaps generally have been carefully washed free of any remaining alkali used in the saponification procedure. As much glycerol as possible is usually left in the soap, and perfumes and medicinal agents are sometimes added. Floating soaps are produced by blowing air into the soap as it solidifies. Softer soaps are made by using potassium hydroxide, yielding potassium salts rather than the sodium salts of the acids. They are used in shaving creams and liquid soaps. Scouring soaps have abrasives added such as fine sand or pumice.

SPECIAL INSTRUCTIONS

Read the essay which precedes this experiment. In addition it is necessary to have read Techniques 1 and 2 before starting this experiment. This experiment is short and can easily be scheduled with another experiment.

PROCEDURE

Prepare a solution of 10 g of sodium hydroxide dissolved in a mixture of 18 ml of water and 18 ml of 95% ethanol. Place 10 g of lard (or other fat or oil) in a 250 ml beaker and add the solution to it. Heat the mixture on a steam bath for at least 30 minutes. Prepare another 40 ml of a 50-50 solution of ethanol–water. Small portions of this solution should be added during the 30 minutes whenever necessary to prevent foaming. Stir the mixture constantly.

Prepare a solution of 50 g of sodium chloride in 150 ml water in a 400 ml beaker. If it is necessary to heat the solution to dissolve the salt, it should be cooled before proceeding. Quickly pour the saponification mixture into the cooled salt solution. Stir the mixture thoroughly for several minutes and then cool it to room temperature in an ice bath. Collect the precipitated soap by vacuum filtration using a Büchner funnel (Technique 2, Section 2.3). Wash the soap with two portions of ice-cold water. Continue to draw air through the soap to partially dry the product. Allow the soap to dry overnight. Weigh the product.

If you are preparing the detergent in Experiment 16, save a small sample for the short comparison tests. Submit the remainder of the product to your laboratory instructor in a labeled vial.

QUESTIONS

1. Why should the potassium salts of fatty acids yield softer soaps?

2. Why is the soap derived from coconut oil so soluble?

3. Why does the addition of a salt solution cause the soap to precipitate?

4. Why do you suppose a mixture of ethanol and water is used for the saponification rather than simply water itself?

Experiment 16
PREPARATION OF A DETERGENT

$$CH_3(CH_2)_{10}CH_2O-\overset{\overset{\displaystyle O}{\|}}{\underset{\underset{\displaystyle O}{\|}}{S}}-O^-Na^+$$

We will prepare a detergent, sodium lauryl sulfate, in this experiment. A detergent is usually defined as a synthetic cleaning agent, whereas a soap is derived from a naturally occurring source, a fat or oil. The differences between the two basic types of cleaning agents are discussed in the essay which precedes Experiment 15. In the second part of this experiment, the properties of soap will be compared to those of the prepared detergent.

The reactions used to prepare the detergent are as follows:

$$CH_3(CH_2)_{10}CH_2OH + Cl\!-\!\overset{\displaystyle O}{\underset{\displaystyle O}{\overset{\|}{\underset{\|}{S}}}}\!-\!OH \longrightarrow$$

  Lauryl Alcohol   **Chlorosulfonic
                              Acid**

$$CH_3(CH_2)_{10}CH_2O\!-\!\overset{\displaystyle O}{\underset{\displaystyle O}{\overset{\|}{\underset{\|}{S}}}}\!-\!OH + HCl$$

                              **Lauryl Ester of
                              Sulfuric Acid**

$$2\,CH_3(CH_2)_{10}CH_2\!-\!O\!-\!\overset{\displaystyle O}{\underset{\displaystyle O}{\overset{\|}{\underset{\|}{S}}}}\!-\!OH + Na_2CO_3 \longrightarrow$$

$$2\,CH_3(CH_2)_{10}CH_2O\!-\!\overset{\displaystyle O}{\underset{\displaystyle O}{\overset{\|}{\underset{\|}{S}}}}\!-\!O^- Na^+ + H_2O + CO_2$$

                      Sodium Lauryl Sulfate

In the first step, lauryl alcohol is allowed to react with chlorosulfonic acid to give the lauryl ester of sulfuric acid. Aqueous sodium carbonate is then added to produce the sodium salt (detergent). The aqueous mixture is saturated with solid sodium carbonate and extracted with 1-butanol. The sodium carbonate must be added to give phase separation, otherwise 1-butanol would be soluble in water. The sodium salt (detergent) is more soluble in the 1-butanol (organic layer) than the aqueous layer because of the long hydrocarbon chain which gives the salt considerable organic (nonpolar) character.

SPECIAL INSTRUCTIONS

Read the essay which precedes Experiment 15, especially the sections dealing with the behavior of soaps and detergents in hard water. In addition, it is necessary to have read Technique 5. Handle chlorosulfonic acid with extreme caution.

NOTE TO THE INSTRUCTOR: The lauryl alcohol is usually supplied in a narrow mouth bottle. Melt the alcohol (mp 24 to 27°) and pour the liquid into a wide mouth bottle. When it crystallizes, it is easily broken into small pieces with a spatula.

PROCEDURE

Add 9.5 ml of glacial acetic acid to a **dry** 250 ml beaker. Cool the acetic acid to about 5° in an ice bath. Slowly add 3.5 ml (d = 1.77 g/ml) of chlorosulfonic acid from a buret (in the fume hood), directly into the beaker containing the acetic acid. Allow the contents of the beaker to cool in the ice bath **in the hood.** Be careful not to allow any water to enter the beaker.

> **Use chlorosulfonic acid with extreme caution. It is an extremely strong acid similar to concentrated sulfuric acid. It will cause immediate burns on the skin. Use rubber gloves when dispensing the corrosive liquid.**

While stirring, slowly add 10 g of lauryl alcohol (1-dodecanol), in small pieces, to the beaker over a 5 minute period. Stir the mixture until all of the lauryl alcohol is dissolved and has reacted (about 30 minutes). Pour the reaction mixture into a 250 ml beaker containing 30 g of cracked ice.

Add 30 ml of 1-butanol to the beaker containing the reaction mixture and the cracked ice. Stir the mixture thoroughly for 3 minutes. While stirring, slowly add a saturated sodium carbonate solution in 3 ml portions until the solution is neutral or slightly basic to litmus. Add 10 g of solid sodium carbonate to the mixture to aid in layer separation. Allow the layers to separate in the beaker and then pour the upper 1-butanol layer into a separatory funnel. Some of the aqueous layer may unavoidably be transferred to the separatory funnel with the organic phase. Add another 20 ml portion of 1-butanol to the aqueous layer remaining in the beaker. Stir the solution well for 3 minutes and allow the layers to separate. Again pour the upper layer into the separatory funnel containing the first extract. Discard the remaining aqueous layer.

Allow the phases to separate in the separatory funnel for about 5 minutes. Drain the lower aqueous phase until the upper layer begins to enter the stopcock and discard it (Technique 5, Section 5.4). Pour the organic layer from the top of the separatory funnel into a large beaker (400 ml or larger). Evaporate the 1-butanol solvent (bp 117°) in a hood on a hot plate adjusted to a low temperature. The detergent precipitates as the solvent is removed. Stir the mixture occasionally to prevent the product from decomposing. This is especially important near the end of the evaporation process. When most of the 1-butanol has been removed, place the moist solid in an oven at 80° (or a steam cabinet) in order to completely dry the solid. The drying procedure should be completed by the next laboratory period. It is important not to dry the sample at a high temperature (>80°) because of the danger of decomposition. When the sample has dried, remove the colorless product from the beaker. Weigh the detergent and calculate the percent yield. If the detergent is not totally free of 1-butanol, the yield may exceed 100%. If necessary, continue to dry the sample. Part of the sample is used for the tests which follow. Submit the remaining detergent to the instructor in a labeled vial.

TESTS ON SOAPS AND DETERGENTS

Soap

Add 10 ml of a soap solution[1] to a 125 ml Erlenmeyer flask. Stopper the flask securely with a rubber stopper and shake it vigorously for about 15 seconds. Allow the solution to stand for 30 seconds and observe the level of the foam. Add 4 drops of 4% calcium chloride solution from an eye dropper. Shake the mixture for 15 seconds, and allow it to stand for about 30 seconds. Observe the effect of the calcium chloride on the foam. Do you observe anything else? Add 1 g of trisodium phosphate and shake the mixture again for about 15 seconds. Allow the solution to stand for 30 seconds. What do you observe? Explain the results of these tests in your laboratory report.

Detergent

Dissolve 0.75 g of your prepared detergent in 50 ml of water. Add 10 ml of this solution to a 125 ml Erlenmeyer flask. Stopper the flask securely with a rubber stopper and shake it vigorously for 15 seconds. Allow the solution to stand for about 30 seconds and observe the level of the foam. Add 4 drops of 4% calcium chloride solution. Shake the mixture for about 15 seconds and allow it to stand for 30 seconds. What do you observe? Explain the results of these tests in your laboratory report.

QUESTIONS

1. Draw a mechanism for the reaction of lauryl alcohol with chlorosulfonic acid.

2. Why do you suppose sodium carbonate, rather than some other base, is used for neutralization?

---

[1] The soap solution is prepared as follows: Add one bar of Ivory soap to 1 liter of distilled or soft water. Stir the solution occasionally and allow the mixture to stand overnight. Remove the remainder of the bar. The mixture can be used directly. A large batch should be prepared by the laboratory instructor. Alternatively, a 0.15 g sample of soap from Experiment 15 may be added to 10 ml of distilled or soft water.

Pheromones: Insect Attractants and Repellents

For humans, who are accustomed to heavy reliance on visual and verbal forms of communication, it is difficult to imagine that there are forms of life that depend primarily on the release and perception of **odors** to communicate with one another. Among insects, however, this is perhaps the major form of communication. Many species of insects have developed a virtual "language" based on the exchange of odors. These insects have well-developed scent glands, often of several different types, which have as their sole purpose the synthesis and release of chemical substances. These chemical substances, known as **pheromones,** are secreted by insects and, when detected by other members of the same species, induce a specific and characteristic response. Pheromones are usually of two distinct types: releaser pheromones and primer pheromones. **Releaser pheromones** produce an immediate **behavioral** response in the recipient insect; **primer pheromones** trigger a series of **physiological** changes in the recipient. Some pheromones, however, combine both releaser and primer effects.

Among the most important types of releaser pheromones are the sex attractants. **Sex attractants** are pheromones secreted by either the female or, less commonly, the male of the species to attract the opposite member for the purposes of mating. Since, in larger concentrations, sex pheromones also induce a physiological response in the recipient, for example the changes necessary to the mating act, they also have a primer effect and are, therefore, misnamed.

Anyone who has owned a female cat or dog will know that sex pheromones are not limited to insects. Female cats or dogs widely advertise, by odor, their sexual availability when they are "in heat." This type of pheromone is not uncommon to mammals. Some persons even believe that there are human pheromones which are responsible for attracting certain sensitive males and females to one another. This idea is, of course, responsible for many of the perfumes which are now widely available. Whether or not the idea is correct cannot yet be established, but there are proven sexual differences in the ability of humans to smell certain substances. For instance, Exaltolide, a synthetic lactone of 14-hydroxytetradecanoic acid, can be perceived only by females or, in some cases, by males after they have been injected with an estrogen. Exaltolide is very similar in overall structure to civetone (civet cat) and muskone (musk deer), which are two naturally occurring compounds believed to be mammalian sex pheromones. Whether or not human males emit pheromones has never been established. Curiously, exaltolide is used in perfumes intended for female as well as male use! But while the odor may lead a woman to believe that she smells pleasant, it cannot possibly have any effect on the male. The "musk oils," civetone and muskone, have also been long used in expensive perfumes.

One of the first identified insect sex attractants was that of the gypsy moth. This moth is a common agricultural pest, and it was hoped that the sex attractant that females emitted could be used to lure and trap males. Such a method of insect control would be preferable to inundating large areas with DDT and would be species specific. Nearly thirty years of work were expended in identifying the chemical substance responsible. Early in this period workers had found that an extract from the tail sections of female gypsy moths would attract males, even from a great distance. In subsequent work, 20 milligrams of a pure chemical substance, called **gyplure,** were isolated from solvent extracts of the two extreme tail segments collected from 500,000 female gypsy moths (about 0.1 μg/moth). This points out the fact that pheromones are effective in very minute amounts and that chemists must work with very small amounts to isolate them and prove their structures. It is not unusual to have to process thousands of insects to get even a very small sample of these substances. Very sophisticated analytical and instrumental methods, like spectroscopy, must be used to determine the structure of a pheromone.

In experiments with gyplure, it was found that the male gypsy moth has an almost unbelievable ability to detect extremely small amounts of the substance. He can detect it in concentrations of less than a few hundred **molecules** per cubic centimeter (about 10^{-19} to 10^{-20} g/cc)! When a male moth encounters a small concentration of gyplure, he immediately turns into the wind and flies upwind in search of higher concentrations and the female. In only a mild breeze, a continuously emitting female can activate a space 300 feet high, 700 feet wide, and almost 14,000 feet (nearly 3 miles) long! Many other insects also release sex pheromones. The structures of several insect pheromones are given in this essay.

MAMMALIAN PHEROMONES (?)

Exaltolide
(Synthetic)

Civetone
(Civet Cat)

Muskone
(Musk Deer)

INSECT SEX ATTRACTANTS

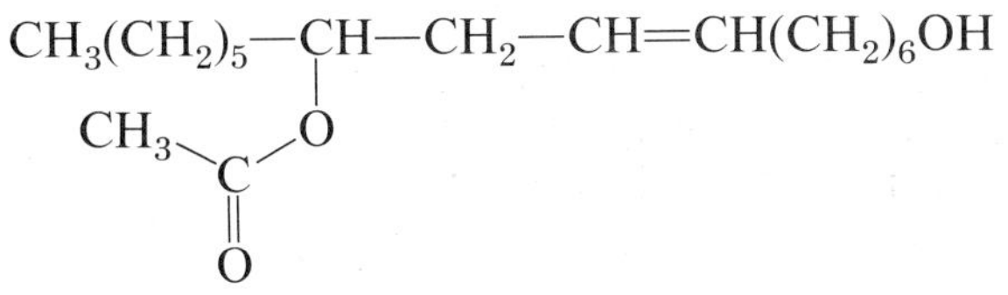

$$CH_3(CH_2)_5-\underset{\underset{\underset{O}{\overset{\|}{C}}}{\overset{|}{\underset{O}{\text{—}}}}{CH}-CH_2-CH=CH(CH_2)_6OH$$

Gyplure
(Gypsy Moth)

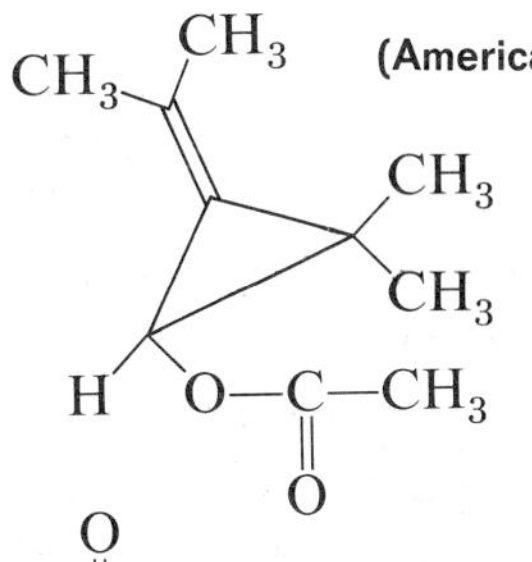

(American Cockroach)

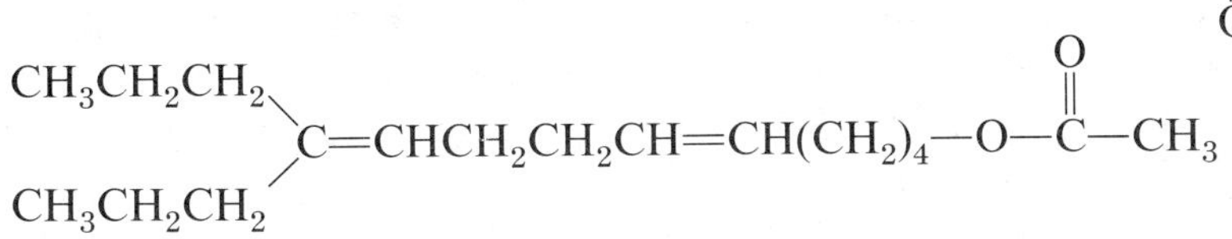

$$\underset{CH_3CH_2CH_2}{\overset{CH_3CH_2CH_2}{\diagdown}}C=CHCH_2CH_2CH=CH(CH_2)_4-O-\overset{\overset{O}{\|}}{C}-CH_3$$

Propylure
(Pink Bollworm Moth)

RECRUITING PHEROMONES

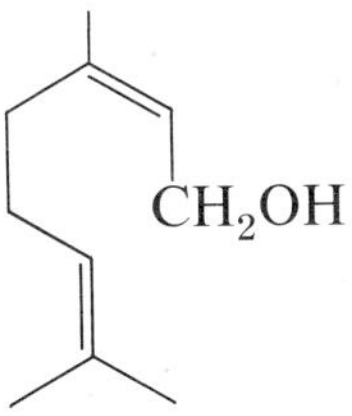

Geraniol
(Honey Bee)

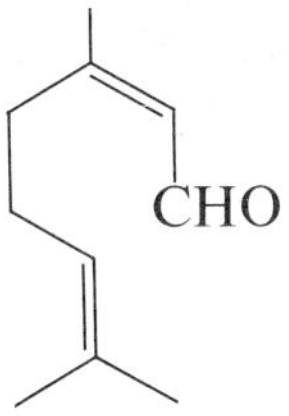

Citral
(Honey Bee)

PRIMER PHEROMONE

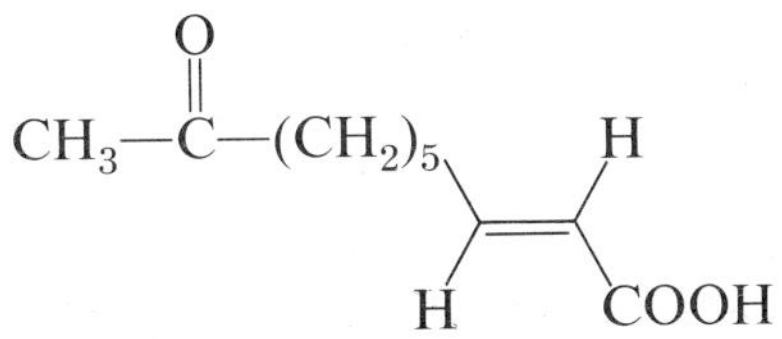

$$CH_3-\overset{\overset{O}{\|}}{C}-(CH_2)_5\underset{H}{\overset{H}{\diagup}}C=C\underset{COOH}{\overset{}{\diagdown}}$$

Queen Substance
(Honey Bee)

ALARM PHEROMONES

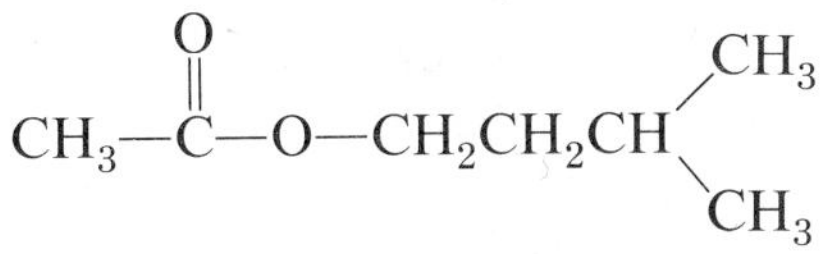

$$CH_3-\overset{\overset{O}{\|}}{C}-O-CH_2CH_2CH\underset{CH_3}{\overset{CH_3}{\diagup}}$$

Isoamyl Acetate
(Honey Bee)

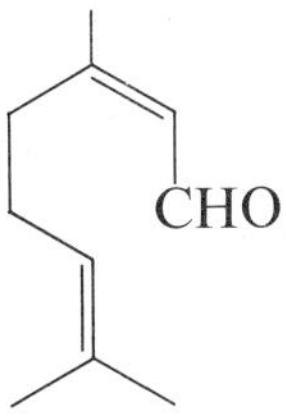

Citral

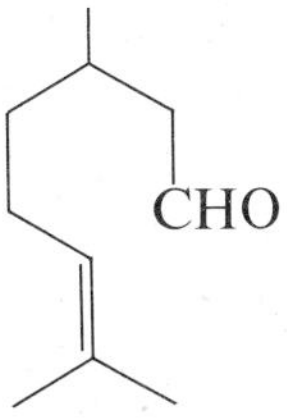

Citronellal
(Ant Species)

The most important example of a primer pheromone is found in honey bees. A bee colony consists of only one queen bee, several hundred male drones, and thousands of worker bees, which are undeveloped females. It has recently been found that the queen, the only female that has achieved full development and reproductive capacity, secretes a primer pheromone called the "queen substance." The worker females, while tending the queen bee, continuously ingest quantities of this queen substance. This pheromone prevents the workers from rearing any competitive queens and also prevents the development of ovaries in all other females in the hive. The substance is also active as a sex attractant, as it attracts drones to the queen during her "nuptial flight."

Honey bees also produce several other important types of pheromones. It has long been known that bees will "swarm" after an intruder. It has also been known that isoamyl acetate would induce a similar type of behavior on the part of bees. Isoamyl acetate (Experiment 11) is an **alarm pheromone.** When an angry worker bee stings an intruder, she discharges, along with the sting venom, a mixture of pheromones that incite the other bees to swarm upon and attack the intruder. Isoamyl acetate is a major component of the alarm pheromone mixture. Alarm pheromones have also been identified in many other insects. In less aggressive insects than bees or ants, the alarm pheromone may take the form of a **repellent** which induces the insects to go into hiding or leave the immediate vicinity.

Honey bees also release **recruiting** or **trail pheromones.** These pheromones attract others to a source of food. Honey bees secrete recruiting pheromones when they locate flowers where large amounts of sugar syrup are available. Although the recruiting pheromone is a complex mixture, both geraniol and citral have been identified as components. In a similar fashion, when ants locate a source of food, they drag their tails along the ground on their way back to the nest and continuously secrete a trail pheromone. Other ants will follow this trail to the source of food.

In some species of insects, **recognition pheromones** have been identified. In carpenter ants, a caste-specific secretion has been found in the mandibular glands of the males of five different species. These secretions have several functions, one of which is to allow members of the same species to recognize one another. Insects not having the correct recognition odor are immediately attacked and expelled from the nest. In one species of carpenter ant, the recognition pheromone has been shown to have methyl anthranilate as a major component.

Although it is clear that insect attractants are pheromones, the case of insect repellents is less straightforward. We simply do not yet know enough about these substances or about how they exert their effect. Currently, the most widely used insect repellent is the synthetic substance N,N-diethyl-m-toluamide (Experiment 17), also called Deet. It is effective against fleas, mosquitoes, chiggers, ticks, deer flies, sand flies, and biting gnats. It has the widest spectrum of activity of any known repellent, but there are other repellents that are effective against each of these individual insects. Why these substances repel insects is unknown. They may simply be compounds that provide unpleasant odor or taste to a wide variety of insects. They may be alarm pheromones for the species affected. Or, they may be alarm pheromones of a hostile species. In time, perhaps the mystery will be solved.

Recently, N,N-diethyl-m-toluamide has been found in a natural insect source. Workers investigating the sex attractant from the pink bollworm moth identified the major chemical component of the pheromone as a substance called propylure. Synthetic propylure, however, failed to attract male pink bollworm moths when used to bait traps. It was subsequently found that propylure is activated when mixed with 10 percent Deet, although it was not quite as active as propylure extracted from the natural source. Soon Deet was found in large amounts in the adult female moth. It is characteristic of pheromones that the individual components are frequently inactive until combined with several minor substances which activate them.

We do not yet know all the types of pheromones that any given species of insect may use, but it seems that as few as ten or twelve pheromones could constitute a "language" that could adequately regulate the entire life cycle of a colony of social insects.

REFERENCES

Beroza, M., Ed. *Chemicals Controlling Insect Behavior.* New York: Academic Press, 1970.
Beroza, M., and Jacobson, M. "Trapping Insects by Their Scents." *Chemistry, 38* (June, 1965), 6.
Jacobson, M., and Beroza, M. "Insect Attractants." *Scientific American, 211* (August, 1964), 20.
Sondheimer, E., and Simeone, J. B. *Chemical Ecology.* New York: Academic Press, 1970.
Vollmer, J. J., and Gordon, S. A. "Chemical Communication." *Chemistry, 48* (April, 1975), 6.
Wilson, E. O. "Pheromones." *Scientific American, 208* (May, 1963), 100.
Wright, R. H. "Why Mosquito Repellents Repel," *Scientific American, 233* (July, 1975), 105.

Experiment **17**

N,N-DIETHYL-*m*-TOLUAMIDE: THE INSECT REPELLENT "OFF"

Preparation Of An Amide
Addition Funnel, Extraction
Column Chromatography
(Vacuum Distillation)

In this experiment, we will synthesize the active ingredient of the insect repellent "OFF." *N,N*-Diethyl-*m*-toluamide belongs to the class of chemical compounds called **amides.** Amides have the generalized

structure $R-\overset{\overset{\displaystyle O}{\displaystyle \|}}{C}-NH_2$. The amide to be prepared in this experiment is a **disubstituted** amide, i.e., two of the hydrogens on the amide $-NH_2$ group have been replaced with ethyl groups.

The synthesis requires **two** reaction steps. These are shown below.

First, *m*-toluic acid will be converted to its acid chloride derivative. Second, the acid chloride will be allowed to react with diethylamine. The acid chloride will not be isolated and purified, but will be reacted **in situ** (in the solution used to form it). This is a common way to prepare

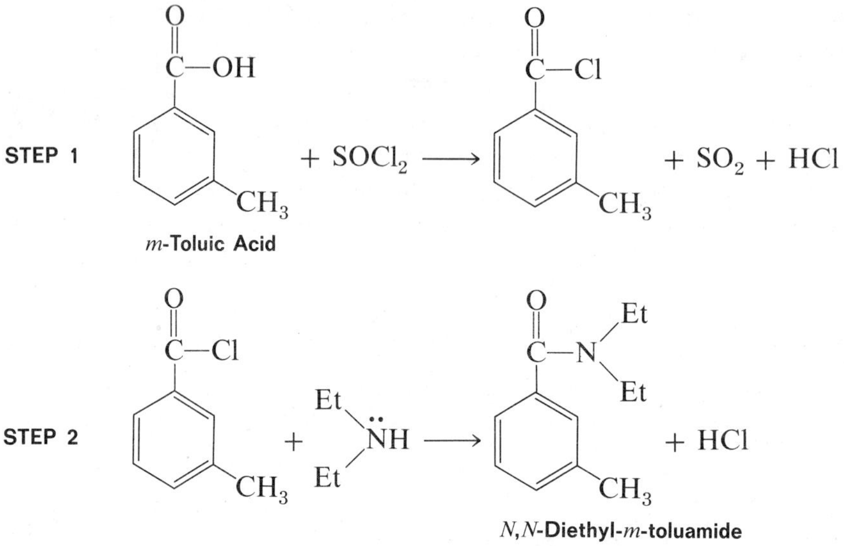

STEP 1

$$\text{m-Toluic Acid} + SOCl_2 \longrightarrow + SO_2 + HCl$$

m-Toluic Acid

STEP 2

$$+ \underset{Et}{\overset{Et}{\underset{\cdot\cdot}{N}H}} \longrightarrow + HCl$$

N,N-**Diethyl-*m*-toluamide**

amides. It should be noted that amides cannot be prepared directly from the admixture of an acid and an amine. If an acid and an amine are mixed, an **acid-base** reaction occurs, giving the conjugate base of the acid which will not react further while in solution.

$$RCOOH + R_2NH \longrightarrow RCOO^-R_2NH_2^+$$

If the amine salt is isolated as a crystalline solid and strongly heated, the amide can be prepared.

$$RCOO^-R_2NH_2^+ \xrightarrow{\text{heat}} RCONR_2 + H_2O$$

This is not a convenient laboratory method. Amides are usually prepared via the acid chloride, as in this experiment.

SPECIAL INSTRUCTIONS

As an introduction, read the essay which precedes this experiment. When measuring reagents keep the following in mind: Since thionyl chloride reacts with water to liberate HCl and SO_2, all equipment should be dry so that decomposition of the reagent is avoided. Likewise, in the second step of the reaction, **anhydrous** ether should be used because water will react with both thionyl chloride and the intermediate acid chloride. It should also be noted that thionyl chloride is a **noxious** and **corrosive** chemical and should be handled with care. If it is spilled on the skin, serious burns will result. Thionyl chloride will also react with any moisture in the air and decompose to give hydrogen chloride gas. Therefore, the reagent should be dispensed **in the hood** from a bottle which should be kept tightly closed. The reagent should not be measured and left to stand on your desk until it is used because it will hydrolyze. Diethylamine, used in the second step of the reaction, is also noxious and corrosive. In addition, it is quite volatile (bp 56°). It should be handled carefully, in the hood, using the same caution as for thionyl chloride.

PROCEDURE

Place 4.1 g of *m*-toluic acid (3-methylbenzoic acid) and 4.5 ml of thionyl chloride (d = 1.65 g/ml) in a dry 500 ml three neck round bottom flask. Refer to the figure below, and equip the flask with a reflux

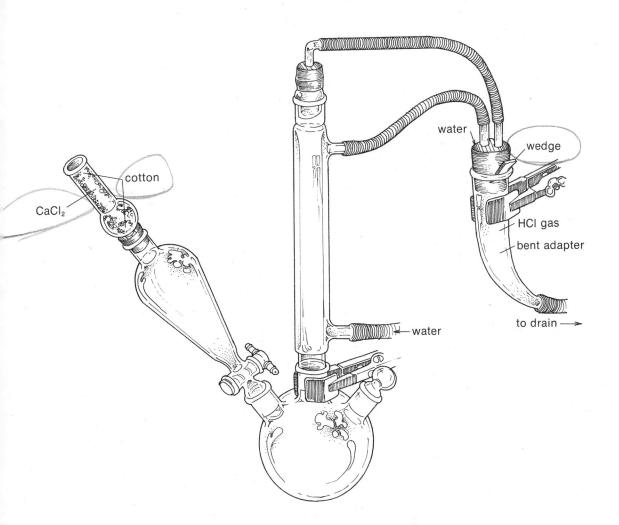

condenser and a gas trap of the type shown. Hydrogen chloride is evolved in the reaction and the gas trap will prevent it from escaping into the room. Add a separatory funnel fitted with a drying tube (see the figure again) and stopper the unused neck of the flask. Add a boiling stone, start the circulation of water in the reflux condenser, and heat the mixture **gently,** using either a steam bath or a heating mantle, until the evolution of gases stops (about 20 minutes).

Allow the flask to cool and add 50 ml of dry ether (bp 35°). Pour a solution of 10 ml of diethylamine (d = 0.7108 g/ml) dissolved in 20 ml of anhydrous ether into the separatory funnel and replace the drying tube. Open the stopcock in the separatory funnel and add the diethylamine solution, a little at a time, over a period of about 20 to 25 minutes. Watch the boiling action and do not allow it to become too vigorous. When the diethylamine is added, a voluminous white cloud will form in the flask. Allow the cloud to settle each time before more diethylamine is added. Also, take care not to add the diethylamine in such large portions that the white material rises up in the arm of the flask and clogs the opening in the addition (separatory) funnel.

After adding the diethylamine, wash the contents of the flask into a separatory funnel, using about 20 ml of a 5% sodium hydroxide solution. If required, use a little water to wash any solid that has accumulated in the condenser into the flask, and then also add this solution to the material in the separatory funnel. Shake the separatory funnel and allow the layers to separate (Technique 5, Section 5.4). If there is not a distinct ether layer, much of the ether may have evaporated during the course of the reaction and some additional ether should be added. Following the sodium hydroxide wash, wash the solution again with another 20 ml portion of 5% sodium hydroxide, then with 20 ml of 10% hydrochloric acid, and finally with 20 ml of water. Dry the ether layer over anhydrous sodium sulfate (Technique 5, Section 5.6). Decant the solution from the sodium sulfate and evaporate the ether on a steam bath in the hood.

The product may be purified by a reduced pressure distillation (Technique 6, Sections 6.6 and 6.7) or by column chromatography (Technique 10). Column chromatography is the easier procedure. If the product is distilled, several (3 or 4) students should combine their crude products and distill cooperatively. If it is purified by column chromatography, alumina should be used as the adsorbent (~25 g alumina in a 25 mm diameter column), and the elution solvent should be petroleum ether or ligroin. N,N-Diethyl-m-toluamide will be the first compound to elute. Evaporation of the petroleum ether should give the purified amide as a pure colorless or light tan liquid (bp 160 to 163° at 20 mm).

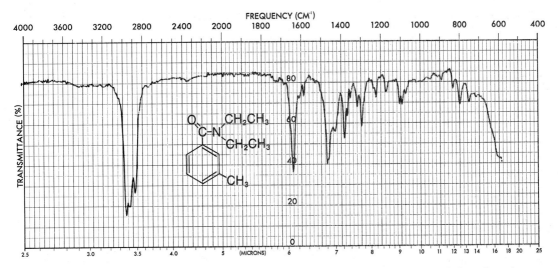

Infrared absorption spectrum of N,N-Diethyl-m-toluamide, neat

At your instructor's option, determine the infrared spectrum of your product.

REFERENCE

Wang, B. J-S. "An Interesting and Successful Organic Experiment." *Journal of Chemical Education, 51* (October, 1974), 631. (The synthesis of *N,N*-diethyl-*m*-toluamide.)

QUESTIONS

1. Write an equation that describes the reaction of thionyl chloride with water.

2. What reaction would take place if the acid chloride of *m*-toluic acid were mixed with water?

3. What is the substance responsible for the voluminous white cloud that forms when diethylamine is added to the reaction mixture? Write an equation that describes its formation.

4. Why is the final reaction mixture extracted with 5% sodium hydroxide?

5. If *N,N*-diethyl-*m*-toluamide distills at 165° at 20 mm, what should its boiling point be at 760 mm?

6. Write a mechanism for each of the steps in the preparation of *N,N*-diethyl-*m*-toluamide.

7. Interpret each of the major peaks in the infrared spectrum of *N,N*-diethyl-*m*-toluamide.

Petroleum

Crude petroleum is a liquid which consists of hydrocarbons, as well as some related sulfur, oxygen, and nitrogen compounds. There may also be other elements, including metals, present in trace amounts. Crude oil is formed by the decay of marine animal and plant organisms which existed millions of years ago. Over many millions of years, under the influence of temperature, pressure, catalysts, radioactivity, and bacteria, the decayed matter was converted into what we now know as crude oil. The crude oil is trapped in pools beneath the ground by various geological formations.

Most crude oils have a density of between 0.78 and 1.00 g/ml. As a liquid, crude oil may be as thick and black as melted tar or as thin and colorless as water. Its characteristics depend upon the particular oil field from which it comes. Pennsylvania crude oils are high in straight-chain alkane compounds (called **paraffins** in the petroleum industry), which makes those crude oils useful in the manufacture of lubricating oils. Oil fields in California and Texas produce crude oil with a higher percentage of cycloalkanes (also called **naphthenes** by the petroleum industry). Some Middle East fields produce crude oil containing up to 90 percent cyclic hydrocarbons. Petroleum contains molecules in which numbers of carbons range from one to sixty.

When petroleum is refined to convert it into a variety of useable products, it is initially subjected to a fractional distillation. The following table lists the various fractions which are obtained from fractional distillation. Each of these fractions has its own particular uses. Each fraction may be subjected to further purification, depending upon the desired application.

GASOLINE

The gasoline which is obtained directly from fractional distillation of crude oil is called **straight run gasoline.** An average barrel of crude

PETROLEUM FRACTION	COMPOSITION	COMMERCIAL USE
Natural Gas	C_1 to C_4	Fuel for heating
Gasoline	C_5 to C_{10}	Motor fuel
Kerosine	C_{11} and C_{12}	Jet fuel and heating
Light Gas Oil	C_{13} to C_{17}	Furnaces, diesel engines
Heavy Gas Oil	C_{18} to C_{25}	Motor oil, paraffin wax, petroleum jelly
Residuum	C_{26} to C_{60}	Asphalt, residual oils, waxes

oil will yield about 19% straight run gasoline. This yield presents two immediate problems. First, there is not enough gasoline contained in crude oil to satisfy man's needs for fuel to power automobile engines. Second, the straight run gasoline obtained from crude oil is a rather poor fuel for modern engines. It must be "refined" at a chemical refinery.

The initial problem of the small quantity of gasoline available from crude oil may be solved by **cracking** and **polymerization.** Cracking is a refinery process by which large hydrocarbon molecules are broken down into smaller molecules. Heat and pressure are required for cracking, and a catalyst must also be used. Gaseous hydrogen may also be present during the cracking process, in which case only saturated hydrocarbons are produced. Silica-alumina and silica-magnesia are among

$$C_{16}H_{34} + H_2 \xrightarrow[\text{heat}]{\text{catalyst}} 2\ C_8H_{18} \quad \textbf{CRACKING}$$

the more effective cracking catalysts. The alkane mixtures which are produced tend to have a fairly high proportion of branched-chain isomers. These branched isomers improve the quality of the fuel.

In the polymerization process, also carried out at a refinery, small molecules of alkenes are reacted with each other to form larger molecules, which are also alkenes. The newly-formed alkenes may be hydro-

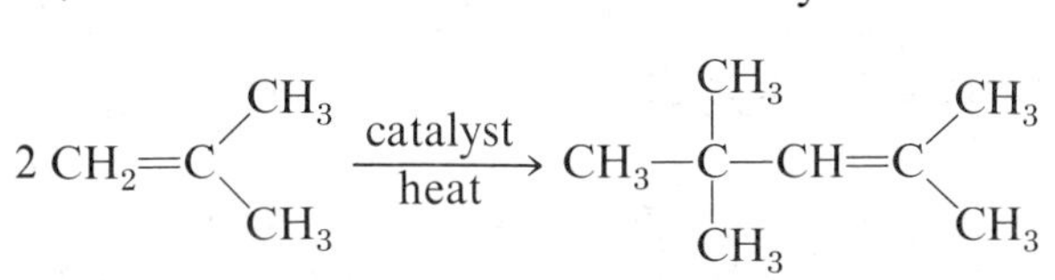

2-Methylpropene
(Isobutylene)

2,4,4-Trimethyl-2-pentene

POLYMERIZATION

genated to form alkenes. The reaction sequence shown here is a very

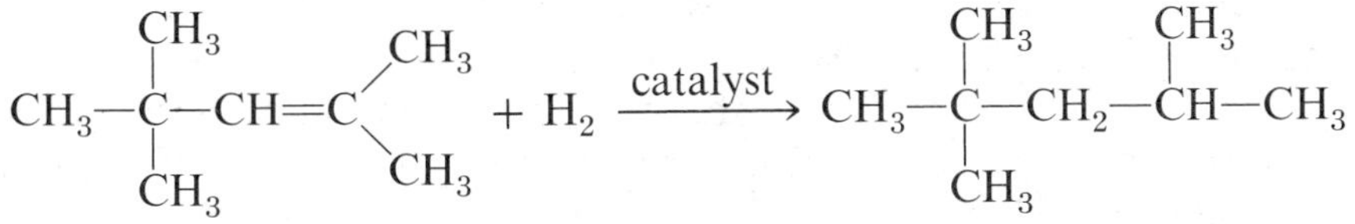

2,2,4-Trimethylpentane
(Iso-octane)

common and important one in petroleum refining, since the product, 2,2,4-trimethylpentane (or "Iso-octane") forms the basis for determining the quality of gasoline. Using these refining methods, the percentage of gasoline which may be obtained from a barrel of crude oil may rise to as much as 45 or 50 percent.

KNOCKING

The internal combustion engine, as found in most automobiles, operates in four cycles or strokes. The four strokes are intake, compres-

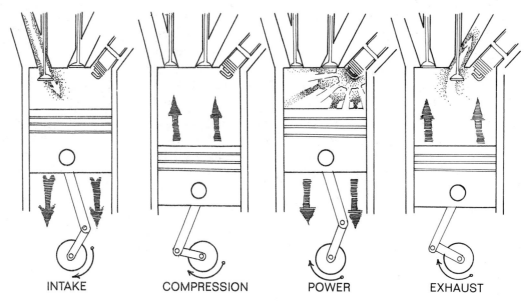

| INTAKE | COMPRESSION | POWER | EXHAUST |

Operation of a four cycle engine

sion, power and exhaust. They are illustrated in the figure. It is the power stroke which is of greatest interest to us from a chemical point of view, since combustion occurs during this stroke.

When the fuel-air mixture is ignited, it does not explode. Rather, it burns at a controlled, uniform rate. The gases closest to the spark are ignited first; then they in turn ignite the molecules farther from the spark, and so on. The combustion proceeds in a wave of flame, or **flame front,** which starts at the spark plug and proceeds uniformly outward from that point until all of the gases in the cylinder have ignited. Because a certain time is required for this burning to occur, the initial spark is timed to ignite just before the piston has reached the top of its travel. In this way, the piston will be at the very top of its travel at the precise instant that the flame front, and the increased pressure which accompanies it, reaches the piston. The result is a smoothly applied force to the piston, driving it downward.

If heat and compression should cause some of the fuel-air mixture to ignite before the flame front has reached it, the timing of the combustion sequence would be disturbed. In such a case, the gases would explode, rather than burn smoothly. The result is **knocking** or **detonation** (sometimes called "pinging"), where the fuel explodes in a violent and uncontrolled manner. The transfer of power to the piston under these conditions is much less efficient than in normal combustion. The wasted energy is merely transferred to the engine block. The explosive force may even damage the engine.

It has been found that the tendency of a fuel to knock is a function of the structures of the molecules composing the fuel. Normal hydrocarbons, those with straight carbon chains, have a greater tendency to lead to knocking than do those alkanes with highly branched chains.

The quality of a gasoline, then, is a measure of its resistance to knocking, and this quality is improved by increasing the proportion of branched-chain alkanes in the mixture. Such chemical refining processes as **reforming** and **isomerization** are used to convert normal alkanes to branch-chain alkanes, thus improving the knock-resistance of gasoline.

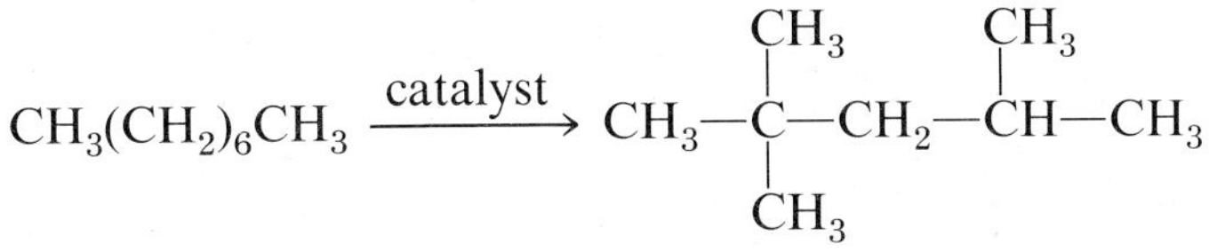

$$CH_3(CH_2)_6CH_3 \xrightarrow{\text{catalyst}} \underset{\underset{CH_3}{|}}{\overset{\overset{CH_3}{|}}{CH_3-C-CH_2-}}\overset{\overset{CH_3}{|}}{CH}-CH_3 \qquad \text{REFORMING}$$

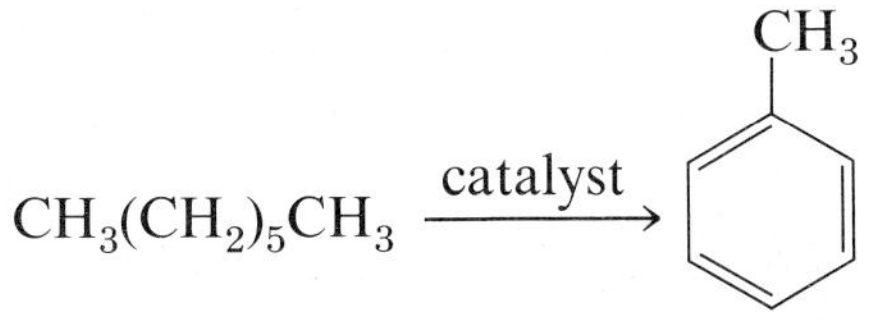

$$CH_3(CH_2)_5CH_3 \xrightarrow{\text{catalyst}} \qquad \text{REFORMING}$$

$$CH_3-CH_2-CH_2-CH_2-CH_3 \xrightarrow{\text{AlBr}_3}$$

$$\underset{\underset{CH_3}{|}}{\overset{\overset{CH_3}{|}}{CH_3-CH-}}CH_2-CH_3 \qquad \text{ISOMERIZATION}$$

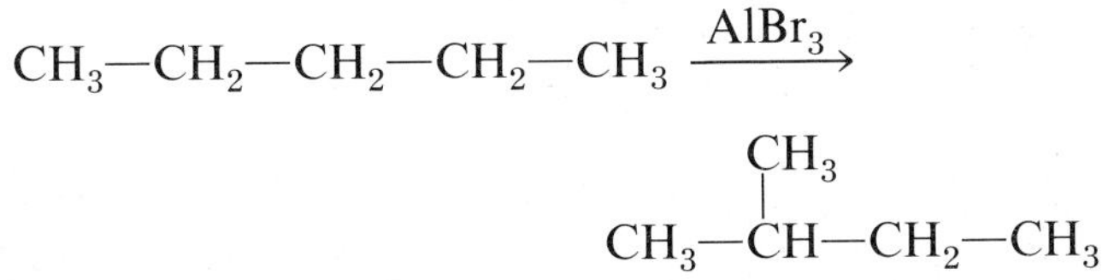

None of these processes converts all of the normal hydrocarbons into branch-chain isomers; consequently, additives are also added to gasoline to improve the knock-resistance of the fuel. Aromatic hydrocarbons are effective in improving the knock-resistance of gasoline, and they are often used in unleaded gasoline. The most common additives used to reduce knocking are **tetraethyllead** and **tetramethyllead.** These substances are added in quantities which by law must not exceed 4.0 ml of tetraalkyllead per gallon of gasoline. The usual amount of tetra-

$$\underset{\underset{CH_2-CH_3}{|}}{\overset{\overset{CH_2-CH_3}{|}}{CH_3-CH_2-Pb-CH_2-CH_3}} \qquad \underset{\underset{CH_3}{|}}{\overset{\overset{CH_3}{|}}{CH_3-Pb-CH_3}}$$

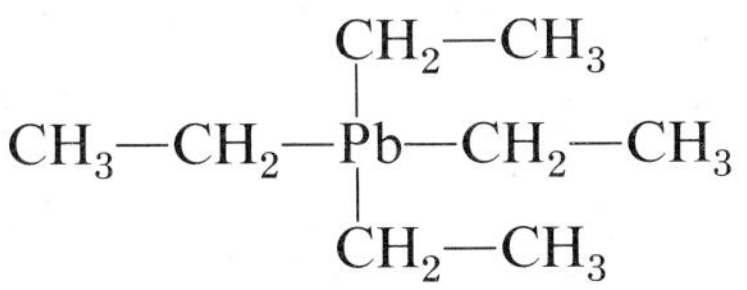

Tetraethyllead   **Tetramethyllead**

alkyllead added to gasoline is 3.0 ml per gallon. The United States Surgeon General has imposed the limits of tetraalkyllead which may be used, since an excessive amount of that substance might create a health hazard. The combination of increased proportions of branch-chain hydrocarbons, increased quantities of aromatic hydrocarbons, and added tetraalkyllead creates a gasoline which burns cleanly, without knocking, in modern high-compression automobile engines. Different manufacturers adjust the proportion of each of these in their gasoline to achieve the desired octane rating.

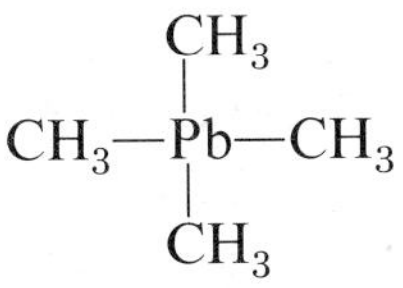

OCTANE RATINGS OF ORGANIC COMPOUNDS

COMPOUND	RESEARCH OCTANE NUMBER	COMPOUND	RESEARCH OCTANE NUMBER
Octane	− 19	1-Pentene	91
Heptane	0	2-Hexene	93
Hexane	25	Butane	94
Pentane	62	Propane	97
Cyclohexane	83	1-Butene	97
2,2,4-Trimethylpentane	100	Methanol	106
Cyclopentane	101	m-Xylene	118
Benzene	106	Toluene	120

OCTANE RATINGS

A fuel may be classified according to its anti-knock characteristics. The most important rating system is the **octane rating** of gasoline. In this method of classification, the anti-knock properties of a fuel are compared in a hot engine to the anti-knock properties of a standard mixture of heptane and 2,2,4-trimethylpentane. This latter compound is called "iso-octane," hence the name, "octane rating." A fuel which has the same anti-knock properties as a given mixture of heptane and "iso-octane" would have an octane rating numerically equal to the percentage of "iso-octane" in that reference mixture. Today's 91 octane (Research method) regular gasoline is a mixture of compounds which, when taken together, have the same anti-knock characteristics as a test fuel composed of 9% heptane and 91% "iso-octane." Other substances besides hydrocarbons may have high resistance to knocking. A table of various substances with their octane ratings is given above.

POLLUTION PROBLEMS

The number of grams of air required for the complete combustion of one mole of gasoline (assuming a formula of C_8H_{18}) is 1735 g. This gives rise to a theoretical air-fuel ratio of 15.1 to 1 for complete combustion. For a variety of reasons, however, it is neither easy nor advisable to supply each cylinder with a theoretically correct air-fuel mixture. The power and performance of an engine improve with a slightly richer (lower air-fuel ratio) mixture. Maximum power is obtained from an engine when the air-fuel ratio is near 12.5 to 1, while maximum economy is obtained when the air-fuel ratio is near 16 to 1. Under conditions of idling or full load (i.e., acceleration), the air-fuel ratio is lower than would be theoretically correct. As a result, complete combustion does not take place in an internal combustion engine, and carbon monoxide (CO) is produced in the exhaust gases. Other types of non-ideal combustion behavior give rise to the presence of unburned hydrocarbons in the exhaust. The high combustion temperatures cause the nitrogen and oxygen of the air to react, forming a variety of nitrogen oxides in the exhaust. Each of these materials contributes to air pollution. Under the influence of sunlight, which is of sufficient energy to break covalent bonds, these materials may react with each other and with air to produce **smog.** Smog consists of **ozone,** which deteriorates rubber and damages plant life; **particulate matter,** which produces haze; **oxides of nitrogen,** which produce a brownish color in the atmosphere; and a variety of eye irritants, such as **peroxyacetyl nitrate** (PAN). Lead particles, from tetraalkyllead, may also cause problems, since they are toxic. Sulfur compounds in the gasoline may lead to the production of noxious gases in the exhaust.

$$CH_3-\overset{\overset{\displaystyle O}{\displaystyle \|}}{C}-O-O-NO_2$$

PAN
Peroxyacetyl Nitrate

Current efforts to reverse the trend of deteriorating air quality caused by automotive exhaust have taken many forms. Initial efforts at modifying the air-fuel mixture of engines produced some improvements in emissions of carbon monoxide, but at the cost of increased nitrogen oxide emissions. With more stringent air-quality standards being imposed by the Environmental Protection Agency, attention has turned to alternate sources of power, such as the Wankel (rotary) engine, electric motors, and steam or gas turbine engines.

In the meantime, since the standard internal combustion engine remains the most attractive powerplant because of its great flexibility and reliability, efforts at improving its emissions continue. The advent of **catalytic converters,** which are muffler-like devices containing catalysts capable of converting carbon monoxide, unburned hydrocarbons, and nitrogen oxides into harmless gases, has resulted from such efforts. Unfortunately, the catalysts are rendered inactive by the lead additives in gasoline. Unleaded gasoline must be used, but it takes more crude oil to make a gallon of unleaded gasoline than is required for the production of leaded gasoline. Other hydrocarbons must be added to play the role of antiknock agent formerly played by tetraalkyllead. Also, a fine mist of sulfuric acid may be discharged in the exhaust as a result of the action of the catalysts. Consequently, catalytic converters are not the final answer to emission control.

Some promise seems to be shown by modifications in the basic design of the combustion chambers of internal combustion engines. Such modified engines as the stratified-charge engines produced by Honda Motor Company of Japan appear to meet the presently mandated exhaust emission standards without requiring catalytic converters and without the large losses in fuel economy which are found with other emission control measures. Methanol has been proposed as an alternative to gasoline as a fuel. Some preliminary tests have indicated that the amounts of the principal air pollutants in automobile exhaust are greatly lowered when methanol is used instead of gasoline in a typical automobile. It will be interesting to see what further developments the dual problems of air quality and shortage of petroleum will bring to the design of automotive engines.

REFERENCES

"Facts About Oil." New York: American Petroleum Institute, 1970.

Greek, B. F. "Gasoline." *Chemical and Engineering News, 52* (November 9, 1970).

Kerr, J. A., Calvert, J. G., and Demerjian, K. L. "The Mechanism of Photochemical Smog Formation." Chemistry in Britain, *8* (1972), 252.

Lane, J. C. "Gasoline and Other Motor Fuels." *Encyclopedia of Chemical Technology, 10* (1966), 463. John Wiley & Sons, Inc.

Pierce, J. R. "The Fuel Consumption of Automobiles." *Scientific American, 232* (January, 1975), 34.

Reed, T. B., and Lerner, R. M. "Methanol: A Versatile Fuel for Immediate Use." Science, *182* (1973), 1299.

Wildeman, T. R. "The Automobile and Air Pollution." *Journal of Chemical Education, 51* (1974), 290.

Experiment **18**

GAS CHROMATOGRAPHIC ANALYSIS OF GASOLINES

Gasoline
Gas Chromatography

In this experiment you will analyze samples of gasoline by means of gas chromatography. From this analysis you should learn something about the composition of these fuels. Although all gasolines are compounded from the same basic hydrocarbon components, different companies blend these components in different proportions in order to obtain a gasoline with similar properties. Sometimes the composition of the gasoline may vary depending upon the composition of the crude petroleum from which the gasoline was derived. Frequently, refineries vary the composition of gasoline in response to differences in climate or to seasonal changes. In the winter or in cold climates, the relative proportion of butane and pentane isomers is increased in order to increase the volatility of the fuel. This increased volatility permits easier starting. In the summer or in warm climates, the relative proportion of these volatile hydrocarbons is reduced. The decreased volatility thus achieved reduces the possibility of vapor lock formation. Occasionally, differences in composition may be detected by examining the gas chromatograms of a particular gasoline over a period of several months. In this experiment, we will not attempt to detect such differences.

There are different octane rating requirements for "regular" and "premium" gasolines. You will be able to observe differences in the composition of these two types of fuels. You should pay particular attention to increases in the proportions of those hydrocarbons which raise octane ratings in the premium fuels. If you analyze an unleaded gasoline, you should be able to observe differences in the composition of this type of gasoline as compared to leaded fuels of a similar octane rating.

You will be asked to analyze a sample of a premium and a regular gasoline, preferably from the same company. If an unleaded gasoline from that company is also available, you may be required to analyze it as well. Other students in the class may be analyzing gasolines from other companies. If so, a comparison of equivalent grades of gasoline from one company to another should be made.

Discount service stations usually purchase their gasoline from one of the large petroleum refining companies. If you analyze gasoline from a discount service station, a comparison of that gasoline to an equivalent grade from a major supplier may prove to be interesting, particularly in terms of the similarities.

SPECIAL INSTRUCTIONS

Before performing this experiment, you should read the essay which precedes this experiment and Technique 12. Your instructor may wish to have each student in the class collect samples of gasoline from different service stations. A list should be compiled of all the different gasoline companies represented in the nearby area, and each student should be assigned to collect a sample from a different company. You should collect the gasoline sample in a labeled screw cap jar. It may be handy to take a funnel along with you as well. An easy way to collect a gasoline sample for this experiment is to drain the excess gasoline from the nozzle and hose of the pump into the jar immediately after the gasoline tank of a car has been filled. The collection of gasoline in this manner must be done **immediately after** the gas pump has been used. If not, the volatile components of the gasoline may evaporate, thus changing the composition of the gasoline. Only a very small sample (a few **milli**liters) of gasoline is required, since the gas chromatographic analysis requires no more than a few **micro**liters of material. Be certain to close the cap of the sample jar tightly, in order to prevent the selective evaporation of the more volatile components. The label on the jar should list the brand of gasoline and the grade.

If you live in a state where the collection of gasoline in glass containers is illegal, your instructor will supply you with a fireproof metal container in which to collect the gasoline sample. Alternatively, the instructor may provide you with gasoline samples.

Always remember that gasoline contains many highly volatile and flammable components. Do not breathe the vapors and do not use open flames near gasoline. Also recall that gasoline contains tetraethyllead and is, therefore, toxic.

This experiment may be assigned along with another short one, since it requires only a few minutes of each student's time to carry out the actual gas chromatography. In order to proceed as efficiently as possible, it may be convenient to arrange an appointment schedule for use of the gas chromatograph.

> NOTE TO THE INSTRUCTOR: The gas chromatograph should be prepared as follows: column temperature, 110 to 115°; injection port temperature, 110 to 115°; carrier gas flow rate, 40 to 50 ml/min. The column should be approximately 12 feet in length and should contain a rather nonpolar stationary phase similar to silicone oil (SE-30) on Chromosorb W or some other stationary phase which separates components principally according to their boiling points.

PROCEDURE

Depending on the instructor's wishes, analysis of a series of standard materials may be required. Reference substances which should be used include pentane, hexane, cyclohexane, benzene, heptane, toluene, and m-xylene, although others may appropriately be included. If the samples are each analyzed individually, a sample size of about 0.5 microliters will be adequate for each reference substance. A better alternative is to analyze a reference mixture which contains all of these standard substances and to compare it to a similar gas chromatogram previously recorded by the instructor. The instructor may post a copy of this chromatogram (with the peaks identified), or he may provide each student with a copy. An example of such a determination is provided at the end of this experiment. If a reference mixture is to be analyzed, a reasonable sample size is 1 microliter. The sample is injected into the gas chromatograph, and the retention times of each of the components are measured and recorded (Technique 12, Sections 12.7 and 12.8).

The instructor may prefer to perform the sample injections himself or to have a laboratory assistant perform them. The sample injection procedure requires careful technique, and the special microliter syringes that are required are very delicate and expensive. If students are to perform the sample injections themselves, it is **essential** that adequate instruction be obtained beforehand.

Inject the sample of regular gasoline onto the gas chromatography column and wait for the gas chromatogram to be recorded. Next inject the sample of premium gasoline, and obtain its gas chromatogram.

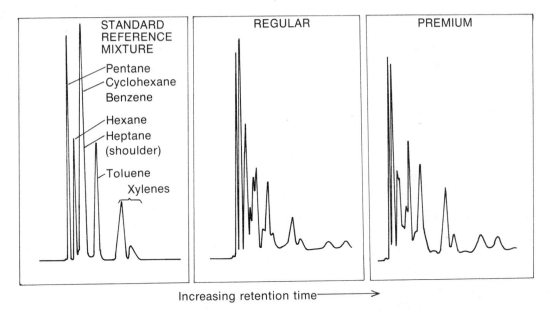

Gas chromatograms of a standard reference mixture, a regular gasoline, and a premium gasoline

Compare these chromatograms to each other and identify as many of the components as possible. For comparison, sample gas chromatograms of a regular gasoline and a premium gasoline are provided at the end of this experiment. Determine the differences between the two grades of fuel. If a sample of unleaded gasoline was also obtained, record its gas chromatogram and compare it to the chromatograms of the other two fuels. Be certain to compare the retention times of the components in each gasoline sample to those of the standards in the reference mixture very carefully. Retention times of compounds vary with the conditions under which they were determined. It is best to analyze the reference mixture and each of the gasoline samples in succession, in order to reduce the variations in retention times that may occur over a period of time. If there are unidentified peaks in the gas chromatograms of the gasoline samples, try to guess their probable identity. Compare the gas chromatograms with those of students who have analyzed gasolines from another dealer.

The report to the instructor should include the actual gas chromatograms as well as an identification of as many of the components in each grade of fuel as possible.

QUESTIONS

1. How do regular and premium grades of gasoline differ in this analysis?

2. Explain the relative retention times of benzene, toluene, and *m*-xylene.

3. What would you expect the analysis of an unleaded gasoline to reveal in this experiment?

4. Is it possible to detect tetraethyllead in this experiment?

5. If you were a forensic chemist working for the police department, and the fire marshall brought you a sample of gasoline found at the scene of an arson attempt, do you think you could identify the service station where the arsonist purchased the gasoline? Explain.

ESSAY
Terpenes and Phenylpropanes

Distillation of plant materials by gentle heating will often yield "essential oils." These essential oils contain the odoriferous constituents of the plant in highly concentrated form. The methods of isolating essential oils have been known for a long while. By the end of the Middle Ages and the beginning of the Renaissance, more than 60 essential oils were already known. Today the list is much longer. Many of these oils are now isolated by steam distillation (see Experiment 19) or by solvent extraction of the appropriate plant parts.

Early investigations of the chemical composition of essential oils showed them to have major constituents of one of two types: terpenes or aromatic compounds. Many of them were found to be isomeric hydrocarbons of formula $C_{10}H_{16}$. These isomeric hydrocarbons of ten carbons were called **terpenes.** Subsequent investigations showed that there were also oxygenated derivatives of the terpenes in essential oils. These were found to be mainly alcohols and ketones, although a few were shown to be acids and esters. The oxygenated compounds were called **terpenoids.** Eventually, it was found that there are also minor and less-volatile plant constituents having carbon skeletons composed of 15, 20, 30, and 40 carbon atoms. These minor constituents were also called terpenes.

The recurrence in plants of compounds composed of C_5 units, in multiples of two or more, was an intriguing coincidence to chemists. This novel fact, along with the popular interest in many of the essential oils which are common flavoring and odor principles, led to an extensive investigation of these chemicals. By 1877 two circumstances allowed Otto Wallach (1847–1931) to postulate a rule concerning these compounds which subsequently became very famous.

First, the thermal "cracking" of natural rubber was found to give copious quantities of a hydrocarbon of formula C_5H_8, called **isoprene.**

Natural Rubber   **Isoprene**

Secondly, by 1877 a sufficient number of terpenoid compounds had had their structures elucidated to allow Wallach to recognize that, in addition to having formulas based on multiples of a C_5 unit, many of them had structural features in common with isoprene. He formulated a diagnostic rule for terpenes called the **isoprene rule.** This rule stated that if a molecule were a terpene, it should be divisible, at least formally, into isoprene units. The structures of two terpenes which follow the isoprene rule are shown below, along with a diagrammatic division of their structures into **isoprene units.**

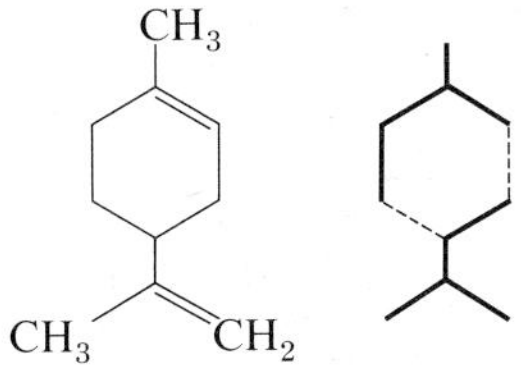

Limonene
(Turpentine)

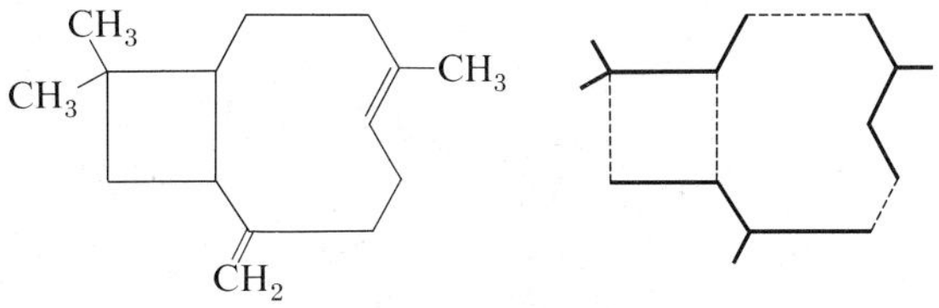

Caryophyllene
(Oil of Cloves)

Since compounds of formula $C_{10}H_{16}$ were originally called "terpenes," these compounds (C_{10}) are now called **monoterpenes** and the other terpenes are classified in the following way:

CLASS	NO. OF CARBONS	CLASS	NO. OF CARBONS
Hemiterpenes	5	Diterpenes	20
Monoterpenes	10	Triterpenes	30
Sesquiterpenes	15	Tetraterpenes	40

In the examples given above, limonene (Experiment 21) is a monoterpene, and caryophyllene (Experiment 19) is a sesquiterpene. β-Carotene (Experiment 9) is a tetraterpene. Camphor, borneol, and isoborneol (Experiment 20), and carvone (Experiment 21), are monoterpen**oid** compounds. Terpenes having numbers of carbons other than 10, 15, 20, 30, or 40 carbons are quite rare.

Aromatic compounds, those containing a benzene ring, are the second major type of compound found in essential oils. Some of these compounds, like p-cymene, are actually cyclic terpenes that have been aromatized, but most are of a different origin, and are based on a

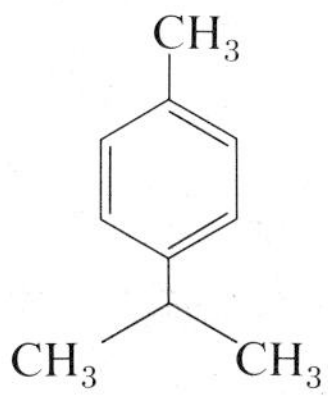

p-Cymene

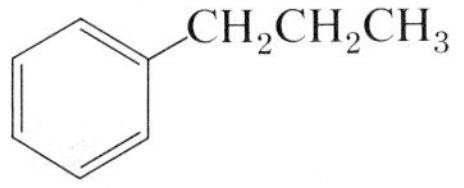

Phenylpropane

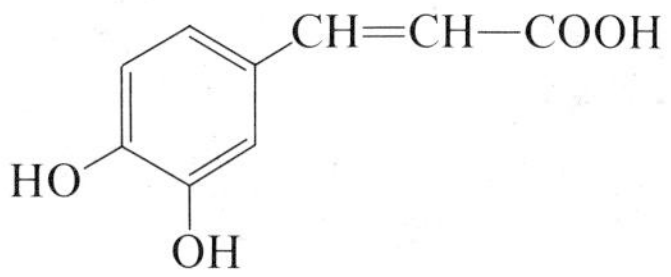

Caffeic Acid
(Coffee)

phenylpropane skeleton. Frequently, the phenyl ring of a phenylpropane compound will have several hydroxyl or methoxyl groups. It is also common to find compounds of phenylpropane origin which have had the three carbon side chain cleaved. Thus, phenylmethane derivatives are also common in plants.

Eugenol
(Cloves)

Vanillin
(Vanilla Bean)

THE BIOCHEMICAL ORIGINS

All of the above-mentioned compounds are secondary metabolites of plants. The primary metabolic processes which plants carry out are those which are crucial or central to the plant's existence. These processes include photosynthesis, glycolysis, respiration, and amino acid synthesis. When sunlight is available, plants carry out **photosynthesis** and convert carbon dioxide from the air into glucose. Glucose molecules form a sort of fuel and basic raw material for the plant. Glucose and its stored form, starch, can both serve as the source of all biosynthetic intermediates and as a source of chemical energy in the plant. As needed, the plant can break down glucose, a six carbon molecule, into two three carbon units. A major intermediate derived from this process, called **glycolysis,** is phosphoenol pyruvate (see chart). Phosphoenol pyruvate is the source, via decarboxylation, of an important two carbon unit, acetyl coenzyme A.

Acetyl coenzyme A is a central compound in plant metabolism because it can be used to provide energy for the plant. In the process called **respiration,** acetyl coenzyme A molecules are converted back to carbon dioxide and water. The specific metabolic sequence for this is called the **citric acid cycle.** The plant harnesses the thermodynamic energy derived from this process and stores it (as ATP) until it is needed to drive other processes. Thus, the citric acid cycle is like the motor of the plant and glucose is the fuel. Photosynthesis provides the fuel, and respiration "burns" it. These processes are **primary** metabolic processes in the plant.

Glucose also serves as a source of carbons for the plant to synthesize other needed materials. For instance, one of the chemical intermediates from the citric acid cycle has the ability to incorporate nitrogen from the soil into its structure. The amino acids and the nucleic acids are derived from this intermediate. The amino acids are used to form essential proteins and enzymes, and the nucleic acids are needed to form RNA and DNA, the genetic material of the plant. The synthesis of these materials also constitutes primary metabolism.

Compounds which are not in these mainstream processes are called **secondary metabolites.** Terpenes and phenylpropanes constitute two important classes of secondary plant metabolites. Terpenes are synthesized by a metabolic pathway called the **mevalonate pathway,** named after the key intermediate of the pathway, mevalonic acid (see chart). Mevalonic acid is formed by the condensation of three acetyl coenzyme A units. Decarboxylation of mevalonic acid gives rise to two isomeric five carbon compounds, **isopentenyl pyrophosphate** and **3,3-dimethylallyl pyrophosphate.** These two compounds are precursors of all the terpenes. They have the same basic carbon skeleton as isoprene. Dimerization of these two units leads to monoterpenoids (C_{10}) via the metabolic intermediate geranyl pyrophosphate.

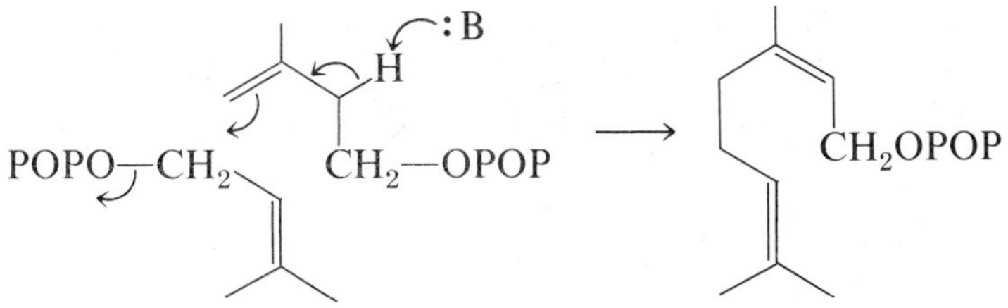

$(C_5) \times 2$ [POPO— = pyrophosphate]   Geranyl Pyrophosphate

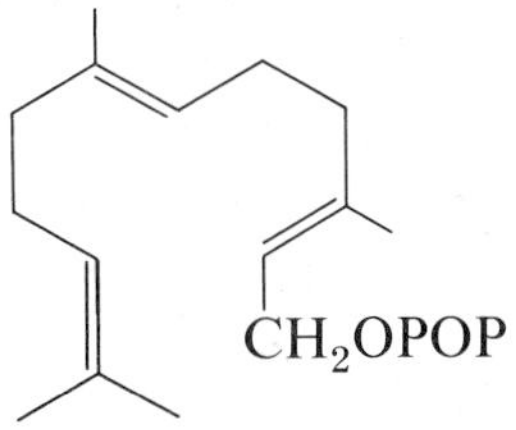

Farnesyl Pyrophosphate

Trimerization leads to farnesyl pyrophosphate and the sesquiterpenoid (C_{15}) compounds. Tetramerization (via geranyl-geranyl pyrophosphate) leads to diterpenoid (C_{20}) compounds. The only other processes that seem to occur regularly are the formation of C_{30} and C_{40} units, which are formed by dimerizations of C_{15} and C_{20} units rather than by the linkage of six or eight successive isoprene (C_5) units.

The phenylpropane compounds are formed by another sequence of secondary plant metabolism, the **shikimic acid pathway.** Plants have the ability to convert glucose into other sugars (pentose phosphate pathway). If glucose is converted to the 4-phosphate derivative of erythrose, and this is condensed with phosphoenol pyruvate, another important metabolic intermediate in plants is produced, **shikimic acid** (see chart). Condensation of shikimic acid with a second molecule of phosphoenol pyruvate, followed by a rearrangement, leads to **prephenic acid** (see chart) which is the precursor of all the phenylpropanoid compounds. As this intermediate is also a precursor to the amino acids phenylalanine and tyrosine, it is an important pathway in plants. Aside from these two amino acids, the other compounds produced by this pathway are secondary metabolites not crucial to the plant's existence.

Although both the shikimic acid and mevalonate pathways are common to plants, only the mevalonate pathway is common in animals, where it has a limited scope. The absence of the shikimic acid pathway in animals means that they are unable to synthesize either phenylalanine or tyrosine. These two amino acids are essential to the diet.

Mammals do not synthesize most of the smaller terpenes found in plants, but the mevalonate pathway does lead to steroids in mammals. Steroids are important hormones (see the essay preceding Experiment 8). For more than 20 years, chemists concerned with natural products suspected that there was a relationship between steroids and terpenoids. It was not until after World War II, when the use of radioactive isotopes became common, that the suspicion was confirmed. Using C^{14} labeled tracer compounds, chemists proved that enzymes in the liver could synthesize cholesterol from acetyl coenzyme A units. The missing link was squalene, a compound found in shark's liver, the only animal that accumulates large quantities of this substance.

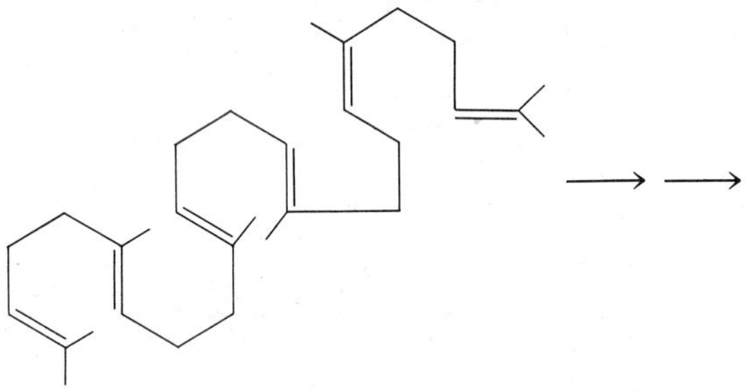

Squalene

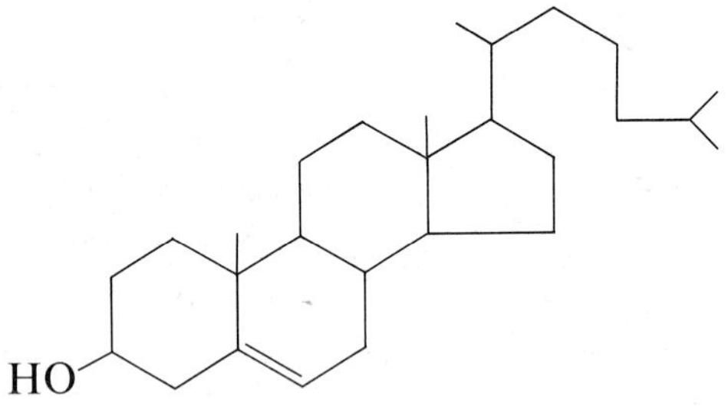

Cholesterol

Squalene is a triterpenoid and is the precursor of all mammalian steroids. They are formed by an extensive series of cyclization and rearrangement reactions starting with squalene. Most plants (although there are some significant exceptions) do not synthesize steroids, but they do produce a great many terpenoid compounds as the end products of the metabolism of mevalonic acid. For many of these secondary metabolites of plants, the function that they serve in the plant, if any, is still unknown. However, many of them are of great service to man, for whom they add a little spice and essence to life.

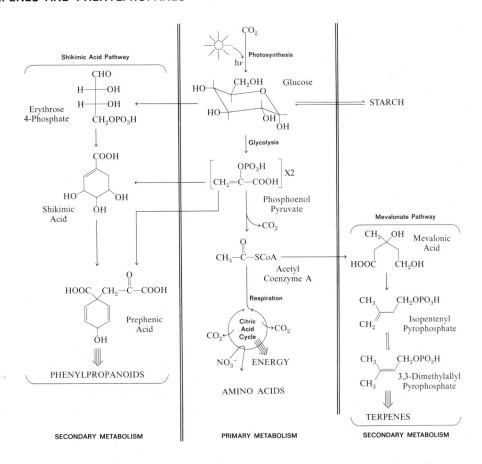

REFERENCES

Cornforth, J. W. "Terpene Biosynthesis." *Chemistry in Britain, 4* (1968) 102.

Geissman, T. A., and Crout, D. H. G. *Organic Chemistry of Secondary Plant Metabolism.* San Francisco: Freeman, Cooper and Company, 1969.

Goodwin, T. W., and Mercer, E. I. *Introduction to Plant Biochemistry.* New York: Pergamon Press, 1972.

Hendrickson, J. B. *The Molecules of Nature.* New York: W. A. Benjamin, 1965.

Pinder, A. R. *The Chemistry of the Terpenes.* New York: Wiley, 1960.

Ruzicka, L. "History of the Isoprene Rule." *Proceedings of the Chemical Society* (*London*), (1959) 341.

Sterrett, F. S. "The Nature of Essential Oils, Part I: Production." *Journal of Chemical Education, 39* (1962), 203.

Sterrett, F. S. "The Nature of Essential Oils. Part II: Chemical Constituents. Analysis." *Journal of Chemical Education, 39* (1962), 246.

Experiment 19

ESSENTIAL OILS FROM SPICES

Isolation Of A Natural Product
Steam Distillation
Derivative Formation

Anyone who has walked through a pine or cedar forest, or anyone who loves flowers and spices, knows that many plants and trees have distinctly pleasant odors. The essences or aromas of plants are due to volatile or **essential oils,** many of which have been valued since antiquity for their characteristic odors (e.g., frankincense and myrrh). A list of the commercially important essential oils would run to over 200 entries. Allspice, almond, anise, basil, bay, caraway, cinnamon, clove, cumin, dill, eucalyptus, garlic, jasmine, juniper, orange, peppermint, rose, sassafras, sandalwood, spearmint, thyme, violet, and wintergreen are but a few familiar examples of such valuable essential oils. Essential oils are used mainly for their pleasant odors and flavors in perfumes, incense, scents, spices, and as flavoring agents in foods. A few are also valued for their antibacterial and antifungicidal action. Some are used medicinally (e.g., camphor and eucalyptus) and others as insect repellents (citronella). Chaulmoogra oil represents one of the few known curative agents for leprosy. Turpentine is used as a solvent for many paint products.

Essential oil components are often found in the glands or intercellular spaces in plant tissue. They may occur in all parts of the plant, but they are often concentrated in the seeds or flowers. Many of the components of essential oils are steam volatile and may be isolated by steam distillation. Other methods of isolating essential oils include solvent extraction and pressing (expression) methods. From an earlier essay we saw that esters were frequently responsible for characteristic odors and flavors of fruits and flowers. In addition to the esters, essential oils may be composed of complex mixtures of hydrocarbons, alcohols, and carbonyl compounds. These latter components usually belong to one of the two groups of natural products called **terpenes** or **phenylpropanoids.** For details about these types of materials, see the essay which precedes this experiment.

The major constituents of the essential oils from cloves, cinnamon, cumin, and allspice are aromatic and volatile with steam. In this experiment we shall isolate the major component of the essential oil derived from these spices by steam distillation. At your instructor's option, the conversion of these oils into a solid derivative may also be required.

A derivative is a new compound to which the original compound is easily converted by a simple reaction. The original compound may be difficult to purify or difficult to characterize accurately by its physical properties since it is a liquid or oil. A derivative is a solid crystalline compound of definite melting point. The derivative is easy to characterize or identify. Its identity, once established, serves as circumstantial evidence to help identify the material from which it was formed.

The technique of steam distillation permits the separation of volatile components from nonvolatile materials without the need for raising the temperature of the distillation above 100°. Steam distillation provides a means of isolation of the essential oils without the risk of decomposing them thermally.

SPECIAL INSTRUCTIONS

Before beginning this experiment read the essay "Terpenes and Phenylpropanoids" and Techniques 3, 8 and 17. Your instructor may choose to make the formation of a derivative an optional part of the experiment. As the instructor indicates, you are to perform only one of the following three experiments: 19A, B, or C.

Experiment 19A
OILS OF CLOVE OR ALLSPICE

Both oil of cloves (from **Eugenia caryophyllata**) and oil of allspice (from **Pimenta officinalis**) are rich in **eugenol** (4-allyl-2-methoxyphenol). Caryophyllene is present in small amounts along with other terpenes. Eugenol (bp 250°) is a phenol or aromatic hydroxy compound.

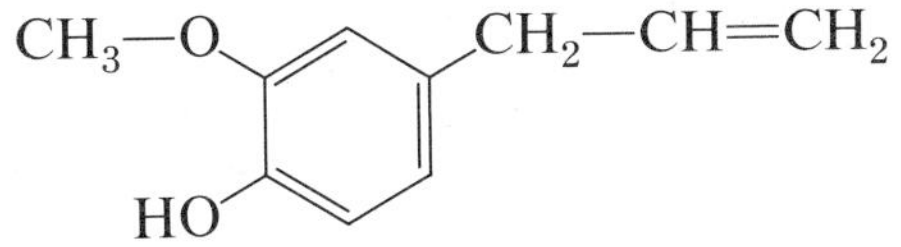

Eugenol 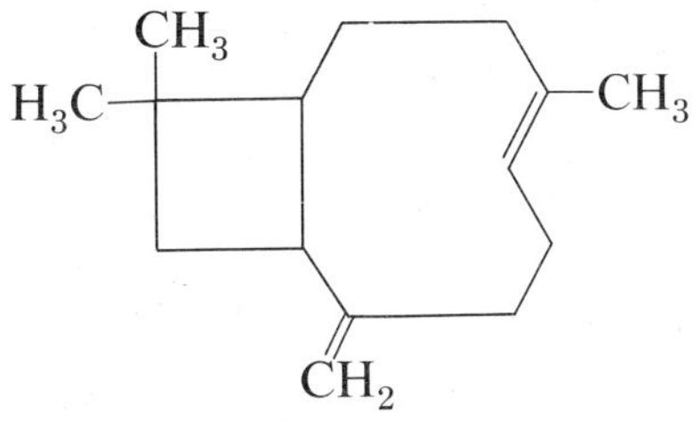   Caryophyllene

In this experiment you will be asked to isolate the eugenol and to characterize it by means of infrared and nuclear magnetic resonance spectroscopy. At your instructor's option, you may also be asked to convert eugenol to its benzoate derivative by reaction with benzoyl chloride.

PROCEDURE

Assemble an apparatus for steam distillation using a 500 ml three neck round bottom flask (Technique 8, Section 8.3B, Direct Method). The collection flask may be a 125 ml Erlenmeyer flask. The heat source will be a Bunsen burner flame. Place 15 g of ground allspice or 10 g of clove buds into the flask and add 150 ml of water. Begin heating the liquid in the flask so as to provide a slow, but steady rate of distillation. During the distillation continue to add water from the addition (separatory) funnel at a rate which will just maintain the original level of the liquid in the distilling flask. Continue the steam distillation until about 100 ml of distillate has been collected.

Empty the water from the addition funnel and place the distillate in it. Extract the distillate with two 10 ml portions of methylene chloride. Separate the layers and discard the aqueous phase. If the methylene chloride layer is separated carefully, it will not need to be dried, although drying with a small amount of anhydrous sodium sulfate may be advisable. Decant the solution from any drying agent and evaporate most of the solvent on a steam bath in the hood. Transfer the remaining liquid to a previously weighed test tube. Concentrate the contents of the test tube by further careful heating on a steam bath until nothing but an oily residue remains. Dry the outside of the test tube and weigh

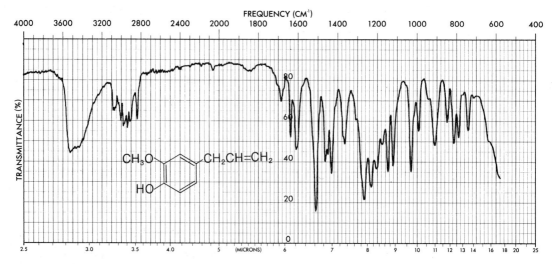

Infrared spectrum of eugenol, neat

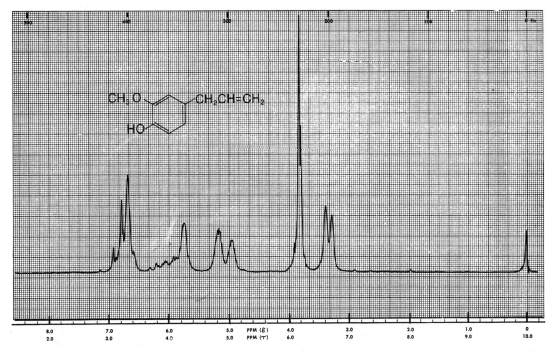

NMR of eugenol, CDCl₃

it. Calculate the weight percent recovery of the oil based on the original amount of allspice or cloves used.

Determine the infrared spectrum and, at the instructor's option, the nuclear magnetic resonance spectrum of the oil. The report should include the spectra which were recorded. The major absorptions in the spectra should be interpreted.

PREPARATION OF A DERIVATIVE (OPTIONAL)

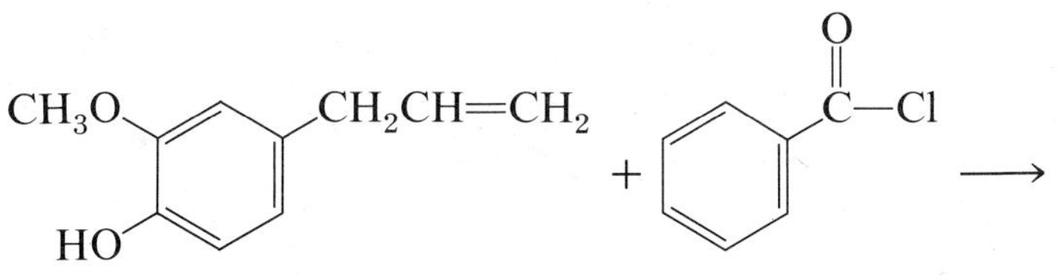

Eugenol   +   **Benzoyl Chloride**   ⟶

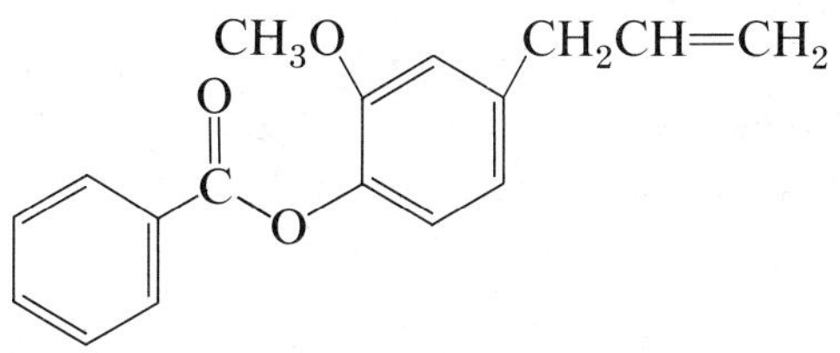

Eugenol Benzoate

Place about 0.2 ml of the crude eugenol in a small test tube and add 1 ml of water. Add drops of a 1M solution of sodium hydroxide to the test tube until the oil just dissolves. The final solution may be cloudy, but there should be no large oil droplets apparent. Carefully add 0.1 ml (4 or 5 drops) of benzoyl chloride. An excess of benzoyl chloride should be avoided since it will make the crystallization of the final product impossible to achieve. Heat the mixture on the steam bath for 5 or 10 minutes. Cool the mixture, and scratch the inside of the test tube with a glass rod until the mixture solidifies. If it does not solidify, cool the mixture in an ice bath. If the oil still will not solidify, decant the aqueous layer, add a few drops of methanol, and continue scratching and cooling the oil. Collect the solid on a Hirsch funnel and wash it carefully with a small volume of cold water. Recrystallize the solid from a **minimum** amount of boiling methanol. If an oil separates from the solution, reheat the solution and cool it more slowly while scratching it and adding seed crystals. Collect the crystals on a Hirsch funnel. Dry the crystals and record their melting point. The melting point of eugenol benzoate is 70°. Submit the crystals in a labeled vial to the instructor.

Experiment **19B**
OIL OF CUMIN

The major portion of the volatile oil from cumin seeds (**Cumin cymium**) is **cuminaledehyde** (p-isopropylbenzaldehyde). Cuminaldehyde has a boiling point of 235–236°.

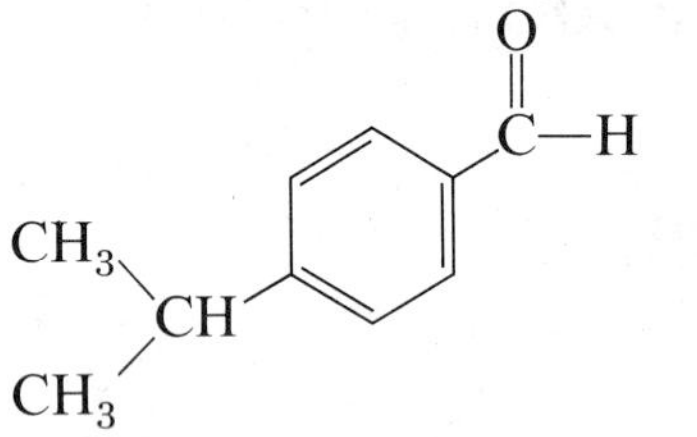

Cuminaldehyde

In this experiment you will be asked to isolate cuminaldehyde from cumin and to characterize it by means of infrared and nuclear magnetic resonance spectroscopy. At your instructor's option, you may also be asked to convert cuminaldehyde to its semicarbazone derivative by reaction with semicarbazide hydrochloride.

PROCEDURE

Assemble an apparatus for steam distillation using a 500 ml three neck round bottom flask (Technique 8, Section 8.3B, Direct Method).

The collection flask may be a 125 ml Erlenmeyer flask. The heat source will be a Bunsen burner flame or a heating mantle. Place 15 g of ground cumin seeds into the flask and add 150 ml of water. Begin heating the liquid in the flask so as to provide a slow, but steady, rate of distillation. During the distillation continue to add water from the addition (separatory) funnel at a rate which will just maintain the original level of the liquid in the distilling flask. Continue the steam distillation until about 100 ml of distillate has been collected.

Empty the water from the addition funnel and place the distillate in it. Extract the distillate with two 10 ml portions of methylene chloride. Separate the layers and discard the aqueous phase. If the methylene chloride layer is separated carefully, it will not need to be dried, although drying with a small amount of anhydrous sodium sulfate may be advisable. Decant the solution from any drying agent and evaporate most of the solvent on a steam bath in the hood. Transfer the remaining liquid to a previously weighed test tube. Concentrate the contents of the test tube by further careful heating on a steam bath until nothing but an oily residue remains. Dry the outside of the test tube and weigh it. Calculate the weight percent recovery of the oil based on the original amount of cumin used.

Determine the infrared spectrum and, at the instructor's option, the nuclear magnetic resonance spectrum of the oil. The report should include the spectra which were recorded, and an interpretation of the major peaks in the spectra.

PREPARATION OF A DERIVATIVE (OPTIONAL)

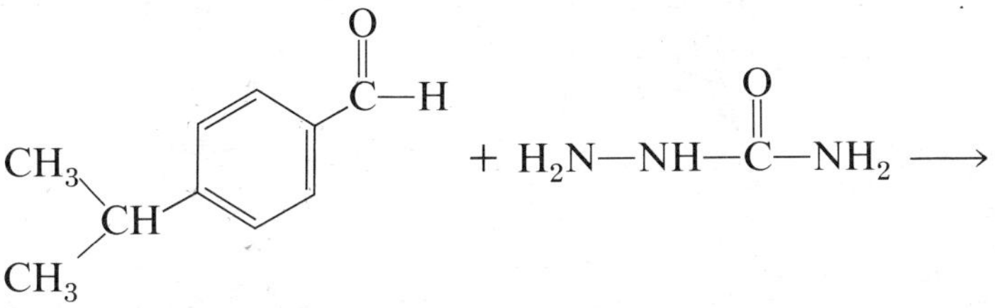

Cuminaldehyde   Semicarbazide

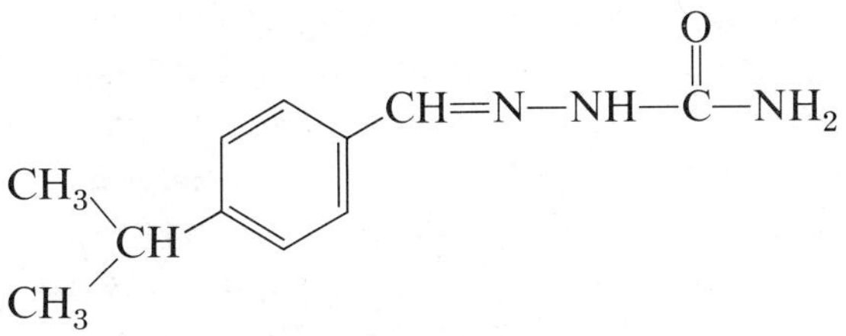

Cuminaldehyde Semicarbazone

Dissolve 0.20 g of semicarbazide hydrochloride and 0.30 g of anhydrous sodium acetate, which will serve as a buffer, in 2 ml of water. To this mixture add 3 ml of absolute ethanol. Add this solution to the cumin oil and warm the mixture on the steam bath for five minutes. Cool the mixture and allow the cuminaldehyde semicarbazone to crystallize. Collect the crystals in a Hirsch funnel and recrystallize them from methanol. Allow the crystals to dry and determine their melting point. The melting point of cuminaldehyde semicarbazone is 216°. Submit the derivative in a labeled vial to the instructor.

Experiment 19C
OIL OF CINNAMON

The principal component of cinnamon oil (from **Cinnamomum zeylanicum**) is cinnamaldehyde (trans-3-phenylpropenal). Cinnamaldehyde has a boiling point of 252°.

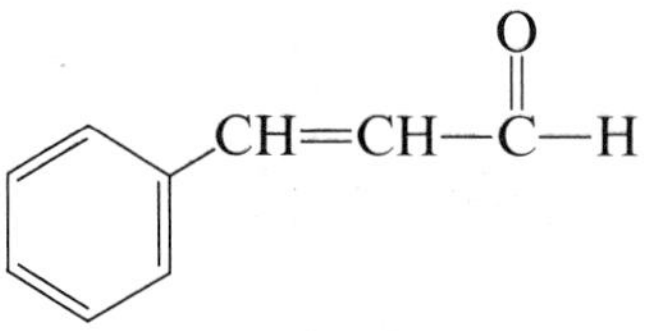

Cinnamaldehyde

In this experiment you will be asked to isolate cinnamaldehyde from cinnamon and to characterize it by means of infrared and nuclear magnetic resonance spectroscopy. At your instructor's option, you may also be asked to convert cinnamaldehyde to its semicarbazone derivative by reaction with semicarbazide hydrochloride.

PROCEDURE

Assemble an apparatus for steam distillation using a 500 ml three neck round bottom flask (Technique 8, Section 8.3A, Live Steam Method). The collection flask may be a 125 ml Erlenmeyer flask. Place 15 g of ground cinnamon bark in the flask and add 100 ml of hot water. Admit steam from the steam line so as to provide a slow, but steady rate of distillation. Continue the steam distillation until about 100 ml of distillate has been collected.

Place the distillate in a separatory funnel and extract it with two 10 ml portions of methylene chloride. Separate the layers and discard the aqueous phase. If the methylene chloride layer is separated carefully, it will not need to be dried, although drying with a small amount of anhydrous sodium sulfate may be advisable. Decant the solution from any drying agent and evaporate most of the solvent on a steam bath in the hood. Transfer the remaining liquid to a previously weighed test

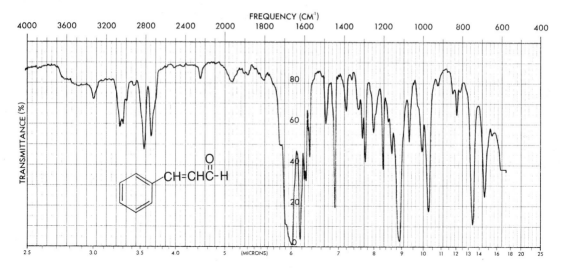

Infrared spectrum of cinnamaldehyde, neat

tube. Concentrate the contents of the test tube by further careful heating on a steam bath until nothing but an oily residue remains. Dry the outside of the test tube and weigh it. Calculate the weight percent recovery of the oil based on the original amount of cinnamon bark used.

Determine the infrared spectrum and, at the instructor's option, the nuclear magnetic resonance spectrum of the oil. The report should include the spectra which were recorded, and an interpretation of the major peaks in the spectra.

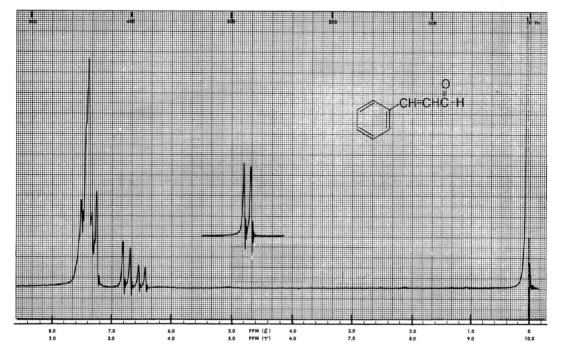

NMR of cinnamaldehyde, offset peaks, 300 HZ

PREPARATION OF A DERIVATIVE (OPTIONAL)

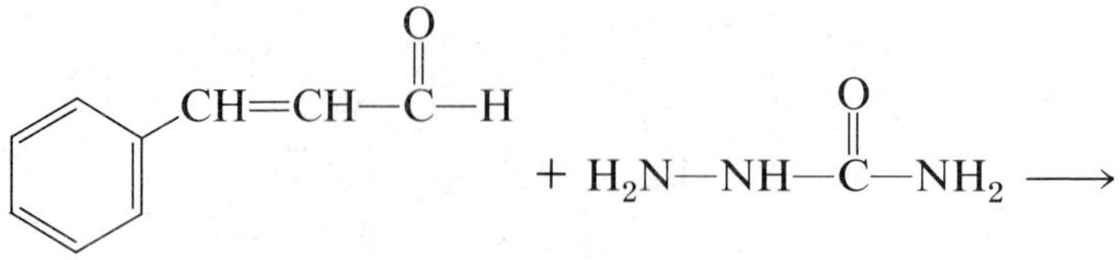

$$\text{Cinnamaldehyde} \quad \text{CH=CH—C—H} \quad + \quad H_2N—NH—C—NH_2 \quad \longrightarrow$$

Cinnamaldehyde	Semicarbazide

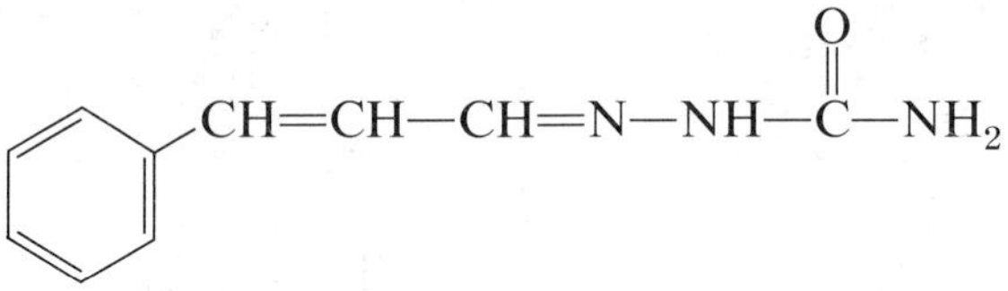

$$\text{CH=CH—CH=N—NH—C—NH}_2$$

Cinnamaldehyde Semicarbazone

Dissolve 0.20 g of semicarbazide hydrochloride and 0.30 g of anhydrous sodium acetate, which will serve as a buffer, in 2 ml of water. To this mixture add 3 ml of absolute ethanol. Add this solution to the cinnamon oil and warm the mixture on the steam bath for five minutes. Cool the mixture and allow the cinnamaldehyde semicarbazone to crystallize. Collect the crystals in a Hirsch funnel and recrystallize them from methanol. Allow the crystals to dry and determine their melting point. The melting point of cinnamaldehyde semicarbazone is 215°. Submit the derivative in a labeled vial to the instructor.

Experiment **20**

AN OXIDATION-REDUCTION SCHEME: BORNEOL, CAMPHOR, ISOBORNEOL

Sodium Dichromate Oxidation
Sodium Borohydride Reduction
Stereochemistry
Sublimation

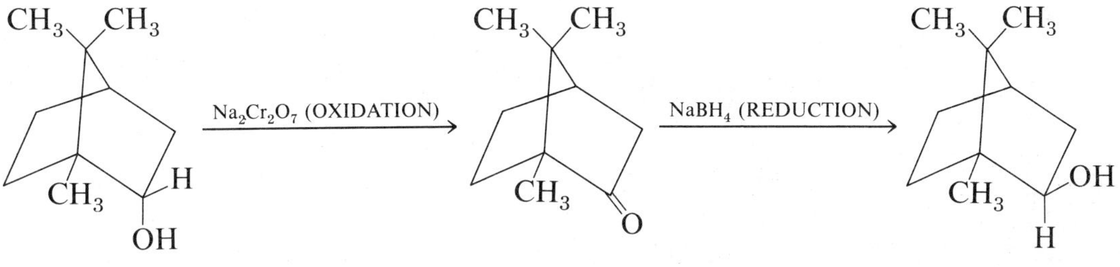

Borneol   Camphor   Isoborneol

This experiment illustrates the use of an oxidizing agent (sodium dichromate) for conversion of a secondary alcohol (borneol) to a ketone (camphor). The camphor is purified by sublimation and then is reduced by sodium borohydride to give the **isomeric** alcohol, isoborneol. The isoborneol is also purified by sublimation. The spectra of borneol, camphor, and isoborneol are compared in order to detect structural differences and to determine the extent to which the final step produces a pure alcohol isomeric with the starting material.

OXIDATION OF BORNEOL WITH SODIUM DICHROMATE

Secondary alcohols are easily oxidized to ketones by the use of sodium dichromate or other chromium VI compounds, such as chromium trioxide or sodium chromate.

The half-reactions and overall reaction employed in the oxidation of borneol with dichromate are as follows:

Half-reaction: $Cr_2O_7^= + 14 H^+ + 6 e^- \longrightarrow 2 Cr^{+++} + 7 H_2O$

Half-reaction: 3 $\longrightarrow$ 3 $+ 6 H^+ + 6 e^-$

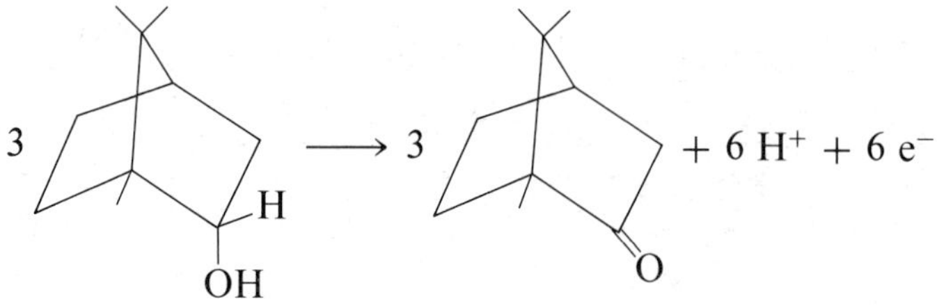

Net reaction:

3 $+ Cr_2O_7^= + 8 H^+ \longrightarrow$ 3 $+ 2 Cr^{+++} + 7 H_2O$

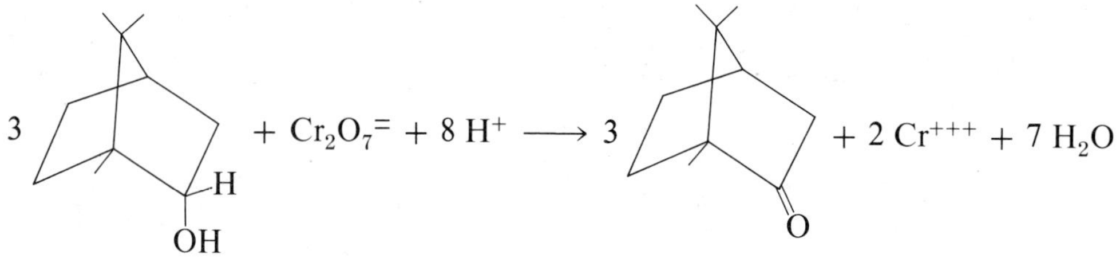

The mechanism for the oxidation is as follows:

$Cr_2O_7^= + H_2O \rightleftharpoons 2 CrO_4^= + 2 H^-$

Dichromate   **Chromate**

No net change
in the oxidation
number of chromium

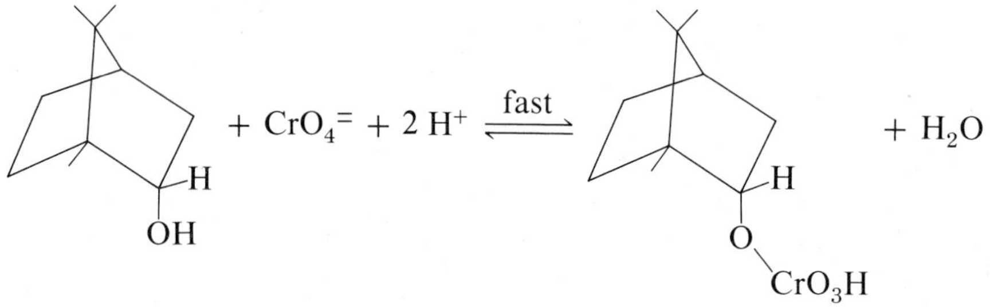

$+ CrO_4^= + 2 H^+ \xrightarrow{\text{fast}}$ $+ H_2O$

**Borneol ester of
chromic acid**

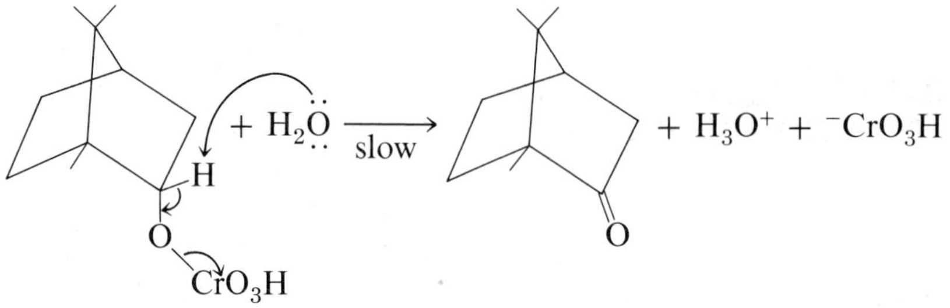

$+ H_2\ddot{O} \xrightarrow{\text{slow}}$ $+ H_3O^+ + {}^-CrO_3H$

REDUCTION OF CAMPHOR WITH SODIUM BOROHYDRIDE

Metal hydrides (sources of $H:^-$) of the Group III elements such as lithium aluminum hydride ($LiAlH_4$) and sodium borohydride ($NaBH_4$) are widely used in reducing carbonyl groups. Lithium aluminum hydride, for example, reduces many compounds containing carbonyl groups, such as aldehydes, ketones, carboxylic acids, esters, or amides, while sodium borohydride only reduces aldehydes and ketones. The reduced reactivity of borohydride even allows it to be used in alcohol and water solvents, whereas lithium aluminum hydride reacts violently with these solvents to produce hydrogen gas, and thus must be used in nonhydroxylic solvents. In the present experiment, sodium borohydride is used because it is easily handled, and the results of reductions using the two reagents are essentially the same. Great care need not be taken in keeping it away from water as is required with lithium aluminum hydride.

The mechanism of action of sodium borohydride in reducing a ketone is as follows:

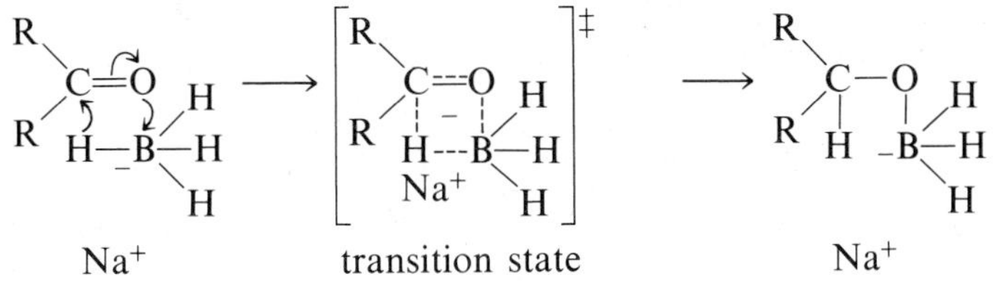

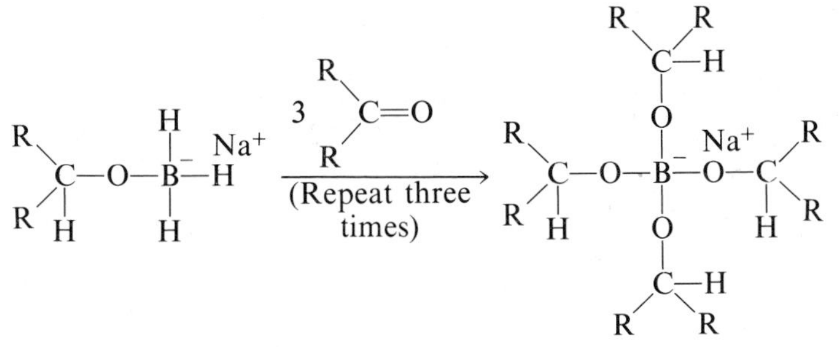

One notes in this mechanism that all four hydrogen atoms are available as hydrides (H^-), and thus one mole of hydride can reduce four moles of ketone. All of the steps are irreversible. Usually an excess of hydride is used because there is uncertainty regarding the purity of the material.

Once the final tetraalkoxyboron compound (I) is produced, it can be decomposed (along with excess hydride) at elevated temperatures as shown below:

$$(R_2CH-O)_4B^-\ Na^+ + 4\ R'OH \longrightarrow 4\ R_2CHOH + (R'O)_4B^-\ Na^+$$
(I)

The stereochemistry of the reduction is of considerable interest. The hydride can approach the camphor molecule more easily from the bottom side (**endo**-approach) than from the top side (**exo**-approach).

If attack occurs at the top, there is a large steric repulsion created by one of the two **geminal** methyl groups. Attack at the bottom avoids this steric interaction.

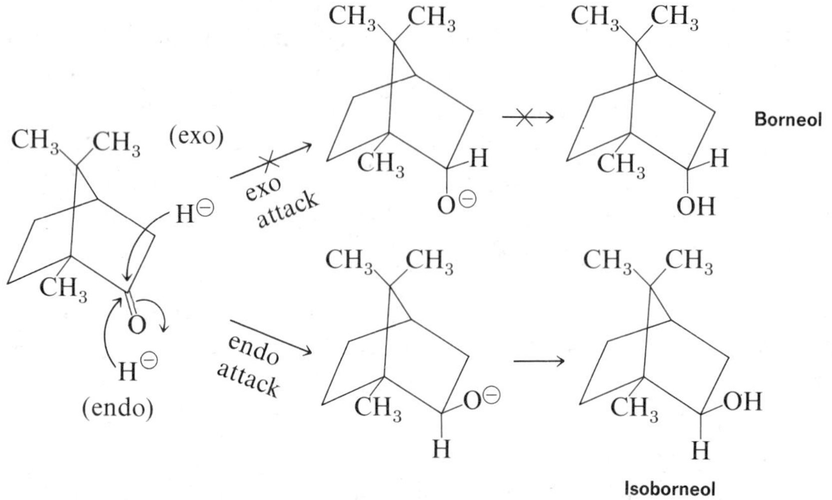

It is expected, therefore, that **isoborneol,** the alcohol produced from the attack at the **least** hindered position, will **predominate but will not be the exclusive product** in the final reaction mixture. The precentage composition of the mixture can be determined by spectroscopy.

It is interesting to note that when the methyl groups are removed (as in 2-norbornanone), the top side (**exo**-approach) is favored, and the opposite stereochemical result is obtained. Again, the reaction does not give **exclusively** one product.

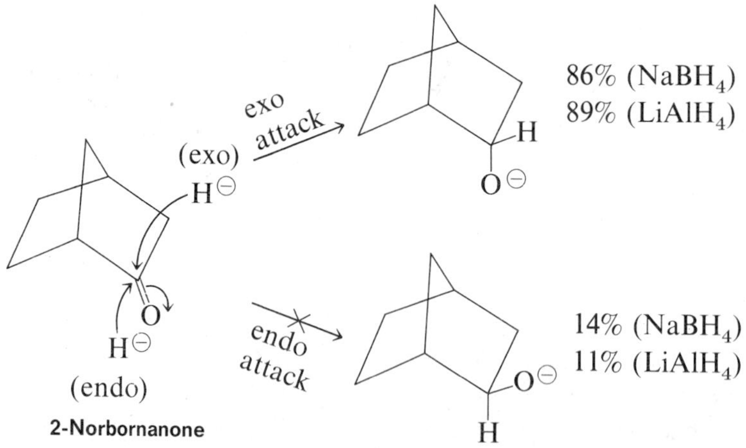

Bicyclic systems such as camphor and 2-norbornanone react in a predictable way according to steric influences. This effect has been termed "steric approach control." However, in the reduction of simple acyclic and monocyclic ketones, the reaction appears to be influenced primarily by thermodynamic factors. This latter effect has been termed "product development control." In the reduction of 4-*t*-butylcyclohexanone, the thermodynamically more stable product is produced by "product development control."

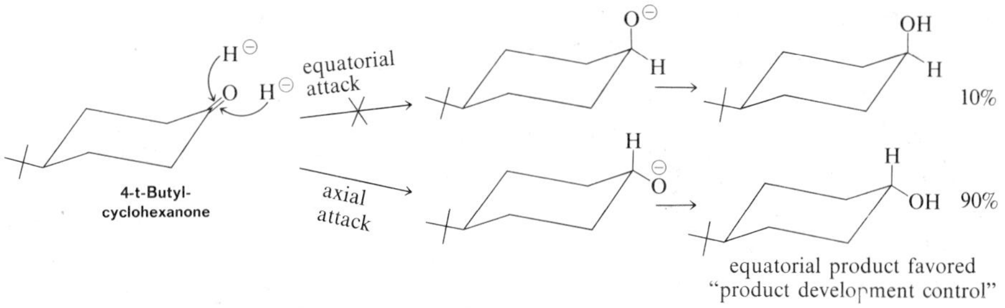

4-t-Butyl-cyclohexanone

equatorial attack

axial attack

equatorial product favored
"product development control"

10%

90%

SPECIAL INSTRUCTIONS

General background information on terpenes is given in the essay which precedes Experiment 19. Borneol, camphor and isoborneol are examples of terpenes. It is necessary to have read Techniques 1 through 5 and 13 before starting this experiment. You may also need to consult Technique 17 and Appendices Three and Four. The reactants and products are all highly volatile and must be stored in tightly closed containers.

PROCEDURE

OXIDATION OF BORNEOL TO CAMPHOR

Dissolve 2.0 g of sodium dichromate dihydrate in 8 ml of water, and **carefully** add 1.6 ml of concentrated sulfuric acid with an eyedropper. Place the oxidizing solution in an ice bath. While this solution is cooling, dissolve 1.0 g of racemic borneol in 4 ml of ether in a 25 ml Erlenmeyer flask and cool it in an ice bath. Remove 6 ml of the sodium dichromate oxidizing mixture prepared above, and **slowly** add it with an eyedropper to the **cold** ether solution over a period of 10 minutes. **Swirl** the reaction mixture in the ice bath between additions and continue swirling for an additional 5 minutes following the final addition of oxidant. Pour the mixture into a separatory funnel and rinse the Erlenmeyer flask, first with a 10 ml portion of ether, and then with 10 ml of water. **Add both** of the rinsings to the separatory funnel.

The entire mixture will be very dark, and it may be difficult to see the interface between the aqueous and organic phases. In this case, use a light source such as a lamp or flashlight in order to detect the interface. If there still is difficulty, begin to drain the lower aqueous phase until the interface is observed. When the lighter colored ether phase passes into the narrow neck of the separatory funnel, it should become visible.

Complete the removal of the aqueous phase, and pour the ether layer into a storage vessel. Return the aqueous layer to the separatory funnel and extract it with two successive 10 ml portions of ether. Each time add the ether phase to the storage vessel and return the aqueous layer to the separatory funnel.

Return the combined ether extracts to the separatory funnel and extract them with 10 ml of 5% sodium bicarbonate. A small amount of solid material may be produced at the interface. Carefully remove the lower aqueous layer and as much of this solid as possible without losing the ether layer. Finally, wash the ether layer with 10 ml of water, and drain the lower aqueous layer. The ether layer contains the desired camphor. Pour it out of the top of the separatory funnel, decanting it away from any solids. Dry the ether layer thoroughly with a small amount (1 or 2 g) of anhydrous magnesium sulfate in a stoppered Erlenmeyer flask. Swirl the flask gently until the ether phase is **clear.** Decant the dry ether phase into a beaker and evaporate the solvent in the hood on a steam bath (using a boiling stone) or with a stream of **dry** air. When the ether has evaporated and a solid has appeared, remove the flask from the heat source **immediately,** otherwise the product may prematurely sublime and be lost. Weigh the product. Approximately 0.6 to 0.7 g of crude camphor should be obtained. Purify all the material by vacuum (aspirator) sublimation (Technique 13, Figure 13–2a or 13–2b). Camphor sublimes rapidly at about 115° under reduced pressure. Scrape the purified material from the cold finger, weigh it, and calculate the percentage yield. Determine the melting point in a sealed capillary tube to prevent sublimation. The melting point of pure racemic camphor is 174°. Save a small amount of purified camphor for an infrared spectrum. The remainder of the camphor is reduced in the next step to isoborneol. Store the compounds in tightly closed containers until needed. For the ir spectrum, dissolve the sample in carbon tetrachloride, place the solution between the salt plates, and mount the plates in a holder (see Technique 17, Part A). An ir spectrum for camphor is shown at the end of the experiment.

REDUCTION OF CAMPHOR TO ISOBORNEOL

In a 25 ml Erlenmeyer flask dissolve 0.5 g of the racemic camphor obtained above in 2 ml of methanol. Cautiously and intermittently add portions of 0.3 g of sodium borohydride to the above solution. If necessary, cool the flask in an ice bath to keep the reaction mixture at room temperature. When all the borohydride is added, boil the contents of the flask on a steam bath for one minute.

Pour the hot reaction mixture into about 15 g of chipped ice. When the ice melts, collect the white solid by suction and dry it thoroughly. Purify the crude isoborneol by vacuum (aspirator) sublimation as before. Isoborneol rapidly sublimes at about 110 to 130° under reduced pressure. Weigh the purified material and calculate the percentage yield. Determine the melting point (sealed tube); **pure** racemic isoborneol melts at 212°. Determine the infrared spectrum of the purified product by the method given above. Compare it to the spectra for borneol and isoborneol given below.

PERCENTAGE OF ISOBORNEOL AND BORNEOL OBTAINED FROM THE REDUCTION OF CAMPHOR

The percentage of each of the isomeric alcohols in the borohydride reduction mixture can be determined from the nmr spectrum[1] (see Technique 17 Part B and Appendix Four). The nmr spectra of the pure alcohols are shown below. The hydrogen on the carbon bearing the hydroxyl group appears at $\delta 4.0$ (6.0τ) for borneol and $\delta 3.6$ (6.4τ) for isoborneol. One can obtain the product ratio by integrating these peaks (using an expanded presentation) in the nmr spectrum of "isoborneol" obtained after the borohydride reduction. In the spectrum shown below, an isoborneol to borneol ratio of 5:1 was obtained. The percentages obtained are 83% isoborneol and 17% borneol.

REFERENCES

Brown, H. C., and Muzzio, J. "Rates of Reaction of Sodium Borohydride with Bicyclic Ketones." *Journal of the American Chemical Society, 88* (1966), 2811.

Dauben, W. G., Fonken, G. J., and Noyce, D. S. "Stereochemistry of Hydride Reductions." *Journal of the American Chemical Society, 78* (1956), 2579.

Flautt, T. J., and Erman, W. F. "The Nuclear Magnetic Resonance Spectra and Stereochemistry of Substituted Boranes." *Journal of the American Chemical Society, 85* (1963), 3212.

Markgraf, J. H. "Stereochemical Correlations in the Camphor Series." *Journal of Chemical Education, 44* (1967), 36.

QUESTIONS

1. Interpret the major absorption bands in the infrared spectra of camphor, borneol and isoborneol.

2. Explain why the **gem**-dimethyl groups appear as separate peaks in the nmr of isoborneol while they are not resolved in borneol.

3. A sample of isoborneol prepared by reduction of camphor was analyzed by infrared spectroscopy and showed a strong band at 1760 cm^{-1} (5.7μ). This result was unexpected. Why?

---

[1] The percentages can also be obtained by gas chromatography, using a Varian-Aerograph, Model A-90-P instrument. Use a 5 ft column of 5% Carbowax 4000 on acid-washed firebrick, and operate the device at 110° with a 50 ml/min flow rate. The compounds are dissolved in a low-boiling solvent for analysis. The retention times for camphor, isoborneol, and borneol are 6, 10, and 12 min, respectively.

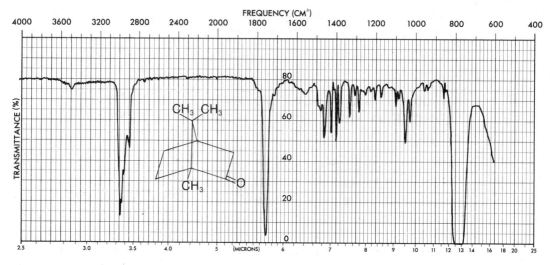

Infrared spectrum of camphor, CCl₄

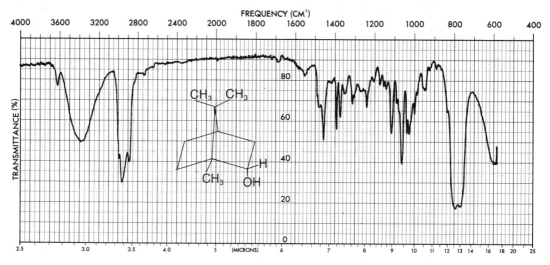

Infrared spectrum of borneol, CCl₄

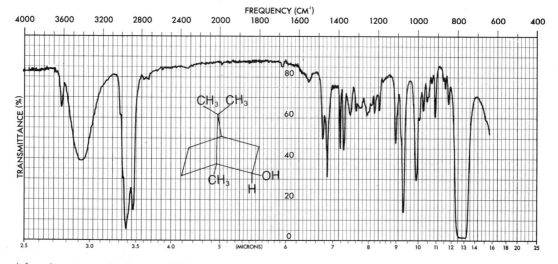

Infrared spectrum of isoborneol, CCl₄

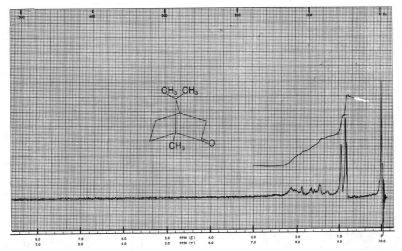

NMR of camphor, CCl$_4$

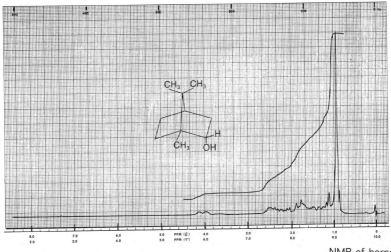

NMR of borneol, CDCl$_3$

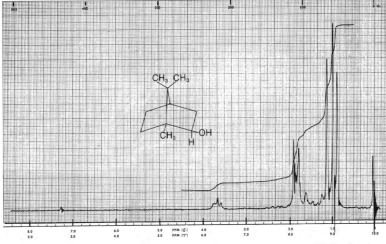

NMR of isoborneol, CDCl$_3$

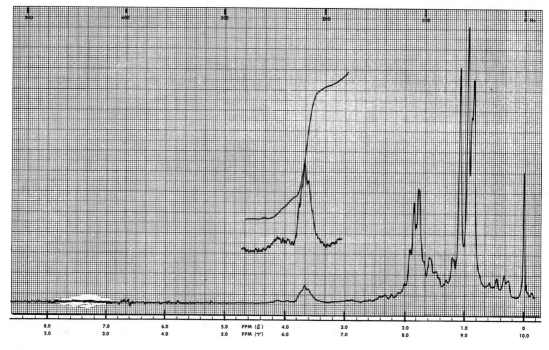

NMR of borohydride reduction product, CDCl$_3$

ESSAY

The Stereochemical Theory of Odor

The human nose has an almost unbelievable ability to distinguish odors. Just consider for a few moments the different substances which you are able to recognize by odor alone. Your list should be a very long one. A person with a trained nose, a perfumer, for instance, can often recognize even individual components in a mixture. Who has not met at least one cook who could sniff almost any culinary dish and identify the seasonings and spices that were used? The olfactory centers in the nose can identify odorous substances even in very small amount. With some substances, studies have shown that as little as one ten millionth of a gram (10^{-7} gram) can be perceived. Of course, many animals (for example, dogs) and insects (see the essay on pheromones that precedes Experiment 17) have an even lower threshold of smell than humans.

There have been many theories of odor, but very few have persisted for any length of time. Strangely enough, one of the oldest theories, although in modern dress, is still the most current theory. Lucretius, one of the early Greek atomists, suggested that substances having odor gave off a vapor of tiny "atoms," all of the same shape and size, and that they gave rise to the perception of odor when they entered pores in the nose. The pores would have to be of various shapes and the odor perceived would depend on which pores the atoms were able to enter. We now have many similar theories regarding the action of drugs (receptor site theory), and the interaction of enzymes with their substrates (the lock and key hypothesis).

A substance must have certain physical characteristics to have the property of odor. First, it must be volatile enough to give off a vapor which can reach the nostrils. Second, once it reaches the nostrils, it must display at least a small, if minute, water solubility so that it can pass through the layer of moisture (mucous) that covers the nerve endings in the olfactory area. Third, it must also have lipid solubility to allow it to penetrate the lipid (fat) layers that form the surface membranes of the nerve cell endings.

Once we pass these criteria, we come to the heart of the question. Why do substances have different odors? In 1949 R. W. Moncrieff, a Scot, resurrected Lucretius' hypothesis. He proposed that in the olfactory area of the nose there is a system of receptor cells of several different types and shapes. He further suggested that each receptor site corresponded to a different type of primary odor. Molecules which would fit these receptor sites would display the characteristics of that primary odor. It would not be necessary for the entire molecule to fit into the receptor, so that for larger molecules, any portion might fit into and activate the receptor. Molecules exhibiting complex odors would presumably be able to activate several different types of receptors.

Moncrieff's hypothesis has been substantially strengthened by the work of J. E. Amoore, which he began as an undergraduate at Oxford in 1952. After an extensive search of the chemical literature, Amoore concluded that there were only seven basic primary odors. By sorting molecules with similar odor types, he even formulated possible shapes for the seven necessary receptors. For instance, from the literature, he culled more than 100 compounds that were described as having a "camphoraceous" odor. Comparing the sizes and shapes of all of these molecules, he postulated a three dimensional shape for a camphoraceous receptor site. Similarly, he derived shapes for the other six receptor sites. The seven primary receptor sites which he formulated are shown in the figure, along with a typical prototype molecule of the appropriate shape to fit the receptor. The shapes of the sites are shown in perspective. Pungent and putrid odors are not thought to require a particular shape in the odorous molecule, but rather to require a particular charge distribution type.

That compounds of roughly similar shape exhibit similar odors can

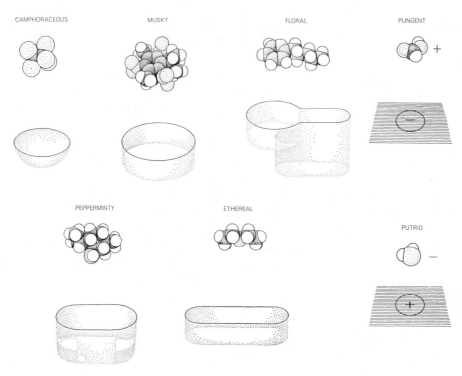

CAMPHORACEOUS   MUSKY   FLORAL   PUNGENT

PEPPERMINTY   ETHEREAL   PUTRID

From "The Stereochemical Theory of Odor," by J. E. Amore, J. W. Johnston, Jr., and M. Rubin. Copyright © by Scientific American, Inc. All rights reserved.

quickly be verified by the reader by comparing the odors of nitro-benzene, acetophenone and benzaldehyde on the one hand, and of d-camphor, hexachloroethane and cyclooctane on the other. Each group of substances has the same basic odor **type** (primary), but the individual molecules differ in **quality** of the odor. Some are sharp, some pungent, other sweet, etc. The latter group all have a camphoraceous odor, and they are all of approximately the same shape.

An interesting corollary to the Amoore theory would be the postulate that, if the receptor sites are chiral, then optical isomers (enantiomers) of a given substance might exhibit **different** odors. This circumstance proves to be true in a number of cases. It occurs in the cases of (+)- and (−)-carvone, and is investigated in Experiment 21 in this text.

Several workers have tested Amoore's hypothesis by experiment. The results of these studies are generally favorable to the hypothesis—so favorable that some chemists now elevate the hypothesis to the level of a theory. In several cases, researchers have been able to "synthesize" odors almost indistinguishable from the real thing by the proper blending of primary odor substances. The primary odor substances used were unrelated to the chemical substances composing the natural odor. These experiments, and others, are described in the articles listed at the end of this essay.

REFERENCES

Amoore, J. E. *The Molecular Basis of Odor*. American Lecture Series Publication No. 773. Springfield, Illinois: Thomas, 1970.

Amoore, J. E., Johnston, J. W., Jr., and Rubin, M. "The Stereochemical Theory of Odor." *Scientific American, 210* (February, 1964).

Amoore, J. E., Rubin, M., and Johnston, J. W., Jr. "The Stereochemical Theory of Olefaction." *Proceedings of the Scientific Section of the Toilet Goods Association* (Special Supplement to No. 37) (October, 1962), 1–47.

Moncrieff, R. W. *The Chemical Senses*. London: Leonard Hill, 1951.

Experiment **21**

SPEARMINT AND CARAWAY OIL: (−) and (+)-CARVONES

Stereochemistry
Vacuum Fractional Distillation
Gas Chromatography, Spectroscopy
Optical Rotation, Refractometry

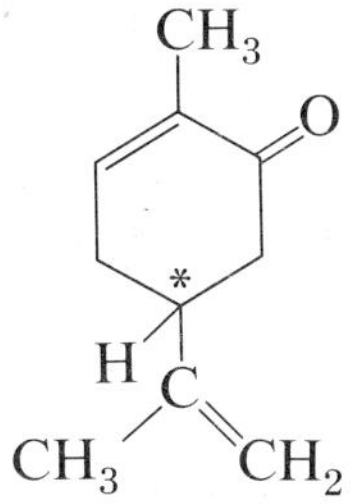

In this experiment, we will isolate (+)-carvone from caraway seed oil and (−)-carvone from spearmint oil by vacuum fractional distillation. The odors of these optical isomers are distinctly different. The presence of one or the other of these isomers is responsible for the characteristic odors of each of the two oils. The difference in their odors is to be expected since the odor receptors in the nose are chiral (see the essay which precedes this experiment). Although we should expect the optical rotations of the isomers (enantiomers) to be of opposite sign, the other physical properties should be identical. Thus, for both (+) and (−)-carvone we predict that the infrared and nuclear magnetic resonance spectra, the gas chromatographic retention times, and the refractive indices should all be identical, within experimental error.

Hence, the only difference in properties one may observe for the two carvones will be the odors and the signs of rotation in a polarimeter.

Caraway seed oil contains mainly limonene and (+)-carvone. The gas chromatogram for this oil is shown in the figure below. Limonene (left hand peak) has the lower retention time. The (+)-carvone (bp 230°) can easily be separated from the lower boiling limonene (bp 177°) by vacuum fractional distillation. The separation is relatively easy due to the large boiling point difference. **Spearmint oil** contains mainly (−)-carvone with a smaller amount of limonene and very small amounts of the lower boiling terpenes, α- and β-phellandrene. The gas chromatogram for this oil is also shown in the figure below. The (−)-carvone can also easily be separated in this oil from the three lower boiling components, the α- and β-phellandrenes, and limonene, by vacuum fractional distillation. These latter three terpenes, however, are

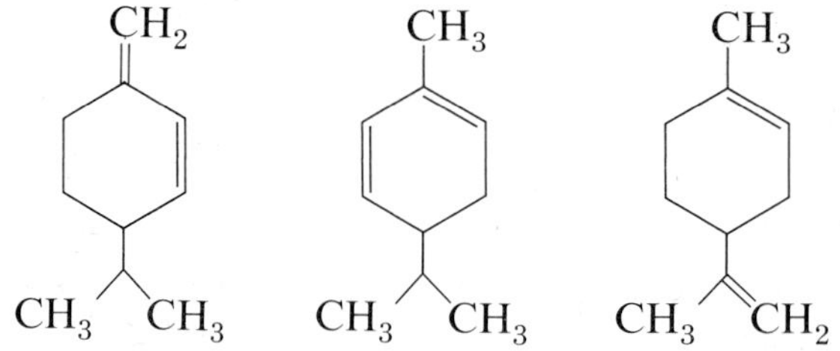

CH₂   CH₃   CH₃

α-Phellandrene   β-Phellandrene   Limonene

CH₃ CH₃   CH₃ CH₃   CH₃ CH₂

only difference you have lost the ketone group

not easily separated because of their similar boiling points.

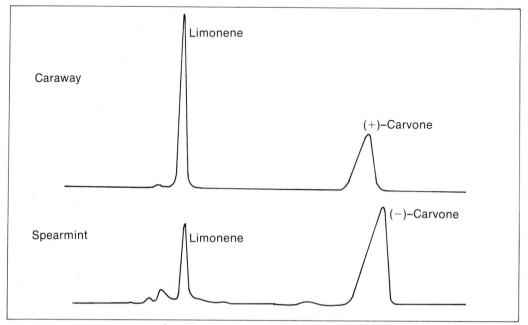

Gas chromatograms of caraway and spearmint oil

SPECIAL INSTRUCTIONS

Read the essay, The Stereochemical Theory of Odor, which precedes this experiment. In addition, it will be necessary to have read Techniques 6, 7, 9, 12, 15, and 16. In order to conduct successfully a fractional distillation under reduced pressure, one must be certain that the aspirator and hoses are in good condition. It is best to use a manometer in the system during the entire distillation.

NOTE TO THE INSTRUCTOR: This experiment may be scheduled along with another experiment. One-half of the class should conduct this experiment while the other half conducts an experiment not requiring the use of aspirators. In this way, the water pressure will be high enough to obtain adequate pressures at the aspirator. Alternatively, or in addition, the students may work in pairs. The column in the gas chromatograph must be heated and equilibrated well in advance of the laboratory period.

PROCEDURE

VACUUM FRACTIONAL DISTILLATION

Place 30 ml of spearmint or caraway seed oil in a 100 ml round bottom flask. Assemble the vacuum fractional distillation apparatus as shown in the figure below. Use a trap arrangement and manometer as shown in Technique 6, Figure 6–7 or Technique 9, Figure 9–6. The distillation apparatus is similar to that given in Technique 6, Figure 6–7, except that a fractionating column, filled with stainless steel sponge (Technique 7, Figure 7–7c), is inserted between the Claisen head and the distilling head. As explained in Technique 6, Section 6.6, it is very important that the apparatus be assembled carefully. All of the rubber tubing must be free of cracks. The ebulliator tube must fit securely in an adapter or rubber stopper. Wrap the fractionating column and the distilling head with glass wool for insulation.

When the apparatus has been assembled, conduct the distillation as outlined in Technique 6, Section 6.7 and Technique 7, Section 7.6. Use a heating mantle or oil bath to heat the distilling flask. It is important that a good vacuum be obtained, preferably about 20 mm pressure. If a number of students are using the aspirators on a given bench, the obtainable pressure may not be low enough to proceed with the distillation. See Technique 6, Section 6.1 for the effect of pressure on the boiling point when a pressure other than 20 mm is obtained.

With spearmint oil, the first fraction distills up to a temperature of about 82° at 20 mm pressure. When the low boiling fraction has been removed, the temperature will drop somewhat, and the rate of distilla-

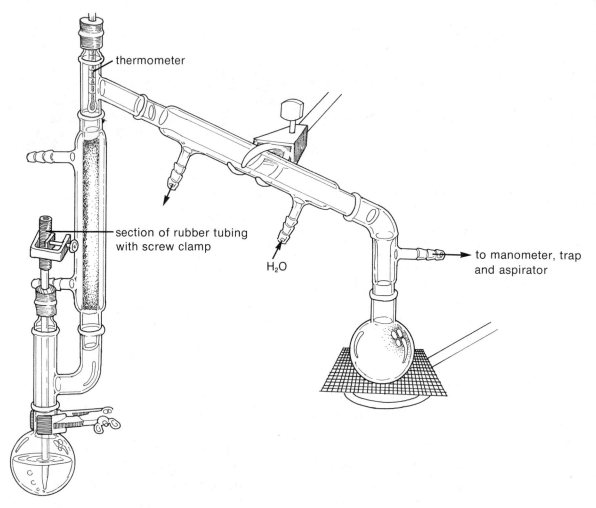

thermometer

section of rubber tubing
with screw clamp

H₂O

to manometer, trap
and aspirator

Apparatus for vacuum fractional distillation

tion take-off in the receiver will slow considerably. The temperature of the heating source should then be raised, and the temperature at the thermometer should once again begin to rise. At that point, the distillation should be stopped, and the heating source lowered. About 4 or 5 g of distillate may be collected in this first fraction.

To change receiving flasks, open the screw clamps on the trap assembly and ebulliator, and turn off the aspirator. Remove the receiver, and replace it with a clean one. Following this, continue the distillation procedure (Technique 6, Section 6.7). Place the heating source back under the distilling flask and raise its temperature. Collect the second fraction in the new receiver. This fraction, which contains the carvone, will boil from about 82° up to about 113° at 20 mm. Most of this material will boil from about 105 to 113° at 20 mm. About 16 or 17 g of distillate may be obtained. Even though some low boiling material is present in this fraction, it should contain nearly pure carvone.

The caraway oil[1] can be distilled in the same manner as the spearmint oil, and a larger first fraction (limonene) is obtained (10 or 11 g). Again the temperature at the thermometer drops somewhat near the end of the first fraction. The temperature of the heating source is raised as before. When the temperature of the vapor again rises, a second fraction (9 or 10 g) of (+)-carvone is collected in another receiver.

ANALYSIS OF THE CARVONES

The samples obtained by distillation should be analyzed using the following methods. The instructor will indicate which methods should be used. Compare your results to those obtained by someone who used a different oil. After the analyses are completed, submit the carvone to the instructor in a labeled **glass** vial.

Odor. About 8 to 10% of the population cannot detect the difference in the odors of the optical isomers. For most people, however, the difference is quite obvious.

Gas Chromatography. See Technique 12. With the help of the instructor or assistant, analyze each of the fractions on any relatively nonpolar column such as a 12 ft × 1/8 in 30% GE SE-30 on a Chromasorb W column.[2] Determine the retention times of the components. Calculate the purity of the carvone sample by the method explained in Technique 12, Section 12.10.

Polarimetry. See Technique 15. With the help of the instructor or assistant, obtain the observed optical rotation, α, of the (−)-carvone from spearmint, and both the (+)-limonene and (+)-carvone from caraway oil. The sample or samples may all be analyzed directly, without dilution, in the smallest cell available, which may have a path length, l, of 0.5 dm. The specific rotation is calculated from the relationship, $[\alpha] = \alpha/(c)(l)$, given in Technique 15. The concentration, c, will be equal to the density of the substances analyzed at 20°. The values are: (+)-carvone (0.9608 g/ml), (+)-limonene (0.8411 g/ml), and (−)-carvone (0.9593 g/ml). The literature values for the specific rotations are as follows: $[\alpha]_{20}^{D} = +58.8°$ for (+)-carvone, $−56.7°$ for (−)-carvone, and $+125°$ for (+)-limonene. They are not identical because of the presence of trace amounts of impurities.

---

[1] It has been noted in the literature that caraway oil may foam in certain instances, especially at higher pressures. At pressures below 40 mm, foaming should not be expected. If foaming is experienced, it will usually stop once the limonene fraction (lower bp) has been removed.

[2] A 6 ft × 1/8 in column of 15% QF-1 on Chromasorb W 80/100 support may also be used.

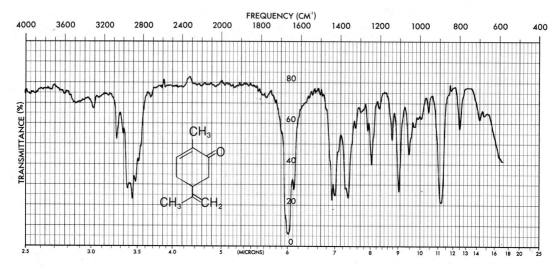

Infrared spectrum of (+)-carvone from caraway oil, neat

Refractive Index. See Technique 16. Obtain the refractive index for the carvone sample. The (+) and (−)-carvones each have refractive indices at 23° equal to 1.4950.

Spectroscopy. See Technique 17. Obtain the infrared spectrum of the (−)-carvone sample from spearmint and the (+)-carvone and (+)-limonene samples from caraway. Compare the carvone and limonene spectra to those shown. The nmr spectra for (−)-carvone and (+)-limonene are shown on the next page. At the option of the instructor, obtain an nmr spectrum of the carvone for comparison.

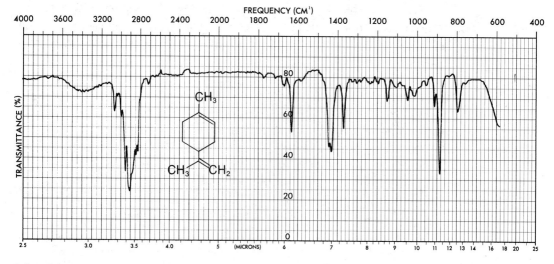

Infrared spectrum of (+)-limonene, neat

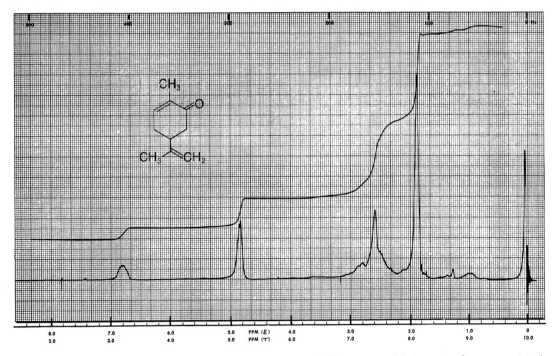

NMR spectrum of (−)-carvone from spearmint oil

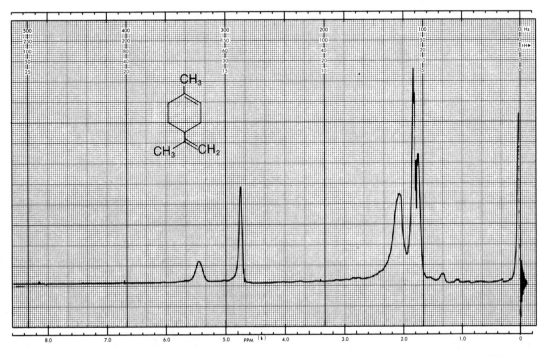

NMR spectrum of (+)-limonene

REFERENCES

Friedman, L., and Miller, J. G. "Odor Incongruity and Chirality." *Science, 172* (1971), 1044.
Murov, S. L., and Pickering, M. "The Odor of Optical Isomers." *Journal of Chemical Education, 50*
        (1973), 74.
Russell, G. F., and Hills, J. I. "Odor Differences between Enantiomeric Isomers." *Science, 172*
        (1971), 1043.

QUESTIONS

1. Interpret the ir and the nmr spectra for carvone and the ir spectrum for limonene.

2. How could you establish that the difference in odor of the distilled carvones was **not** due to a trace component? Assume that the impurity is not abundant enough to appear in the spectra. In considering your answer, read Technique 12, Section 12.9.

3. Assign the chiral (asymmetric) center in α-phellandrene, β-phellandrene, and limonene.

Experiment **22**
REACTIVITIES OF SOME ALKYL HALIDES

S_N1/S_N2 Reactions

The reactivities of alkyl halides in nucleophilic substitution reactions depend upon two important factors. These factors are reaction conditions and substrate structure.

SODIUM OR POTASSIUM IODIDE IN ACETONE

A reagent composed of sodium or potassium iodide dissolved in acetone is quite useful in classifying alkyl halides according to their reactivity in an S_N2 reaction. Iodide ion is an excellent nucleophile, and acetone is a rather non-polar solvent. The tendency to form a precipitate increases the completeness of the reaction. Sodium and potassium iodide are soluble in acetone, but the corresponding bromides and chlorides are not soluble. Consequently, as bromide ion or chloride ion is produced, it is precipitated from the solution. According to LeChatelier's principle, the precipitation of a product from the reaction solution drives the equilibrium toward the right; such is the case in the reaction described here:

$$R—Cl + Na^+I^- \longrightarrow RI + NaCl\downarrow$$

$$R—Br + Na^+I^- \longrightarrow RI + NaBr\downarrow$$

SILVER NITRATE IN ETHANOL

A reagent composed of silver nitrate dissolved in ethanol is useful in classifying alkyl halides according to their reactivity in an S_N1 reaction. Nitrate ion is a rather poor nucleophile, but ethanol is a moderately powerful ionizing solvent. The silver ion, because of its ability to coordinate the leaving halide ion to form a silver halide precipitate, greatly assists the ionization of the alkyl halide. Again, the formation of a precipitate as one of the reaction products also serves to enhance the reaction.

$$R—Cl + Ag^+NO_3^- \longrightarrow R^+ NO_3^- + AgCl\downarrow$$

$$R—Br + Ag^+NO_3^- \longrightarrow R^+ NO_3^- + AgBr\downarrow$$

SPECIAL INSTRUCTIONS

Before beginning this experiment review the chapters dealing with nucleophilic substitution in your textbook. This experiment requires very little time, and it may be performed along with another longer experiment. Some of the compounds used in this experiment are lachrymators. They must be disposed of in a hood sink.

PROCEDURE

SODIUM IODIDE IN ACETONE

Label a series of 10 clean dry test tubes (10 × 75 mm test tubes may be used) from 1 to 10. In each test tube place 0.2 ml of one of the following halides: (1) 2-chlorobutane, (2) 2-bromobutane, (3) t-butyl chloride, (4) 1-chlorobutane, (5) crotyl chloride (CH_3CH=$CHCH_2Cl$), (6) chloroacetone ($ClCH_2COCH_3$), (7) benzyl chloride, (8) bromobenzene, (9) bromocyclohexane, and (10) bromocyclopentane.

Into each test tube add 2 ml of a 15% NaI in acetone solution, noting the time of each addition. Record the time required for the formation of any precipitates. After about 5 minutes, place any test tubes that do not contain a precipitate in a 50° water bath. Be careful not to allow the temperature of the water bath to exceed 50° because the acetone will evaporate and/or boil out of the test tube. At the end of 6 minutes, cool the test tubes to room temperature, and note whether a reaction has occurred. Record the results and explain why each compound possesses the reactivity which was observed. Explain the reactivities in terms of structure.

Generally speaking, reactive halides give a precipitate within 3 minutes, moderately reactive halides will give a precipitate when heated, and unreactive halides do not yield a precipitate even after heating.

SILVER NITRATE IN ETHANOL

Label a series of ten clean dry test tubes from 1 to 10, as described in the previous experiment. Place 0.2 ml of the appropriate halide into each test tube, as was described for the sodium iodide test.

Add 2 ml of a 1% ethanolic silver nitrate solution to each test tube, noting the time of each addition. Record the time required for the formation of any precipitates.

After about 5 minutes, heat each solution which has not yielded any precipitate to boiling on the steam bath. Note whether a precipitate has formed.

Record the results and explain why each compound possesses the reactivity which was observed. Explain the reactivities in terms of structural effects, as before.

Again, reactive halides will give a precipitate within 3 minutes, moderately reactive halides will give a precipitate when heated, and unreactive halides do not yield a precipitate even after heating.

QUESTIONS

1. In both the sodium iodide in acetone and silver nitrate in ethanol tests, why should 2-bromobutane react faster than 2-chlorobutane?

2. In the silver nitrate in ethanol test why should the cyclopentyl compound react faster than the cyclohexyl compound?

3. When treated with sodium iodide in acetone, benzyl chloride reacts much faster than 1-chlorobutane, even though both compounds are primary alkyl chlorides. Explain this rate difference.

4. How would you expect the four compounds shown below to compare in their behavior in the two tests?

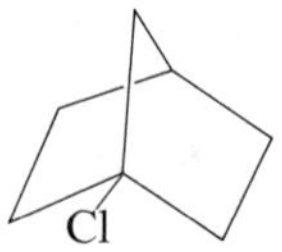

1 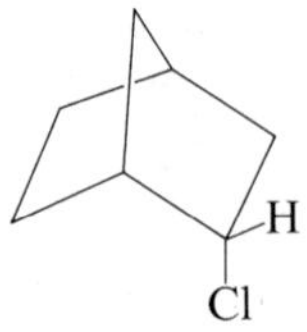   2 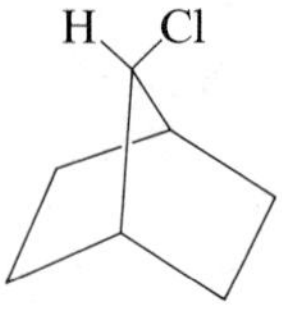   3 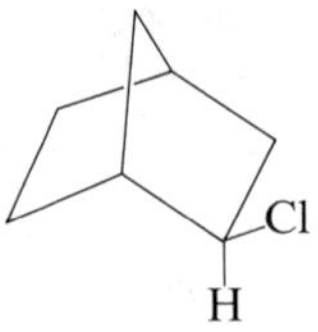   4

5. How would you predict that chlorocyclopropane would behave in each of these tests?

Experiment **23**

SYNTHESIS OF *n*-BUTYL BROMIDE AND *t*-PENTYL CHLORIDE

Synthesis of Alkyl Halides
Extraction
Simple Distillation

Two syntheses of alkyl halides from alcohols are described in this experiment. In the first procedure, a primary alkyl halide, *n*-butyl bromide, is prepared as shown in equation **1**.

$$CH_3CH_2CH_2CH_2OH + NaBr + H_2SO_4 \longrightarrow$$
n-Butyl Alcohol

$$CH_3CH_2CH_2CH_2Br + NaHSO_4 + H_2O \quad \textbf{(1)}$$
n-Butyl Bromide

In the second procedure, a tertiary alkyl halide, *t*-pentyl chloride (*t*-amyl chloride), is prepared as shown in equation **2**.

$$
\begin{array}{ccc}
& \overset{\displaystyle CH_3}{\underset{\displaystyle OH}{CH_3CH_2\overset{|}{\underset{|}{C}}CH_3}} + HCl \longrightarrow & \overset{\displaystyle CH_3}{\underset{\displaystyle Cl}{CH_3CH_2\overset{|}{\underset{|}{C}}CH_3}} + H_2O \quad \textbf{(2)}
\end{array}
$$

t-Pentyl Alcohol   *t*-Pentyl Chloride

These reactions serve as an interesting contrast in mechanisms. The *n*-butyl bromide synthesis proceeds by an S_N2 mechanism, while *t*-pentyl chloride is prepared by an S_N1 reaction.

n-BUTYL BROMIDE

This alkyl halide can easily be prepared by allowing *n*-butyl alcohol to react with sodium bromide and sulfuric acid by equation **1**. The sodium bromide reacts with sulfuric acid to produce hydrobromic acid. An excess of sulfuric acid serves to shift the equilibrium and thus to speed the reaction by producing a higher concentration of hydrobromic acid. The sulfuric acid also protonates the hydroxyl group of *n*-butyl alcohol so that water is displaced rather than the hydroxide ion (OH^-). The acid also protonates the water as it is produced in the reaction, and thus deactivates this nucleophile. This deactivation prevents the alkyl halide from being converted back to the alcohol by nucleophilic attack of water.

The reaction proceeds **via** an S_N2 mechanism as follows:

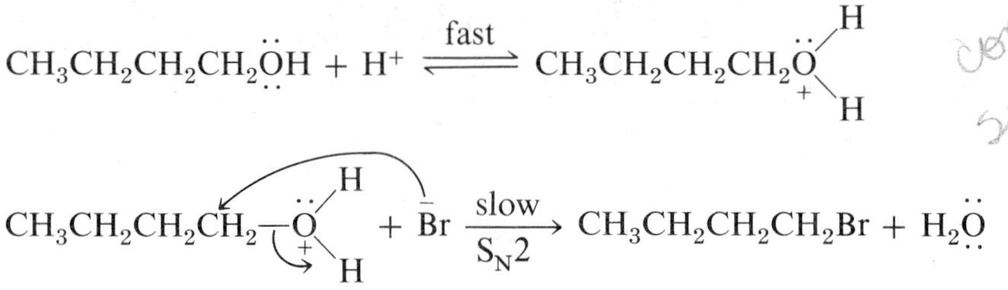

$$CH_3CH_2CH_2CH_2\ddot{O}H + H^+ \underset{}{\overset{fast}{\rightleftharpoons}} CH_3CH_2CH_2CH_2\overset{..}{\underset{+}{O}}\overset{H}{\diagdown_H}$$

$$CH_3CH_2CH_2CH_2-\overset{..}{\underset{+}{O}}\diagup^{H}_{\diagdown H} + \bar{B}r \xrightarrow[S_N2]{slow} CH_3CH_2CH_2CH_2Br + H_2\ddot{O}$$

Primary substrates such as *n*-butyl alcohol usually react by the S_N2 mechanism.

During the isolation of the *n*-butyl bromide, the crude product is washed with sulfuric acid to remove any remaining *n*-butyl alcohol. The sulfuric acid also can remove the by-products of the reaction, such as 1-butene and dibutyl ether, by making use of the basic character of these by-products. Alkyl halides are, however, not basic.

t-PENTYL CHLORIDE

This alkyl halide can easily be prepared by allowing *t*-pentyl alcohol to react with hydrochloric acid by equation **2**. The reaction is conducted in a separatory funnel. As the reaction proceeds, the insoluble alkyl halide forms as an upper phase.

The reaction proceeds **via** an S_N1 mechanism as follows:

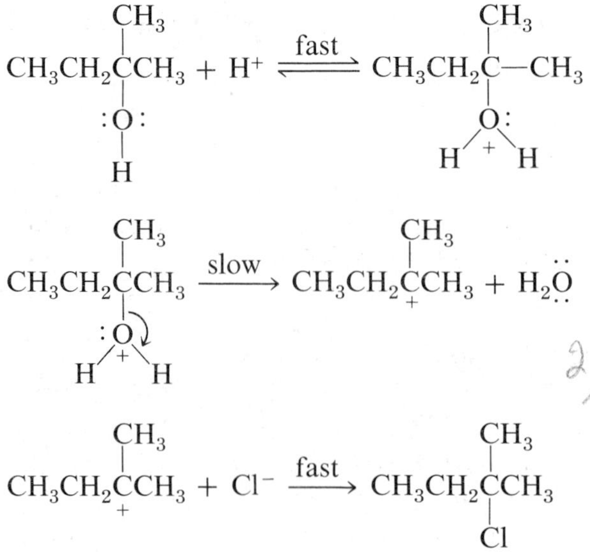

$$\underset{\underset{\underset{H}{|}}{\underset{:\ddot{O}:}{|}}{\overset{\overset{CH_3}{|}}{CH_3CH_2CCH_3}} + H^+ \overset{fast}{\rightleftharpoons} \underset{\underset{H\diagup\underset{+}{O}:\diagdown H}{}}{\overset{\overset{CH_3}{|}}{CH_3CH_2C-CH_3}}$$

$$\underset{\underset{H\diagup\underset{+}{\ddot{O}}\diagdown H}{}}{\overset{\overset{CH_3}{|}}{CH_3CH_2CCH_3}} \xrightarrow{slow} \underset{+}{\overset{\overset{CH_3}{|}}{CH_3CH_2\overset{..}{C}CH_3}} + H_2\ddot{O}$$

$$\underset{+}{\overset{\overset{CH_3}{|}}{CH_3CH_2CCH_3}} + Cl^- \xrightarrow{fast} \underset{\underset{Cl}{|}}{\overset{\overset{CH_3}{|}}{CH_3CH_2CCH_3}}$$

Tertiary substrates such as *t*-pentyl alcohol usually react by the S_N1 mechanism.

A small amount of an alkene, 2-methyl-2-butene, is produced as a by-product in this reaction. If sulfuric acid had been used as for *n*-butyl bromide, a considerable amount of this alkene would have been produced.

SPECIAL INSTRUCTIONS

Before starting, review the appropriate chapters in your textbook. In addition, it is necessary to have read Techniques 1, 5, and 6 before starting this experiment. As your instructor indicates, perform either the *n*-butyl bromide or the *t*-pentyl chloride procedure.

Procedure 23A

n-BUTYL BROMIDE

Place 24.0 g of sodium bromide in a 250 ml round bottom flask, and add 25 ml of water and 17 ml (d = 0.81 g/ml) of *n*-butyl alcohol (1-butanol). Cool the mixture in an ice bath and slowly add 20 ml (d = 1.84 g/ml) of concentrated sulfuric acid with continuous swirling in the ice bath. Add several boiling stones to the mixture. Assemble the apparatus as shown in the figure. The inverted funnel in the beaker serves as a trap to absorb the hydrogen bromide gas which is evolved during the reaction period. An alternate trap arrangement is shown in Experiment 17. Place some water and sodium hydroxide pellets in the beaker. The funnel must be adjusted so that it projects only **slightly** below the surface of the liquid in the beaker. Heat the mixture with a heating mantle, oil bath, or flame until the mixture begins to reflux gently. Reflux the mixture for 30 minutes. Two layers will form during this time.

At the end of this reflux period remove the heating source and allow the mixture to cool. Remove the condenser and reassemble the apparatus for simple distillation (Technique 6, Figure 6–6). Add several new boiling stones to the round bottom flask. Distill the mixture and collect the distillate in a receiver cooled in an ice bath. The alkyl halide co-distills with water and then separates into two phases in the receiver. Distill the mixture until the distillate appears to be clear. The temperature should reach 110 to 115° by that time. While the distillation is still

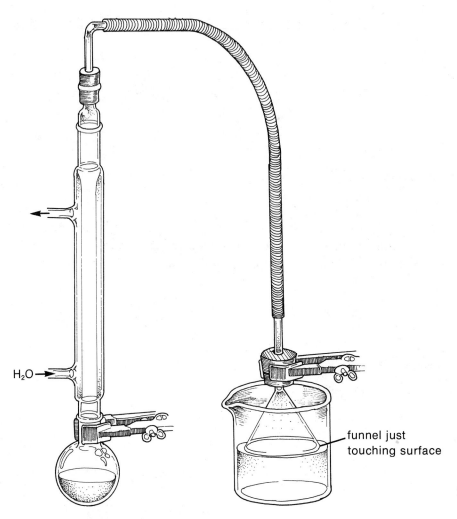

Apparatus for preparing *n*-butyl bromide

proceeding, remove the receiver and collect a few drops of distillate in a test tube containing some water. Check to see if the distillate is completely soluble (miscible). If it is, no alkyl halide is present and the distillation may be stopped. If the distillate produces insoluble droplets in the water then the distillation must be continued until the distillate becomes completely water soluble. The distillate collected in the receiver contains principally *n*-butyl bromide and water, with smaller amounts of sulfuric acid and hydrogen bromide.

Transfer the distillate to a separatory funnel, add 25 ml of water to it, and shake the mixture (Technique 5, Section 5.4). Drain the lower layer, which contains *n*-butyl bromide (d = 1.27 g/ml), from the funnel. Discard the aqueous layer after making certain that the correct layer has been saved. Return the alkyl halide to the funnel and add to it 15 ml of **cold** concentrated sulfuric acid. Swirl the mixture (unstoppered) until it is thoroughly mixed. Then stopper the funnel and shake it thoroughly, but carefully, to insure that sulfuric acid does not leak out of the stopper or stopcock. Allow several minutes for the phases to separate. Sulfuric acid has a density of 1.84 g/ml. Which phase contains *n*-butyl bromide? If there is some doubt about the identity of the phases, read Technique 5, Section 5.4. Drain the lower layer from the funnel. Allow several minutes for further layer separation and again drain the remaining lower layer. Wash the *n*-butyl bromide with 15 ml of 10% sodium hydroxide solution. Carefully separate the layers and be certain to save the organic layer. Dry the crude *n*-butyl bromide over 1.5 g of anhydrous calcium chloride in a small Erlenmeyer flask (Technique 5, Section 5.6). Stopper the flask and swirl the contents until the liquid is **clear.** The drying process may be accelerated by **gently** warming the mixture on a steam bath.

Decant the **clear** liquid into a **dry** distilling flask. Add a boiling stone and distill the crude *n*-butyl bromide in a **dry** apparatus (Technique 6, Section 6.4, Figure 6–6). Collect the material that boils between 98 and 102°. Weigh the product and calculate the percent yield. Submit the sample in a labeled vial to the instructor.

Procedure 23B

t-PENTYL CHLORIDE

In a 125 ml separatory funnel, place 22 ml (d = 0.805 g/ml) of *t*-pentyl alcohol (*t*-amyl alcohol or 2-methyl-2-butanol) and 50 ml (d = 1.18 g/ml, 37.3% HCl) of concentrated hydrochloric acid. Do not stopper the funnel. Gently swirl the mixture in the separatory funnel for about one minute. After this period of swirling, stopper the separatory funnel, and carefully invert it. Without shaking the separatory funnel, immediately open the stopcock to release the pressure. Close the stopcock, shake the funnel several times, and again release the pressure through the stopcock (Technique 5, Section 5.4). Shake the funnel for 2 to 3 minutes, with occasional venting. Allow the mixture to stand in the separatory funnel until the two layers have completely separated. The *t*-pentyl chloride has a density of 0.865 g/ml. Which layer contains the alkyl halide? Separate the layers.

The operations in this paragraph should be conducted as rapidly as possible since the *t*-pentyl chloride is somewhat unstable in water and sodium bicarbonate solution. Wash (swirl and shake) the organic layer with one 25 ml portion of water. Again, separate the layers and discard the aqueous phase after making certain that the proper layer has been saved (Technique 5, Section 5.4). Wash the organic layer with a 25 ml portion of 5% aqueous sodium bicarbonate. Gently swirl the funnel (unstoppered) until the contents are thoroughly mixed. Stopper the funnel, and carefully invert it. Release the excess pressure through the stopcock. Gently shake the separatory funnel, with frequent release of pressure. Following this, vigorously shake the funnel, again with release of pressure, for about one minute. Allow the layers to separate, and drain the lower aqueous bicarbonate layer. Wash (swirl and shake) the organic layer with one 25 ml portion of water, and again drain the lower aqueous layer.

Transfer the organic layer to a small dry Erlenmeyer flask. Pour it from the top of the separatory funnel. Dry the crude *t*-pentyl chloride over anhydrous calcium chloride until it is clear (Technique 5, Section 5.6). Swirl the alkyl halide with the drying agent to aid the drying process. Decant the **clear** material into a small **dry** distilling flask. Add a boiling stone and distill the crude *t*-pentyl chloride in a **dry** apparatus (Technique 6, Section 6.4, Figure 6–6) by using a steam bath. Collect the pure *t*-pentyl chloride in a receiver, cooled in ice. Collect the material which boils between 79 and 84°. Weigh the product and calculate the percent yield. Submit the sample to the instructor in a labeled vial.

QUESTIONS

n-Butyl Bromide

1. Sulfuric acid was used to remove unreacted alcohol from the crude alkyl halide. Explain how it removes the alcohol. Write the equation.

2. Dibutyl ether and 1-butene can be formed as by-products in this reaction. Explain how they are removed by sulfuric acid. Give the reactions.

3. Look up the density of *n*-butyl chloride. Assume that this alkyl halide had been prepared instead of the bromide. Decide whether the alkyl halide would appear as the upper or lower phase in the separatory funnel at each stage of the isolation: after the reflux; after the co-distillation; after addition of water to the distillate; after washing with sulfuric acid; after washing with sodium hydroxide.

4. Why must the crude alkyl halide be dried carefully with calcium chloride before the final distillation? See Technique 7, Section 7.7.

t-Pentyl Chloride

1. Aqueous sodium bicarbonate was used to wash the crude *t*-pentyl chloride. Why would it be undesirable to wash the halide with aqueous sodium hydroxide?

2. Some 2-methyl-2-butene may be produced in the reaction as a by-product. Give the mechanism for its production. How can it be removed during the purification procedure?

3. How is the unreacted *t*-pentyl alcohol removed in this experiment? Look up the solubility of the alcohol and alkyl halide in water.

4. Why must the crude alkyl halide be dried carefully with calcium chloride before the final distillation? See Technique 7, Section 7.7.

NUCLEOPHILIC SUBSTITUTION REACTIONS: COMPETING NUCLEOPHILES

Nucleophilic Substitution
Refractometry
Gas Chromatography
NMR Spectroscopy

The purpose of this experiment is to compare the relative nucleophilicities of chloride and bromide ions toward *n*-butyl alcohol (1-butanol) on one hand, and toward *t*-butyl alcohol (2-methyl-2-propanol) on the other. The two nucleophiles will be present at the same time in each reaction, and they will be competing with each other for substrate.

In general, alcohols do not react readily in simple nucleophilic displacement reactions. If they are attacked by nucleophiles directly, hydroxide ion, a strong base, must be displaced. Such a displacement is not energetically favorable, and it cannot occur to any reasonable extent.

$$X^- + ROH \; \not\!\!\longrightarrow \; R-X + OH^-$$

In order to avoid this problem, nucleophilic displacement reactions with alcohols as substrates are carried out in acidic media. In a rapid initial step the alcohol is protonated, then water, a very stable molecule, is displaced. This displacement is energetically very favorable, and the reaction proceeds in high yield.

$$ROH + H^+ \rightleftharpoons R-O^+{\Large\langle}^{H}_{H}$$

$$X^- + R-O^+{\Large\langle}^{H}_{H} \longrightarrow R-X + H_2O$$

Once the alcohol is protonated, the substrate reacts by either the S_N1 or the S_N2 mechanism, depending upon the structure of the alkyl group of the alcohol. For a brief review of these mechanisms, you should consult the chapters on nucleophilic substitution in your textbook.

You will analyze the products of the reaction by a variety of techniques in order to determine the relative amounts of alkyl chloride and alkyl bromide formed in each reaction. You will be asked to develop an explanation of the results you obtain. Half the class will use *n*-butyl alcohol as a substrate, and the other half will use *t*-butyl alcohol. You will need to compare your results with those of a student who used the substrate complementary to yours. The ammonium halides (NH_4Cl and NH_4Br) are used as sources of halide ions in this experiment, since they are more soluble at the concentrations used than are the corresponding sodium or potassium salts.

SPECIAL INSTRUCTIONS

Before beginning this experiment, review the appropriate chapters in your textbook. Also read Techniques 1, 5, 12, 16, 17 and Appendix Four, Nuclear Magnetic Resonance. Before beginning this experiment, you should decide which alcohol to use as a substrate. Concentrated sulfuric acid is very corrosive; be careful when handling it.

> NOTE TO THE INSTRUCTOR: Be certain that the *t*-butyl alcohol has been melted before the beginning of the laboratory period. Prepare the gas chromatograph as follows: column temperature, 85°; injection port temperature, 85°; carrier gas flow rate, 50 ml/min. The column should be approximately twelve feet in length and should contain a stationary phase similar to silicone oil (SE-30).

PROCEDURE

Cautiously pour 50 ml of concentrated sulfuric acid over 60 g of ice in a beaker. When this is mixed, pour the mixture into a 500 ml round-bottom, three-neck flask, as shown in the figure. Add 13.5 g (0.25 mole) of ammonium chloride, 24.5 g (0.25 mole) of ammonium bromide, and a boiling stone. Be certain that no salts remain adhering to the ground glass joints, or they will not fit tightly. Attach a reflux condenser to this flask and fit it with a trap for acidic gases (see figure). Heat the mixture until all of the solids dissolve. It may be necessary to swirl the flask to achieve complete solution. Stop heating the solution and, if the solvent is boiling, wait for the boiling action to stop. Some solids may precipitate. Don't worry about these. Choose either 1-butanol or 2-methyl-2-propanol (*t*-butyl alcohol). Slowly add 14.8 g of the alcohol through the addition funnel. **Gently** reflux the resulting solution for about 1.5 hours for the 1-butanol and 0.5 hours for the *t*-butyl alcohol. If the mixture is heated too strongly, all of the organic products will be lost from the top of the condenser.

Allow the reaction mixture to cool with the reflux condenser still in place. When the solution has reached a temperature near room temperature, cool the mixture in an ice bath. Some solids may precipitate. Ignore these and decant the solution away from them into a

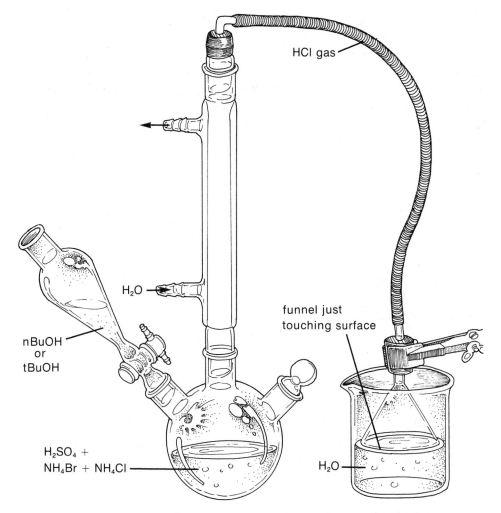

HCl gas

H_2O

nBuOH
or
tBuOH

funnel just
touching surface

H_2SO_4 +
NH_4Br + NH_4Cl

H_2O

Apparatus for S_N1/S_N2 experiment

separatory funnel. The organic product layer should separate. There may be **three** layers if there is still some unreacted alcohol. Treat the alcohol layer as if it were part of the halide layer. Discard the water layer (be sure you know which it is) and wash the halide mixture (and alcohol, if present) successively with two 20 ml portions of concentrated sulfuric acid, one 100 ml portion of water, and one 100 ml portion of 5% aqueous $NaHCO_3$.

CAUTION: The last extraction should be done as quickly as possible. With the *t*-butyl halides, particularly, the evolution of CO_2 may *never* stop. The halides hydrolyze readily to produce HCl or HBr. For this reason, you should *not* continue to extract beyond about 5 minutes.

Transfer the organic layer to a 50 ml Erlenmeyer flask and remove traces of water by adding 2 or 3 g of anhydrous calcium chloride. The alkyl halides are volatile, so stopper the flask tightly. When the liquid is clear, decant it into a dry, ground glass stoppered flask, and stopper the flask **tightly.** If this material is to be stored for any period of time, the flask should be well sealed by the use of stopcock grease. Do not store the liquid in a container with a cork or a rubber stopper, because these will absorb the halides.

ANALYSIS PROCEDURES

The ratio of 1-chlorobutane to 1-bromobutane, or *t*-butyl chloride to *t*-butyl bromide, must be determined. At your instructor's option you may do this by one of three methods: gas chromatography, refractive index, or nuclear magnetic resonance spectroscopy.

GAS CHROMATOGRAPHY

The instructor or laboratory assistant may either choose to perform the sample injections himself or to allow the students in the class to perform them. In the latter case, it is **essential** that adequate instruction be obtained beforehand. A reasonable sample size is 1 microliter. The sample is injected into the gas chromatograph, and the gas chromatogram is recorded. The alkyl chloride, because of its greater volatility, has a shorter retention time than the alkyl bromide.

Once the gas chromatogram has been obtained, determine the relative areas of the two peaks (Technique 12, Section 12.10). While the peaks may be cut out and weighed on an analytical balance, as a method of determining areas, the method of triangulation is preferred. Record the percentages of alkyl chloride and alkyl bromide in the reaction mixture.

REFRACTIVE INDEX

Measure the refractive index of the product mixture (Technique 16). In order to determine the composition of the mixture, assume a linear relationship between the refractive index and the molar composition of the mixture. At 20°C the refractive indices of the alkyl halides are:

1-chlorobutane	1.4015	2-chloro-2-methylpropane	1.3877
1-bromobutane	1.4398	2-bromo-2-methylpropane	1.4280

If the temperature of the laboratory room is not 20°C, the refractive index must be corrected. Add 0.0004 refractive index units to the observed reading for each degree above 20°C, and subtract the same amount for each degree below this temperature. Record the percentages of alkyl chloride and alkyl bromide in the reaction mixture.

NUCLEAR MAGNETIC RESONANCE

The instructor or a laboratory assistant will record the nmr spectrum of the reaction mixture. Submit a labeled sample vial containing the mixture for this spectral determination. The spectrum will also contain integration of the important peaks (Appendix Four, Nuclear Magnetic Resonance). If the substrate alcohol was 1-butanol, the resulting halide mixture will give rise to a rather complicated spectrum. Each alkyl halide will show a downfield triplet, caused by the CH_2 group nearest the halogen. This triplet will appear further downfield in the alkyl chloride than in the alkyl bromide. These triplets will overlap each other, but one branch of each triplet will be available for comparison. Compare the integral of the **downfield** branch of the triplet for 1-chlorobutane to the **upfield** branch of the triplet for 1-bromobutane. The spectrum shown below provides an example. The relative heights of these integrals correspond to the relative amounts of each halide in the mixture.

If the substrate alcohol was 2-methyl-2-propanol, the resulting halide mixture will show two peaks in the nmr spectrum. Each halide

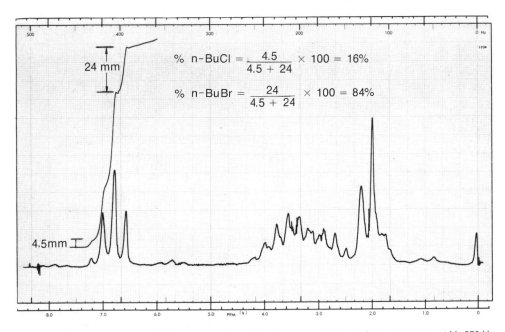

500
400
300
200
100
0 Hz
24 mm
% n-BuCl = 4.5 / (4.5 + 24) × 100 = 16%
% n-BuBr = 24 / (4.5 + 24) × 100 = 84%
4.5mm
8.0
7.0
6.0
5.0
PPM
4.0
3.0
2.0
1.0
0

NMR spectrum of 1-chlorobutane and 1-bromobutane, sweep width 250 Hz

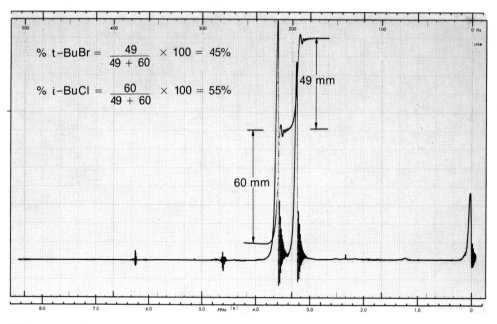

% t-BuBr = 49 / 49 + 60 × 100 = 45%
% t-BuCl = 60 / 49 + 60 × 100 = 55%
49 mm
60 mm
500
400
300
200
100
0 Hz
8.0
7.0
6.0
5.0
PPM
4.0
3.0
2.0
1.0
0

NMR spectrum of *t*-butyl chloride and *t*-butyl bromide, sweep width 250 Hz.

will show a singlet, since all of the CH_3 groups are equivalent and not coupled to each other. In the reaction mixture, the downfield peak is due to *t*-butyl chloride, while the upfield peak is caused by *t*-butyl bromide. Compare the integrals of these peaks. The spectrum shown below provides an example. The relative heights of these integrals correspond to the relative amounts of each halide in the mixture.

THE REPORT

Record the percentages of alkyl chloride and alkyl bromide in the reaction mixture. The report must include the percentage of each alkyl halide determined by each method used in this experiment. In addition, the report should include a discussion of the mechanism of the reaction studied. The student should compare his results to those of a student who used the alternative substrate. Any differences in product distribution and mechanism should be discussed in the report. All gas chromatograms, refractive index data, and spectra should be attached to the report.

QUESTIONS

1. Which is a better nucleophile, chloride ion or bromide ion?

2. What is the purpose of the extraction with concentrated sulfuric acid after the reaction has been completed?

3. What would be the product distribution if 2-butanol had been the substrate?

4. Why does the alkyl chloride show nmr peaks further downfield than the corresponding peaks in the alkyl bromide?

5. What is the principal by-product of these reactions?

6. A student left his alkyl halides (RCl and RBr) in an open container for several hours. What happens to the composition of the halide mixture during that time?

7. In the $NaHCO_3$ extraction of the t-butyl halides, carbon dioxide evolution may never cease. Develop an explanation, with equations, for this observation. How does the RCl/RBr ratio change during this process?

ESSAY

The Rates of Chemical Reactions

In a study of chemical kinetics we are concerned with the question, "How fast does a reaction proceed?" In the experiment which follows this essay the answer to that question for two particular chemical reactions will be provided by the experimental results which you will obtain.

The chief object of a kinetic study is the determination of the relationship between the rate at which a chemical reaction occurs at a given temperature and the concentrations of the reactants. In addition to this determination, such experiments as the study of variations in the reaction velocity with temperature provide hints which can lead to an elucidation of the **reaction mechanism,** or the pathway by which the reaction proceeds. The majority of useful organic reactions are complex, taking place in a series of steps. One of these steps must be slower than the others, and this step determines the rate of the overall reaction. The slow step acts as a sort of "kinetic bottleneck." The kinetic study of a reaction will often allow us to say, with a reasonable degree of confidence, just which species participate in this **rate-determining step.**

The rate of a reaction can be defined as the change in the concentration of a reactant per time unit, or mathematically, for a reaction:

$$A + B \longrightarrow Products$$

The rate equation may be written in any of three equivalent ways:

$$Rate = -\frac{d[A]}{dt} = -\frac{d[B]}{dt} = +\frac{d[Products]}{dt}$$

So a kinetic experiment is one in which the concentrations of reactants or products are determined at certain increments of time.

Chemical reactions occur by means of a series of collisions between reactant molecules or atoms. Let us consider the following hypothetical gas phase reaction:

$$A_2 + B_2 \longrightarrow 2\ AB$$

What this equation implies is that a molecule of A_2 collides with a molecule of B_2, and this collision produces two molecules of AB. Since this collision involves two bodies, or particles, and is therefore a simple system to consider, we can reach some useful conclusions about this reaction. First of all, if we should double the number of molecules of A_2, we should expect the number of collisions also to double. If the number of collisions is doubled, we should expect the rate of the reaction to double, since twice the number of A_2 molecules are reacting. Similarly, if we double the number of B_2 molecules, we should expect the number of collisions and the reaction rate to double, also.

It should be pointed out, however, that not every collision of A_2 with B_2 leads to the formation of two molecules of AB. Some collisions do not lead to any reaction at all, but simply result in the molecules of A_2 and B_2 bouncing off of each other. This lack of reaction with a collision can come about if a collision occurs with insufficient energy to bring about reaction, or if a collision occurs where the molecules of A_2 and B_2 do not collide with the proper orientation.

THE EFFECT OF TEMPERATURE

Since the kinetic energy of a system increases as the temperature of that system increases, we would expect the number of collisions and the energy with which those collisions occur to increase as the temperature of a reaction is increased. The minimum energy for a reaction to occur, E_a, is called the **activation energy** for the reaction.

ENERGY DIAGRAMS

If we plot the potential energy of the system **versus reaction coordinate,** which is defined as a measure of the progress of the reaction for each molecule or atom, we see that the activation energy is really the amount of energy required for the reaction to pass over the maximum point of potential energy. This maximum point of potential energy is called the **transition state.** It can be seen that the energy required to create the transition state represents an energy barrier which the reaction must overcome in order to proceed on to products. The difference in potential energies between the reactant state ($A_2 + B_2$) and the

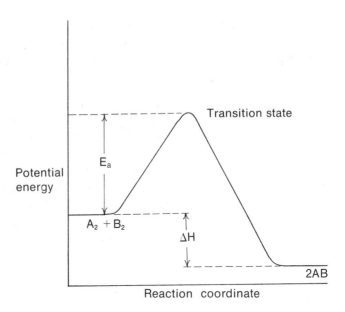

Reaction coordinate

product state (2 AB) is the net heat (enthalpy) of the reaction, ΔH. The net heat of the reaction is independent of the height of the energy barrier, hence it is independent of the value of the activation energy. The specie, or the arrangement of molecules and atoms, found at the top of the energy barrier, or at the transition state, is properly called the **activated complex,** although it is also commonly called the **transition state.** This arrangement of atoms is the most unstable arrangement encountered during the course of the reaction.

If we begin with an A_2 molecule and a B_2 molecule, we have particles which are moving, and hence possess kinetic energy in addition to the potential energy shown on the above graph. When these molecules collide, their kinetic energy is converted into potential energy, the reaction begins, and the system begins to move up the energy "hill." If enough kinetic energy is converted into potential energy, the system reaches the top of the hill and starts down the far side. During the descent, potential energy is converted back into kinetic energy, and the product molecules leave with some velocity. Since the potential energy of the products is less than that of the reactants, there is a net release of energy in this reaction, and it is **exothermic.**

According to this diagram, it is possible for this reaction to run backwards. But if it does, it will be an **endothermic** process, and its activation energy will be larger than for the forward reaction, so the reverse process will be slower.

KINETIC EQUATIONS

Since the number of collisions, and therefore the reaction rate, is proportional to the number of molecules or atoms present in the system, or to the concentration, we can express the reaction rate in terms of concentrations of the reactants (or pressure, in the case of gases). If we assume that the reaction of A_2 and B_2 gives two molecules of AB in only one step, we can write the rate equation

$$-\frac{d[A_2]}{dt} = k[A_2][B_2]$$

where k is a proportionality constant, and is called the **rate constant.**

The reaction rate will be proportional to the concentration of some (but not necessarily all) of the reactants, raised to certain powers (or exponents). In a kinetic study, the manner in which the concentrations of these reactants influence the rate of a reaction is examined. For example, in a reaction:

$$A + 2B + C \longrightarrow D$$

if doubling the concentration of A doubles the rate, then

$$Rate \propto [A]$$

If doubling the concentration of B raises the rate by a factor of four, then

$$Rate \propto [B]^2$$

If doubling the concentration of C does not change the rate of the reaction, then

$$Rate = [C]^0 = 1$$

Putting these together, for this hypothetical example, we have:

$$Rate \propto [A][B]^2$$

or

$$Rate = k[A][B]^2$$

This latter equation is called the **rate equation** or **kinetic equation.** It describes the dependence of the rate upon the concentrations of the various reactants, and the rate constant provides some measure of how sensitive the reaction rate is to changes in the concentrations.

Notice that [C] does **not** appear in the rate equation. It is not necessary that all of the species listed in the balanced **stoichiometric** equation appear in the **kinetic** equation. Furthermore, the coefficients in the stoichiometric equation need not be reflected as exponents in the kinetic equation. It must be emphasized that the rate equation, and therefore the order of the reaction, cannot be determined by examining the stoichiometric equation. Frequently the true rate-equation is somewhat different than what we would predict from the stoichiometry of a reaction. The only way that a rate equation can be determined is **by experiment.**

The **reaction order** designates the dependence of the rate upon the concentrations of the various reactants. The reaction order is the sum of the exponents in the concentration terms for each of the reactants.

ZERO-ORDER REACTIONS

The rate equation for a **zero-order reaction** is

$$-\frac{d[A]}{dt} = k$$

If we integrate this expression, we obtain:

$$[A]_0 - [A]_t = kt$$

$[A]_0 = $ initial concentration of A

$[A]_t = $ concentration of A at some value, t, of time.

A plot of $(A_0 - A_t)$ versus time provides a straight line whose slope is equal to the value of the rate constant, k. Zero-order reactions are most often encountered in cases of gas phase reactions at metal surfaces or with certain enzyme-catalyzed reactions in living systems. They do not occur very often in organic reactions in solution.

FIRST ORDER REACTIONS

For a **first-order reaction,** which depends upon the instability and dissociation of a single type of particle, the rate must be proportional to the concentration of this one particle; that is, if we double the concentration, the rate will be doubled. We may express this fact, for the generalized reaction

$$A \longrightarrow \text{Products}$$

by the rate equation

$$-\frac{d[A]}{dt} = k[A]$$

If we integrate this expression, we obtain:

$$\ln\left(\frac{[A]_0}{[A]_t}\right) = kt$$

$[A]_0 =$ initial concentration of A
$[A]_t =$ concentration of A at some value, t, of time

A plot of $\ln\dfrac{[A]_0}{[A]_t}$ versus time provides a straight line whose slope is equal to the value of the rate constant, k. First-order reactions are quite common in organic chemistry. The S_N1 nucleophilic substitution very closely follows first-order behavior (see the section dealing with pseudo-first-order reactions).

SECOND ORDER REACTIONS

A **second-order reaction,** which results from the collision of two particles in the rate-determining step of the mechanism, may have either of the following forms:

$$-\frac{d[A]}{dt} = k[A][B]$$

$$-\frac{d[A]}{dt} = k[A]^2$$

The integrated rate expressions are, respectively:

$$\frac{1}{[B]_0 - [A]_0}\ln\frac{[A]_0[B]_t}{[B]_0[A]_t} = kt$$

$[A]_0$ and $[B]_0 =$ initial concentrations of A or B, respectively

$$\frac{1}{[A]_t} - \frac{1}{[A]_0} = kt$$

$[A]_t$ and $[B]_t =$ concentrations of A or B, respectively, at some value, t, of time.

In either case, a plot of the left-hand side of the equation versus time provides a straight line whose slope provides the rate constant. The second-order reaction is by far the most common type of reaction, since most collisions are two-body processes. Common examples include the S_N2 nucleophilic substitution reaction, the E2 elimination reaction, the substitution reactions of carboxylic derivatives, the addition to alkenes, and the carbonyl addition reactions.

Third-order reactions depend upon some combination of three species. Since a three-particle collision is much less likely than a two-body collision, these types of reactions do not occur very often.

PSEUDO FIRST ORDER REACTIONS

If one or more of the reactants is present in very large excess, their relative concentrations will change very little as compared to the corresponding concentration change for another reactant present in a smaller amount. The reaction, under these conditions, behaves as if it were of a lower kinetic order than it actually is. For example, in the second-order reaction,

$$A + B \longrightarrow Products$$

$$-\frac{d[A]}{dt} = k[A][B]$$

if we make the concentration of A so high that the amount of A used in the course of the reaction represents a negligible fraction of the total amount of A present initially, then

$$k[A] \approx k'$$

and the reaction reduces to a **pseudo-first-order reaction,** whose kinetic equation is

$$-\frac{d[A]}{dt} = k'[B]$$

Very often a complex reaction is simplified by using a large excess of one or more reactants. The resulting simplified kinetic equation is much easier to handle mathematically. This method is very useful to the kineticist. Common types of pseudo-first-order reactions include a wide variety of enzyme-catalyzed reactions and nucleophilic substitution reactions where the nucleophile is the solvent (e.g., hydrolysis reactions, where water is the nucleophile as well as the solvent).

REFERENCES

Frost, A. A., and Pearson, R. G. *Kinetics and Mechanism* (2nd edition). New York: Wiley, 1961. Chapters 1, 2, and 3.

Gilliom, R. D. *Introduction to Physical Organic Chemistry*. Reading, Mass: Addison-Wesley, 1970. Chapters 6 and 7.

King, E. L. *How Chemical Reactions Occur*. New York: W. A. Benjamin, 1963. Chapters 1 through 6.

Laidler, K. J. *Chemical Kinetics* (2nd edition). New York: McGraw-Hill, 1965. Chapters 1, 2, and 3.

THE HYDROLYSIS OF SOME ALKYL CHLORIDES

Synthesis of an Alkyl Halide
Kinetics

There are two chemical reactions of interest in this experiment. The first is the preparation of the alkyl chlorides whose hydrolysis rates are to be measured. The chloride formation is a simple nucleophilic substitution reaction, carried out in a separatory funnel. Because the concentration of the initial alkyl chloride does not need to be determined for the kinetic experiment, isolation and purification of the alkyl chloride is not required.

$$ROH + HCl \longrightarrow R\!-\!Cl + H_2O$$

The second reaction is the actual hydrolysis, and the rate of this reaction will be measured. Under the conditions of this experiment, the reaction proceeds by an S_N1 pathway. The reaction rate is monitored by measuring the rate of appearance of hydrochloric acid. The concentration of hydrochloric acid is determined by titration with aqueous sodium hydroxide.

$$R\!-\!Cl + H_2O \xrightarrow{\text{solvent}} R\!-\!OH + HCl$$

The rate equation for the S_N1 hydrolysis of an alkyl chloride is:

$$+\frac{d[\text{HCl}]}{dt} = k[\text{RCl}]$$

Let c equal the initial concentration of RCl. At some time, t, x moles/liter of alkyl chloride will have decomposed and x moles/liter of HCl will have been produced. The remaining concentration of alkyl chloride at that value of time is equal to $(c - x)$. The rate equation becomes:

$$+\frac{dx}{dt} = k(c - x)$$

When integrated, this gives

$$\ln\left(\frac{c}{c-x}\right) = kt$$

converting to base 10 logarithms:

$$2.303 \log\left(\frac{c}{c-x}\right) = kt$$

This equation is of the form appropriate for a straight line ($y = mx + b$) with slope m, and with intercept b equal to zero. If the reaction is indeed first order, a plot of $\log(c/c - x)$ versus t will provide a straight line whose slope is $k/2.303$.

Evaluation of the term $(c/c - x)$ remains a problem, since it is experimentally difficult to determine the concentration of alkyl chloride. We can, however, determine the concentration of hydrochloric acid produced by titration with base. Because the stoichiometry of the reaction indicates that the number of moles of alkyl chloride consumed equals the number of moles of hydrochloric acid produced, c must also equal the number of moles of HCl produced when the reaction has gone to completion (the so-called "infinity concentration" of HCl), and x equals the number of moles of HCl produced at some particular value at time t. From these equalities, we can re-write the integrated rate expression in terms of volume of base used in the titration. At the end-point of the titration:

$$\# \text{ of moles HCl} = \# \text{ of moles NaOH}$$

or

$$x = \# \text{ of moles NaOH at time } t$$

and

$$c = \# \text{ of moles NaOH at time } \infty.$$
$$\# \text{ of moles NaOH} = [\text{NaOH}] \, V \qquad V = \text{Volume}$$

Substituting and cancelling:

$$\left(\frac{c}{c-x}\right) = \frac{[\text{NaOH}]V_\infty}{[\text{NaOH}](V_\infty - V_t)} \qquad \begin{array}{l} V_\infty = \text{Volume of NaOH used when the} \\ \qquad \text{reaction is complete} \\ V_t = \text{Volume of NaOH used at time, } t \end{array}$$

This integrated rate equation becomes:

$$2.303 \log\left(\frac{V_\infty}{V_\infty - V_t}\right) = kt$$

It should be noticed that the concentration of the base used in the titration cancels out of this equation, so that it is not necessary to know the concentration of base nor the amount of alkyl chloride used in the experiment.

A plot of $\log (V_\infty / V_\infty - V_t)$ **versus** t will provide a straight line whose slope is equal to $k/2.303$. The slope is determined according to the following figure.

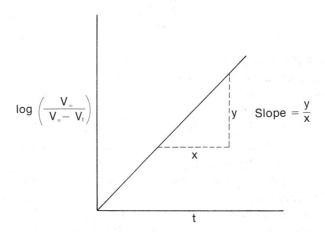

If the time is measured in minutes, the units of k are min^{-1}. The experimental points plotted on the graph may contain a certain amount of scatter, but the line which is drawn is the best **straight** line. The line should pass through the origin of the graph. With some reactions, competing processes may cause the line to contain a certain amount of curvature. In these cases, the slope of the initial portion of the line is used, before the curvature becomes too important.

One other value which is often cited in kinetic studies is the **half-life** of the reaction τ. The half-life is the time required for one half of the reactant to undergo conversion to products. During the first half-life, 50% of the available reactant is consumed. At the end of the second half-life, 75% of the reactant has been consumed. For a first-order reaction, the half-life is calculated by:

$$\tau = \frac{\ln 2}{k} = \frac{0.69315}{k}$$

Two alkyl chlorides will be studied by the class in a variety of solvents. The class data will be compared to determine relative reactivities of the alkyl chlorides.

SPECIAL INSTRUCTIONS

Before beginning this experiment, you should read the essay which precedes it, dealing with the methods of kinetics. Concentrated hydrochloric acid is quite corrosive. Care should be exercised in handling it. Avoid breathing the vapors. Some of the alkyl chlorides hydrolyze rather rapidly. It is necessary for students to work in pairs in this experiment in order to perform the measurements rapidly.

PROCEDURES

THE PREPARATION OF THE ALKYL CHLORIDES

An alcohol should be selected by the student. The choices include *t*-butyl alcohol (2-methyl-2-propanol) and α-phenylethyl alcohol (1-phenylethanol). Place the alcohol (11 ml) in a separatory funnel along with 25 ml of cold, concentrated hydrochloric acid (specific gravity 1.18, 37.3% hydrogen chloride). Shake the separatory funnel vigorously, with frequent venting in order to relieve any excess pressure, over a thirty minute period. Remove the aqueous layer. Wash the organic phase quickly with three 5 ml portions of cold water, followed by a washing with 5 ml of 5% sodium bicarbonate solution. Place the organic product in a small Erlenmeyer flask over three to four grams of anhydrous calcium chloride. Shake the flask occasionally over a five minute period. Carefully decant the alkyl chloride from the drying agent into a small Erlenmeyer flask which can be stoppered tightly. The alkyl chloride is used in this experiment without prior distillation. Because the true concentration of alkyl chloride introduced into the hydrolysis reaction is determined by titration, it is not necessary to purify the product prepared in this part of the experiment.

THE KINETIC STUDY OF THE HYDROLYSIS OF AN ALKYL CHLORIDE

Because these chlorides hydrolyze rather rapidly under the conditions used in this experiment, it is necessary that students perform the kinetic studies working in pairs. One student will perform the titrations, while the other student will measure the time and record the data.

Prepare a stock solution of alkyl chloride by dissolving about 0.6 grams of alkyl chloride in 50 ml of dry, reagent-grade acetone. Store this solution in a stoppered container to protect it from moisture. Use a 125 ml Erlenmeyer flask to carry out the hydrolysis reaction. The flask should contain a magnetic stirring bar, 50 ml of solvent (see table below for the appropriate solvent), and two or three drops of bromthymol blue indicator. Use absolute ethanol in preparing the aqueous ethanol solvent. Do not use denatured ethanol, as the denaturing agents may interfere with the reactions being studied. Bromthymol blue has a yellow color in acid solution and a blue color in alkaline solution.

Place a 50 ml buret filled with approximately 0.01 N sodium hydroxide above the flask. The exact concentration of sodium hydroxide used need not be known. Record the initial volume of sodium hydroxide at time, t, equal to 0.0 minutes. Add about 2 ml of sodium hydroxide from the buret to the Erlenmeyer flask, and precisely record the new volume in the buret. Start the stirrer. At time = 0.0 minutes, **rapidly** add 1.0 ml of the acetone solution of the alkyl chloride from a pipet. Start the timer when the pipet is about half empty. The indicator will undergo a color change, passing from blue through green to yellow when sufficient hydrogen chloride has been formed in the reaction to neutralize the sodium hydroxide in the flask. Record the time at which the color change occurred. This color change may not be rapid. One should attempt to use the same color as the end-point each time. Add another 2 ml of sodium hydroxide from the buret, precisely record the volume, and also record the time at which this second volume of sodium hydroxide is consumed. Repeat this sodium hydroxide addition two more times (four total). Finally, allow the reaction to go to completion for an hour without excess sodium hydroxide being present. Stopper the Erlenmeyer flask during this period.

After the reaction has gone to completion, **accurately** titrate the amount of hydrogen chloride in solution to the end point. The end point is reached when the color of the solution remains constant for at least 0.5 minute. The time corresponding to this final volume is infinity $(t = \infty)$. Repeat this process in the other two solvent mixtures indicated in the table below. These experiments may be carried out while waiting for the infinity titration of the previous experiments, provided that a separate buret is used for each run, thereby permitting an accurate determination of the infinity concentrations of hydrogen chloride produced.

EXPERIMENTAL CONDITIONS

COMPOUND	SOLVENT MIXTURES (volume percent of organic phase in water)
t-butyl chloride	40% ethanol
	25% acetone
	10% acetone
α-phenylethyl chloride	50% ethanol
	40% ethanol
	35% ethanol

THE HYDROLYSIS OF α-PHENYLETHYL CHLORIDE
IN 50% ETHANOL

TIME MIN.	VOL. NaOH RECORDED	VOL. NaOH USED	$V_\infty - V_t$	$\dfrac{V_\infty}{V_\infty - V_t}$	$\ln\left(\dfrac{V_\infty}{V_\infty - V_t}\right)$
0.00	0.2	0.0	6.9	1.00	0.000
8.46	2.2	2.0	4.9	1.41	0.343
18.25	4.2	4.0	2.9	2.37	0.863
31.80	5.9	5.7	1.2	5.75	1.750
47.72	6.8	6.6	0.3	23.00	3.136
100 (∞)	7.1	6.9	0.0	–	–

The data should be plotted according to the method described in the introductory section of this experiment. The rate constant, k, and the half-life, τ, must be reported. The report to the instructor should include the plot of the data as well as a table of data. A sample table of data is shown above. Explain your results especially with regard to the effect of changing the water content of the solvent on the rate of the reaction. At the option of the instructor, the results from the entire class may be compared.

QUESTIONS

1. Plot the data given in the table above. Determine the rate constant and the half-life for this example.

2. What are the principal by-products of these reactions? Give the rate equations for these competing reactions. Should the production of these by-products proceed at the same rate as the hydrolysis reactions? Explain.

3. Compare the energy diagrams for an $S_N 1$ reaction in solvents with two different percentages of water. Explain any differences in the diagrams and their effect on the reaction rate.

4. Compare the rate of hydrolysis of t-cumyl chloride (2-chloro-2-phenylpropane) to that of α-phenylethyl chloride (1-chloro-1-phenylethane) in the same solvent. Explain any differences which might be expected.

CYCLOHEXENE *no bunsen burner*

Preparation of an Alkene
Dehydration of an Alcohol
Bromine and Permanganate
Tests for Unsaturation

(handwritten notes: due 10; dates; brown to clear; over the double bond; purple to clear; CH₂; reaction scheme sketches with H₃PO₄/Δ, Br₂, KMnO₄)

Cyclohexanol   Cyclohexene

$$\text{Cyclohexanol} \xrightarrow[\Delta]{H_3PO_4} \text{Cyclohexene} + H_2O$$

Alcohol dehydration is an acid-catalyzed reaction performed by strong, concentrated mineral acids, such as sulfuric and phosphoric acids. The acids protonate the alcoholic hydroxyl group permitting it to dissociate as water. Loss of a proton from the intermediate (elimination) results in an alkene. Since sulfuric acid often causes extensive charring in this reaction, phosphoric acid, which is comparatively free of this problem, will be used.

The equilibrium that attends this reaction will be shifted in favor of the product, cyclohexene, by distilling it from the reaction mixture as it is formed. The cyclohexene (bp 83°) will co-distill with the water that is also formed. By continuously removing the products a high yield of cyclohexene will be obtained. Since the starting material, cyclohexanol, is also rather low boiling (bp 161°), the distillation must be done carefully, not allowing the temperature to rise much above 100°.

Unavoidably, a small amount of phosphoric acid co-distills with the products. It is removed by washing the distillate mixture with aqueous sodium carbonate. To remove the water which co-distills with cyclohexene, and any traces of water introduced in the base extraction, the product is dried over anhydrous calcium chloride.

Compounds containing double bonds will react with a bromine solution (red) to decolorize it. Similarly, they will react with a solution of potassium permanganate (purple) to discharge its color and produce a brown precipitate (MnO_2). These tests are often used as qualitative

$$\underset{\substack{Br \quad Br \\ \text{(colorless)}}}{} \xleftarrow{Br_2\text{(red)}} \quad \xrightarrow{KMnO_4\text{(purple)}} \underset{\substack{OH \quad OH \\ \text{(colorless)}}}{} + \underset{\text{(brown)}}{MnO_2}$$

tests to determine the presence of a double bond in an organic molecule (see Experiment 50C). Both tests will be performed on the cyclohexene formed in this experiment.

SPECIAL INSTRUCTIONS

Phosphoric acid is very corrosive. Do not allow any acid to come in contact with your skin. Since a flame is used to carry out the distillation, keep in mind that cyclohexene will burn and is quite volatile. Be careful not to allow your flame to get near the collection flask. Also, do not leave your product in an open container on your desk top, because a neighbor's flame may ignite the vapors from your product. Finally, you should not flush your distillation residues down an open laboratory sink. This could fill the laboratory with dangerous vapors. Dispose of all residues in a hood sink. Before proceeding, you should have read Techniques 1, 5, and 6.

PROCEDURE

Assemble a distillation apparatus as illustrated in Figure 6–6, Technique 6. Use a 100 ml distilling flask and a 50 ml collection flask. The collection flask should be immersed up to its neck in an ice water bath to minimize the possibility of cyclohexene vapors escaping into the laboratory.

Place 20 ml of cyclohexanol (d = 0.96) and 5 ml of 85% phosphoric acid in the distilling flask and mix the solution thoroughly. Add several boiling stones, start circulating the cooling water in the condenser and heat the mixture until the product begins to distill. The temperature of the distilling vapor should be regulated so that it does not exceed 100°. This can best be done if the mixture is heated with a Bunsen burner. Hold the burner by its base and apply the heat so as to maintain a slow, but steady, rate of distillation. If the temperature rises above 100°, remove the burner for a few seconds before continuing to heat.

When only a few milliliters of residue remain in the distilling flask, stop the distillation. Saturate the distillate with solid sodium chloride. Add the salt, little by little, and shake the flask gently. When no more salt will dissolve, add enough 10% aqueous sodium carbonate solution to make the distilled solution basic to litmus. Pour the neutralized mixture into a separatory funnel and separate the two layers. Drain the aqueous layer through the stopcock and then pour the upper layer (cyclohexene) through the neck of the separatory funnel into a 125 ml Erlenmeyer flask. Add about 2 or 3 g of anhydrous calcium chloride to the flask and swirl the mixture occasionally until the solution is dry and clear (about 10 to 15 minutes). If there is not sufficient time to complete the experiment, it may be stopped here. Store the cyclohexene over calcium chloride in a lightly greased and stoppered standard taper flask.

Reassemble a distillation apparatus as before, but this time use a 50 ml flask for the distilling flask. Again, cool the receiver in an ice

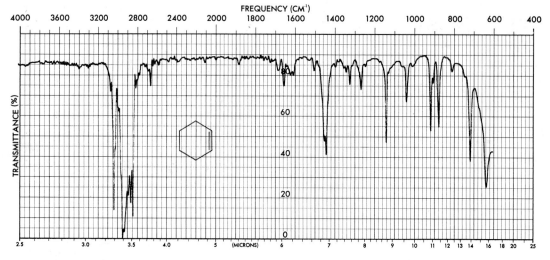

Infrared spectrum of cyclohexene, neat

water bath. Decant the dry cyclohexene solution into the distilling flask and add a boiling stone. Distill the cyclohexene and collect the material that boils over a range of 80 to 85°. Weigh the product and calculate the yield. After performing the tests below, place the sample in a labeled, glass, screw cap vial and submit it, along with the report, to the instructor.

UNSATURATION TESTS

Place 8 to 10 drops of cyclohexanol in each of two small test tubes. In each of another pair of small test tubes, place 8 to 10 drops of the

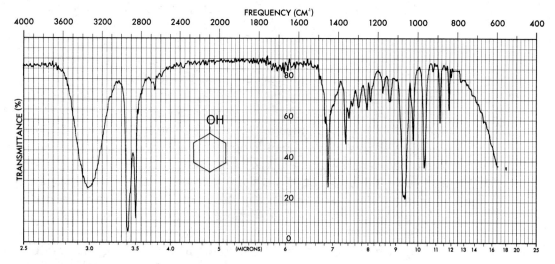

Infrared spectrum of cyclohexanol, neat

cyclohexene you prepared. Do not confuse the test tubes. Take one test tube from each group and add to each a solution of bromine in carbon tetrachloride, drop by drop, until the red color is no longer discharged. Record the result in each case. Test the remaining two test tubes in a similar fashion with a solution of potassium permanganate. Since aqueous potassium permanganate is not miscible with organic compounds, you will have to add about 0.5 ml of dioxane to each test tube before performing the test. Record your results and explain them.

QUESTIONS

1. Draw a mechanism for the dehydration of cyclohexanol catalyzed by phosphoric acid.

2. What would be the alkene produced on dehydration of each of the following alcohols?

a) 1-methylcyclohexanol   d) 2,2-dimethylcyclohexanol
b) 2-methylcyclohexanol   e) 1,2-cyclohexanediol
c) 4-methylcyclohexanol

3. In the workup procedure for cyclohexene, why is salt added before the layers are neutralized and separated?

4. What is the purpose of addition of the sodium carbonate solution? Give an equation.

5. Compare and interpret the infrared spectra of cyclohexene and of cyclohexanol (see above).

Experiment 27
CAMPHENE FROM BORNEOL OR ISOBORNEOL

Elimination (Dehydration)
Rearrangement

In this experiment we will prepare camphene from either borneol or isoborneol by the acid catalyzed elimination of water (dehydration). The straightforward dehydration product, bornylene, is not produced in this reaction. Rather, the rearranged alkene, camphene, is obtained. In this experiment the proton is supplied by the catalyst potassium bisulfate ($KHSO_4$).

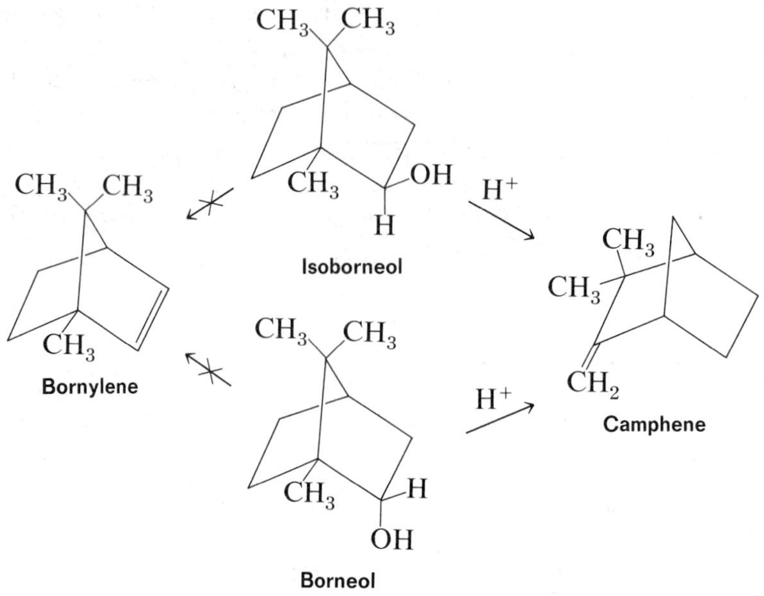

Isoborneol

Bornylene

Camphene

Borneol

The mechanism of the elimination/rearrangement reaction for isoborneol is as follows:

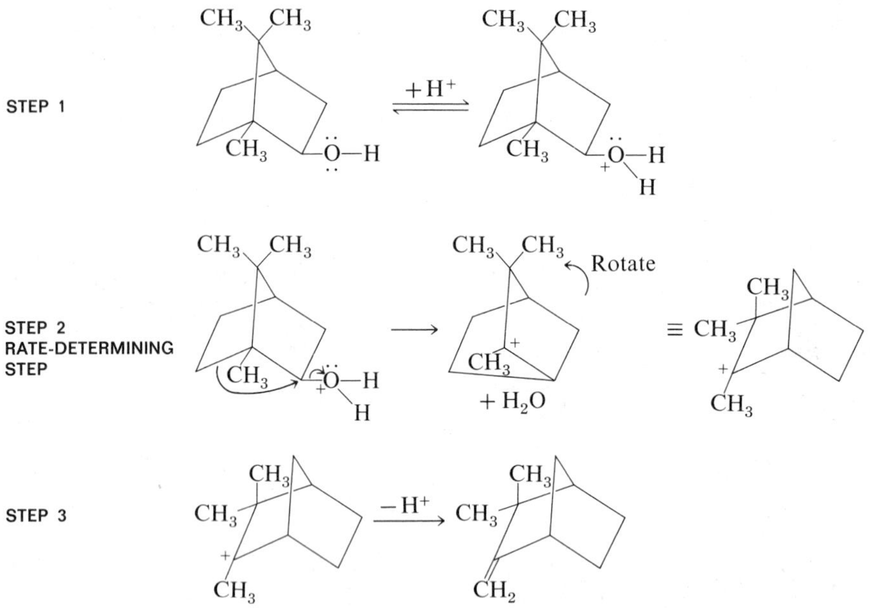

STEP 1

STEP 2
RATE-DETERMINING
STEP

STEP 3

In step one, the alcohol is protonated in order to provide a good leaving group (water). In step two, the rate-determining step, the rearrangement occurs by migration of the ring simultaneously with displacement of water. The tertiary (3°), carbonium ion intermediate produced at that point can lose a proton to give the alkene, camphene (step three).

Isoborneol is more reactive towards dehydration than borneol. The reactivity order can be predicted by looking at the stereochemistry of the two isomeric alcohols. In protonated isoborneol, the migrating methylene bridge and leaving group are constrained to be **trans** to one another (see structure below). The **trans** stereochemistry is ideal for participation by the methylene because backside attack can occur easily (step two). In protonated borneol, the neighboring methylene group and the leaving group are **cis** related (see structure below) and no participation can occur. Participation must enhance the rate of reaction. Since no participation can occur with borneol, its rate of dehydration is slower than isoborneol.

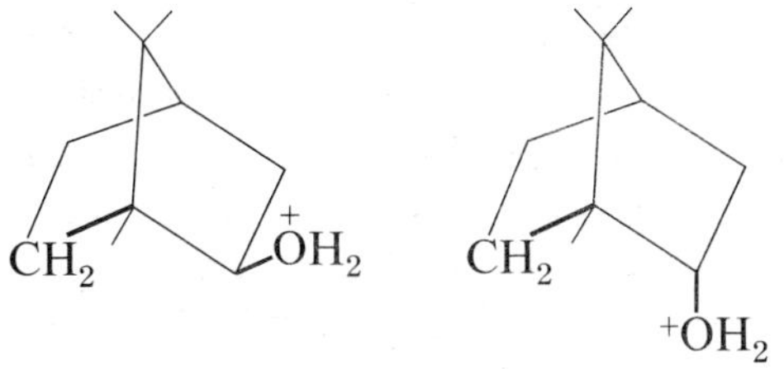

trans- **stereochemistry**	*cis-* **stereochemistry**
(isoborneol)	**(borneol)**

The mechanism of the elimination/rearrangement reaction for borneol is shown below. Steps 1 (OH down) and 3 are identical to the mechanism shown for isoborneol. Step 2a, the rate-determining step, shows the formation of a secondary (2°) carbonium ion **without** participation. Since participation does not occur, the overall rate of the reaction will be slowed relative to isoborneol. The secondary (2°) carbonium ion then rearranges to a tertiary (3°) carbonium ion in step 2b. This provides the driving force for the reaction. The elimination of a proton occurs in the last step (step 3).

STEP 2a
RATE-DETERMINING
STEP

STEP 2b

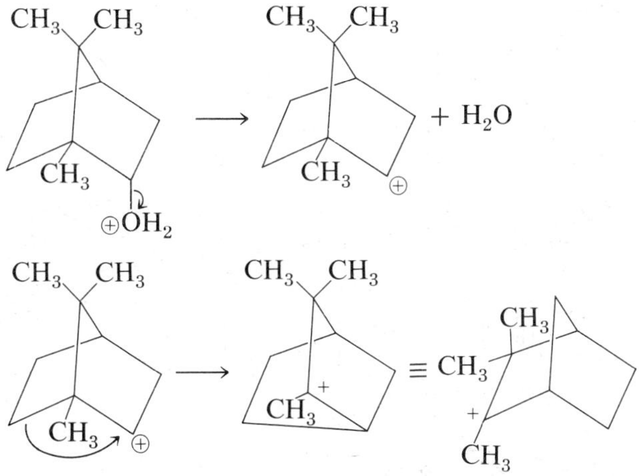

SPECIAL INSTRUCTIONS

It is helpful to have read Techniques 1, 5 (Section 5.6), and 6 (Section 6.4) before performing this experiment. The camphene sublimes readily. Be sure to store it in a stoppered container.

PROCEDURE

Grind 5 g of borneol or isoborneol together with 10 g of anhydrous potassium bisulfate, using a mortar and pestle, until they are thoroughly pulverized. Transfer this powder to a 25 ml distilling flask with the aid of a powder funnel (wide neck). If a powder funnel is not available, a makeshift funnel can be made by rolling a piece of smooth notebook paper into a cone and fastening it together with tape.

Assemble the apparatus as shown in the figure. Wrap the distilling head with glass wool for insulation. Place mineral oil or another type of bath oil in a beaker to serve as a heating bath. Mount a thermometer in the bath as shown in the figure. Pass steam through the condenser in order to keep the product from solidifying in the condenser.

Heat the beaker containing the oil with a Bunsen burner (not a microburner) until the temperature rises to about 180° in the bath. At that point, the product should begin to distill as the dehydration proceeds. Record the time at which the dehydration begins to occur and the temperature of the bath at that time. It is essential that no solidification occur in the condenser; if the system closes, the apparatus may explode. To prevent a closed system, it may be necessary to remove

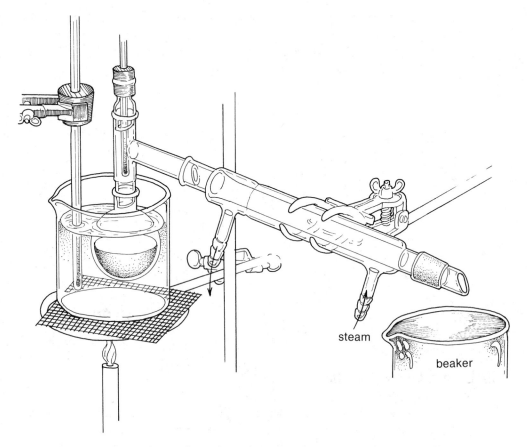

Apparatus for dehydration of borneol or isoborneol

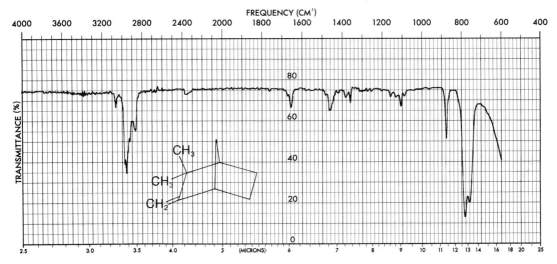

Infrared spectrum of camphene, CCl₄

the solid product, which collects at the end of the condenser, with a spatula. If necessary, the temperature of the oil may be raised to about 220° to complete the distillation of the product. Watch the temperature indicated by the thermometer inserted in the distilling head. When the reaction is completed, the temperature will drop significantly. Borneol is converted to camphene in about 30 minutes, while isoborneol requires a shorter reaction time. Record the time at which the distillation is complete. Also record the temperature of the bath at the end of the distillation. Clean and dry the distilling flask and condenser. Pour the mineral oil back into a storage vessel.

Melt the crude camphene on the steam bath and remove any significant amount of water (lower phase) with a capillary pipet. Since the temperature of the condenser is about 100° during the dehydration reaction, only a small quantity of water is usually collected in the receiver (beaker). With a pipet carefully transfer the crude product back to the 25 ml distilling flask. Add 1 g of Drierite (anhydrous calcium sulfate) and a boiling stone to the flask. Reassemble the apparatus and again pass steam through the condenser. Use a Bunsen burner to distill the crude product from the drying agent, but do not use the oil bath. Collect the product in a pre-weighed beaker. It may again be necessary to use a spatula to remove the solid from the end of the condenser during the distillation. The observed boiling range is usually from 154 to 158°. Determine the weight of the colorless camphene and calculate the percentage yield.

Determine the melting point of the product. Pure racemic camphene has a melting point of 52 to 54°. Store the camphene in a stoppered container since it sublimes readily. At the instructor's option, determine the infrared spectrum of camphene. The camphene may be dissolved in carbon tetrachloride. Place a few drops of the solution on the salt plates and mount the plates in the holder. Compare the spectrum that is obtained to the one given above. In the report to the instructor, give the reaction time and temperature range at which the dehydration was conducted. Submit the camphene to the instructor in a labeled vial.

QUESTIONS

1. Interpret the infrared spectrum of camphene.

2. What is the solid which remains in the distilling flask after the dehydration reaction?

3. Give the mechanism and the expected product which would be obtained upon dehydration of neopentyl alcohol (2,2-dimethyl-1-propanol).

4. Repeat question three, for cyclobutylcarbinol (cyclobutylmethanol).

5. Why is the reaction time for borneol longer than for isoborneol?

Experiment **28**
TRIPHENYLMETHANOL AND BENZOIC ACID

Grignard Reactions
Extraction

In this experiment a Grignard reagent, or organomagnesium reagent, will be prepared. The reagent to be prepared is phenylmagnesium bromide.

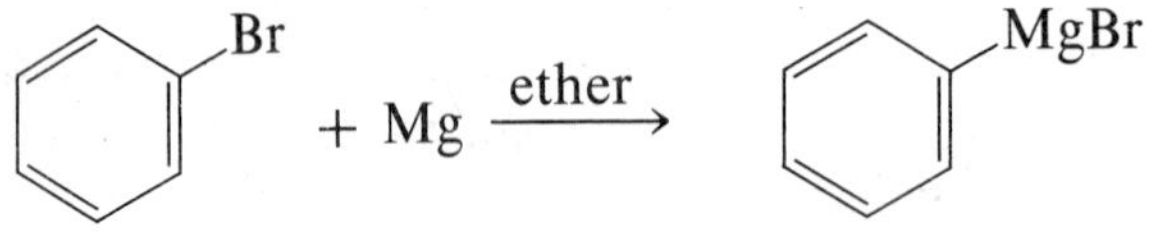

Bromobenzene   **Phenylmagnesium Bromide**

This reagent will be converted to a tertiary alcohol or to a carboxylic acid, depending upon the procedure selected.

PROCEDURE 28A:

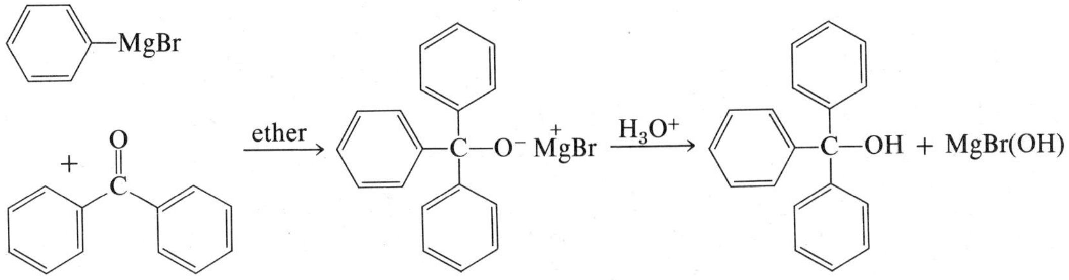

Benzophenone   Triphenylmethanol

PROCEDURE 28B:

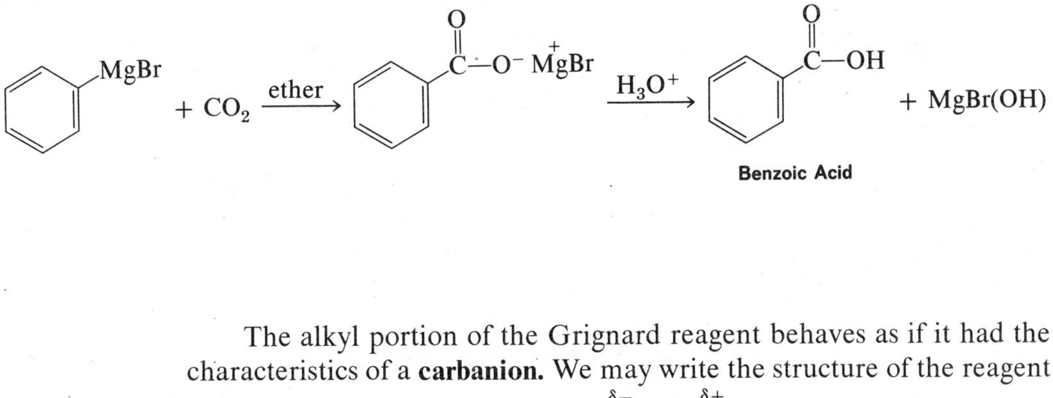

Benzoic Acid

The alkyl portion of the Grignard reagent behaves as if it had the characteristics of a **carbanion.** We may write the structure of the reagent as a partially ionic compound: $\overset{\delta-}{R} \dots \overset{\delta+}{MgX}$. This partially bonded carbanion is a Lewis base. It reacts with strong acids, as one would expect, to give an alkane.

$$\overset{\delta-}{R} \dots \overset{\delta+}{MgX} + HX \longrightarrow R-H + MgX_2$$

Any compound with a suitably acidic hydrogen will donate a proton to destroy the reagent. Water, alcohols, terminal acetylenes, phenols, and carboxylic acids are all sufficiently acidic to bring about this reaction.

The Grignard reagent will also function as a good nucleophile in nucleophilic addition reactions of the carbonyl group. The carbonyl group has electrophilic character at its carbon atom (due to resonance), and a good nucleophile will seek out this center for addition.

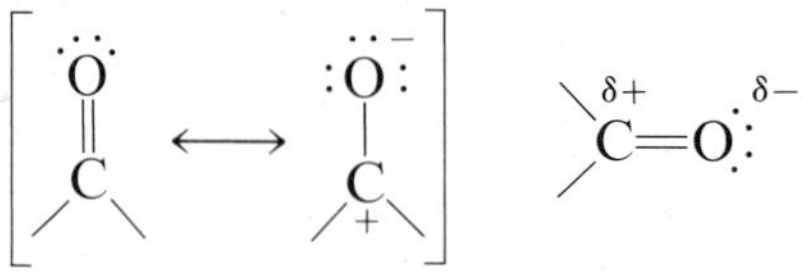

The magnesium salts produced form a complex with the addition product, an alkoxide salt, and in a second step of the reaction, these must be protonated by addition of dilute aqueous acid.

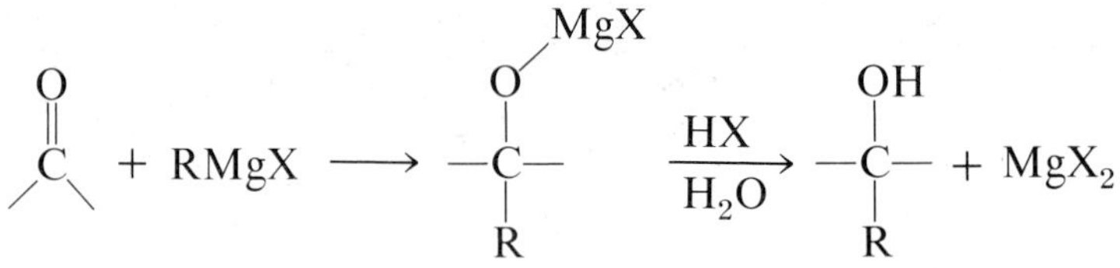

STEP I   STEP II

The Grignard reaction is used synthetically to prepare secondary and tertiary alcohols from aldehydes and ketones, respectively. It will react with esters twice to give tertiary alcohols. Synthetically, it also can be allowed to react with carbon dioxide to give carboxylic acids,

$$RMgX + O{=}C{=}O \longrightarrow RC\overset{O}{\overset{\|}{-}}OMgX \xrightarrow[H_2O]{HX} R\overset{O}{\overset{\|}{-}}C{-}OH$$

and with oxygen to give hydroperoxides.

$$RMgX + O_2 \longrightarrow ROOMgX \xrightarrow[H_2O]{HX} ROOH$$

Because of its reactions with water, carbon dioxide, and oxygen, when the Grignard reagent is used, it must be protected from air and moisture. The apparatus in which the reaction is to be conducted must be scrupulously dry (recall that 18 ml of H_2O is one mole), and the solvent must be free of water, or anhydrous. During the reaction the flask must be protected by a calcium chloride drying tube. Oxygen should also be excluded. In practice this can be done by allowing the solvent ether to reflux. This blanket of solvent vapor keeps air from the surface of the reaction mixture.

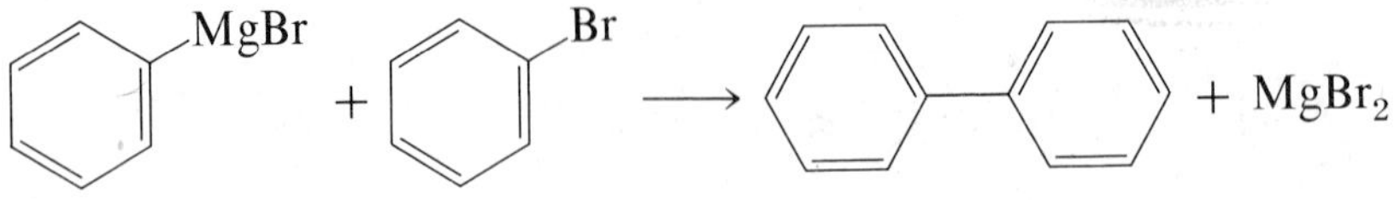

Biphenyl

In the experiment being described here, the principal impurity is **biphenyl,** which is formed by a heat- or light-catalyzed coupling reaction of the Grignard reagent and unreacted bromobenzene. A high reaction temperature will favor the formation of this product. Biphenyl is highly soluble in petroleum ether, and it is easily separated from triphenyl-methanol by crystallization from that solvent.

SPECIAL INSTRUCTIONS

Before beginning this experiment, read Techniques 1 and 5. This reaction must be conducted in one laboratory period to the point where either benzophenone is added (Procedure 28A) or the Grignard reagent is poured over Dry Ice (Procedure 28B). The Grignard reagent cannot be stored.

FIRE SAFETY

This reaction involves the use of large quantities of diethyl ether, which is extremely flammable. Be certain that no open flames of any sort are in your vicinity when you are using ether. It is recommended that you reread the introductory chapter, "Laboratory Safety."

THE APPARATUS

All glassware used in a Grignard reaction must be **scrupulously** dried. Surprisingly large amounts of water adhere to the walls of glass-ware, even glassware which is apparently dry. It is important to perform a flaming operation since truly dry glassware is required in this reaction. The magnesium turnings must also be dried carefully. The glassware can be dried in a drying oven before use, but the best procedure is to "flame" the assembled apparatus. The apparatus is assembled as shown in the figure, with drying tubes attached in the proper locations. At this point, no hoses are attached, and no water is in the cooling jacket of the condenser. The glassware should appear dry, with no droplets of water visible on the inside of the apparatus. The magnesium which is to be used for the reaction is placed in the 500 ml flask, but no other reagents are included. The entire apparatus, particularly the 500 ml flask with the magnesium, is assembled on a ring stand, taken to a **hood** (ether is being used in the laboratory), and **gently** heated with a Bunsen burner flame for about 5 minutes. Do not heat the magnesium strongly. The apparatus is then allowed to cool. When it has cooled, the remaining reagents may be added and the reaction may be begun.

In subsequent operations, all other equipment, including the gradu-ated cylinders used for measurements, must be dry. Furthermore, **anhydrous** ether must be used as a solvent. It will absorb moisture from the air if ether containers are not kept tightly closed.

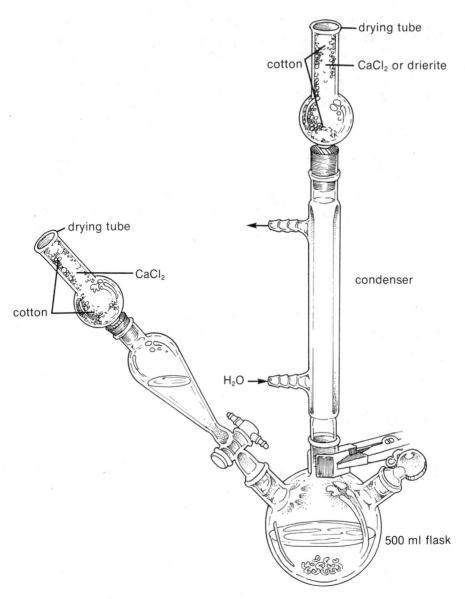

Apparatus for Grignard reactions

STARTING A GRIGNARD REACTION

Starting a Grignard reaction can often be a frustrating experience. Failure to form the Grignard reagent is the most common difficulty. This difficulty can arise from several causes: 1) the halide may be too unreactive, 2) the magnesium may have an oxide coating, or 3) traces of water may be present in the reagents or the equipment.

Bromobenzene is fairly reactive, so the problem of an unreactive halide should not be an important factor in this experiment. The oxide coating on the magnesium can be removed, at least partially, by rubbing or crushing the magnesium chips with a glass rod. If this is done in the flask, care should be taken not to punch a hole in the bottom of the flask. Crushing the magnesium exposes a fresh surface area at which the reaction may begin. The addition of a small crystal of iodine will also sometimes help to clean the surface of the magnesium. If traces of water are present, it is frequently helpful to prepare a small sample of the Grignard reagent in a test tube. When this reaction is started, it is added to the main reaction mixture. This "extra" reagent will quickly remove the trace of water by reacting with it.

PROCEDURE

PHENYLMAGNESIUM BROMIDE

Have both a steam bath and an ice bath ready for use, since it may be necessary to control the rate of the reaction by alternately heating and cooling the reaction mixture. Fit a clean, dry 500 ml round-bottom, three-neck flask with a dry reflux condenser and an addition (separatory) funnel. Place drying tubes on both the addition funnel and condenser. Add 2.64 g of magnesium turnings and flame the apparatus according to the procedure described previously. Obtain 15 ml (density = 1.49 g/ml) of bromobenzene. Add 10 ml of **anhydrous** ether (contained in metal cans) and 1 ml of the previously measured bromobenzene to the reaction flask. Dissolve the rest of the bromobenzene in 75 ml of anhydrous ether, stir the solution, stopper it, and store it for later addition to the reaction. Use a stirring rod to crush two or three chips of the magnesium against the side of the reaction flask. Be careful not to punch a hole in the flask. Replace the stopper and note if any reaction begins to take place. If not, heat the flask gently with a steam bath for a minute or two. Only a very small amount of steam is needed to heat the ether to its boiling point. Remember that the presence of water vapor in the reaction will make starting the reaction very difficult. Stop the heating, remove the flask from the steam bath, and check the reaction mixture to see if any reaction is taking place. If not, try crushing some more of the magnesium chips with a stirring rod. Evidence of reaction will be the formation of a brown-gray, cloudy material in the solution. Also, the ether will continue to reflux when no heat is being applied. The formation of small bubbles at the surface of certain pieces of magnesium will be noticeable. If the reaction does not commence within 2 to 3 minutes, repeat the heating procedure and continue crushing the magnesium with the stirring rod. If, after two or three attempts, the reaction still will not start, refer to the "Starting a Grignard Reaction" section in the Special Instructions for additional advice.

While attempting to start the reaction, place the remainder of the bromobenzene-ether solution in the addition (separatory) funnel. Replace the drying tube. When the reaction has started, as evidenced by the bubbling action which proceeds in the absence of heat, add the bromobenzene solution to the magnesium suspension. This addition should be dropwise at a rate of 1 to 3 drops per second. If the condensation ring does not remain in the lower third of the condenser, slow the rate of addition. Heating should not be necessary as the reaction is sufficiently exothermic to boil the solvent without the use of additional heat. If the reaction begins to reflux too vigorously, moderate it with an ice bath. Do not overcool the reaction, as this will slow the reaction excessively. When all the bromobenzene has been added, reflux the ether gently for at least 15 minutes on the steam bath. As indicated by your instructor, proceed to either Procedure 28A or 28B.

Procedure 28A

TRIPHENYLMETHANOL

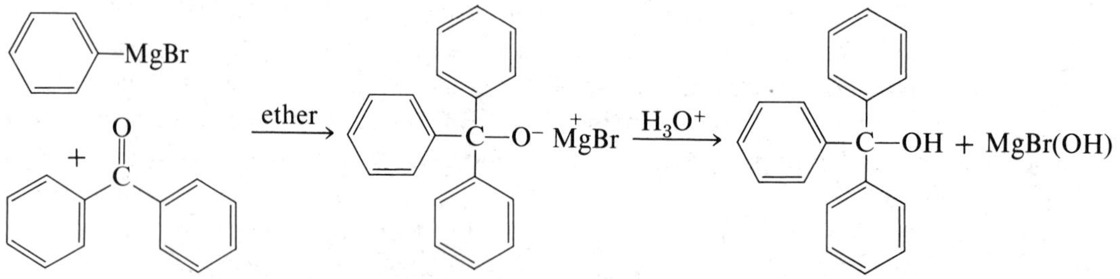

While the phenylmagnesium bromide solution is refluxing, make a solution of 18.1 g of benzophenone in 50 ml of anhydrous ether. Place the solution in the addition (separatory) funnel. After the 15 minute reflux period, add the benzophenone solution to the phenylmagnesium bromide solution as rapidly as possible, again keeping the condensation ring in the lower third of the condenser. The ice bath may be used to moderate the reaction. Swirl the flask during the addition to ensure proper mixing. Cool the solution to room temperature and add 6 M hydrochloric acid cautiously **(dropwise at first)** until the aqueous solution is acidic to litmus. Use a stirring rod to take solution from the lower layer, since litmus does not work in nonaqueous solvents. It may be necessary to add some more ether if it evaporates. Separate the layers using a separatory funnel. Wash the water layer with 25 ml of **solvent grade** ether and combine the two ether fractions. Solvent grade ether is cheaper than the anhydrous ether that was used above. The principal difference is that solvent grade ether contains some water. Water can no longer interfere at this point of the reaction. Dry the ether solution with 2 to 3 g of anhydrous powdered sodium sulfate, and gravity filter (through a fluted filter paper) the suspension to remove the drying agent. Evaporate the ether in a hood. Alternatively, the ether may be removed by distillation, using a steam bath as the heat source.

The material which remains after all the ether has evaporated has the appearance of a brown oil. The crude product contains the desired triphenylmethanol as well as the by-product, biphenyl. Most of the biphenyl can be removed by adding 100 ml low-boiling **petroleum** ether or ligroin. These latter solvents are mixtures of hydocarbons which easily dissolve the hydrocarbon, biphenyl, and leave behind the alcohol, triphenylmethanol. They are not to be confused with diethyl ether ("ether"). Swirl the crude product with the solvent for several minutes, and then suction filter the mixture through a Büchner funnel. Take 0.5 g of the remaining solid and recrystallize it from a minimum amount of hot petroleum ether or ligroin. This recrystallization requires rather large quantities of solvent. An alternative procedure is to recrystallize the crude triphenylmethanol from the minimum amount of benzene. Report the melting point of the purified material (literature

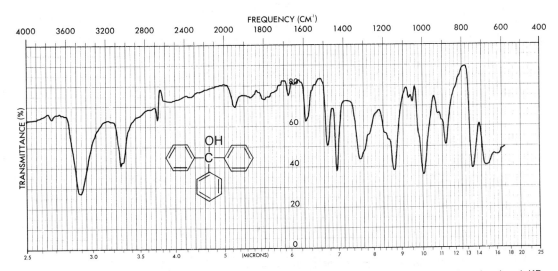

Infrared spectrum of triphenylmethanol, KBr

value = 164°) and the yield in grams, as well as the percent yield. Place both the crude and recrystallized products in properly labeled vials and submit them to the instructor. At the option of the instructor, determine the infrared spectrum, as a KBr mull, of the purified triphenylmethanol (Technique 17, Part A).

Procedure 28B

BENZOIC ACID

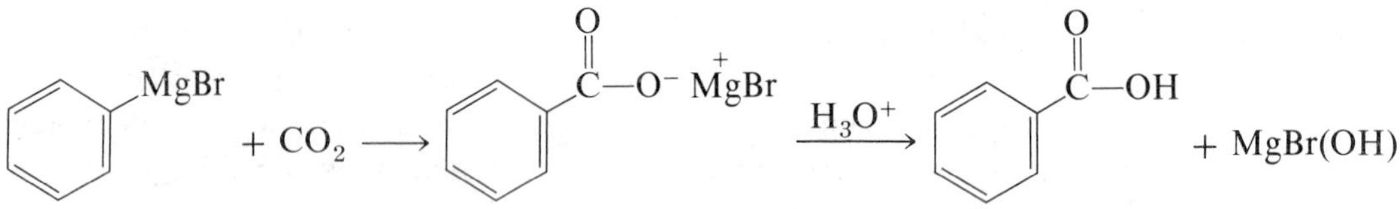

When the phenylmagnesium bromide has been prepared and the mixture has stopped refluxing, remove the condenser from the flask and pour the entire contents slowly but steadily over 35 g of crushed Dry Ice contained in a 600 ml beaker. Exercise caution in handling Dry Ice. Contact with the skin can cause severe frostbite. Always use cotton gloves or tongs. The Dry Ice is best crushed by wrapping one lump in a clean, dry towel and striking it on the bench top or the floor. Dry Ice condenses moisture from the atmosphere to form a coating of ice over its surface. Use the Dry Ice immediately after crushing it to avoid side reactions caused by this condensed water. Cover the reaction mixture with a watch glass and allow it to stand until the excess Dry Ice has completely sublimed. The Grignard addition compound will appear as a viscous glassy mass. If the mass is too viscous to stir, add an additional 50 ml of ether.

Hydrolyze the Grignard addition product by adding a mixture of 50 g of crushed ice to which 15 ml of concentrated hydrochloric acid has been added. Stir the mixture until two layers appear, then transfer the mixture to a separatory funnel. Remove the lower aqueous layer and discard it. Wash the upper ether layer once with 50 ml of water. If the ether layer appears dark yellow or brown due to the presence of iodine, add 10 ml of a saturated solution of sodium bisulfite in water

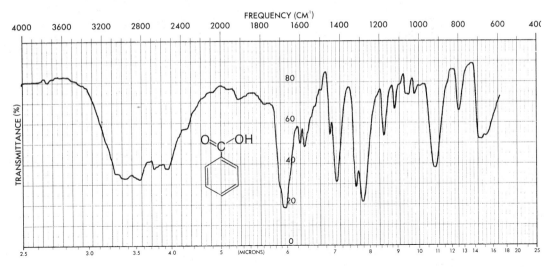

Infrared spectrum of benzoic acid, KBr

to the 50 ml of water used to wash the ether layer. Separate and discard the water layer. Extract the ether layer three times with successive 50 ml portions of 5% sodium hydroxide solution to remove the benzoic acid. Combine the basic extracts and discard the ether layer, which contains the by-product, biphenyl. Place the combined basic extracts in a 400 ml beaker and remove the ether by evaporation in the hood. Stir the mixture until the ether dissolved in the alkaline solution has been completely removed. Ether is soluble in water to the extent of 7%. Unless the ether is removed before the benzoic acid is precipitated, the product may appear as a waxy solid, rather than as crystals. Cool the alkaline solution and precipitate the benzoic acid by the addition of 20 ml of concentrated hydrochloric acid. Collect the precipitated benzoic acid on a Büchner funnel by vacuum filtration. Wash the collected crystals with several small portions of cold water and allow the crystals to dry. Recrystallize a 0.5 g sample of benzoic acid from water. Determine the melting point of this recrystallized material (literature value = 122°). At the option of the instructor, determine the infrared spectrum, as a KBr mull, of the purified benzoic acid (Technique 17, Part A). Weigh the product and submit both the crude and recrystallized products in properly labeled vials to the instructor.

QUESTIONS

1. Benzene is often produced as a side product during Grignard reactions using phenylmagnesium bromide. How can its formation be explained? Give a balanced equation for its formation.

2. Why is ligroin or petroleum ether used to recrystallize triphenylmethanol, rather than a solvent like ethanol?

3. Write a balanced equation for the reaction of benzoic acid with hydroxide ion. Why is it necessary to extract the ether layer with sodium hydroxide?

4. Interpret the major peaks in the infrared spectrum of either triphenylmethanol or benzoic acid, depending upon the procedure used in this experiment.

Experiment **29**
NITROBENZENE

Aromatic Substitution
Simple Distillation

The nitration of benzene to prepare nitrobenzene is an example of an electrophilic aromatic substitution reaction where a proton of the aromatic ring is replaced by a nitro group.

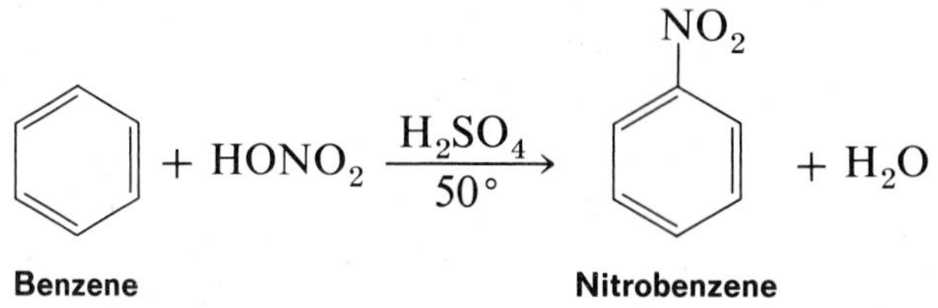

Benzene   **Nitrobenzene**

Many such aromatic substitution reactions are known to occur when an aromatic substrate is allowed to react with a suitable electrophilic reagent, and many other groups besides nitro may be introduced into the ring.

You may recall that alkenes (which are electron rich due to an excess of electrons in the π system) can react with an electrophilic reagent. An intermediate is formed which is electron deficient. It reacts with the nucleophile to complete the reaction. The overall sequence is called **electrophilic addition.** Addition of HX to cyclohexene is an example.

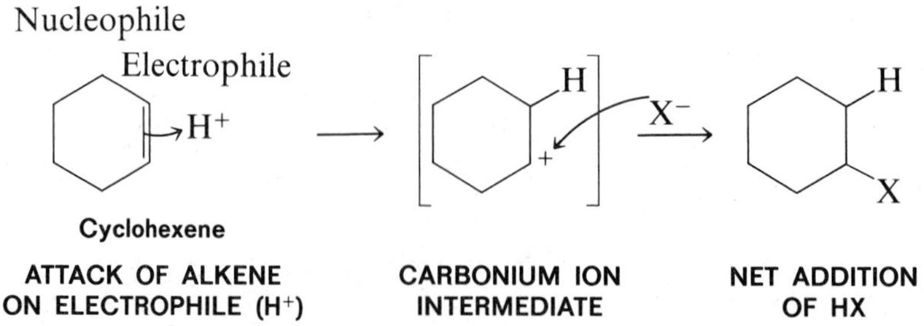

Nucleophile
Electrophile

Cyclohexene

| ATTACK OF ALKENE | CARBONIUM ION | NET ADDITION |
| ON ELECTROPHILE (H⁺) | INTERMEDIATE | OF HX |

Aromatic compounds are not fundamentally different from cyclohexene. They can also react with electrophiles. However, due to resonance in the ring, the electrons of the π system are generally less available for addition reactions since this would mean the loss of the stabilization that resonance provides. In practice this means that aromatic compounds will only react with **powerfully electrophilic reagents** and usually at somewhat elevated temperatures.

Benzene, for example, can be nitrated at 50° with a mixture of concentrated nitric and sulfuric acids; the electrophile is NO_2^+ (nitronium ion), the formation of which is promoted by action of the concentrated sulfuric acid on nitric acid:

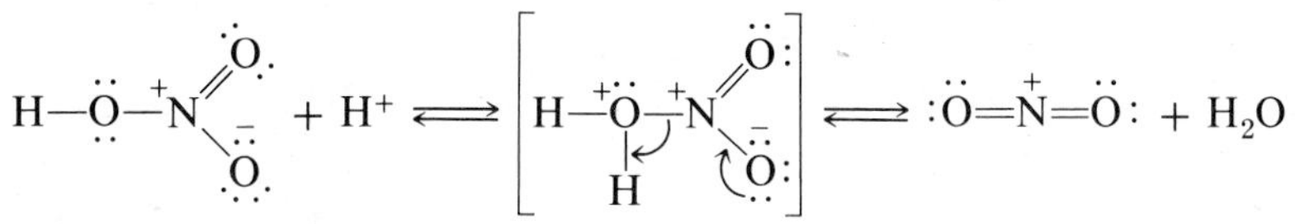

Nitric Acid   **Nitronium Ion**

The nitronium ion thus formed is sufficiently electrophilic to add to the benzene ring, **temporarily** interrupting ring resonance.

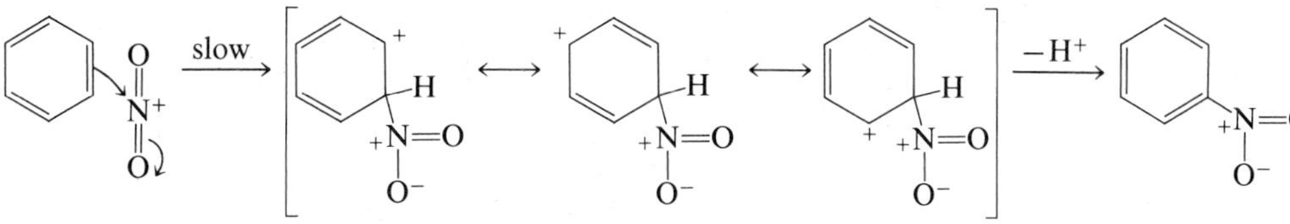

The intermediate first formed is somewhat stabilized by resonance and does not rapidly undergo reaction with a nucleophile as in the case of the unstabilized carbonium ion formed from cyclohexene plus an electrophile. In fact, aromaticity can be restored to the ring if **elimination** occurs instead. (Recall that elimination is often a reaction of carbonium ions.) Removal of a proton, probably by HSO_4^-, from the sp^3 ring carbon **restores the aromatic system** and yields a net **substitution** wherein a hydrogen has been replaced by a nitro group. Many similar reactions are known, and they are called **electrophilic aromatic substitution reactions.**

In principle, every hydrogen of the ring could subsequently be replaced by a nitro group. For reasons beyond the scope of our discussion here, this does not occur. Each electron-withdrawing nitro group introduced **deactivates** or makes the ring less susceptible to further substitution by rendering the aromatic system less nucleophilic. The formation of m-dinitrobenzene from nitrobenzene occurs at a rate about 10^4 times slower than that of the formation of mononitrobenzene from benzene. Nevertheless, small amounts of dinitrobenzene and trinitrobenzene are formed as side products. At temperatures higher than 50°, proportionately more of these products will be formed. Since these two compounds are explosives, it is important to maintain proper temperature control. Also, when distilling the final product it is important not to distill to dryness since the higher boiling di- and tri-nitrated products may explode when excessively heated. In the reaction which follows we shall attempt to limit the formation of higher nitration products both by controlling the temperature and by limiting the amount of sulfuric acid used.

Water has a retarding effect on the nitration reaction since it interferes with the nitric-sulfuric acid equilibria that form the nitronium ions. The smaller the amount of water present, the more active is the nitrating mixture. Also, the reactivity of the nitrating mixture can be controlled by varying the amount of sulfuric acid used. This acid must protonate nitric acid, which is a **weak** base, and the more acid available, the more of the protonated species (and hence NO_2^+) in the solution. Water interferes since it is a stronger base than H_2SO_4 or HNO_3.

SPECIAL INSTRUCTIONS

Before starting this experiment you should have read Techniques 5 and 6. It is important that the temperature of the reaction mixture be maintained below 60°. Nitrobenzene is toxic. Do not inhale the vapor. Avoid contact with the skin.

PROCEDURE

Before starting the nitration reaction, prepare an ice bath of sufficient size for cooling the reaction vessel.

To 60 ml (d = 1.84 g/ml) of concentrated sulfuric acid in a dry 500 ml Erlenmeyer flask, cautiously add, with swirling, 50 ml (d = 1.41 g/ml) of concentrated nitric acid. Place a thermometer in the flask and let the mixture cool to near room temperature.

While the nitrating mixture is cooling, obtain 44 ml (d = 0.88 g/ml) of benzene. Add about 5 ml of the benzene to the mixed acids with swirling, and note the temperature rise. It is necessary here and throughout the experiment to swirl the contents of the flask continuously because benzene is only slightly soluble in the mixed acids. If the temperature should exceed 60°, immerse the flask immediately in the ice bath (continue swirling the contents). Add the rest of the benzene, in about 5 ml portions, at such a rate that the temperature remains between 50° and 60°, using the ice bath only when necessary. After all the benzene has been added and the strongly exothermic reaction has subsided, maintain a temperature of 50 to 60° for an additional 15 minutes with a steam or hot water bath (remember to swirl the mixture).

Cool the reaction mixture in the ice bath and then transfer it to a separatory funnel. Remove the lower phase (mixed acids), and carefully pour it down the drain, to be followed by a large amount of water. Wash the organic phase successively with 25 ml portions of water, 5% aqueous sodium hydroxide, and water (Technique 5, Section 5.4). The density of nitrobenzene is 1.20 g/ml. Be certain that the correct layer is saved after each of the washing steps. Transfer the washed nitrobenzene to a 125 ml Erlenmeyer flask containing about 5 g of anhydrous calcium chloride. Swirl the mixture until the cloudiness has disappeared (Technique 5, Section 5.6). It may be necessary to warm the mixture gently on the steam bath for a few minutes. Transfer the dry material to a distilling flask (use decantation rather than filtration, if possible).

Assemble a simple distillation apparatus, but do not pass water through the condenser (Technique 6, Section 6.4). Be sure to grease the joints in the apparatus lightly or they may freeze. Distill the nitrobenzene at atmospheric pressure, using a flame, and collect the material which boils over a 10° range. According to the literature, the boiling point for nitrobenzene is 210° at 760 mm. Be sure to distill the material slowly, but with evenly distributed heating. This heating procedure can be accomplished by holding the Bunsen burner in the hand and slowly passing it back and forth under the distilling flask. CAUTION: Do not distill to the point of dryness. Determine the weight of the product and calculate the percent yield. At the option of the instructor, determine the infrared spectrum of the product and compare it to the one given below. Submit the nitrobenzene to the instructor in a labeled vial, unless it is to be used in the next experiment to prepare aniline.

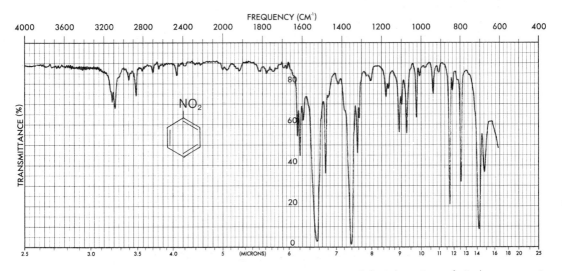

Infrared spectrum of nitrobenzene, neat

QUESTIONS

1. Why is *m*-dinitrobenzene produced as a by-product rather than the *o*- or *p*-dinitrobenzenes?

2. Why does the amount of *m*-dinitrobenzene increase at higher temperatures?

3. Interpret the infrared spectrum of nitrobenzene.

4. The boiling point that one observes for nitrobenzene is often low. Why? (See Technique 6, Section 6.8.) Estimate a correction and add it to the observed value.

5. Suppose that one desired to distill the nitrobenzene (bp 210°) under reduced pressure at 30 mm. What would be the estimated boiling point at 30 mm pressure?

6. Indicate the products formed on nitration of each of the following compounds: toluene, chlorobenzene, and benzoic acid.

Experiment **30**

ANILINE

Reduction of a Nitro Group
Steam Distillation
Simple Distillation

Amino and hydroxyl groups cannot be introduced into an aromatic ring system in any direct fashion (at least when given the typical substrate). They are usually formed **via** another functional group. The conversion of a nitro group to an amino group is typical of this approach. Aniline may be prepared in this way.

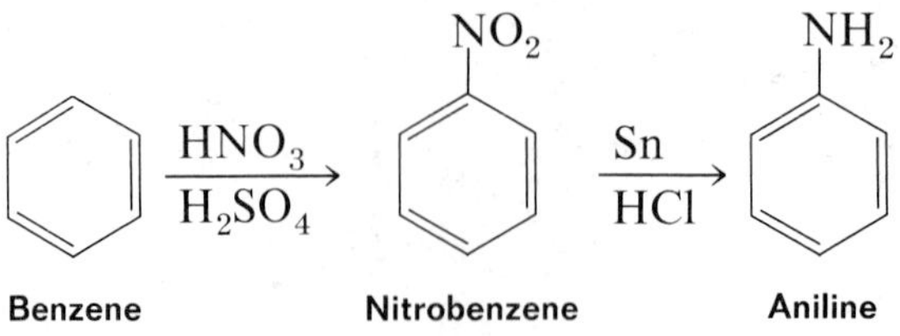

$$\text{Benzene} \xrightarrow[\text{H}_2\text{SO}_4]{\text{HNO}_3} \text{Nitrobenzene} \xrightarrow[\text{HCl}]{\text{Sn}} \text{Aniline}$$

In this experiment the nitrobenzene which was prepared in Experiment 29 may be used to carry out the second step of the above reaction sequence.

The conversion of a nitro group to an amino group is accomplished through the use of reducing agents. Catalytic reduction can be used, although an aromatic nitro group is most easily reduced by use of a moderately active metal and an acid. Tin and hydrochloric acid are the most commonly used components for reducing nitro groups, although iron is also used.

$$2 \;\langle\!\!\!\;\rangle\!\!-\!NO_2 + 3\,Sn + 14\,HCl \longrightarrow 2 \;\langle\!\!\!\;\rangle\!\!-\!NH_3{}^+\,Cl^- + 3\,SnCl_4 + 4\,H_2O$$

Tin, however, is relatively cheap, provides a moderate reaction rate, and does not lead to difficulties that a metal like copper (which forms amine complexes) would generate. The oxidation of tin metal provides a large enough redox potential to accomplish the reduction of the nitro group.

$$R\!-\!NO_2 + 6\,H^+ + 6\,e^- \longrightarrow R\!-\!NH_2 + 2\,H_2O$$

The exact stepwise mechanism for this process is not known. The oxygen atoms of the nitro group, however, end up in water molecules, and clearly the hydrochloric acid is the source of hydrogen atoms which replace the oxygen atoms.

Since the reaction is carried out in hydrochloric acid solution, the product aniline is rapidly converted to its hydrochloride salt.

$$\langle\!\!\!\;\rangle\!\!-\!NH_2 + HCl \longrightarrow \langle\!\!\!\;\rangle\!\!-\!NH_3{}^+\,Cl^-$$

Aniline **Aniline Hydrochloride**

To generate "free" aniline (i.e., aniline not converted to the hydrochloride salt), aqueous base (NaOH) must be used when the reduction reaction has been completed.

$$\langle\!\!\!\;\rangle\!\!-\!NH_3{}^+\,Cl^- + NaOH \longrightarrow \langle\!\!\!\;\rangle\!\!-\!NH_2 + H_2O + NaCl$$

Along with free aniline one also obtains large amounts of tin hydroxides. These hydroxides form an emulsion with aniline. Thus, one cannot extract the mixture (mess!) with organic solvents. Instead, the aniline can be isolated from the reaction mixture by steam distillation. The aniline co-distills with water, while the inorganic tin hydroxides remain behind in the distilling flask.

Aniline is a very useful compound. For example, it may be converted to sulfanilic acid (Experiment 31), *p*-nitroaniline (Experiment 32) **via** acetanilide (Experiment 2), or to sulfa drugs (Experiment 41) **via** acetanilide (Experiment 2).

SPECIAL INSTRUCTIONS

It is necessary to have read Techniques 5, 6, and 8 (especially Section 8.3A) before starting this experiment. You should be prepared to carry this reaction sequence through the steam distillation stage by the end of one laboratory period. This is a suitable stopping point. Alternatively, the methylene chloride extract may be stored over anhydrous sodium sulfate. Your instructor may ask you to distill the aniline by vacuum distillation, in which case it is best for two or more students to combine their crude aniline. Aniline is somewhat toxic and should not be spilled on the skin. In addition, it should not be inhaled.

PROCEDURE

Before starting, prepare an ice bath for use if the reaction becomes too vigorous. Combine 15 g of nitrobenzene (Experiment 29) and 30 g of granulated tin in a 500 ml three neck, round bottom flask fitted with a reflux condenser. Stopper the two unused necks.

Measure 70 ml (d = 1.18 g/ml, 37.3% HCl) of concentrated hydrochloric acid into a 125 ml Erlenmeyer flask. Add about 10 ml of the acid through the reflux condenser and swirl the reaction mixture. A noticeable reaction should be observed, and the reaction mixture may begin to boil. Add the rest of the acid in small portions (5 ml) over about a 30 min period. Continue to swirl the contents of the reaction flask during that time. If the reaction becomes too vigorous, it can be moderated by the use of the ice bath. It may also be necessary to add smaller portions of acid at less frequent intervals. After all of the acid has been added and all noticeable reaction has ceased, heat the mixture on a steam bath for 30 minutes. Swirl it occasionally.

Cool the mixture and add 25 ml of water to it. Slowly add, with stirring, 50 ml of aqueous sodium hydroxide (75 g/100 ml) to the solution. The resulting solution should be basic to litmus paper. If it is not basic, add more of the sodium hydroxide solution.

Arrange an apparatus for steam distillation using live steam (Technique 8, Section 8.3A, Figure 8–2). Be sure to grease the joints lightly so that they will not freeze together in the presence of a base. Pass live steam through the hot liquid but do not distill it at too rapid a rate. Check to see that the condenser is actually cooling the distillate by touching the lower part of the condenser. It should not be hot. An ice bath should be used to cool the receiving flask. If the flow of steam is stopped at any time, the screw clamp on the steam trap must be opened first; otherwise the condensing steam will create a vacuum and draw the distillation mixture out of the distilling flask. Distill the mixture until the distillate loses its milky appearance. Since aniline has moderate solubility in water, the loss of a milky appearance does not necessarily mean that all the aniline has distilled. Hence, continue the distillation and collect an extra 50 ml of distillate. If the distilling flask begins to fill with water during the steam distillation, heat the flask with a Bunsen burner to distill some of the water.

Estimate the volume of liquid and add about 20 g NaCl/100 ml of distillate. This procedure nearly saturates the distillate with sodium chloride and forces the separation of the water and aniline layers. Extract the mixture with two successive 30 ml portions of methylene chloride. Gently invert the separatory funnel, rather than shaking it vigorously, to avoid the formation of an emulsion. Combine the methylene chloride extracts in an Erlenmeyer flask, and add 5 g of anhydrous sodium sulfate to dry the solution. Stopper the flask and swirl it until the solution is clear. Decant the solution away from the anhydrous sodium sulfate and remove the methylene chloride by distillation. Use a simple distilling apparatus,[1] add boiling stones, and heat with a steam bath (Technique 6, Figure 6–6). Methylene chloride boils at 41°. Pour the methylene chloride distillate in a container provided for it. After all of the methylene chloride has been removed, continue the distillation using a flame as a source of heat. When the temperature reaches 150°, stop the water flow in the condenser, drain it by disconnecting the water inlet, and continue the distillation. Collect the product that boils over a 10° range. The literature value for the boiling point of aniline is 184° at 760 mm. Determine the weight of the product and calculate the percent yield. At the option of the instructor, determine the infrared spectrum of the aniline sample and compare it to the one given below. Submit the aniline to the instructor in a labeled vial.

QUESTIONS

1. Derive the balanced equation, shown in the introduction to this experiment, for the reduction of nitrobenzene to aniline hydrochloride with tin and hydrochloric acid. Use the half-reaction method.

---

[1]Alternatively, a vacuum distillation may be conducted at the option of the instructor (see Technique 6, Section 6.6 and 6.7, Figure 6–7). Two or more students should combine their crude aniline.

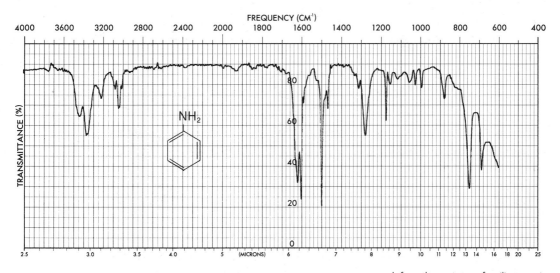

Infrared spectrum of aniline, neat

2. What is the limiting reagent in this experiment? How many moles of each of the other reagents are required to react with this limiting reagent? How many moles of each are in excess?

3. Explain why aniline can be steam distilled, but tin hydroxide cannot.

4. What would happen if one accidentally forgot to make the solution basic with sodium hydroxide before steam distilling the mixture?

5. Why does sodium chloride cause the aniline to separate from the steam distillate?

6. This text describes the preparation of sulfanilic acid, acetanilide, p-nitroaniline, and sulfanilamide (a sulfa drug). Give the reactions which are used to prepare each of these compounds, starting from benzene. Once an intermediate has been prepared, it need not be prepared again. You will also need to use acetic anhydride and inorganic compounds.

7. Interpret the infrared spectrum of aniline.

Experiment 31

SULFANILIC ACID

Aromatic Substitution
Sulfonation
Sulfonamides

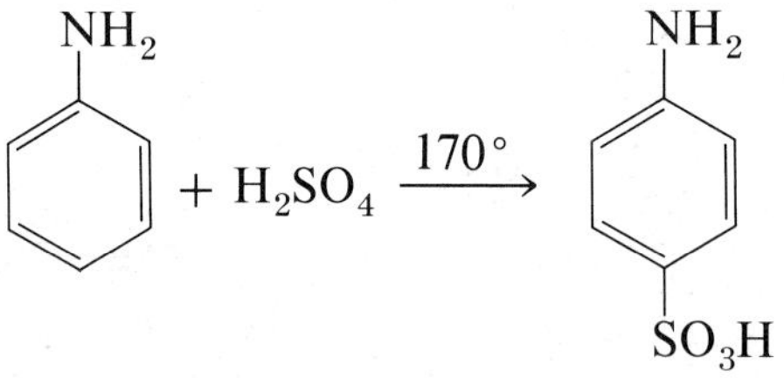

$$\text{Aniline} + H_2SO_4 \xrightarrow{170°} \text{Sulfanilic Acid}$$

Aniline   Sulfanilic Acid

Sulfanilic acid (4-aminobenzenesulfonic acid) will be prepared by the sulfonation of aniline (Experiment 30). The amino group is normally an ortho, para directing group owing to its electron-releasing resonance with the aromatic ring. In acid solutions, however, the amino group is often protonated, giving rise to an anilinium ion. Since the protonated amino group carries a positive formal charge ($-NH_3^+$), and since the resonance possibilities become lost on protonation, the amino group often becomes a meta directing group in substitution reactions that require acidic reaction conditions. Thus, while bromination of aniline gives bromine substitution at the ortho and para ring positions, nitration, by acid solution, results in a nitro group predominantly at the meta position.

In this experiment, three factors lead to predominantly para substitution. First, under the high temperature conditions of this reaction, aniline is not simply protonated; it is converted to its sulfonamide derivative.

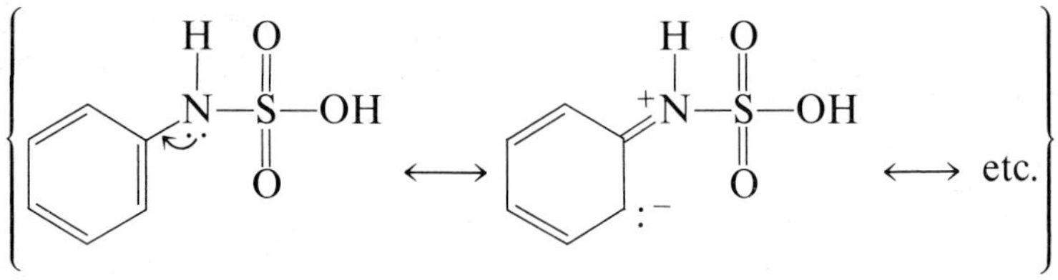

Aniline Sulfonamide

The sulfonamide is an **amide** of sulfuric acid and, thus, the nitrogen atom is rendered less basic than in aniline. It does not, therefore, become protonated to a large extent. Secondly, since the amino nitrogen still retains a pair of electrons after formation of the sulfonamide, these electrons can participate in resonance with the aromatic ring ($+R$ effect). This resonance renders the sulfonamide group an ortho, para directing group. Finally, as the sulfonamide group is quite large, the amount of ortho substitution is much diminished due to steric hindrance. Sulfonation of aniline (aniline sulfonamide) at 170°, therefore, proceeds mainly at the para position.

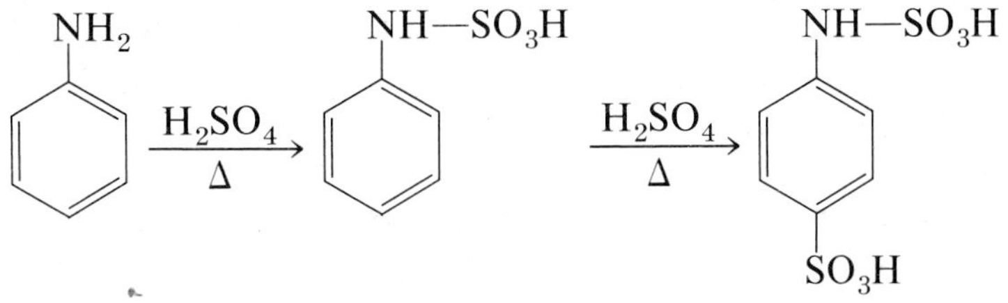

At the end of the reaction, the sulfonyl group is removed by hydrolysis, to give sulfanilic acid.

Since sulfanilic acid is a bifunctional compound with both an acidic and a basic functional group, it commonly exists in the form of a doubly charged ion, a zwitterion. In fact, depending on the pH of the solution, sulfanilic acid may adopt one of several different charge forms:

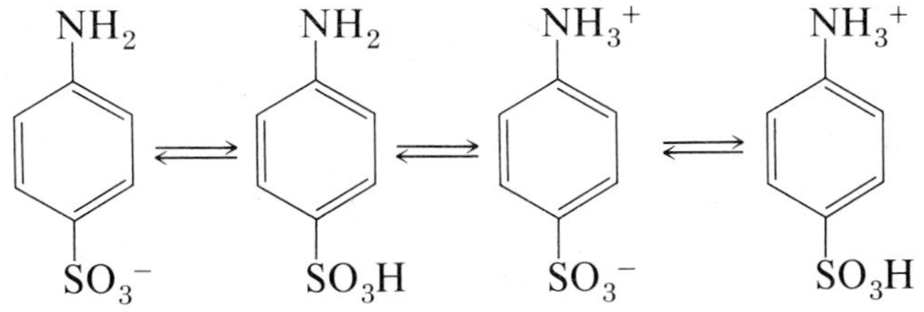

The product isolated in this experiment is the internal salt (zwitterion). Since it is precipitated from an aqueous solution, the salt is hydrated. The waters of hydration can be removed by drying the compound in an oven at 100°. Because sulfanilic acid is a salt, it does not give a definite melting point. Rather, it **chars** at a temperature between 280° and 290° (mp 288° d).

SPECIAL INSTRUCTIONS

Aniline is somewhat toxic and should not be spilled on the skin. In addition, it should not be inhaled. Concentrated sulfuric acid is corrosive and should be handled with care, especially when it is hot. The reaction itself is quite sensitive to temperature. If the reaction mixture is heated much above 180°, a wonderful, black, viscous tar will be the sole product. Before proceeding, you should have read Techniques 1 through 4.

PROCEDURE

Place 10 g of aniline in a 500 ml three neck flask fitted with a reflux condenser and a thermometer. Using the thermometer adapter, the thermometer should be placed in one of the side necks and positioned so that it does not quite touch the bottom of the flask. The unused neck of the flask should be stoppered. Slowly and cautiously, add 17.4 ml of concentrated sulfuric acid (d = 1.84 g/ml) to the flask through the unused neck. Replace the stopper and heat the mixture to 170° for 2 to 2.5 hours. A heating mantle or an oil bath should be used to heat the mixture. The temperature should be controlled rather carefully.

Allow the mixture to cool to about 50°. Below this temperature, the reaction mixture may become viscous and solidify, making it rather difficult to remove from the flask. If the mixture solidifies, it should be warmed on a steam bath until it is liquid again. The presence of sulfuric acid requires careful pouring of the reaction mixture into a beaker containing 100 ml of ice water. Using a glass rod, stir the resulting viscous mass vigorously. The sulfanilic acid should separate as a gray crystalline solid. Pour some of the ice water from the beaker into the reaction flask and, using a small bent spatula, scrape the sides of the flask and wash them with the water to free any material that was not transferred to the beaker. After another short period of stirring, collect the product by vacuum filtration and rinse it with a little cold water.

Recrystallize the crude sulfanilic acid from boiling water. The solubilities of sulfanilic acid in water are: 6.67 g/100 ml at 100° and 1.08 g/100 ml at 20°. Decolorization will probably be necessary. Collect the crystals by vacuum filtration using a Büchner funnel and allow them to dry in the air. Since sulfanilic acid is still quite soluble in water at 20°, a good deal of material remains in the mother liquors from the crystallization. These should be concentrated in volume, decolorized, and crystallized to provide a second crop of crystals. Do not combine this second batch of crystals with the first, but weigh them separately, when dry, and use the combined weight of both batches of crystals to calculate the yield. Compare the color of the second crop of material to that of the first crop. Determine the melting points of both crops of material and compare them. If the product is to be used to prepare methyl orange, proceed to Experiment 33; if not, submit both samples, in labeled vials, to the instructor.

QUESTIONS

1. Draw a mechanism for the formation of the sulfonamide from aniline and sulfuric acid.

2. Draw a mechanism for the sulfonation of aniline sulfonamide.

3. Draw a mechanism for the hydrolysis of the final intermediate in the reaction sequence.

4. Explain why aniline sulfonamide undergoes aromatic substitution only once (monosubstitution), while aniline often undergoes substitution several times.

5. Compare the features of this reaction to those of the one which follows (Experiment 32, *p*-Nitroaniline). What are the similarities?

6. Although sulfanilic acid is quite soluble in water, it is not appreciably soluble in benzene or ether. Explain.

7. How might you increase the yield of material recovered from the first crystallization of sulfanilic acid without resorting to the collection of a second crop of crystals?

Experiment 32

p-NITROANILINE

Electrophilic Aromatic Substitution
Hydrolysis of an Amide
Protective Groups
Crystallization
Thin Layer Chromatography

In this experiment, we shall convert acetanilide to *p*-nitroaniline. The sequence of reactions, beginning with aniline, is shown below. The conversion of aniline to acetanilide, the first step, was performed in Experiment 2.

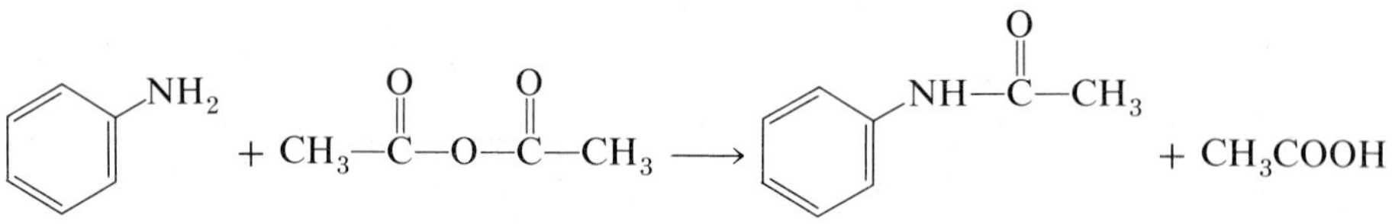

Aniline + $CH_3-\overset{O}{\overset{\|}{C}}-O-\overset{O}{\overset{\|}{C}}-CH_3$ → Acetanilide + CH_3COOH

Acetanilide + HNO_3 $\xrightarrow{H_2SO_4}$ *p*-Nitroacetanilide + ortho isomer + H_2O

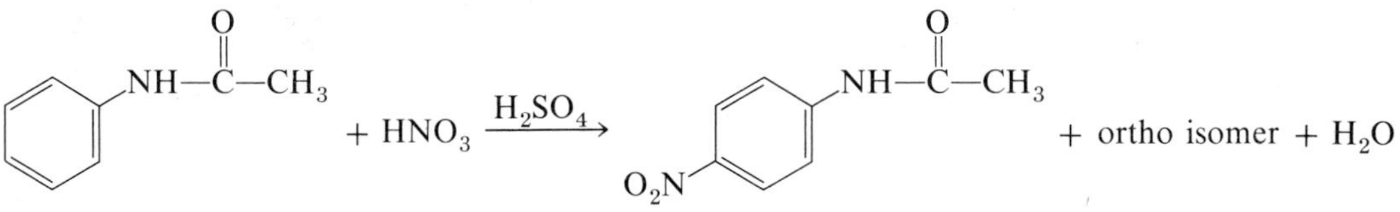

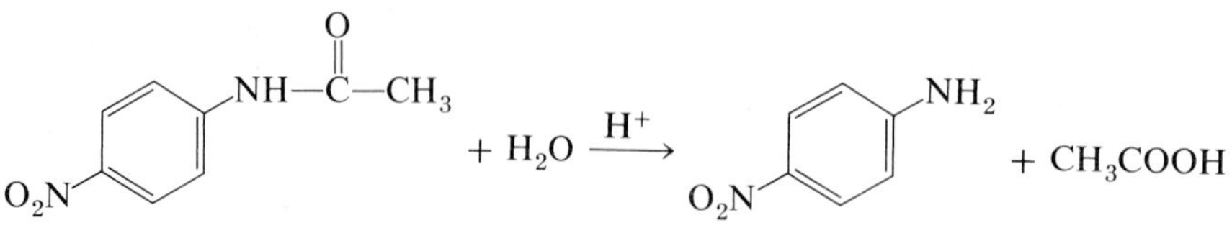

+ H_2O $\xrightarrow{H^+}$ *p*-Nitroaniline + CH_3COOH

The mechanism for the nitration reaction is essentially identical to that given for the nitration of benzene in Experiment 29. The nitronium ion is directed to the positions ortho and para to the acetamido ($-NHCOCH_3$) group. This occurs because the resonance electron-releasing effect of that group increases the electron density at those positions, helping to stabilize the intermediates which are formed.

Substitution para to the acetamido group is favored over substitution ortho to that group, because the great bulk of the acetamido group shields the ortho positions from approach by reagents. This steric hindrance makes ortho substitution much less likely than para substitution, where the bulk of the acetamido group has no influence. The ortho substitution product is formed in small quantities in this reaction, but the isolation and purification of this product will not be attempted in this experiment.

If one wished to convert aniline to *p*-nitroaniline, it would seem reasonable, at first glance, to carry out the nitration reaction directly on aniline, without passing through the amide intermediates. The amino group, due to its strong resonance electron-releasing effect would, in theory, direct substitution to the positions ortho and para to itself. In the usual reaction, the free amino group would also activate the benzene ring so greatly toward electrophilic aromatic substitution that substitution would occur at all three ortho and para positions. It would be difficult to obtain the monosubstituted aniline from this type of reaction. In the nitration of aniline there is the possibility that products with two or three nitro groups would be formed. However, since most electrophilic aromatic substitution reactions occur in acidic media, the basic amino group is converted to the cationic ammonium group ($-NH_3^+$). This latter group is electron-withdrawing, meta directing, and deactivating toward further substitution. Since a large proportion of the aniline molecules in acidic solution are protonated, there is a slow formation of the meta substituted aniline, along with some ortho and para substitution. These competing reactions are illustrated for the case of nitration:

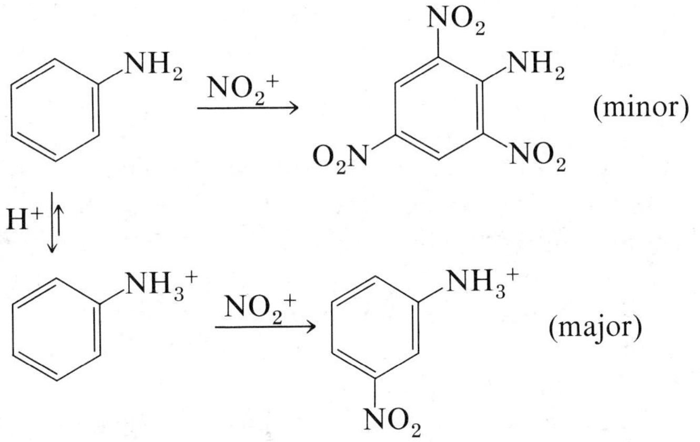

In the case of nitration of anilines, there is the additional complication that the anilines are susceptible to oxidation. This oxidation caused by the nitric acid, which is a powerful oxidizing agent, further reduces the yields of desired products in this reaction. It can be seen that it is very difficult to obtain reasonable yields of *p*-nitroaniline from the direct nitration of aniline.

For this reason, the amino group will initially be converted to the acetamido group with acetic anhydride. The acetyl group, which is thus attached to the amino group, serves as a **protective group.** The acetyl group reduces the reactivity of the amino group with acids, since the nitrogen, being in the form of an amide, is no longer basic. The acetyl group also protects the amino function against oxidation. The acetyl group also reduces the electron-releasing resonance of the amino group. As a result, the substituent no longer activates the benzene ring toward multiple substitution as strongly as it did before the acetyl group was attached. Monosubstitution is now possible, so products with only one nitro group may now be isolated from the reaction mixture.

A protective group must fulfill three important requirements. First, it must be easy to attach to the molecule. The reaction to install the protective group must proceed in high yield, in order not to waste starting material, and it must not require reaction conditions which would result in the decomposition of the molecule being protected. Secondly, the protective group must be stable under the conditions of the reaction in which it is expected to perform its protective function. Obviously, if the protective group were to come off the molecule in the course of a reaction, it would no longer be able to protect the functional group of interest. Thirdly, the protective group must be one which may be removed easily once it has fulfilled its protective function. This last requirement is very important. Obviously, if one wished to restore the original functional group at the end of a reaction sequence, a protective group which could not be removed would not fulfill its role satisfactorily. In this experiment, the acetyl group is an example of a useful protective group. The original acetylation of the amino group proceeds in high yield and under mild reaction conditions. The acetyl group is stable under the nitration conditions. Finally, it may be removed easily by hydrolysis to regenerate the original amino group.

SPECIAL INSTRUCTIONS

Before performing this experiment, you should read the introductory material in Experiments 2 and 29, as well as Techniques 5 and 11. Concentrated sulfuric and nitric acids, in combination, form a very hazardous and corrosive mixture. These acids should be poured together carefully, and this procedure should be carried out in a hood, since noxious vapors are produced. At the option of the instructor, you may begin with either a sample of acetanilide obtained from the stockroom or the sample of acetanilide which you prepared in Experiment 2. If the acetanilide is obtained from Experiment 2, be certain that it has been thoroughly dried before weighing it.

PROCEDURE

Place 3.0 g of acetanilide in a 125 ml Erlenmeyer flask. Add about 5 ml of concentrated sulfuric acid to the acetanilide. Dissolve most of the solid by swirling and stirring the mixture. Do not be concerned if a small amount of undissolved solid remains. It will dissolve in later stages of this procedure. Place the flask in an ice bath. Place 1.8 ml of concentrated nitric acid in a small flask, and add about 5 ml of concentrated sulfuric acid to it. CAUTION: This is a hazardous mixture. This mixing should be done carefully in a hood. Mix the acids thoroughly.

Using a disposable capillary pipet, add small portions of the mixed acids to the cooled sulfuric acid solution of acetanilide. After each addition of acids, swirl the mixture thoroughly in the ice bath. Do not allow the flask to become warm to the touch. After 20 minutes, including the time required for the addition of the nitric-sulfuric acid mixture, add 25 ml of an ice-water mixture to the reaction. A suspension of nitro-acetanilide isomers will result. Allow this mixture to stand for five minutes, with occasional stirring.

In order to hydrolyze the nitroacetanilides to the corresponding nitroanilines, the material in the flask will be heated, using the dilute sulfuric acid already present in the flask as the hydrolyzing medium. Add a boiling stone to the Erlenmeyer flask and heat the flask using a Bunsen burner. A wire gauze will serve to disperse the burner flame. Heat the mixture to a gentle boil and continue heating for 15 minutes. During this period the color of the mixture may darken, and the solid will dissolve. Cool the flask in an ice bath, and when it is cool, add 25 ml of concentrated aqueous ammonium hydroxide, in 5 or 6 portions, to the flask. This addition must be conducted in a hood because noxious fumes are evolved during the addition. Swirl the flask in the ice bath after each portion of ammonium hydroxide is added. The nitroaniline isomers will precipitate during this latter addition.

Collect the precipitated nitroanilines on a Büchner funnel using vacuum filtration. Wash the solid thoroughly with small portions of water. While continuing the vacuum, allow the solid to air-dry on the Büchner funnel for several minutes.

Scrape the solid material from the filter paper into a 25 × 10 mm test tube and add sufficient ethanol to dissolve the solid when the ethanol is boiling. Use the steam bath to heat the solution. When sufficient ethanol has been added to just dissolve the solid while the ethanol is boiling, allow the solution to cool. When the first crystals appear, place the test tube in an ice bath to bring the crystallization to completion. Filter the crystals of *p*-nitroaniline by vacuum filtration using a Hirsch funnel and a small filter flask. **Save the filtrate for a later tlc analysis.** Wash the crystals with a minimum amount of **cold** ethanol and allow them to dry by drawing air through them on the filter for a few minutes. **Save and dry a small sample of the crystals** for a later determination of the melting point and a tlc analysis.

Dissolve the crude *p*-nitroaniline in ethanol, using 15 ml of ethanol for each gram of *p*-nitroaniline. Warm the solution to dissolve the solid. Add about 1 g of activated charcoal to the solution and swirl it for a few minutes. Filter the charcoal from the solution by gravity, using a fluted filter paper. It probably will be necessary to repeat this gravity filtration an additional time. Concentrate the filtrate to about one-third of its original volume on the steam bath. Allow the solution to cool. When the first crystals appear, place the flask in an ice bath. After crystallization is complete, collect the crystals on a Hirsch funnel by vacuum filtration. Allow the crystals to air dry on the Hirsch funnel by continuing to draw a vacuum on them. When the crystals are dry, weigh them. Determine the melting points of the two samples of crystals obtained before and after this final crystallization (literature value, mp = 149°). **Save a small sample of this purified *p*-nitroaniline for tlc analysis.** A labeled vial of the product should be submitted to the instructor.

For the thin layer chromatographic analysis, the three samples (filtrate from first crystallization, crude *p*-nitroaniline, and purified product) are each dissolved in a few drops of ethanol. Each sample is spotted on a single tlc plate which has been prepared from a microscope slide using silica gel G (Technique 11, Sections 11.2A, 11.3, and 11.4). Using chloroform as the solvent, develop the plate and compare the pattern of spots obtained for each sample. The materials being studied are colored, so they should be visible. However, it may be necessary to intensify the colors of the spots by briefly exposing the tlc plate to iodine vapors (Technique 11, Section 11.6).

The report should contain the melting points of the two samples of *p*-nitroaniline with a discussion of the differences observed in melting point behavior. The percentage yield should also be reported. The report should also include a discussion of the tlc results obtained with the three samples analyzed, with particular emphasis on the differences between the pattern of spots for each sample.

QUESTIONS

1. Explain why the acetamido group is an ortho, para directing group. Why should it be less effective in activating the aromatic ring toward further substitution than an amino group?

2. Outline the mechanism of the acid-catalyzed hydrolysis of *p*-nitroacetanilide to yield *p*-nitroaniline.

3. *o*-Nitroaniline is more soluble in ethanol than *p*-nitroaniline. Propose a scheme by which a pure sample of *o*-nitroaniline might be obtained from this reaction.

4. If you were to use column chromatography to separate *o*-nitroaniline from the filtrate, explain how you would do this. What adsorbent and solvent(s) would you use? Which compound would elute first?

Synthetic Dyes

The practice of using dyes is an ancient art. Substantial evidence exists that plant dyestuffs were known long before man began to keep written history. Prior to this century, practically all dyes were obtained from natural plant or animal sources. Dyeing was a complicated and secret art passed from one generation to the next. Dyes were extracted from plants mainly by macerating the roots, leaves, or berries in water. The extract was often boiled and then strained before use. In some cases, it was necessary to make the extraction mixture acidic or basic before the dye could be liberated from the plant tissues. Applying the dyes to cloth was also a complicated process. **Mordants** were used to fix the dye to the cloth, or even to modify its color (see the essay preceding Experiment 34).

Madder is one of the oldest known dyes. Alexander the Great was reputed to have used the dye to trick the Persians into overconfidence during a major battle. Using madder, a root bearing a brilliant red dye, he simulated blood stains on the tunics of his soldiers. The Persians, seeing the apparently incapacitated Greek army, became overconfident and, much to their surprise, were overwhelmingly defeated. Through modern chemical analysis, we now know the structure of the dye found in madder root. It is called **alizarin** (see structures) and is very similar in structure to another ancient dye, **henna,** which has been responsible for a long line of synthetic redheads. Madder is obtained from the plant **Rubia tinctorum.** Henna is a dye preparation prepared from the leaves of the Indian henna plant **(Lawsonia alba)** and an extract of **Acacia catechu.**

Indigo is another plant dyestuff with a long history. This dye, obtained from the plant **Indigofera tinctoria,** has been known in Asia for more than 4000 years. To produce indigo by the ancient process, the leaves of the indigo plant are cut and allowed to ferment in water. During the fermentation process, **indican** (see chart) is extracted into the solution, and the attached glucose molecule is split off to produce **indoxyl.** The fermented mixture is transferred to large open vats where the liquid is beaten with bambo sticks. During this process, the indoxyl is air-oxidized to indigo. Indigo, a strong blue dye, is insoluble in water and precipitates. Today, indigo is made synthetically, and its principal use is in the dyeing of denim to produce "blue jeans" material.

Many plants yield dyestuffs that will dye wool or silk, but there are few of these that will dye cotton well. Most will not dye synthetic fibers like polyester or rayon. In addition, the natural dyes, with a few exceptions, do not cover a wide range of colors, nor do they yield "brilliant" colors. Even though some people prefer the softness of the "homespun" colors from natural dyes, the **synthetic dyes,** which give rise to deep, brilliant colors, are much preferred today. Also, synthetic

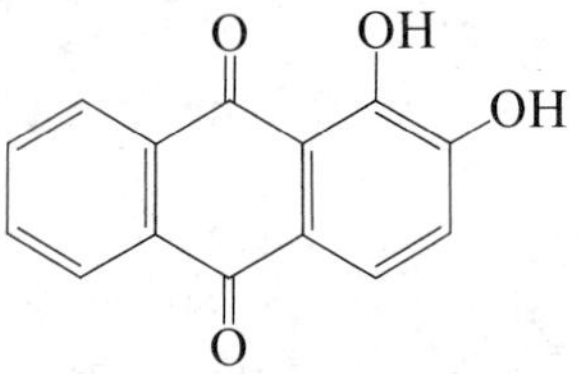

Alizarin 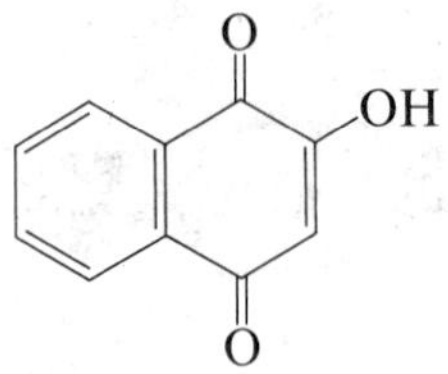   Henna

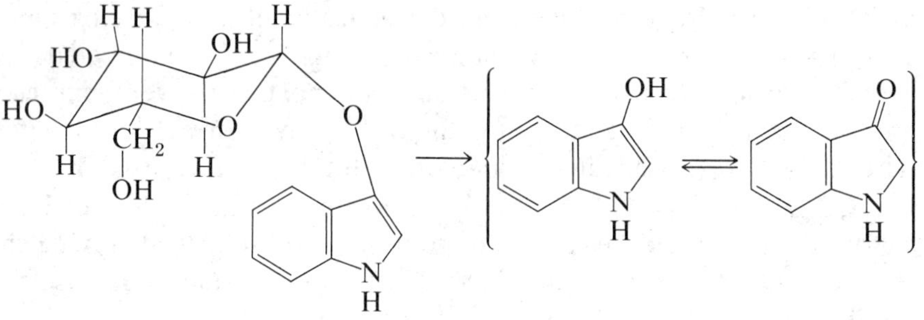

Indican   Indoxyl

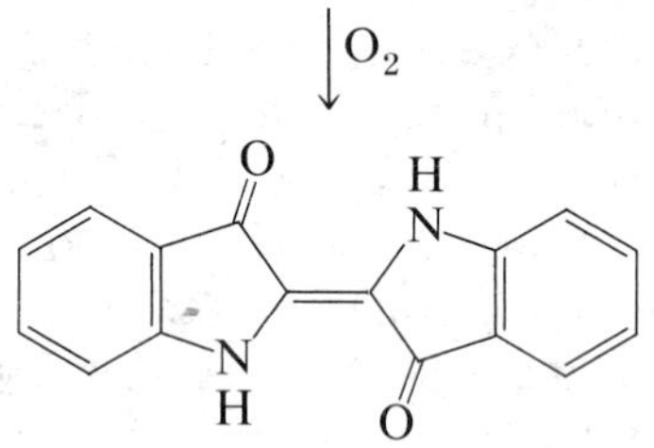

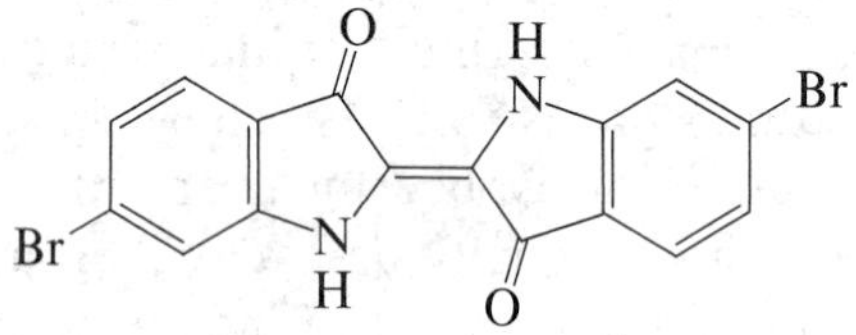

Tyrian Purple   Indigo

dyes that will dye the popular synthetic fibers can now be manufactured. Thus, today we have available an almost infinite variety of colors, as well as dyes to dye any type of fabric.

Before 1856, all dyes in use were obtained from natural sources. However, an accidental discovery by W. H. Perkin, an English chemist, started the development of a huge synthetic dye industry, mostly in England and Germany. Perkin, then only age 18, was trying to synthesize quinine. Structural organic chemistry was not very well developed at that time, and the chief guide to the structure of a compound was its molecular formula. Based on formulas, Perkin thought that it might be possible to synthesize quinine by the oxidation of allyltoluidine:

$$2\ C_{10}H_{13}N + 3[O] \longrightarrow C_{20}H_{24}N_2O_2 + H_2O$$

Allyltoluidine   Quinine

He proceeded to make allyltoluidine and to oxidize it with potassium dichromate. Of course, the reaction was unsuccessful because allyltoluidine bore no structural relationship to quinine. He obtained no quinine, but he did recover a reddish-brown precipitate with properties that interested him. He decided to try the reaction with a simpler base, aniline. On treating aniline sulfate with potassium dichromate, he obtained a black precipitate which could be extracted with ethanol to give a beautiful purple solution. This purple solution subsequently proved to be a good dye for fabrics. After receiving favorable comments from dyers, Perkin resigned his post at the Royal College and went on to found the British coal tar dye industry. He became a very successful industrialist and retired at age 36 (!) to devote full time to research. The dye he synthesized became known as **mauve**. The structure of mauve was not proved until much later. From the structure (see figures) it is clear that the aniline which Perkin used was not pure and that it contained the *o*-, *m*-, and *p*-toluidines as well.

Mauve was the first synthetic dye, but soon (1859) the triphenylmethyl dyes, pararosaniline, malachite green, and crystal violet (see figures) were discovered in France. These dyes were produced by treating mixtures of aniline and/or the toluidines with nitrobenzene, an oxidizing agent, and in a second step, with concentrated hydrochloric acid. The triphenylmethyl dyes were soon joined by **synthetic** alizarin (Lieberman, 1868), **synthetic** indigo (Baeyer, 1879), and the azo dyes (Griess, 1862). The azo dyes, also manufactured from aromatic amines, revolutionized the dye industry.

TRIPHENYLMETHANE DYES

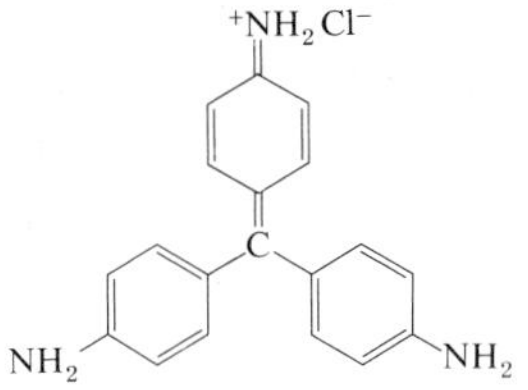

Pararosaniline

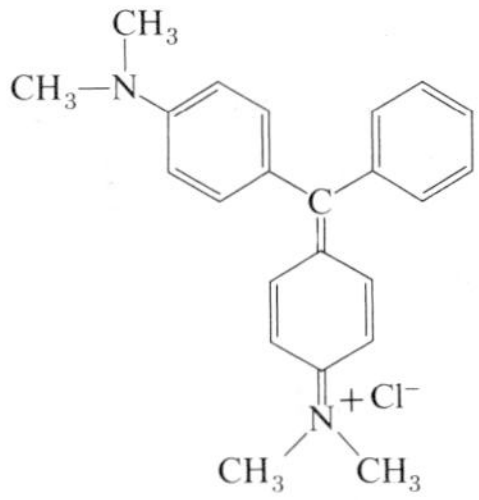

Malachite Green

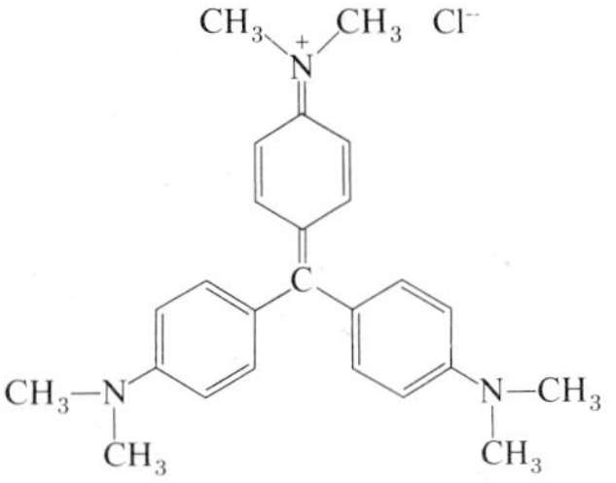

Crystal Violet

AZO DYES

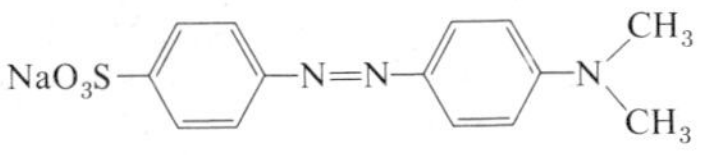

Methyl orange

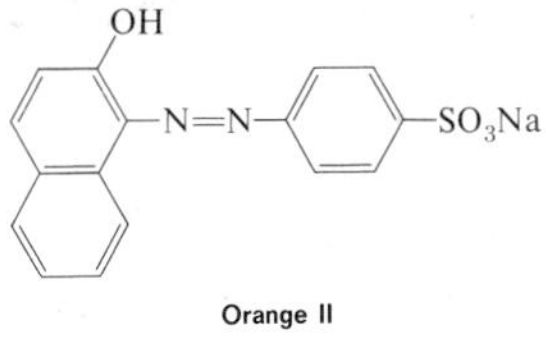

Orange II

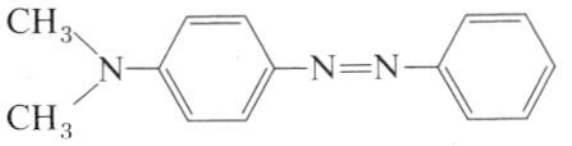

Butter Yellow

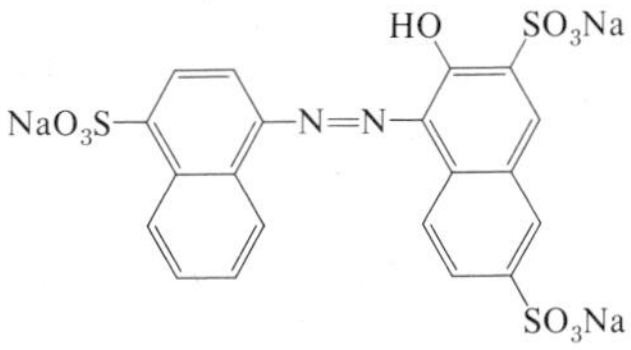

Amaranth Red

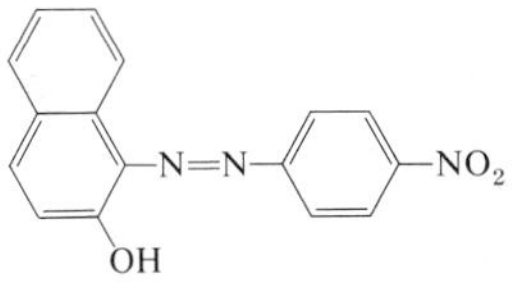

Para Red

MAUVE

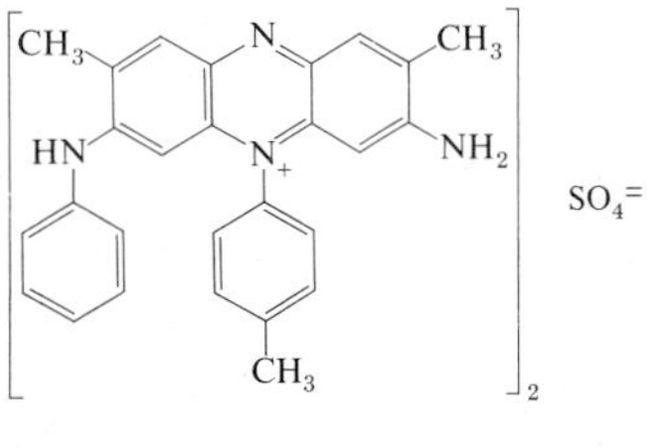

The azo dyes are one of the most common types of dye still in use today. They are used as dyes for clothing (see essay preceding Experiment 34), as food dyes (see essay preceding Experiment 35), and as pigments in paints. In addition, they are used in printing inks and in certain color printing processes. Azo dyes have the following basic structure:

$$Ar—N=N—Ar'$$

Several of these dyes are illustrated in the accompanying figures. The unit containing the nitrogen–nitrogen bond is called an **azo** group, a strong chromophore that imparts a brilliant color to these compounds.

To produce an azo dye, an aromatic amine is treated with nitrous acid to give a **diazonium ion** intermediate. This process, which is shown below, is called **diazotization.**

$$Ar—NH_2 + HNO_2 + HCl \longrightarrow Ar—\overset{+}{N}{\equiv}N\colon Cl^- + 2\,H_2O$$

The diazonium ion is an electron deficient (electrophilic) intermediate, and an aromatic compound, suitably rich in electrons (nucleophilic), will add to it. The most commonly used species are aromatic amines and phenols. Both of these types of compounds are usually more nucleophilic at a ring carbon than at either nitrogen or oxygen. This is due to resonance of the following types:

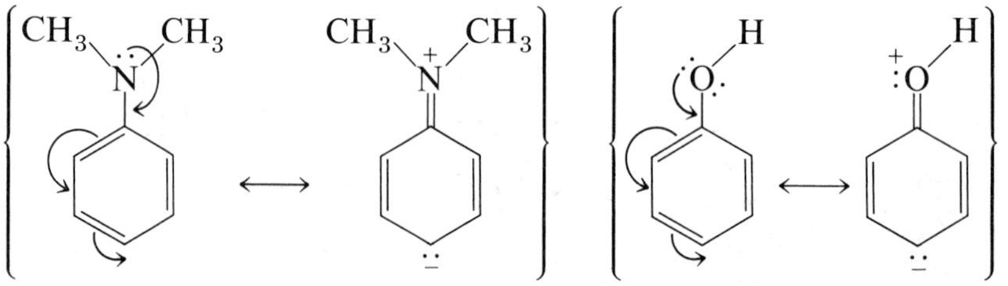

The addition of the amine or the phenol to the diazonium ion is called the **diazonium coupling** reaction, and it takes place as shown below.

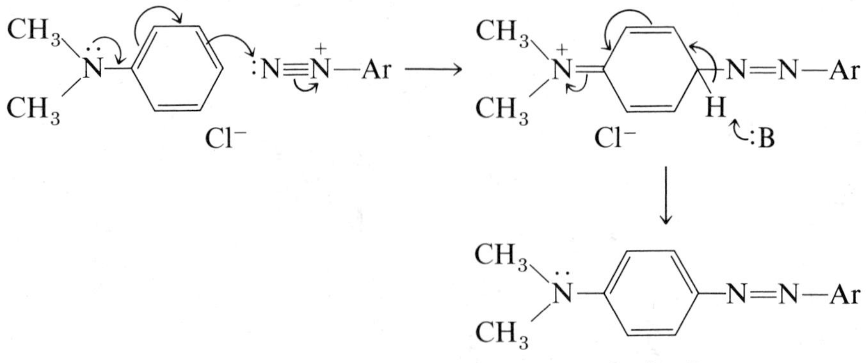

An Azo Dye

Azo dyes are both the largest and the most important group of synthetic dyes. In making the azo linkage, many combinations of $ArNH_2$ and $Ar'NH_2$ (or $Ar'OH$) are possible. These combinations give rise to dyes with a broad range of colors, encompassing yellows, oranges, reds, browns, and blues. The preparation of an azo dye is given in Experiment 33.

The azo dyes, the triphenylmethyl dyes, and mauve are all synthesized from the anilines (Aniline, o-, m-, and p-toluidine) and aromatic substances (benzene, naphthalene, anthracene). All of these substances can be found in **coal tar,** a crude material that is obtained by distilling coal. Perkin's discovery led to the formation of a multimillion dollar industry based on coal tar, a material that was once widely regarded as a foul smelling nuisance. Today, these same materials can be recovered from crude oil or from petroleum as by-products in the refining of gasoline. Although we no longer utilize coal tar, many of the dyes are still in wide use.

REFERENCES

Abrahart, E. N. *Dyes and their Intermediates.* London: Pergamon Press, 1968.

Allen, R. L. M. *Colour Chemistry.* New York: Appleton-Century Crofts, 1971.

Jaffee, H. E., and Marshall, W. J. "The Origin of Color in Organic Compounds." *Chemistry, 37* (December, 1964), 6.

Juster, N. J. "Color and Chemical Constitution." *Journal of Chemical Education, 39* (1962), 596.

Sementsov, A. "The Medical Heritage from Dyes." *Chemistry, 39* (November, 1966), 20.

Shaar, B. E. "Chance Favors the Prepared Mind. Part One: Aniline Dyes," *Chemistry, 39* (January, 1966), 12.

Experiment 33
METHYL ORANGE

Diazotization
Diazonium Coupling
Azo Dyes

In this experiment the azo dye, **methyl orange,** will be prepared by the diazo coupling reaction. It is prepared from sulfanilic acid (Experiment 31) and *N,N*-dimethylaniline. The first product obtained from the coupling is the bright red acid form of methyl orange, called **helianthin.** In base, helianthin is converted to the orange sodium salt, called methyl orange.

Methyl Orange

Helianthin (red)

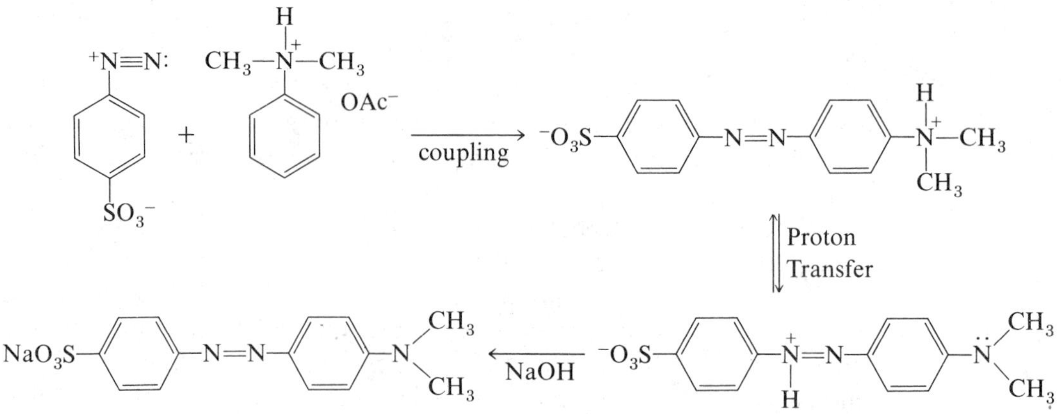

Although sulfanilic acid is insoluble in acid solutions, it is nevertheless necessary to carry out diazotization reaction in an acid (HNO_2) solution. To avoid this problem, sulfanilic acid is precipitated from a solution in which it is initially soluble. The precipitate is a fine suspension and reacts instantly with nitrous acid. To accomplish this, sulfanilic acid is first dissolved in basic solution.

Sulfanilic Acid

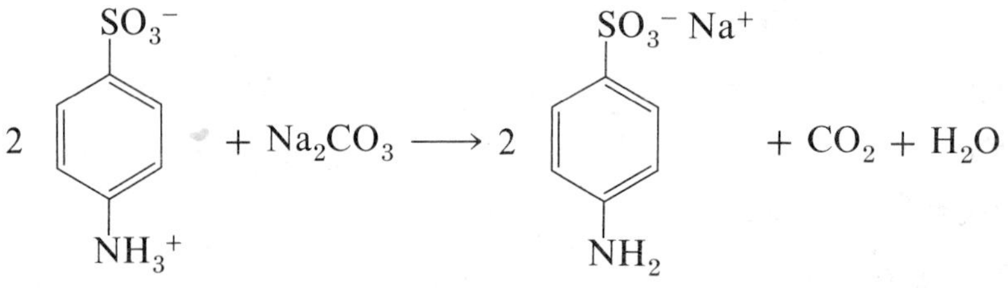

When the solution is acidified during the diazotization to form nitrous acid,

$$NaNO_2 + HCl \longrightarrow NaCl + HNO_2$$

the sulfanilic acid is precipitated out of solution as a finely divided solid which is immediately diazotized. The finely divided diazonium salt is reacted immediately with dimethylaniline in the solution in which it was precipitated.

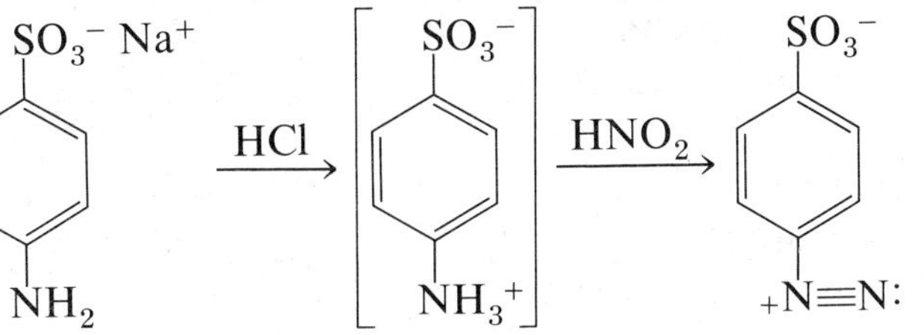

Methyl orange is often used as an acid-base indicator. In solutions which are more basic than pH 4.4, methyl orange exists almost entirely as the **yellow** negative ion. In solutions which are more acidic than pH 3.2, it is protonated to form a **red** dipolar ion.

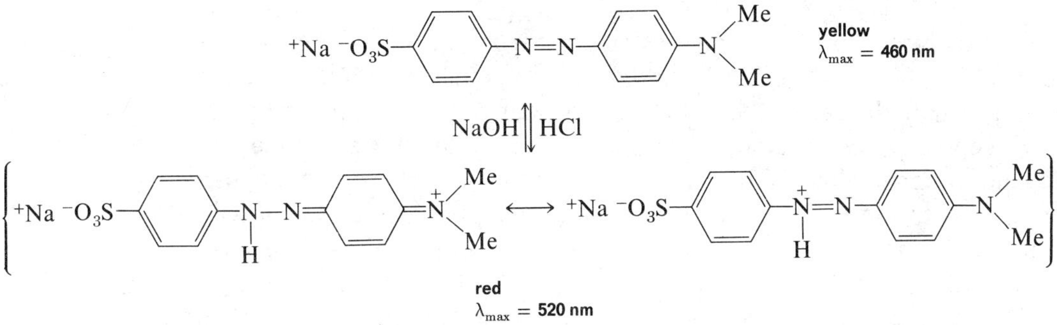

Thus, methyl orange can be used as an indicator for titrations that have their endpoints in the pH 3.2 to 4.4 region. The indicator is usually prepared as a 0.01% solution in water. In higher concentrations in basic solution, of course, methyl orange appears **orange** in color.

Azo compounds are easily reduced at the nitrogen–nitrogen double bond by reducing agents. Sodium hydrosulfite ($Na_2S_2O_4$) is often used to bleach azo compounds.

$$R'-N=N-R \xrightarrow{Na_2S_2O_4} R'-NH_2 + H_2N-R$$

Other good reducing agents, such as stannous chloride in concentrated hydrochloric acid, will also work.

SPECIAL INSTRUCTIONS

As an introduction to this experiment, you should read the essay which precedes this experiment. You will also find of interest the essays preceding Experiments 34 and 35. To perform the experiment, you should have read and familiarized yourself with Techniques 1 through 4.

Temperature is important in this reaction. You must keep the temperature of the reaction mixture below 10° when the diazonium salt is formed. If the diazonium solution is stored for even a few minutes, it should be kept in an ice bath. If the aqueous diazonium salt solution is allowed to rise in temperature, the diazonium salt will be hydrolyzed to a phenol, thus reducing the yield of the desired product.

If you have performed Experiment 31 (Sulfanilic Acid) as an earlier experiment, your instructor may wish to have you use the material already prepared for this experiment.

PROCEDURE

DIAZOTIZED SULFANILIC ACID

Dissolve 1.1 g of anhydrous sodium carbonate in 50 ml of water. Use a 125 ml Erlenmeyer flask. Add 4.0 g of sulfanilic acid monohydrate (3.6 g, if anhydrous) to the solution and heat it on a steam bath until it dissolves. A small amount of suspended material may make the solution appear cloudy. To remedy this, a small portion of Norit can be added to the solution; the still-hot solution should then be gravity filtered using fluted paper moistened with hot water. Rinse the filter paper with a little (2 to 5 ml) hot water. Cool the filtrate to **room temperature,** add 1.5 g of sodium nitrite, and stir until it dissolves. Pour this solution, with stirring, into a 400 ml beaker containing 25 ml of ice water to which 5 ml of concentrated hydrochloric acid has been added. In a short time the diazonium salt of sulfanilic acid should separate as a finely divided white precipitate. Keep this suspension cooled in an ice bath until it is to be used.

METHYL ORANGE

In a test tube mix together 2.7 ml of dimethylaniline and 2.0 ml of glacial acetic acid. Add this solution to the cooled suspension of diazotized sulfanilic acid in the 400 ml beaker. Stir the mixture vigorously with a stirring rod. In a few minutes a red precipitate of helianthin should form. Keep the mixture cooled in an ice bath for about 15 minutes to assure completion of the coupling reaction. Next, add 30 ml of a 10% aqueous sodium hydroxide solution. Do this slowly and with stirring while still cooling the beaker in an ice bath. Check with litmus or pH paper to make sure the solution is basic. If not, add extra base. Heat the mixture to boiling with a Bunsen burner for 10 to 15 minutes to dissolve most of the newly formed methyl orange. When all (or most) of the dye is dissolved, add 10 g of sodium chloride, and cool the mixture in an ice bath. The methyl orange should recrystallize. When the solution has cooled and the precipitation appears complete, collect the product by vacuum filtration using a Büchner funnel. Rinse the beaker with two cold portions of a saturated aqueous sodium chloride solution (10 ml each) and wash the filter cake with these rinse solutions.

To purify the product, transfer the filter cake and paper to a large beaker containing about 150 ml of boiling water. Maintain the solution at a gentle boil for a few minutes, stirring it constantly with a glass stirring rod. Not all of the dye will dissolve, but the salts with which it is contaminated will dissolve. Remove the filter paper and allow the solution to cool to room temperature. Cool the mixture in an ice bath and, when it is cold, collect the product by vacuum filtration using a Büchner funnel. Allow the product to dry, weigh it, and calculate the percentage yield. The product is a salt. Since salts do not generally have well-defined melting points, the melting point determination should not be attempted.

At your instructor's option, you may be asked to recrystallize a 1 or 2 g portion of the product derived from the boiling water. Prepare a saturated solution of the dye in boiling water. It may not be possible to get all of the sample to dissolve. Remove the insoluble material by filtering the hot solution by gravity through a fluted filter placed in a funnel which has been preheated on a steam bath. Set the filtered solution aside to cool and crystallize. Collect the product by vacuum filtration using a Büchner funnel. Submit the purified sample, along with your report, to the instructor.

TESTS (OPTIONAL)

Obtain a square of Multifiber Fabric 10A from your instructor. This cloth contains alternate bands of acetate rayon, cotton, nylon, acrylic, polyester, and wool (see Experiment 34). Prepare a dye bath by dissolving 0.5 g of methyl orange (your crude material will suffice) in 300 ml of water, to which 10 ml of a 15% aqueous sodium sulfate solution and 5 drops of concentrated sulfuric acid have been added. Heat the solution to just below its boiling point. Immerse the fabric in the bath for about 5 to 10 minutes. Remove the fabric, rinse it well, and note the results.

Make the dye bath basic by adding sodium carbonate. Then, add a solution of sodium dithionite (sodium hydrosulfite) until the color of the bath is discharged. Add a slight excess. Now place the very end portion of the dyed fabric in the bath for a few minutes. Note the result.

INDICATOR ACTION (OPTIONAL)

Dissolve a few crystals of methyl orange in a small amount of water in a test tube. Alternately add a few drops of dilute hydrochloric acid, and then a few drops of dilute sodium hydroxide solution, until the color changes are apparent in each case.

QUESTIONS

1. Why does the dimethylaniline couple with the diazonium salt at the para position of the ring?

2. Give a mechanism for the production of a phenol from the diazonium salt which was prepared from sulfanilic acid.

3. What would be the result if cuprous chloride were added to the diazonium salt prepared in this reaction?

4. The diazonium coupling reaction is an electrophilic aromatic substitution reaction. Give a mechanism that clearly indicates this fact.

5. In the essay on food colors that precedes Experiment 35, the structures of several azo food colors are given. Indicate how you would synthesize each of these dyes by means of the diazo coupling reaction.

6. Immediately after the coupling reaction in this experiment, a proton transfer occurs. The proton is transferred from the dimethylamino group to the azo linkage. Why is this latter protonated form lower in energy than the one which is formed initially?

ESSAY

Dyes and Fabrics

Not all dyes are suitable for use with every type of fabric. A dye must have the property of **fastness.** It must bond strongly to the fibers of the material and remain there even after repeated washings. It must not fade on exposure to light. Additionally, a dye must dye the fabric evenly if it is to be a commercially important dye. **Levelness** is the term used to refer to the uniformity of the dye on the fiber. A level dye, then, is one that colors the fabric evenly after its application. Just how a dye bonds to given fiber is a subject of high complexity because not all dyes and not all fibers are equivalent. A dye which is good for wool or silk may not dye cotton at all. Conversely, a dye that is fast and level on cotton may not be satisfactory for wool.

To understand the mechanism by which a dye bonds to a fiber, it is necessary to know both the structure of the dye and the structure of the fiber. The main types of fabric fiber are illustrated in Table One.

The two natural fibers, wool and silk, are very similar in structure. They are both polypeptides, i.e., polymers made of amino acid units.

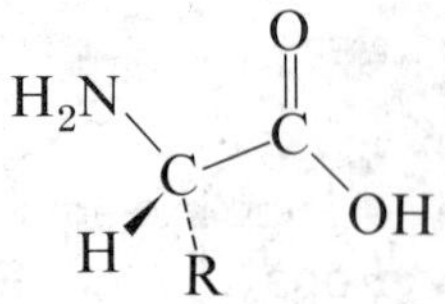

BASIC AMINO
ACID STRUCTURE

All amino acids have the same basic structure, but differ in the nature of the substituent group, R. In naturally occurring amino acids the

NATURAL FIBERS

Wool (Polypeptide)
Silk

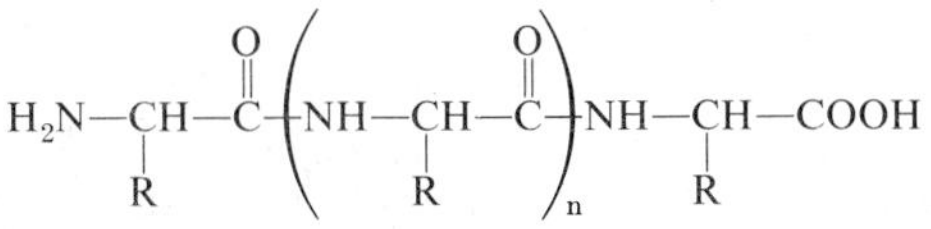

Cotton (Cellulose)

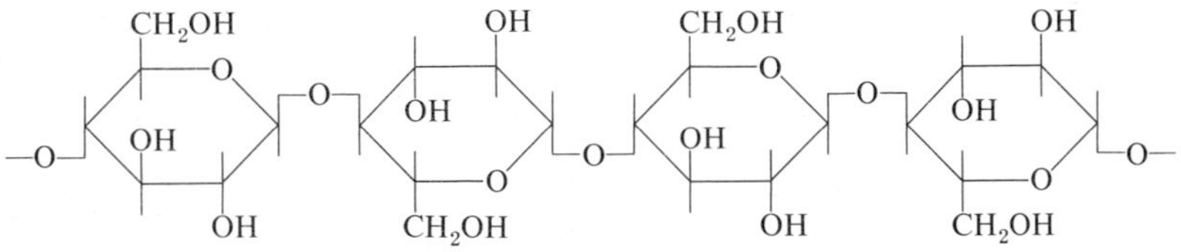

SYNTHETICS

Acetate (Cellulose Acetate)

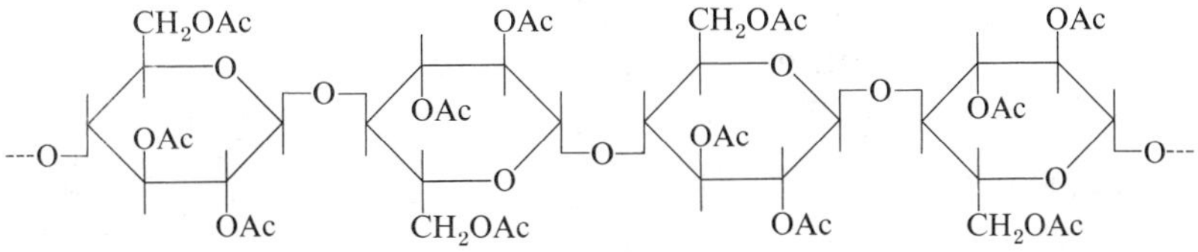

Nylon (Polyamide)

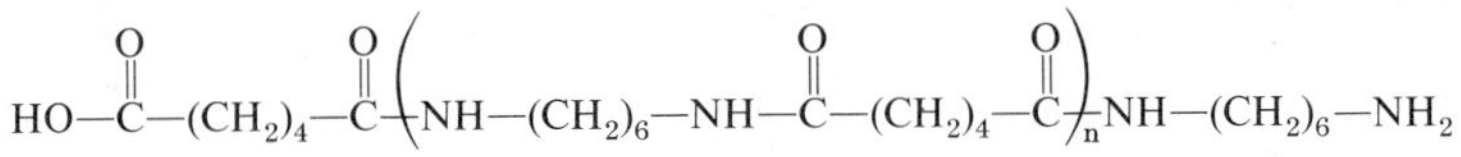

Dacron (Polyester)

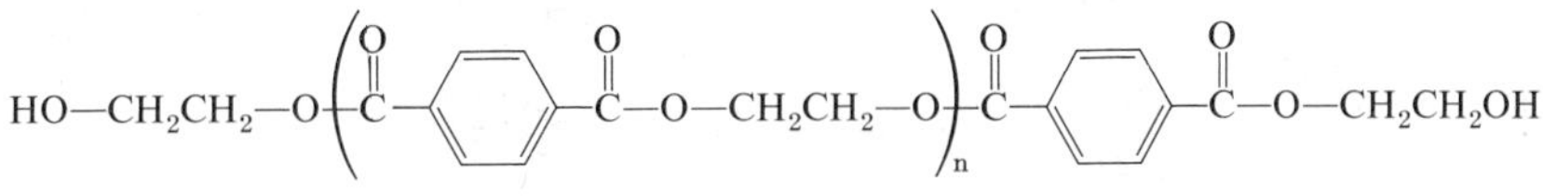

Orlon (Acrylic, Polyacrylonitrile)

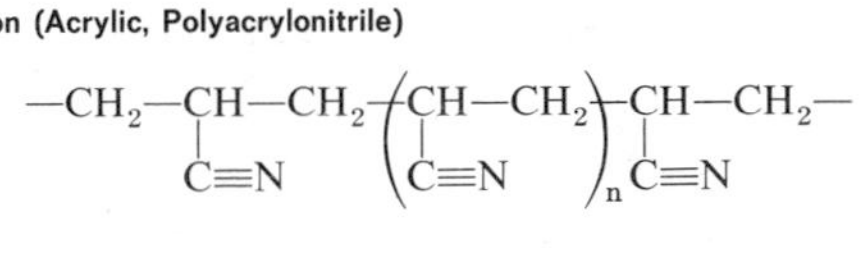

Table One. FIBER TYPES

various possible side chain R groups consist of substituents which vary from basic (—NH_2), to neutral (—CH_2OH, —CH_3), to acidic (—COOH).

In wool and silk, there are a number of ways in which a dye may bind to the fiber. It may make chemical links either to the terminal —NH_2 or —COOH groups of the polypeptide chain, or to the functional groups present in the side chains of the component amino acids.

Dyes that attach themselves to a fiber by direct chemical interaction are called **direct dyes.** Picric acid is a direct dye for wool or silk. Since it is a strong acid, it interacts with the basic end groups or side chains in wool or silk to form a **salt linkage** between itself and the fiber. The picric acid gives up its proton to some basic group on the fiber and becomes an anion which strongly bonds to a cationic group on the fiber by ionic interaction (salt formation). This interaction is illustrated in Figure 1.

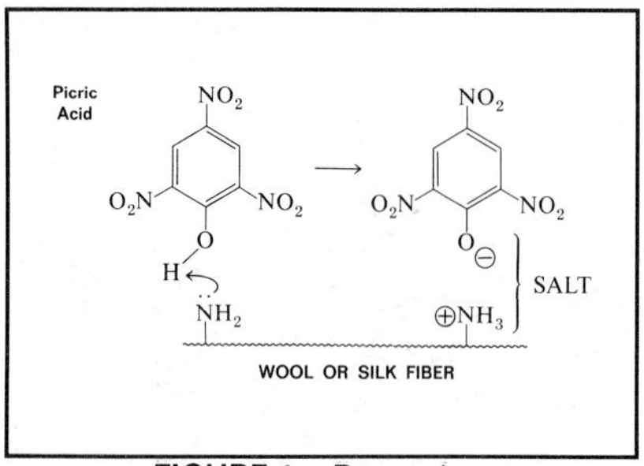

FIGURE 1. Direct dye.

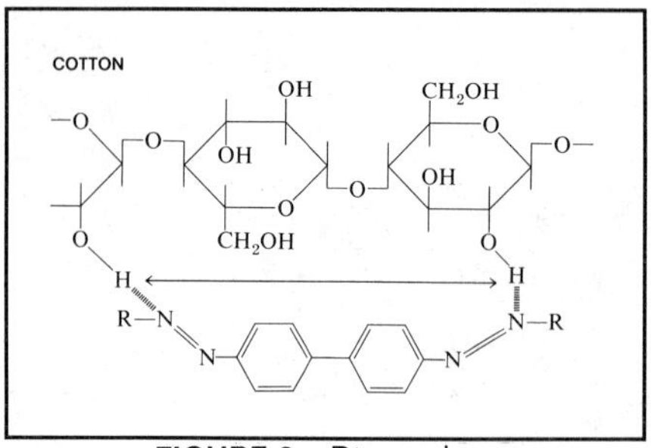

FIGURE 2. Disazo dye.

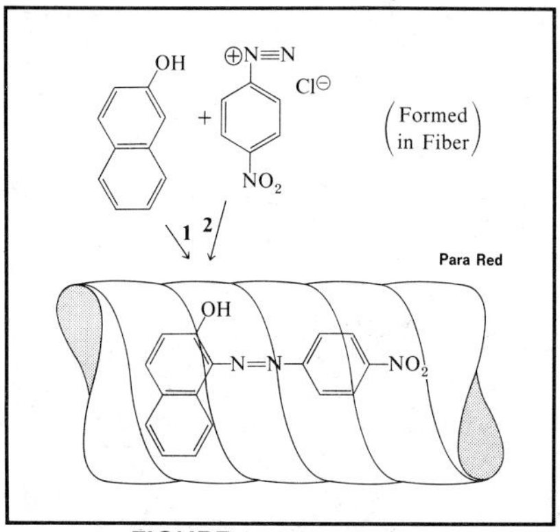

FIGURE 3. Ingrain dye.

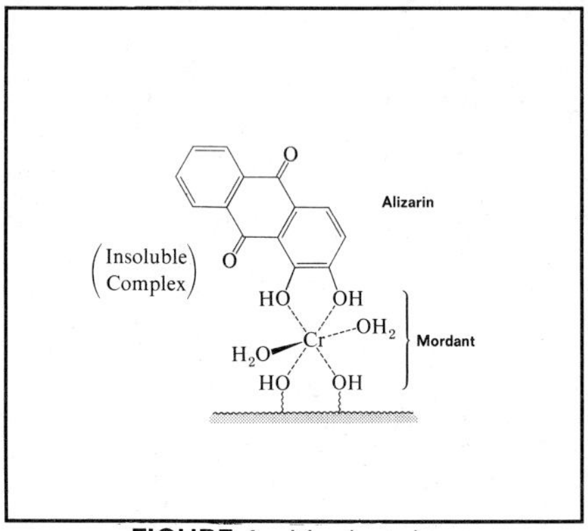

FIGURE 4. Mordant dye.

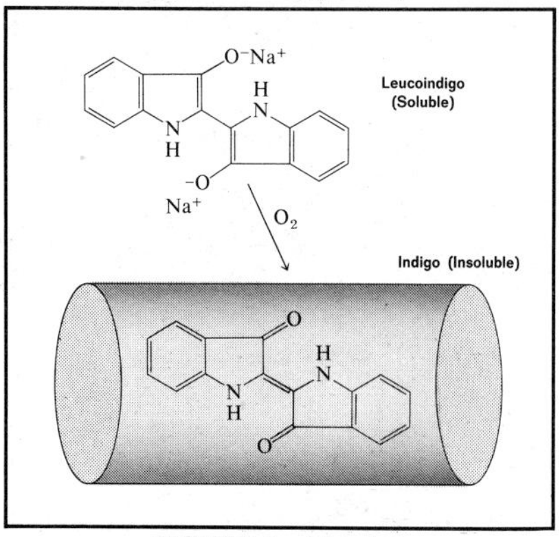

FIGURE 5. Vat dye.

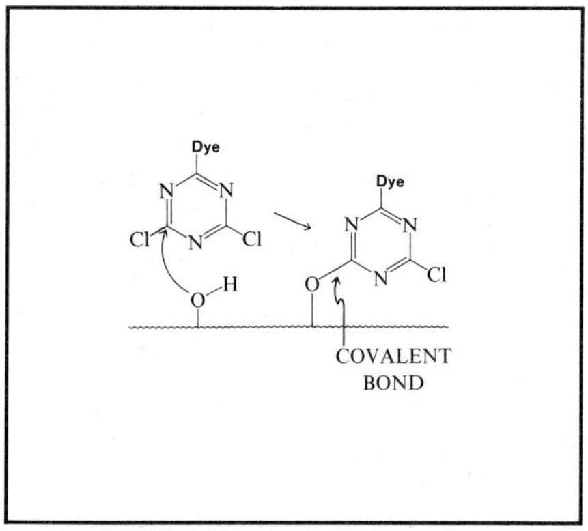

FIGURE 6. Fiber reactive dye.

At biological pH, most amino acids exist in their zwitterionic form:

$$H_3\overset{+}{N}-CH-COO^-$$
$$|$$
$$R$$

This means that in wool or silk there are $-NH_3^+$ and $-COO^-$ groups as well as $-NH_2$ and $-COOH$ groups. The $-NH_2$ and $-COO^-$ groups are basic, and the $-COOH$ and $-NH_3^+$ groups are acidic. Thus, in an alternate mechanism to that given in Figure 1, picric acid might give up its proton to a $-COO^-$ group to form its anion. The picrate anion would then bond to some $-NH_3^+$ group already present on the chain.

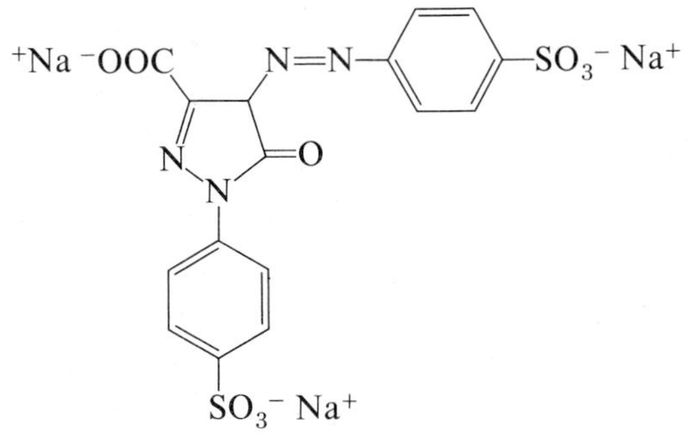

Tartrazine

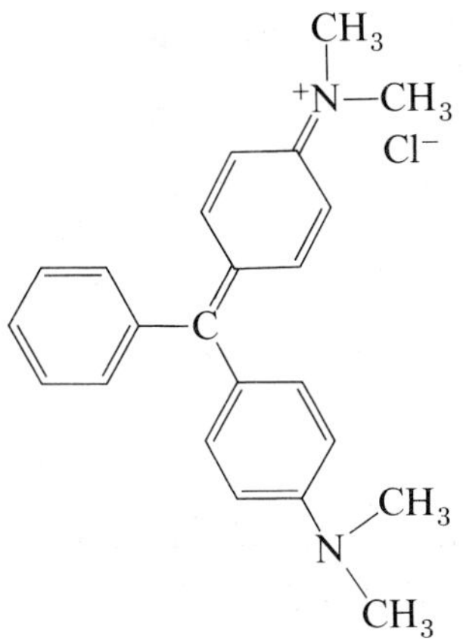

Malachite
Green

Two other types of direct dyes for wool or silk are the **anionic** and **cationic** dyes. Anionic dyes are also called **acid dyes**; cationic dyes are also called **basic dyes**. Tartrazine, an azo dye, is an important yellow acid dye (anionic) for wool and silk. Malachite green, a triphenylmethane dye, is an important basic dye (cationic) for wool and silk. Anionic dyes interact with the acidic $-NH_3^+$ groups in the fiber.

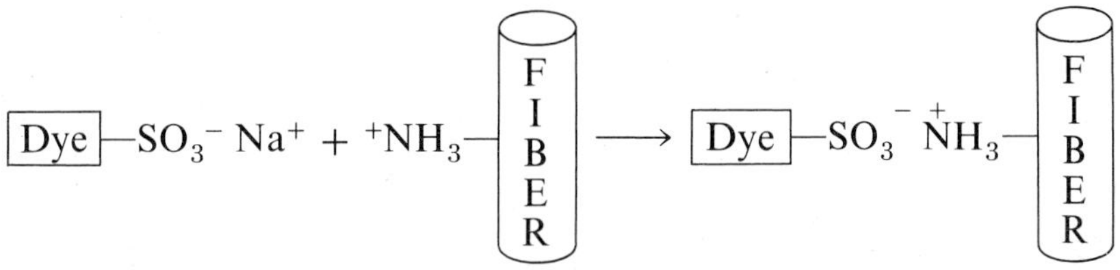

ANIONIC DYE

Cationic dyes interact with the $-COO^-$ groups in the fiber.

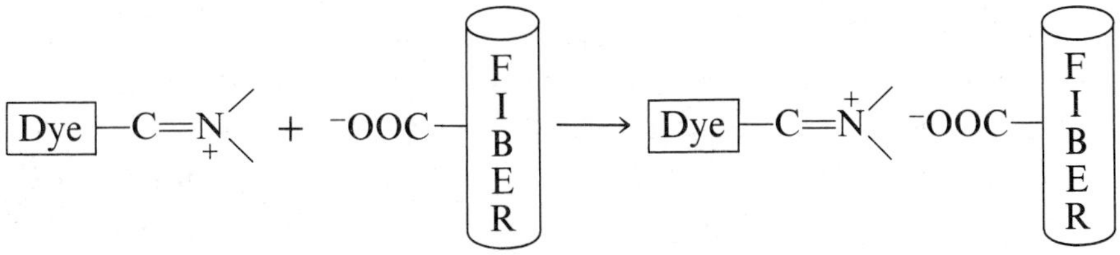

CATIONIC DYE

Of the synthetic fibers, nylon is most like wool and silk. It is constructed with "peptide" or amide linkages ($-NH-\overset{\underset{\|}{O}}{C}-$), being formed from a diamine ($NH_2-(CH_2)_6-NH_2$) and a dicarboxylic acid ($HOOC-(CH_2)_4-COOH$). Nylon can be synthesized so that either $-NH_3^+$ or $-COO^-$ groups predominate at the ends of the chains. It is more usual for nylon fabrics to contain an excess of $-NH_3^+$ groups. Therefore, nylon is usually dyed with anionic (acid) dyes.

Dacron (a polyester fiber), Orlon (an acrylic fiber) and cotton (cellulose) do not contain many anionic or cationic groups in their structures and are not easily dyed by picric acid or by dyes of the anionic and cationic types. Orlon is sometimes synthesized with $-SO_3^-$ groups incorporated into the chain:

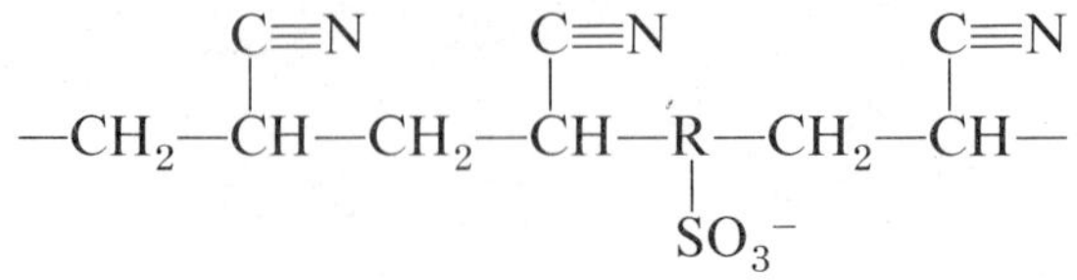

$$-CH_2-\overset{\overset{\displaystyle C\equiv N}{|}}{CH}-CH_2-\overset{\overset{\displaystyle C\equiv N}{|}}{CH}-R-CH_2-\overset{\overset{\displaystyle C\equiv N}{|}}{\underset{\underset{\displaystyle SO_3^-}{|}}{CH}}-$$

When modified in this way, Orlon can be dyed with cationic dyes.

Azo dyes are important for dyeing cotton fibers. Two such dyes are Congo Red and Para Red. Congo Red is a direct dye for cotton. Para Red is a **developed dye** for cotton. We shall explain both types in detail.

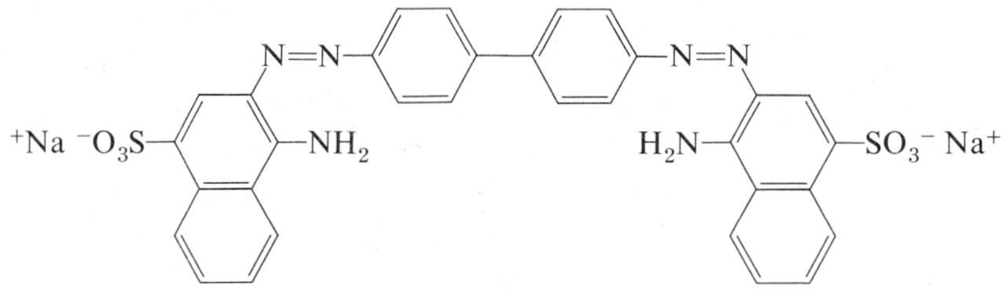

Congo Red (A Disazo Dye)

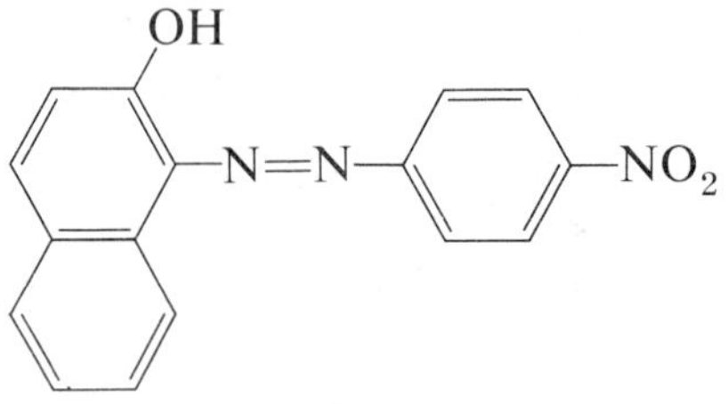

Para Red (An Azo Dye)

Cotton, which has only hydroxylic groups in its structure, does not dye well with picric acid or with anionic and cationic dyes. Simple azo dyes do not bond well to cotton, but **disazo** dyes are direct dyes for cotton. Congo Red is a disazo dye. A disazo dye has two azo linkages $(R-N{=}N{\sim}N{=}N-R)$. When the two azo groups are separated by about 10.2 to 10.8 Å, a disazo dye is a fast, direct dye for cotton. The mode of attachment for these dyes is not well understood. Models show that the repeating length between structurally similar hydroxyl groups in cellulose is about the same distance (10.3 Å) as that required for a disazo dye to be a direct dye. In addition, acetate fibers do not dye with disazo dyes. Since the hydroxyl groups are absent in the acetate fiber, it is thought that hydrogen bonding between the hydroxyl groups and the azo linkages accounts for the attachment of a disazo dye to cotton. This is shown in Figure 2.

Although simple azo dyes are not direct dyes for cotton, they can nonetheless be used to dye cotton as **developed dyes,** also called **ingrain dyes.** Ingrain dyes are synthesized right inside the fiber. Individually, the two components used to synthesize the dye will diffuse into the pores and spaces between the fibers in the fabric. The fully formed dye would be too large a molecule to do this and could only bond to the surface of the fibers. When the individual components react to form the dye, it is trapped **inside** or "in the grain" of the fibers. Azo dyes are used as ingrain dyes for cotton. Para Red is a typical ingrain azo dye. To form para red, the fabric is soaked first in a solution of the coupling component β-naphthol, and then soaked in a solution of diazotized p-nitroaniline. This process is depicted in Figure 3.

A third type of dye for cotton is the **mordant dye.** Alizarin, shown in Figure 4, is a typical mordant dye. Mordant dyes usually have incorporated into their structures groups capable of forming chelate complexes with heavy metals such as copper, chromium, tin, iron, and aluminum. Cotton, which has many hydroxyl groups, can also coordinate with these metals. To dye a fabric with a mordant dye, the fabric is first treated with a **mordant.** Mordants are heavy metal salts which can complex with the cotton fibers. Typical mordants are alum, copper sulfate, ferric chloride, stannous chloride, and potassium dichromate. After the fiber is treated with the mordant, it is dyed. The dye also complexes with the mordant, and the mordanting metal links the dye to the fabric. This process is depicted in Figure 4.

For a dye to be fast when using a mordant, the complex formed between the dye, the fiber, and the mordanting metal must be very stable and **insoluble.** Different mordants (metals) often lead to different colors with the same dye.

When cationic dyes, such as malachite green, are used on cotton, the fabric must be treated with tannic acid as "mordant." In this case, the mordant (substance which binds the dye to the fiber) is not a metal. It is simply a substance which forms an insoluble precipitate when mixed with the dye. Tannic acid forms insoluble complexes with cationic dyes. In this process, the cotton is first impregnated with the mordant (tannic acid) and then treated with the dye.

A type of dye that can be used for all fibers, both natural and synthetic, is the **vat dye.** Vat dyes are normally water soluble in their reduced form, but when oxidized become insoluble. Indigo is a vat dye. Indigo is an insoluble blue dye. It can be reduced by sodium dithionite (sodium hydrosulfite) to leucoindigo, a soluble form which is colorless.

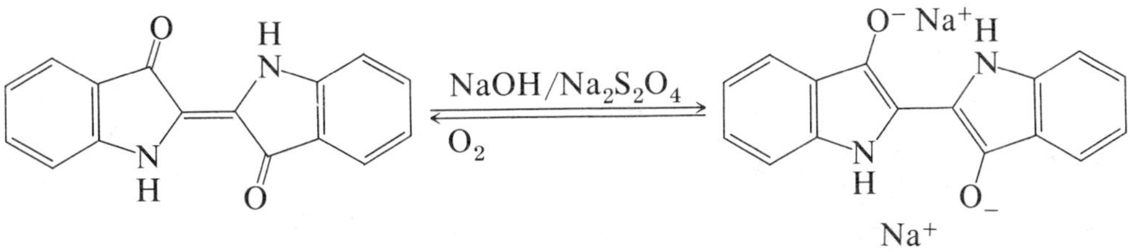

Indigo
(blue, insoluble)

Leucoindigo
(colorless, soluble)

In the vat process for indigo, the fabric is impregnated with the soluble leucoindigo in a hot dye bath or **vat.** Then the fabric is removed from the vat, and the leucoindigo is allowed to air oxidize to the insoluble indigo. The indigo precipitates both inside and on the surface of the fabric fibers. Since the indigo is insoluble in aqueous solutions, the dye is fast. The process is illustrated in Figure 5. Indigo is used for "blue jeans."

On fibers like Dacron (Polyester) and Acetate, no groups are present that allow any direct dye–fiber interactions. These fibers are hard to dye. **Disperse dyes,** however, can be used. In the disperse process, a dye which is water insoluble is used. It is allowed to penetrate the fiber by use of an organic solvent as a "carrier." Often the dye is suspended in an aqueous solution and a small amount of the carrier solvent or substance is added. The carrier dissolves the dye and carries it into the fiber where it is "dispersed." The dye becomes trapped within the fiber only because of water insolubility. It is not bonded to the fibers, but merely dispersed among them.

The most recent types of dyes are the **fiber reactive dyes.** The main class of these dyes, the Procion dyes, are based on cyanuric chloride. The chlorides in this compound may be replaced by nucleophilic substitution reactions. If a dye with a good nucleophilic group is allowed to react with cyanuric chloride in basic solution, the dye becomes bound to the cyanuric chloride ring by a **covalent** bond and a Procion dye is produced.

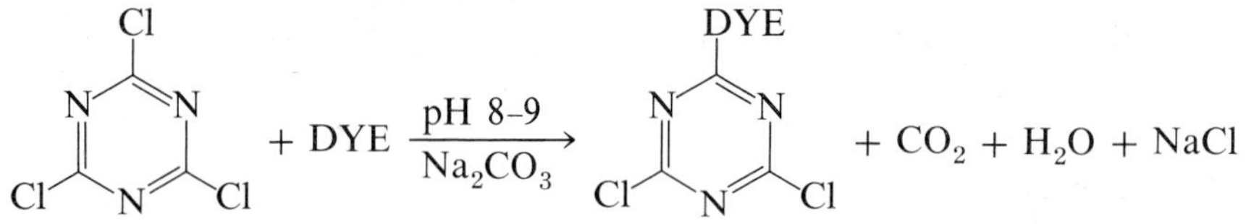

Cyanuric Chloride   Procion Dye

Generally, the dye molecule must contain an —OH or an —NH_2 group to replace the chlorine in cyanuric chloride. If the dye replaces only one of the chlorines, two chlorines still remain which can be replaced. If the fiber to be dyed has —OH or —NH_2 groups, these can displace the remaining chlorines thus forming a **covalent** bond between the cyanuric chloride and the fiber. The net result is that the dye is attached to the fiber, through cyanuric chloride, by covalent bonds. This process is shown in Figure 6.

REFERENCES

Abrahart, E. N. *Dyes and Their Intermediates.* London: Pergamon Press, 1968.
Allen, R. L. M. *Colour Chemistry.* New York: Appleton-Century Crofts, 1971.
Davidson, M. F. *The Dye Pot.* Published by the author, Route 1, Gatlinburg, Tenn. 37738.
Editorial Committee of the Brooklyn Botanic Garden. *Dye Plants and Dyeing—A Handbook.*

Published by the Brooklyn Botanic Garden, 100 Washington Avenue, Brooklyn, New York 11225. A special printing of *Plants and Gardens,* Vol. 20, No. 3.
Giles, C. H. *A Laboratory Course in Dyeing.* Society of Dyers and Colorists Publication, 1957.
Maille, A. *Tie and Dye.* New York: Ballantine Books, 1971.
Nea, S. *Tie-Dye.* New York: Van Nostrand, 1971.

Experiment **34**

DYES, FABRICS AND DYEING

Dyes and Fibers
Azo Dyes, Direct Dyes, Vat
  Dyes, Developed Dyes, Disperse
  Dyes, Substantive Dyes
Mordants. Wool, Cotton,
  Silk, Nylon, Polyester

In this experiment, we shall examine how dyes adhere to fibers in several fabrics; wool, cotton, polyester, and nylon (or silk). The various classes of dyes and their fastness are examined: direct (anionic and cationic), disazo, vat, ingrain (developed), disperse, and mordant. Finally, the action of mordants on several types of dyes is investigated. The theory of the experiment is discussed in the previous essay "Dyes and Fabrics."

SPECIAL INSTRUCTIONS

This entire experiment can be completed in one lab period, but you will need to organize your time efficiently. It will be necessary to perform several experiments simultaneously. Since all of the experiments involve waiting periods of as long as 15 to 20 minutes, these waiting periods allow for time to begin the next experiment. If your instructor so chooses, the experiment may be extended to two periods. In this case, both the mordanting and Part A should be completed in the first period. Part B may then be completed in the subsequent period. To understand the experiment, it is necessary to have read the essay which precedes it.

Care should be taken not to immerse your hands in either the mordant or the dye baths. Many of these materials are toxic. In addition, all azo dyes should be suspected of having possible carcinogenic activity. You will also find it socially uncomfortable to sport yellow, green, red, and blue digits.

PROCEDURES

Obtain from the instructor or laboratory assistant a kit which contains the following materials:

DYES (in waxed envelopes)

Picric Acid	(0.5 g)	Alizarin	(0.1 g)
Indigo	(0.1 g)	Methyl Orange	(0.1 g)
Congo Red	(0.1 g)	Malachite Green	(0.1 g)
Eosin	(0.1 g)		

FABRIC

5 Rectangles of wool   (2″ × 3″) ⎫ If available, six 2 1/2 inch
5 Squares of cotton   (3″ × 3″) ⎪   squares of Multifiber
3 Squares of polyester   (2″ × 2″) ⎬   Fabric 10A may be
3 Squares of silk   (2″ × 2″) ⎪   more conveniently
        (or nylon)   (2″ × 2″) ⎭   used

YARNS   OTHER
    36 10″ Pieces of cotton yarn   6 Safety pins, string
    72 5″ Pieces of wool yarn   An index card
                                                    One 12 inch square
                                                        aluminum foil

Before beginning the experiments, tie the cotton and wool yarns into 18 hanks of each by the following methods.

COTTON: Take two 10″ strands of cotton yarn, fold them in half, and tie an overhand knot in the center of the folded lengths (Figure 1A). Prepare 18 such hanks. Tie the knot loosely or the yarn will resist dyeing.

WOOL: Take four 5″ strands of wool yarn and tie an overhand knot in the center of them (Figure 1B). Prepare 18 of these hanks also.

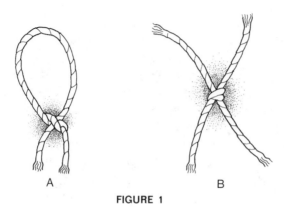

A   B

FIGURE 1

MORDANTING

Since mordanting requires some time, it should be done before proceeding to the initial dyeing experiments in Part A. The mordanted yarns will be used in Part B. Five large (2 l) mordant baths will be found in the laboratory. They should be kept warm on steam baths or hot plates. The five mordants to be used are alum (potassium aluminum sulfate), copper (cupric sulfate), chrome (potassium dichromate), tin (stannous chloride), and tannic acid. A fixing bath of tartar emetic (potassium antimony tartrate) will also be found. It is to be used after the tannic acid mordant. All the solutions are 0.1 M in the mordants. The samples should be mordanted 15 to 20 minutes in the mordant baths. The tin mordant is especially hard on wool, so keep the time brief for this mordant. On the other hand, if possible, the tannic acid mordant should be left for a longer time.

It will be convenient to mordant several samples at once in the following way. On three safety pins, fix 3 hanks each of wool and cotton yarns by passing the point of each pin through the knots in the centers of the hanks. On a fourth safety pin, stick 4 hanks each of cotton and wool yarns, using the same method. On a fifth pin, fix one hank each of wool and cotton. Tie a 10 inch length of string through the spring eye of each pin. Make five labels from an index card. Punch a hole in each label and tie one label to the unfastened end of the string on each pin (Figure 2). Print your name on each label. Also print the mordant to be used for each set of samples. The first three sets of hanks (3 each of wool and cotton) will be mordanted in the copper, chrome, and tin baths. The fourth set (4 each of wool and cotton) will be mordanted in alum. The last set (1 each of wool and cotton) will be mordanted in the tannic acid bath.

Place each set of samples in the appropriate bath. Use a long spatula or forceps (not your fingers) to make sure the samples are fully immersed and wetted. The label should not be immersed, but should hang over the edge of the bath. Mordant the samples for 15 minutes in the hot baths. They should then be removed from the baths so as

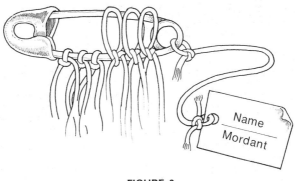

FIGURE 2

to allow excess mordant to drain back into the bath. In addition, squeeze out any excess mordant by pressing the sample against the side of the beaker with a spatula or stirring rod. Press each of the samples between a double thickness of paper toweling to remove any remaining mordant. After pressing, place all the samples, except that from the tannic acid bath, on a hot steam bath which has its top covered with a sheet of aluminum foil. This will help to dry the samples somewhat. The last sample (from the tannic acid bath) should be fixed in a bath of tartar emetic for 5 minutes and then dried. Leave the samples to dry and proceed to the next set of experiments. Experiments in Parts A and B may be done simultaneously since there are waiting periods in each set of procedures.

A. EXPERIMENTS WITH MATERIAL PATCHES
(Or With Multifiber Fabric)

Chart 1 outlines the experiments to be performed with patches of material. The entire set of experiments may be simplified if Multifiber Fabric is available. Multifiber Fabric 10A[1] has six fibers, wool, acrylic, polyester, nylon, cotton, and acetate rayon, woven in sequence. None of the experiments in this part will involve mordants. All materials (fabrics) should have been laundered to remove sizing. If not, they should be washed in soapy water and thoroughly rinsed before dyeing.

1. Picric Acid (Direct to Wool, Silk, and Nylon)

Dissolve the picric acid (about 0.5 g) in about 60 ml of water to which 2 or 3 drops of concentrated sulfuric acid have been added. Use a small beaker and place it on a steam bath for warming. Immerse

---

[1] Available from: Testfabrics, Inc., P.O. Box 118, 200 Blackford Ave., Middlesex, New Jersey 08846.

CHART 1. EXPERIMENTS WITH MATERIAL PATCHES

DYE TYPE	DYE	WOOL	COTTON	POLYESTER	SILK OR NYLON
DIRECT	Picric Acid				
VAT	Indigo			///////	
DISAZO (ANIONIC)	Congo Red			///////	
AZO	Para Red				///////
INGRAIN	Para Red			///////	///////
DISPERSE	Para Red	///////	///////		///////

one piece each of wool, cotton, polyester, and silk (or nylon) fabric. Be sure that each piece is thoroughly wetted in the dye solution. Add a little more water if necessary. After about 2 minutes, remove the patches with a forceps and transfer them to another beaker containing warm water. Rinse the samples well and then discard the water. Repeat the rinse process until the dye no longer runs (colors the rinse water). Note the results.

2. Indigo (Vat Dye)

Place the indigo powder (about 0.1 g) in a small Erlenmeyer flask containing a solution of 0.2 g of sodium dithionite (sodium hydrosulfite) and one sodium hydroxide pellet in 40 ml of warm water. Stopper the flask and shake it gently for several minutes until the indigo is dissolved and bleached. The leucoindigo solution will not be colorless, but will appear somewhat green in color. Add more sodium dithionite if the indigo powder did not dissolve. Heat the solution on a steam bath. Unstopper the flask and, using a forceps or a glass rod, insert a patch of wool into the leucoindigo solution. Stopper the flask and shake the solution to be sure that the material is thoroughly wetted. After about 30 seconds, remove the sample from the flask and squeeze it between a double thickness of paper towelling to remove the excess dye solution. Hang the sample to dry in the air and allow the leucoindigo to become oxidized. If a deeper blue is desired, repeat the dyeing process. It may be necessary to add a bit more sodium dithionite to the indigo solution if it is not still bleached. Repeat the dyeing process using cotton and silk (or nylon). Note and compare the results.

3. Congo Red (Disazo Dye for Cotton)

Dissolve the Congo Red (about 0.1 g) in a beaker containing 100 ml of hot water to which 1 ml of 10% aqueous sodium carbonate and 1 ml of 10% aqueous sodium sulfate have been added. Keep the solution warm on a steam bath. Introduce pieces of wool, cotton, and silk (or nylon). After 10 minutes in the hot dye bath, remove the samples and wash them in warm water until no more dye is removed. Note the results.

4. Para Red (Developed or Ingrain Dye)

Solution 1—Diazotized *p*-Nitroaniline. Place 1.4 g *p*-nitroaniline, 25 ml of water, and 5 ml of 10% hydrochloric acid in a small beaker. Heat the solution until most of the *p*-nitroaniline dissolves (add a bit more acid if necessary). Cool the solution to about 5°C in an ice bath. Add, all at once, a solution of 0.7 g of sodium nitrite dissolved in 10 ml of water. Stir the solution well, keeping it in the ice bath. If solid remains, the solution may be filtered by vacuum filtration using a Büchner funnel, but filtration is not necessary. Keep this solution in the ice bath, and prepare solution 2.

Solution 2—β-Naphthol. Place 0.5 g of β-naphthol in a small beaker containing 100 ml of hot water. While stirring, add a solution of 10% sodium hydroxide **by drops** until most of the β-naphthol dissolves. Not all of the β-naphthol will dissolve, and it is important not to add a large amount of base because the cotton will disintegrate in a strongly basic solution. The wool will fare even worse.

Place a sample of cotton in solution 2 and allow it to soak for only 2 or 3 minutes. Remove the sample from the bath and pat it dry between paper towels. Dilute solution 1 with about 100 ml of cold water and place into it the sample removed from solution 2. After several minutes, remove the sample and rinse it well in water. Note the result.

Next, place a sample of wool in solution 2 for 2 or 3 minutes. Remove it and dry it between paper towels. After this, immerse the sample in solution 1 for several minutes, and then remove it and rinse it well. Note the result.

After completing the ingrain dyeings, mix the two ingrain solutions (1 and 2) together and stir them well. Split the resulting dye mixture into two parts and proceed to parts 5 and 6.

5. Para Red (Azo Dye)

Use half of the Para Red solution prepared by mixing the two ingrain solutions (part 4). Add enough sulfuric acid to make the solution acidic to litmus or pH paper (about 3 drops of concentrated sulfuric acid). Immerse pieces of wool, cotton, and polyester in the hot dye bath for about five minutes. Remove the samples and rinse them well. Note the results.

6. Para Red (Disperse)

In its insoluble, finely-dispersed form, Para Red may be used as a disperse dye. Use half of the suspension of Para Red produced above by mixing the two ingrain solutions (part 4). Add to this solution 0.1 g of biphenyl (the carrier) and several drops of a surfactant (liquid detergent). Immerse a piece of the polyester cloth in the solution and heat it on a steam bath for 15 to 20 minutes. Stir the solution frequently. Remove the sample and note the results.

B. EXPERIMENTS WITH YARN

Prepare four dye baths (use beakers) by dissolving each of the following dyes (0.1 g each) in 200 ml of hot water: Eosin, Alizarin,

Methyl Orange, and Malachite Green. Add 0.1 g of sodium carbonate to the Alizarin bath. Keep the dye baths hot on a steam bath. The experiments to be done are outlined in Chart Two. Prepare a grid similar to that in Chart Two on a piece of notebook paper. You will be able to keep your samples straight by placing them on this sheet. Perform the dyeing operations in the sequence given. This is necessary as some of the mordants have a tendency to precipitate the dyes.

1. Immerse one hank each of **unmordanted** cotton and wool in each bath. After about 20 minutes, remove the samples and rinse them thoroughly in a beaker of warm water. Change the rinse water for each sample. Blot the samples dry between paper towels. Note the results.

2. Place one hank each of the wool and cotton yarns mordanted in alum in each dye bath. After about 15 minutes, remove the samples from the dye baths and rinse them well in warm water. Blot them dry between paper towels. Note the result.

3. Place the two samples (wool and cotton) mordanted in tannic acid and fixed in tartar emetic in the Malachite Green bath. After 10 to 15 minutes, remove the samples and wash them well in warm water. Blot them dry between paper towels. Note the results. The Malachite Green bath will not be used for any further experiments and may be discarded.

4. Place one hank each of the wool and cotton yarns mordanted in copper in each of the three remaining dye baths: Eosin, Alizarin, and Methyl Orange. After 10 to 15 minutes, remove and dry these samples, rinsing as before.

5. In each of the three dye baths (Eosin, Alizarin, and Methyl Orange) place one hank each of the wool and cotton yarns mordanted in chrome. After 10 to 15 minutes, remove and dry these samples.

6. In each of the three dye baths (Eosin, Alizarin, and Methyl Orange) place one hank each of the wool and cotton yarns mordanted in tin. After 10 to 15 minutes, remove and dry these samples.

CHART 2. EXPERIMENTS WITH YARN

DYE TYPE	DYE	COTTON					WOOL				
		UN	AL	CPR	CHR	TIN	UN	AL	CPR	CHR	TIN
MORDANT	Eosin										
	Alizarin										
AZO (ANIONIC)	Methyl Orange										
		UN	AL	TAR			UN	AL	TAR		
TRIPHENYL-METHANE (CATIONIC)	Malachite Green										

UN = unmordanted   AL = Alum   CPR = Copper   CHR = Chrome
TIN = Tin   TAR = Tannic Acid + Tartar Emetic

ESSAY
Food Colors

Before 1850, the majority of the colors added to foods were derived from natural biological sources. Some of these natural colors are listed in the table that follows.

Red	Alkanet root	Yellow	Annato seed (bixin)
	Beets (betanin)		Carrots (β-carotene)
	Cochineal insects		Crocus stigmas
	(carminic acid)		(saffron)
	Sandalwood		Tumeric (rhizome)
Orange	Brazilwood	Green	Chlorophyll
Brown	Caramel	Blue	Purple Grape Skins
	(charred sugar)		(oenin)

A wide variety of colors can be obtained from these natural sources, many of which are still used today, but they have been largely supplanted by synthetic dyes.

After 1856, when Perkin succeeded in synthesizing mauve, the first coal tar dye (see the essay preceding Experiment 33), and when many other new synthetic dyes began to be discovered, artificial colors began to find their way into foodstuffs with increasing regularity. Today, better than 90% of the coloring agents added to foods are synthetic.

The synthetic dyes have certain advantages over the natural coloring agents. Many of the natural dyes are sensitive to degradation by light and oxygen, or by bacterial action. Therefore, they are not stable or long lasting. Synthetic colors can be devised which have a much longer shelf life. The synthetic dyes are also stronger, give more intense colors, and they can be used in smaller quantity to achieve a given color than can the natural agents. Often these artificial coloring materials are cheaper than the natural colors. This fact of economics is often especially true when the smaller amounts that are required are taken into account.

Why should artificial colors be added to foods at all? This question is easier to answer from the manufacturer's point of view than it is from that of the consumer. The manufacturer knows that, to a certain extent, the eye appeal of a product will affect its sales. As an example, a consumer is more likely to select for purchase an orange which has a bright orange skin than to purchase one with a mottled green and yellow skin. This is true even though the flavor and nutritive value of the orange may not be affected at all by the color of the skin. Sometimes, more than eye appeal is involved. The consumer is a creature of habit and is accustomed to have certain foodstuffs a particular color. Consider for a moment how you might react to green margarine or blue steak! For obvious reasons, they would not sell very well. Both butter and margarine are artificially colored yellow. Natural butter has a yellow color only in the summer; in the winter it is colorless, and manufacturers customarily add yellow coloring. Margarine must always be colored yellow.

Thus, the colors that are added to foodstuffs are added for quite a different reason than that which justifies the use of other types of food additives. Other additives may be added to foods for either nutritional or technological reasons. Some of these additives can be justified by good arguments. For instance, during the processing of many foods, valuable vitamins and minerals are lost. Many manufacturers replace these lost nutrients by "enriching" their product. In another instance, preservatives are sometimes added to food to forestall spoilage due to oxidation or the growth of bacteria, yeasts, and molds. With modern marketing practices, which involve the shipping and warehousing of products over long distances and time periods, the use of preservatives is a virtual necessity in many instances. Other additives, such as thickeners and emulsifiers, are often added for technological reasons, for example, to improve the texture of the foodstuff.

With food colors, however, there is no nutritional or technological necessity for their use. In fact, in some cases, dyes have been used to deceive customers. For instance, yellow dyes have been used in both cake mixes and egg noodles to suggest a higher egg content than was actually present. On the grounds that they are unnecessary, and perhaps dangerous, many persons have advocated that the use of all synthetic food dyes be abandoned.

Of all the food additives, dyes have come under the heaviest attack. As early as 1906, the government took steps to protect the consumer. At the turn of the century, there were more than 90 dyes used in foods. There were no governmental regulations, and the same dyes that were used for dyeing clothes could be used to color foodstuffs. The first legislation governing the use of dyes was passed in 1906, when food colors known to be harmful were removed from the market. At that time only 7 colors were approved for use in food. In 1938 the law was extended, and any batch of dye destined for use in food was required to be **certified** as to its chemical purity. Previously, certification was voluntary on the part of the manufacturer. At that time there were 15 food colors in general use and each was given a color and an F, D and C number designation rather than a chemical name. In 1950, when the number of dyes in use had expanded to 19, an unfortunate incident led to the discontinuation of three of the dyes: two were F, D & C Oranges Number 1 and Number 2, and the other was F, D & C Red Number 32. These dyes were removed when several children became seriously ill after eating popcorn colored by them.

Since that time, research has revealed that many of these dyes are either toxic, that they can cause birth defects, that they can cause heart trouble, or that they are carcinogenic (cancer inducing). Because of experimental evidence, mainly with chick embryos, rats, and dogs, Reds

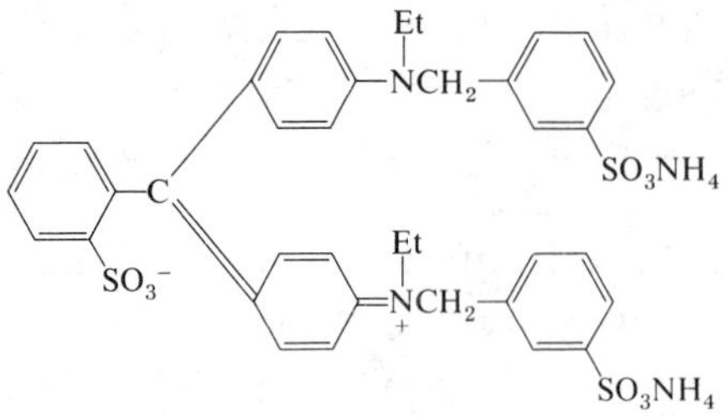

F, D&C Blue No. 1
(Brilliant Blue FCF)

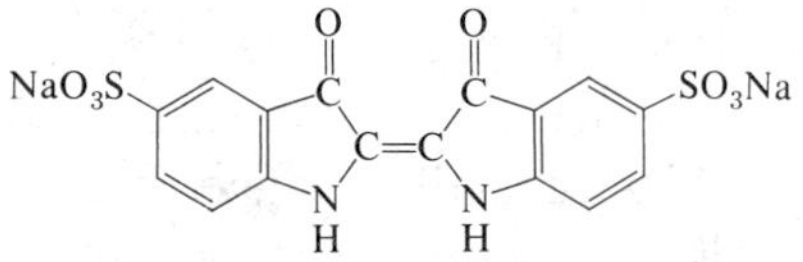

F, D&C Blue No. 2
(Indigo Carmine)

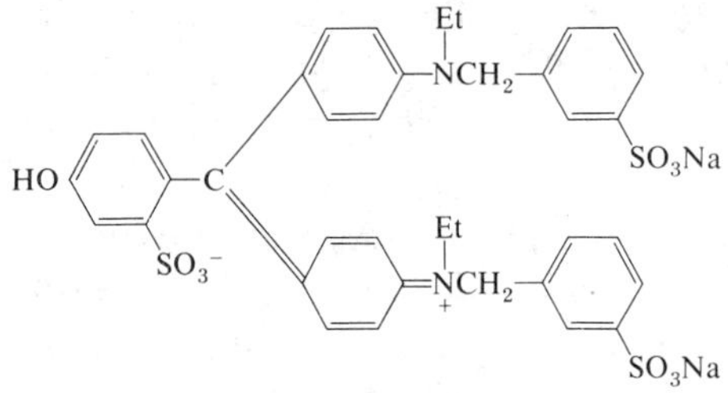

F, D&C Green No. 3
(Fast Green FCF)

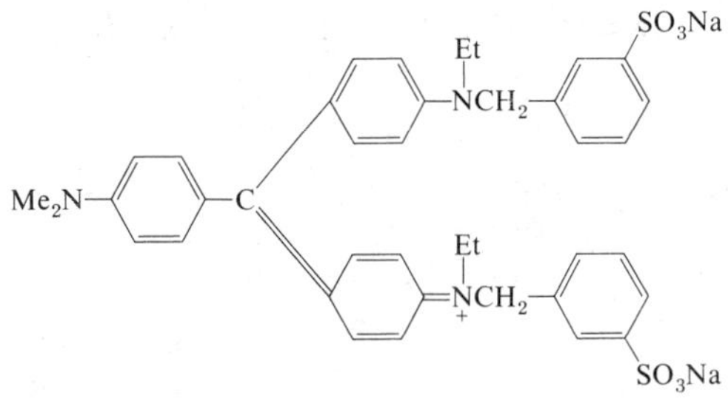

F, D&C Violet No. 1
(Benzyl Violet)

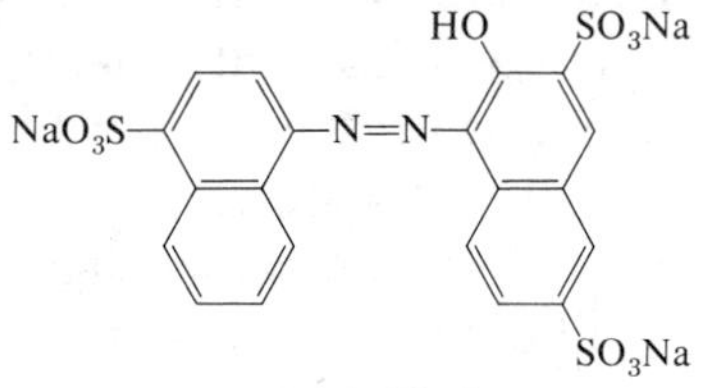

F, D&C Red No. 2
(Amaranth)

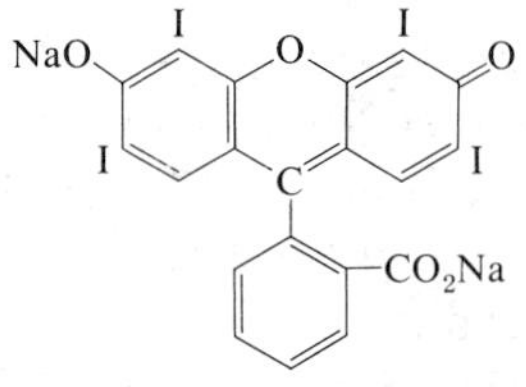

F, D&C Red No. 3
(Erythrosine)

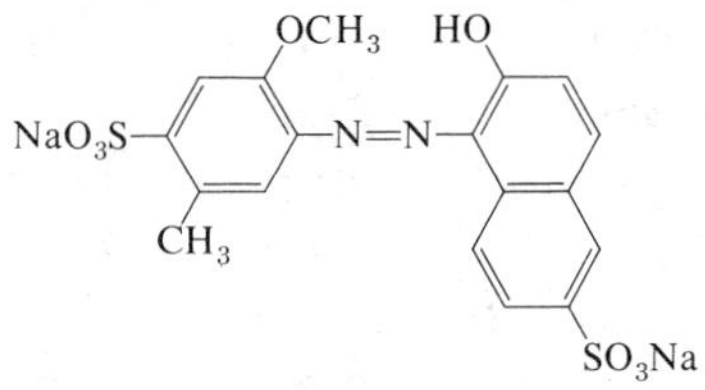

F, D&C Red No. 40
(Allura Red)

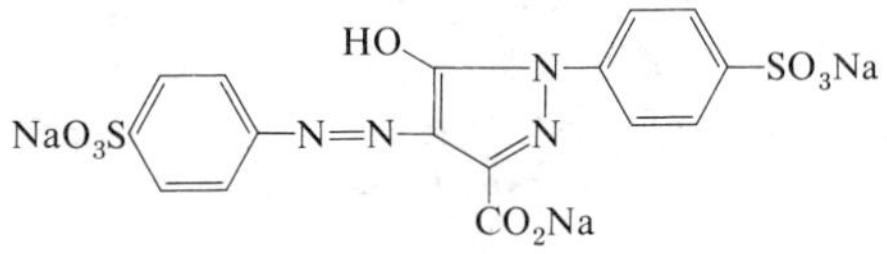

F, D&C Yellow No. 5
(Tartrazine)

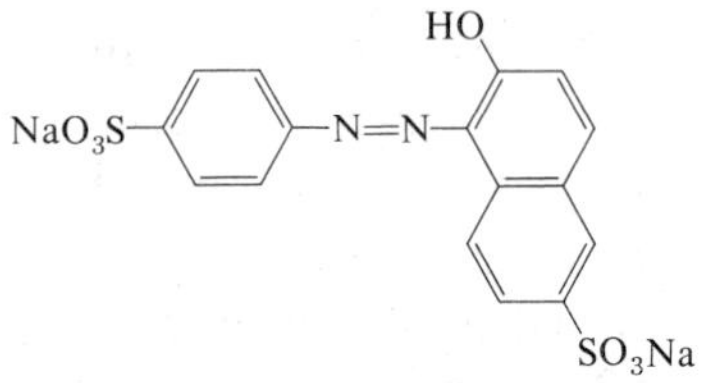

F, D&C Yellow No. 6
(Sunset Yellow)

The nine food colors currently approved by the Food and Drug Administration (1975)

Numbers 1 and 4, and Yellows Numbers 1, 2, 3, and 4 were also removed from the approved list in 1960. All of these are still proscribed except Reds Numbers 4 and 32, which are restricted to particular uses. In 1965, the ban on Red Number 4 was partially lifted to allow it to be used to color maraschino cherries. This use was allowed because there was no substitute dye available which would dye cherries, and it was thought that since maraschino cherries are mainly decorative, they are not properly considered a foodstuff. This use of Red Number 4 was considered to be a minor use. Similarly, Red Number 32, which may not be used to color food to be eaten, is now called Citrus Red Number 2, and is allowed only for the purpose of dyeing the skins of oranges.

The structures of the major food dyes are shown in the figure. Note that a good many of them are azo dyes. Since many of the dyes with the azo linkage have been shown to be carcinogens, many persons suspect all such dyes. In 1960, the law was amended to require that any new dyes submitted for approval be submitted to extensive scientific testing before they could be approved. They must be shown to be free from causing birth defects, organic dysfunction, and cancer. Old dyes may be subject to reconsideration if experimental evidence suggests that this is necessary.

The most controversial case right now is that of Red Number 2, also called Amaranth. In many tests, some even performed by FDA chemists, there is mounting evidence that this dye may be harmful. In some studies it has been concluded that Red Number 2 may cause birth defects, natural abortion of fetuses, and possibly cancer. Red Number 2 is the most widely used food dye in the industry, being used for everything from ice cream to cherry soda. The FDA has not yet activated a ban on this dye even though its own chemists and biochemists have made the recommendation that it be proscribed. Proscription of Red Number 2 would not be disastrous, since a ready substitute is available in Red Number 40.

Red Number 40, the most recently accepted food dye, was approved in 1971. Before gaining approval, the Allied Chemical Corporation, which holds exclusive patent rights to the dye, carried out the most thorough and expensive testing program ever given to a food dye. These tests even included a study of possible birth defects. Red Number 40, called Allura Red, seems destined to replace Red Number 2, since it can be used for an extremely wide variety of applications, including the dyeing of maraschino cherries, where it could replace the provisionally listed Red Number 4.

There are currently nine allowed dyes for food use. The structures of these nine approved food dyes are given in the accompanying chart. The use of these nine dyes is unrestricted. In addition, there are three other dyes that are approved for restricted uses. Red Number 4 may be used to color maraschino cherries, Citrus Red Number 2 (old Red Number 32) may be used to color the skins of oranges, and Orange B may be used to color the skins of sausages.

REFERENCES

Jacobson, M. F. *Eater's Digest: The Consumer's Factbook of Food Additives*. New York: Doubleday and Company (Anchor No. A 861), 1972.
"Red Food Coloring: How Safe Is It?" *Consumer's Reports,* 130 (February, 1973).
Sanders, H. J. "Food Additives." *Chemical and Engineering News,* Part I, (October 10, 1966), 100; Part II, (October 17, 1966), 108.
Turner, J. *The Chemical Feast*. New York: Grossman, 1970.
Winter, R. *A Consumer's Dictionary of Food Additives*. New York: Crown Publishers, 1972.

CHROMATOGRAPHY OF SOME DYE MIXTURES

Thin Layer Chromatography
  Pre-prepared Plates
  Hand-dipped Plates
Paper Chromatography

In this experiment, three different types of chromatography are used to separate mixtures of dyes. Two types of dye mixtures will be used. The first type of mixture will be that represented by the commercial food colors which can be purchased in any grocery store. These are usually available in small packages containing bottles of red, yellow, blue, and green food dye mixtures. As will be seen in the experiment, each of these colors is rarely compounded of only a single dye. For instance, the blue dye usually has a small admixture of a red dye to make it more brilliant in color. The most commonly used dyes in these food color packages are F, D & C Blue Number 1, F, D & C Red Number 2, and F, D & C Yellows Numbers 5 and 6 (see the essay which precedes this experiment for the structures).

The second type of dye mixture to be used is one compounded from three dyes not approved for food use. This mixture contains a red dye (Sudan Red G), a blue dye (Indophenol Blue), and a yellow dye (Butter Yellow). The structures of these three dyes are shown below.

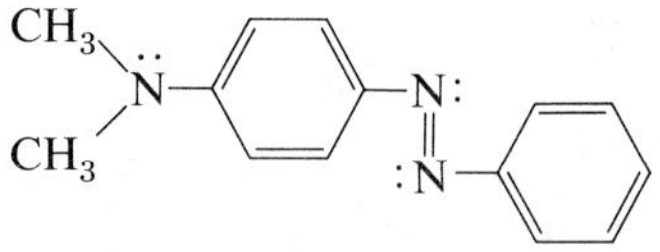

p-Dimethylaminoazobenzene
(Butter Yellow)

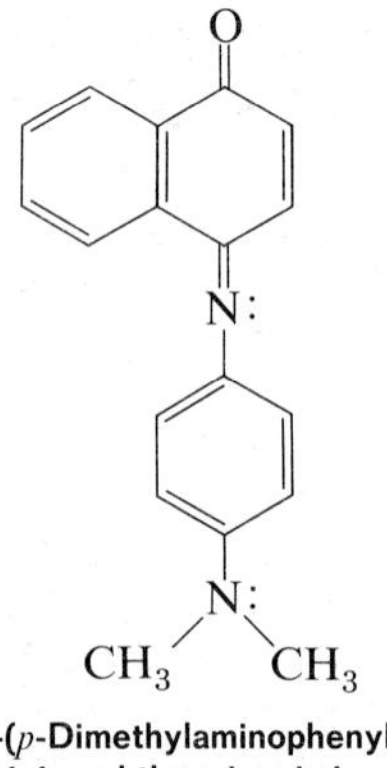

N-(*p*-Dimethylaminophenyl)
-1,4-naphthoquinonimine
(Indophenol Blue)

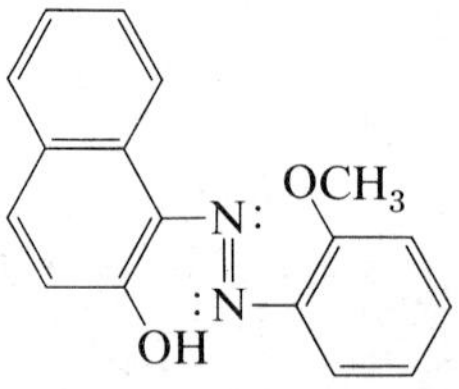

1-(*o*-Methoxyphenylazo)-
2-naphthol
(Sudan Red G)

SPECIAL INSTRUCTIONS

To understand this experiment, it will be necessary to read the essay, "Food Colors," which precedes it. In addition, it is necessary to have read Technique 10 (Column Chromatography), Sections 10.1 through 10.4, and Technique 11 (Thin Layer Chromatography).

The instructor may choose to do only a part of this experiment or, perhaps, to do all three parts. If all three parts are to be completed, the paper chromatography section should be started first because it requires the longest development time. The other experiments can be completed during the period of time that the solvent is ascending the paper chromatogram.

PROCEDURES

PAPER CHROMATOGRAPHY OF FOOD COLORS

At least 12 capillary micropipets will be required for the experiment. Prepare these according to the method described and illustrated in Technique 11, Section 11.3.

Prepare about 90 ml of a development solvent consisting of:

> 30 ml 2 N NH_4OH ($=$ 4 ml conc. NH_4OH + 26 ml H_2O)
> 30 ml 1-Pentanol ($= n$-Amyl or n-Pentyl Alcohol)
> 30 ml Absolute Ethanol

The entire mixture may be prepared in a 100 ml graduated cylinder. Mix the solvent well and pour it into the development chamber for storage. A 32 oz wide mouth screw cap jar (or a Mason jar) will serve as an appropriate development chamber. Cap the jar tightly to prevent losses of solvent due to evaporation.

Next, obtain a 12 cm $\times$ 24 cm sheet of Whatman Number 1 paper. Using a pencil (not a pen), lightly draw a 24 cm long line about 2 cm up from the long edge of the sheet. Using a centimeter ruler and the pencil, measure and mark off two dashed lines each about 2 cm from each short end of the paper. Then make 9 small marks at 2 cm intervals along the line on the long axis of the paper. These are the positions at which the samples will be spotted (see the illustration).

If they are available, starting from left to right, spot the reference dyes F, D & C Red Number 2 (Amaranth), F, D & C Blue Number 2 (Indigo Carmine), and F, D & C Yellow Number 5 (Tartrazine). These should be available in 2% aqueous solutions. It may be wise to practice the spotting technique on a small piece of Whatman Number 1 filter paper before attempting to spot the actual chromatogram. The correct method of spotting is described in Technique 11, Section 11.3. It is important that the spots be made as small as possible and that the paper not be overloaded. If either of these errors is made, the spots

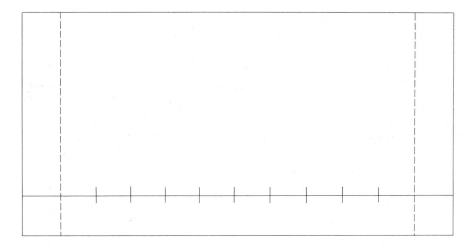

will tail and overlap one another after development. The applied spot should be 1 or 2 mm (1/16 in) in diameter.

On the remaining 6 positions (9 if standards are not used) you may spot any dyes of your choice. Use of a red, blue, green, and yellow dye from a single manufacturer is suggested. These four dyes could then be compared to two (say a yellow and a green) from another manufacturer. If the dyes are supplied in screw cap bottles, the pipets can be filled simply by dipping them into the bottle. However, if the dyes are supplied in squeeze bottles it will be easiest to place a drop of the dye on a microscope slide and to insert the pipet into the drop. One microscope slide should suffice for all the samples.

When the samples have been spotted, hold the paper upright with the spots at the bottom and coil it into a cylinder. Line up the two dashed lines and fasten the cylinder together with a paper clip or a staple. When the spots have dried, place the cylinder, spotted edge down, in the development chamber. The solvent level should be below the spots or they will dissolve in the solvent. Cap the jar and wait until the solvent ascends to the top of the paper. This will take about 40 minutes, and the remaining parts of the experiment (if required) may be performed while waiting.

When the solvent has ascended to within about 1 cm from the top of the paper, remove the cylinder, open it quickly, and, using a pencil, mark the level of the solvent. This uppermost level is the solvent front. Allow the chromatogram to dry. Then, using a ruler marked in millimeters, measure the distance that each spot has travelled relative to the solvent front and calculate its R_f value (see Technique 11, Section 11.8). Using the list of approved food dyes in the "Food Colors" essay, and the reference dyes (if used), try to determine which particular dyes were used to formulate the food colors you tested. Be sure to examine the dye package (or the bottles) to see if the desired information is given. What conclusions can be reached? Include your chromatogram along with your report.

SEPARATION OF FOOD COLORS USING PRE-PREPARED TLC PLATES

Obtain from the instructor a 5 cm $\times$ 10 cm sheet of a pre-prepared silica gel tlc plate (Eastman Chromagram Sheet No. 13180 or No. 13181). These plates have a flexible backing, but they should not be bent excessively. They should be handled carefully or the adsorbent may flake off of them. In addition, they should be handled only by the edges. The surface should not be touched.

Using a lead pencil (not a pen), **lightly** draw a line across the short dimension of the plate about 1 cm from the bottom. Using a centimeter ruler, mark off four 1 cm intervals on the line (see figure). These are the points at which the samples will be spotted.

Prepare at least four capillary micropipets as described and illustrated in Technique 11, Section 11.3. Starting from left to right, spot first a red color food dye, then a blue dye, a green dye, and a yellow dye. The correct method of spotting a tlc slide is described in Technique 11, Section 11.3. It is important that the spots be made as small as possible and that the slide not be overloaded. If either of these errors is made, the spots will tail and will overlap one another after development. The applied spot should be about 1 or 2 mm (1/16 in) in diameter. If small scrap pieces of the tlc plates are available, it would be a good idea to practice spotting on these prior to preparing the actual sample plate.

Prepare a development chamber from an 8 oz wide mouth screw cap jar. It **should not** have the filter paper liner which is described in Technique 11, Section 11.4. These plates are very thin and, if they touch a liner at any point, solvent will begin to diffuse onto the plate from that point. The development solvent, which can be prepared in a 10 ml graduated cylinder, should be a 60/40 mixture of methanol and acetone. Mix the solvent well and pour enough into the development chamber to give a solvent depth of about 0.5 cm (or less). If the solvent level is too high, it will cover the spotted substances, and they will dissolve into the solvent reservoir.

Place the spotted tlc slide in the development chamber, cap the jar tightly, and wait for the solvent to rise to almost the top of the slide. When the solvent is close to the top edge, remove the slide and, using a pencil (not a pen), quickly mark the position of the solvent front. Allow the plate to dry. Using a ruler marked in millimeters, measure the distance that each spot has travelled relative to the solvent front and calculate its R_f value (see Technique 11, Section 11.8).

At your instructor's option, and if the dyes are available, you may be asked to prepare a second plate spotted with a set of reference dyes. The reference dyes will include F, D & C Red Number 2 (Amaranth), F, D & C Blue Number 2 (Indigo Carmine), and F, D & C Yellow Number 5 (Tartrazine) and Number 6 (Sunset Yellow). If this second set of dyes is analyzed, it should be possible (using the list of approved dyes in the "Food Colors" essay) to determine the identity of the dyes used to formulate the food colors tested on the first plate. Be sure to examine the package (or bottles) of the food dyes to determine if the desired information is given.

Since the plates are rather fragile, a sketch of the slide(s) should be included along with your report.

SEPARATION OF A DYE MIXTURE USING HAND-DIPPED TLC SLIDES

Using a silica gel slurry, prepare three hand-dipped microscope slide tlc plates by the method described in Technique 11, Section 11.2 A. Prepare a developing chamber from a 4 oz wide mouth screw cap jar as described in Technique 11, Section 11.4. Finally, prepare several capillary micropipets as described in Technique 11, Section 11.3.

Next, obtain from the reagent shelf a bottle containing a mixture of the three dyes, Sudan Red G, Indophenol Blue, and Butter Yellow.[1] Using a capillary micropipet, spot the dye mixture twice on each plate. The correct method of spotting a tlc plate is described in Technique 11, Section 11.3. It is important that the spots be made as small as possible and that the plates not be overloaded. If either of these errors is made, the spots will tail and will overlap one another after development. The applied spot should be 1 or 2 mm (1/16 in) in diameter.

Develop the first slide using chloroform as the development solvent. When the solvent has risen to within about 0.5 cm from the top of the slide, remove it and, using a lead pencil, quickly mark the position of the solvent front. Set the slide aside to dry. Empty the development chamber and allow it to dry. Refill the development chamber with a solvent consisting of 8.5 ml of cyclohexane and 1.5 ml of ethyl acetate. Develop the second slide in this solvent. Be sure to mark the solvent front when the slide is removed. Finally, using benzene as the solvent,

---

[1] Available from the Mallinckrodt Chemical Company. Mallinckrodt No. 3082 TLC Dye Mixture (0.1% W/V in benzene).

develop the third slide. Using a ruler calibrated in millimeters, measure the distance that each spot travelled relative to its solvent front and calculate an R_f value for that spot (Technique 11, Section 11.8). Include sketches of each plate in your report and explain the results. Note that ethyl acetate is capable of hydrogen bonding. Also note that Sudan Red is capable of internal hydrogen bonding and that oxygen forms stronger hydrogen bonds than nitrogen. A solvent capable of hydrogen bonding can interfere with internal hydrogen bonding.

ESSAY

Polymers and Plastics

Chemically, plastics are composed of chainlike molecules of high molecular weight, called polymers. Polymers have been built up from simpler chemicals, called monomers. A different monomer or combination of monomers is used to manufacture each different type or family of polymers. There are many polymers around us that are familiar. Examples of manmade polymers are Teflon, nylon, Dacron, polyethylene, polyester, Orlon, epoxy, vinyl, polyurethan, silicones, Lucite, and boat resin. Examples of natural polymers are starch and cellulose (from glucose), rubber (from isoprene) and proteins (from amino acids). Certainly, polymers have made a great impact on our society. They are rapidly replacing many metals for making objects, and man-made polymeric textiles are replacing natural fibers for making cloth. Of course, in the process of creating these materials, a problem has arisen in the disposal of them since many are not biodegradable.

CHEMICAL STRUCTURES OF POLYMERS

Basically a polymer is made up of many repeating molecular units formed by sequential addition of monomer molecules to one another. Many monomer molecules of A, say 1000 to 1 million, can be linked together to form a gigantic polymeric molecule:

$$\text{Many A} \longrightarrow \text{etc.---A-A-A-A-A---etc. } \textbf{or } \left(\text{A}\right)_n$$

Monomer   **Polymer**
Molecules   **Molecule**

One may also link together monomers which are different, to produce a polymer with an alternating structure. This type of polymer is called a co-polymer.

Many A + many B $\longrightarrow$ etc.—A-B-A-B-A-B—etc. **or** $\dashleftarrow\text{A-B}\dashrightarrow_n$

Monomer   Polymer
Molecules   Molecule

TYPES OF POLYMERS

For convenience, chemists divide polymers into several main classes, depending on their method of synthesis.

1. **Addition polymers** are formed by a reaction in which monomer units simply add to one another to form a long chain (generally linear or branched) polymer. The monomers usually contain carbon–carbon double bonds. Familiar examples of addition polymers are: polyethylene and Teflon. The process may be represented as follows:

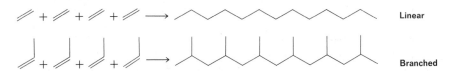

2. **Condensation polymers** are formed by reactions of bi- or polyfunctional molecules, with the elimination of some small molecule (such as water, ammonia, or hydrogen chloride) as a by-product. Familiar examples of condensation polymers are: nylon, Dacron, and polyurethan. The process may be represented as follows:

$$\text{H}-\boxed{}-\text{X} + \text{H}-\boxed{}-\text{X} \longrightarrow \text{H}-\boxed{}-\boxed{}-\text{X} + \text{HX}$$

3. **Crosslinked polymers** are formed by the linking together of long chains into one gigantic, 3-dimensional structure with tremendous rigidity. Addition and condensation polymers can exist with a crosslinked network, depending on the monomers used in the synthesis. Familiar examples of crosslinked polymers are: Bakelite, rubber, and casting (boat) resin. The process may be represented as follows:

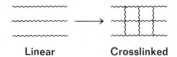

Linear   Crosslinked

Industrialists and technologists often divide polymers into other classes, as well.

1. **Thermoplastics** are materials which may be softened (melted) by heat and reformed (molded) into another shape. Weaker, non-covalent bonds are broken in the heating process. Technically, thermoplastics are the materials which we refer to as "plastics." Both addition and condensation polymers can fall

in this classification scheme. Familiar examples include polyethylene (addition polymer) and nylon (condensation polymer).

2. **Thermoset plastics** are materials which can be melted initially,
but upon further heating become permanently hardened. They
cannot be softened and remolded into another shape without
destroying the polymer by breaking the covalent bonds. Chemically, thermoset plastics are crosslinked polymers. Bakelite is
an example of a thermoset plastic.

Polymers can also be classified in other ways, for example, many
varieties of rubber are often referred to as elastomers, Dacron is a fiber,
and polyvinyl acetate is an adhesive. The addition and condensation
classification will be used in this essay.

ADDITION POLYMERS

Most of the polymers made are of the addition type. The monomers
generally contain a carbon–carbon double bond. The most important
example of an addition polymer is the well known polyethylene. In
this case the monomer is ethylene. Countless numbers of ethylene
molecules are linked together into long chain polymeric molecules by
breaking the pi bond and creating two new single bonds between the
monomer units. The number of recurring units may be large or small,
depending on the polymerization conditions.

Ethylene
Monomer

Polyethylene
Polymer

This reaction may be promoted by heat, pressure, and a chemical
catalyst. The molecules produced in a typical reaction vary in the
number of carbon atoms in their chains. In other words, a mixture of
polymers of varying length is produced, rather than a pure compound.

Polyethylenes, with linear structures, can pack together easily and
are referred to as high density polyethylenes. They are fairly rigid
materials. Low density polyethylenes consist of branched chain molecules, with some crosslinking in the chains. They are more flexible than
the high density polyethylenes. The reaction conditions and the catalysts
which produce low and high density polyethylenes are quite different.
The monomer, however, is the same in each case.

Another example of an addition polymer is polypropylene. In this
case, the monomer is propylene. The polymer which results has a
branched methyl on alternate carbon atoms of the chain.

Propylene
Monomer

Polypropylene
Polymer

TABLE 1. ADDITION POLYMERS

EXAMPLE	MONOMER(S)	POLYMER	USE
Polyethylene	$CH_2{=}CH_2$	$—CH_2—CH_2—$	Most common and important polymer: bags, insulation for wires, squeeze bottles
Polypropylene	$CH_2{=}CH$ $\quad\ CH_3$	$—CH_2—CH—$ $\qquad\ CH_3$	Fibers, indoor-outdoor carpets, bottles
Polystyrene	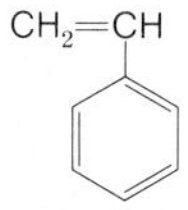 $CH_2{=}CH$ (phenyl)	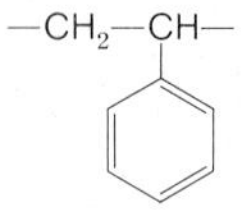 $—CH_2—CH—$ (phenyl)	Styrofoam, inexpensive household goods, inexpensive molded objects
Polyvinyl Chloride (PVC)	$CH_2{=}CH$ $\qquad Cl$	$—CH_2—CH—$ $\qquad\quad Cl$	Synthetic leather, clear bottles, floor covering, phonograph records, water pipe
Polytetrafluoroethylene (Teflon)	$CF_2{=}CF_2$	$—CF_2—CF_2—$	Non-stick surfaces, chemically resistant films
Polymethyl Methacrylate (Lucite, Plexiglas)	$\qquad CO_2CH_3$ $CH_2{=}C$ $\qquad CH_3$	$\qquad CO_2CH_3$ $—CH_2—C—$ $\qquad\ CH_3$	Unbreakable "glass", latex paints
Polyacrylonitrile (Orlon, Acrilan, Creslan)	$CH_2{=}CH$ $\qquad CN$	$—CH_2—CH—$ $\qquad\quad CN$	Fiber used in sweaters, blankets, carpets
Polyvinyl Acetate (PVA)	$CH_2{=}CH$ $\qquad OCCH_3$ $\qquad\ \ \ O$	$—CH_2—CH—$ $\qquad\quad OCCH_3$ $\qquad\qquad O$	Adhesives, latex paints, chewing gum, textile coatings
Natural Rubber	$\qquad\ CH_3$ $CH_2{=}CCH{=}CH_2$	$\qquad\quad CH_3$ $—CH_2—C{=}CH—CH_2—$	The polymer is crosslinked with sulfur (vulcanization)
Polychloroprene (Neoprene Rubber)	$\qquad\ Cl$ $CH_2{=}CCH{=}CH_2$	$\qquad\quad Cl$ $—CH_2—C{=}CH—CH_2—$	Crosslinked with ZnO, resistant to oil, gasoline
Styrene Butadiene Rubber (SBR)	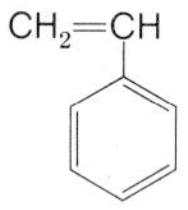 $CH_2{=}CH$ (phenyl) $CH_2{=}CHCH{=}CH_2$	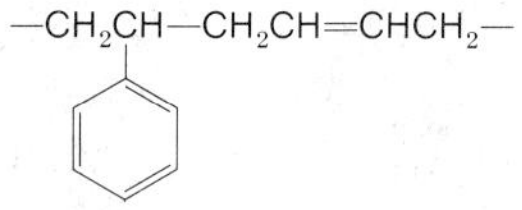 $—CH_2CH—CH_2CH{=}CHCH_2—$ (phenyl)	Crosslinked with peroxides: Most common rubber, used for tires. 25% styrene, 75% butadiene

A number of common addition polymers are shown in Table 1. Some of their principal uses are also listed. The last three entries in the table all have a carbon–carbon double bond remaining after the polymer is formed. These bonds activate or participate in a further reaction to form crosslinked polymers, called elastomers, a term almost synonymous with rubber, because of their common characteristics.

CONDENSATION POLYMERS

Condensation polymers are more complex than addition polymers because the monomers contain more than one type of functional group. In addition, most are copolymers made from more than one type of monomer. You will recall, in contrast, that addition polymers are all prepared from substituted ethylene molecules. The single functional group in each case is a double bond(s), and a single type of monomer is generally used.

Dacron, a polyester, may be prepared by the reaction of a dicarboxylic acid with a bifunctional alcohol (a diol):

$$HO-\overset{O}{\underset{}{C}}-\langle\rangle-\overset{O}{\underset{}{C}}\boxed{OH \quad H}OCH_2CH_2OH \longrightarrow -\overset{O}{\underset{}{C}}-\langle\rangle-\overset{O}{\underset{}{C}}-OCH_2CH_2-O- + H_2O$$

Terephthalic Acid	Ethylene Glycol	Dacron

Nylon 6-6, a polyamide, may be prepared by the reaction of a dicarboxylic acid with a bifunctional amine:

$$HO-\overset{O}{\underset{}{C}}(CH_2)_4\overset{O}{\underset{}{C}}\boxed{OH \quad H}\overset{H}{\underset{}{N}}(CH_2)_6\overset{H}{\underset{}{N}}H \longrightarrow -\overset{O}{\underset{}{C}}(CH_2)_4\overset{O}{\underset{}{C}}-\underset{H}{\overset{}{N}}(CH_2)_6\underset{H}{\overset{}{N}}- + H_2O$$

Adipic Acid	Hexamethylene-diamine	Nylon

Notice, in each case, that a small molecule, water, is eliminated as a product of the reaction. A number of other condensation polymers are listed in Table 2. Linear (or branched) chain polymers as well as crosslinked polymers are produced in condensation reactions.

The nylon structure contains the amide linkage at regular intervals, $-\overset{O}{\underset{}{C}}-\overset{H}{\underset{}{N}}-$. This type of linkage is extremely important in nature because of its presence in proteins or polypeptides. Proteins are gigantic polymeric substances made up of monomer units of amino acids. They are linked together by the peptide (amide) bond. Proteins and amino acids are discussed in the essay preceding Experiment 54.

Other important condensation polymers which occur naturally are starch and cellulose. They are polymeric materials made up of the sugar monomer, glucose. These polymers are discussed in the essay which precedes Experiment 51. Another important condensation polymer which occurs naturally is the DNA molecule. A DNA molecule is made up of deoxyribose sugar linked together with phosphates to form the backbone of the molecule. A portion of a DNA molecule is shown in the essay which precedes Experiment 6.

TABLE 2.   CONDENSATION POLYMERS

EXAMPLE	MONOMERS	POLYMER	USE
Polyamides (Nylon)	$HOOC(CH_2)_n$—$COOH$ $H_2N(CH_2)_nNH_2$	—$CO(CH_2)_nCO$—$NH(CH_2)_nNH$—	Fibers, molded objects
Polyesters (Dacron, Mylar, Fortrel)	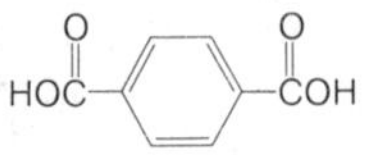 $HOOC$—⬡—$COOH$ $HO(CH_2)_nOH$	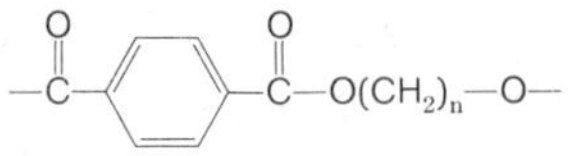 —CO—⬡—CO—$O(CH_2)_n$—O—	Linear polyesters: fibers, recording tape
Polyesters (Glyptal Resin)	phthalic anhydride $HOCH_2CHCH_2OH$ $\quad\quad\;|$ $\quad\quad OH$	—CO—⬡—$COOCH_2CHCH_2O$—	Crosslinked polyester: paints
Polyesters (Casting Resin)	$HOOCCH{=}CHCOOH$ $HO(CH_2)_nOH$	—$COCH{=}CHCO$—$O(CH_2)_n$—O—	Crosslinked with styrene and peroxide: fiber glass boat resin
Phenol-Formaldehyde Resin (Bakelite)	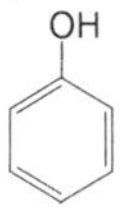 phenol (⬡OH) $CH_2{=}O$	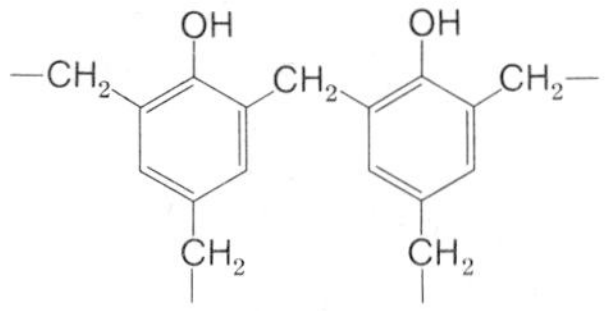 phenol-formaldehyde crosslinked structure	Mixed with fillers: molded electrical goods, adhesives, laminates, varnishes
Cellulose Acetate*	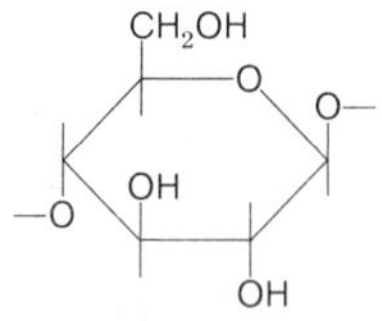 glucose unit, OH groups CH_3COOH	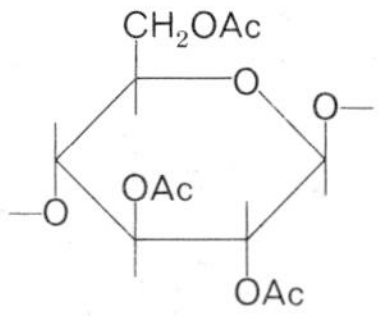 glucose unit with CH_2OAc, OAc groups	Photographic film
Silicones	$\begin{array}{c}CH_3\\ Cl{-}Si{-}Cl\\ CH_3\end{array}$ H_2O	$\begin{array}{c}CH_3\\ {-}O{-}Si{-}O{-}\\ CH_3\end{array}$	Water-repellent coatings, temperature-resistant fluids and rubbers (CH_3SiCl_3 crosslinks in water)
Polyurethans	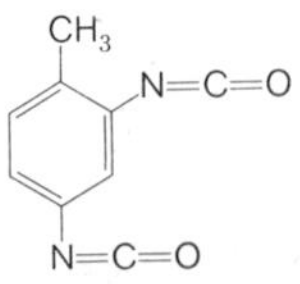 CH_3—⬡—$N{=}C{=}O$ $N{=}C{=}O$ $HO(CH_2)_nOH$	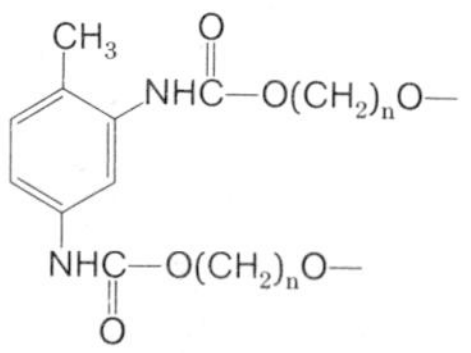 CH_3—⬡—$NHCO$—$O(CH_2)_nO$— $NHCO$—$O(CH_2)_nO$—	Rigid and flexible foams, fibers

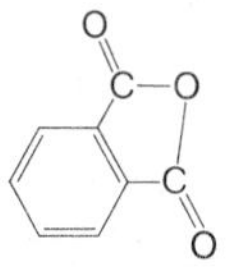

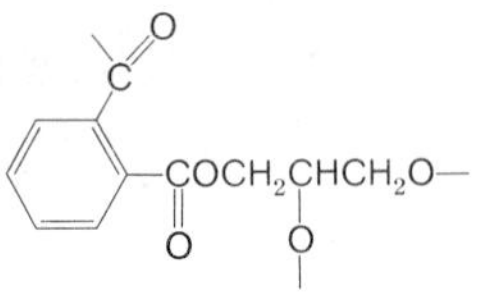

* Cellulose, a polymer of glucose, is used as the monomer.

PROBLEMS WITH PLASTICS

Plastics have certainly become very common in our society. However, they are not without their problems. There are disposal problems, health hazards, littering problems, fire hazards, and energy shortages associated with their manufacture and use.

Plasticizers and Health Hazards. Certain types of plastics such as polyvinyl chloride (PVC) are mixed with plasticizers which soften the plastic so that it is more pliable. If plasticizers were not added, the plastic would be hard and brittle. Some of the plasticizers used in vinyl plastics are phthalate esters. The structure of a phthalate ester is shown below. The esters are volatile, low molecular weight compounds. Part of the new car "smell" comes from the odor of these esters as they evaporate from the vinyl upholstery. The vapor often condenses on the windshield as an oily, insoluble film. After some time the vinyl material may lose enough plasticizer to cause it to crack. Phthalate esters may constitute a health hazard. Sometimes vinyl containers incorporating phthalate plasticizers are used to store blood. The esters are leached from blood bags made of PVC and may be partly responsible for shock lung, a condition that sometimes leads to death during a blood transfusion. The long term effects of these plasticizers are, however, not known.

Recently, a rare and fatal form of liver cancer (angiosarcoma) was discovered among small numbers of workers in chemical companies making polyvinyl chloride (PVC). The monomer used in making PVC is vinyl chloride, a gas. The structure is shown below. Currently, industry is required to eliminate this health hazard by reducing or eliminating vinyl chloride from the atmosphere.

Other types of plasticizers once used were the polychlorinated biphenyls (PCB). These compounds have similar physiological effects to DDT, and are even more persistent in the environment! The PCBs are actually a mixture of compounds where the hydrogens on the basic hydrocarbon structure are replaced with chlorines (from one to ten hydrogens can be replaced). One typical PCB which may be present in a plasticizer mixture is shown below. PCBs are no longer being sold except for use in closed systems, where they cannot leak into the environment.

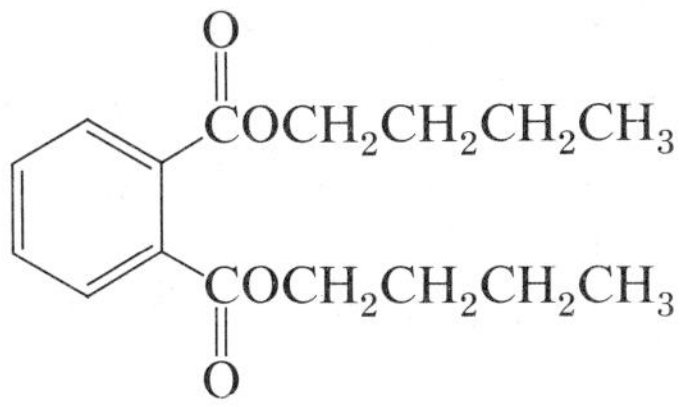

Dibutyl Phthalate	Vinyl Chloride	A Polychlorinated Biphenyl (PCB)

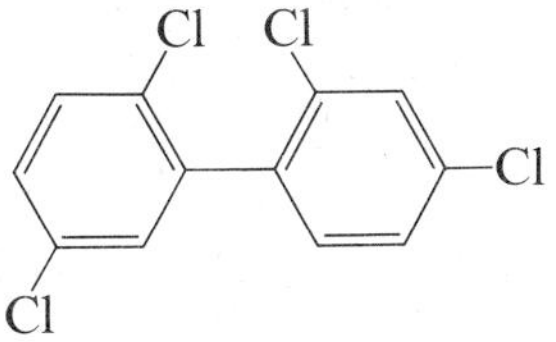

Disposability Problems. What do we do with all of our wastes? Currently, the most popular method is to bury our garbage in sanitary landfills. However, as we run out of good places to bury our garbage, incineration appears to be an attractive method for solving the solid waste problem. Plastics, which consist of about 2% of our garbage, burn readily. The new high temperature incinerators are extremely efficient and can be operated with very little air pollution. In addition, it should be possible to burn our garbage and generate electrical power from it.

Ideally, we should either recycle all of our wastes, or not produce the waste in the first place. Plastic waste consists of about 55% polyethylene and polypropylene, 20% polystyrene, and 11% PVC. All of these polymers are thermoplastics and can be recycled. They can be resoftened and remolded into new goods. Unfortunately, thermosetting plastics (crosslinked polymers) cannot be remelted. They decompose upon high temperature heating. Thus, thermosetting plastics should not be used for "disposable" purposes. In order to recycle plastics effectively, we must sort the materials according to the various types. This requires will power as well as knowing what the plastics are that we are discarding. Neither task is easy to perform!

Littering Problems. Plastics, if they are well made, will not corrode, or rust, and will last almost indefinitely. Unfortunately, these desirable properties also lead to a problem when plastics are buried in a landfill or thrown on the landscape—they do not decompose. Currently, research is being undertaken to discover plastics which are biodegradable or photodegradable so that either microorganisms or light from the sun can decompose our litter and garbage. Some success has been achieved.

Fire Hazards. Numerous injuries are caused by clothing made of polymers, especially children's clothing. Many of these organic fibers burn readily. To combat this problem, flame retardant fabrics have been developed recently, especially for children's sleepwear.

Toxic gases are sometimes liberated when plastics burn. For example, hydrogen chloride and hydrogen cyanide are generated when PVC and polyacrylonitriles, respectively, are burned. This presents a problem that compounds the fire danger.

Energy Shortage. The demand for energy has increased at an alarming rate, leading to the energy crisis. The production of polymers requires petroleum as a raw material and as a source of energy to operate the manufacturing process. Unfortunately, fossil fuels are a nonrenewable resource, and as their availability decreases, we will have an even greater problem. On the other hand, natural substances, such as cotton, are renewable resources; perhaps for some uses they would actually be better, and less costly than the synthesized polymers. There are many plastics, however, which are superior to natural materials. The answer lies in using plastics wisely.

REFERENCES

"Biodegradability: Lofty Goal for Plastics." *Chemical and Engineering News,* (September 11, 1972), 37–38.

"Degradable Plastic Lids are now Commercial." *Chemical and Engineering News,* (June 19, 1972), 32.

Flory, P. J. "Understanding Unruly Molecules." *Chemistry,* (May, 1964), 6–13.

Heckert, W. W. "Synthetic Fibers," *Journal of Chemical Education, 30* (1953), 166.

Kaufman, M. *Giant Molecules.* New York: Doubleday, 1968.

Keller, E. "Nylon—from Test Tube to Counter," *Chemistry,* (September. 1964), 8–23.

Mark, H. F. "Giant Molecules." *Scientific American, 197* (1957), 80.

Mark, H. F. "The Nature of Polymeric Materials." *Scientific American, 217* (1967), 149.

McGrew, F. C. "Structure of Synthetic High Polymers." *Journal of Chemical Education, 35* (1958), 178.

"More Trouble Brewing for Vinyl Chloride." *Chemical and Engineering News,* (September 2, 1974), 12–13.

Morton, M. "Big Molecules." *Chemistry,* (January, 1964), 13–17.

Morton, M. "Design and Formation of Long Chain Polymers." *Chemistry,* (March, 1964), 6–11.

Oster, B. "Polyethylene." *Scientific American, 197* (1957), 139.

Natta, G. "How Giant Molecules are Made." *Scientific American, 197* (1957), 98.

"Plasticizers: Found in Heart Cells." *Chemical and Engineering News,* (November 8, 1971), 7–8.

"Plastics Face Growing Pressure from Ecologists." *Chemical and Engineering News,* (March 29, 1971), 12–13.

"PVC Makers Hit with Health Hazard Suit." *Chemical and Engineering News,* (September 2, 1974), 9–10.

Experiment **36**

PREPARATION OF POLYMERS: POLYESTER, NYLON, POLYSTYRENE AND POLYURETHAN

Condensation Polymers
Addition Polymers
Crosslinked Polymers

In this experiment the syntheses of two polyesters (Procedure 36A), nylon (Procedure 36B), polystyrene (Procedure 36C), and polyurethan (Procedure 36D) are described. These polymers are representative of the more important commercial plastics. They are also representative of the main classes of polymers: condensation (linear polyester, nylon), addition (polystyrene), and crosslinked (Glyptal polyester, polyurethan).

SPECIAL INSTRUCTIONS

Before performing this experiment, it is necessary to have read the essay which precedes this experiment. Toluene diisocyanate (TDI) is toxic. It will irritate the skin and eyes. Avoid breathing the vapor.

POLYESTERS

Linear and crosslinked polyesters are prepared in this experiment. Both are examples of condensation polymers.

The linear polyester is prepared as follows:

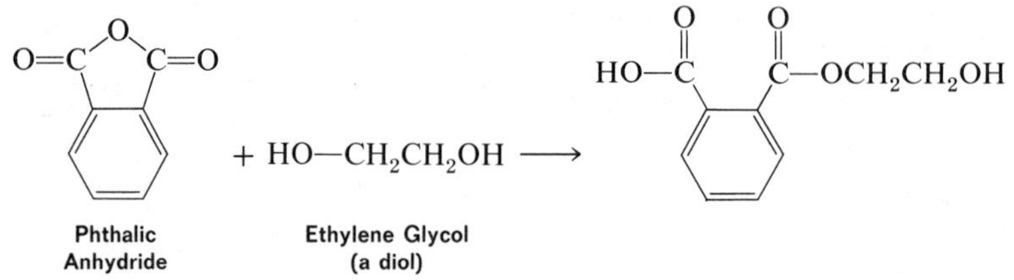

Phthalic Anhydride	Ethylene Glycol (a diol)	

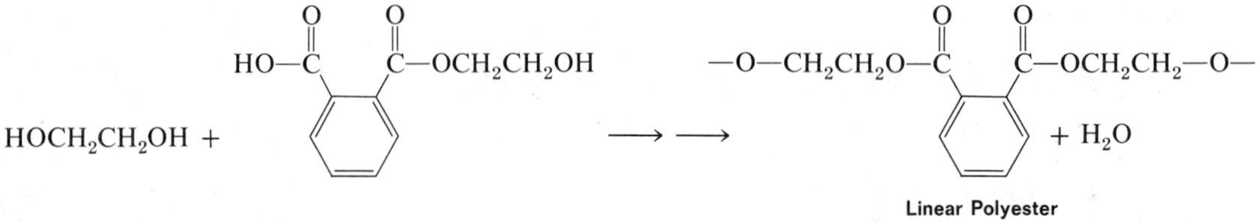

This linear polyester is isomeric with Dacron, which is prepared from terephthalic acid and ethylene glycol (see the preceding essay). Dacron and the linear polyester made in this experiment are both thermoplastics.

If more than two functional groups are present in one of the monomers, the polymer chains may be linked to one another (crosslinked) to form a three-dimensional network. Such structures are usually more rigid than linear structures and are useful in making paints and coatings. They may be classified as thermosetting plastics. The polyester, Glyptal, is prepared as follows:

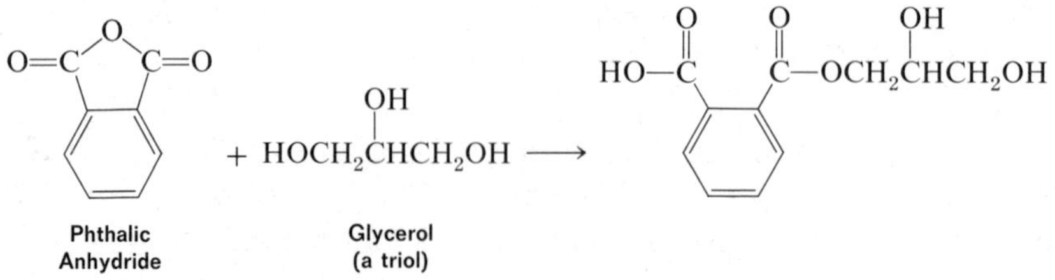

Phthalic Anhydride

Glycerol (a triol)

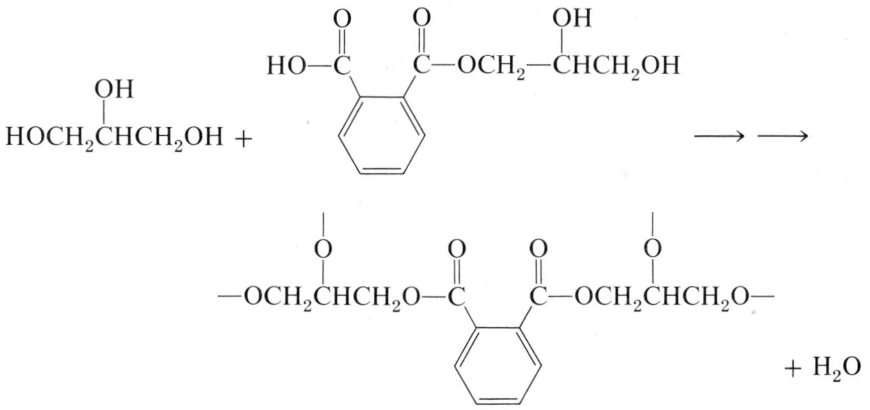

$$HOCH_2\overset{\overset{\displaystyle OH}{|}}{C}HCH_2OH \ + \ \underset{\text{(phthalic ring)}}{HO-\overset{\overset{\displaystyle O}{\|}}{C} \quad \overset{\overset{\displaystyle O}{\|}}{C}-OCH_2-\overset{\overset{\displaystyle OH}{|}}{C}HCH_2OH} \longrightarrow \longrightarrow$$

$$-OCH_2\overset{\overset{\displaystyle O}{|}}{C}HCH_2O-\overset{\overset{\displaystyle O}{\|}}{C} \quad \overset{\overset{\displaystyle O}{\|}}{C}-OCH_2\overset{\overset{\displaystyle O}{|}}{C}HCH_2O-$$

$$+ \ H_2O$$

Crosslinked Polyester
(Glyptal Resin)

The reaction of phthalic anhydride with a diol (ethylene glycol) is described in the following procedure. This linear polyester will be compared to the crosslinked polyester (Glyptal) prepared from phthalic anhydride and a triol (glycerol).

PROCEDURE

Place 2 g of phthalic anhydride and 0.1 g of sodium acetate in each of two test tubes. To one tube add 0.8 ml of ethylene glycol, and to the other add 0.8 ml of glycerol. Clamp both tubes so that they may be heated with a flame simultaneously. Heat the tubes gently until the solutions appear to boil (due to the elimination of water during the esterification), then continue the heating for five minutes. Allow the tubes to cool and compare the viscosity and brittleness of the two polymers.

Procedure **36**B

POLYAMIDE (NYLON)

Reaction of a dicarboxylic acid, or one of its derivatives, with a diamine leads to a linear polyamide through a condensation reaction. Commercially, nylon 6-6 (so called because each monomer has 6 carbons) is made from adipic acid and hexamethylenediamine. In this experiment, you will use the acid chloride instead of adipic acid.

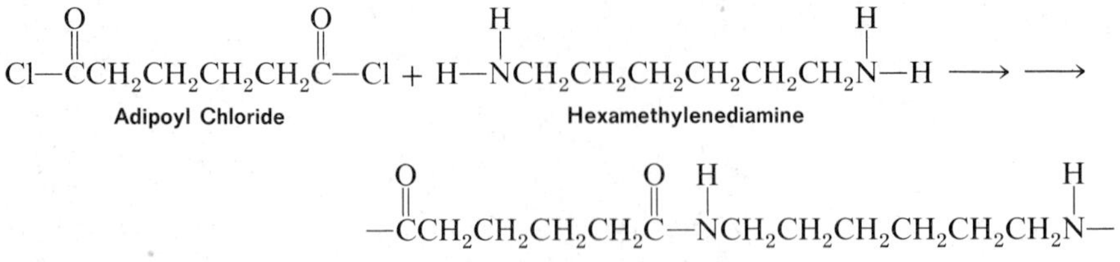

$$Cl-\underset{\underset{}{\overset{O}{\parallel}}}{C}CH_2CH_2CH_2CH_2\underset{\underset{}{\overset{O}{\parallel}}}{C}-Cl \ + \ H-\underset{\overset{H}{\mid}}{N}CH_2CH_2CH_2CH_2CH_2CH_2\underset{\overset{H}{\mid}}{N}-H \ \longrightarrow \ \longrightarrow$$

Adipoyl Chloride   Hexamethylenediamine

$$-\underset{\underset{}{\overset{O}{\parallel}}}{C}CH_2CH_2CH_2CH_2\underset{\underset{}{\overset{O}{\parallel}}}{C}-\underset{\overset{H}{\mid}}{N}CH_2CH_2CH_2CH_2CH_2CH_2\underset{\overset{H}{\mid}}{N}-$$

Nylon 6-6

The acid chloride is dissolved in cyclohexane and this is added **carefully** to hexamethylenediamine dissolved in water. These liquids will not mix and two layers will form. At the point of contact between the layers (interface), the nylon will form. It can then be drawn out continuously to form a long strand of nylon. Imagine how many molecules have been linked together in this long strand! It is a fantastic number.

PROCEDURE

Pour 10 ml of a 5% aqueous solution of hexamethylenediamine (1,6-hexanediamine) into a 50 ml beaker. Add 10 drops of 20% sodium hydroxide solution. Carefully add 10 ml of a 5% solution of adipoyl

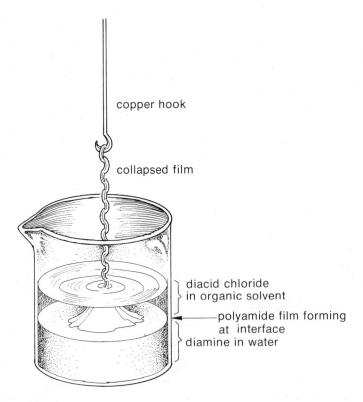

copper hook

collapsed film

diacid chloride in organic solvent

polyamide film forming at interface

diamine in water

Preparation of nylon

chloride in cyclohexane to the solution by pouring it down the wall of the slightly tilted beaker. Two layers will form (see figure), and there will be an immediate formation of a polymer film at the liquid–liquid interface. Using a copper wire hook (a six-inch piece of wire bent at one end) gently free the walls of the beaker from polymer strings. Then hook the mass at the center, and slowly raise the wire so that polyamide forms continuously, producing a rope which can be drawn out for many feet. The strand may be broken by pulling it at a faster rate. Rinse the rope several times with water, and lay it on a paper towel to dry. With the piece of wire, vigorously stir the remainder of the two-phase system to form additional polymer. Decant the liquid, wash the polymer thoroughly with water. Allow the polymer to dry. Do not discard the nylon in the sink. Use a waste container.

Procedure 36c

POLYSTYRENE

An addition polymer, polystyrene, is prepared in this experiment. Reaction may be brought about by free radical, cationic, or anionic catalysts, the first of these being most common. In this experiment, polystyrene is prepared by free radical-catalyzed polymerization.

The reaction is initiated by a free radical source. The initiator used will be benzoyl peroxide, a relatively unstable molecule which, at 80 to 90°, decomposes with homolytic cleavage of the oxygen–oxygen bond.

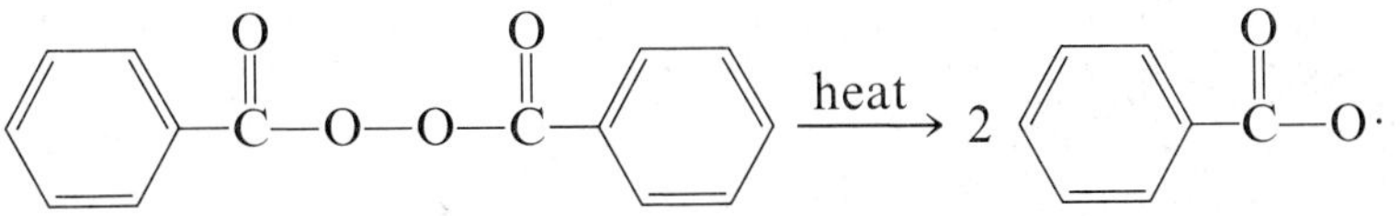

Benzoyl peroxide   Benzoyl radical

If an unsaturated monomer is present, the catalyst radical adds to it, initiating a chain reaction by producing a new free-radical. If we let R stand for the catalyst radical, the reaction with styrene can be represented as:

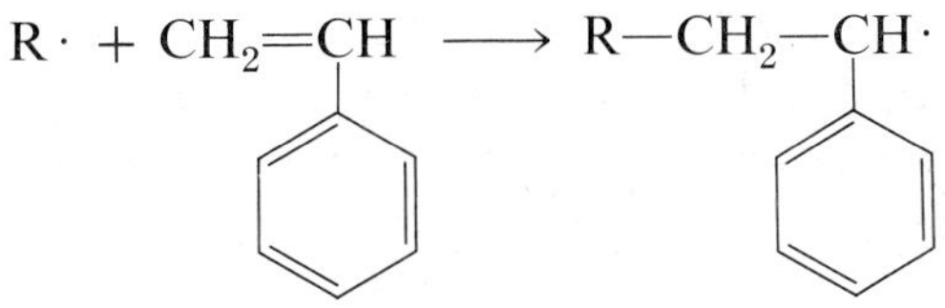

$$R \cdot + CH_2 {=} CH \longrightarrow R{-}CH_2{-}CH \cdot$$

The chain continues to grow:

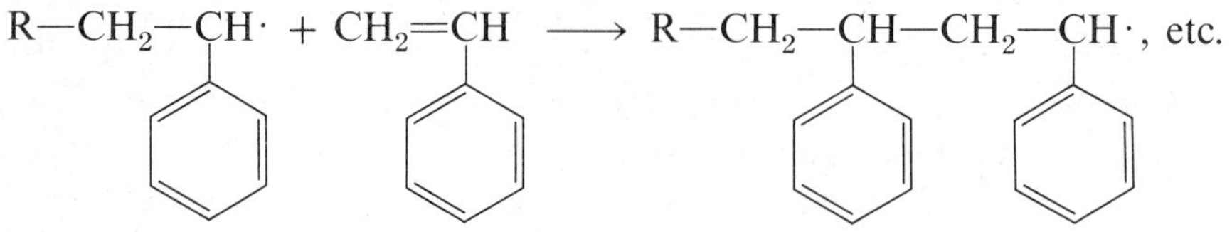

$$R-CH_2-CH\cdot \ + \ CH_2=CH \ \longrightarrow \ R-CH_2-CH-CH_2-CH\cdot, \ etc.$$

The chain may be terminated by combination of two radicals (either both polymer radicals, or one polymer radical and one initiator radical) or by abstraction of a hydrogen atom from another molecule.

PROCEDURE

Since the glassware is difficult to clean, this experiment is best performed by the laboratory instructor. One large batch should be made for the entire class (at least 10 times the amounts given). Perform the experiment in a hood. Place several thicknesses of newspaper in the hood.

> **CAUTION:** Benzoyl peroxide is flammable and may detonate upon impact or on heating (friction). It should be weighed on glassine (glazed not ordinary) paper. Clean up *all* spills with water. Wash the glassine paper with water before discarding it.

Place 25 to 30 ml of styrene monomer in a 150 ml beaker and add 0.7 g of benzoyl peroxide. Heat the mixture on a hot plate until the mixture develops a yellow color. When the color disappears and bubbles begin to appear, immediately take the beaker of styrene off the hot plate since the reaction is exothermic. After the reaction subsides, put the beaker of styrene back on the hot plate and continue heating it until the liquid becomes very syrupy. With a stirring rod, draw out a long filament of material from the beaker. If this filament can be cleanly snapped after a few seconds of cooling, the polystyrene is ready to be poured. If the filament does not break, continue heating the mixture and repeat the above process until the filament breaks easily. Pour the syrupy liquid on a watch glass. After cooling, the polystyrene can be lifted from the glass surface by gently prying it with a spatula.

POLYURETHAN FOAM

A crosslinked polymer, polyurethan foam, is prepared in this experiment from a diisocyanate and a triol. The main reaction is the addition of the alcohol across the $-N=C-$ bond of an isocyanate:

$$R-N=C=O \longrightarrow R-N-C=O$$
$$H-OR \qquad\qquad H \quad OR$$

Isocyanate + Alcohol   Urethan

With a diisocyanate, and a triol, the reaction can proceed in **three** directions leading to a large molecule which is rigidly held into a three dimensional structure.

The foaming is caused by the evolution of carbon dioxide in a manner strongly resembling the baking of bread, where carbon dioxide is evolved by fermentation of sugars with yeast, causing the bread to rise. In the present preparation, the carbon dioxide is produced by the small amount of water present which decomposes a small amount of the isocyanate:

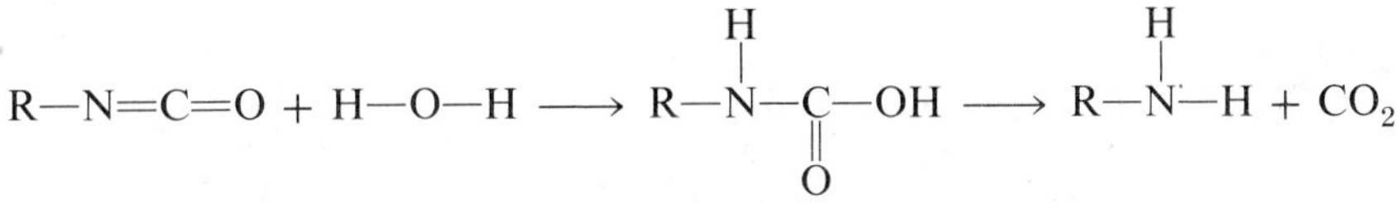

$$R-N=C=O + H-O-H \longrightarrow R-N-C-OH \longrightarrow R-N-H + CO_2$$

The evolution of carbon dioxide bubbles creates pores in the viscous mixture as the foam sets into a rigid mass. Thus, the foam has excellent buoyant properties. The cell size and structure of the foams are controlled by adding silicone oil.

The structure of the polymer is as follows:

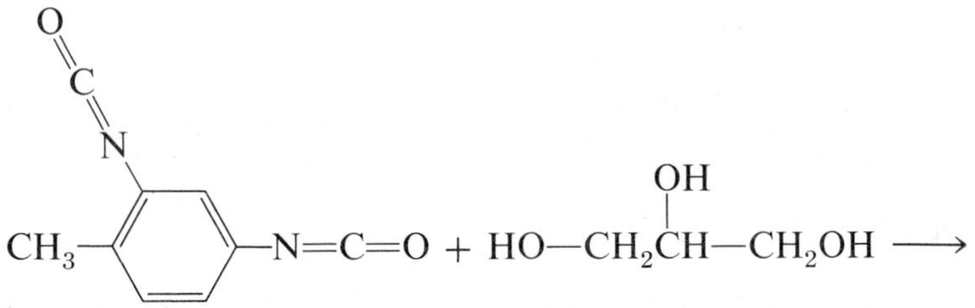

$$CH_3 - \quad -N=C=O + HO-CH_2\overset{OH}{CH}-CH_2OH \longrightarrow$$

Toluene Diisocyanate   Glycerol

The polymer can grow in all of the indicated directions.

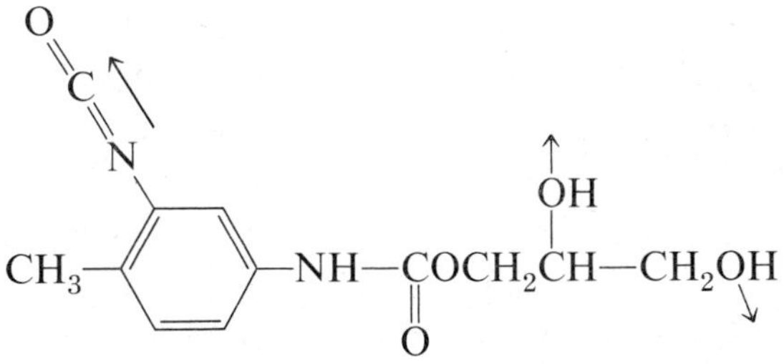

$$CH_3 - \quad -NH-COCH_2\overset{OH}{CH}-CH_2OH$$

Castor oil, a triol, can also react in the same way. It is a triglyceride (fat) of ricinoleic acid (see the essay which precedes Experiment 14):

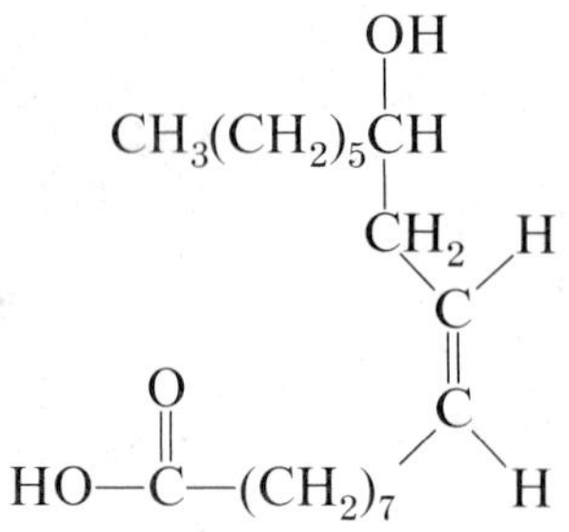

$$CH_3(CH_2)_5\overset{\overset{\displaystyle OH}{|}}{C}H$$

Ricinoleic Acid

In the present experiment, glycerol, castor oil, small amounts of water, silicone oil (a foaming agent), and stannous octoate (a catalyst) are mixed together. The diisocyanate (TDI) is added to this mixture. The mixture is stirred and foaming begins. Commercial foams are not usually prepared from these simple materials. A polymeric diol or triol is usually used instead of glycerol and castor oil.

PROCEDURE

> **CAUTION:** Toluene diisocyanate (TDI) is toxic. It will irritate the skin and eyes. Avoid breathing the vapor. It may cause an allergic respiratory response. Work in a hood or in an area with adequate ventilation. Keep the container tightly closed when it is not in use (TDI reacts with moisture in the air). After handling TDI, wash your hands thoroughly.

Pour 17 ml of mixture A (shake well before using) into a waxed soft drink cold cup.[1] Then add 10 ml of toluene diisocyanate (tolylene-2,4-diisocyanate). Stir the mixture rapidly and thoroughly with a stirring rod until a smooth and creamy mixture is obtained. The mixture should become warm and should begin to evolve bubbles of carbon dioxide after about 1 minute. When the gas begins to evolve, immediately stop stirring the mixture (foaming will be spontaneous). Do not breathe the evolved vapors. Place the mixture in the hood. After the foaming has ceased, allow the material to cool and set thoroughly. The polyurethan is initially sticky but after several hours it will become quite firm. The

---

[1] Mixture A is prepared as follows: Place 350 g of castor oil, 100 g of glycerol, 50 drops of stannous octoate (stannous 2-ethylhexanoate) 50 drops of Dow-Corning 200 silicone oil (this is estimated since it is difficult to measure) and 150 drops of water in a bottle. Cap the bottle and shake it thoroughly. Allow this mixture to stand no more than 12 hours before use.

paper container can then be removed. The material will shrink noticeably upon standing.

QUESTIONS

1. Ethylene dichloride ($ClCH_2CH_2Cl$) and sodium polysulfide (Na_2S_4) react to form a chemically resistant rubber, Thiokol A. Give the structure for the rubber.

2. Vinylidene chloride ($CH_2=CCl_2$) is polymerized with vinyl chloride to make Saran. Give a structure, which includes at least two units, for the copolymer formed.

3. Isobutylene ($CH_2=C(CH_3)_2$) is used to prepare cold-flow rubber. Give a structure of the addition polymer formed from this alkene.

4. Kel-F is an addition polymer with the following partial structure. What is the monomer used to prepare it?

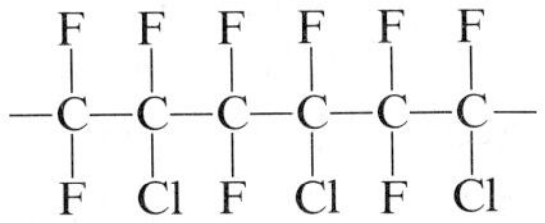

$$-\overset{\overset{\displaystyle F}{|}}{\underset{\underset{\displaystyle F}{|}}{C}}-\overset{\overset{\displaystyle F}{|}}{\underset{\underset{\displaystyle Cl}{|}}{C}}-\overset{\overset{\displaystyle F}{|}}{\underset{\underset{\displaystyle F}{|}}{C}}-\overset{\overset{\displaystyle F}{|}}{\underset{\underset{\displaystyle Cl}{|}}{C}}-\overset{\overset{\displaystyle F}{|}}{\underset{\underset{\displaystyle F}{|}}{C}}-\overset{\overset{\displaystyle F}{|}}{\underset{\underset{\displaystyle Cl}{|}}{C}}-$$

5. Maleic anhydride reacts with ethylene glycol to produce an alkyd resin. Give a structure of the condensation polymer produced.

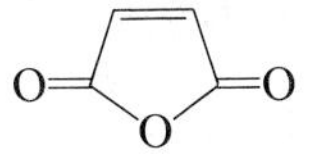

Maleic anhydride

6. Kodel is a condensation polymer made from terephthalic acid and 1,4-cyclohexanedimethanol. Give the structure of the resulting polymer.

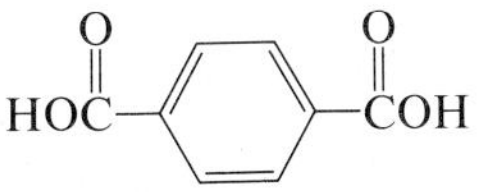

Terephthalic acid 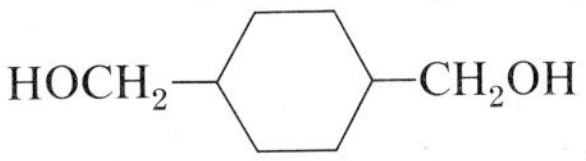   1,4-Cyclohexanedimethanol

ESSAY

Cyanohydrins in Nature

The addition of hydrogen cyanide to aldehydes is a typical nucleophilic addition reaction of the aldehyde functional group. The adduct is called a **cyanohydrin.**

$$R-CHO + HCN \rightleftharpoons R-CH\overset{\displaystyle OH}{\underset{\displaystyle CN}{<}}$$ A Cyanohydrin

Because cyanide ion is required to be the attacking nucleophile, the reaction is catalyzed by base. It is usually carried out with sodium or potassium cyanide by adding only that amount of acid required to keep the pH near 7 or 8. In the presence of excess acid, the cyanide ion, which is a fairly strong base, is protonated and its effectiveness as a nucleophile is diminished.

In the cyanohydrin, the hydrogen derived from the original alde-hyde is made acidic due to its new position **alpha** to the cyano group which provides resonance stabilization.

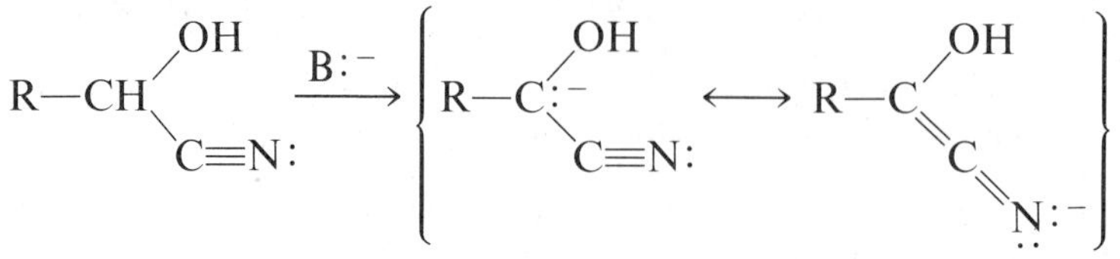

In a slightly basic KCN solution, the removal of this proton promotes the formation of the cyanohydrin by immediately converting it to its resonance-stabilized conjugate anion (Le Chatelier's principle). This anion can also behave as a nucleophile toward an unreacted molecule of benzaldehyde. An example of this type of behavior can be found in the benzoin condensation as is described in Experiment 37.

The cyanohydrin formed from benzaldehyde is called mandeloni-trile since, when hydrolyzed in strongly **basic** solutions, it yields man-delic acid.

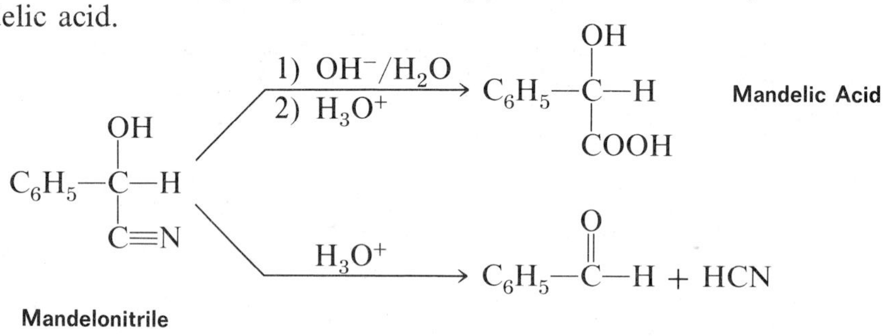

In contrast, when hydrolyzed under **acidic** conditions, mandelonitrile decomposes to regenerate benzaldehyde and hydrogen cyanide.

Surprisingly, mandelonitrile is found in many forms in nature. Perhaps the most interesting occurrence of this compound is in the millipede **Apheloria corrugata**. This millipede utilizes mandelonitrile as a part of its protective apparatus. It synthesizes and stores the cyano-hydrin in a series of 22 glands which are arranged in pairs on several of the body sections just above the legs. Each gland has two compart-ments. The inner compartment is a large saclike reservoir lined with cells that secrete mandelonitrile as an aqueous emulsion. The inner compartment is separated from the outer compartment by a muscular valve. The second compartment, the vestibule, is lined with cells that secrete an enzyme that decomposes mandelonitrile into benzaldehyde and hydrogen cyanide, a poisonous gas. When the millipede is alarmed by a predator, it opens the valve separating the two compartments and, by contraction of the storage reservoir, forces mandelonitrile into the

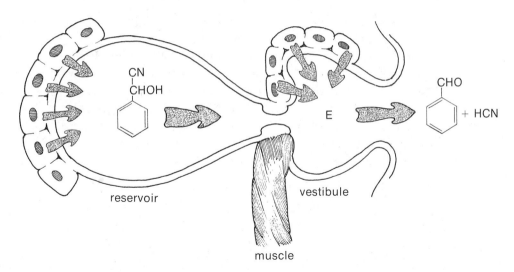

The reactor gland of *Ampheloria corrugata*

vestibule where it is mixed with the enzyme and forced outside. The products of the dissociation either kill or repel the predator. A single Apheloria can secrete enough hydrogen cyanide to kill a small bird or small mouse. The benzaldehyde is also an effective repellant for many of the predators.

Mandelonitrile is also found in several plants of the Rosaceae family. However, it is not found in the free form, but as a **glycoside,** that is, with its hydroxyl group attached through an acetal linkage to a sugar. The most commonly occurring glycoside is **amygdalin,** which is found in the seeds (pits), leaves, and bark of many plants such as bitter almond **(Prunus amydalus),** apricot, wild cherry, peach, and plum.

Amygdalin

Two similar glycosides, **prunlaurasin** and **sambunigrin,** are found in cherry laurel and **Sambucus nigra,** respectively. These glycosides differ from amygdalin only in their stereochemistry. On hydrolysis, amygdalin yields two molecules of glucose and D-mandelonitrile. Sambunigrin yields L-mandelonitrile, and prunlaurasin yields racemic (D,L) mandelonitrile.

The seeds of most of these plants also contain two enzymes called **emulsin** and **prunase.** Emulsin will hydrolyze one glucose molecule from amygdalin to produce **prunasin,** or mandelonitrile monoglucoside. Prunasin is often found along with amydalin. The second enzyme, prunase, cleaves the second glucose molecule from amygdalin (pruna-sin) to give mandelonitrile. The occurrence of the two enzymes, along with the glycoside, assures that free mandelonitrile will be produced when the contents of the seed are digested in the stomach of a predator. The free mandelonitrile is then rapidly hydrolyzed to benzaldehyde and hydrogen cyanide in the acidic stomach medium. Thus, amygdalin, prunlaurasin, and sambunigrin probably constitute protective mecha-nisms for the plants in which they are found. Many of these plants produce an otherwise luscious and attractive fruit.

REFERENCES

Claus, E. P., Tyler, V. E., and Brady, L. R. *Pharmacognosy.* "Cyanophore Glycosides." Philadelphia: Lea and Febiger, 1970, pp. 114–118.
Sondheimer, E., and Simeone, J. B., (editors). *Chemical Ecology.* Chapter 8, "Chemical Defense against Predation in Arthropods." New York: Academic Press, 1970.

Experiment 37
THE BENZOIN CONDENSATION

Condensation Reaction

Benzaldehyde does not possess alpha hydrogens and, therefore, will not undergo a self aldol condensation. In fact, in strongly basic solution, benzaldehyde, like other aldehydes which lack alpha hydrogens, un-dergoes the Cannizzaro reaction, yielding benzyl alcohol and sodium benzoate.

$$2\ C_6H_5CHO + NaOH \longrightarrow C_6H_5CH_2OH + C_6H_5COO^-Na^+$$

However, in the presence of cyanide ion, benzaldehyde will undergo a unique self condensation reaction, called the **benzoin condensation,** to yield an α-hydroxy ketone called **benzoin.** In this experiment, we shall use the benzoin condensation to synthesize benzoin.

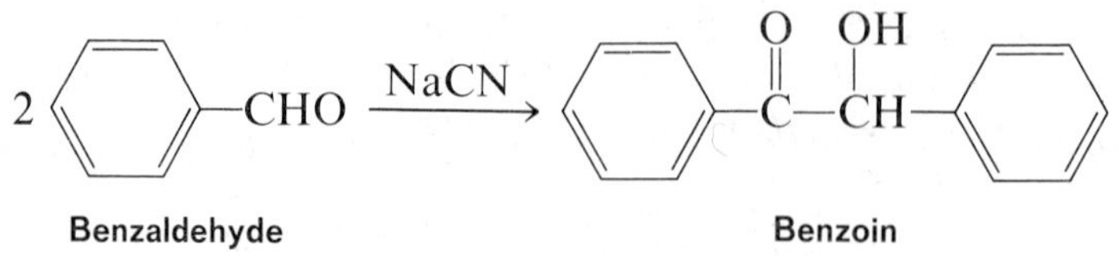

Benzaldehyde   Benzoin

The complete mechanism for this reaction is shown in the accompany-ing chart. The first step is the formation of the cyanohydrin **1,** which in the basic reaction medium immediately forms its conjugate anion **2.** The conjugate anion, **2,** is stabilized by resonance which involves both the cyano group and the aromatic ring. In a second step of the reaction, the cyanohydrin anion makes a nucleophilic addition to a second molecule of benzaldehyde to give the adduct **3**. After a proton transfer to form the anion, **4**, cyanide is expelled forming benzoin, an α-hydroxy ketone, **(5)**.

The reaction is carried out in 95% aqueous ethanol, and the product, which is sparingly soluble crystallizes from the reaction mixture on cooling. The product is collected by vacuum filtration and recrystallized from 95% ethanol.

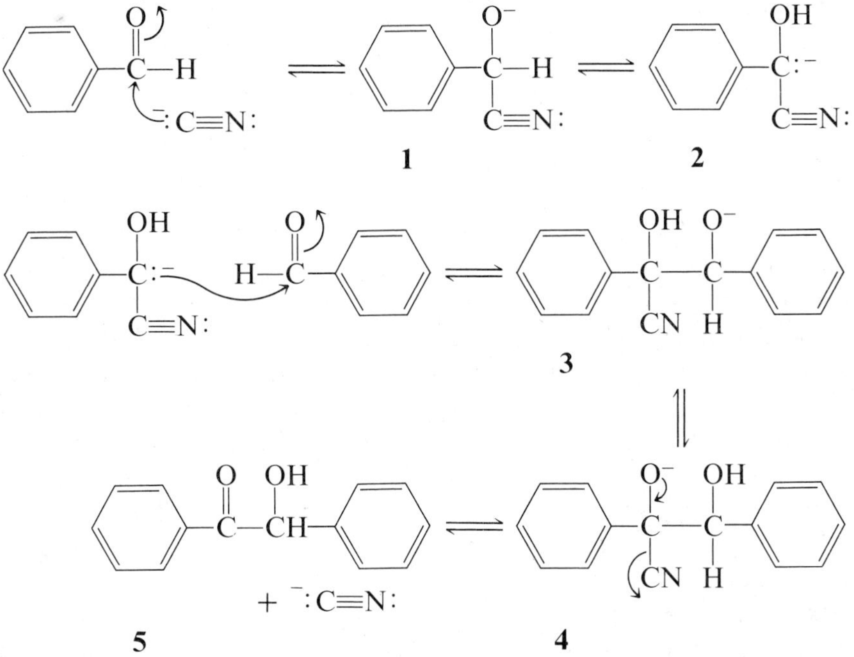

SPECIAL INSTRUCTIONS

A knowledge of Techniques 1 through 4 is required for this experiment. In addition, the essay on cyanohydrins which precedes the experiment should be read.

CAUTION: Sodium cyanide is extremely hazardous and toxic. When weighing this substance be careful not to spill any of it, and do not allow it to come into contact with your skin. Do not breathe the dust. Any spilled material should be cleaned up immediately and disposed of down a drain in the hood. Any area of the skin which has come into contact with cyanide should be washed thoroughly with water. Do not allow sodium cyanide to come into contact with acid because hydrogen cyanide, a toxic gas, will be generated.

Be sure to take all the cautionary measures mentioned in the note above, and to flush all cyanide residues down the sink with **large** volumes of water. If possible, the reaction should be carried out in a hood. If a hood is not available, it is wise to attach a trap like that shown in Experiment 23 in order to control any escaping vapors. The beaker should be filled with 15% aqueous sodium hydroxide.

The benzaldehyde used for this experiment should be pure. It is easiest to use a fresh bottle that has no benzoic acid, a solid white precipitate, evident in the bottom of the bottle. If benzoic acid is present, the benzaldehyde should be distilled before use.

PROCEDURE

Using a 100 ml round bottom flask, assemble an apparatus for heating under reflux (Technique 1, Figure 1–4). Place 20 ml of 95% ethanol, 15.0 g of pure benzaldehyde, and a solution of 1.5 g of sodium cyanide in 15 ml of water in the flask. Reflux the mixture gently for 0.5 hour and then cool the flask in an ice bath (leave the condenser attached). The product should precipitate. Collect the crude benzoin by vacuum filtration using a Büchner funnel (Technique 2, Section 2.3). Wash the product well with several portions of cold water to remove all the sodium cyanide (see the caution above). Set the benzoin aside to dry. A second batch of benzoin can be obtained by concentrating the filtrate. This should be done in a beaker in the hood. It will be necessary to use a hot plate to boil the solution. The crystals obtained should be colorless or pale yellow. Weigh the crude material and record the weight.

Recrystallize the crude benzoin from 95% ethanol (about 8 ml per

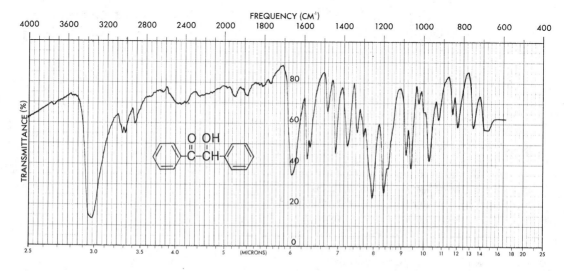

Infrared spectrum of benzoin, KBr

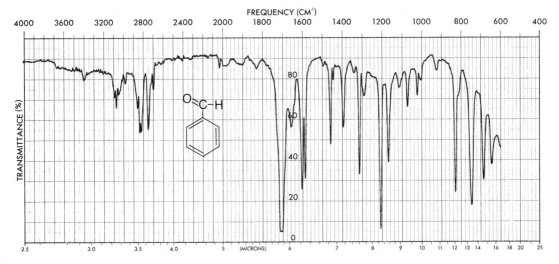

Infrared spectrum of benzaldehyde, neat

gram of crude crystals) to yield pure benzoin (mp 134 to 135°). Weigh the purified material, calculate the yield, and determine its melting point. At your instructor's option, determine the infrared spectrum of the benzoin as a KBr mull. If the sample is to be used to prepare benzil (Experiment 39), proceed to that experiment. If not, place the sample in a labeled vial and submit it with your report.

QUESTIONS

1. Give the structure of the product that would be formed by the action of cyanide ion on acetaldehyde. Will the reaction be the same as for benzaldehyde?

2. Interpret the major peaks in the infrared spectra of benzoin and benzaldehyde.

3. Give all of the possible resonance structures of the conjugate base of benzaldehyde cyanohydrin (2).

ESSAY

Thiamine as a Coenzyme

Vitamin B_1, or thiamine, as its pyrophosphate derivative, thiamine pyrophosphate, is a coenzyme of universal occurrence in all living systems. It was originally discovered as a required nutritional factor (vitamin) in man by its link with the disease beriberi. **Beriberi** is a disease of the peripheral nervous system caused by a deficiency of vitamin B_1 in the diet. Symptoms include pain and paralysis of the extremities, emaciation, or swelling of the body. The disease is most common in the Far East.

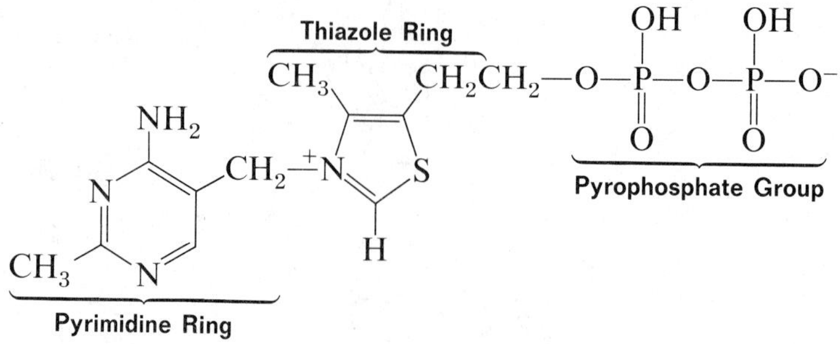

THIAMINE PYROPHOSPHATE

Thiamine serves as a coenzyme (defined later) for three important types of enzymatic reactions:

1. non-oxidative decarboxylations of α-keto acids,

$$R-\overset{\overset{\textstyle O}{\|}}{C}-COOH \xrightarrow{B_1} R-\overset{\overset{\textstyle O}{\|}}{C}-H + CO_2$$

2. oxidative decarboxylations of α-keto acids, and

$$R-\overset{\overset{\textstyle O}{\|}}{C}-COOH \xrightarrow{B_1/O_2} R-\overset{\overset{\textstyle O}{\|}}{C}-OH + CO_2$$

3. formation of acyloins (α-hydroxy ketones).

$$R-\overset{\overset{\textstyle O}{\|}}{C}-COOH + R-\overset{\overset{\textstyle O}{\|}}{C}-H \xrightarrow{B_1} R-\overset{\overset{\textstyle O}{\|}}{C}-\overset{\overset{\textstyle OH}{|}}{C}H-R + CO_2$$

or

$$2\,R-\overset{\overset{\textstyle O}{\|}}{C}-COOH \xrightarrow{B_1} R-\overset{\overset{\textstyle O}{\|}}{C}-\overset{\overset{\textstyle OH}{|}}{C}H-R + 2\,CO_2$$

or

$$2\,R-\overset{\overset{\textstyle O}{\|}}{C}-H \xrightarrow{B_1} R-\overset{\overset{\textstyle O}{\|}}{C}-\overset{\overset{\textstyle OH}{|}}{C}H-R$$

Most biochemical processes are no more than **organic chemical reactions** carried out under special conditions. It is easy to lose sight of this fact. However, most of the steps of the ubiquitous metabolic pathways can, where they have been studied sufficiently, be explained mechanistically. There is a simple organic reaction that serves as a model for almost every biological process. These reactions, however, are modified ingeniously through the intervention of a protein molecule ("enzyme") to make them more efficient (greater yield), more selective in choice of substrate (molecule being acted upon), more stereospecific in their result, and able to occur under milder conditions (pH) than would normally be possible.

Experiment 38 is designed to illustrate this last fact. As a biological reagent, the coenzyme thiamine is used to carry out an organic reaction **without** resorting to the use of an enzyme. The reaction is an acyloin condensation (see above) of benzaldehyde:

$$2\ C_6H_5\text{—CHO} \longrightarrow C_6H_5\text{—}\overset{\overset{\displaystyle O}{\|}}{C}\text{—}\overset{\overset{\displaystyle OH}{|}}{CH}\text{—}C_6H_5$$

In Experiment 37, a similar condensation is described, where sodium cyanide is used as the catalyst. Nature would clearly prefer to use a reactant that is somewhat milder and less toxic than cyanide ion to carry out the acyloin condensations necessary to everyday metabolism. Thiamine constitutes just such a reagent.

Chemically speaking, the most important part of the entire thiamine molecule is the central ring—the thiazole ring—which contains nitrogen and sulfur. This ring constitutes the **reagent** portion of the coenzyme. The other portions of the molecule, although important in a biological sense, are not necessary to the **chemistry** that thiamine initiates. Undoubtedly the pyrimidine ring and the pyrophosphate group have important ancillary functions, such as enabling the coenzyme to make the correct attachment to its associated protein molecule ("enzyme"), or enabling it to achieve the correct degree of polarity and the correct solubility properties necessary to allow free passage of the coenzyme across the cell membrane boundary (i.e., to allow it to get to its site of action). These properties of thiamine are no less important to its biological functioning than its chemical reagent capabilities; however, only the latter is our concern here.

Experiments with the model compound 3,4-dimethylthiazolium bromide have given an explanation for the mechanisms of thiamine catalyzed reactions. It was found that this model thiazolium compound rapidly exchanged the C-2 proton for deuterium in D_2O solution. At pD = 7 (no pH here!) this proton was completely exchanged in a matter of seconds!

3,4-Dimethylthiazolium
Bromide

This indicates that the C-2 proton is more acidic than one would have expected. It is apparently easily removed because the conjugate base is a highly stabilized **ylid.** An ylid is a compound or intermediate with positive and negative formal charges on adjacent atoms.

The sulfur atom plays an important role in stabilizing this ylid.

This was shown by comparing the rate of exchange of 1,3-dimethyl-imidazolium ion to that of the thiazolium ion shown above. The di-nitrogen compound exchanged its C-2 proton more slowly than the sulfur containing ion. Sulfur, being in the third row of the periodic chart, has d-orbitals available for bonding to adjacent atoms. Thus, it has fewer geometrical restrictions than do carbon and nitrogen atoms, and can form carbon–sulfur multiple bonds in situations where carbon and nitrogen normally would not.

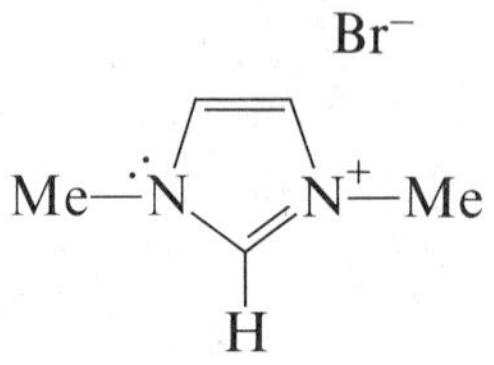

1,3-Dimethylimidazolium Bromide

DECARBOXYLATION OF α-KETO ACIDS

On the basis of the above, it is now thought that the active form of thiamine is its ylid. The system is interestingly constructed, as is seen in the decarboxylation of pyruvic acid by thiamine.

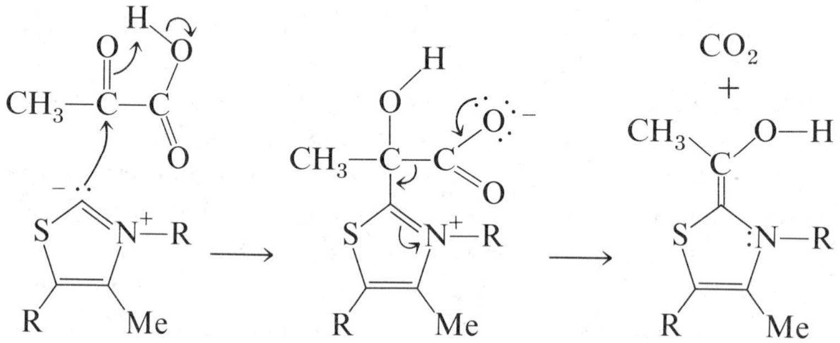

Notice especially how the positively charged nitrogen provides a site to accommodate the electron pair which is released on decarboxylation. Thiamine is regenerated by use of this same pair of electrons which become protonated in vinylogous fashion on carbon. The other product is the protonated form of acetaldehyde, the decarboxylation product of pyruvic acid.

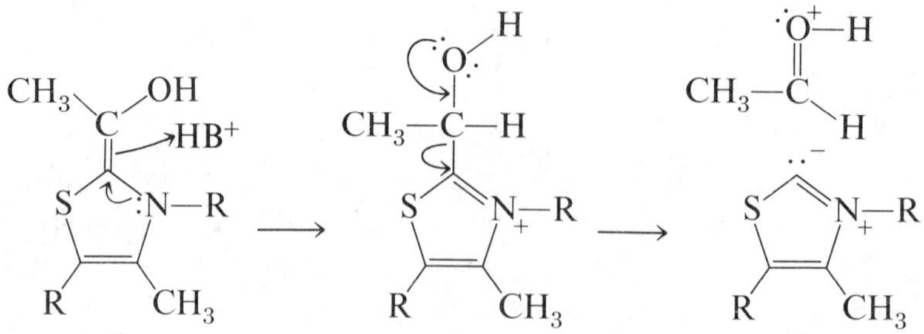

OXIDATIVE DECARBOXYLATION OF α-KETO ACIDS

In oxidative decarboxylations two additional coenzymes, lipoic acid and coenzyme A, are involved. An example of this type of process, which occurs in all living organisms, is found in the metabolic process called **glycolysis**. It is found in the steps which convert pyruvic acid to acetyl coenzyme A, which then enters the citric acid cycle (Krebs cycle, tricarboxylic acid cycle) to provide an energy source for the organism. In this process, the enamine intermediate, **3** (shown above), is first oxidized by lipoic acid and then transesterified by coenzyme A.

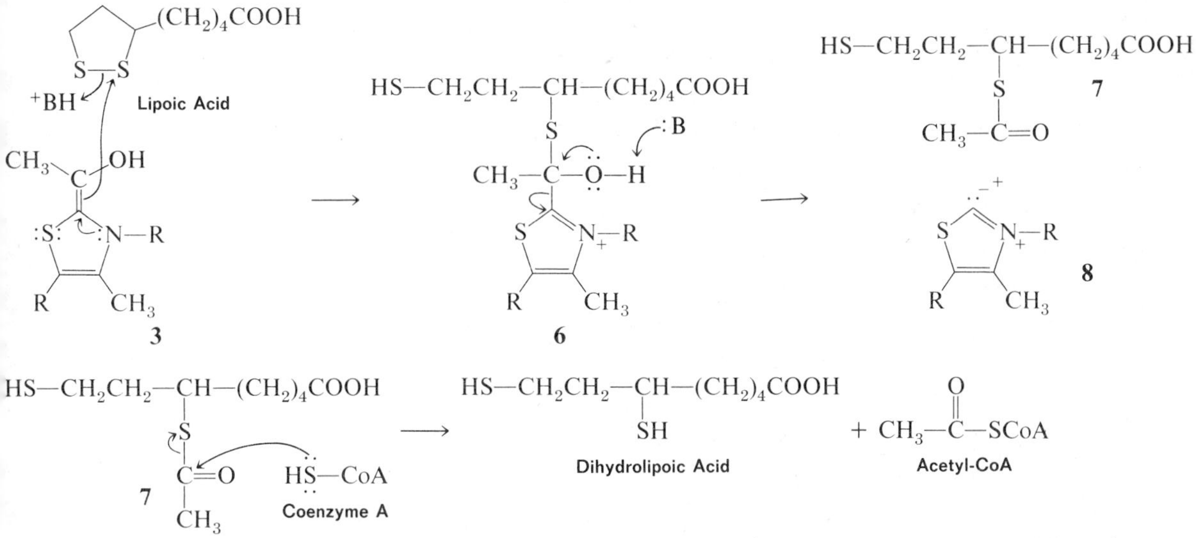

Lipoic Acid

$HS-CH_2CH_2-CH-(CH_2)_4COOH$

$HS-CH_2CH_2-CH-(CH_2)_4COOH$ **7**

3   **6**   **8**

$HS-CH_2CH_2-CH-(CH_2)_4COOH$ **7**

HS—CoA   Coenzyme A

$HS-CH_2CH_2-CH-(CH_2)_4COOH$ SH   Dihydrolipoic Acid

$+ CH_3-\overset{O}{\overset{\|}{C}}-SCoA$ Acetyl-CoA

Following this sequence of events, the dihydrolipoic acid is oxidized (through a chain of events involving molecular oxygen) back to lipoic acid, and the acetyl coenzyme A is condensed with oxaloacetic acid to form citric acid. The formation of citric acid begins the citric acid cycle. Notice that acetyl coenzyme A is a thioester of acetic acid and could be hydrolyzed to give acetic acid, not an aldehyde. Thus, an oxidation has taken place in this sequence of events.

ACYLOIN CONDENSATIONS

The enamine intermediate **3** can also function much like the enolate partner in an acid catalyzed aldol condensation. It can condense with a suitable carbonyl-containing acceptor to form a new carbon–

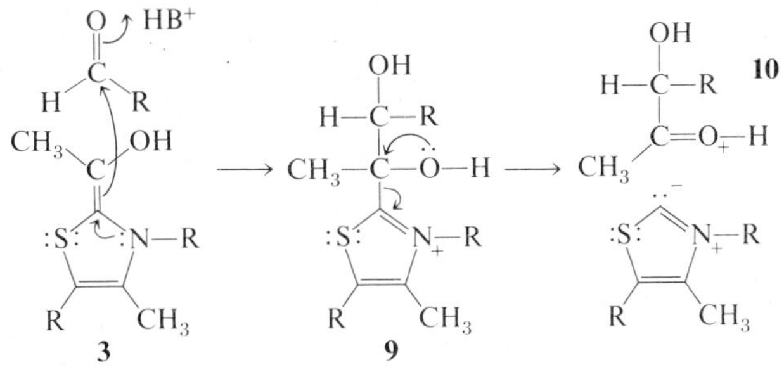

3   **9**   **10**

carbon bond. Decomposition of the adduct 9 to regenerate the thiamine ylid, yields the protonated acyloin **10.**

THE FUNCTION OF A COENZYME

In biological terminology, thiamine is called a **coenzyme.** It must bind to an enzyme before the enzyme is activated. The enzyme also binds the substrate. Reaction between the coenzyme and the substrate occurs while they are both bound to the enzyme (a large protein). Without the coenzyme thiamine, no chemical reaction would occur. The coenzyme is the **chemical reagent.** The protein molecule (the enzyme) facilitates and mediates the reaction by controlling stereochemical, energetic, and entropic factors, but in this case, is non-essential to the overall result (see Experiment 38). A special name is given to coenzymes that are essential to the nutrition of an organism. They are called

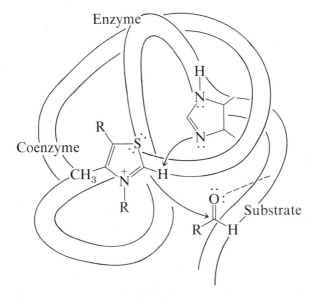

Thiamine (the coenzyme) and the substrate aldehyde are bound to the protein molecule, here called an enzyme. A possible catalytic group (imidazole) is also shown.

vitamins. Many biological reactions are of this type, where a chemical reagent (coenzyme) and a substrate are bound to an enzyme for reaction and, after the reaction, are again released into the medium.

REFERENCES

Bernhard, S. *The Structure and Function of Enzymes,* Chapter 7, "Coenzymes and Cofactors." New York: W. A. Benjamin, 1968.

Bruice, T. C., and Benkovic, S. *Bioorganic Mechanisms.* Volume 2, Chapter 8, "Thiamine Pyrophosphate and Pyridoxal-5′-Phosphate." New York: W. A. Benjamin, 1966.

Lowe, J. N., and Ingraham, L. L. *An Introduction to Biochemical Reaction Mechanisms.* Chapter 5, "Coenzyme Function and Design." Englewood Cliffs, New Jersey: Prentice-Hall, 1974.

COENZYME SYNTHESIS OF BENZOIN

Coenzyme Chemistry
Benzoin Condensation

In this experiment, a benzoin condensation of benzaldehyde will be carried out using a biological coenzyme, thiamine hydrochloride, as the catalyst.

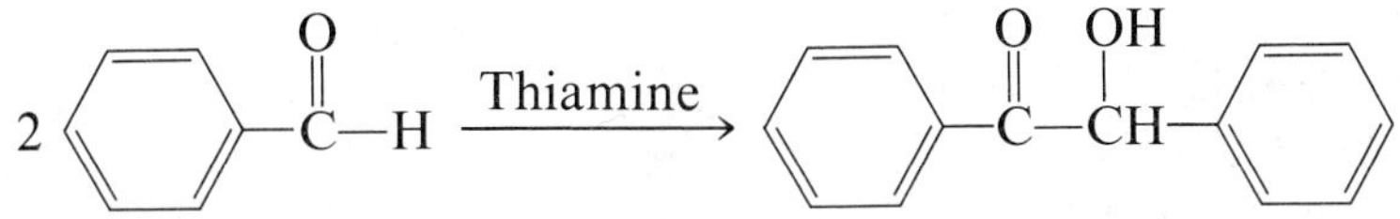

Benzaldehyde   Benzoin

The same reaction can be accomplished using cyanide ion, an inorganic reagent, as the catalyst. A mechanism for the cyanide catalyzed benzoin condensation is given in Experiment 37. The mechanistic information necessary to understand how thiamine accomplishes this same reaction is given in the essay which precedes this experiment.

SPECIAL INSTRUCTIONS

A familiarity with the material in Techniques 1 through 4 is necessary to perform this experiment. A careful reading of the essay "Thiamine as a Coenzyme" is also essential because it explains the mechanism of action of thiamine. It will also be helpful to compare the mechanism of the cyanide-catalyzed benzoin condensation (Experiment 37) with that of the reaction in this experiment.

The benzaldehyde used for this experiment **must** be free of benzoic acid. Benzaldehyde is oxidized easily in air, and crystals of benzoic acid are often visible in the bottom of the reagent bottle. If solid is evident in the bottle of reagent, it **must** be redistilled. If the lab assistant or instructor has not redistilled the benzoin provided for this experiment, you will have to distill it. However, if a **newly opened** bottle of benzaldehyde is available, it will usually be unnecessary to redistill it. If in doubt, consult the instructor.

Thiamine hydrochloride is a heat sensitive reagent. It should be stored in a refrigerator when not in use. Since it may decompose on heating, you should take care not to heat the reaction mixture too vigorously. It is best to use a fresh bottle of this reagent.

PROCEDURE

Dissolve 3.5 g of thiamine hydrochloride in about 10 ml of water in a 100 ml round bottom flask equipped with a condenser for reflux (Technique 1, Figure 1–4). Add 30 ml of 95% ethanol and cool the solution by swirling the flask in an ice-water bath. Meanwhile, place about 10 ml of 2 M sodium hydroxide solution in a small Erlenmeyer flask. Cool this solution in the ice bath also. Then, over a period of about 10 minutes, add the cold sodium hydroxide solution, through the condenser, to the thiamine solution. Measure 20 ml of benzaldehyde and add it, also through the condenser, to the reaction mixture. Add a boiling stone and heat the mixture gently on a steam bath for about 90 minutes. Do not heat the mixture under vigorous reflux. Allow the mixture to cool to room temperature, and then induce crystallization of the benzoin (it may already have begun) by cooling the mixture in an ice-water bath. If the product separates as an oil, reheat the mixture until it is once again homogeneous, and then allow it to cool more slowly than before. Scratching of the flask with a glass rod may be required.

Collect the product by vacuum filtration using a Büchner funnel. Wash the product with two 50 ml portions of **cold** water. Weigh the crude product and then recrystallize it from 95% ethanol. The solubility of benzoin in boiling 95% ethanol is about 12 to 14 g per 100 ml. Weigh the product, calculate the percentage yield, and determine its melting point (mp 134 to 136°).

At your instructor's option, determine the infrared spectrum of the benzoin as a KBr mull. A spectrum of benzoin may be found below for comparison.

The benzoin may be converted to benzil (Experiment 39). However,

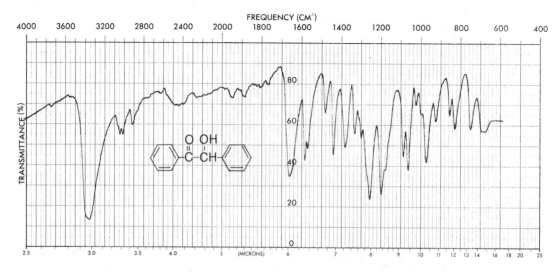

Infrared spectrum of benzoin, KBr

if you are not scheduled to perform this experiment, submit the sample of benzoin, along with your report, to the instructor.

QUESTIONS

1. The infrared spectrum of benzoin is given in Experiment 37. Interpret the major peaks in the spectrum.

2. Why is sodium hydroxide added to the solution of thiamine hydrochloride?

3. Using the information given in the essay that precedes this experiment, formulate a complete mechanism for the thiamine-catalyzed conversion of benzaldehyde to benzoin.

4. How do you think the appropriate enzyme would have affected the reaction yield?

5. What modifications of conditions would be appropriate if the enzyme were to be used?

6. Refer to the essay which precedes this experiment. It gives a structure for thiamine pyrophosphate. Using this structure as a guide, draw a structure for **thiamine hydrochloride.** The pyrophosphate group is absent in the latter compound.

Experiment **39**
BENZIL

Oxidation
Crystallization

In this experiment an α-diketone, benzil, will be prepared by the oxidation of an α-hydroxyketone, benzoin (Experiment 37 or 38).

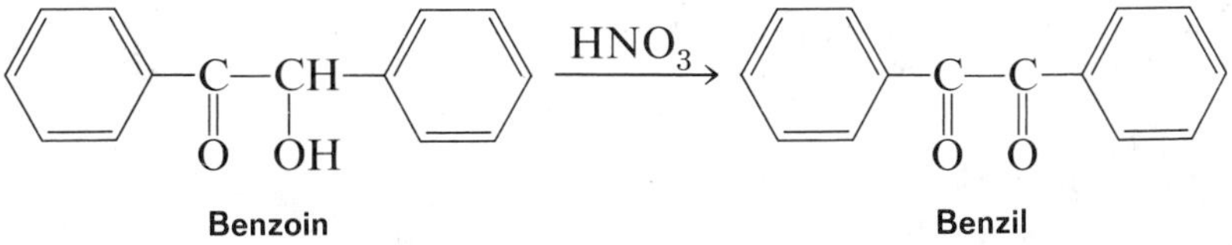

Benzoin $\xrightarrow{\text{HNO}_3}$ **Benzil**

This oxidation may easily be performed with mild oxidizing agents such as Fehling's solution (alkaline cupric tartrate complex) or copper sulfate in pyridine. In this experiment, the oxidation will be performed with nitric acid.

SPECIAL INSTRUCTIONS

To perform this experiment, you should have read Techniques 1 through 4. The benzoin prepared in Experiment 37 or 38 may be used in this experiment.

PROCEDURE

Place 10.0 g of benzoin (Experiment 37 or 38) in a 250 ml Erlenmeyer flask with 50 ml of concentrated nitric acid. Heat the mixture on a steam bath in the hood. Shake the mixture occasionally until nitrogen oxide gases (red) are no longer evolved (about 1 hour). Alternatively, the mixture may be heated at the desk if a trap arrangement such as that shown in Experiment 23 is used. Sodium hydroxide is added to the trap to react with the nitrogen oxide gases. Pour the reaction mixture into 150 ml of cool tap water and stir it vigorously until the oil crystallizes completely as a yellow solid. Vacuum filter the crude benzil and wash it well with cold water to remove the nitric acid.

Recrystallize the product from 95% ethanol (4 ml/g). Scratch the solution with a stirring rod as it cools. The solution will become supersaturated unless this is done. Yellow, needle-like crystals are formed. Cool the mixture in an ice bath to complete the crystallization. Vacuum filter the crystals using a Büchner funnel. Press the crystals with a clean stopper or cork to remove the excess solvent. Dry the product either in air overnight or in an oven at 75° for about 10 minutes. Weigh the benzil and calculate the percent yield, then determine the melting point. The melting point of pure benzil is 95°. The value obtained is

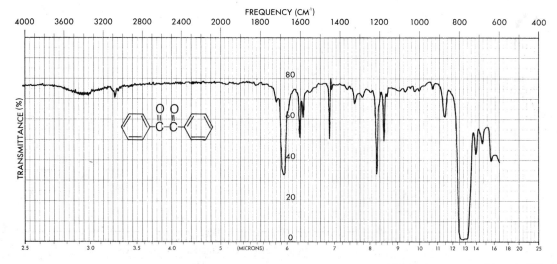

Infrared spectrum of benzil, CCl_4

often lower than this, ranging from a low value of 84° to a high value of 92°. Material in this range of melting points is of sufficient purity for conversion to benzilic acid (Experiment 40) or tetraphenylcyclopentadienone (Experiment 45). Submit the benzil to the instructor unless it is to be used to prepare benzilic acid or tetraphenylcyclopentadienone. At the instructor's option, obtain the infrared spectrum of benzil in carbon tetrachloride or chloroform. Compare it to the infrared spectrum of benzoin shown in Experiment 37.

Experiment 40
BENZILIC ACID

Anionic Rearrangement

In this experiment, benzilic acid will be prepared via the rearrangement of the α-diketone, benzil. A preparation of benzil is given in Experiment 39. The reaction proceeds in the following way:

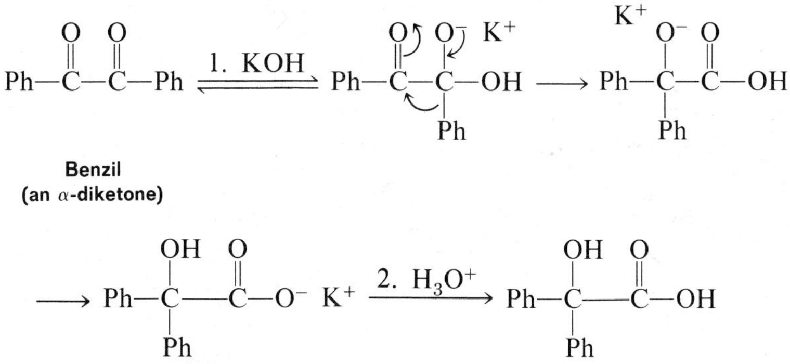

Benzil
(an α-diketone)

Potassium
Benzilate

Benzilic acid
(an α-hydroxyacid)

The driving force for the reaction is provided by the formation of a stable carboxylate salt (potassium benzilate). Once this salt is produced, acidification yields benzilic acid. The reaction can generally be used to convert aromatic α-diketones to aromatic α-hydroxyacids. Other compounds, however, also will undergo a benzilic acid type rearrangement (see questions).

SPECIAL INSTRUCTIONS

This reaction involves a 15 minute reflux period, followed by an overnight crystallization period. The reaction may be scheduled with another experiment.

PROCEDURE

Dissolve 5.5 g of potassium hydroxide in 12 ml of water in an Erlenmeyer flask. The mixture may be heated to dissolve the base, then cooled again to room temperature. In a 100 ml round bottom flask, dissolve 5.5 g of benzil (Experiment 39) in 17 ml of 95% ethanol, heating slightly, if necessary, to dissolve the solid. Add the potassium hydroxide solution to the round bottom flask and swirl the contents of the flask. Attach a reflux condenser to the round bottom flask. Reflux the mixture on a steam bath for 15 minutes. During this period, the initial blue-black coloration will be replaced by a brown color. Transfer the contents of the flask to a beaker or an evaporating dish and cover it with a watch glass. Allow the mixture to stand until the next laboratory period. The potassium salt of benzilic acid will crystallize during this period. Collect the crystals by vacuum filtration and wash the crystals with 2 ml of ice cold 95% ethanol.

Dissolve the potassium benzilate in a minimum amount of hot water in a 250 ml Erlenmeyer flask. More than 100 ml of hot water will be needed to dissolve this solid. Add a small amount of decolorizing carbon and shake or stir the mixture for a few minutes. Filter the **hot** solution by gravity using a fluted filter. Acidify the filtrate with concentrated hydrochloric acid to a pH of about 2. Allow the mixture to cool slowly to room temperature and then complete the cooling in an ice bath. Collect the benzilic acid by vacuum filtration using a Büchner funnel. Wash the crystals thoroughly with water to remove salts, and remove the wash water by drawing air through the filter. Dry the product thoroughly by allowing it to stand until the next laboratory period.

Determine the melting point of the product. Pure benzilic acid melts at 150°. If necessary, recrystallize the product from benzene, using 11 ml/0.5 g of crude product. Heat the mixture to boiling on a steam bath (some impurities will remain undissolved) and gravity filter it through a fast fluted filter paper. Cool the filtrate and scratch the solution to induce crystallization. Allow the mixture to stand at room temperature until crystallization is complete (about 15 minutes). Cool the mixture in an ice bath to 10° (benzene freezes at 5°!) and vacuum filter the crystals. Wash them with a small amount of **cold** benzene and dry them by drawing air through the filter. Determine the melting point of the crystallized product.

At the instructor's option, determine the infrared spectrum of the benzilic acid in potassium bromide. Calculate the percent yield. Submit the sample(s) to your laboratory instructor in labeled vials.

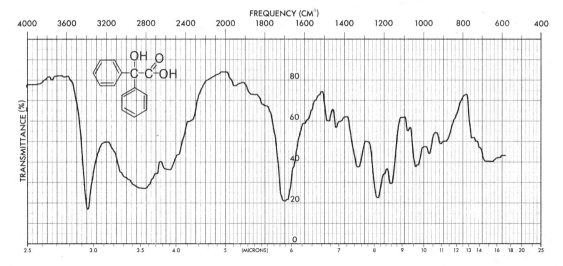

Infrared spectrum of benzilic acid, KBr

QUESTIONS

1. Show how to prepare the following compounds starting from the appropriate aldehyde (see Experiments 37 and 39).

a. CH₃O— (structure with OH, C—CO₂H, OCH₃)

b. (furan structure with OH, C—CO₂H)

2. Give the mechanisms for the following transformations:

a. (phenanthrenequinone) $\xrightarrow[\text{2. H}^+]{\text{1. KOH/ALC.}}$ (fluorene OH CO₂H)

b. HO—C—CH₂—C—C—CH₂C—OH $\xrightarrow[\text{2. H}^+]{\text{1. KOH/H}_2\text{O}}$ HO—CCH₂—C—CH₂C—OH with CO₂H

Citric Acid

c. Ph—C—C—Ph $\xrightarrow[\text{CH}_3\text{OH}]{^-\text{OCH}_3}$ PhC—COCH₃ with OH, Ph, O

3. Interpret the infrared spectrum of benzilic acid.

The Sulfa Drugs

The history of **chemotherapy** extends back as far as 1909, when Paul Ehrlich first used the term. Although Ehrlich's original definition of the word was somewhat limited, he is, nevertheless, recognized as one of the giants of medicinal chemistry. Chemotherapy might be defined as the treatment of disease by chemical reagents. It is preferable that these chemical reagents exhibit a toxicity toward only the pathogenic organism, rather than toward the organism and the host as well. A chemotherapeutic agent would not be useful if it poisoned the patient at the same time that it cured the patient's disease!

In 1932, the German dye manufacturing firm, I. G. Farbenindustrie, patented a new drug, Prontosil. Prontosil is a red azo dye, and it was first prepared for its dye properties. Remarkably, it was discovered that Prontosil exhibited antibacterial action when it was used to dye wool. This discovery led to studies of Prontosil as a drug capable of inhibiting the growth of bacteria. The following year, Prontosil was successfully used against staphylococcal septicemia, a blood infection. In 1935, Gerhard Domagk published the results of his research which indicated that Prontosil was capable of curing streptococcic infections of mice and rabbits. Prontosil was shown to be active against a wide variety of bacteria in later work. This important discovery, which paved the way for a tremendous amount of research on the chemotherapy of bacterial infections, earned for Domagk the 1939 Nobel Prize in Medicine, but an order from Hitler prevented Domagk from accepting this honor.

$$H_2N-\underset{\underset{NH_2}{|}}{\bigcirc}-N=N-\bigcirc-SO_2NH_2 \qquad H_2N-\bigcirc-SO_2NH_2$$

Prontosil   Sulfanilamide

Prontosil is an effective antibacterial substance **in vivo,** that is, when injected into a living animal. The medicinal activity of Prontosil does not occur when the drug is tested **in vitro,** that is, on a bacterial culture grown in the laboratory. In 1935, the research group at the Pasteur Institute in Paris, headed by J. Tréfouël, learned that Prontosil is metabolized in animals to **sulfanilamide.** Sulfanilamide had been known since 1908. Experiments with sulfanilamide showed that it exhibited the same action as Prontosil **in vivo,** but moreover, it was also active **in vitro** where Prontosil was known to be inactive. It was concluded that the active portion of the Prontosil molecule was the sulfanilamide moiety. This discovery led to an explosion of interest in sulfonamide derivatives. Well over a thousand sulfonamide substances were prepared within a few years of these discoveries.

Although a great number of sulfonamide compounds were prepared, only a relative few exhibited useful anti-bacterial properties. These few medicinally active sulfonamides, or **sulfa drugs,** as the first useful antibacterial drugs became the wonder drugs of their day. An antibacterial drug may be either **bacteriostatic** or **bacteriocidal.** A bacteriostatic drug suppresses the growth of bacteria; a bacteriocidal drug kills bacteria. Strictly speaking, the sulfa drugs are bacteriostatic. Some of the most common sulfa drugs are given in the following table.

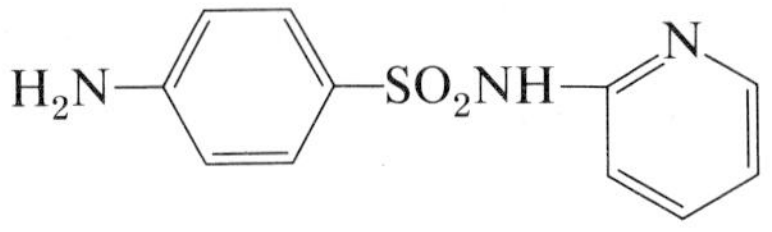

Sulfapyridine

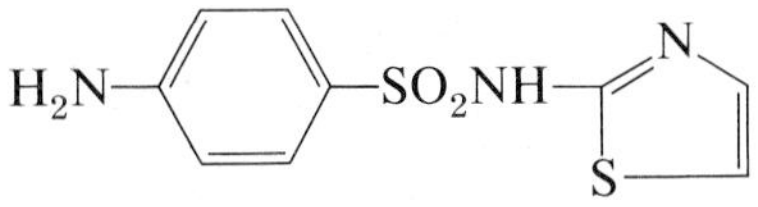

Sulfathiazole

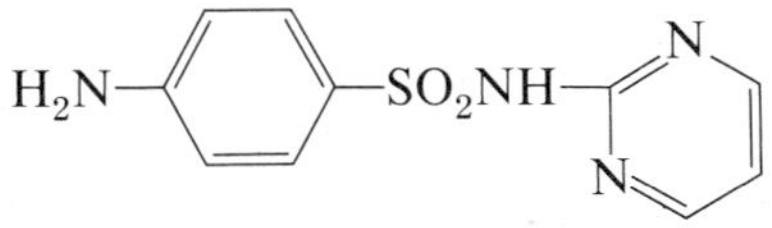

Sulfadiazine

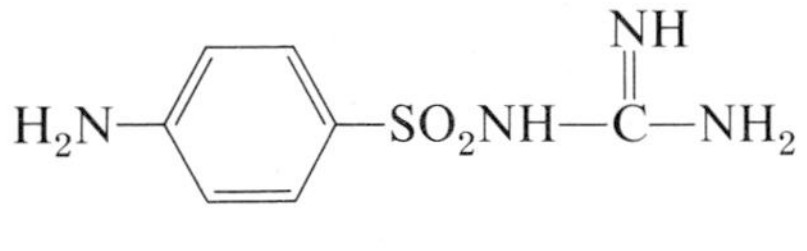

Sulfaguanidine

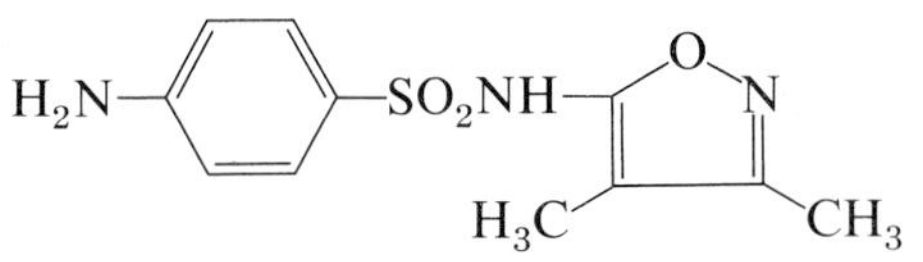

Sulfisoxazole

These more complex sulfa drugs have a variety of important applications. Although they do not possess the simple structure characteristic of sulfanilamide, they tend to be less toxic than the simpler compound.

Sulfa drugs began to lose their importance as generalized antibacterial agents when antibiotics began to be produced in large quantities. In 1929, Sir Alexander Fleming made his famous discovery of **penicillin.** In 1941, penicillin was first used successfully in the treatment of humans. Since that time, the field of study of antibiotics has spread to molecules which bear little or no structural similarity to the sulfonamides. Besides penicillin derivatives, antibiotics which are derivatives of **tetracycline,** including Aureomycin and Terramycin, were also discovered. These new antibiotics possess high activities against bacteria, and they do not usually have the severe unpleasant side effects which many of the sulfa drugs possess. Nevertheless, the sulfa drugs still find wide application in the treatment of such diseases as malaria, tuberculosis, leprosy, meningitis, pneumonia, scarlet fever, plague, respiratory infections, and infections of the intestinal and urinary tracts.

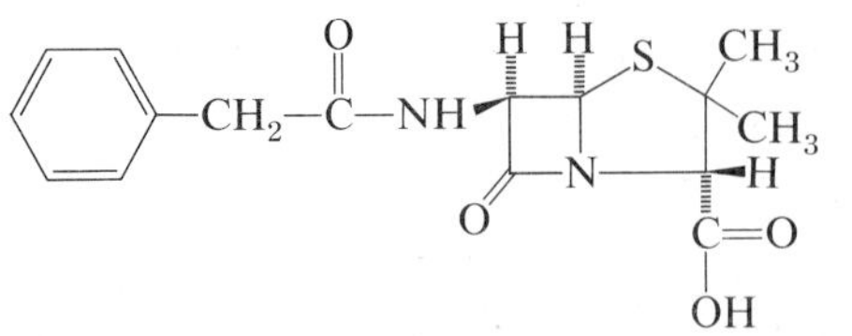

Penicillin G 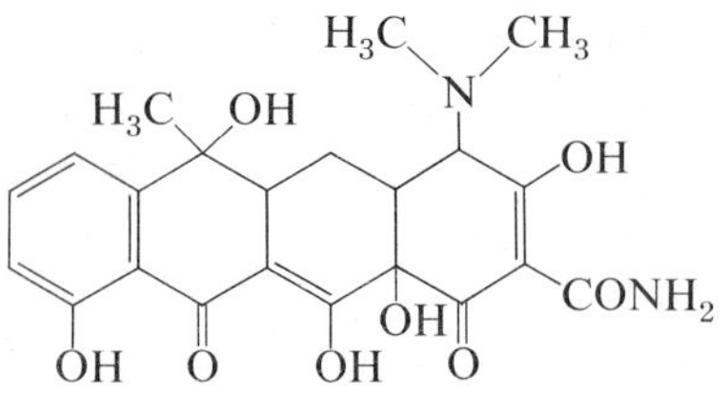   Tetracycline

Even though the importance of sulfa drugs has declined, the studies of the mechanism of action of these materials provide a very interesting insight into how chemotherapeutic substances might behave. In 1940, Woods and Fildes discovered that *p*-aminobenzoic acid (PABA) inhibits the action of sulfanilamide. They concluded that sulfanilamide and PABA, because of their structural similarity, must compete with each other within the organism even though they cannot carry out the same chemical function. Further studies indicated that sulfanilamide does not kill bacteria, but inhibits their growth. In order to grow, bacteria require an enzyme-catalyzed reaction which uses **folic acid** as a cofactor. Bacteria synthesize folic acid using PABA as one of the components. When sulfanilamide is introduced into the bacterial cell, it competes with PABA for the active site of the enzyme which carries out the incorporation of PABA into the molecule of folic acid. Because sulfanilamide and PABA compete for an active site due to their structural similarity, and because sulfanilamide is incapable of carrying out the chemical transformations characteristic of PABA once it has formed a complex with the enzyme, sulfanilamide is called a **competitive inhibitor** of the enzyme. Once the enzyme has formed a complex with sulfanilamide, it is incapable of catalyzing the reaction required for the synthesis of folic acid. Without folic acid, the bacteria cannot synthesize the nucleic acids required for growth. As a result, the bacterial growth is arrested until the body's immune system can respond and kill the bacteria.

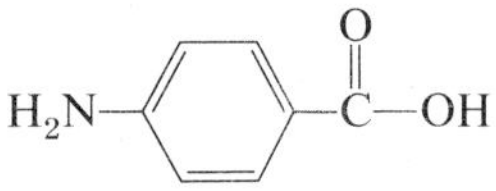

p-Aminobenzoic Acid
(PABA)

PABA residue

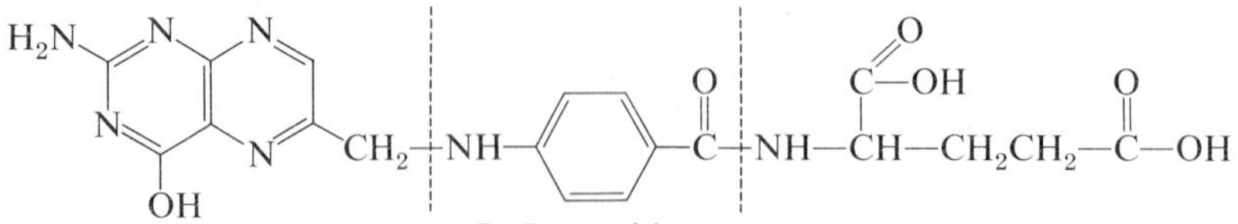

Folic Acid

One might well ask the question, "Why, when someone takes sulfanilamide as a drug, doesn't it inhibit the growth of **all** cells, bacterial and human alike?" The answer is very simple. Animal cells cannot synthesize folic acid. Folic acid must be a part of the diet of animals, which makes it an essential vitamin. Since animal cells receive their fully synthesized folic acid molecules through the diet, only the bacterial cells are affected by the sulfanilamide, and only their growth is inhibited.

For the vast majority of drugs, a detailed picture of the mechanism of their action is unavailable. The case of the sulfa drugs, however, provides a rare glimpse into how many other therapeutic agents might carry out their medicinal activity.

REFERENCES

Amundsen, L. H. "Sulfanilamide and Related Chemotherapeutic Agents." *Journal of Chemical Education, 19* (1942), 167.
Evans, R. M. *The Chemistry of Antibiotics used in Medicine.* London: Pergamon Press, 1965.
Fieser, L. F., and Fieser, M. *Topics in Organic Chemistry.* Chapter 7, "Chemotherapy." New York: Reinhold, 1963.
Garrod, L. P., and O'Grady, F. *Antibiotic and Chemotherapy.* Edinburgh: E. and S. Livingstone, Ltd., 1968.
Goodman, L. S. and Gilman, A. *The Pharmacological Basis of Therapeutics* (4th edition). Chapter 56. "The Sulfonamides," by L. Weinstein. New York: Macmillan, 1970.
Sementsov, A. "The Medical Heritage from Dyes." *Chemistry, 39* (November, 1966), 20.
Zahner, H., and Maas, W. K. *Biology of Antibiotics.* Berlin: Springer-Verlag, 1972.

Experiment **41**

SULFA DRUGS: SULFANILAMIDE, SULFAPYRIDINE, AND SULFATHIAZOLE

In this experiment you will prepare one of these sulfa drugs: sulfanilamide (Procedure 41A), sulfapyridine (Procedure 41B), or sulfathiazole (Procedure 41C) by one of the following synthetic schemes:

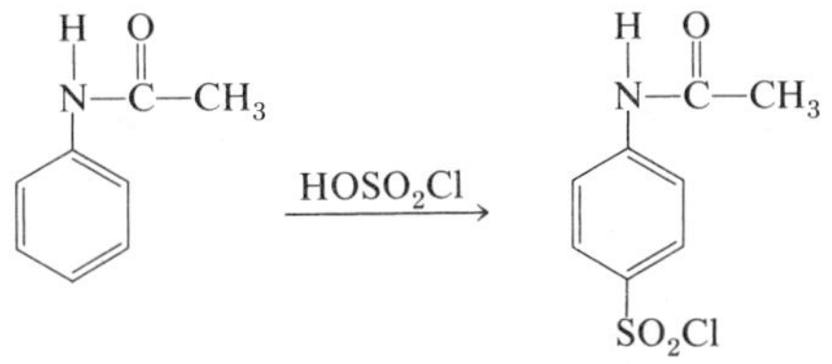

Acetanilide (1)   *p*-Acetamidobenzenesulfonyl Chloride (2)

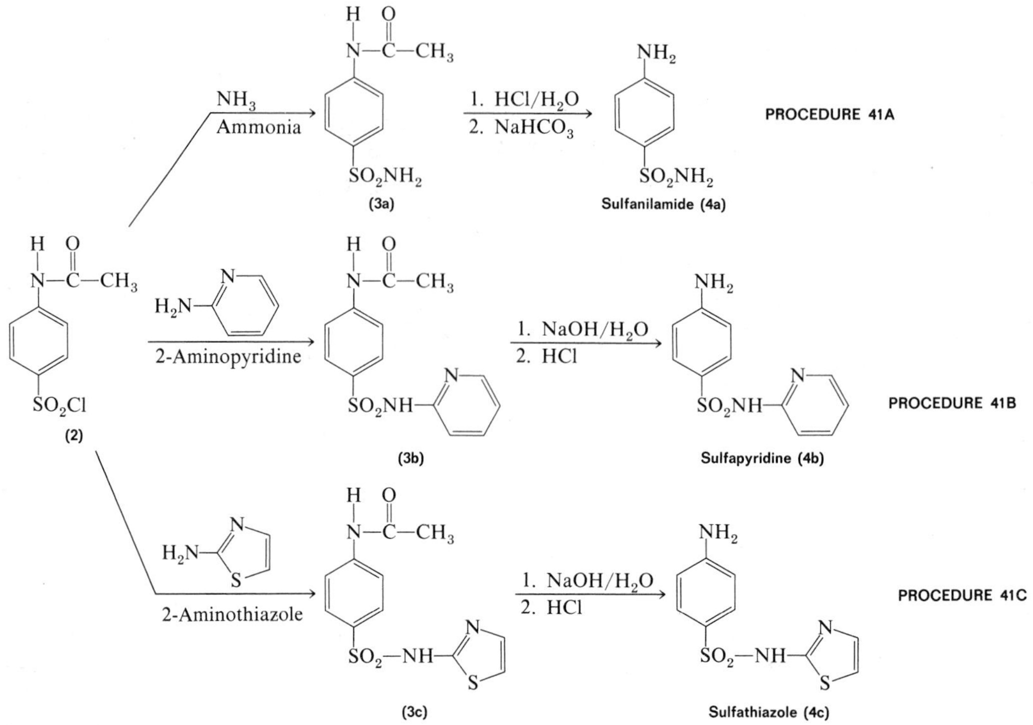

The syntheses each involve the conversion of acetanilide (**1**) to *p*-acetamidobenzenesulfonyl chloride (**2**). This intermediate, which is prepared by everyone, is then converted to one of three possible acetyl derivatives of sulfa drugs (**3a, 3b,** or **3c**). From these, you will then obtain the sulfa drugs (**4a, 4b,** or **4c**) upon hydrolysis. At the option of the instructor, you may test the prepared sulfa drug on several different kinds of bacteria (Procedure 41D).

p-ACETAMIDOBENZENESULFONYL CHLORIDE (2)

Acetanilide (**1**) which may be easily prepared from aniline (see Experiment 2), is allowed to react with chlorosulfonic acid to yield *p*-acetamidobenzenesulfonyl chloride (**2**). The acetamido group directs

Acetanilide (1)

p-Acetamidobenzenesulfonyl Chloride (2)

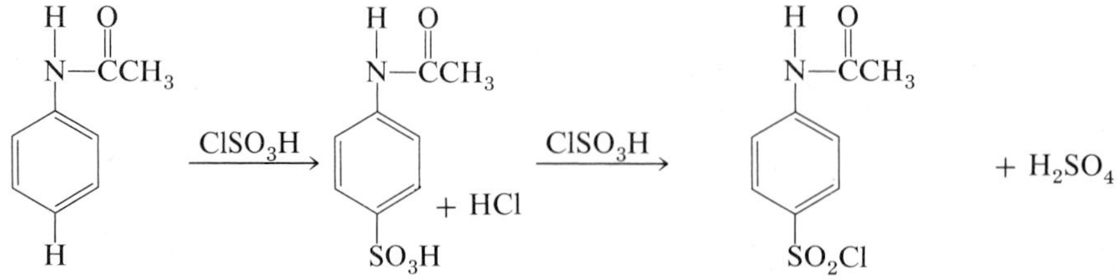

substitution almost totally to the **para** position. The reaction is an example of a typical electrophilic aromatic substitution reaction. Two problems would result if aniline itself were used in the reaction. First, the amino group in aniline would be protonated in strong acid to become a **meta** director, and secondly, the chlorosulfonic acid would react with the amino group, rather than with the ring, to give C_6H_5—$NHSO_3H$. For these reasons, the amino group has been "protected" by acetylation. It will be removed in the final step, after it is no longer needed, to regenerate the free amino group present in sulfa drugs. The product is isolated by pouring the reaction mixture into ice water, which decomposes the excess chlorosulfonic acid. The product, p-acetamidobenzenesulfonyl chloride (**2**) is fairly stable in water; nevertheless, it is converted slowly into the corresponding sulfonic acid (Ar—SO_3H). Thus, water must be removed as soon as possible after the preparation of this intermediate. This is done by dissolving the sulfonyl chloride (**2**) in chloroform, which separates the immiscible water. Anhydrous material is obtained from the chloroform upon crystallization.

SULFANILAMIDE (4a)

The intermediate sulfonyl chloride (**2**) is converted to the amide (**3a**) by reaction with aqueous ammonia. Excess ammonia neutralizes the hydrogen chloride produced. The only side reaction is the hydrolysis of the sulfonyl chloride, in the presence of water, to the sulfonic acid. The next step is the acid catalyzed hydrolysis of the protecting acetyl group to generate the protonated amino group. Note that of the two amide linkages present, **only** the carboxylic acid amide (acetamido group) is cleaved and not the sulfonic acid amide (sulfonamide). The salt of the sulfa drug which results is converted to sulfanilamide (**4a**) upon addition of the base, sodium bicarbonate.

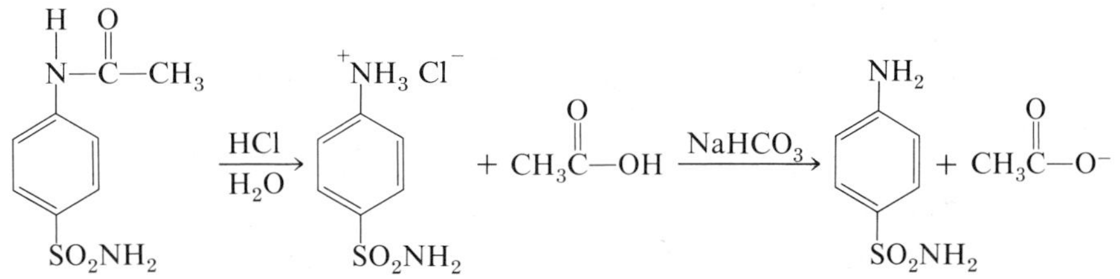

(3a)   Sulfanilamide (4a)

SULFAPYRIDINE (4b)

The intermediate sulfonyl chloride (**2**) is converted to the amide (**3b**) by reaction with 2-aminopyridine. The other product of the reaction, hydrogen chloride, reacts with the solvent pyridine to form a

salt. If pyridine were not present, then the 2-aminopyridine would be
inactivated as a nucleophile by salt formation, and this would reduce
the yield of the reaction. The pyridine must be very dry since the
sulfonyl chloride reacts rapidly with water in pyridine solution.

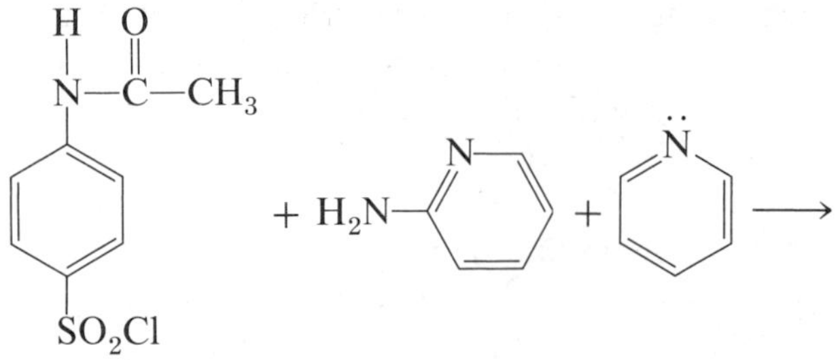

| (2) | 2-Aminopyridine | Pyridine |

| (3b) | Pyridine
Hydrochloride |

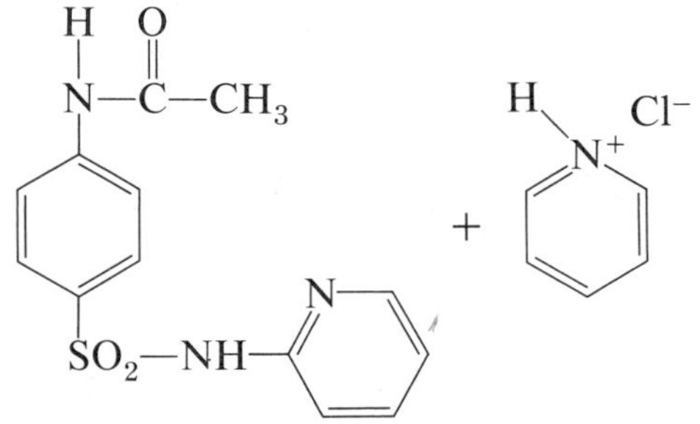

The second step (**3b** to **4b**) is a base-catalyzed hydrolysis which preferentially cleaves the acetamido group (see p. 313). The salt of the sulfa
drug is converted to sulfapyridine (**4b**) upon addition of acid. The salt is
formed because sulfonamides are reasonably acidic compounds.

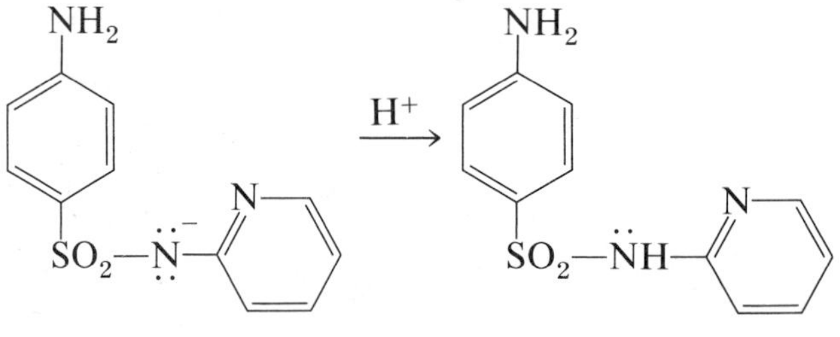

| Salt | Sulfapyridine (4b) |

SULFATHIAZOLE (4c)

The intermediate sulfonyl chloride (**2**) is converted to the amide
(**3c**) by reaction with 2-aminothiazole in the presence of pyridine. The
reaction proceeds as for sulfapyridine above. The side reaction which
may occur results from possible tautomerism of the 2-aminothiazole.

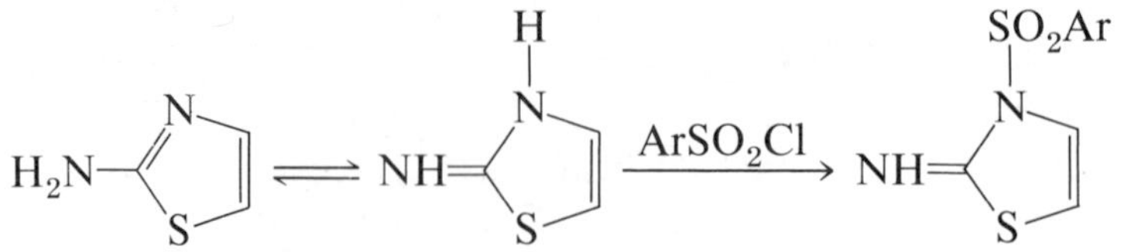

| A | B | C |

The tautomer **B** can react with the sulfonyl chloride (**2**) to give an isomer of sulfapyridine (**C**). Fortunately, little of this isomer (**C**) is produced in this experiment. The second step (**3c** to **4c**) is a base catalyzed hydrolysis which is discussed above under sulfapyridine.

SPECIAL INSTRUCTIONS

Read the essay which precedes this experiment. The chlorosulfonic acid must be handled with care since it is a corrosive liquid. The *p*-acetamidobenzenesulfonyl chloride (**2**) should be used during the same laboratory period in which it was prepared, or during the very next period. It is unstable and will not survive long periods of storage.

The instructor may divide the class into groups of three students. Each student in the group may then pick a different sulfa drug and synthesize it. The three drugs (sulfanilamide, sulfapyridine, and sulfathiazole) may then be tested on several kinds of bacteria (Procedure 41D).

PROCEDURES

p-ACETAMIDOBENZENESULFONYL CHLORIDE (2)

Assemble an apparatus as shown in the figure to help trap the hydrogen chloride gas that forms as a by-product of this reaction. Insert a short section of glass tubing into a one hole rubber stopper. Insert the stopper into the flask after the acetanilide and chlorosulfonic acid have been added. Use a piece of rubber tubing to lead the vapors away from the reaction flask to an inverted funnel placed **just below** the surface of a beaker of water containing about 1 g of sodium hydroxide. As the gas is evolved, it dissolves in the water and reacts with the sodium hydroxide.

Place 12 g of dry acetanilide (**1**) in a **dry** 250 ml Erlenmeyer flask. Melt the acetanilide by gentle heating with a flame. Swirl the heavy oil so that it is deposited uniformly on the lower wall and bottom of the flask. Cool the flask in an ice bath. To the solidified material, add 33 ml of chlorosulfonic acid, $ClSO_2OH$ (d = 1.77 g/ml) in one portion.

CAUTION: Chlorosulfonic acid is an extremely noxious and corrosive chemical and should be handled with care. Use only dry glassware with this reagent. Should the chlorosulfonic acid be spilled on the skin, wash it off immediately with water. Wear safety glasses.

Apparatus for making *p*-acetamidobenzenesulfonyl chloride

The funnel must
be placed just
below the surface
of the water

Remove the flask from the ice bath and swirl it. Hydrogen chloride gas is evolved vigorously, so be certain that the rubber stopper is securely placed in the neck of the flask. The reaction mixture usually will not have to be cooled. If the reaction becomes too vigorous, however, slight cooling may be necessary. After ten minutes, the reaction should have subsided and only a small amount of acetanilide should remain. Heat the flask for an additional 10 minutes on the steam bath to complete the reaction (continue to use the trap). After this time, remove the trap assembly, and cool the flask in an ice bath in the hood.

The operations in this paragraph should be conducted as rapidly as possible since the *p*-acetamidobenzenesulfonyl chloride reacts with water. **Slowly** pour the cooled mixture with vigorous stirring (it will splatter somewhat) into a beaker containing 200 ml of crushed ice. Rinse the flask with some cold water and transfer the contents to the beaker containing the ice. Stir the precipitate to break up the lumps and then vacuum filter the mixture. Wash the crude *p*-acetamidoben-zenesulfonyl chloride with a small amount of cold water. Dissolve the solids in 150 ml of boiling chloroform in the hood, but be certain to avoid the toxic vapors. The solid may dissolve slowly. Replace the chloroform solvent as it evaporates. Transfer the mixture to a **warm** separatory funnel. Drain the lower chloroform layer rapidly, but carefully, from the funnel and away from the upper water layer.

Cool the mixture in an ice bath and collect the crystalline *p*-aceta-midobenzenesulfonyl chloride by vacuum filtration using a Büchner funnel. Draw air through the funnel to dry the product. Weigh the material and calculate the percent yield. Determine the melting point of the product. The anhydrous compound melts at 149°. The *p*-acetamidobenzenesulfonyl chloride should be used during the same laboratory period or the very next one. If it must be stored, the dry product must be placed in a tightly stoppered bottle. Convert the *p*-acetamidobenzenesulfonyl chloride (**2**) into one of three possible sulfa drugs by one of the following procedures, 41A, 41B, or 41C.

Procedure **41** A

SULFANILAMIDE (4a)

Place 5 g of *p*-acetamidobenzenesulfonyl chloride (**2**) in a 125 ml Erlenmeyer flask and add (in the hood) 15 ml of concentrated ammonium hydroxide. Stir the mixture well with a stirring rod. A reaction usually begins immediately and the mixture becomes warm. Heat the mixture on a steam bath in the hood for 15 minutes with frequent stirring. During this time the material becomes a more pasty suspension. Remove the flask and place it in an ice bath. When the mixture is well cooled, add 6 N hydrochloric acid (5 to 10 ml) until the mixture is acidic to litmus paper. Continue cooling the mixture in the ice bath until it is thoroughly cold, and filter the product on the Büchner funnel with the vacuum. Wash the product (compound **3a**) with 50 ml of cold water. The material may be used directly in the next step or it can be allowed to dry in air.

Transfer the crude product (**3a**) to a small round bottom flask and add 3 ml of concentrated hydrochloric acid, 6 ml of water, and a boiling stone. Attach a reflux condenser to the flask. Allow the mixture to reflux until the solid has dissolved (about 10 min) and then reflux for an additional 10 minutes. Cool the mixture to room temperature. If a solid which is unreacted starting material appears, bring the mixture to a boil again for several minutes. Upon cooling to room temperature, no further solids should appear. To this cooled solution, add 5 ml of water and a small amount of decolorizing carbon. Shake the mixture and then filter it by gravity into a 200 ml or larger beaker. Rinse the flask with 10 ml of water and pour the rinsings through the filter. To the filtrate, cautiously add a solution of 4 g of sodium bicarbonate (dissolved in a minimum amount of water) with stirring, until the solution is neutral to litmus. Foaming will occur after each addition of the bicarbonate solution because of carbon dioxide evolution. The sulfanilamide will precipitate during the neutralization process. Cool the mixture thoroughly in an ice bath and filter the product by vacuum filtration through a Büchner funnel. Dry the sulfanilamide as much as possible by drawing air through the filter. Recrystallize the solid from water, using about

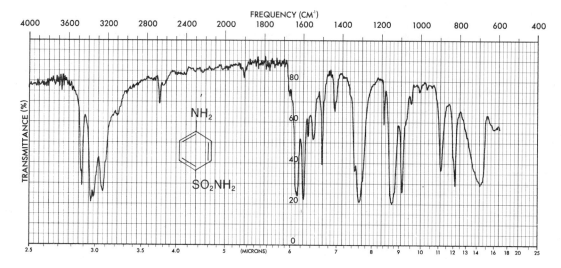

Infrared spectrum of sulfanilamide, KBr

10 to 12 ml of water per gram of crude product. Allow the material to dry until the next laboratory period. Determine the melting point. Pure sulfanilamide melts from 163 to 164°. Weigh the product and calculate the percent yield. Submit the sulfanilamide to the instructor in a labeled vial or save it for the tests with bacteria in Procedure 41D. At the instructor's option, determine the infrared spectrum in potassium bromide.

Procedure **41** B

SULFAPYRIDINE (4b)

Dissolve 2.4 g of 2-aminopyridine in 10 ml of anhydrous pyridine (dried over KOH pellets) in a small round bottom flask. Add 6 g of dry p-acetamidobenzenesulfonyl chloride (2) to the mixture. Attach a reflux condenser to the flask. Allow the mixture to reflux for 15 minutes. Cool the mixture and pour it into 50 ml of water. Add some water to the flask to aid in the transfer operation. Stir the mixture in an ice bath until the oil crystallizes. Filter the solid (3b) by vacuum filtration through a Büchner funnel.

Transfer the crude product (3b) to a round bottom flask and dissolve it in 20 ml of 10% sodium hydroxide solution. Allow the resulting solution to reflux for 40 minutes. After cooling the mixture, carefully neutralize it with 6N hydrochloric acid. Filter the precipitated sulfa-

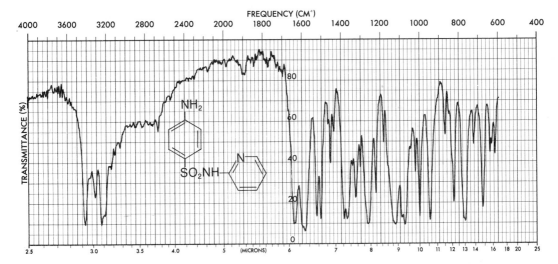

FREQUENCY (CM⁻¹)

Infrared spectrum of sulfapyridine, KBr

pyridine by vacuum filtration and recrystallize it from 95% ethyl alcohol. About 100 ml of alcohol will be needed to dissolve the sulfapyridine. Watch it carefully as it dissolves. When the point is reached where the remaining solid no longer dissolves, remove the remaining solid by gravity filtration. Cool the filtrate slowly to room temperature. Filter the mixture by vacuum filtration on a Büchner funnel. Dry the sulfapyridine in air and determine its melting point. The recorded melting point is 190 to 193°. Weigh the product and calculate the percent yield. Submit the sulfapyridine to the instructor in a labeled vial or save it for the tests with bacteria in Procedure 41D. At the instructor's option, determine the infrared spectrum in potassium bromide.

Procedure 41 c

SULFATHIAZOLE (4c)

Dissolve 2.5 g of 2-aminothiazole in 10 ml of anhydrous pyridine (dried over KOH pellets) in an Erlenmeyer flask. To this mixture, add in small portions 6.5 g of dry p-acetamidobenzenesulfonyl chloride (2) with swirling. The rate of addition should be such that the temperature does not rise above 40°. Cool the mixture slightly if the temperature exceeds this value. After the addition, complete the reaction by heating it for 30 minutes on a steam bath. Cool the mixture and then pour it into 75 ml of water. Add some water to the flask to aid in the transfer operation. The oil (3c) should solidify when stirred with a glass rod.

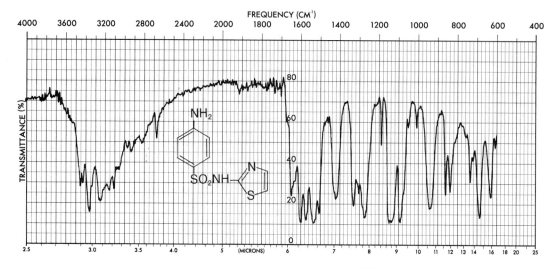

Infrared spectrum of sulfathiazole, KBr

Isolate the crystals by vacuum filtration. Wash them with cold water, and then press them dry.

Weigh the solid (**3c**) and dissolve it in 10% sodium hydroxide (10 ml per gram of solid) in a round bottom flask. Attach a reflux condenser and heat the mixture under reflux for 1 hour. Cool the solution and add concentrated hydrochloric acid until the mixture is at pH 6. If too much acid is added, adjust the pH to 6 by adding 10% sodium hydroxide. Then add solid sodium acetate until the solution is just basic to **litmus.** Heat the mixture to a boil and cool it in an ice bath. Filter the sulfathiazole by vacuum filtration and recrystallize it from water (the solubility of sulfathiazole in hot water is quite low). Cool the solution slowly to room temperature. Filter the mixture by vacuum filtration. Allow the sulfathiazole to dry until the next laboratory period. Determine the melting point. Pure sulfathiazole melts from 201 to 202°. Weigh the product and calculate the percentage yield. Submit the sulfathiazole in a labeled vial to the instructor or save it for the bacteria tests in Procedure 41D. At the instructor's option, determine the infrared spectrum in potassium bromide.

Procedure **41**D

THE TESTING OF SULFA DRUGS ON BACTERIA (OPTIONAL)

At the option of the instructor, test the prepared sulfa drug on two types of bacteria; **Aerobacter aerogenes** and **Bacillus subtilis.** A group of three students should team up so that all three drugs are included on each plate. A specific area in the laboratory should be equipped for this experiment. Dissolve 0.25 g of sulfanilamide or sulfathiazole in 70 ml of boiling distilled water. Dissolve 0.25 g of sulfapyridine in a boiling mixture of 50 ml of water and 20 ml of 95% ethanol.

While the drugs are dissolving, obtain agar plates which have been pre-inoculated with the bacteria.[1] The name of the bacteria has been written on the top of the Petri dish. With a grease pencil divide the bottom of the Petri dish into quadrants, as shown in the figure. In each quadrant print one of the following symbols: SA (sulfanilamide), SP (sulfapyridine), ST (sulfathiazole), or C (control). Place the names or initials of the students in the group on the top of the Petri dish.

Dip a pair of tweezers in 95% ethanol and place them in a Bunsen burner flame to sterilize them. It is essential to maintain sterile conditions at all times. The attitude that contaminating bacteria are everywhere is a good one. The bacteria tend to drop into solutions from above, but never move up or sideways. After the tweezers have been sterilized, dip two tabs into one of the sulfa drug solutions while it is still boiling, and remove the tabs. Immediately place them under a watch glass on a piece of filter paper to dry for a few minutes. Again sterilize the tweezers and repeat the above steps in succession with the two remaining solutions of sulfa drugs. Finally, dip one tab in a beaker of boiling distilled water. This tab serves as a control, since it does not contain any sulfa drug. Lift the lid of a Petri dish containing one of the bacteria, **straight up** so that it is directly above the bottom (see figure). It should be lifted just high enough to allow room for the tweezers. Insert the tweezers with the control tab. Place this tab in the center of the quadrant that was previously marked. Sterilize the tweezers and transfer the tabs, in succession, to the center of the appropriate quadrants on the agar plate. Sterilize the tweezers each time a new sulfa drug is being transferred. Repeat this operation with another agar plate containing the second bacteria to be investigated. Again, care must be taken to avoid contamination of the bacteria.

Once the tabs have been placed on the plates, store the agar plates in one general area. Do not place them in your drawer. The temperature should be maintained at 25°C. The bacteria will then begin to grow. The plates should be observed 24 and 48 hours after the introduction of the tabs. During the first 24 hours, the drugs may have the greatest growth-inhibiting effect. Thus, it is important to observe the results at that time or before.

Holding the plates approximately level, observe the plates from the bottom. The originally clear plates will be cloudy in the area where bacteria has grown. Growth should be observed around the control tab. The agar will be clear in the area where the sulfa drug has inhibited the growth of the bacteria. One usually observes a clear circular area

---

[1]Instructions for preparing the agar plates are in the instructor's manual or are available from the authors.

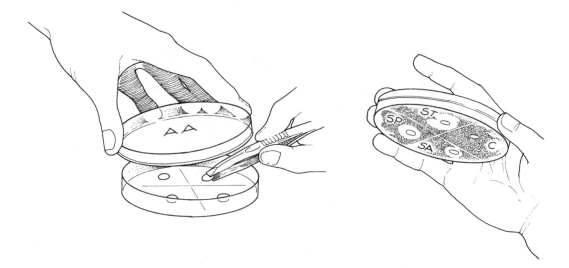

where inhibition has occurred (see figure). The larger the circle, the greater the effect of the sulfa drug in killing the bacteria. Inspect **all** of the plates, not just your own group's plates. Sometimes the bacteria will not grow at all. If this has happened, the plate will be totally clear, even around the control tab. Assess the relative effect of the three sulfa drugs on each of the bacteria. Report the results to the instructor in your report.

QUESTIONS

1. Give an equation showing how excess chlorosulfonic acid is decomposed.

2. In the preparation of sulfanilamide, why was aqueous sodium bicarbonate used, rather than aqueous sodium hydroxide, to neutralize the solution in the final step?

3. In the final steps of the preparation of sulfapyridine and sulfathiazole, the acidity must be carefully controlled. Why? Give equations to show the behavior of these sulfa drugs in both acidic and basic solution.

4. At first glance, it might appear possible to prepare sulfanilamide from sulfanilic acid (Experiment 31) by the following set of reactions:

$$
\underset{\underset{SO_3H}{}}{\overset{NH_2}{\bigcirc}} \xrightarrow{PCl_5} \underset{\underset{SO_2Cl}{}}{\overset{NH_2}{\bigcirc}} \xrightarrow{NH_3} \underset{\underset{SO_2NH_2}{}}{\overset{NH_2}{\bigcirc}}
$$

However, when the reaction is conducted in this way, a polymeric product is produced after the first step. What is the structure of the polymer? Why does *p*-acetamidobenzenesulfonyl chloride not produce a polymer?

5. In the preparation of sulfathiazole, an impurity is sometimes produced (shown

below). How could you separate this impurity from sulfathiazole by using acids and/or bases?

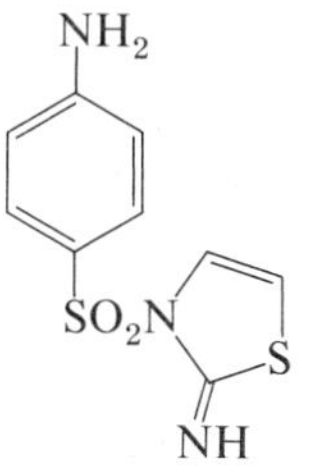

$$NH_2$$

$$SO_2N \diagdown S \diagdown NH$$

Experiment **42**

p-AMINOBENZOIC ACID

Synthesis Of Vitamin
Amide Formation
Permanganate Oxidation
Amide Hydrolysis

The object of this set of chemical reactions is to prepare a vitamin (for bacteria) **p-aminobenzoic acid** (or PABA). For a description of the importance of PABA in biological processes, read the covering essay immediately preceding Experiment 41.

The synthesis of *p*-aminobenzoic acid involves three reactions which will be outlined in the following paragraphs. The first of these reactions is the conversion of the commercially available **p-toluidine** into **N-acetyl-p-toluidine** (or *p*-acetotoluidide), the corresponding amide. The reaction is carried out by the treatment of the amine, *p*-toluidine, with acetic anhydride. Such a procedure is a standard method for the prepa-

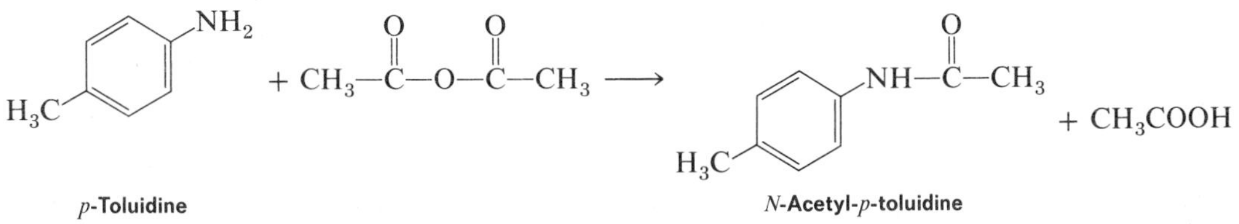

$$H_3C \diagdown NH_2 \quad + \quad CH_3-\overset{O}{\underset{\|}{C}}-O-\overset{O}{\underset{\|}{C}}-CH_3 \longrightarrow \quad H_3C \diagdown NH-\overset{O}{\underset{\|}{C}}-CH_3 \quad + \quad CH_3COOH$$

p-Toluidine   *N*-**Acetyl-*p*-toluidine**

ration of amides. The reason for acetylating the amine in the initial step is to protect it during the second step, which is a permanganate oxidation. If the oxidation of the methyl group to the corresponding carboxyl group were carried out directly with *p*-toluidine, the highly reactive amino group would also be oxidized in the reaction. To prevent this undesired oxidation, a **protective group** is employed. A protective group is a functional group that is added during the course of a reaction sequence to shield a particular position on a molecule from undesired reactions. A good protective group is one that is easily added to the substrate molecule, does not permit the protected group from undergoing reactions, and is easily removed after the protective role has been played. This last point is quite important, since if the protective group cannot be removed, the original functional group cannot be utilized. In the example used in this experiment, the protective group is the **acetyl group.** The amide which is formed in this initial reaction is stable toward the oxidation conditions used in the second step. As a result, the methyl group may be oxidized without destroying the amino function in the process.

The second step of this reaction sequence is the oxidation of the methyl group to the corresponding carboxyl group, with potassium permanganate serving as the oxidizing agent. Because of the great

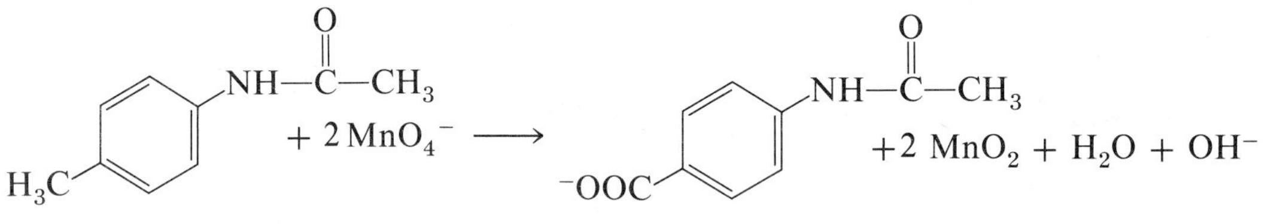

stability of substituted benzoic acids, alkyl groups attached to aromatic rings may be oxidized rather easily to the corresponding benzoic acids. During the course of the oxidation, the violet solution of permanganate ion is converted to a brown precipitate, manganese dioxide, as Mn(VII) is reduced to Mn(IV). A small amount of magnesium sulfate is added to serve as a buffer in order that the solution does not become excessively basic as hydroxide ion is produced. The product of the reaction is not the carboxylic acid, but rather it is the salt which is produced directly from the reaction. Acidification of the reaction mixture yields the carboxylic acid, which precipitates from solution.

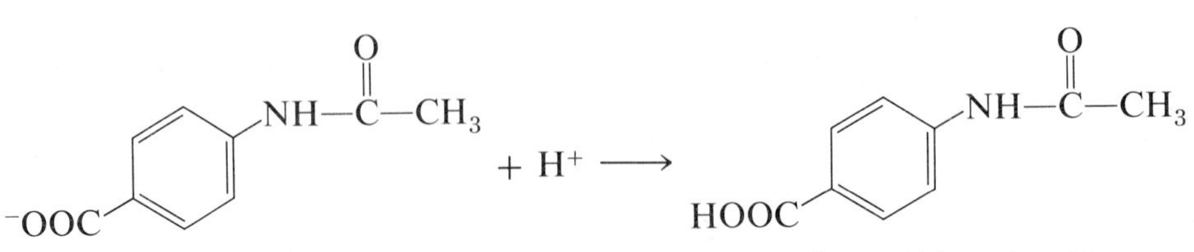

p-Acetamidobenzoic acid

The final step of this procedure is the hydrolysis of the amide functional group to remove the protecting acetyl group, thus yielding *p*-aminobenzoic acid. The reaction proceeds easily in dilute aqueous

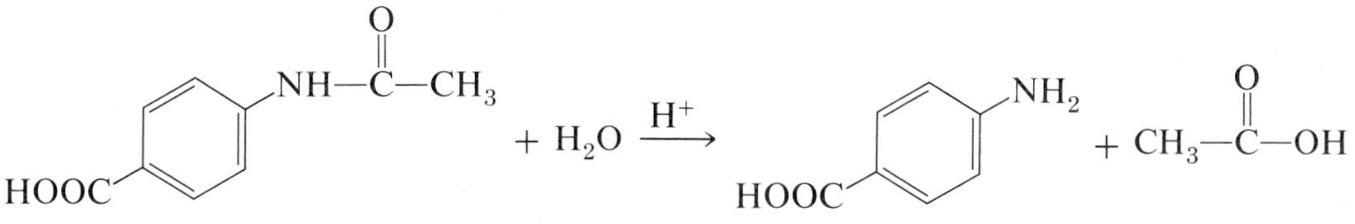

p-Aminobenzoic acid

acid, and it is a quite characteristic reaction of amides in general. The product is crystallized from dilute aqueous acetic acid.

SPECIAL INSTRUCTIONS

Before performing this experiment, read Techniques 1, 2, 3, and 4. This experiment should be conducted in two parts. It is essential that the procedure up through the placing of the crystals of *p*-acetamido-benzoic acid in the drying oven be carried out in the first laboratory period. Between one and one-and-a-half hours of reaction time are required to reach this point, along with considerable time spent in performing other operations.

PROCEDURE

N-ACETYL-*p*-TOLUIDINE

Place 16 g of powdered *p*-toluidine in a 500 ml Erlenmeyer flask. Add 400 ml of water and 15 ml of concentrated hydrochloric acid. If necessary, warm the mixture on a steam bath, with stirring, to facilitate solution. If the solution is dark-colored, add 0.5 to 1.0 g of decolorizing charcoal, stir it for several minutes, and filter it by gravity.

Prepare a solution of 24 g of sodium acetate trihydrate in 40 ml of water. If necessary, use a steam bath to warm the solution until all of the solid has dissolved.

Warm the decolorized solution of *p*-toluidine hydrochloride to 50°C. Add 16.7 ml (d = 1.08) of acetic anhydride, stir rapidly, and immediately add the previously prepared sodium acetate solution. Mix the solution thoroughly, and cool the mixture in an ice bath. A white solid should appear at this point. Filter the mixture by vacuum using a Büchner funnel, wash the crystals three times with cold water, and allow the crystals to stand in the filter to air dry, while maintaining the vacuum. These crystals will not be isolated and dried, but will be used directly in the next step.

p-ACETAMIDOBENZOIC ACID

Place the previously prepared, wet *N*-acetyl-*p*-toluidine in a one liter beaker, along with 50 g of magnesium sulfate hydrate and 750 ml of water. Place the flask on a steam bath, and adjust the steam flow to a gentle rate. Add a sludge made up of 60 g potassium permanganate and a small amount of water to the reaction in approximately teaspoon quantities. Increase the flow rate of the steam and allow the reaction to proceed for one hour. It is important to heat the reaction mixture thoroughly. It may be helpful to place the beaker down **into** the steam cone in order to heat the beaker more effectively. During the reaction period, the mixture must be stirred every few minutes. After one hour, the mixture should be quite brown in color. Vacuum filter the **hot** solution through a bed of Celite. Wash the precipitated manganese dioxide with a small amount of hot water. If the filtrate shows the presence of excess permanganate (by its purple color), add **not more than** one milliliter of ethanol, heat the solution in the steam bath for another 30 minutes and filter the **hot** solution through fluted filter paper once more. Cool the colorless filtrate and acidify with excess 20% sulfuric acid solution. A white solid should form at this point. Filter the solid by vacuum and dry it in an oven. The yield based on *p*-toluidine and the melting point should be determined at this point. The melting point of the pure material is 250 to 252°C. In some cases an inorganic salt may be produced. If a salt is produced, simply proceed to the next step of the procedure.

p-AMINOBENZOIC ACID

Prepare a dilute solution of hydrochloric acid by mixing 47 ml of concentrated hydrochloric acid and 47 ml of water. Place the previously prepared *p*-acetamidobenzoic acid in a 300 ml round bottom flask which has a reflux condenser attached to it. Add the hydrochloric acid solution. Reflux the mixture gently for 30 minutes. Allow the reaction mixture to cool, transfer it to a 500 ml Erlenmeyer flask, add 94 ml of water, and make the reaction mixture just alkaline to litmus paper with dilute ammonia solution. For each 30 ml of the final solution, add one milliliter of glacial acetic acid, chill the solution in an ice bath, and initiate the crystallization, if necessary, by scratching the inside of the flask with a glass rod. Filter the crystals by vacuum and allow them to dry. Determine the yield for this step, basing the overall yield on *p*-toluidine.

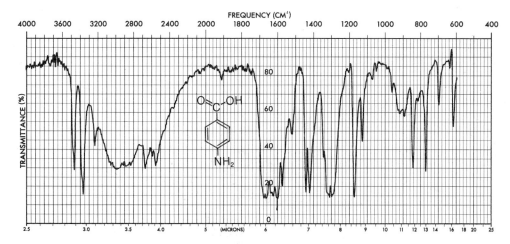

Infrared spectrum of *p*-aminobenzoic acid, KBr

Determine the melting point of the product. The melting point of the pure *p*-aminobenzoic acid is 186 to 187°. Frequently the melting point of the product is somewhat lower. Attempts to recrystallize the product are not sufficiently successful to be recommended. Recrystallization is not necessary before the *p*-aminobenzoic acid is used in Experiment 43. At the option of the instructor, an infrared spectrum of the *p*-aminobenzoic acid may be determined (Technique 17, Part A, and Appendix Three).

The *p*-aminobenzoic acid may be saved for use in Experiment 43, the preparation of either benzocaine or procaine. If it is not to be used for a later experiment, submit the sample of *p*-aminobenzoic acid, in a labeled vial, to the instructor.

REFERENCE

Kremer, C. B. "The Laboratory Preparation of a Simple Vitamin: *p*-Aminobenzoic Acid." *Journal of Chemical Education, 33* (1956), 71.

QUESTIONS

1. Write a mechanism for the reaction of *p*-toluidine with acetic anhydride. What is the purpose of the sodium acetate added to this reaction?

2. In the oxidation step, if excess permanganate remains after the reaction period, a small amount of ethanol is added to discharge the purple color. Write a chemical equation which describes the reaction of permanganate with ethanol.

3. Write a mechanism for the acid-catalyzed hydrolysis of *p*-acetamidobenzoic acid to form *p*-aminobenzoic acid.

4. Interpret the major absorption bands in the infrared spectrum of *p*-aminobenzoic acid.

Local Anesthetics

Local anesthetics or "painkillers" are a well-studied class of compounds with which chemists have shown their ability to study the essential features of a naturally occurring drug and to improve upon them by substituting totally new synthetic surrogates. Often such substitutes are superior with respect to their desired medical effects as well as in their lack of unwanted side effects or hazards.

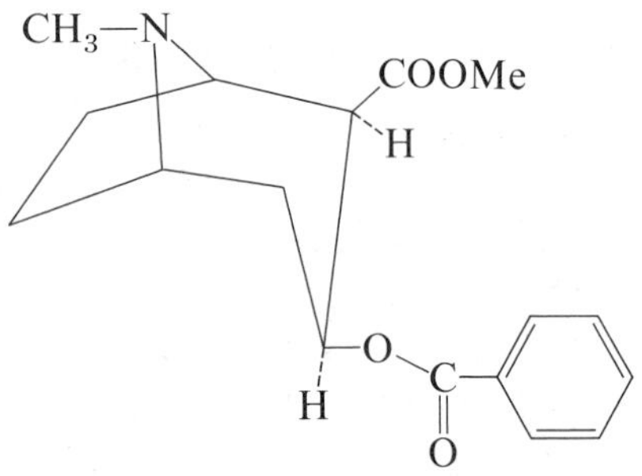

Cocaine 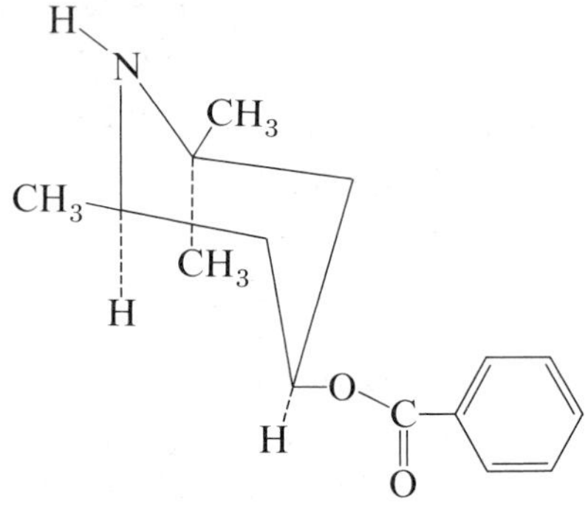   Eucaine

The coca shrub (*Erythroxylon coca*) grows wild in Peru, specifically in the Andes Mountains, at elevations from about 1500 to 6000 feet above sea level. The natives of South America have long chewed these leaves for their stimulant effects. Leaves of the coca shrub have even been found in pre-Inca Peruvian burial urns. The leaves bring about a definite sense of mental and physical well-being and have the power to increase endurance. For chewing, the Indians smear the coca leaves with lime and roll them. The lime [$Ca(OH)_2$] apparently releases the free alkaloid components; it is remarkable that the Indians learned this subtlety long ago by some empirical means. The pure alkaloid responsible for the properties of the coca leaves is **cocaine.**

Moderate indulgence, as practiced by the coca chewing Indians, probably produces no more ill effects than does moderate tobacco smoking. The amounts of cocaine consumed in this manner by the Indians are extremely small. Without such a crutch of central nervous system stimulation, the natives of the Andes would probably find it more difficult to perform the nearly Herculean tasks of their daily lives, such as carrying heavy loads over the rugged mountainous terrain. Unfortunately, over-indulgence can lead to mental and physical deterioration and eventually an unpleasant death.

The pure alkaloid in large quantities is a common drug of addiction, which is psychological if not physical. Sigmund Freud first made a detailed study of cocaine in 1884. He was particularly impressed by the ability of the drug to stimulate the central nervous system and used it as a replacement drug to wean one of his addicted colleagues from morphine. This attempt was successful but, unhappily, the colleague became the world's first known cocaine addict.

An extract from coca leaves was one of the original igredients in Coca-Cola. However, early in the present century, government officials, with much legal difficulty, forced the manufacturer to omit coca from its beverage. The company has managed to this day to maintain the **coca** in its trademarked title even though "Coke" contains none!

Our interest in cocaine, however, lies in its anesthetic properties. The pure alkaloid was isolated in 1862 by Niemann who noted that it had a bitter taste and produced a queer numbing sensation on the tongue rendering it almost devoid of sensation. (Oh, those brave, but foolish chemists of yore who used to **taste** everything!) In 1880 Von Anrep found that the skin was made numb and insensitive to the prick of a pin when cocaine was injected subcutaneously. Having failed at attempts to reconstitute morphine addicts, Freud and his assistant, Karl Koller, turned to a study of the anesthetizing properties of cocaine. Eye surgery is made difficult by involuntary reflex movements of the eye in response to even the slightest touch. Koller found that a few drops of a solution of cocaine would overcome this problem. Not only can cocaine serve as a local anesthetic, but it can also be used to produce mydriasis (dilation of the pupil). The ability of cocaine to block signal conduction in nerves (particularly of pain) led to its rapid medical use in spite of its dangers. It soon found use as a "local" in both dentistry (1884) and in surgery (1885). In this type of application it was injected directly into the particular nerves it was intended to deaden.

Soon after the structure of cocaine was established, chemists began to search for a substitute. Cocaine has several drawbacks for wide medical use as an anesthetic. In eye surgery, as mentioned above, it also produces mydriasis. It can also become a drug of addiction. And, finally, it has a dangerous effect on the central nervous system.

The first totally synthetic substitute was eucaine (see above). This was synthesized by Harries in 1859 and retains many of the essential skeletal features of the cocaine molecule. The development of this new anesthetic confirmed, in part, the portion of the cocaine structure essential for local anesthetic action. The advantage of eucaine over cocaine is that it does not produce mydriasis and is not habit-forming. Unfortunately, it has a high toxicity.

A further attempt at simplification led to piperocaine. The molecular portion common to cocaine and eucaine is outlined by dotted lines. Piperocaine is only 1/3 as toxic as cocaine itself.

Piperocaine

The most successful synthetic for many years was the drug procaine, also known more commonly by its trade name Novocain (see table). Novocain is only 1/4 as toxic as cocaine, giving a better margin of safety in its use. The toxic dose is also almost ten times the effective amount, and it is not a habit-forming drug.

Over the years hundreds of new local anesthetics have been synthesized and tested. For one reason or another most of them have not come into general use. In fact, the search for the perfect local anesthetic is still under way. All of the drugs which have been found to be active have the following structural features in common. At one end of the molecule is an aromatic ring. At the other is a secondary or tertiary amine. These two essential features are separated by a central chain of atoms usually 1 to 4 units long. The aromatic part is usually an ester

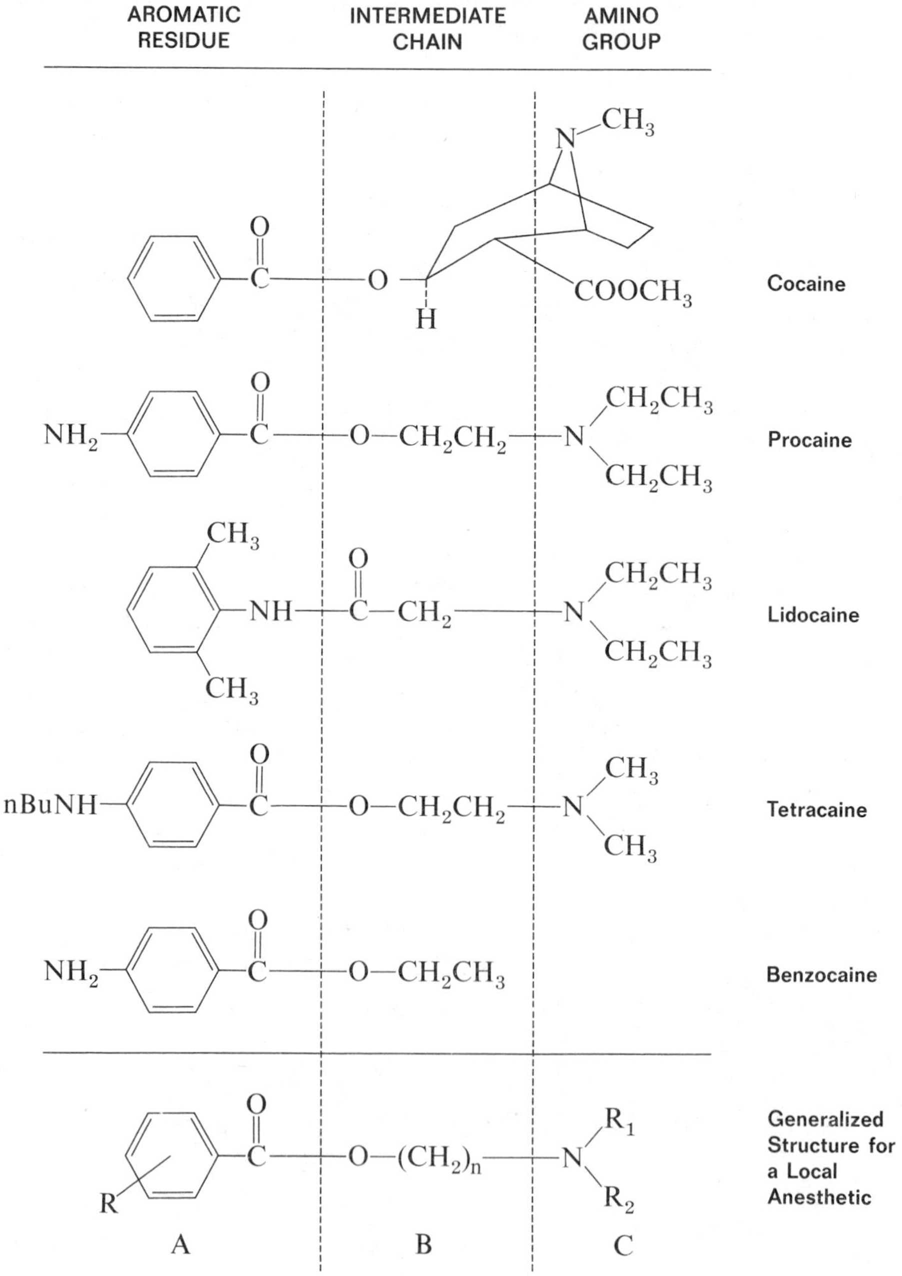

AROMATIC RESIDUE	INTERMEDIATE CHAIN	AMINO GROUP	

Local Anesthetics

of an aromatic acid. In fact, the ester group is important to the bodily detoxification of these compounds. In deactivating them, the first step is a hydrolysis of this ester linkage, a process which occurs in the blood stream. Compounds which do not have the ester link are both longer lasting in their effect and, in general, more toxic. An exception is lidocaine, which is an amide. The tertiary amino group is apparently necessary to enhance the solubility of the compounds in the injection solvent. Most of these compounds are used in their hydrochloride salt

$$-\ddot{N}\Big\langle\begin{smallmatrix}R\\[2pt]R\end{smallmatrix}\ +\ HCl\ \longrightarrow\ -\overset{\overset{\displaystyle R}{|}}{\underset{\underset{\displaystyle R}{|}}{N}}\!\overset{\oplus}{-}H\ Cl^{\ominus}$$

forms which can be dissolved in water for injection. Benzocaine, in contrast, is active as a local anesthetic but is not used for injection. It does not suffuse well into tissue and is not water soluble. It is used primarily in skin preparations where it can be applied directly in an ointment or salve. It is an ingredient of many sunburn preparations.

How these drugs act to stop pain conduction is not currently well understood. Their main site of action is at the nerve membrane. They seem to compete with calcium at some receptor site altering the permeability of the membrane and keeping the nerve slightly depolarized electrically.

REFERENCES

Foye, W. O *Principles of Medicinal Chemistry.* Chapter 14, "Local Anesthetics." Philadelphia: Lea and Febiger, 1974.

Goodman, L. S., and Gilman, A. *The Pharmacological Basis of Therapeutics* (4th edition). Chapter 20, "Cocaine; Procaine and Other Synthetic Local Anesthetics," J. M. Ritchie, et al. New York: Macmillan, 1970.

Ray, O. S. *Drugs, Society, and Human Behavior.* Chapter 11, "Stimulants and Depressants." St. Louis: Mosby, 1972.

Taylor, N. *Narcotics: Nature's Dangerous Gifts.* Chapter 3, "The Divine Plant of the Incas." New York: Dell 1970. Paperbound revision of *Flight from Reality.*

Taylor, N. *Plant Drugs that Changed the World.* New York: Dodd, Mead, and Company, 1965. Pp. 14–18.

Wilson, C. O., Gisvold, O., and Doerge, R. F. *Textbook of Organic Medicinal and Pharmaceutical Chemistry* (6th edition). Chapter 22, "Local Anesthetic Agents," R. F. Doerge. Philadelphia: J. B. Lippincott, 1971.

BENZOCAINE AND PROCAINE

Esterification

In this experiment, procedures are given for the preparation of two local anesthetics, benzocaine (Procedure 43A) and procaine (Procedure 43B). Procaine hydrochloride is often referred to by its trade name, Novocain. In both of these experiments, the products are made by direct esterification of *p*-aminobenzoic acid with ethanol or 2-diethylaminoethanol, respectively. At the instructor's option, you may test the prepared anesthetics on a frog's leg muscle.

PROCEDURE 43A:

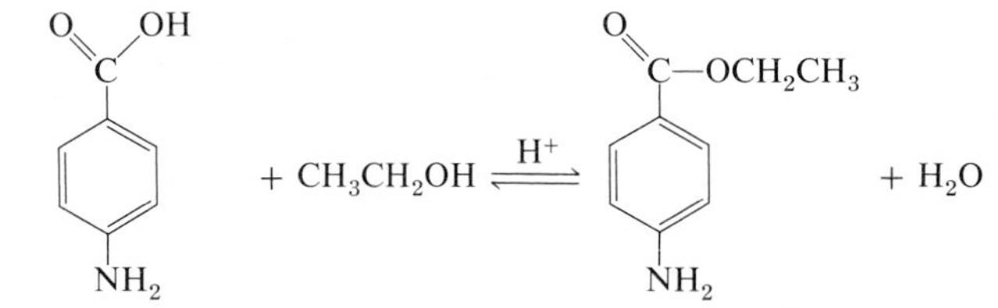

p-Aminobenzoic Acid   Ethyl *p*-Aminobenzoate
(Benzocaine)

PROCEDURE 43B:

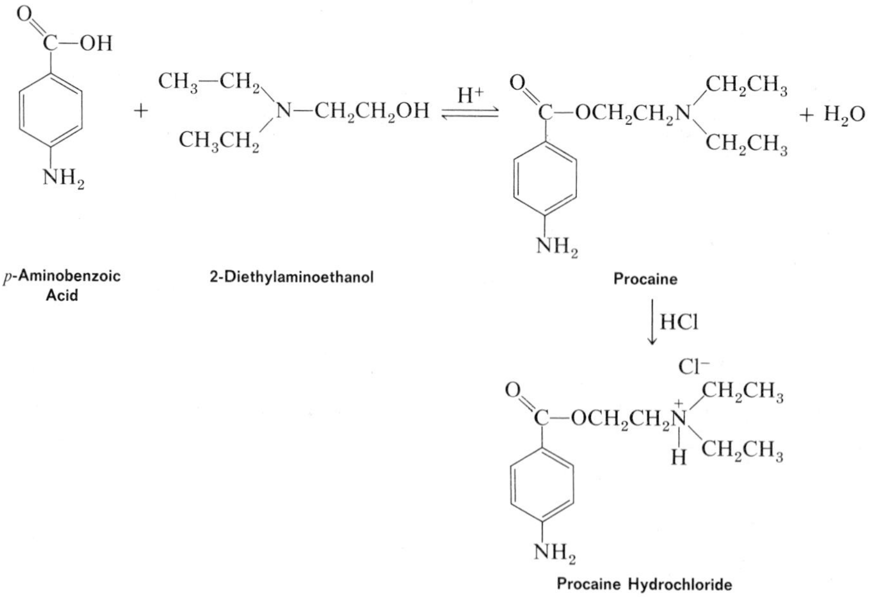

p-Aminobenzoic
Acid   2-Diethylaminoethanol   Procaine

Procaine Hydrochloride
(Novocain)

SPECIAL INSTRUCTIONS

Read the essay which precedes this experiment. In Procedure 43B, you should start the reaction of *p*-aminobenzoic acid with 2-diethylaminoethanol one laboratory period in advance because the rate of esterification is quite slow (1 to 2 days). Since the reflux is untended, other work may be performed during the period of heating.

NOTE TO THE INSTRUCTOR. If desired, instructions for testing either procaine or benzocaine on a frog's leg muscle are given in the Instructor's Guide.

Procedure **43A**

BENZOCAINE

Add 5.0 g of *p*-aminobenzoic acid (Experiment 42), 15 ml of 95% ethanol, and 1 ml of concentrated sulfuric acid to a 100 ml round bottom flask. Attach a reflux condenser. Heat the mixture under reflux for 2 hours. Cool the mixture and neutralize the acid with 10% aqueous sodium carbonate. Extract the aqueous layer with ether and evaporate the ether using a steam bath in a hood. Recrystallize the crude solid from ethanol and water using the mixed solvent method (Technique 3, Section 3.7). Collect the benzocaine by vacuum filtration using a Büchner funnel. After the solid is dry, weigh it, calculate the percentage yield, and determine its melting point. The melting point of pure benzocaine is 92°. At the option of the instructor, obtain the infrared spectrum in chloroform, and the nmr spectrum in carbon tetrachloride. Submit the sample in a labeled vial to the instructor.

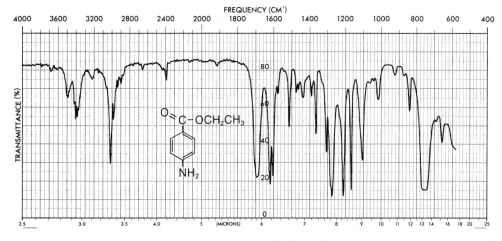

Infrared spectrum of benzocaine, CHCl$_3$ (CHCl$_3$ solvent: 3030, 1220 and 750 cm^{-1})

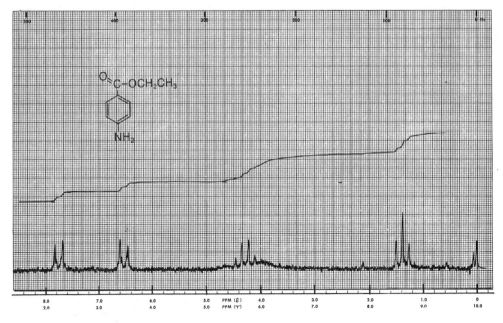

NMR of benzocaine, CCl$_4$

Procedure **43**B
PROCAINE

PROCAINE DIHYDRATE

If predistilled material is not available, purify a quantity of 2-di-ethylaminoethanol by two successive simple distillations (bp 160 to 162°) in order to remove oxidation products. A Claisen head is required because of foaming difficulties. Add 5.0 g of *p*-aminobenzoic acid (Experiment 42) and 16 ml of the clear, purified 2-diethylaminoethanol to a 250 ml Erlenmeyer flask. **Slowly** and **carefully** add 40 ml of **cold** concentrated sulfuric acid to the mixture while cooling the flask in an ice bath. With a spatula, break up the large pieces of material in the mixture as well as possible and stopper the flask with a cork. The name of the student should be written on the cork. Heat the mixture on a steam bath in a hood or in an oven at 100° for 1 or 2 days (until the next laboratory period). Check the mixture after the first half hour of heating to see if all of the large pieces have dissolved. If they have not dissolved, crush them with a spatula and swirl the solution in order to mix it adequately.

Cool the solution in an ice bath after the heating period is completed. Add 60 ml of cold water to the solution. This will help reduce the exothermic effect during the neutralization with concentrated ammonium hydroxide. Approximately 100 ml of concentrated (28%) ammonium hydroxide should be required to neutralize the 40 ml of concentrated sulfuric acid originally used. Make the solution basic by **slowly** adding 100 ml of **cold** concentrated ammonium hydroxide. At this point, monitor the pH of the solution and add more base until a pH of 8 is attained. A brown oil will form as an upper layer in the flask. It will solidify upon cooling. It is important that the solution be cooled in an **ice-salt bath** until this oil crystallizes. The ice-salt mixture must nearly cover the Erlenmeyer flask in order to cool adequately the upper oily layer. Scratch the sides of the flask with a glass rod to help induce crystallization. This crystallization process may take up to one hour. Some white ammonium sulfate may also be produced at the bottom of the flask during the neutralization process. Carefully remove the brown solid at the **top** of the flask with a spatula or spoon and place it in a Büchner funnel under vacuum. Wash the crude solid with about 10 ml of cold water to remove any ammonium sulfate which may have been occluded in the crude procaine dihydrate.

Dissolve the crude procaine dihydrate in 5 to 10 ml of hot 95% ethanol and add boiling water with stirring until the solution becomes permanently cloudy (Technique 3, Section 3.7). Keep the procaine alcohol solution on a steam bath during this process. The product should crystallize in about 10 minutes in an ice bath. If the purified procaine dihydrate does not begin to crystallize during that time period, scratch the brown oil with a glass rod to induce crystallization. Collect and dry the procaine dihydrate by vacuum filtration. Do not leave the product exposed to air for very long because it is hygroscopic. Weigh the material and calculate the percent yield. Save a small sample of the purified procaine dihydrate in an airtight bottle for a melting point determination (pure dihydrate, mp 51°).

PROCAINE HYDROCHLORIDE

Calculate the amount of 0.736 M hydrochloric acid solution required (5 ml/g) for each gram of purified procaine dihydrate obtained from the previous step.[1] This calculated volume corresponds to procaine of **maximum** purity. If the product is pure, it will take 5 ml of 0.736 M hydrochloric acid to dissolve each gram of procaine and to yield a solution with a pH of 5. If the procaine dihydrate is not pure the resulting procaine hydrochloride solution may be too acidic, leading to product loss.

To ensure a more rapid reaction, break up the large pieces of procaine dihydrate with a spatula before mixing them with the acid. Use a transfer pipet to add 4/5 of the calculated volume of 0.736 M hydrochloric acid solution to the powdered procaine, and dissolve as

---

[1] Preparation of 0.736 M HCl: Add 7.24 g of concentrated hydrochloric acid to a 100 ml volumetric flask and dilute it to volume.

much solid as possible by **gently** heating the mixture on a steam bath. Check the pH with pH paper. The solution will be basic if undissolved solid remains. If solids remain or the pH is too high, **carefully** add more of the remaining acid solution, while continuously stirring and monitoring the pH, until all the solid dissolves and the pH reaches 5. Use a steam bath with a strong stream of air blowing on the surface of the mixture to evaporate the solution to dryness. As the liquid evaporates, triturate (stir, mix and grind) the crude product until it crystallizes and is broken into small pieces. Dry the crude procaine hydrochloride in an oven at 100° for 10 minutes, or allow it to dry in air until the next laboratory period.

Purify the crude product by dissolving it in boiling isopropyl alcohol, using at least 50 ml of solvent for each gram of procaine dihydrate originally used in this step. Decolorize the solution by boiling it with decolorizing carbon for a few minutes. Clean and dry a large filter flask and large Büchner funnel. Prepare a pad of Filter Aid (Celite) with isopropyl alcohol (Technique 2, Section 2.4). Discard the slurry solvent which is collected in the filter flask. Now, vacuum filter the hot procaine solution through the layer of Celite. Rinse the Celite with two 40 ml portions of boiling isopropyl alcohol to remove any product which may have remained in the Celite pad. If the filtrate is still quite colored, repeat the above procedure with more decolorizing carbon. Reduce the volume of the isopropyl alcohol solution (filtrate) to 20 ml per gram of procaine dihydrate originally used. Crystallize the procaine hydrochloride in an ice bath. Collect the crystalline product by vacuum filtration, dry it for 20 minutes in an oven at 100°. Weigh the product and calculate the percent yield for this step and the overall yield from p-aminobenzoic acid. Determine the melting point (mp 155 to 156°). Submit the sample in a labeled vial to the instructor. At the instructor's option, obtain the infrared spectrum in potassium bromide.

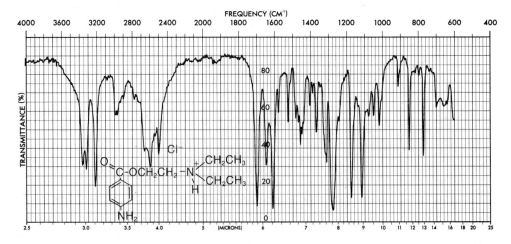

Infrared spectrum of procaine hydrochloride, KBr

ESSAY

Barbiturates

Barbituric acid results from the condensation of malonic acid and urea. It was first prepared in 1864 by Adolph von Baeyer, a young research assistant to Kekulé at the University of Ghent. One story has it that von Baeyer celebrated the advent of the new compound by visiting a nearby tavern. The tavern was one also frequented by artillery officers of the region, and as it happened, it was the day of their patron saint, St. Barbara. Somehow, in ensuing festivities, the name of Barbara was joined to that of urea, thereby christening the new compound, barbituric acid.

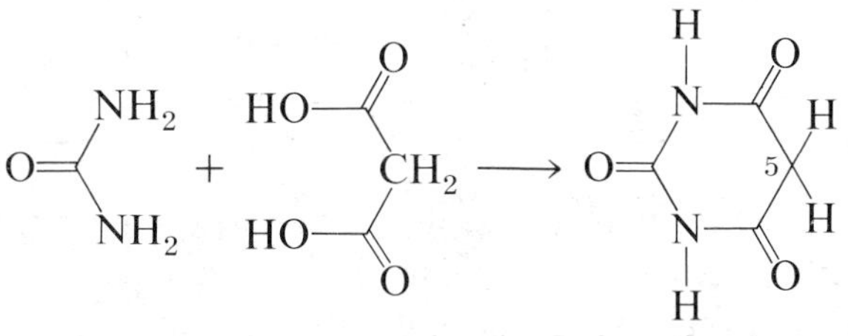

Urea   Malonic Acid   Barbituric Acid

The derivatives of barbituric acid are today called barbiturates and many of them are among the most widely used of the sedative-hypnotic drugs. The first physiologically active drug, Barbital or Veronal, was introduced in 1903; the method of synthesis of this compound and its many later analogs has undergone little change. The usual method begins with diethylmalonate. This diester has two acidic hydrogens adjacent to the ester groups, and they may be removed with base. The resultant anion is a good nucleophile and may be **alkylated** with a suitable substrate.

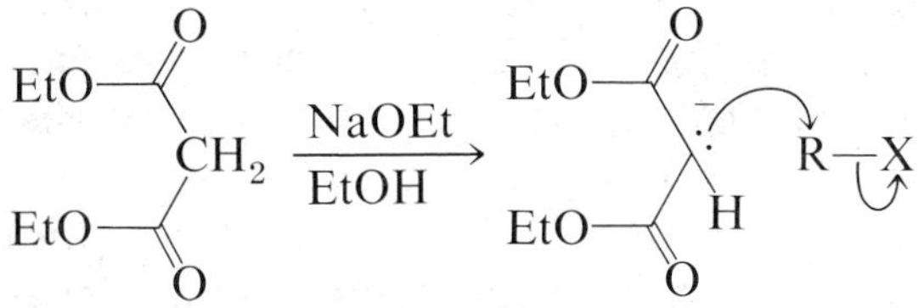

Diethylmalonate

337

STRUCTURES AND TIME FACTORS FOR DIFFERENT BARBITURATES

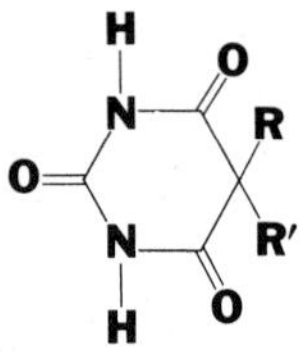

GENERIC NAME	BRAND NAME	R	R'
		Long Acting	
Barbital	Veronal	$-CH_2CH_3$	$-CH_2CH_3$
Phenobarbital	Luminal	$-CH_2CH_3$	phenyl
		Intermediate Acting	
Amobarbital	Amytal	$-CH_2CH_3$	$-CH_2CH_2-CH \Big\langle \begin{smallmatrix} CH_3 \\ CH_3 \end{smallmatrix}$
		Short Acting	
Pentobarbital	Nembutal	$-CH_2CH_3$	$-\underset{\underset{CH_3}{\vert}}{CH}-CH_2CH_2CH_3$
Secobarbital	Seconal	$-CH_2-CH{=}CH_2$	$-\underset{\underset{CH_3}{\vert}}{CH}-CH_2CH_2CH_3$

	TIME TO TAKE EFFECT	DURATION OF ACTION
Long Acting	1 hr	6–10 hr
Intermediate	1/2 hr	5–6 hr
Short Acting	1/4 hr	2–3 hr

Since there are two α-hydrogens, this process may be repeated to give a dialkyl diethylmalonate derivative. This product is condensed with urea to give a barbiturate, a 5,5-dialkylbarbituric acid.

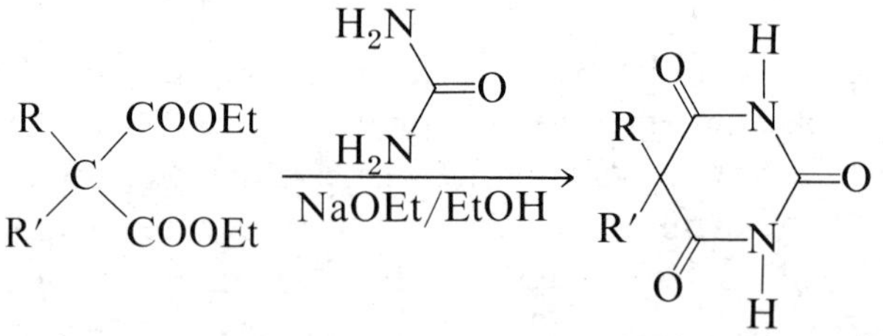

Both hydrogens must be replaced by alkyl groups for the compound to exhibit sedative-hypnotic character as a drug. This is probably due to the susceptibility of the α-hydrogens to rapid metabolic attack within the body and a subsequent degradation of the compound.

Chemists have synthesized and pharmacologists have tested many of these drugs. Some of the more common ones are listed in the accompanying table. The barbiturates exhibit a wide variety of responses in the body depending mainly on the identity of the substituted groups. Some generalizations can be made. Increasing the length of an alkyl chain up to 5 or 6 carbon atoms in the 5-position enhances the sedative action; beyond that, depressant action decreases, and the drugs become more effective as anticonvulsants for control of epileptic seizures. Branched or unsaturated chains in the 5-position generally produce a briefer duration of action. In fact, barbiturates are classified into three categories with respect to the time required for them to take effect and the time of duration of their activity (see table). Compounds with phenyl or ethyl groups in the 5 position seem to have the longest time of duration.

The medical and physiological use of barbiturates depends on the dose size in a manner typical of all these drugs. In small doses the drugs are mild sedatives, acting to relieve tension and anxiety. In this use they have been replaced by the more modern tranquilizer drugs. At 3 to 5 times the sedative dose, sleep is produced. In large doses barbiturates act as anesthetics. Sodium pentothal, the sodium salt of thiopental, is one of the most widely used anesthetics for surgery. It has an ultra-short action.

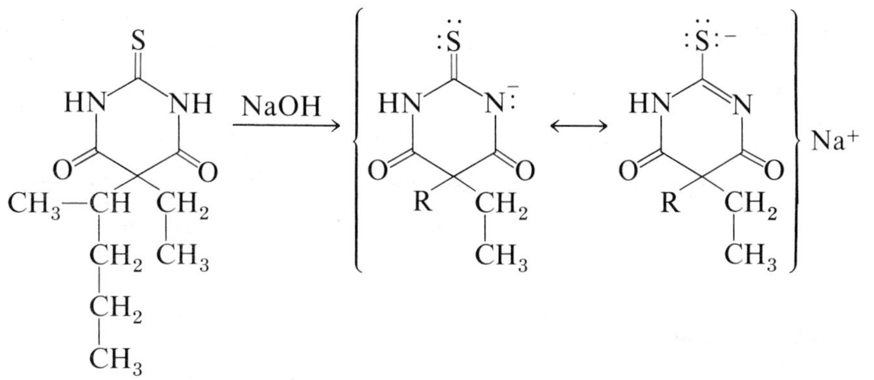

Sodium Pentothal

At lower dose levels, sodium pentothal was used during WW II as a "truth drug." With the right dose level, a kind of twilight sleep or hypnosis could be induced with the patient only half conscious. At this level of narcosis (sleep), a subject had little self control or will power and was very susceptible to suggestion. He could not hide truthful answers, even if he so wished. Such drug treatment was also an integral part of "brain washing," a practice of the Communist Chinese that gained notoriety during the Korean War. A combination of drugs and psychological warfare on the patient was used to totally change his perspectives. Sodium pentothal and another drug, scopolamine (not a barbiturate), were widely used for these purposes.

The barbiturates are widely used as prescription sleeping pills. Many persons find sleep brought about by these drugs as refreshing as natural sleep. Many, however, awake with a hangover, dizziness, drowsiness and a headache. Tests have proved that whether or not a person experiences this hangover, his level of efficiency is reduced. Experimental subjects consistently score lower than average on mental and memory tests.

High levels of barbiturates, of course, cause death. An overdose of sleeping pills is one of the most common forms of suicide, both successful and unsuccessful. Barbiturate use can also lead to addiction and chronic intoxication. The addiction is real and withdrawal symptoms of nausea, tremors, and vomiting are experienced. From the person trying to avoid anxiety to the chronic sleeping pill user, legal barbiturate misuse is second only to alcohol abuse; in addition, a large illegal market is also extant. Both alcohol and barbiturates are difficult drugs to use if one wishes to achieve the initial euphoria produced. In each case the user inevitably slips into sedation and sleep. In combination, of course, these two central nervous system depressants, alcohol and barbiturates, can be and nearly always are fatal. Unfortunately, just how the barbiturates bring about narcosis, sedation, and anesthesia is not currently well understood.

REFERENCES

Adams, E. "Barbiturates," *Scientific American, 198* (1958), 60.

Foye, W. O. *Principles of Medicinal Chemistry*. "Barbiturates." Philadelphia: Lea and Febiger, 1974. Pp. 165–171.

Goodman, L. S., and Gilman, A. *The Pharmacological Basis of Therapeutics* (4th edition). Chapter 9, "Hypnotics and Sedatives. I. The Barbiturates," S. K. Sharpless. New York: Macmillan, 1970.

Ray, O. S. *Drugs, Society, and Human Behavior*. Chapter 11, "Stimulants and Depressants." St. Louis: Mosby, 1972.

Experiment 44

5-*n*-BUTYLBARBITURIC ACID

Condensation Reactions
Amide Formation, Alkylation
Handling Sodium Metal

A derivative of barbituric acid that has little or no sedative-hypnotic potential and is relatively ineffective on humans will be synthesized in this experiment. The product is, however, toxic and should be treated with respect. The synthetic target is 5-*n*-butylbarbituric acid, a monoalkyl barbiturate. The synthetic scheme is as follows:

$$2 \ C_2H_5OH + 2 \ Na \xrightarrow{C_2H_5OH} 2 \ NaOC_2H_5 + H_2$$

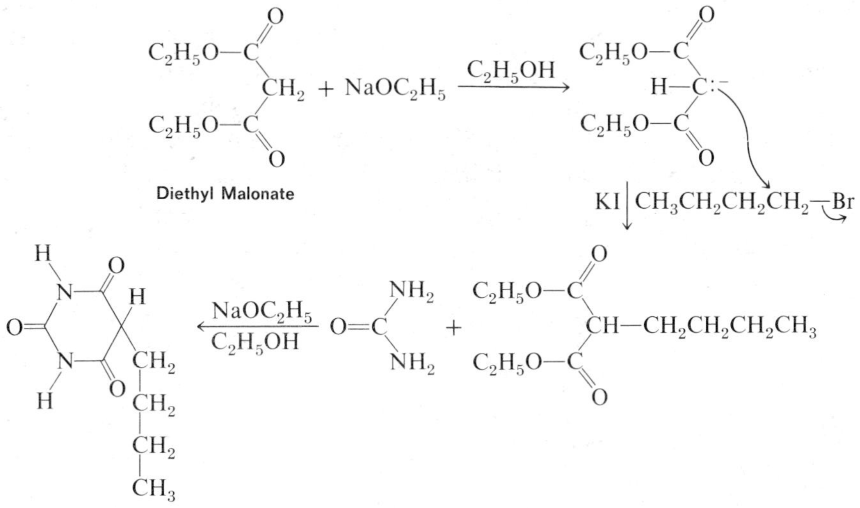

Diethyl Malonate

5-*n*-Butylbarbituric Acid   Urea   Diethyl *n*-Butylmalonate

The first step is the alkylation of diethylmalonate with *n*-butyl bromide. The high boiling point of diethyl butylmalonate dictates the use of reduced pressure distillation in its isolation. In doing the reaction itself, you will learn how to handle sodium metal and how to use a manometer in conjunction with a reduced pressure distillation. Note that potassium iodide is used to catalyze the alkylation reaction. Presumably *n*-butyl iodide is formed **in situ** (in the solution) and iodide ion is more easily displaced by malonate ion than is bromide ion.

$$I^- + R-Br \rightleftharpoons R-I + Br^-$$

Note also that during the alkylation step ethanol is used as the solvent. This is dictated by the fact that it is the diethyl ester of malonic acid being used. Another alcohol would lead to ester interchange and a more complex mixture of products.

The final step is a condensation reaction between urea and diethyl butylmalonate. It should be noted that the product, 5-*n*-butylbarbituric acid, is capable of two tautomeric forms, one having aromatic resonance.

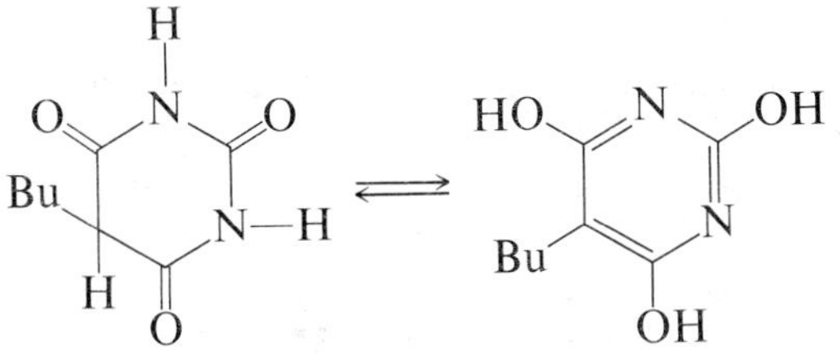

SPECIAL INSTRUCTIONS

Before beginning this experiment you should read the essay which precedes it. It is important to read Techniques 1 through 6 and, especially, 14. When sodium reacts with ethanol, hydrogen gas is evolved in large quantities. Be certain that all flames in the laboratory are extinguished before dissolving sodium in ethanol. Alternatively, this operation may be conducted in a hood. The reaction must be carried out under strict anhydrous conditions. The apparatus and the alcohol must be dry. The formation of diethyl *n*-butylmalonate requires at least 45 minutes of heating under reflux. The formation of 5-*n*-butylbarbituric acid requires at least 2 hours of heating under reflux.

PROCEDURE

DIETHYL *n*-BUTYLMALONATE

Working in a hood, add 1.32 g of sodium metal (cut into small pieces; see Technique 14, Section 14.4) cautiously and piece by piece to a 250 ml round bottom flask fitted with a reflux condenser and containing 25 ml of dry **absolute** ethanol.[1] Use forceps to handle the sodium metal. It is important that the amount of sodium used be no more or no less than the amount specified. Be certain that any unused necks of the flask are stoppered and that all ground glass joints are greased. The heat evolved when the sodium dissolves in the ethanol will cause the ethanol to boil. If necessary, cool the flask in an ice bath to keep the reaction from becoming too vigorous. After the sodium has dissolved and the boiling action has subsided, attach a calcium chloride drying tube to the top of the reflux condenser. Heat the mixture on a steam cone, if necessary, to dissolve any particles of sodium that remain. Add 0.75 g of anhydrous **powdered** KI down the condenser and continue heating the mixture on the steam cone until all the solid has dissolved (or nearly so). Add 8.8 g of dry diethylmalonate. Loosen the clamp holding the apparatus and shake it gently or swirl it to mix the ingredients. Heat the mixture for an additional 5 to 10 minutes and then add 7 g of *n*-butyl bromide in three equal portions, down the condenser, allowing the reaction to subside after each portion is added. Heat the mixture for an additional 45 minutes on the steam cone. The product forms as a yellow oil on the top of the mixture.

Cool the mixture to about room temperature and decant as much liquid as possible away from the inorganic salts that form as a white

---

[1] The ethanol used in this experiment must be anhydrous. A freshly opened bottle of **absolute** (100%) ethanol works well. Do not use denatured or 95% ethanol.

solid at the bottom of the flask. Evaporate as much ethanol as possible on a steam bath in the hood, using a stream of air directed at the surface of the solution to accelerate the evaporation. Add 20 ml of water and 0.5 ml of concentrated hydrochloric acid to the resulting concentrated yellow liquid and pour the mixture back into the 250 ml flask. After all the solid has dissolved and the mixture has cooled, add 20 ml of ether, swirl the solution, and transfer it to a separatory funnel. Shake and separate the layers. Wash the organic layer once with 10 ml of water, once with 15 ml of 5% sodium bicarbonate solution, and finally with about 15 ml of water. When washing the organic layer with water, an emulsion may form which can be broken with 2 or 3 ml of fresh ether. Dry the ether layer with 1 or 2 g of sodium sulfate until the solution appears clear. This should take about 2 minutes. Gravity filter the solution and place the filtrate in a 50 ml flask. Remove the ether by distillation on a steam bath in the hood (use a boiling stone). Assemble an apparatus for distillation under reduced pressure, using a manometer if possible (Technique 6, Sections 6.6 and 6.7 and Technique 9, Section 9.4). Be certain that all ground glass joints are greased, but avoid an excess of grease. Make certain that the rubber tubing used has no cracks in it. Good rubber seals for the thermometer and air bleeder are absolutely necessary in order to obtain a high vacuum. Collect the fraction boiling from 120 to 138° at 21 to 25 mm Hg (115 to 130° at 17 mm Hg).[2] Weigh the clear, colorless product and calculate the percentage yield. At the option of the instructor, obtain the infrared spectrum of diethyl *n*-butylmalonate as a neat liquid.

5-*n*-BUTYLBARBITURIC ACID

Prepare a sodium ethoxide solution in a 250 ml round bottom flask, as in the previous step. Use one mole of sodium metal per mole of diethyl *n*-butylmalonate. Dissolve the sodium in absolute ethanol at a ratio of 47 ml of ethanol per gram of sodium used. The last few pieces can be dissolved by placing the flask on a steam bath, taking care not to introduce water vapor into the reaction (use a calcium chloride drying tube). After all the sodium has dissolved, add the diethyl *n*-butylmalonate, followed by a warm (about 70°) solution of **dry** urea dissolved in **absolute** ethanol at a ratio of 18 ml of ethanol for each gram of urea used.[3] Use one mole of urea for each mole of diethyl *n*-butylmalonate. Swirl the mixture well and heat it under reflux using a heating mantle or an oil bath at 110 to 115° for 2 hours. This heating period may be conducted over more than one laboratory period. At the beginning of

---

[2] Some unreacted diethyl malonate may be obtained in this reaction. If an appreciable amount of this lower boiling material is obtained, change receiver flasks when the correct boiling point is reached.

[3] The urea should be dried for about 45 minutes at 105 to 110°. The laboratory instructor may prepare a large quantity for use by the entire class.

the reflux period, place a Claisen adapter (with the central joint stoppered) on top of the condenser. Attach a calcium chloride drying tube to the side arm of the Claisen adapter, using a thermometer adapter. All the pieces of the apparatus must be firmly clamped in case bumping should occur. It is advisable to tape a wire gauze over the top of the drying tube to prevent spillage of the drying agent when the mixture bumps. It is a normal expectation that the reaction mixture will bump occasionally, but these precautions will avoid the loss of product. Once the mixture begins to boil smoothly, careful monitoring of the reaction is no longer necessary.

After completion of the reflux period, add 44 ml of warm (50°) water per gram of sodium used. Acidify the resulting solution with 4 ml of concentrated hydrochloric acid per gram of sodium used in the reaction. Concentrate the solution to about 50 ml, using a hot plate (add a boiling stone) and an air stream. Cool the solution in an ice bath for 15 minutes and collect the crude product using vacuum filtration. If a rubbery, white solid is formed, smelling of diethyl n-butylmalonate, wash it with 75 ml of petroleum ether while it is still in the Büchner funnel and set the solution aside. Dry the solid and recrystallize it from boiling water, using a hot plate to heat the water. The recrystallization is conducted using 20 ml of water per gram of diethyl n-butylmalonate initially used. Crystallize the product by placing it in an ice-salt bath for 30 minutes, stirring the solution occasionally after the first 20 minutes of cooling. Collect the product by vacuum filtration using a large Büchner funnel and air dry the solid for 20 to 30 minutes. Occasionally turn the crystals with a spatula to expose the wet surfaces. Air dry the product overnight or in an oven at 105 to 110° for three hours. Weigh the solid and determine the melting point of the pure 5-n-butylbarbituric acid (literature mp = 209 to 210°). Calculate the percentage yield for both the second step and for the overall reaction sequence.

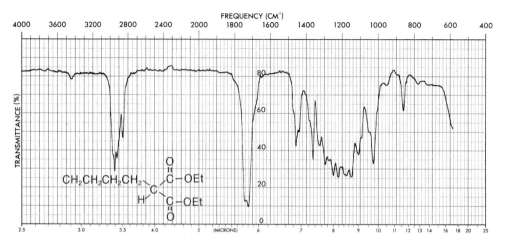

Infrared spectrum of diethyl n-butylmalonate, neat

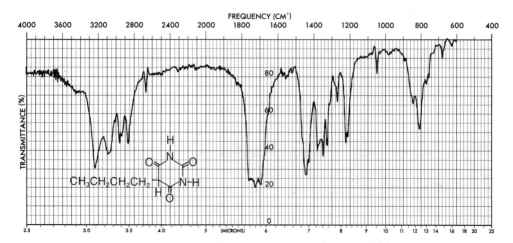

Infrared spectrum of 5-*n*-butylbarbituric acid, KBr

Dry the petroleum ether extract obtained earlier, remove the solvent by distillation using a steam bath in the hood, and weigh the remaining liquid. This residue is mostly unreacted diethyl *n*-butylmalonate. Calculate the percent of starting material recovered. Recalculate the percentage yield for the second step from the amount of reagent actually consumed. At the option of the instructor, determine the infrared spectrum of *n*-butylbarbituric acid as a KBr mull.

QUESTIONS

1. Diethylmalonate has two acidic hydrogens. Bearing this in mind, what by-product would you expect in the first step of this reaction sequence?

2. The presence of water poses a serious problem in both steps of this reaction sequence. Write equations which explain this.

3. After the reaction of urea with diethyl *n*-butylmalonate, the reaction mixture must be neutralized with dilute hydrochloric acid. Why? Give the equations for the reaction.

4. Interpret the major absorption bands in the infrared spectra of diethyl *n*-butyl-malonate and *n*-butylbarbituric acid.

ESSAY

Aldol Condensations—Aldolases

The aldol condensation is a common organic reaction. It is one of the few reactions, and perhaps the most useful reaction, used to form new carbon–carbon bonds. It is, therefore, a mainstay reaction for the synthetic organic chemist. Nature has not eschewed this reaction either. Many biological systems utilize the aldol condensation or one of its variants to build up the carbon skeletons of many complex naturally occurring compounds.

In thinking of the aldol condensation, one usually thinks first of the base-catalyzed aldol condensation. In basic solution, the hydrogen atoms attached to a carbon which is alpha to a carbonyl group, may be removed to produce an enolate ion. The enolate ion is quite nucleophilic (at carbon) and will add across the carbonyl group of a second molecule of the original carbonyl compound to give an **"aldol."**

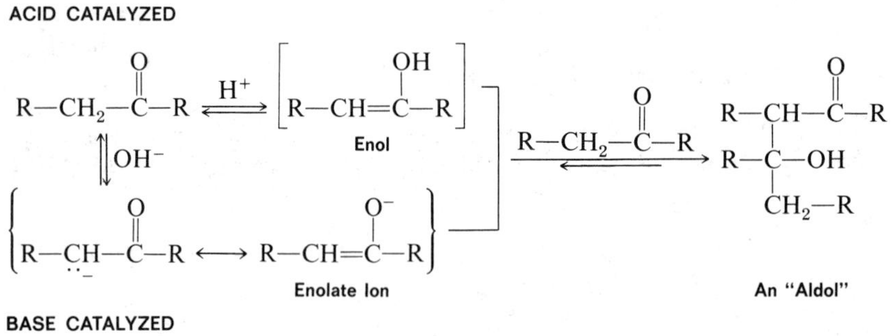

ACID CATALYZED

BASE CATALYZED

The reaction may also be catalyzed by acid, proceeding through the enol form of the carbonyl compound. Both reactions are equilibrium processes, and quite often the position of the equilibrium does not lie in favor of the aldol product, but rather in favor of the reactants. In such cases, when the pure product is subjected to the reaction conditions, the aldol reaction will reverse to a large extent to reestablish equilibrium concentrations of the products and reactants. Such backward reactions are called "reverse aldol" reactions, and are quite common in organic chemistry. Reverse aldol reactions are also found in biochemical systems. In many of these reactions it is necessary to shift the equilibrium position of the reaction by one of the usual methods (excess of reactants, removal of product, etc.) in order to achieve the desired result.

Both aldol condensations and reverse aldol condensations occur in biochemical systems. A biochemical reaction sequence that occurs quite commonly in all living systems is **glycolysis.** In the glycolysis process, glucose, a six-carbon sugar, is converted first to an isomeric sugar, fructose. Fructose is then converted to its diphosphate derivative, fructose-1,6-diphosphate, which is degraded by a **reverse** aldol reaction to two three-carbon fragments, dihydroxyacetone phosphate and glyceraldehyde-3-phosphate.

GLYCOLYSIS (MITOCHONDRIA)

Glucose $\rightleftharpoons$ Fructose $\rightleftharpoons$ Fructose-1,6-diphosphate $\rightleftharpoons$ Glyceraldehyde-3-phosphate + Dihydroxyacetone Phosphate $\rightleftharpoons$ CO_2

P = Phosphate (PO_3H^-)

PHOTOSYNTHESIS (CHLOROPLASTS)

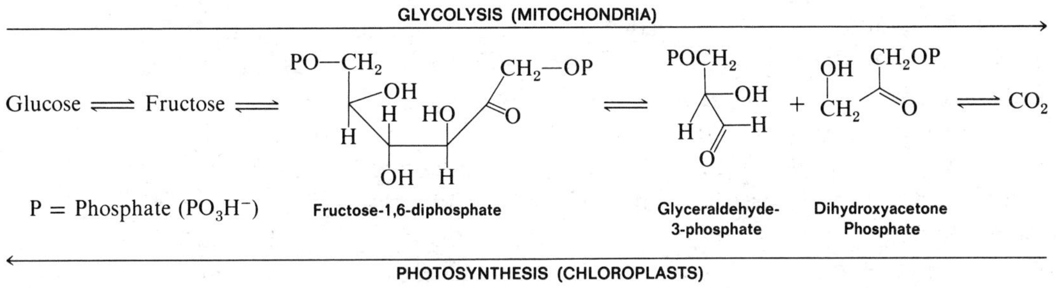

Both of the three carbon compounds produced by the cleavage of fructose-1,6-diphosphate can be converted to phosphoenol pyruvate which may be used for many purposes. It can provide acetyl coenzyme A units which may be used either to derive energy for the cell (citric acid cycle) or for synthesis (see the essay which precedes Experiment 19).

The reverse aldol condensation of fructose-1,6-diphosphate is catalyzed by an enzyme called **aldolase.** The same enzyme also has the ability to catalyze the forward reaction if it is used under different circumstances. For instance, in **photosynthesis** this enzyme is utilized to **form** fructose-1,6-diphosphate from dihydroxyacetone phosphate and glyceraldehyde-3-phosphate. Of course, these two reactions take place under different circumstances and, in fact, in different cellular organelles. Glycolysis occurs mainly in mitochondria, whereas photosynthesis occurs mainly in chloroplasts. Actually the equilibrium for this aldol process lies in favor of the fructose moiety. In mitochondria, however, the products are efficiently removed by other chemical processes as quickly as they are formed, thereby shifting the position of the equilibrium.

Although the aldolase discussed here is specific for dihydroxyace-tone, it can condense dihydroxyacetone phosphate with almost any available aldehyde. Thus, many sugars other than fructose may be synthesized using this enzyme. For instance, condensation with hydroxy-acetaldehyde (R=OH) produces D-xylulose-1-phosphate, and conden-sation with acetaldehyde (R=H) produces 5-deoxy-D-xylulose-1-phos-phate:

$$
HOCH_2-\overset{\overset{\displaystyle O}{\|}}{C}-CH_2OP \;+\; \underset{\underset{\displaystyle CH_2R}{|}}{CHO} \;\overset{Aldolase}{\rightleftharpoons}\;
\begin{array}{c}
CH_2OP \\
| \\
C=O \\
| \\
HO-C-H \\
| \\
H-C-OH \\
| \\
CH_2R
\end{array}
$$

Aldol condensations occurring in biological systems cannot, of course, occur under strongly acidic or strongly basic conditions. There-fore, the enzyme must achieve its result by some other mechanism. Aldolase has been studied sufficiently for some knowledge of its mech-anism to be known. When dihydroxyacetone is mixed with the enzyme, aldolase, in D_2O solution, the alpha hydrogens of the dihydroxyacetone are exchanged for deuterium. No such change occurs with glyceral-dehyde. Hence, the enzyme complexes strongly with dihydroxyacetone. The workers studying this system theorized that the enzyme aldolase contained an amino group at its active site and that the complex be-tween dihydroxyacetone and the enzyme was in fact due to the forma-tion of an imine (C=N) linkage. To prove this, they mixed the enzyme and dihydroxyacetone to form the complex, and then added a solution of sodium borohydride to reduce the imine. Following the reduction, the reduced enzyme-substrate complex was hydrolyzed. On hydrolysis, N-glycyllysine was obtained, which both proved the theory and identi-fied lysine as the amino acid involved in the complex formation.

$$
ENZYME-NH_2 \;+\; HOCH_2-\overset{\overset{\displaystyle O}{\|}}{C}-CH_2OP \;\rightleftharpoons\; ENZYME-N=C\overset{\displaystyle CH_2OP}{\underset{\displaystyle CH_2OH}{<}}
$$

Aldolase   **Imine**

$$\Big\downarrow NaBH_4$$

$$
\underset{\underset{\displaystyle N\text{-Glycyllysine}}{}}{HOOC-\underset{\underset{\displaystyle NH_2}{|}}{CH}(CH_2)_4NH-CH\overset{\displaystyle CH_2OH}{\underset{\displaystyle CH_2OH}{<}}} \;\overset{H_3O^+}{\longleftarrow}\; ENZYME-NH-CH\overset{\displaystyle CH_2OP}{\underset{\displaystyle CH_2OH}{<}}
$$

Based on these experiments, the following mechanism was pro-posed for the action of aldolase.

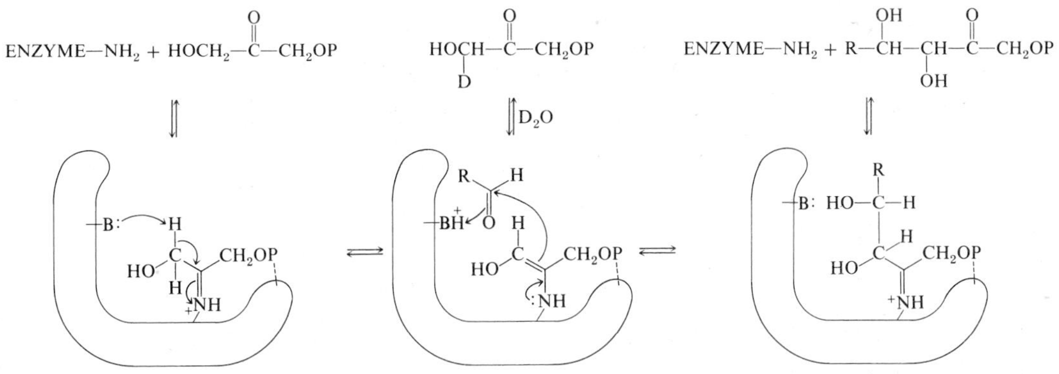

The isomerization of the imine **1** to the enamine **2** is, of course, an important and well-known modification of the aldol condensation, the **enamine reaction.** A similar condensation can be carried out by mixing a carbonyl compound with a secondary amine to form an enamine intermediate which is sufficiently nucleophilic to condense with another carbonyl acceptor.

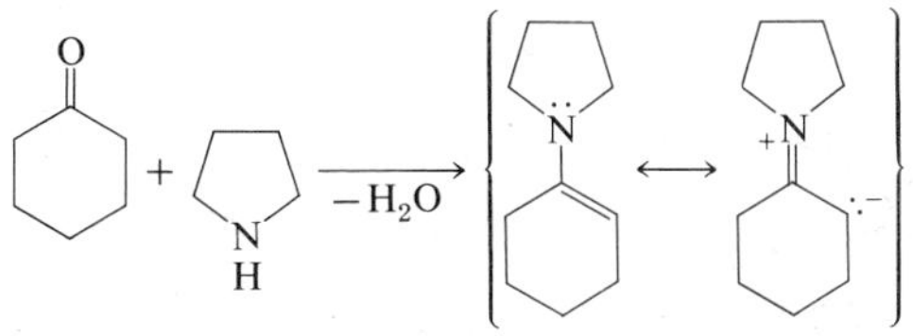

Aldolases which operate via the enamine mechanism described above are called Class I Aldolases. They occur frequently in both vertebrate and invertebrate animals and in green plants. Another type of aldolase, the Class II Aldolase, is found in bacteria, fungi, blue green algae, and certain protozoans, like the Euglena, which also has Class I type aldolases. The Class II Aldolases utilize a metal (M^+) to complex with the α-hydroxyketone, causing it to enolize. The enolized complex is then capable of condensing with aldehydes. In different enzymes,

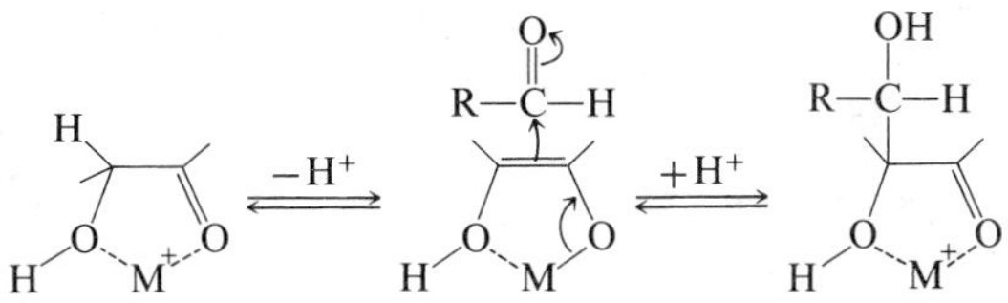

various metals may be involved, but Zn^{+2}, Co^{+2}, and Fe^{+2} are the most common ions.

Another device sometimes used to form an enol capable of reaction is the decarboxylation of a β-ketoacid. For instance, in order to condense acetyl coenzyme A units, acetyl coenzyme A is converted to malonyl coenzyme A. On decarboxylation, malonyl coenzyme A yields an enol which will condense with acetyl coenzyme A units which have been bound to the appropriate enzyme.

Acetyl Coenzyme A Malonyl Coenzyme A

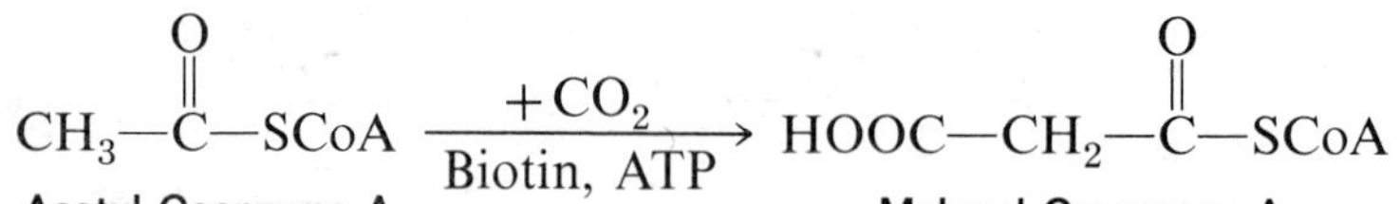

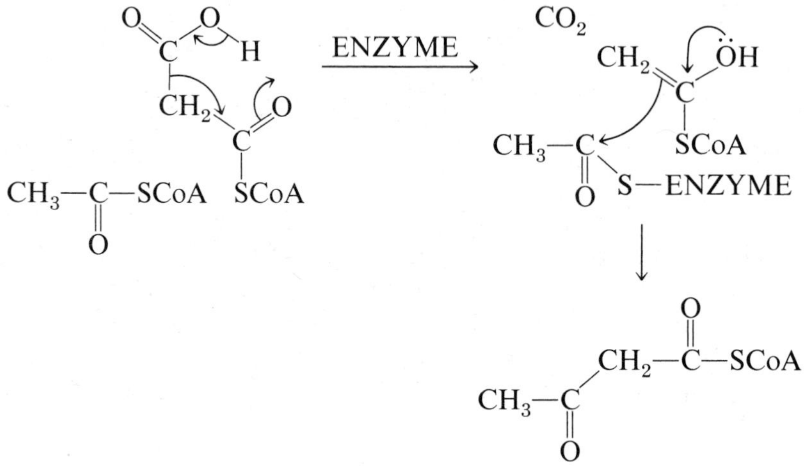

This reaction, which is actually a Claisen ester condensation, is important to the synthesis of the fatty acids, terpenes, and acetogenins. Long chains of acetyl units may be constructed which when reduced give rise to the fatty acids, all of which have even numbers of carbon atoms due to their method of synthesis from these two carbon units.

Finally, probably the most well known aldol condensation occurring in biological systems is that which forms citric acid from acetyl coenzyme A and oxaloacetic acid, thus beginning the citric acid cycle.

Citric Acid

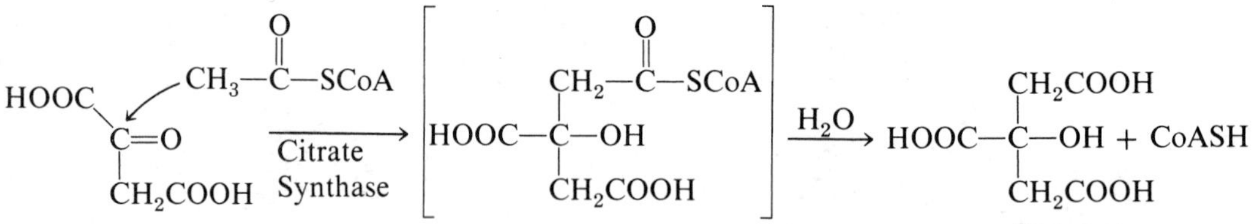

Unfortunately, the details of the mechanism of this reaction are at present not well understood.

REFERENCES

Lehninger, A. L. *Biochemistry: The Molecular Basis of Cell Structure and Function.* New York: Worth Publishers, 1970.

Mahler, H. R., and Cordes, E. H. *Biological Chemistry* (2nd edition). New York: Harper and Row, 1971.

TETRAPHENYLCYCLOPENTADIENONE

Aldol Condensation

In this experiment tetraphenylcyclopentadienone will be prepared by the reaction of dibenzyl ketone (1,3-diphenyl-2-propanone) with benzil (Experiment 39) in the presence of base.

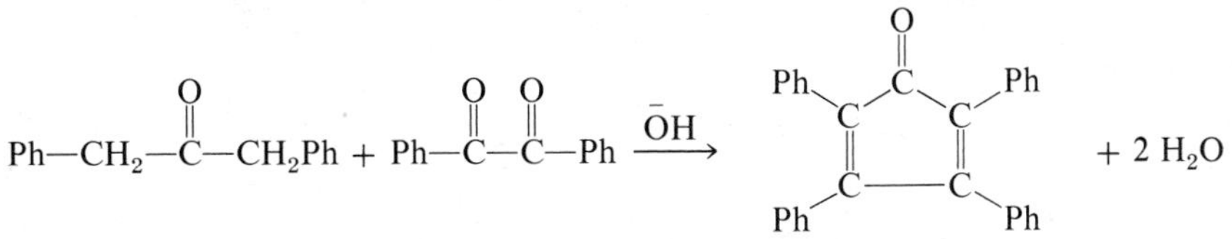

$$Ph-CH_2-\overset{O}{\underset{\|}{C}}-CH_2Ph + Ph-\overset{O}{\underset{\|}{C}}-\overset{O}{\underset{\|}{C}}-Ph \xrightarrow{\bar{O}H} \text{(Tetraphenylcyclopentadienone)} + 2\ H_2O$$

Dibenzyl Ketone   Benzil   Tetraphenylcyclopentadienone

This reaction proceeds via an aldol condensation reaction, with dehydration giving the purple unsaturated cyclic ketone. A stepwise mechanism for the reaction may proceed as follows:

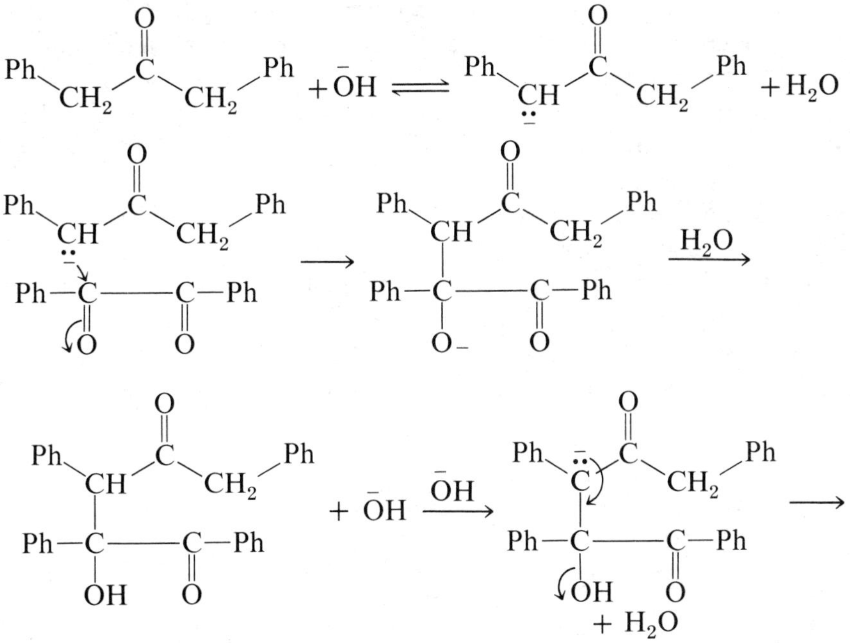

Aldol Intermediate
(A)

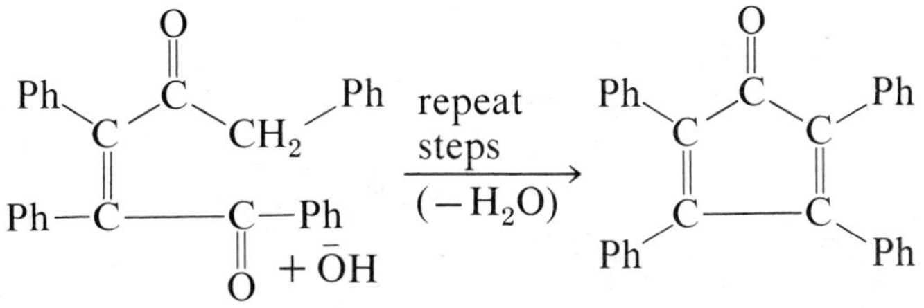

(B)

The aldol intermediate, A, readily loses water to give the highly conjugated system, B, which reacts further by intramolecular aldol condensation to give the dienone product following dehydration.

SPECIAL INSTRUCTIONS

This reaction can be completed in about one hour. The product can then be converted to tetraphenylphthalic anhydride (Experiment 46). Read the essay which precedes this experiment.

PROCEDURE

Add 4.2 g of benzil (Experiment 39), 4.2 g of dibenzyl ketone (1,3-diphenyl-2-propanone, 1,3-diphenylacetone), and 30 ml of **absolute** ethanol to a 100 ml round bottom flask. Attach a reflux condenser to the flask. Heat the mixture on a steam bath until the solids dissolve. Also dissolve 0.6 g of potassium hydroxide in 6 ml of absolute ethanol in another container. As the solid dissolves, crush the pieces with a spatula to aid in the solution process.

Raise the temperature of the benzil and dibenzyl ketone solution until it is just below its boiling point and slowly add the solution of potassium hydroxide through the top of the condenser in two portions (CAUTION: foaming may occur). The mixture will immediately turn a deep purple color. Once the potassium hydroxide has been added, allow the mixture to reflux for 15 minutes. Shake the flask several times during this period. Cool the mixture below 5° in an ice bath. Vacuum filter the deep purple crystals, wash them with three 10 ml portions of cold 95% ethanol, and dry them in an oven for 30 minutes or in air overnight.

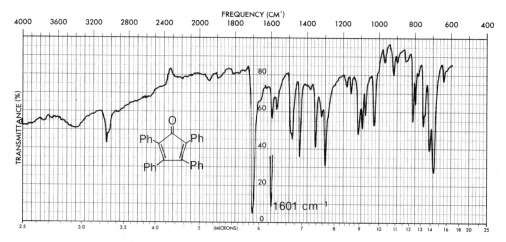

Infrared spectrum of tetraphenylcyclopentadienone, KBr

The crude product is of sufficient purity (mp 218 to 220°) for the preparation of tetraphenylphthalic anhydride (Experiment 46). Weigh the product and calculate the percentage yield. Determine the melting point. A small portion may be recrystallized, if desired, from a 1:1 mixture of ethanol and benzene (16 ml/0.5 g), mp 219 to 220°. At the instructor's option, determine the infrared spectrum of tetraphenyl-cyclopentadienone in potassium bromide. Submit the product to the instructor in a labeled vial or save it for Experiment 46.

QUESTIONS

1. Interpret the infrared spectrum of tetraphenylcyclopentadienone.

2. Draw the structure of the product that you would expect upon reaction of a mixture of benzaldehyde and acetophenone with base.

3. Suggest several possible by-products of this reaction.

ESSAY

The Diels-Alder Reaction and Insecticides

That the addition of an unsaturated molecule across a diene system forms a substituted cyclohexene has been known since the 1930's. The original research dealing with this type of reaction was performed by Otto Diels and Kurt Alder in Germany, and the reaction has become known as the **Diels-Alder reaction.** The Diels-Alder reaction is the reaction of a **diene** with a specie capable of reacting with the diene, the **dienophile.**

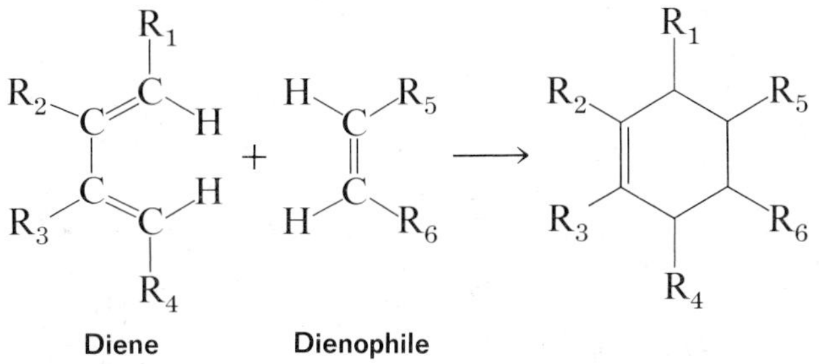

Diene   Dienophile

The product of the Diels-Alder reaction is usually a structure that contains a cyclohexene ring system. If the substituents shown above are simply alkyl groups or hydrogen atoms, the reaction proceeds only under rather extreme conditions of temperature and pressure. However, with more complex substituents, Diels-Alder reaction may proceed at low temperatures and under fairly mild conditions. The reaction of tetraphenylcyclopentadienone with maleic anhydride to form tetraphenyl-1,2-dihydrophthalic anhydride, described in Experiment 46, is an example of a Diels-Alder reaction carried out under reasonably mild conditions.

A commercially important use of the Diels-Alder reaction involves the use of hexachlorocyclopentadiene as the diene. Depending upon the dienophile, a variety of chlorine-containing addition products may be synthesized. Nearly all of these products are powerful **insecticides.** Three insecticides which are synthesized by the Diels-Alder reaction are shown below.

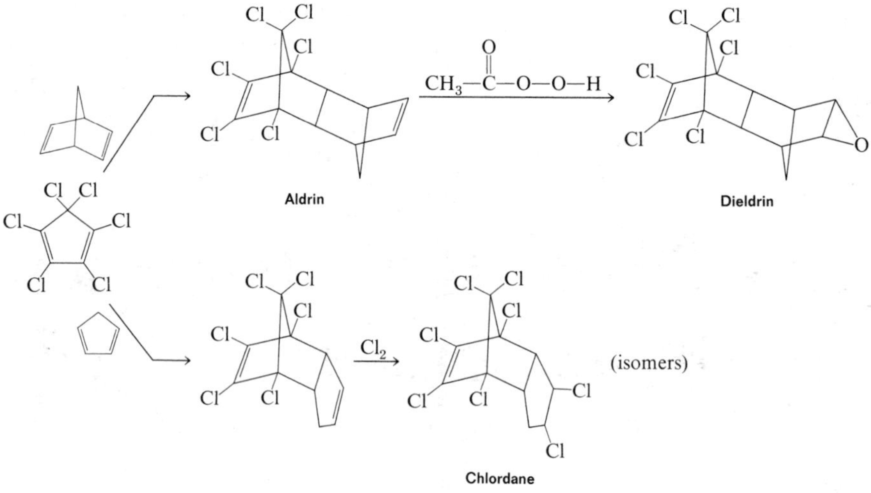

Dieldrin and Aldrin are named after Diels and Alder. They are used against the insect pests of fruits, vegetables, and cotton, against soil insects, termites, and moths, and in the treatment of seeds. Chlordane is used in veterinary medicine against insect pests of animals, including fleas, ticks, and lice.

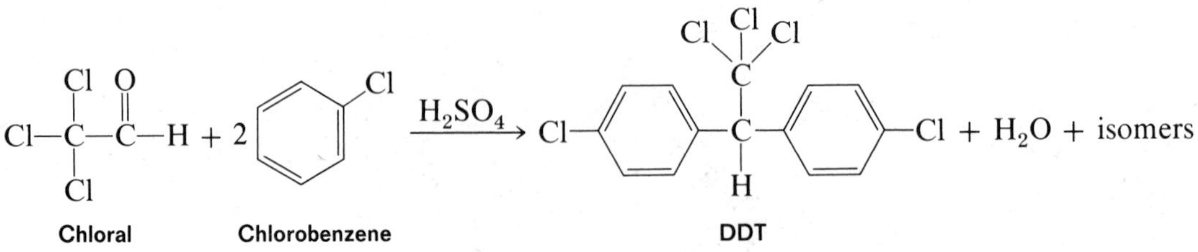

The best known of the insecticides, DDT, is not prepared **via** the Diels-Alder reaction, but nevertheless is the best illustration of difficulties which are encountered when insecticides are used indiscriminantly. DDT was first synthesized in 1874, and its insecticidal properties were first demonstrated in 1939. Its commercial synthesis is performed quite easily, using inexpensive reagents.

At the time of its introduction, DDT was an important boon to mankind. It was effective in the control of lice, fleas, and malaria-carrying mosquitoes, and thus served to aid in the control of human and animal disease. The use of DDT rapidly spread to the control of hundreds of insects which damage fruit, vegetable, and grain crops.

Pesticides which persist in the environment for a long while after application are called "hard" pesticides. Beginning in the 1960's, some of the harmful effects of such hard pesticides as DDT and the other chlorocarbon materials became known. DDT is a fat-soluble material, and it is therefore likely to collect in the fat, nerve, and brain tissues of animals. The concentration of DDT in tissues increases in animals further up the food chain. Thus, birds which eat poisoned insects accumulate large quantities of DDT. Animals which feed on the birds accumulate even more DDT. In birds, at least two undesirable effects of DDT have been recognized. First, birds whose tissues contain large amounts of DDT have been observed to lay eggs whose shells are too thin to survive until young birds are hatched. Second, large quantities of DDT in the tissues appear to interfere with normal reproductive cycles. The massive destruction of bird populations that sometimes occurs after heavy spraying with DDT has become an issue of great concern. Today the brown pelican and the bald eagle are in danger of extinction. The use of chlorocarbon insecticides has been identified as a principal reason for the decline in the numbers of these birds.

Because DDT is chemically quite inert, it persists in the environment without decomposing to harmless materials. It can decompose very slowly, but the decomposition products are every bit as harmful as DDT itself is. Consequently, each application of DDT means that still more DDT will pass from specie to specie, from food source to predator, until it concentrates in the higher animals, possibly endangering their existence. Even man may be threatened. As a result of evidence of the harmful effects of DDT, in the early 1970's the Environmental Protection Agency banned the use of DDT. Nevertheless, DDT may still be used for certain purposes, although permission of the Environmental Protection Agency is required. In 1974, such permission was granted for the use of DDT against the tussock moth in the forests of Washington and Oregon.

Because the life cycles of insects are quite short, they have the ability to evolve an immunity to insecticides quite rapidly. As early as 1948, several strains of DDT-resistant insects were identified. Today, the malaria-bearing mosquitoes are almost completely resistant to DDT, a rather ironic development. Other chlorocarbon insecticides have been used as alternatives to DDT against resistant insects. Examples of other chlorocarbon materials include Dieldrin, Aldrin, Chlordane, and the substances shown below. Heptachlor and Mirex are also prepared using Diels-Alder reactions.

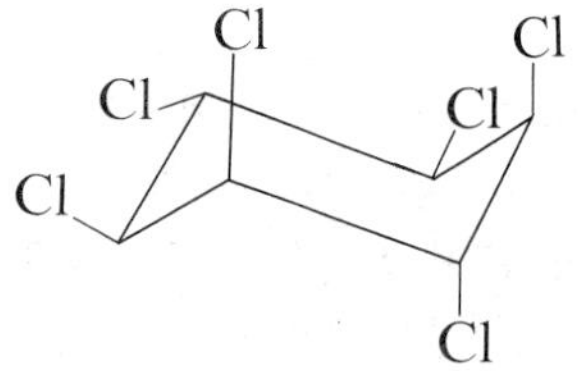

Lindane 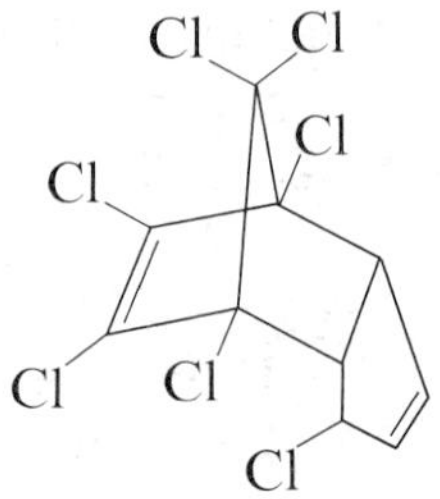   Heptachlor 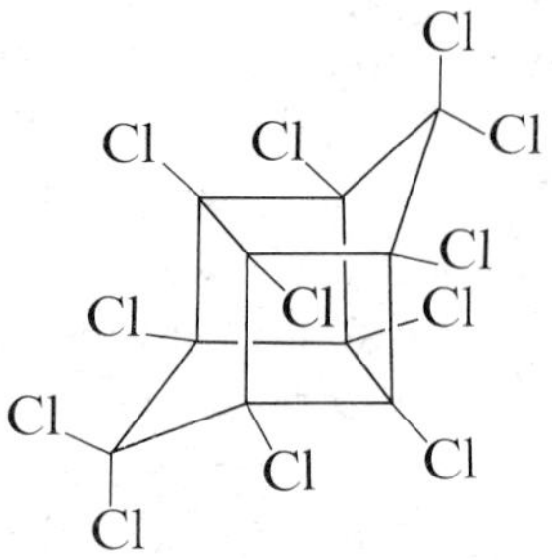   Mirex

In spite of their structural similarity, Chlordane and Heptachlor exhibit different behavior. Unlike Heptachlor, Chlordane is rather short-lived and is less toxic to mammals. Nevertheless, all of the chlorocarbon insecticides have been the objects of a great deal of suspicion. A ban on the use of Dieldrin and Aldrin also has been ordered by the Environmental Protection Agency. In addition, strains of insects resistant to Dieldrin, Aldrin, and other materials have been observed. Some of these insects actually become addicted to the chlorocarbon insecticide and thrive on it!

The problems associated with the chlorocarbon materials have led to the development of the "soft" insecticides. These usually are organophosphorus or carbamate derivatives, and they are characterized by a short lifetime before they are decomposed to harmless materials in the environment.

Examples of organophosphorus insecticides are shown below.

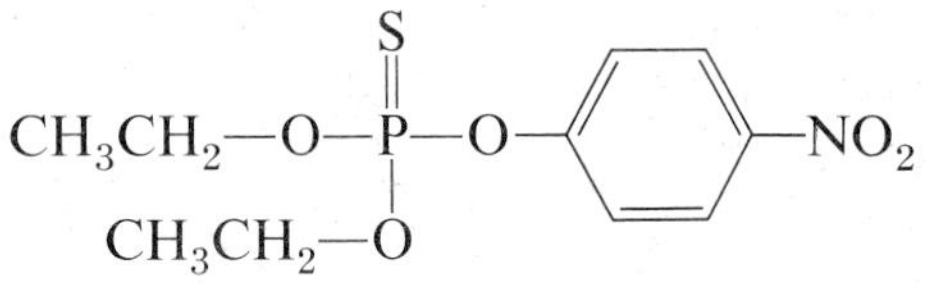

Parathion

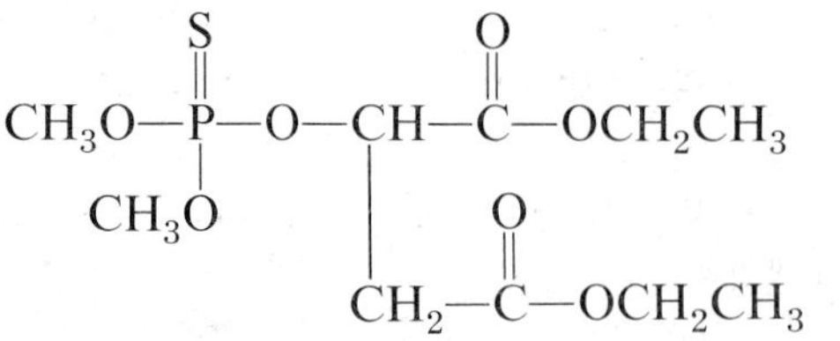

Malathion 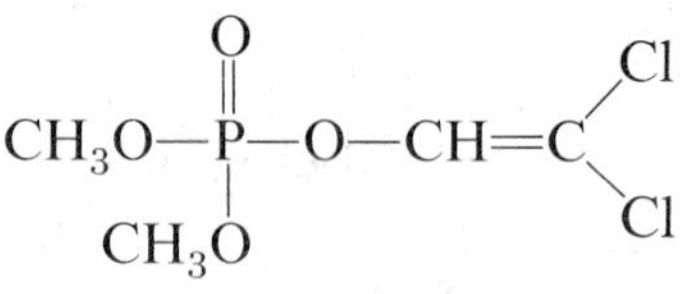   DDVP or Dichlorvos

Parathion and Malathion are used widely in agriculture. DDVP is used in "pest strips" which are used to combat household insect pests. The organophosphorus materials are not persistent in the environment, and so they are not passed from specie to specie along the food chain, as the chlorocarbon compounds are. However, the organophosphorus compounds are highly toxic to humans. Some loss of life among migrant and other agricultural workers has been caused by accidents involving the application of these materials. Stringent safety precautions must be applied when the organophosphorus insecticides are being used.

The carbamate derivatives, including Carbaryl, tend to be less toxic than the organophosphorus compounds. They are also readily degraded to harmless materials. Nevertheless, insects resistant to the "soft" insecticides have also been observed. Furthermore, the organophosphorus and carbamate derivatives are much more destructive of nontarget pests than the chlorocarbon compounds. The danger to earthworms, mammals, and birds is very high.

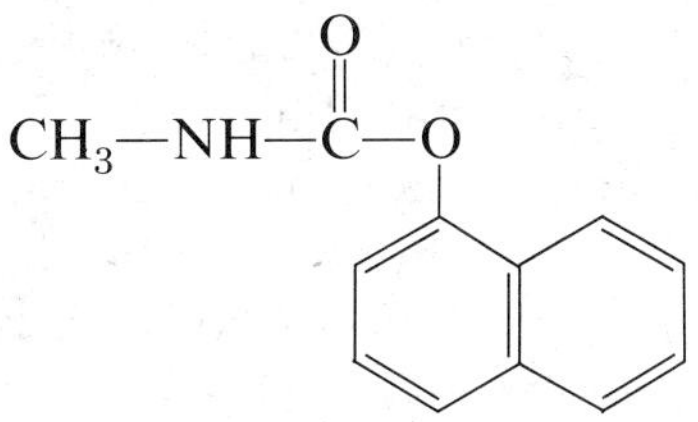

$$\text{CH}_3\text{—NH—}\overset{\displaystyle\overset{O}{\|}}{\text{C}}\text{—O}$$

Carbaryl

ALTERNATIVES TO INSECTICIDES

Several alternatives to the massive application of insecticides have recently been explored. Insect attractants, including the pheromones (see the essay preceding Experiment 17), have been used in localized traps. Such methods have been effective against the gypsy moth. A "confusion technique," whereby a pheromone is sprayed into the air in such high concentrations that male insects are no longer able to locate females, has been studied. These methods are specific for the target pest and do not cause repercussions in the general environment.

Recent research has been focused on the use of an insect's own biochemical processes as a means to control pests. Experiments with **juvenile hormone** have shown promise. Juvenile hormone is one of three internal secretions which are used by insects to regulate growth and metamorphosis from the larval stage to the pupa and hence to the adult stage. At certain stages in the metamorphosis from larva to pupa, juvenile hormone must be secreted; at other stages it must be absent, or the insect will either develop abnormally or fail to mature. Juvenile hormone plays a role in maintaining the juvenile or larval stage of the growing insect. The male **Cecropia** moth, which is the mature form of the silkworm, has been used as a source of juvenile hormone. The structure of the Cecropia juvenile hormone is shown below. This material has been shown to prevent the maturation of yellow fever mosquitoes and human body lice. In as much as insects are not expected to develop a resistance to their own hormones, it is hoped that insects will not be likely to develop a resistance to juvenile hormone.

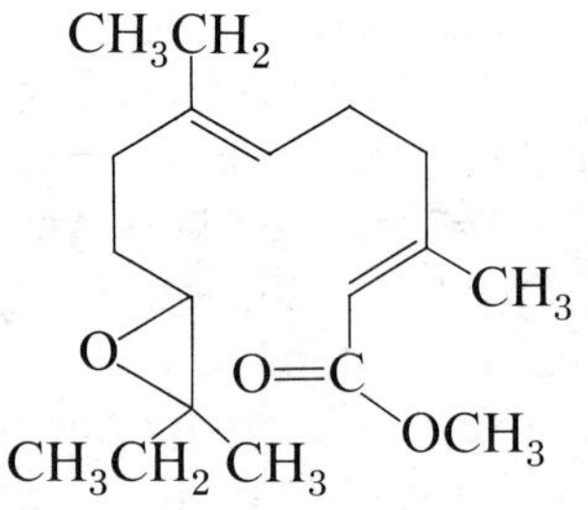

Cecropia Juvenile Hormone 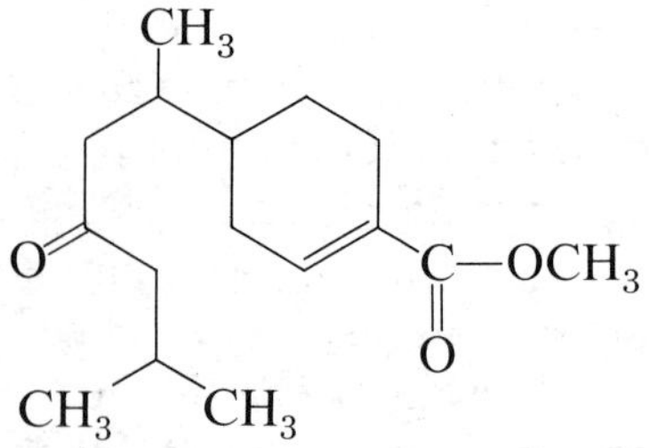   Paper Factor

While the natural substance is very difficult to obtain in sufficient quantities for use in agriculture, synthetic analogs have been prepared, and they have been shown to have similar properties and effectiveness to the natural substance. Williams, Sláma, and Bowers have identified and characterized a substance found in the American balsam fir (**Abies balsamea**), known as **Paper Factor**, which is active against the linden bug, **Pyrrhocoris apterus**, a European cotton pest. This substance is merely one out of thousands of terpenoid materials synthesized by the fir tree. The investigation of other terpenoid substances as potential juvenile hormone analogs is being conducted.

Certain plants are capable of synthesizing substances which protect them against insects. Included among these natural insecticides are the **pyrethrins** and derivatives of **nicotine** (see the essay preceding Experiment 5).

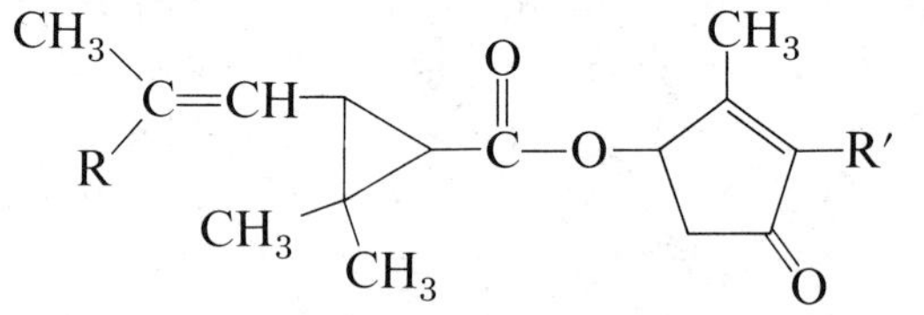

Pyrethrin

$$R = CH_3 \quad or \quad COOCH_3$$
$$R' = CH_2CH=CHCH=CH_2 \quad or \quad CH_2CH=CHCH_3$$
$$or \quad CH_2CH=CHCH_2CH_3$$

The search for environmentally suitable means of controlling agricultural pests continues with a great sense of urgency. Insects cause billions of dollars of damage to food crops each year. With food becoming increasingly scarce and with the world's population growing at an exponential rate, the need to prevent such losses to food crops is absolutely essential.

REFERENCES

Berkoff, C. E. "Insect Hormones and Insect Control." *Journal of Chemical Education, 48* (1971), 577.

Carson, R. *Silent Spring.* Boston: Houghton-Mifflin, 1962.

Keller, E. "The DDT Story." *Chemistry, 43* (February, 1970), 8.

O'Brien, R. D. *Insecticides: Action and Metabolism.* New York: Academic Press, 1967.

Peakall, D. B. "Pesticides and the Reproduction of Birds." *Scientific American, 222* (April, 1970), 72.

Pryde, L. T. *Environmental Chemistry: An Introduction.* Menlo Park, California: Cummings, 1973. Chapter 7.

Pyle, J. L. *Chemistry and the Technological Backlash.* Englewood Cliffs, New Jersey: Prentice-Hall, 1974. Chapter 4.

Rudd, R. L. *Pesticides and the Living Landscape.* Madison, Wisconsin: University of Wisconsin Press, 1964.

Williams, C. M. "Third-Generation Pesticides." *Scientific American, 217* (July, 1967), 13.

Experiment **46**

TETRAPHENYLPHTHALIC ANHYDRIDE

Diels-Alder Reaction

In this experiment you will prepare tetraphenylphthalic anhydride. The first step is a Diels-Alder reaction between tetraphenylcyclopentadienone (Experiment 45) and maleic anhydride to give an adduct (A) which then loses carbon monoxide to give tetraphenyl-1,2-dihydrophthalic anhydride.

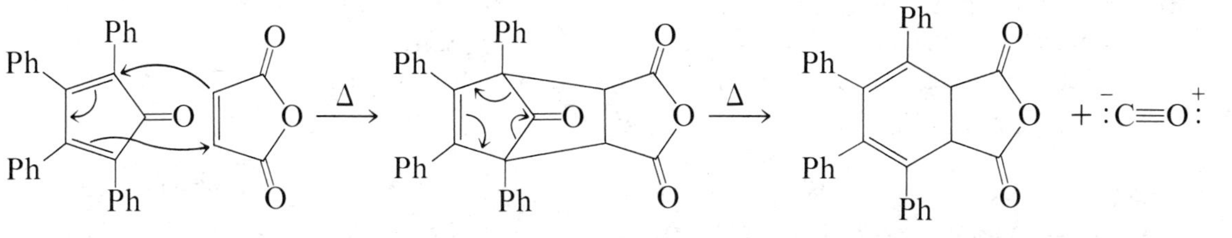

Tetraphenylcyclo-
pentadienone   (A)   Tetraphenyl-1,2-
dihydrophthalic Anhydride

The adduct (A) has the **endo** configuration shown below rather than the **exo** configuration.

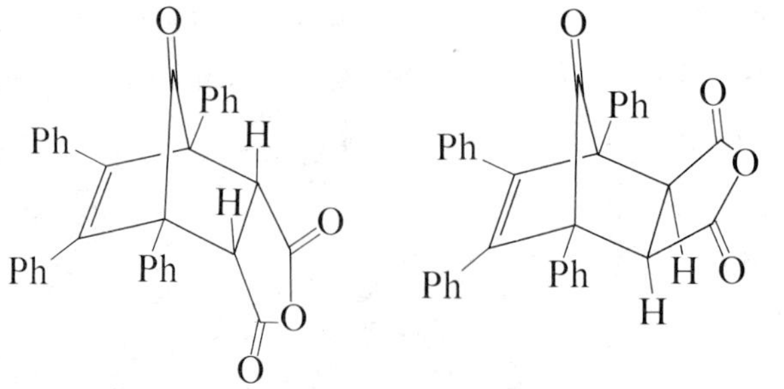

endo   exo

Although the tetraphenyl-1,2-dihydrophthalic anhydride may be isolated, it is, instead, converted to the fully aromatized product in the second step of the reaction sequence. In effect, the tetraphenyl-1,2-dihydrophthalic anhydride must be dehydrogenated to produce the aromatic ring in tetraphenylphthalic anhydride. This is done by bromination which produces the intermediate bromide (B), which easily loses hydrogen bromide to form the aromatic ring.

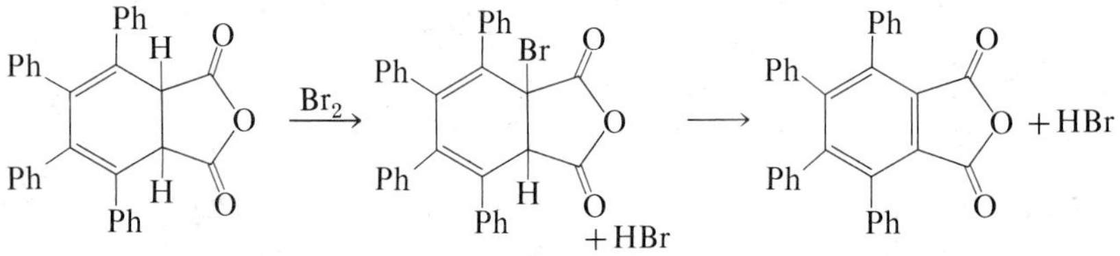

Tetraphenyl-1,2-   (B)   Tetraphenylphthalic
dihydrophthalic Anhydride   Anhydride

The bromination proceeds smoothly because the proton removed is both allylic and alpha to the carbonyl. The reaction probably proceeds **via** the enol which is easily brominated.

SPECIAL INSTRUCTIONS

Read the essay which precedes this experiment. There are two successive three-hour reflux periods in this experiment. The time may be used efficiently by completing shorter experiments during these reflux periods. Special care must be taken to avoid breathing the carbon monoxide gas which is produced in the first step of this synthesis.

PROCEDURE

Thoroughly mix 7.0 g of tetraphenylcyclopentadienone (Experiment 45) and 1.9 g of maleic anhydride[1] and place the mixture in a 100 ml

---

[1]The maleic anhydride may need to be purified by the instructor or assistant. The crude anhydride, which may contain some acid, is added to chloroform. The anhydride is soluble in this solvent, while the acid is not soluble. After filtering the solid, the chloroform is removed to yield the purified anhydride (mp 48 to 52°).

round bottom flask. Add 5 ml of bromobenzene to the flask to produce a thick mixture of solids. Attach a reflux condenser to the flask and heat the mixture under gentle reflux in the hood (carbon monoxide is evolved!) or at a desk if a trap is used. A trap for carbon monoxide is made by filling a large test tube one-half full with a solution of cuprous chloride and ammonium chloride in aqueous ammonia.[2] If a trap is to be used, secure a short section of glass tubing with a clamp so that the end is immersed several centimeters below the surface of the solution in the test tube. Fasten the other end of the glass tubing to an aspirator trap with rubber tubing (Technique 2, Figure 2-4). Connect this latter trap to the top of the condenser with a piece of rubber tubing. The aspirator trap serves as a safety device to prevent the copper solution from accidently backing up into the reaction mixture.

Reflux the mixture for a total of 3 hours during one or two laboratory periods or reflux until the dark brown color of the solution disappears and a light yellow mass of solid appears. Shake the mixture occasionally during this reflux period. During the latter part of the reflux period, prepare a solution of 1.4 ml of bromine in 2 ml of bromobenzene.

CAUTION: Handle bromine with extreme caution; it is a strong oxidant. Do not breathe the vapors or have it come in contact with the skin. Use an eye dropper to transfer this corrosive liquid.

Cool the mixture, which contains tetraphenyl-1,2-dihydrophthalic anhydride, following the reflux period,[3] and add the bromine solution through the top of the condenser. Shake the mixture thoroughly. After the first exothermic reaction subsides, gently reflux the mixture for 3 hours in a hood, or with a trap connected to the condenser. The trap (Experiment 23) is used to prevent bromine vapors from entering the room. The solution in the trap consists of 1 to 2 g of sodium bisulfite added to 100 ml of water. This solution will react with any bromine evolved from the reaction mixture.

Cool the flask in an ice bath for about 10 minutes after the reflux period. Add about 7 ml of ligroin to the cooled mixture. Break up any

---

[2] A large amount of trapping agent may be prepared as follows: Dissolve 200 g of cuprous chloride and 250 g of ammonium chloride in 700 ml of water. Add to the solution 1/3 its volume of concentrated (28%) ammonium hydroxide.

[3] The tetraphenyl-1,2-dihydrophthalic anhydride may be isolated at this point. Cool the flask to room temperature and add 50 ml of ligroin (bp 63 to 75°) to the precipitated solid. Break up the solid and filter it by vacuum. Wash the crude product twice with 10 ml portions of ligroin and dry the solid. The melting point of the crude product is 237 to 240°. Dissolve the crude crystals in 120 ml of hot benzene, filter the solution by gravity and add 120 ml of ligroin to the filtrate. Cool the mixture thoroughly in an ice bath and filter the solid that separates. Determine the melting point (mp 240 to 243°). The infrared spectrum in potassium bromide is shown at the end of the experimental procedure. Occasionally, some of this material may be partially aromatized after the lengthy reflux. If this is the case, the melting point will be high and the carbonyl absorption will be shifted in the infrared spectrum.

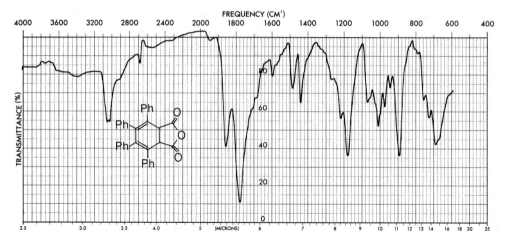

Infrared spectrum of tetraphenyl-1,2-dihydrophthalic anhydride, KBr

lumps and vacuum filter the mixture. Wash the solid with 3 portions of ligroin (2 to 3 ml each). Transfer the product to a beaker and crush the solid thoroughly. Add about 15 ml of ligroin and heat it for a few minutes with stirring. Cool the mixture in an ice bath and decant the liquid away from the solid. Add another 15 ml portion of ligroin and heat it for a few minutes. Cool the mixture and vacuum filter it using a Büchner funnel.

Recrystallize the crude tetraphenylphthalic anhydride from benzene using 8 or 9 ml of solvent per gram of solid. Heat the mixture to boiling and gravity filter it through fast fluted filter paper. Slowly cool the mixture to room temperature and vacuum filter the purified tetraphenylphthalic anhydride. Dry the solid for one hour in an oven at 110° to remove the residual benzene which crystallizes with the product.

Weigh the tetraphenylphthalic anhydride and calculate the percent yield. Determine the melting point (288 to 290°). At the option of the

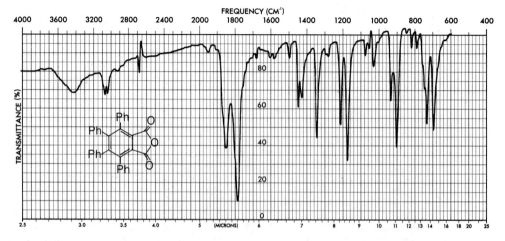

Infrared spectrum of tetraphenylphthalic anhydride, KBr

instructor, determine the infrared spectrum of tetraphenylphthalic anhydride in potassium bromide and compare it to the spectrum given above for tetraphenyl-1,2-dihydrophthalic anhydride. Submit the sample to the instructor in a labeled vial.

QUESTIONS

1. Give the structure of the product that would be produced from the reaction of tetraphenylcyclopentadienone with diphenylacetylene.

Experiment **47**

PHOTOREDUCTION OF BENZOPHENONE

Photochemistry
Photoreduction
Energy Transfer

The photoreduction of benzophenone is one of the oldest and most thoroughly studied photochemical reactions. It was discovered early in the history of photochemistry that solutions of benzophenone are unstable to light when certain solvents are used. If benzophenone is dissolved in a "hydrogen donor" solvent, such as 2-propanol, and exposed to ultraviolet light, an insoluble dimeric product, benzpinacol, will form.

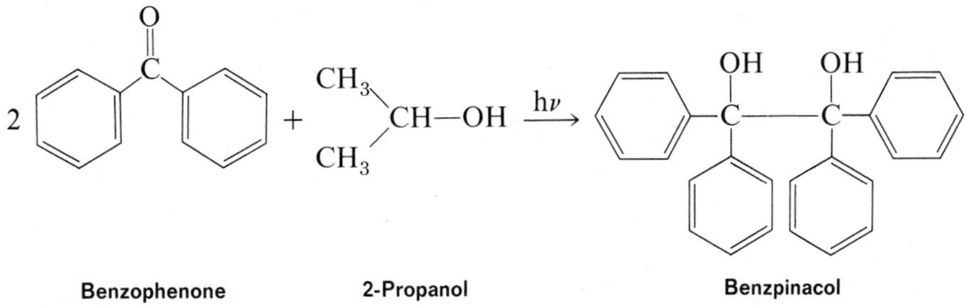

| Benzophenone | 2-Propanol | Benzpinacol |

To understand this reaction, it is necessary to review some simple photochemistry as it relates to aromatic ketones. In the typical organic molecule, all of the electrons are paired in the occupied orbitals. When such a molecule absorbs ultraviolet light of the appropriate wavelength, an electron from one of the occupied orbitals, usually the one of highest energy, is excited to an unoccupied molecular orbital, usually to the one of lowest energy. During this transition, the electron must retain its spin value, because a change of spin is a quantum mechanically forbidden process during an electronic transition. Thus, just as the two electrons in the highest occupied orbital of the molecule originally had their spins paired (opposite), so will they retain paired spins in the first electronically excited state of the molecule. This is true even though the two electrons will be in **different** orbitals after the transition. This first excited state of a molecule is called a **singlet state** (S_1), since its spin multiplicity ($2S + 1$) is one. The original unexcited state of the molecule is also a singlet state, since its electrons are paired, and it is called the **ground state** singlet state (S_0) of the molecule.

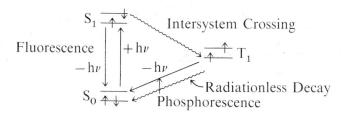

The electronic states of a typical molecule and the interconversion processes that may occur. In each state (S_0, S_1, T_1), the lower line represents the highest occupied orbital and the upper line represents the lowest unoccupied orbital of the unexcited molecule. Straight lines represent processes in which a photon is absorbed or emitted. Wavy lines represent radiationless processes—those which occur **without** emission or absorption of a photon.

The excited state singlet, S_1, may return to the ground state, S_0, by reemission of the absorbed photon of energy. This process is called **fluorescence.** Alternatively, the excited electron may undergo a change of spin to give a state of higher multiplicity, the excited **triplet state,** so called because its spin multiplicity ($2S + 1$) is three. The process of conversion from the first excited singlet state to the triplet state is called **intersystem crossing.** Because it has a higher multiplicity, the triplet state will inevitably be a lower energy state than the excited singlet state (Hund's rule). Normally this change of spin (intersystem crossing) is a quantum mechanically forbidden process, just as a direct excitation of the ground state (S_0) to the triplet state (T_1) is forbidden. However, in those molecules in which the singlet and triplet states lie close to one another in energy, the two states inevitably have a number of overlapping vibrational states, i.e., states in common, a situation which allows the "forbidden" transition to occur. In many molecules where S_1 and T_1 are similar in energy ($\Delta E < 10$ Kcal/mole), intersystem crossing proceeds at a rate which is faster than the rate of fluorescence, and the molecule is rapidly converted from its excited singlet state to its triplet state. In benzophenone, S_1 undergoes intersystem crossing to T_1 with a rate of $k_{isc} = 10^{10}$ sec^{-1}, meaning that the lifetime of S_1 is only 10^{-10} second. The rate of fluorescence for benzophenone is $k_f = 10^6$ sec^{-1}, meaning that intersystem crossing occurs at a rate which is 10^4 times as fast as fluorescence. Thus, the conversion of S_1 to T_1 in benzophenone is an essentially quantitative process. In molecules which have a wide energy gap between S_1 and T_1, this situation would be reversed. As will be seen shortly, the naphthalene molecule presents a situation of this latter type.

Since the excited triplet state is lower in energy than the singlet state, it cannot easily return to that state. Nor can it easily return to the ground state by returning the excited electron to its original orbital. Once again, the transition, $T_1 \longrightarrow S_0$, would require a change of spin for the electron, and this is a forbidden process. Hence, the triplet excited state usually has a long lifetime (relative to other excited states) since it generally has nowhere that it can easily go. Even though the process is forbidden, the triplet, T_1, may eventually return to the ground state, S_0, by a process called a **radiationless transition.** In this process, the excess energy of the triplet is lost to the surrounding solution as heat, thereby "relaxing" the triplet back to the ground state, S_0. This process is the study of much current research, and is not well understood. In the second process, where a triplet state may revert to the ground state, **phosphorescence,** the excited triplet emits a photon to dissipate the excess energy and returns directly to the ground state. Although this process is "forbidden," it nevertheless occurs when there is no other open pathway by which the molecule may dissipate its excess energy. In benzophenone, radiationless decay is the faster process with rate $k_d = 10^5$ sec^{-1}, and phosphorescence, which is not observed, has a slower rate of $k_p = 10^2$ sec^{-1}.

Benzophenone is a ketone. Ketones have **two** possible excited singlet states, and, in consequence, two excited triplet states as well. This occurs since two relatively low energy transitions are possible in benzophenone. It is possible to excite one of the π electrons in the carbonyl π bond to the lowest energy unoccupied orbital, a π^* orbital. It is also possible to excite one of the non-bonded or n electrons on oxygen to the same orbital. The first type of transition is called a π-π^* transition, while the second is called an n-π^* transition. These transitions and the states which result are illustrated pictorially below.

Spectroscopic studies show that for benzophenone and most other ketones, the n-π* excited states S_1 and T_1 are of lower energy than the π-π* excited states. An energy diagram depicting the excited states of benzophenone (along with one that depicts naphthalene) is shown below.

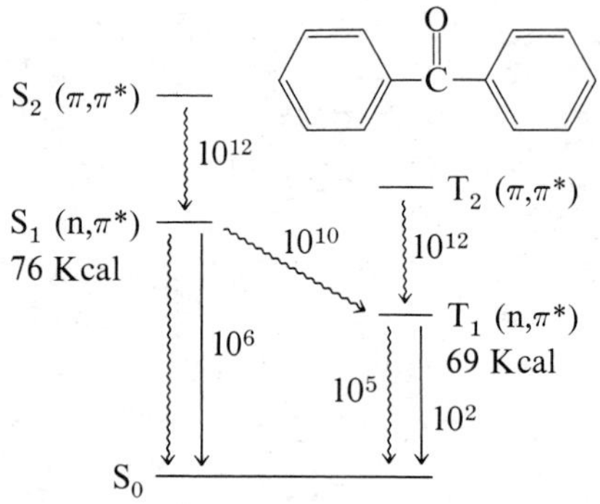

EXCITED STATES OF BENZOPHENONE 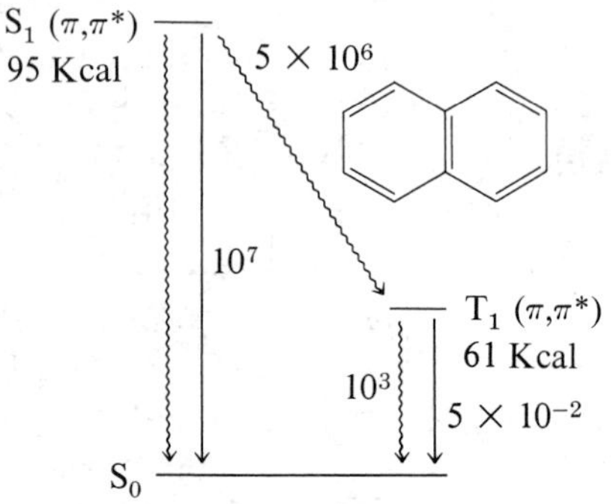   EXCITED STATES OF NAPHTHALENE

It is now known that the photoreduction of benzophenone is a reaction of the n-π* triplet state (T_1) of benzophenone. The n-π* excited states have radical character at the carbonyl oxygen atom due to the unpaired electron in the non-bonding orbital. Thus, the radical-like and energetic T_1-excited state specie can abstract a hydrogen atom from a suitable donor molecule to form the diphenylhydroxymethyl radical. Two of these radicals, once formed, may couple to form benzpinacol. The complete mechanism for photoreduction is outlined below.

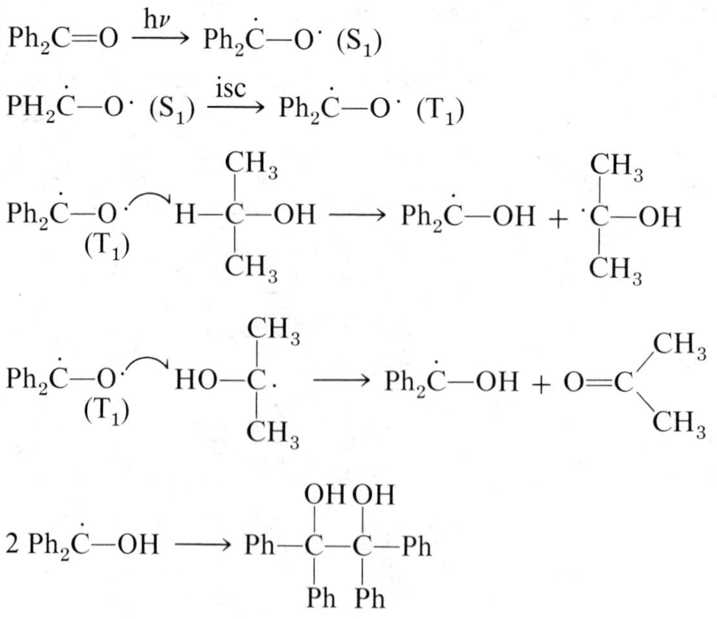

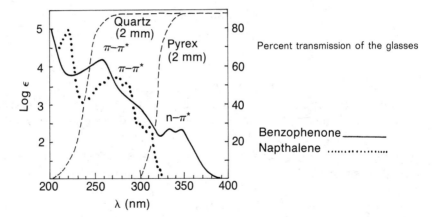

Percent transmission of the glasses

Benzophenone _____
Napthalene ....

Many photochemical reactions must be carried out in a quartz apparatus because they require ultraviolet radiation of shorter wavelengths (higher energy) than those wavelengths which are able to pass through Pyrex glass. Benzophenone, however, requires radiation of approximately 350 nm ($=350$ mμ or 3500 Å) to become excited to its n-π^* singlet state S_1, a wavelength which readily passes through Pyrex glass. In the figure above, the ultraviolet absorption spectra of benzophenone and naphthalene are given. Superimposed on their spectra are two curves which show the wavelengths which may be transmitted by Pyrex and quartz glass, respectively. Pyrex will not allow any radiation of wavelength shorter than approximately 300 nm to pass, whereas quartz will allow wavelengths as short as 200 nm to pass. Thus, when benzophenone is placed in a Pyrex flask, the only electronic transition which may occur is the n-π^* transition which occurs at 350 nm.

However, even if it were possible to supply benzophenone with radiation of the appropriate wavelength to produce the second excited singlet state of the molecule, this singlet would rapidly convert to the lowest singlet state, S_1. S_2 has a lifetime of less than 10^{-12} second. The conversion process, $S_2 \longrightarrow S_1$, is called an **internal conversion.** Internal conversions are processes of conversion which occur between excited states of the same multiplicity (singlet-singlet, or triplet-triplet), and they usually occur very rapidly. Thus, when either S_2 or T_2 are formed, they readily convert to S_1 and T_1, respectively. As a consequence of their very short lifetimes, very little is known about the properties or the exact energies of S_2 and T_2 of benzophenone.

ENERGY TRANSFER

Using a simple **energy transfer** experiment, it is possible to show that the photoreduction of benzophenone proceeds via the T_1 excited state of benzophenone, rather than the S_1 excited state. If naphthalene is added to the reaction, the photoreduction process is stopped because the excitation energy of the benzophenone triplet is transferred to naphthalene. The naphthalene is said to have **quenched** the reaction. This occurs in the following way.

When the excited states of molecules have sufficiently long life-times, it is often possible for them to transfer their excitation energy to another molecule. The mechanisms by which these transfers can occur are complex and cannot be explained here; the essential require-ments, however, can be outlined. First, for two molecules to exchange their respective states of excitation, the process must occur with an overall decrease in energy. Second, the spin multiplicity of the total system must not change. These two features may be illustrated by giving the two most common examples of energy transfer—singlet transfer and triplet transfer. In these two examples, the superscript 1 denotes an excited singlet state, the superscript 3 denotes a triplet state, and the subscript 0 denotes a ground state molecule. A and B are different molecules.

$$A^1 + B_0 \longrightarrow B^1 + A_0 \quad \text{SINGLET ENERGY TRANSFER}$$
$$A^3 + B_0 \longrightarrow B^3 + A_0 \quad \text{TRIPLET ENERGY TRANSFER}$$

In the first process, singlet energy transfer, excitation energy is transferred from the excited singlet state of A to a ground state molecule of B, converting it to its excited singlet state and returning A to its ground state. In the second process, triplet energy transfer, a similar interconversion of excited state and ground state occurs. Singlet energy transfer can occur through space by means of a dipole-dipole coupling mechanism, but triplet energy transfer requires a collision of the two molecules involved in the transfer. In the usual organic medium, about 10^9 collisions occur per second. Thus, if a triplet state A^3 has a lifetime longer than 10^{-9} second, and if an acceptor molecule, B_0, which has a lower triplet energy than that of A^3 is available, energy transfer can be expected to occur. If the triplet A^3 undergoes a reaction (like photo-reduction) at a rate slower than the rate of collisions in the solution, and if an acceptor molecule is added to the solution, the reaction can be **quenched.** The acceptor molecule, which is called a **quencher,** deacti-vates or "quenches" the triplet before it has an opportunity to react. Naphthalene has the ability to quench benzophenone triplets in this way, and to stop the photoreduction reaction.

Naphthalene cannot quench the excited state singlet S_1 of benzo-phenone, since its own singlet has an energy (95 Kcal/mole) which is higher than that of benzophenone (76 Kcal/mole). In addition, the conversion process $S_1 \longrightarrow T_1$ occurs very rapidly (10^{-10} second) in ben-zophenone. Thus, naphthalene can only intercept the triplet state of benzophenone. The triplet excitation energy of benzophenone (69 Kcal/mole) is transferred to naphthalene ($T_1 = 61$ Kcal/mole) in an exothermic collision process. Finally, the naphthalene molecule does not absorb light of the wavelengths transmitted by Pyrex glass (see spectra above) and, therefore, benzophenone is not inhibited from absorbing energy when naphthalene is present in solution. Thus, the fact that naphthalene quenches the photoreduction reaction of benzophenone indicates that this reaction proceeds via the triplet state, T_1, of benzophenone. If naphthalene did not quench the reaction, the singlet state of benzophenone would be indicated as the reactive intermediate. In the experiment that follows, the photoreduction of benzophenone will be attempted both in the presence and in the absence of added naphthalene.

SPECIAL INSTRUCTIONS

This experiment may be performed concurrently with some other experiment. It requires only 15 minutes of time during the first laboratory period, and only about 15 minutes of time in a subsequent laboratory period about one week later.

It is important that the reaction be left in a place where it will receive **direct** sunlight. If not, the reaction will be slow and may require more than a week for completion.

PROCEDURE

Obtain two 20 × 150 mm test tubes and label them No. 1 and No. 2 near the top of the tubes. Place 2.5 g of benzophenone in the first tube. Place 2.5 g of benzophenone and 2.5 g of naphthalene in the second tube. Add about 10 ml of 2-propanol (isopropyl alcohol) to each tube and warm it on the steam bath to dissolve the solids. When the solids have dissolved, add 1 drop of glacial acetic acid to each tube, and then fill each tube nearly to the top with more 2-propanol. Stopper the tubes tightly with rubber stoppers, shake them well, and place them in a beaker on a window sill where they will receive direct sunlight. The reaction will require about one week for completion. If the reaction has occurred during this period, the product will have crystallized from the solution. Observe the result in each test tube. Collect the product by vacuum filtration, dry it, and determine its melting point and percentage yield. At your instructor's option, determine the infrared spectrum of the benzpinacol as a KBr mull. Submit the product to your instructor along with the report.

QUESTIONS

1. Can you think of a way to produce the benzophenone n-π* triplet, T_1, **without** having benzophenone pass through its first singlet state? Explain.

2. A similar reaction to the one here described occurs when benzophenone is treated with magnesium metal (pinacol reduction).

$$2\ Ph_2C{=}O \xrightarrow{\ Mg\ } Ph_2\overset{\displaystyle OH}{\underset{\displaystyle |}{C}}{-}{-}\overset{\displaystyle OH}{\underset{\displaystyle |}{C}}Ph_2$$

Compare the mechanism of this reaction to the photoreduction mechanism. What are the differences?

3. Which of the following molecules would you expect to be useful in quenching benzophenone photoreduction? Explain.

Oxygen   ($S_1 = 22$ Kcal/mole)   Biphenyl ($T_1 = 66$ Kcal/mole)
9,10-Diphenylanthracene ($T_1 = 42$ Kcal/mole)   Toluene ($T_1 = 83$ Kcal/mole)
trans-1,3-Pentadiene   ($T_1 = 59$ Kcal/mole)   Benzene ($T_1 = 84$ Kcal/mole)
Naphthalene   ($T_1 = 61$ Kcal/mole)

4. Acetone ($S_1 = 84$ Kcal/mole, $T_1 = 78$ Kcal/mole) does not undergo photoreduction. Explain

ESSAY

Fireflies and Photochemistry

The production of light as a result of a chemical reaction is called **chemiluminescence.** A chemiluminescent reaction generally produces one of the product molecules in an electronically excited state. The excited state emits a photon, and light is produced. If a reaction which produces light is biochemical, occurring in a living organism, the phenomenon is called **bioluminescence.**

The light produced by fireflies and other bioluminescent organisms has fascinated observers for many years. There are many different organisms which have developed the ability to emit light. They include bacteria, fungi, protozoans, hydras, marine worms, sponges, corals, jellyfishes, crustaceans, clams, snails, squids, fishes, and insects. Curiously, other than fish, this list does not include many of the higher forms of life. Amphibians, reptiles, birds, mammals, and the higher plants are excluded. Among the marine species, none is a freshwater organism. The excellent Scientific American article by McElroy and Seliger (see references) delineates the natural history, characteristics, and habits of many of these bioluminescent organisms.

The first significant studies of a bioluminescent organism were those of the French physiologist Raphael Dubois in 1887. He studied the mollusk **Pholas dactylis,** a bioluminescent clam indigenous to the Mediterranean Sea. Dubois found that a cold water extract of the clam was able to emit light for several minutes following the extraction. When the light emission ceased, he found that it could be restored by a material extracted from the clam by hot water. A hot water extract of the clam did not alone produce the luminescence. Reasoning carefully, Dubois concluded that there was an enzyme in the cold water extract that was destroyed in hot water. The luminescent compound, however, could be extracted without destruction in either hot or cold water. He called the luminescent material **luciferin,** and the enzyme which induced it to emit **luciferase,** both names being derived from that of Lucifer, a Latin name meaning "bearer of light." Today the luminescent materials from **all** organisms are called luciferins, and the associated enzymes are called luciferases.

The most extensively studied bioluminescent organism is the firefly. Fireflies are found in many areas of the world and probably thus represent the most familiar example of bioluminescence. In those areas, on a typical summer evening, fireflies, or "lightning bugs," can frequently be seen to emit flashes of light as they cavort in the lawn or garden. It is now universally accepted that the luminescence of fireflies is a mating device. The male firefly flies at a height of about two feet above the ground and emits flashes of light at regular intervals. The female, who remains stationary on the ground, waits a characteristic interval, and then flashes a response. In return, the male will reorient his direction of flight toward her and flash a signal once again. The entire cycle is rarely repeated more than 5 to 10 times before the male reaches the female. Fireflies of different species can recognize one another by their flash patterns, which vary in number, rate, and duration from one species to another.

Although the total structure of the luciferase enzyme of the American firefly, **Photinus pyralis,** is unknown, the structure of the luciferin has been established. However, in spite of a large amount of experimental work, the complete nature of the chemical reactions that produce the light is still subject to some controversy. Nevertheless, the most salient details of the reaction may be outlined.

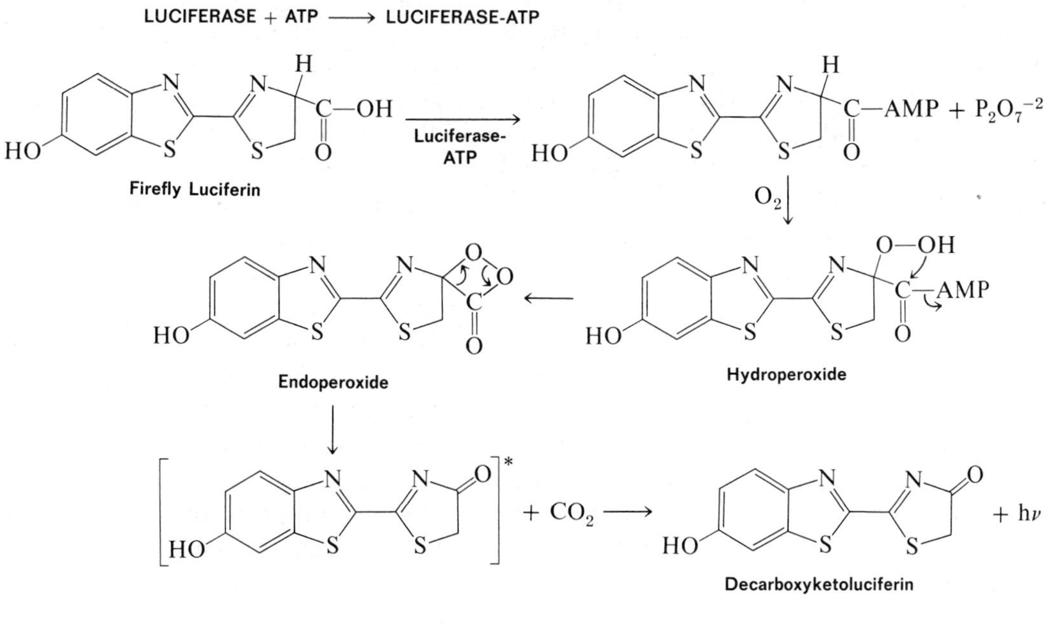

LUCIFERASE + ATP $\longrightarrow$ LUCIFERASE-ATP

Firefly Luciferin

Endoperoxide

Hydroperoxide

Decarboxyketoluciferin

In addition to the luciferin and the luciferase, magnesium (II), ATP (adenosine triphosphate), and molecular oxygen are required to produce the luminescence. In the proposed first step of the reaction, the luciferase complexes with an ATP molecule. In a second step, the luciferin binds to the luciferase enzyme and reacts with the already bound ATP molecule to become "primed." In this reaction, pyrophosphate ion is expelled, and AMP (adenosine monophosphate) becomes attached to the carboxyl group of the luciferin. In a third step, the luciferin-AMP complex is oxidized by molecular oxygen to form a hydroperoxide which cyclizes with the carboxyl group, expelling AMP and forming the cyclic endoperoxide. This reaction would be difficult if the carboxyl group of the luciferin had not been primed with ATP. The endoperoxide is unstable and readily decarboxylates, producing decarboxyketoluciferin in an **electronically excited state,** which is deactivated by the emission of a photon (fluorescence). Thus, it is the cleavage of the four-membered-ring endoperoxide that leads to the electronically excited molecule that emits and produces the bioluminescence.

$$ \ce{>C(O)(O)C<} \longrightarrow \left[\begin{matrix} \ce{>C=O} \\ \ce{>C=O^*} \end{matrix} \right] \longrightarrow 2 \ce{>C=O} + h\nu $$

That one of the two carbonyl groups, either that of the decarboxyketoluciferin or that of the carbon dioxide, should be formed in an excited state is readily predicted on the basis of the orbital symmetry conservation principles of the Woodward-Hoffmann rules. This reaction is formally similar to the decomposition of a cyclobutane ring to yield two ethylene molecules. In analyzing the forward course of that reaction, i.e., 2 ethylene $\rightarrow$ cyclobutane, one can easily show that the reaction, which involves 4 π electrons, is forbidden for two ground state ethylenes, but is allowed only for one ethylene in the ground state and the other in an excited state. This suggests that in the reverse process, one of the ethylene molecules should be formed in an excited state. Extending these arguments to the endoperoxide also suggests that one of the two carbonyl groups should be formed in its excited state.

The emitting molecule, decarboxyketoluciferin, has been isolated and synthesized. When it is excited photochemically by photon absorption in **basic** solution (pH $>$ 7.5 to 8.0), it fluoresces, giving a fluorescence emission spectrum that is identical to the emission spectrum produced by the interaction of firefly luciferin and firefly luciferase. The emitting form of decarboxyketoluciferin has thus been identified as the **enol dianion.** In neutral or acidic solution, the emission spectrum of decarboxyketoluciferin does not match that of the bioluminescent system.

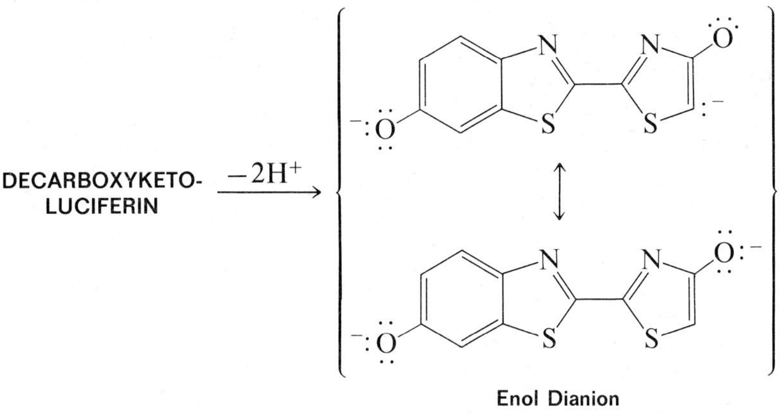

DECARBOXYKETO-
LUCIFERIN $\xrightarrow{-2H^+}$

Enol Dianion

The exact function of the enzyme firefly luciferase is not yet known, but it is clear that all of these reactions occur while luciferin is bound to the enzyme as a substrate. Also, since the enzyme undoubtedly has a number of basic groups ($-COO^-$, $-NH_2$, etc.), the buffering action of those groups would easily explain the fact that it is the enol dianion which is also the emitting form of decarboxyketoluciferin in the biological system.

Most chemiluminescent and most bioluminescent reactions require oxygen. Likewise, most of them produce an electronically excited emitting specie through the decomposition of a **peroxide** of one sort or another. In the experiment which follows, a **chemiluminescent** reaction which also involves the decomposition of a peroxide intermediate is described.

REFERENCES

Clayton, R. K. *Light and Living Matter, Volume 2: The Biological Part*. Chapter 6, "The Luminescence of Fireflies and Other Living Things." New York: McGraw-Hill, 1971.

Harvey, E. N. *Bioluminescence*. New York: Academic Press, 1952.

Hastings, J. W. "Bioluminescence." *Annual Review of Biochemistry, 37* (1968), 597.

McElroy, W. D., and Seliger, H. H. "Biological Luminescence." *Scientific American, 207* (December, 1962), 76.

McElroy, W. D., Seliger, H. H., and White, E. H. "Mechanism of Bioluminescence, Chemiluminescence and Enzyme Function in the Oxidation of Firefly Luciferin." *Photochemistry and Photobiology, 10* (1969), 153.

Seliger, H. H., and McElroy, W. D. *Light: Physical and Biological Action*. New York: Academic Press, 1965.

LUMINOL

Chemiluminescence
Energy Transfer
Reduction Of a Nitro Group
Amide Formation

In this experiment, the chemiluminescent compound luminol, or 5-aminophthalhydrazide, will be synthesized from 3-nitrophthalic acid.

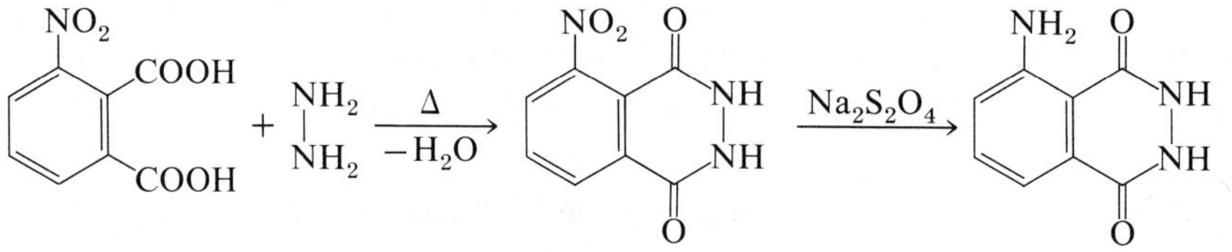

| 3-Nitrophthalic Acid | Hydrazine | | 5-Nitrophthalhydrazide | | Luminol |

The first step of the synthesis is the simple formation of a cyclic diamide, 5-nitrophthalhydrazide, by reaction of 3-nitrophthalic acid with hydrazine. Reduction of the nitro group with sodium dithionite affords luminol.

In neutral solution, luminol exists largely as a dipolar ion (zwitterion). This dipolar ion itself exhibits a weak blue fluorescence after being exposed to light. However, in alkaline solution, luminol is converted to its dianion, which may be oxidized by molecular oxygen to give an intermediate which is chemiluminescent. The reaction is thought to proceed according to the following sequence of events:

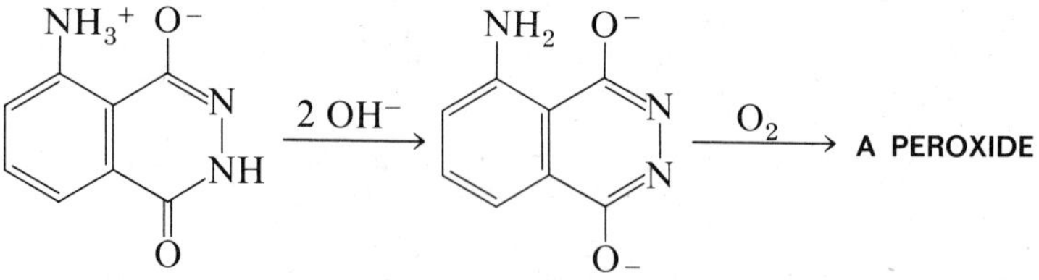

Luminol   Dianion

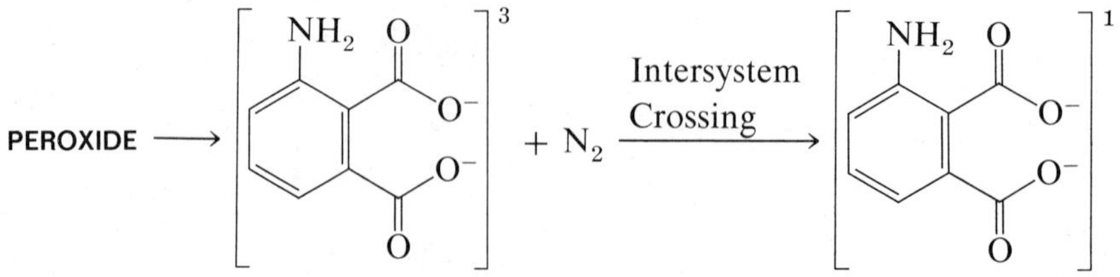

3-Aminophthalate
Triplet Dianion (T_1)

3-Aminophthalate
Singlet Dianion (S_1)

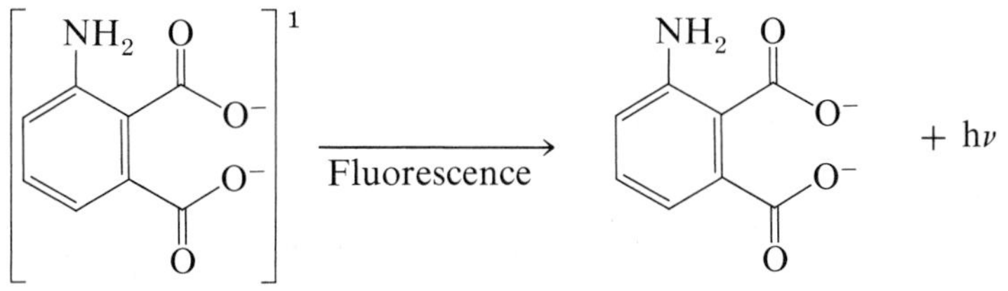

(S[1])

3-Aminophthalate
Ground State Dianion (S[0])

The dianion of luminol undergoes a reaction with molecular oxygen to form a peroxide of unknown structure. This peroxide is unstable and decomposes with the evolution of nitrogen gas, producing the 3-aminophthalate dianion in an electronically excited state. The excited dianion emits a photon which is observed as visible light. One very attractive hypothesis for the structure of the peroxide postulates a cyclic endoperoxide that decomposes by the following mechanism:

A POSTULATE

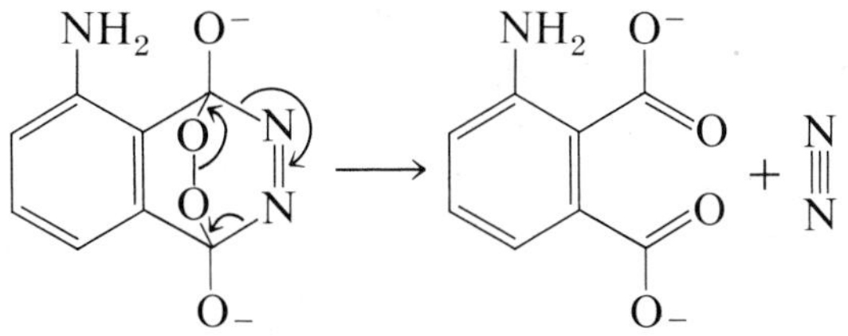

However, certain experimental facts argue against this intermediate. For instance, certain acyclic hydrazides which cannot form a similar intermediate have also been found to be chemiluminescent.

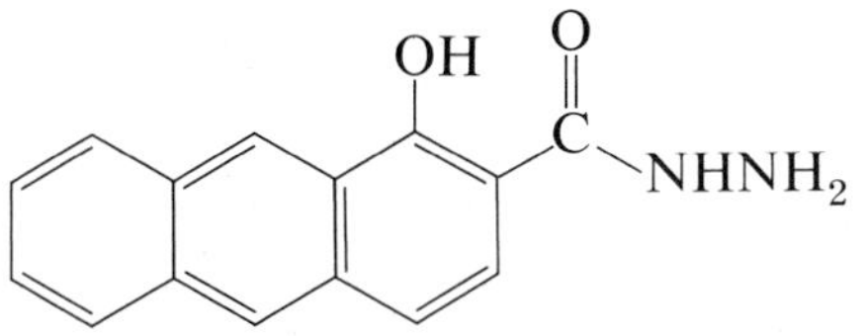

1-Hydroxy-2-anthroic Acid
Hydrazide (Chemiluminescent)

Although the nature of the peroxide is still a matter of debate, the remainder of the reaction is well understood. The chemical products of the reaction have been shown to be the 3-aminophthalate dianion and molecular nitrogen. The intermediate which emits light has been definitely identified as the **excited state singlet**[1] of the 3-aminophthalate dianion. Thus, the fluorescence emission spectrum of the 3-aminophthalate dianion (produced by photon absorption) is identical to the spectrum obtained from the light emitted from the chemiluminescent reaction. However, for a number of reasons which are fairly detailed, it is believed that the 3-aminophthalate dianion is first formed as a vibrationally excited **triplet state** molecule which makes the intersystem crossing to the singlet state before emission of a photon.

---

[1]The terms **singlet, triplet, intersystem crossing, energy transfer,** and **quenching** are explained in the introduction to Experiment 47.

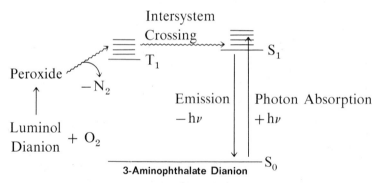

3-Aminophthalate Dianion

The excited state of the 3-aminophthalate dianion may be quenched by suitable acceptor molecules, or the energy (about 50 to 80 Kcal/mole) may be transferred to give emission from the **acceptor** molecules. Several such experiments are described in the experimental procedure of this experiment.

The system chosen for the chemiluminescence studies of luminol in this experiment will employ dimethylsulfoxide $(CH_3)_2SO$ as the solvent, potassium hydroxide as the base required for the formation of the dianion of luminol, and molecular oxygen. Several alternative systems have been used, substituting hydrogen peroxide and an oxidizing agent for molecular oxygen. An aqueous system employing potassium ferricyanide and hydrogen peroxide is an alternative system used frequently.

REFERENCES

Rahaut, M. M. "Chemiluminescence from Concerted Peroxide Decomposition Reactions." *Accounts of Chemical Research, 2* (1969), 80.

White, E. H., and Roswell, D. F. "The Chemiluminescence of Organic Hydrazides." *Accounts of Chemical Research, 3* (1970), 54.

SPECIAL INSTRUCTIONS

Before proceeding with this experiment, it will be helpful to have read the introduction to Experiment 47, which defines many of the photochemical terms. The essay "Fireflies and Photochemistry," which precedes this experiment, also contains many relevant details.

This entire experiment can be completed in less than one hour. When working with hydrazine, one should keep in mind that it is quite toxic and should not be spilled on the skin. Dimethylsulfoxide may also be toxic. One should avoid breathing the vapors or spilling it on the skin.

A darkened room is required to observe adequately the chemiluminescence of luminol. However, a darkened hood which has had its window covered with butcher paper or aluminum foil also works well. Fluorescent dyes other than those mentioned (for instance 9,10-diphenylanthracene) may also be used for the energy transfer experiments. The dyes selected may depend on what is immediately available. The instructor may choose to have each student use one dye for the energy transfer experiments, with one student performing a comparison experiment without a dye.

PROCEDURE

5-Nitrophthalhydrazide. Place 1.3 g of 3-nitrophthalic acid and 2 ml of a 10% aqueous solution of hydrazine[2] in a large side arm test tube. At the same time, heat 20 ml of water in a beaker on a steam bath. Heat the test tube over a microburner until the solid dissolves. Add 4 ml of triethylene glycol and clamp the test tube in an upright position on a ring stand. Insert a thermometer and a boiling stone into the test tube and attach a piece of pressure tubing to the side arm. Connect this tubing to an aspirator (use a trap). Heat the solution with a microburner until the liquid boils vigorously and the refluxing water vapor is drawn away by the aspirator vacuum (the temperature will rise to about 120°). Continue heating and allow the temperature to increase rapidly until it rises above 200°. About five minutes will be required for this heating. Remove the burner briefly when this temperature has been achieved, and then resume gentle heating in order to maintain a fairly constant temperature of 210 to 220° for about two minutes. Allow the test tube to cool to about 100°, add the 20 ml of hot water which was prepared previously, and cool the test tube to room temperature by allowing tap water to flow over the outside of the test tube. Collect the light yellow crystals of 5-nitrophthalhydrazide by vacuum filtration using a Hirsch funnel. It is not necessary to dry the product in order to proceed with the next reaction step.

Luminol (5-Aminophthalhydrazide). Transfer the moist 5-nitrophthalhydrazide to a 20 × 150 mm test tube. Add 6.5 ml of a 10% sodium hydroxide solution, and stir the mixture until the hydrazide dissolves. Add 4 g of sodium dithionite dihydrate (sodium hydrosulfite dihydrate, $Na_2S_2O_4 \cdot 2H_2O$). Using a capillary pipet, add sufficient water to wash the solid from the walls of the test tube. Heat the test tube until the solution boils, stir the solution, and maintain the boiling, with stirring, for 5 minutes. Add 2.6 ml of glacial acetic acid and cool the test tube to room temperature by allowing tap water to flow over the outside of it. Stir the mixture during this cooling step. Collect the light yellow crystals of luminol by vacuum filtration using a Hirsch funnel. Save a small sample of this product, allow it to dry overnight, and determine its melting point (mp 319 to 320°). The remainder of the luminol may be used without drying for the chemiluminescence experiments.

---

[2] A 10% aqueous solution of hydrazine may be prepared by diluting 15.6 g of a commercial 64% hydrazine solution to a volume of 100 ml using water.

FLUORESCENT DYE	COLOR
No Dye	Faint Bluish-White
9-Aminoacridine	Blue-Green
Eosin	Salmon Pink
Fluorescein	Yellow-Green
Dichlorofluorescein	Yellow-Orange
Rhodamine B	Green

Chemiluminescence Experiments.[3] Cover the bottom of a 250 ml Erlenmeyer flask with a layer of potassium hydroxide pellets. Add about 75 ml of dimethylsulfoxide and about 0.2 to 0.3 g of the moist luminol to the flask, stopper it, and shake it vigorously to mix air into the solution. In a dark room, a faint glow of blue-white light will be observed. The intensity of the glow will increase with continued shaking of the flask and occasional removal of the stopper to admit more air.

To observe energy transfer to a fluorescent dye, dissolve a few crystals (1 to 5 mg) of the indicator dye in 2 to 3 ml of water. Add the dye solution to the dimethylsulfoxide solution of luminol, stopper the flask, and shake the mixture vigorously. Observe the intensity and the color of the light which is produced.

A table of some dyes and the colors which are produced when mixed with luminol is given above. Other dyes not included on this list may also be tested in this experiment.

Experiment 49

trans, trans-1,4-DIPHENYL-1,3-BUTADIENE

Wittig Reaction
Handling Sodium Metal

The Wittig reaction is often used to form alkenes from carbonyl compounds. In this experiment a diene, *trans, trans*-1,4-diphenyl-1,3-butadiene, will be formed from cinnamaldehyde and the benzyltri-

---

[3] An alternative method for demonstrating chemiluminescence, using potassium ferricyanide and hydrogen peroxide as oxidizing agents, may be found in E. H. Huntress, L. N. Stanley, and A. S. Parker, *Journal of Chemical Education, 11* (1934), 142.

phenylphosphonium chloride Wittig reagent. The reaction is carried out in two steps. First the Wittig reagent (a salt) is formed by reaction of triphenylphosphine with benzyl chloride:

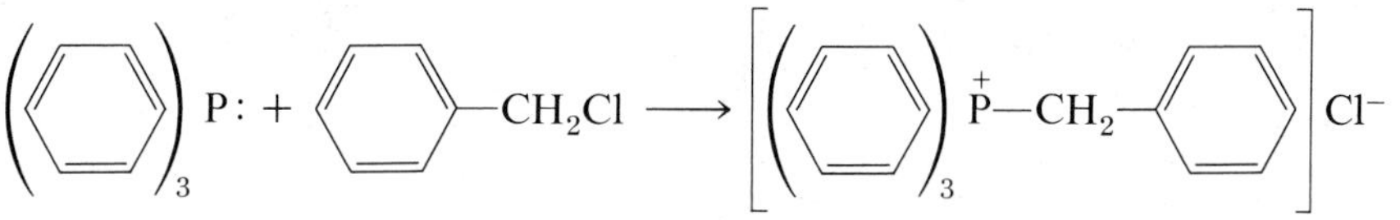

Benzyltriphenylphosphonium Chloride

The reaction is a simple nucleophilic displacement of chloride ion by triphenylphosphine. The salt formed is called the "Wittig reagent" or "Wittig salt."

When treated with base, the Wittig salt forms an **ylid.** An ylid is a specie having adjacent atoms oppositely charged.

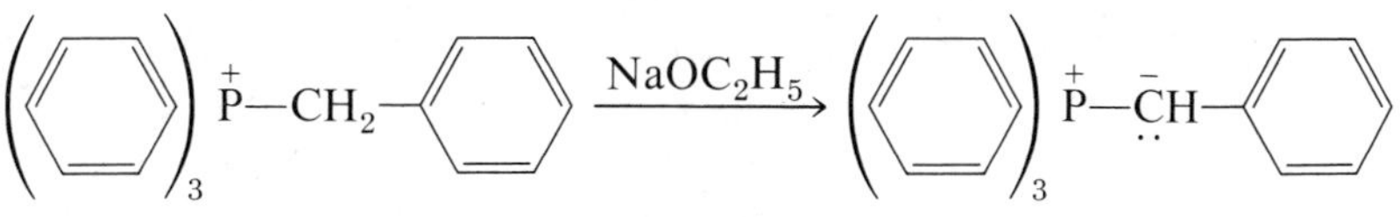

An Ylid

The ylid is stabilized due to the ability of phosphorus to accept more than eight electrons in its valence shell. Thus the carbanion center is stabilized by resonance.

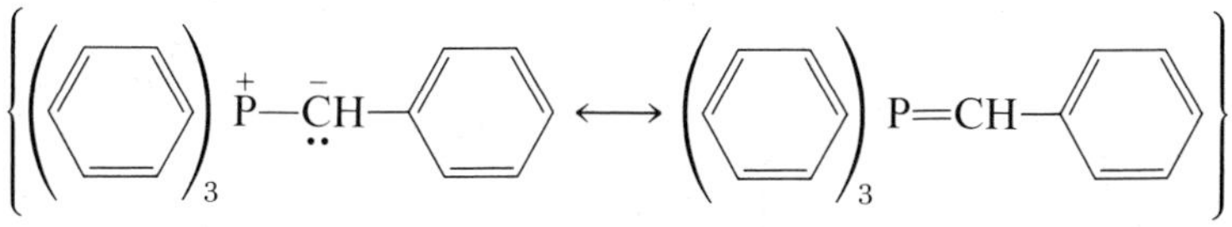

Phosphorus uses its 3d orbitals to form the overlap with the 2p orbital of carbon which is necessary for resonance stabilization.

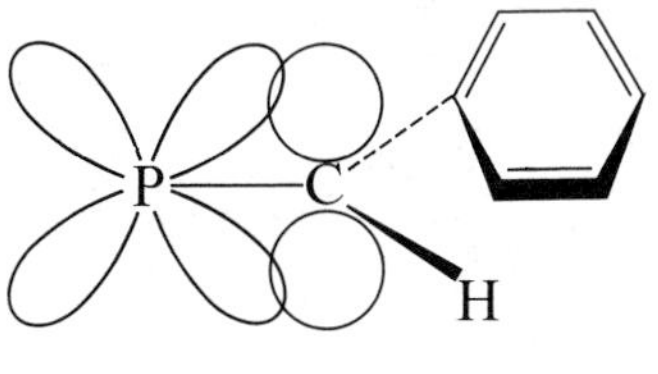

The ylid is a carbanion, and it is nucleophilic toward addition to a carbonyl group.

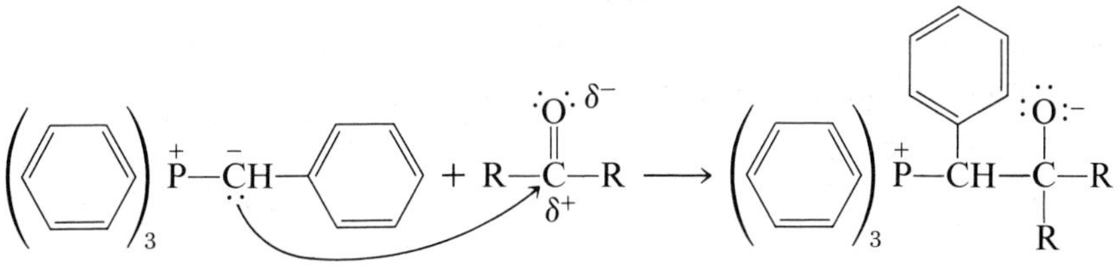

Following this initial nucleophilic addition, a remarkable sequence of events occurs, as outlined in the following mechanism.

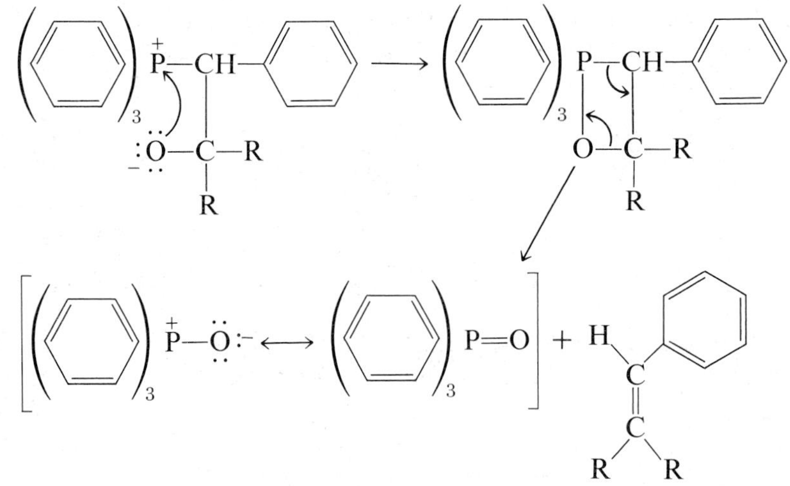

Triphenylphosphine Oxide   **An Alkene**

The addition intermediate, formed from the ylid and the carbonyl compound, cyclizes to form a four-membered-ring intermediate. This new intermediate is unstable and fragments into an alkene and triphenylphosphine oxide. Notice that the ring breaks open in a different way than it was formed. This is due primarily to the fact that triphenylphosphine oxide is thermodynamically a very stable compound. A large decrease in potential energy is achieved upon its formation.

In this experiment, cinnamaldehyde is used as the carbonyl component and gives *trans, trans*-1,4-diphenyl-1,3-butadiene as the product.

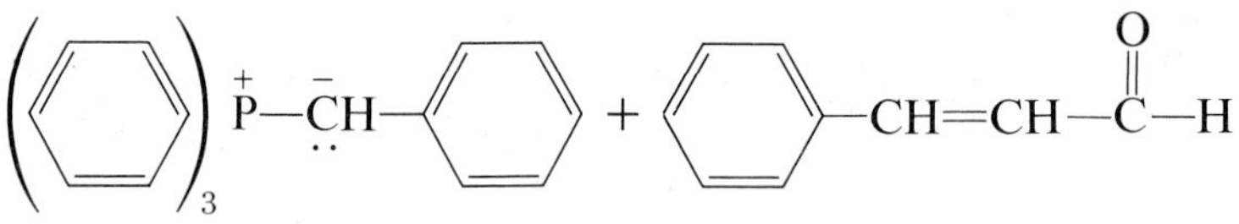

Cinnamaldehyde

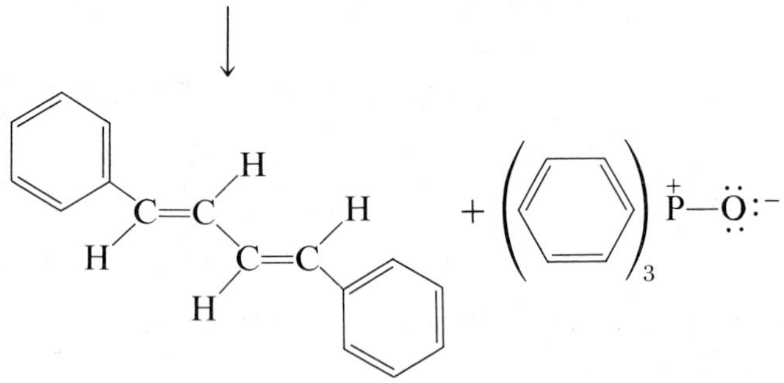

trans, trans-**1,4-Diphenyl-1,3-butadiene**

The *trans, trans* isomer is thermodynamically the most stable isomer of this compound, and it is formed preferentially.

SPECIAL INSTRUCTIONS

Before beginning this experiment it is essential to read Technique 14 thoroughly. Metallic sodium is a very hazardous material, and great care must be taken in handling it. Do not allow sodium to come in contact with water or with moisture. Triphenylphosphine is rather toxic. Be careful not to inhale the dust. Benzyl chloride is a skin irritant and a lachrymator. It should be handled in the hood with care.

PROCEDURE

Benzyltriphenylphosphonium chloride. Place 3.8 g of benzyl chloride, 11.0 g of triphenylphosphine, and 60 ml of *p*-cymene in a 500 ml round bottom, three neck flask fitted with a reflux condenser. Stopper the unused openings. The large flask prevents loss of material resulting from bumping when solid forms during the reaction. Add a boiling stone. Use a heating mantle as a heat source, heat the mixture under reflux for 2 1/2 hours. If a longer reflux period is available, the yield of product will be increased.

Cool the mixture and collect the phosphonium salt by vacuum filtration using a Büchner funnel. Since the *p*-cymene is somewhat expensive, place the filtrate in the container provided for recycling. The *p*-cymene will be recovered later by distillation. Reassemble the filtration apparatus and wash the crystals with a small amount of cold petroleum ether to remove the residual *p*-cymene. Dry the crystals, weigh them, and calculate the percentage yield of phosphonium salt.

trans, trans-**1,4-Diphenyl-1,3-butadiene.** On a ring stand, assemble an apparatus consisting of a dry 500 ml round bottom, three neck flask, a reflux condenser, and an addition (separatory) funnel. Stopper the unused neck of the flask and attach a calcium chloride drying tube to the top of the condenser. An illustration of a similar apparatus may be found in Experiment 28.

Dissolve the sample of benzyltriphenylphosphonium chloride prepared above in 75 ml of absolute ethanol and place the solution in the addition funnel. Stopper the funnel with a second calcium chloride drying tube to prevent the solution from absorbing moisture from the air. Weigh 0.7 g of sodium metal into a beaker of xylene and cut the sodium into small pieces (Technique 14, Section 14.4). Place 75 ml of absolute ethanol in the three neck flask and add the sodium slowly, one piece at a time, through the unused neck of the flask. Replace the stopper after each piece of sodium has been added. Use a tweezers or forceps to handle the sodium. Continue adding the sodium until it has all dissolved. Do not add the sodium at such a rate that it causes the solution to boil vigorously. A slight rate of boiling is acceptable. If the solution becomes warm during this sodium addition, allow it to cool before proceeding to the next step of the procedure.

Once the sodium ethoxide solution has been prepared, add 2.6 g of cinnamaldehyde, dissolved in 25 ml of absolute ethanol, to the flask. Then add the benzyltriphenylphosphonium chloride solution to the flask from the addition funnel. Add the phosphonium salt solution rather rapidly and swirl the contents of the flask by gently moving the ring stand and attached apparatus with a circular motion. Formation of the ylid will be indicated by a transient yellow or orange color. Once the phosphonium salt has been added, allow the reaction mixture to stand for 30 minutes.

After the standing period, crystals of the product should have formed. Collect the crystals by vacuum filtration using a Büchner funnel. Rinse the flask and the crystals with a little **cold** absolute ethanol. Allow the product to dry, weigh it, and calculate the percentage yield of the crude product. Recrystallize the crude material from absolute ethanol and determine the melting point of the pure crystals (literature value = 149 to 150°C). Submit the sample in a labeled vial to the instructor.

QUESTIONS

1. Why should the *trans, trans* isomer be the thermodynamically most stable one?

2. Why is *p*-cymene used for the formation of the phosphonium salt, rather than another solvent, e.g., benzene?

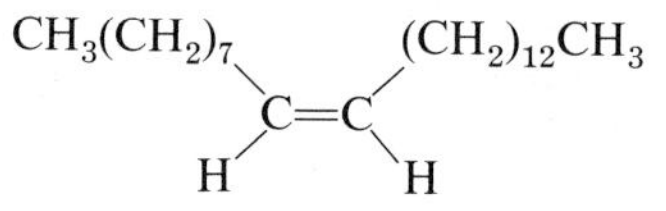

$$CH_3(CH_2)_7\diagdown \qquad \diagup(CH_2)_{12}CH_3$$
$$C{=}C$$
$$H\diagup \qquad \diagdown H$$

Muscalure

3. The sex attractant of the female housefly (**Musca domestica**) is called **muscalure,** and its structure is shown above. Outline a synthesis of muscalure, using the Wittig reaction. Will your synthesis lead to the required **cis** isomer?

Experiment **50**
IDENTIFICATION OF UNKNOWNS

Qualitative organic analysis, the identification and characterization of unknown compounds, is an important part of organic chemistry. Every chemist must learn the appropriate methods used to establish the identity of a compound. In this experiment you will be issued an unknown compound and be asked to identify it through chemical and spectroscopic methods. Your instructor may choose to give you a general unknown or a specific unknown. With a **general unknown,** you must first determine the class of compound to which the unknown belongs, that is, identify its major functional group; and then you must determine the specific compound in that class which corresponds to the unknown. With a **specific unknown,** the class of compound (ketone, alcohol, amine, etc.) will be known in advance, and it will only be necessary to determine which specific member of that class was issued to you as an unknown. This experiment is designed so that the instructor may issue several general unknowns or as many as six successive specific unknowns, each having a different main functional group.

Although there are well over a million organic compounds that an organic chemist might be called upon to identify, the scope of this experiment is more limited. Just over 300 compounds are included in the tables of possible unknowns given for the experiment (see Appendix One). In addition, the experiment is restricted to include only seven major functional groups:

Aldehydes	Amines
Ketones	Alcohols
Carboxylic Acids	Esters
Phenols	

Even though this list of possible functional groups omits some of the important types of compounds (alkyl halides, alkenes, alkynes, aromatics, ethers, amides, mercaptans, nitriles, acid chlorides, acid anhydrides, nitro compounds, etc.), the methods which are introduced here can be applied equally well to these other classes of compounds. The list chosen is sufficiently broad to illustrate all of the principles involved in the identification of an unknown compound.

In addition, although many of the functional groups listed as being excluded will not appear as the **major** functional group in a compound, several of them will frequently appear as **secondary** or subsidiary functional groups. Three examples of this are listed below.

$$Br-\langle\text{C}_6\text{H}_4\rangle-\overset{\displaystyle O}{\overset{\|}{C}}-CH_3 \qquad O_2N-\langle\text{C}_6\text{H}_4\rangle-OH \qquad CH_3O-\langle\text{C}_6\text{H}_4\rangle-CH=CH-CHO$$

MAJOR:	KETONE	PHENOL	ALDEHYDE
SUBSIDIARY:	Halide	Nitro	Alkene Aromatic
	Aromatic	Aromatic	Ether

Those groups which have been included in a subsidiary fashion include:

—Cl	Chloro	—NO$_2$	Nitro	C=C	Double Bond
—Br	Bromo	—C≡N	Cyano	C≡C	Triple Bond
—I	Iodo	—OR	Alkoxy	⬡	Aromatic

The experiment presents all of the major chemical and spectroscopic methods of determining the **major** functional groups, and it includes methods for verifying the presence of the **subsidiary** functional groups as well. In most cases, it will not be necessary to determine the presence of the subsidiary functional groups in order to correctly identify the unknown compound. However, **every** piece of information helps the identification, and, if these groups can be detected easily, one should not hesitate to do so. Finally, complex bifunctional compounds will generally be avoided in this experiment. Only a few of these are included.

HOW TO PROCEED

Fortunately, one can detail a fairly straightforward procedure for determining all the necessary pieces of information. This procedure consists of the following steps:

PART ONE

1. Preliminary classifications as to physical state, color, and odor
2. Melting point or boiling point determination; other physical data
3. Purification, if necessary
4. Determination of solubility behavior in water, and in acids and bases
5. Simple preliminary tests: Beilstein, ignition (combustion)
6. Application of relevant chemical classification tests

PART TWO

7. Determination of infrared and nmr spectra
8. Elemental analysis, if necessary
9. Preparation of derivatives
10. Confirmations of identity

Each of these steps is discussed briefly in the sections below.

PRELIMINARY CLASSIFICATIONS

One should take note of the physical characteristics of the unknown. These include its color, its odor, and its physical state (liquid, solid, crystalline form). Many compounds have characteristic colors or odors, or they crystallize with a specific crystal structure. This type of information can often be found in a handbook, and can be checked later. Compounds with a high degree of conjugation are frequently yellow to red in color. Amines often have a fishlike odor. Esters have a pleasant fruity or floral odor. Acids have a sharp and pungent odor. A part of the training of every good chemist includes a cultivation of the ability to recognize familiar or typical odors. As a note of caution, many compounds have distinctly unpleasant or nauseating odors. Some have corrosive vapors. Any unknown substance should be sniffed with the greatest caution. As a first step, open the container, hold it away from you, and using your hand, carefully waft the vapors toward your nose. If you get past this stage, a closer inspection will be possible.

MELTING POINT OR BOILING POINT DETERMINATION

The single most useful piece of information to have for an unknown compound is its melting point or boiling point. Either piece of data will drastically limit the compounds which are possible. The electric melting point apparatus may be used for a rapid and accurate measurement (see Technique 4, Sections 4.5 and 4.6). To save time, it is often helpful to determine two separate melting points. The first determination may be performed rapidly to obtain an approximate value. Then, the second melting point may be determined more carefully.

The boiling point is easily obtained by a simple distillation of the unknown (Technique 6, Section 6.4) or by a micro boiling point determination (Technique 6, Section 6.2). The simple distillation has the advantage that it will also purify the compound. The smallest distilling flask available should be used if a simple distillation is performed, and one should make certain that the thermometer bulb is fully immersed in the vapor of the distilling liquid. To obtain an accurate boiling point value, the liquid should be distilled fairly rapidly.

If the solid is high melting ($>200°$), or the liquid high boiling ($>200°$), a thermometer correction may be required (Technique 4, Section 4.8 and Technique 6, Section 6.8). In any event, allowance should be made for errors of as large as $\pm5°$ in these values.

PURIFICATION

If the melting point of a solid has a wide range (>4 to $5°$), it should be recrystallized and the melting point redetermined.

If a liquid was highly colored before distillation, if it yielded a wide boiling point range, or if the temperature did not hold constant during the distillation, it should be redistilled a second time to determine a new temperature range.

SOLUBILITY BEHAVIOR

These tests are described fully in Procedure 50A. They are extremely important. The solubility of small amounts of the unknown in water, 5% HCl, 5% $NaHCO_3$, 5% NaOH, concentrated H_2SO_4, and organic solvents is determined. This information reveals whether a compound is an acid, a base, or a neutral substance. The sulfuric acid test reveals if a neutral compound has a functional group that contains an oxygen, a nitrogen, or a sulfur atom that can be protonated. This information allows one to eliminate or to choose various functional group possibilities. The solubility tests must be performed on **all** unknowns.

PRELIMINARY TESTS

The two combustion tests, the Beilstein test (Procedure 50B) and the ignition test (Procedure 50C), are easy and quick to perform, and they often give valuable information. It is recommended that they be performed on all unknowns.

CHEMICAL CLASSIFICATION TESTS

The solubility tests will usually suggest or eliminate several possible functional groups. The chemical classification tests which are listed in Procedures 50D to 50I allow one to distinguish between the possible choices. Choose only those tests which the solubility tests suggest might be meaningful. Time will be wasted doing unnecessary tests. There is no substitute for a first hand thorough knowledge of these tests. Each of the sections should be studied carefully until the significance of each test is understood. In addition, it will be helpful to actually try the tests on **known** substances. In this way, it will be easier to recognize a positive test when it occurs. Appropriate test compounds are listed for many of the tests. When performing a test which is new to you, it is always a good practice to run the test separately on both a known substance and the unknown **at the same time.** This allows for a direct comparison of the results.

Once the melting or boiling point, the solubilities, and the major chemical tests have been performed, it will be possible to identify the major class of the compound. At this stage, using the melting point or boiling point as a guide, it will be possible to compile a list of possible compounds. Inspection of this list will suggest additional tests which must be performed to distinguish among the possibilities. For instance, one compound may be a methyl ketone and the other may not. The iodoform test is called for to distinguish between the two possibilities. The tests for the subsidiary functional groups may also be required. These are described in Procedures 50B and 50C. These tests should also be studied carefully and, again, there is no substitute for a first hand knowledge of them.

One should not perform the chemical tests either willy-nilly or in a methodical comprehensive sequence. One should use the tests selectively. Solubility tests will automatically eliminate the need to perform some of the chemical tests. Each successive test will either eliminate the need for or dictate the use of another test. In addition, one should examine the tables of unknowns carefully. The boiling point or the melting point of the unknown may eliminate the need for many of the tests. For instance, the possible compounds may simply not include one with a double bond. Efficiency is the keyword here. You should not waste time in doing nonsensical or unnecessary tests. Many possibilities may be eliminated on the basis of logic alone.

The manner in which you proceed with the following steps may be limited by your instructor's wishes. Many instructors may wish to restrict your access to infrared and nmr spectra until you have narrowed your choices to a few compounds **all within the same class.** Others may wish you to determine this data routinely. Some instructors may wish students to perform elemental analysis on all unknowns; others may wish to restrict it to only the most essential situations. Most unknowns can be solved without either spectroscopy or elemental analysis. Again, some instructors may require derivatives as a final confirmation of the compound's identity, others may wish not to use them at all.

SPECTROSCOPY

Spectroscopy is probably the most powerful and modern tool available to the chemist for determining the structure of an unknown compound. It is often possible to determine the structure of a compound through the use of spectroscopy alone. On the other hand, there are also situations where spectroscopy is not of much help and where the traditional methods must be relied upon. For this reason, you should not use spectroscopy to the exclusion of the more traditional tests, but rather as a confirmation of those results. Nevertheless, the major functional groups and their immediate environmental features can be determined quickly and accurately with spectroscopy.

ELEMENTAL ANALYSIS

These tests, which allow one to determine the presence of nitrogen, sulfur, or a specific halogen atom (Cl, Br, I) in a compound are often useful. However, many times other information renders these tests unnecessary. A compound identified as an amine on the basis of solubility tests obviously contains nitrogen. Many nitrogen-containing groups (for instance, nitro groups) can be identified by infrared spectroscopy. Finally, it is not usually necessary to identify a specific halogen. The simple information that the compound contains a halogen (any halogen) may be enough information to distinguish between two compounds. A simple Beilstein test will provide this information.

DERIVATIVES

One of the major tests of a correct identification of an unknown compound comes when one tries to convert the compound by means of a chemical reaction to another known compound. This second compound is called a **derivative.** The best derivatives are solid compounds, since the melting point of a solid provides an accurate and reliable identification of most compounds. Solids are also easily purified through crystallization. The derivative can serve to distinguish between two otherwise very similar compounds. Usually they will have derivatives (both prepared by the same reaction) which give different melting points. Tables of unknowns and derivatives are listed in Appendix One at the end of the book. Procedures for preparing derivatives are given in Appendix Two.

CONFIRMATIONS OF IDENTITY

A rigid and final test for the identity of an unknown can be made if an "authentic" sample of the compound is available for comparison. One can compare infrared and nmr spectra of the unknown compound to that of the known compound. If the spectra match, peak for peak, then the identity is probably certain. Other physical and chemical properties may also be compared. If the compound is a solid, a convenient test is the mixed melting point (Technique 4, Section 4.4).

While we cannot be complete in this experiment in terms of the functional groups covered, or the tests described, the experiment should give a good introduction to the methods and the techniques used by chemists to identify unknown compounds. Texts which cover the subject more thoroughly are listed in the references. The student is encouraged to consult these texts for more information, including more specific methods and further classification tests.

REFERENCES

COMPREHENSIVE TEXTBOOKS

Cheronis, N. D., and Entrikin, J. B. *Identification of Organic Compounds*. New York: Wiley-Interscience, 1963.

Pasto, D. J., and Johnson, C. R. *Organic Structure Determination*. Englewood Cliffs, New Jersey: Prentice-Hall, 1969.

Shriner, R. L., Fuson, R. C., and Curtin, D. Y. *The Systematic Identification of Organic Compounds* (5th edition). New York: Wiley, 1964.

SPECTROSCOPY

Bellamy, L. J. *The Infra-red Spectra of Complex Molecules* (2nd edition). New York: Wiley, 1958.

Dyer, J. R. *Applications of Absorption Spectroscopy of Organic Compounds*. Englewood Cliffs, New Jersey: Prentice-Hall, 1965.

Nakanishi, K. *Infrared Absorption Spectroscopy*. San Francisco: Holden-Day, 1962.
Silverstein, R. M., and Bassler, G. C. *Spectrometric Identification of Organic Compounds* (2nd edition). New York: Wiley, 1967.

MORE EXTENSIVE TABLES OF COMPOUNDS AND DERIVATIVES

Rappoport, Z., editor. *Handbook of Tables for Organic Compound Identification*. Cleveland, Ohio: Chemical Rubber Company, 1967.

Procedure **50**A

SOLUBILITY TESTS

Solubility tests should be performed on **every unknown.** They are extremely important in determining the nature of the major functional group present in the unknown compound. The tests are very simple and require only small amounts of the unknown compound. In addition, the solubility tests will reveal whether the compound is a strong base (amine), a weak acid (phenol), a strong acid (carboxylic acid), or a neutral substance (aldehyde, ketone, alcohol, ester). The common solvents used to determine solubility types are:

5% HCl	Concentrated H_2SO_4
5% $NaHCO_3$	Water
5% NaOH	Organic Solvents

The solubility chart on page 389 indicates solvents in which compounds containing the various functional groups are likely to dissolve. The summary charts that are given in sections 50D through 50I repeat this information for each functional group included in this experiment. In this section, the correct procedure for determining whether a compound is soluble in a test solvent is given. Also given is a series of explanations detailing the reasons why compounds having specific functional groups are soluble in only specific solvents. This is accomplished by indicating the type of chemistry or the type of chemical interaction that is possible in each solvent.

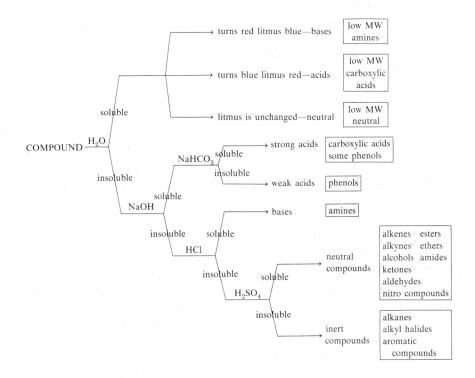

SOLUBILITY TESTS

Procedure. Place about 1 ml of the solvent in a small test tube. Add one drop of an unknown liquid from an eyedropper, or a few crystals of an unknown solid from the end of a spatula, directly into the solvent. Gently tap the test tube with your finger to assure mixing, and then observe whether any mixing lines appear in the solution. The disappearance of the liquid or solid, or the appearance of the mixing lines, indicates that solution is taking place. Add several more drops of the liquid, or a few more crystals of the solid, to determine the extent of the compound's solubility. A common mistake made in determining the solubility of a compound is testing with a quantity of the unknown too large to dissolve in the chosen solvent. Use small amounts. Solids may require several minutes to dissolve. Compounds in the form of large crystals will require more time to dissolve than powders or very small crystals. In some cases it is helpful to pulverize a compound with large crystals in a mortar and pestle. Gentle heating is helpful in some instances, but strong heating is to be discouraged, as it often leads to reaction. When colored compounds dissolve, the solution will often assume the color.

Using the procedure given above, the solubility of the unknown should be determined in each of the following solvents: water, 5% NaOH, 5% $NaHCO_3$, 5% HCl, and concentrated H_2SO_4. With sulfuric acid, a color change may be observed rather than solution. A color change should be regarded as a positive solubility test. Solid unknowns which do not dissolve in any of the test solvents may be inorganic substances. To eliminate this possibility, the solubility of the unknown in several organic solvents, like ether, should be determined. If the compound is organic, a solvent can usually be found which will dissolve it.

If a compound is found to dissolve in water, the pH of the aqueous solution should be estimated by using pH paper or litmus. Compounds soluble in water will usually be soluble in **all** of the aqueous solvents. If a compound is only slightly soluble in water, it may be **more** soluble in another of the aqueous solvents. For instance, a carboxylic acid may be only slightly soluble in water, but very soluble in dilute base. In many cases it will not be necessary to determine the solubility of the unknown in every solvent.

Test Compounds. Five solubility unknowns will be found on the side shelf. The five unknowns include a base, a weak acid, a strong acid, a neutral substance with an oxygen-containing functional group, and a neutral substance which is inert. Using solubility tests, distinguish these unknowns as to type. Verify your answer with the instructor.

SOLUBILITY IN WATER

Compounds with four or less carbons, and which contain oxygen, nitrogen, or sulfur, will often be soluble in water. Almost any functional group containing these elements will lead to water solubility for low molecular weight (C_4) compounds. Compounds having five or six carbons with those elements will often be insoluble in water or have borderline solubility. Branching of the alkyl chain in a compound lowers the intermolecular forces between its molecules. This is usually reflected in a lowered boiling point or melting point and a greater solubility in water for the branched compound than for the corresponding straight chain compound. This occurs simply because the molecules of the branched compound are more easily separated from one another. Thus, *t*-butyl alcohol would be expected to be more soluble in water than *n*-butyl alcohol.

When the ratio of the oxygen, nitrogen or sulfur atoms in a compound to the carbon atoms is increased, the solubility of that compound in water will often increase. This is due to the increased number of polar functional groups. Thus, 1,5-pentanediol would be expected to be more soluble in water than would 1-pentanol.

As the size of the alkyl chain of a compound is increased beyond about four carbons, the influence of a polar functional group is diminished and the water solubility begins to decrease. A few examples of these generalizations are given below.

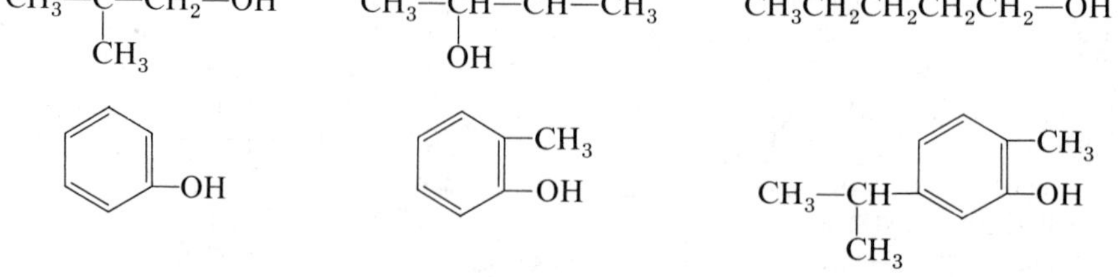

SOLUBILITY IN 5% HCl

The possibility of an amine should be considered immediately if a compound is soluble in dilute acid (5% HCl). Aliphatic amines (RNH_2, R_2NH, R_3N) are basic compounds which readily dissolve in acid because they form hydrochloride salts that are soluble in the aqueous medium.

$$R-NH_2 + HCl \longrightarrow R-NH_3^+ + Cl^-$$

The substitution of an aromatic ring (Ar) for an alkyl group (R) reduces the basicity of an amine somewhat, but the amine will still protonate, and it will still generally be soluble in dilute acid. The reduction in basicity in an aromatic amine is due to the resonance delocalization of the unshared electrons on the amino nitrogen of the free base. The delocalization is lost on protonation, a problem which does not exist for aliphatic amines. The substitution of two or three aromatic rings on an amine nitrogen reduces the basicity of the amine even further. Di- and triaryl amines will not dissolve in dilute HCl since they do not protonate easily. Thus, Ar_2NH and Ar_3N are insoluble in dilute acid. Some very high molecular weight amines, like tribromoaniline (MW 330) may also be insoluble in dilute acid.

| AROMATIC AMINE | Delocalization | No Delocalization |

| ALIPHATIC AMINE | $R-NH_2$ | $R-NH_3^+$ |
| | No Delocalization | No Delocalization |

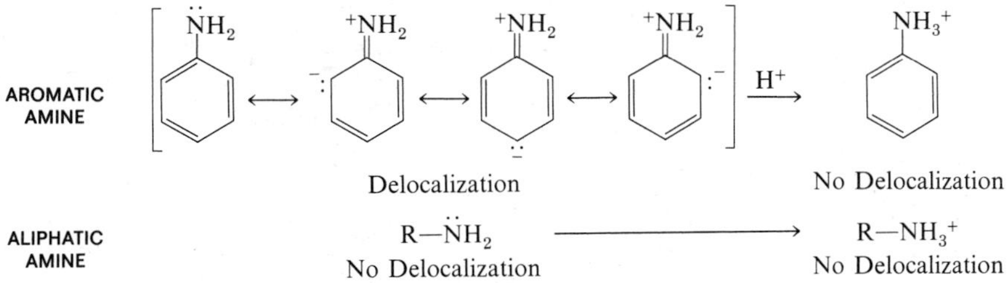

SOLUBILITY IN 5% NaHCO₃ AND 5% NaOH

Compounds which dissolve in sodium bicarbonate, a weak base, are strong acids. Compounds which dissolve in the sodium hydroxide, a strong base, may be either strong or weak acids. Thus, one can distinguish between weak and strong acids by determining their solubility in both strong (NaOH) and weak (NaHCO₃) base. The classification of some functional groups as either weak or strong acids is given in the accompanying table.

STRONG ACIDS soluble in both NaOH and NaHCO₃	WEAK ACIDS soluble in NaOH, but not NaHCO₃
Sulfonic Acids RSO_3H Carboxylic Acids $RCOOH$ Ortho and para substituted di- and trinitrophenols 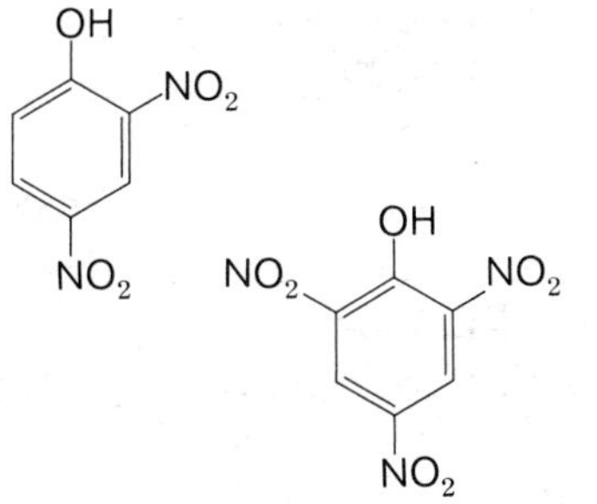	Phenols $ArOH$ Nitroalkanes RCH_2NO_2 R_2CHNO_2 β-Diketones β-Diesters Imides 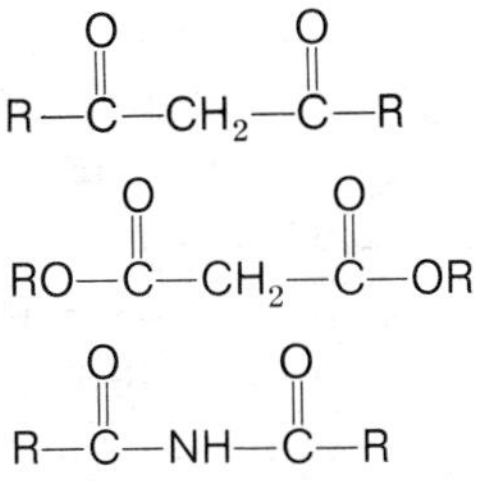 Sulfonamides $ArSO_2NH_2$ $ArSO_2NHR$

For the purposes of this experiment, carboxylic acids ($pK_a \sim 5$) are generally indicated when a compound is soluble in both bases, while phenols ($pK_a \sim 10$) are indicated when it is soluble in NaOH only.

Compounds dissolve in base because they form sodium salts which are soluble in the aqueous medium. However, the salts of some high molecular weight compounds are not soluble and precipitate. The salts of the long chain carboxylic acids, such as myristic (C_{14}), palmitic (C_{16}), and stearic (C_{18}) acids, which form soaps, are in this category. Some phenols also produce insoluble sodium salts, and often these are colored due to resonance in the anion.

Both phenols and carboxylic acids produce resonance stabilized conjugate bases. Thus, bases of the appropriate strength may easily remove their acidic protons to form the sodium salts.

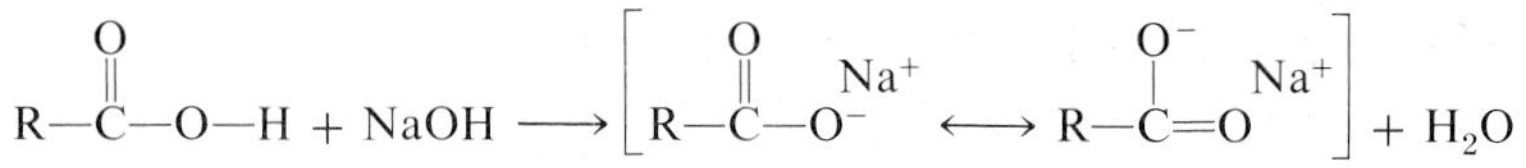

Delocalized Anion

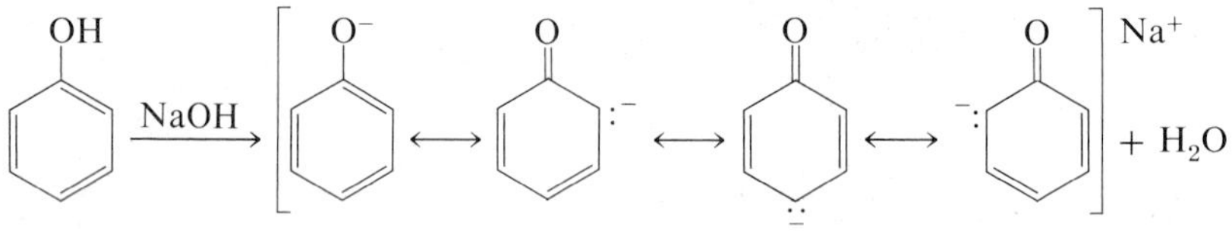

Delocalized Anion

In phenols, substitution of nitro groups in the ortho and para positions of the ring increases the acidity. Nitro groups in these positions provide additional delocalization in the conjugate anion. Phenols which have two or three nitro groups in the ortho and para positions will often dissolve in **both** sodium hydroxide and sodium bicarbonate solutions.

SOLUBILITY IN CONCENTRATED SULFURIC ACID

Many compounds are soluble in cold, concentrated sulfuric acid. Of the compounds included in this experiment, alcohols, ketones, aldehydes, and esters are in this category. Other compounds which also dissolve include alkenes, alkynes, ethers, nitroaromatics, and amides. Since several different types of compounds are soluble in sulfuric acid, further chemical tests and spectroscopy will be required to differentiate between them.

Compounds which are soluble in concentrated sulfuric acid, but not soluble in dilute acid, are extremely weak bases. Almost any compound containing a nitrogen, oxygen, or sulfur atom can be protonated in concentrated sulfuric acid. The ions which are produced are soluble in the medium.

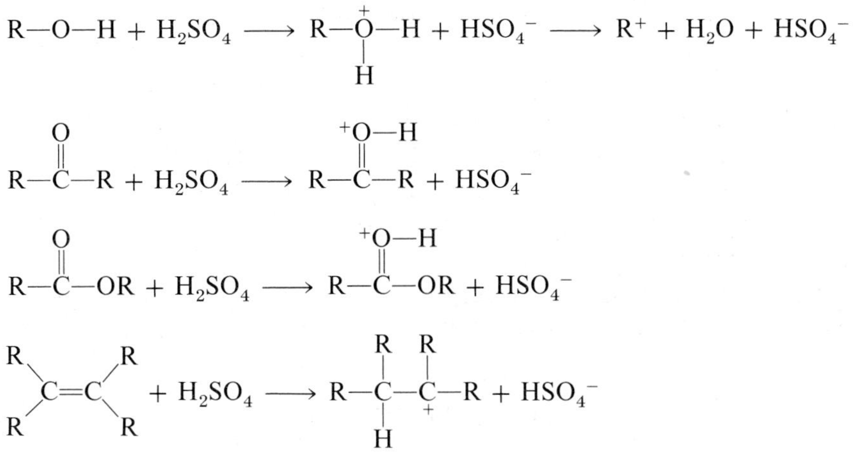

$$R\text{—}O\text{—}H + H_2SO_4 \longrightarrow R\overset{+}{\underset{\underset{H}{|}}{O}}\text{—}H + HSO_4^- \longrightarrow R^+ + H_2O + HSO_4^-$$

$$R\overset{\overset{O}{\parallel}}{\text{—}C}\text{—}R + H_2SO_4 \longrightarrow R\overset{\overset{+O\text{—}H}{\parallel}}{\text{—}C}\text{—}R + HSO_4^-$$

$$R\overset{\overset{O}{\parallel}}{\text{—}C}\text{—}OR + H_2SO_4 \longrightarrow R\overset{\overset{+O\text{—}H}{\parallel}}{\text{—}C}\text{—}OR + HSO_4^-$$

$$\underset{R}{\overset{R}{\diagdown}}C{=}C\underset{R}{\overset{R}{\diagup}} + H_2SO_4 \longrightarrow R\overset{\overset{R}{|}}{\underset{\underset{H}{|}}{C}}\overset{\overset{R}{|}}{\underset{\underset{+}{}}{C}}\text{—}R + HSO_4^-$$

INERT COMPOUNDS

Compounds not soluble in concentrated sulfuric acid or any of the other solvents are said to be **inert.** Compounds which are not soluble in concentrated sulfuric acid include the alkanes, most simple aromatics, and the alkyl halides. Some examples of inert compounds are: hexane, benzene, chlorobenzene, chlorohexane, and toluene.

Procedure **50**B

TESTS FOR THE ELEMENTS (N,S,X)

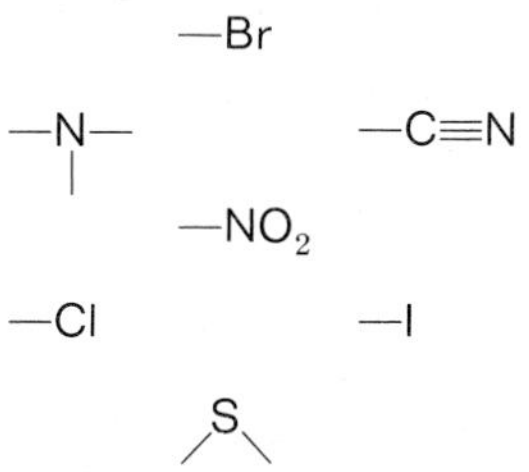

$$\begin{array}{ccc} & \text{—Br} & \\ \text{—N—} & & \text{—C}{\equiv}\text{N} \\ | & \text{—NO}_2 & \\ \text{—Cl} & & \text{—I} \\ & \overset{S}{\diagup\diagdown} & \end{array}$$

Other than amines (Procedure 50G), which are easily detected by their solubility behavior, all other compounds issued in this experiment will contain heteroelements (N,S,Cl,Br,I) only as **secondary** functional groups. They will be subsidiary to some other important functional group. Thus, no alkyl or aryl halides, nitro compounds, thiols, or thioethers will be issued. However, some of the unknowns may contain a halogen or a nitro group. Less frequently, they may contain a sulfur atom or a cyano group.

As an example, consider *p*-bromobenzaldehyde, an **aldehyde** which contains bromine as a ring substituent. The identification of this compound would hinge on the ability of the investigator to identify it as an aldehyde. It could probably be identified **without** proving the existence of bromine in the molecule. That information, however, could make the identification somewhat easier. In this experiment, methods are given for identifying the presence of a halogen or a nitro group in an unknown compound. Also given is a general method (Sodium Fusion) for detecting the principal heteroelements that may exist in organic molecules.

CLASSIFICATION TESTS

HALIDES	NITRO GROUPS	N,S,X (=Cl,Br,I)
Beilstein Test Silver Nitrate Sodium Iodide/Acetone	Ferrous Hydroxide	Sodium Fusion

TESTS FOR THE PRESENCE OF A HALIDE

BEILSTEIN TEST

Procedure. Bend a small loop in the end of a short length of copper wire. Heat the loop end of the wire in a Bunsen burner flame. After cooling, dip the wire directly into a small sample of the unknown. Now, heat the wire in the Bunsen burner flame again. The compound will first burn. After burning, a green flame will be produced if a halogen is present.

Test Compounds. Try this test on bromobenzene and benzoic acid.

The presence of halogen can be detected easily and fairly reliably by the Beilstein test. It is the simplest method to determine the presence of a halogen, but it does not differentiate between chlorine, bromine, and iodine, any one of which will give a positive test. However, in cases where the identity of the unknown has been narrowed to two choices, one of which has a halogen and one which does not, the Beilstein test will often be sufficient to distinguish between the two.

A positive Beilstein test results from the production of a volatile copper halide when an organic halide is heated with copper oxide. The copper halide imparts a blue-green color to the flame.

SILVER NITRATE TEST

Procedure. Add one drop of a liquid or 5 drops of a concentrated ethanolic solution of a solid unknown to 2 ml of a 2% ethanolic silver nitrate solution. If no reaction is observed after 5 minutes at room temperature, heat the solution on a steam bath and note whether a precipitate is formed. If a precipitate is formed, add 2 drops of 5% nitric acid and note whether the precipitate dissolves. Carboxylic acids give a false test by precipitating in silver nitrate, but they dissolve upon addition of nitric acid. Silver halides, on the other hand, do not dissolve in nitric acid.

Test Compounds. Apply this test to benzyl bromide (α-bromotoluene) and bromobenzene. Discard all waste reagents down a hood sink since benzyl bromide is a lachrymator.

This test depends upon the formation of a white or off-white precipitate of silver halide when silver nitrate is allowed to react with a sufficiently reactive halide.

$$RX + Ag^+ NO_3^- \longrightarrow \underset{\text{Precipitate}}{AgX} + R^+ NO_3^-$$

The test does not distinguish between chlorides, bromides, and iodides, but it does distinguish **labile** (reactive) halides, from those which are unreactive. Halides substituted on an aromatic ring will not usually give a positive silver nitrate test; however, alkyl halides of many types will give a positive test.

The most reactive compounds are those able to form stable carbonium ions in solution, and those which are equipped with good leaving groups (X = I, Br, Cl). Benzyl, allyl, and tertiary halides give immediate reaction with silver nitrate. Secondary and primary halides do not react at room temperature, but readily react when heated. Aryl and vinyl halides do not react at all, even at elevated temperatures. This pattern of reactivity fits the stability order for various carbonium ions quite well. Those compounds which produce stable carbonium ions react at faster rates than those which do not.

The fast reaction of benzylic and allylic halides is a result of the resonance stabilization that is available to the intermediate carbonium ions formed. Tertiary halides are more reactive than secondary halides, which are in turn more reactive than primary or methyl halides, due to the fact that alkyl substituents are able to stabilize the intermediate carbonium ions by an electron-releasing effect. The methyl carbonium ion has no alkyl groups, and is the least stable of all those mentioned thus far. Vinyl and aryl carbonium ions are extremely unstable since the charge is localized on an sp^2 hybridized carbon (double bond carbon) rather than on one which is sp^3 hybridized.

SODIUM IODIDE IN ACETONE

Procedure. This test is described in Experiment 22.

THE DETECTION OF NITRO GROUPS

Although nitro compounds will not be issued as distinct unknowns, many of the unknowns may have a nitro group as a secondary functional group. The presence of a nitro group, and hence nitrogen, in an unknown compound is determined most easily by means of infrared spectroscopy. However, many nitro compounds give a positive result in the following test.

FERROUS HYDROXIDE

Procedure. Place 1.5 ml of freshly prepared 5% aqueous ferrous ammonium sulfate in a small test tube and add about 10 mg of the unknown compound. Mix the solution well and then add, first one drop of 3 N sulfuric acid, and then 1 ml of 2 N potassium hydroxide in methanol. Stopper the test tube and shake it vigorously. A positive test is indicated by the formation of a red-brown precipitate, usually within 1 minute.

Most nitro compounds will oxidize ferrous hydroxide to ferric hydroxide, which is a red-brown solid. Formation of the precipitate indicates a positive test.

$$R\text{—}NO_2 + 4\ H_2O + 6\ Fe(OH)_2 \longrightarrow R\text{—}NH_2 + 6\ Fe(OH)_3$$

SPECTROSCOPY

INFRARED

The nitro group gives two strong bands near 1560 and 1350 cm^{-1} (6.4 and 7.4 μ).

DETECTION OF A CYANO GROUP

Although nitriles will not be given as unknowns in this experiment, the cyano group may be a subsidiary functional group whose presence or absence may be important to the final identification of an unknown compound. The cyano group can be hydrolyzed in strong base, with vigorous heating, to give a carboxylic acid and ammonia gas.

$$R\text{—}C\equiv N + 2\ H_2O \xrightarrow[\Delta]{NaOH} R\text{—}COOH + NH_3$$

The ammonia can be detected by its odor or by using moist pH paper. However, this method is somewhat difficult, and the presence of a nitrile group is confirmed most easily by the use of infrared spectroscopy. No other functional groups (except some $C\equiv C$) absorb in the same region of the spectrum as $C\equiv N$.

SPECTROSCOPY

INFRARED

$C\equiv N$ Stretch is a very sharp band of medium intensity near 2250 cm^{-1} (4.5 μ).

SODIUM FUSION TESTS (OPTIONAL)
(Detection of N, S, and X)

When an organic compound containing nitrogen, sulfur, or halide atoms is fused with sodium metal, a reductive decomposition of the compound takes place which converts these atoms to the sodium salts of the inorganic ions CN^-, S^{-2}, and X^-.

$$[N,\ S,\ X] \xrightarrow[\Delta]{Na} NaCN,\ Na_2S,\ NaX$$

When the fusion mixture is dissolved in distilled water, the cyanide, sulfide, and halide ions can be detected by means of standard qualitative inorganic tests.

CAUTION: Read Technique 14 before proceeding. Always remember to manipulate the sodium metal with a knife or a forceps, and not to touch it with your fingers. Keep sodium away from water. Destroy all waste sodium with 1-butanol or ethanol. WEAR SAFETY GLASSES.

SODIUM FUSION

Procedure. Using the procedures outlined in Technique 14, cut a small piece of sodium metal about the size of a small pea (3 mm on a side) and dry it on a paper towel. Place this small piece of sodium in a clean and dry small test tube (10 × 75 mm). Clamp the test tube to a ring stand and heat the bottom of the tube with a microburner until the sodium melts and its metallic vapor can be seen to rise about 1/3 of the way up the tube. The bottom of the tube will probably have a dull red glow. Remove the burner and **immediately** drop the sample directly into the tube. About 10 mg of a solid placed on the end of a spatula or 2 or 3 drops of a liquid should be used. Be sure to drop the sample directly down the center of the tube so that it comes into contact with the hot sodium metal and does not adhere to the side of the test tube. There will usually be a flash or a small explosion if the fusion is successful. If a successful reaction is not observed, the tube should be heated to red heat for a few seconds to assure complete reaction.

Allow the test tube to cool to room temperature and then carefully add 10 drops of methanol, a drop at a time, to the fusion mixture. Using a spatula or a long glass rod, reach into the test tube and stir the mixture to assure complete reaction of any excess sodium metal. The fusion process will have destroyed the test tube for other uses. Thus, the easiest way to recover the fusion mixture is to crush the test tube into a small beaker containing 5 to 10 ml of **distilled** water. The tube is easily crushed if it is placed in the angle of a clamp holder. Tighten the clamp until the tube is securely held near its bottom and then, standing back from the beaker and holding the clamp at its opposite end, continue tightening the clamp until the test tube breaks and the pieces fall into the beaker. Stir the solution well, heat it to boiling, and then filter it by gravity through a fluted filter. Portions of this solution will be used for the tests to detect nitrogen, sulfur, and the halogens.

ALTERNATE METHOD

Procedure.   With some volatile liquids, the method given above will not work. These compounds volatilize before they reach the sodium vapors. For these compounds, place 4 or 5 drops of the pure liquid in the clean, dry test tube, clamp it, and cautiously add the small piece of sodium metal. If there is any reaction, wait until it subsides. Then, heat the test tube to red heat, and proceed according to the instructions given in the second paragraph above.

NITROGEN TEST

Procedure.   Using pH paper and a 10% sodium hydroxide solution, adjust the pH of about 1 ml of the stock solution to pH 13. Add 2 drops of saturated ferrous ammonium sulfate solution and 2 drops of 30% potassium fluoride solution. Boil the solution for about 30 seconds. Then, acidify the hot solution by adding 30% sulfuric acid dropwise until the iron hydroxides dissolve. Avoid using an excess of acid. If nitrogen is present, a dark blue (not green) precipitate of Prussian blue ($NaFe_2(CN)_6$) will form or the solution will assume a dark blue color.

Reagent.   Dissolve 5 g of ferrous ammonium sulfate in 100 ml of water.

SULFUR TEST

Procedure.   Acidify about 1 ml of the test solution with acetic acid and add a few drops of a 1% lead acetate solution. The presence of sulfur is indicated by a black precipitate of lead sulfide (PbS).

HALIDE TESTS

Procedure.   Cyanide and sulfide ions interfere with this test. If they are present, they must be removed. To accomplish this, acidify the solution with dilute nitric acid and boil it for about 2 minutes. This will drive off any HCN or H_2S that is formed. When the solution cools, add a few drops of a 5% silver nitrate solution. A **voluminous** precipitate indicates the presence of a halide. A faint turbidity **does not** indicate a positive test. Silver chloride is white. Silver bromide is off-white. Silver iodide is yellow. Silver chloride will readily dissolve in concentrated ammonium hydroxide, while silver bromide is only slightly soluble.

DIFFERENTIATION OF CHLORIDE, BROMIDE, AND IODIDE

Procedure. Acidify 2 ml of the test solution with 10% sulfuric acid and boil it for about 2 minutes. Cool the solution and add about 0.5 ml of carbon tetrachloride. Add a few drops of chlorine water or 2 to 4 mg of calcium hypochlorite. Check to be sure that the solution is still acidic. Then, stopper the tube, shake it vigorously, and set it aside to allow the layers to separate. An orange to brown color in the carbon tetrachloride layer indicates bromine. A violet color indicates iodine. No color or a **light** yellow color indicates chlorine.

Procedure **50**c

TESTS FOR UNSATURATION

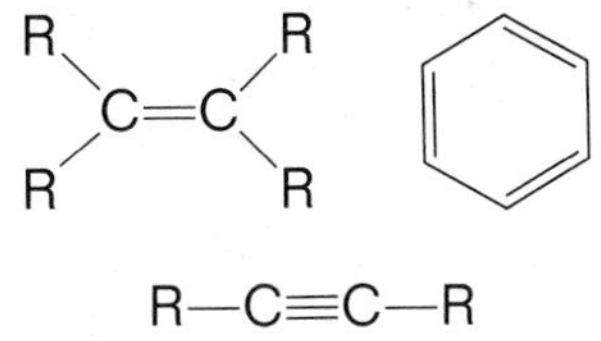

The unknowns issued for this experiment will have neither a double bond nor a triple bond as their **only** functional group. Hence, simple alkenes and alkynes can be ruled out as possible compounds. However, some of the unknowns may have a double or a triple bond **in addition to** another more important functional group. The tests which follow allow one to determine the presence of a double bond or single bond (unsaturation) in such compounds.

CLASSIFICATION TESTS

UNSATURATION	AROMATICITY
Bromine/Carbon Tetrachloride	Ignition Test
Potassium Permanganate	

TESTS FOR SIMPLE MULTIPLE BONDS

BROMINE IN CARBON TETRACHLORIDE

Procedure. Dissolve 50 mg of a solid unknown or two drops of a liquid unknown in 1 ml of carbon tetrachloride (or dioxane). Add a 2% (by volume) solution of bromine in carbon tetrachloride, dropwise with shaking, until the bromine color persists. The test is positive if more than 5 drops of the bromine solution are required to have the color remain for one minute. Usually, a large number of drops of the bromine solution will be required if unsaturation is present. Hydrogen bromide should not be evolved. If hydrogen bromide gas is evolved, one will note a "fog" when blowing across the mouth of the test tube. The HBr may also be detected by using a moistened piece of litmus or pH paper. If hydrogen bromide is evolved, the reaction which occurs is a **substitution reaction** rather than an **addition reaction,** and a double or triple bond is probably not present.

Test Compounds. Try this test with cyclohexene, cyclohexane, benzene, and acetone.

A successful test depends on the addition of bromine, a red liquid, to a double or triple bond to give a colorless dibromide:

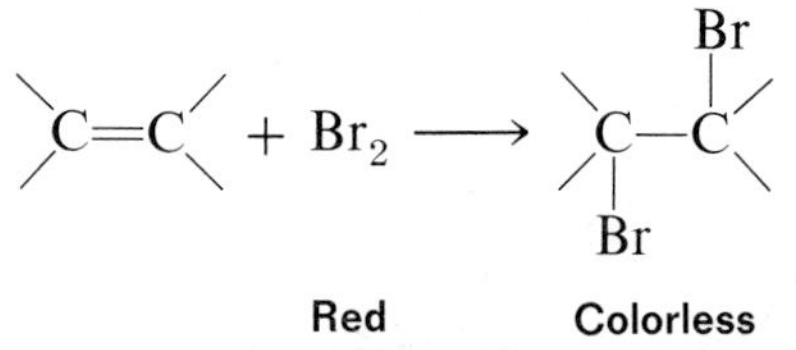

$$\text{\textbackslash C=C / } + \text{Br}_2 \longrightarrow \text{\textbackslash C—C /}$$

Red   Colorless

Not all double bonds will react with bromine-carbon tetrachloride solution. Only those which are electron-rich are sufficiently good nucleophiles to initiate the reaction. A double bond which is substituted by electron withdrawing groups will often fail to react or react quite slowly. Fumaric acid is an example of a compound that fails to give the reaction.

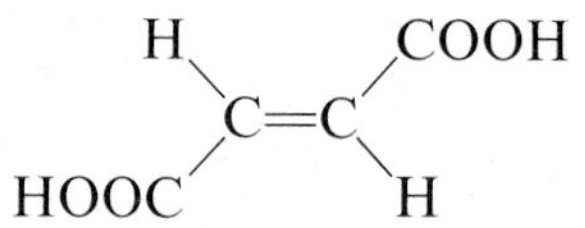

$$\underset{\text{HOOC}}{\overset{H}{\diagdown}} C = C \underset{H}{\overset{COOH}{\diagup}}$$

Fumaric Acid

Aromatic compounds will either not react with bromine/carbon tetrachloride reagent, or they will react by **substitution.** Only those aromatic rings which have activating groups as substituents (—OH, —OR, or —NR$_2$) will give the substitution reaction.

[benzene with OH ring] + Br$_2$ ⟶ [bromophenol ring] + ortho isomers + HBr
etc.

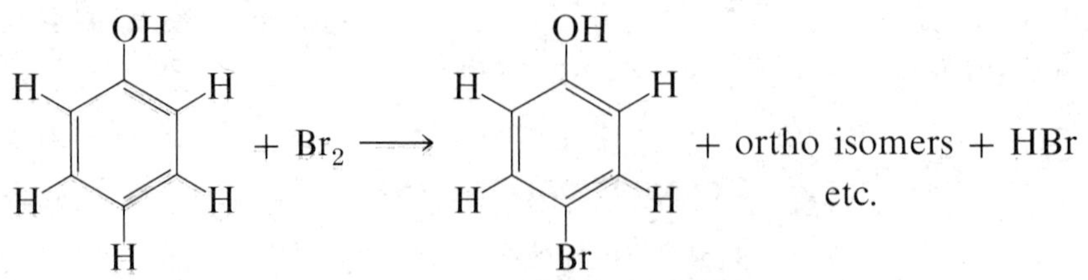

Some ketones and aldehydes will react with bromine to give a **substitution** product, but this reaction is slow except for acetone and some aldehydes and ketones which have a high enol content. When substitution occurs, not only will the bromine color be discharged, but hydrogen bromide gas will also be evolved.

POTASSIUM PERMANGANATE (BAEYER TEST)

Procedure. Dissolve 25 mg of a solid unknown or 2 drops of the liquid unknown in 2 ml of water or 95% ethanol (dioxane may also be used). Slowly add a 1% aqueous solution (weight/volume) of potassium permanganate, drop by drop with shaking, to the unknown. In a positive test, the purple color of the reagent will be discharged, and a brown precipitate of manganese dioxide will form, usually within 1 minute. If alcohol was used as the solvent, the solution should not be allowed to stand for more than 5 minutes, as the alcohol will slowly begin to be oxidized. Since permanganate solutions undergo some decomposition to manganese dioxide on standing, any small amount of precipitate should be interpreted with caution.

Test Compounds. Try this test on cyclohexene and toluene.

This test is positive for double and triple bonds, but not for aromatic rings. It depends on the conversion of the purple MnO_4^- ion to a brown precipitate of MnO_2 following the oxidation of an unsaturated compound.

$$\underset{\text{Purple}}{\overset{\displaystyle }{\scriptstyle\diagdown}C=C\scriptstyle\diagup} + MnO_4^- \longrightarrow \underset{\text{Brown}}{\overset{\displaystyle \underset{OH\ \ OH}{\scriptstyle\diagdown}C-C\scriptstyle\diagup}{}} + MnO_2$$

Other easily oxidized compounds also give a positive test with potassium permanganate solution. These substances include aldehydes, some alcohols, phenols, and aromatic amines. If any of these functional groups are suspected to be present, the test should be interpreted with caution.

SPECTROSCOPY

INFRARED

DOUBLE BONDS (C=C)
C=C Stretch usually occurs near 1680 to 1620 cm^{-1} (5.95 to 6.17 μ)
Symmetrical alkenes may have no absorption

C—H Stretch of vinyl hydrogens occurs >3000 cm^{-1} (3.33 μ), but usually not

TRIPLE BONDS (C≡C)
C≡C Stretch usually occurs near 2250 to 2100 cm^{-1} (4.44 to 4.76 μ)
The peak is usually quite sharp. Symmetrical alkynes show no absorption

C—H Stretch of terminal acetylenes occurs near 3310 to 3200 cm^{-1} (3.02 to 3.12 μ)

higher   than   3150 cm^{-1}
(3.18 μ)
C—H   Out of Plane Bending
occurs near 1000 to 700 cm^{-1}
(10.0 to 14.3 μ)

NUCLEAR MAGNETIC RESONANCE

Vinyl hydrogens have resonance near 5 to 7 δ and have coupling values: J_{trans} = 11 to 18 Hz, J_{cis} = 6 to 15 Hz, $J_{geminal}$ = 0 to 5 Hz. Allylic hydrogens have resonance near 2 δ. Acetylenic hydrogens have resonance near 2.8 to 3.0 δ.

TESTS FOR AROMATICITY

As was the case for unsaturated compounds, none of the unknowns issued in this experiment will be simple aromatic hydrocarbons. All aromatic compounds will have a major functional group as a part of their structure. Nevertheless, in many cases it will be useful to be able to recognize the presence of an aromatic ring. Although spectroscopy provides the easiest method of determining the presence of aromatic systems, often they can be detected by a simple ignition test.

IGNITION TEST

Procedure. Place a small amount of the compound on a spatula and place it in the flame of a Bunsen burner. Observe whether a sooty flame is obtained. Compounds giving a sooty yellow flame have a high degree of unsaturation and may be aromatic.

Test Compound. Try this test with naphthalene.

The presence of an aromatic ring or other centers of unsaturation will lead to the production of a sooty yellow flame in this test. Compounds that contain little oxygen, and which have a high carbon to hydrogen ratio, burn at a low temperature with a yellow flame. A large amount of carbon is produced when they are burned. Compounds which contain oxygen generally burn at a higher temperature with a clean blue flame.

SPECTROSCOPY

INFRARED

C=C   Aromatic Ring Double Bonds
These appear in the 1450 to 1650 cm^{-1} (6 to 7 μ) region. There

are often four sharp absorptions that occur in pairs near 1600 cm^{-1} (6.3 μ) and 1450 cm^{-1} (6.9 μ) that are characteristic of an aromatic ring

Special Ring Absorptions

There are often weak ring absorptions around 2000 to 1600 cm^{-1} (5 to 6 μ). These are often obscured, but when they can be observed, the relative shapes and numbers of these peaks can often be used to ascertain the type of ring substitution (see references)

=C—H　Stretch, Aromatic Ring. The aromatic C—H stretch always occurs at a higher frequency than 3000 cm^{-1} (shorter wavelength than 3.33 μ)

=C—H　Out-of-Plane Bending. These peaks appear in the region 900 to 690 cm^{-1} (11 to 15 μ). The number and position of these peaks can be used to determine the substitution pattern of the ring (see Appendix Three, "Infrared Spectroscopy")

NUCLEAR MAGNETIC RESONANCE

Hydrogens attached to an aromatic ring usually have resonance near 7 δ. Monosubstituted rings not substituted by anisotropic or electronegative groups usually give a single resonance for all the ring hydrogens. Monosubstituted rings with isotropic or electronegative groups usually have the aromatic resonances split into two groups integrating either 3:2 or 2:3. A non-symmetric, para-disubstituted ring has a characteristic four peak splitting pattern (see Appendix Four, "Nuclear Magnetic Resonance").

Procedure **50**D

ALDEHYDES AND KETONES

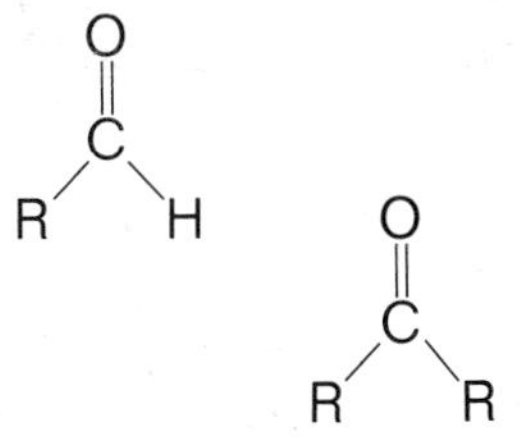

Compounds containing the carbonyl functional group ($>$C$=$O) where it has only hydrogen atoms or alkyl groups as substituents are called aldehydes (RCHO) or ketones (RCOR'). The chemistry of these compounds is primarily due to the chemistry of the carbonyl functional groups. These compounds are identified by the distinctive reactions of the carbonyl function.

SOLUBILITY CHARACTERISTICS	CLASSIFICATION TESTS

HCl NaHCO$_3$ NaOH H$_2$SO$_4$ ETHER
(−)   (−)   (−)   (+)   (+)
WATER: $<$C$_5$ and some C$_6$ (+)
         $>$C$_5$ (−)
	ALDEHYDES AND KETONES
	   2,4-Dinitrophenylhydrazine
	ALDEHYDES ONLY METHYL KETONES
	   Chromic Acid Iodoform Test
	   Tollen's Reagent
	COMPOUNDS WITH HIGH ENOL CONTENT
	   Ferric Chloride Test

CLASSIFICATION TESTS

Most aldehydes and ketones give a solid, orange to red precipitate when mixed with 2,4-dinitrophenylhydrazine. However, only aldehydes will reduce chromium(VI) or silver(I). This difference in behavior allows one to differentiate between aldehydes and ketones.

2,4-DINITROPHENYLHYDRAZINE

Procedure. Place one drop of the liquid unknown in a small test tube and add 1 ml of the 2,4-dinitrophenylhydrazine reagent. If the unknown is a solid, dissolve about 10 mg (estimate) in a minimum amount of 95% ethanol (or dioxane) before adding the reagent. Shake the mixture vigorously. Most aldehydes and ketones will give a yellow to red precipitate immediately. However, some compounds will require up to 15 minutes, or even **gentle** heating, to give a precipitate. The formation of a precipitate indicates a positive test.

Test Compounds. Try this test on cyclohexanone, benzaldehyde, and benzophenone.

Reagent. Dissolve 3.0 g of 2,4-dinitrophenylhydrazine in 15 ml of concentrated sulfuric acid. In a beaker mix together 20 ml of water and 70 ml of 95% ethanol. With vigorous stirring, slowly add the 2,4-dinitrophenylhydrazine solution to the aqueous ethanol mixture. After thorough mixing, filter the solution by gravity through a fluted filter.

Most aldehydes and ketones will give a precipitate, but esters will generally not give this test. Thus, an ester may usually be eliminated on the basis of this test.

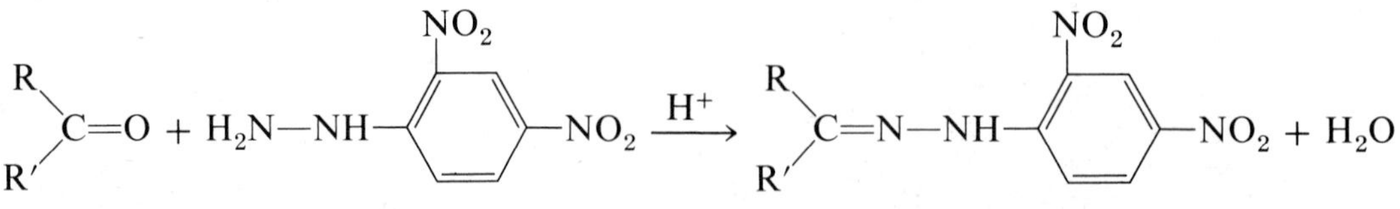

Aldehyde   2,4-Dinitrophenylhydrazine   2,4-Dinitrophenylhydrazone
or Ketone

The color of the 2,4-dinitrophenylhydrazone (precipitate) formed is often a guide to the amount of conjugation in the original aldehyde or ketone. Unconjugated ketones, such as cyclohexanone, give yellow precipitates, while conjugated ketones, such as benzophenone, give orange to red precipitates. Compounds which are highly conjugated give red precipitates. However, the 2,4-dinitrophenylhydrazine reagent is itself orange-red, and the color of any precipitate must be judged with caution. In rare cases, compounds which are either strongly basic or strongly acidic will precipitate the unreacted reagent.

Some allylic and benzylic alcohols give this test because the reagent has the capability to oxidize them to aldehydes and ketones which subsequently react. Also, some alcohols may be contaminated with carbonyl impurities either because of their method of synthesis (reduction) or because they have become air oxidized. A precipitate formed from small amounts of impurity in the solution will be formed in small amount. With some caution, a test which gives only a slight amount of precipitate may usually be ignored. The infrared spectrum of the compound should establish its identity, as well as identify the nature of any impurities present.

CHROMIC ACID TEST

Procedure. Dissolve one drop of a liquid or 10 mg (approximate) of a solid aldehyde in 1 ml of **reagent grade** acetone. Add several drops of the chromic acid reagent, one drop at a time, with shaking. A positive test is indicated by the formation of a green precipitate and a loss of the orange color in the reagent. With aliphatic aldehydes (RCHO), the solution turns cloudy within 5 seconds and a precipitate appears within 30 seconds. Aromatic aldehydes (ArCHO) generally require 30 to 120 seconds for the formation of a precipitate; but some may require even longer.

In a negative test, usually there will be no precipitate. However, in some cases, a precipitate will form, but the solution will remain orange in color.

In performing this test, one should make quite sure that the acetone used for the solvent does not give a positive test with the reagent. Add several drops of the chromic acid reagent to a few drops of the reagent acetone contained in a small test tube. Allow this mixture to stand for 3 to 5 minutes. If no reaction has occurred after this period, the acetone is of sufficient purity to use as a solvent for the test. If a positive test resulted, try another bottle of acetone, or distill some acetone from potassium permanganate to purify it.

Test Compounds. Try this test on benzaldehyde, butanal (butyraldehyde), and cyclohexanone.

Reagent. Dissolve 1.0 g of chromic oxide (CrO_3) in 1 ml of concentrated sulfuric acid. Then, dilute this mixture carefully with 3 ml of water.

This test is a result of the fact that aldehydes are easily oxidized to the corresponding carboxylic acid by chromic acid. The green precipitate is due to chromous sulfate.

$$2\,CrO_3 + 2\,H_2O \overset{H^+}{\rightleftharpoons} 2\,H_2CrO_4 \overset{H^+}{\rightleftharpoons} H_2Cr_2O_7 + H_2O$$

$$\underset{\text{orange}}{3\,RCHO + H_2Cr_2O_7 + 3\,H_2SO_4} \longrightarrow 3\,RCOOH + \underset{\text{green}}{Cr_2(SO_4)_3} + 4\,H_2O$$

Primary and secondary alcohols are also oxidized by this reagent (see Procedure 50H). For this reason, this test is not a useful test for identifying aldehydes **unless** a positive identification of the carbonyl group has already been obtained. Aldehydes will give a 2,4-dinitrophenylhydrazine test, while alcohols will not.

There are numerous other tests used to detect the aldehyde functional group. Most of them are based on an easily detectible oxidation of the aldehyde to a carboxylic acid. The most common of these tests are the Tollen's, Fehling's, and Benedict's tests. The Benedict's test is described in Experiment 51. Only the Tollen's test will be described here.

TOLLEN'S TEST

Procedure.   The reagent must be prepared immediately prior to use. To prepare the reagent, mix 1 ml of Tollen's solution A with 1 ml of Tollen's solution B. A precipitate of silver oxide will form. Add enough dilute ammonia solution (dropwise) to the mixture to **just** dissolve the silver oxide. The reagent so prepared may be used immediately for the test below.

Dissolve one drop of a liquid aldehyde or 10 mg (approximate) of a solid aldehyde in the minimum amount of dioxane. Add this solution, a little at a time, to the 2 or 3 ml of reagent contained in a small test tube. Shake the solution well. If a mirror of silver is deposited on the outer walls of the test tube, the test is positive. In some cases it may be necessary to warm the test tube in a bath of warm water.

CAUTION: The reagent should be prepared immediately prior to use and all residues disposed of immediately after use. Wash any residues down a sink with a large quantity of water. On standing the reagent has a tendency to form silver fulminate, a very explosive **substance. Solutions containing the mixed Tollen's reagent should never be stored.**

Test Compounds.   Try the test on acetone and benzaldehyde.

Reagents.   Solution A. Dissolve 3.0 g of silver nitrate in 30 ml of water. Solution B. Prepare a 10% sodium hydroxide solution.

Most aldehydes will reduce ammoniacal silver nitrate solution to give a precipitate of silver metal. The aldehyde is oxidized to a carboxylic acid.

$$RCHO + 2\ Ag(NH_3)_2OH \longrightarrow 2\ Ag + RCOO^-NH_4^+ + H_2O + NH_3$$

Ordinary ketones do not give a positive result in this test. The test should be used only if it has already been shown that the unknown compound is either an aldehyde or a ketone.

IODOFORM TEST

Procedure. Dissolve 4 drops of a liquid unknown or 0.1 g of a solid in 5 ml of dioxane in a large test tube (18 × 150 mm). Add 1 ml of 10% sodium hydroxide solution. Then, add the iodine-potassium iodide test solution, drop by drop with shaking, until a slight excess gives a definite dark color of iodine. If at this point at least 2 ml of the iodine reagent was added and decolorized, proceed to the directions in the next paragraph. If less than 2 ml of the iodine reagent was used, heat the test tube to 60° in a beaker of water. If the solution is decolorized on heating, add more of the iodine test solution, a little at a time, until the solution no longer becomes decolorized after 2 minutes of heating at 60°. Following this, add a few drops of 10% sodium hydroxide, with shaking, to decolorize any unreacted iodine remaining in the solution. The iodine will be **intensely** colored; the solution will still remain yellow when it is discharged.

Fill the test tube with water and allow it to stand for 15 minutes. A yellow precipitate of iodoform will form if the unknown was a methyl ketone or a compound easily oxidized to a methyl ketone. To prove the identity of the yellow precipitate as iodoform, one should collect, dry, and determine the melting point of the solid (iodoform mp 119 to 121°).

Test Compound. Try the test on acetone.

Reagent. Dissolve 20 g of potassium iodide and 10 g of iodine in 100 ml of water.

The basis of this test is the ability of certain compounds to form a precipitate of iodoform when treated with a basic solution of iodine. Methyl ketones are the most common type of compounds that give a positive result in this test. However, acetaldehyde (CH_3CHO) and alcohols with the hydroxyl group at the 2-position of the chain will also give a precipitate of iodoform. Alcohols of the type described are easily oxidized to methyl ketones under the conditions of the reaction. The other product of the reaction, in addition to iodoform, is the potassium or sodium salt of a carboxylic acid.

$$R-\underset{\underset{\text{OH}}{|}}{C}H-CH_3 \xrightarrow[\text{NaOH}]{I_2} R-\underset{\underset{\text{O}}{\|}}{C}-CH_3 \xrightarrow[\text{NaOH}]{I_2}$$

Alcohol   Methyl Ketone

$$R-\underset{\underset{\text{O}}{\|}}{C}-CI_3 \xrightarrow{OH^-} R-\underset{\underset{\text{O}}{\|}}{C}-O^- + HCI_3$$

Iodoform

FERRIC CHLORIDE TEST

Procedure. Some aldehydes and ketones, those which have a high **enol content,** will give a positive ferric chloride test, as described for phenols in Procedure 50F.

SPECTROSCOPY

INFRARED

The carbonyl group is usually one of the strongest absorbing groups in the infrared spectrum, with a very broad range: 1800 to 1650 cm^{-1} (5.85 to 6.20 μ). The aldehyde functional group has **very characteristic**

CH stretch absorptions: two sharp peaks that lie **far outside** the usual region for —C—H, =C—H, or ≡C—H.

ALDEHYDES

C=O   Stretch at approximately 1725 cm^{-1} (5.80 μ) is normal 1725 to 1685 cm^{-1} (5.80 to 5.95 μ)†

C—H   Stretch (aldehyde
                    —CHO)
   Two weak bands at about 2750 and 2850 cm^{-1} (3.65 and 3.50 μ)

KETONES

C=O   Stretch at approximately 1715 cm^{-1} (5.85 μ) is normal 1780 to 1665 cm^{-1} (5.62 to 6.01 μ)†

NUCLEAR MAGNETIC RESONANCE

Hydrogens which are alpha to a carbonyl group have resonance in the region between 2 to 3 δ. The hydrogen of an aldehyde group has a characteristic resonance between 9 to 10 δ. In aldehydes, there is coupling between the aldehyde hydrogen and any alpha hydrogens (J = 1 to 3 Hz).

DERIVATIVES

The most common derivatives of aldehydes and ketones are the 2,4-diphenylhydrazones, oximes, and semicarbazones.

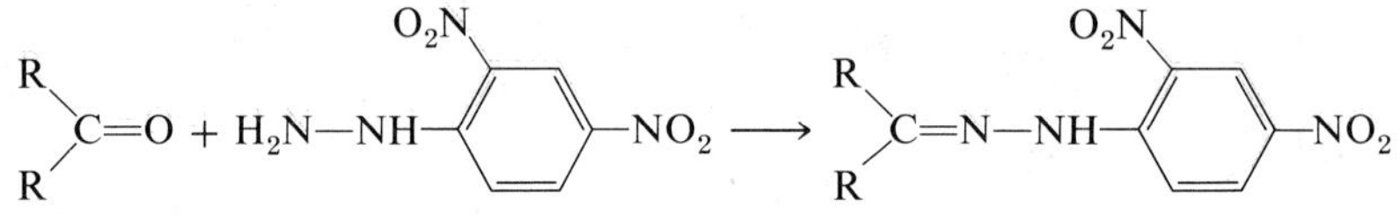

2,4-Dinitrophenylhydrazine   2,4-Dinitrophenylhydrazone

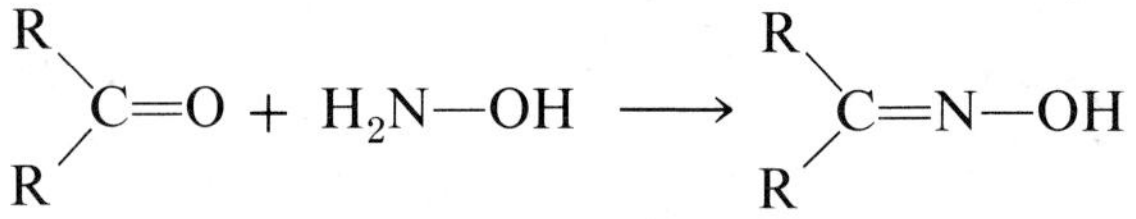

Hydroxylamine   Oxime

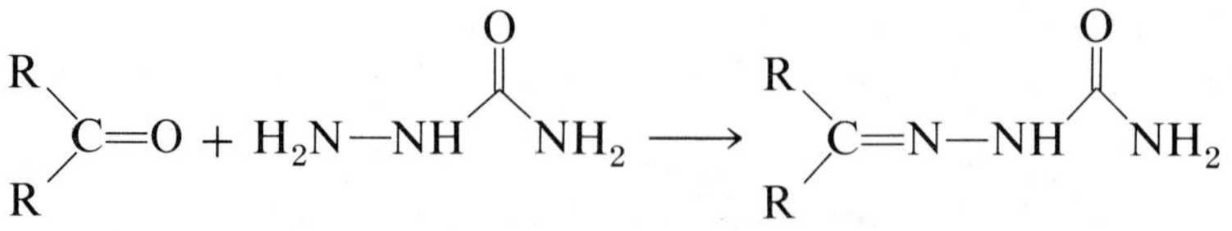

Semicarbazide   Semicarbazone

Procedures for the preparation of these derivatives are given in Appendix Two at the end of the book.

---

†**Conjugation** moves the absorption to lower frequencies (higher wavelength). **Ring Strain** (cyclic ketones) moves the absorption to higher frequencies (lower wavelength).

Procedure 50E

CARBOXYLIC ACIDS

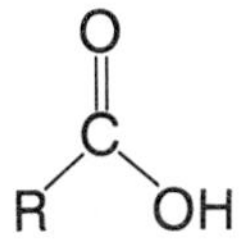

$$\underset{R}{\overset{O}{\underset{\quad}{\|}}}\underset{OH}{C}$$

These compounds are detected mainly by their solubility characteristics. They are soluble in **both** dilute sodium hydroxide and sodium bicarbonate solutions.

SOLUBILITY CHARACTERISTICS	CLASSIFICATION TESTS
HCl NaHCO$_3$ NaOH H$_2$SO$_4$ ETHER (−) (+) (+) (+) (+) WATER: <C$_6$ (+) 　　　　 >C$_6$ (−)	pH of an Aqueous Solution Sodium Bicarbonate Silver Nitrate Neutralization Equivalent

CLASSIFICATION TESTS

pH OF AN AQUEOUS SOLUTION

Procedure. If the compound is soluble in water, simply prepare an aqueous solution and check the pH by using pH paper. If the compound is an acid, the solution will have a low pH.

Compounds which are insoluble in water may be dissolved in ethanol (or methanol) and water. First dissolve the compound in the alcohol and then add water until the solution **just** becomes cloudy. Clarify the solution by adding a few drops of the alcohol, and then determine its pH using pH paper.

SODIUM BICARBONATE

Procedure. Dissolve a small amount of the compound in a 10% aqueous sodium bicarbonate solution. Observe the solution carefully. If the compound is an acid, bubbles of carbon dioxide will be observed to form.

$$RCOOH + NaHCO_3 \longrightarrow RCOO^-Na^+ + H_2CO_3 \text{ (unstable)}$$

$$H_2CO_3 \longrightarrow CO_2 + H_2O$$

SILVER NITRATE

Procedure. Acids give a false silver nitrate test as described in Procedure 50B.

NEUTRALIZATION EQUIVALENT

Procedure. Accurately weigh (3 significant figures) approximately 0.2 g of the acid into a 125 ml Erlenmeyer flask. Dissolve the acid in about 50 ml of water or aqueous ethanol. Titrate the acid using a solution of sodium hydroxide of known normality (about 0.1 N) and a phenolphthalein indicator.

Calculate the neutralization equivalent (N.E.) from the following equation:

$$\text{N.E.} = \frac{\text{mg acid}}{\text{normality of NaOH} \times \text{ml of NaOH added}}$$

The neutralization equivalent is identical to the equivalent weight of the acid. If the acid has only one carboxyl group, the neutralization equivalent and the molecular weight of the acid are identical. If the acid has more than one carboxyl group, the neutralization equivalent is equal to the molecular weight of the acid divided by the number of carboxyl groups, i.e., the equivalent weight. The N.E. can be used much like a derivative to identify a specific acid.

Many phenols are acidic enough to behave very similarly to carboxylic acids. This is especially true of those substituted by electron withdrawing groups at the ortho and para ring positions. These phenols, however, can usually be eliminated either by the ferric chloride test (Experiment 50F) or by spectroscopy (phenols have no carbonyl group).

SPECTROSCOPY

INFRARED

C=O   Stretch is very strong and often broad in the region between 1725 and 1690 cm^{-1} (5.8 to 5.9 μ).

O—H   Stretch is a very broad absorption in the region between 3300 and 2500 cm^{-1} (3.0 to 4.0 μ); usually overlaps the CH stretch region.

NUCLEAR MAGNETIC RESONANCE

The acid proton (—COOH) usually has resonance near 12.0 δ.

DERIVATIVES

Derivatives of acids are usually amides. They are prepared via the corresponding acid chloride.

$$\overset{O}{\overset{\|}{R-C-OH}} + SOCl_2 \longrightarrow \overset{O}{\overset{\|}{R-C-Cl}} + SO_2 + HCl$$

The most common derivatives are the amides, the anilides, and the *p*-toluidides.

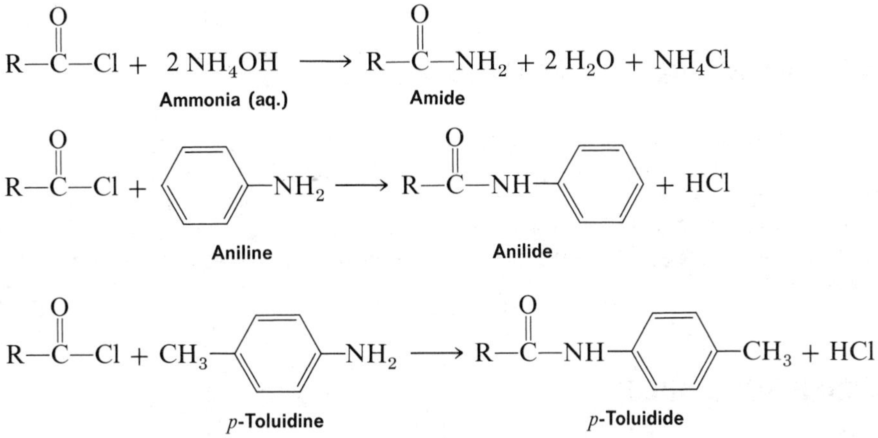

$$R-\overset{\overset{\displaystyle O}{\|}}{C}-Cl + 2\,NH_4OH \longrightarrow R-\overset{\overset{\displaystyle O}{\|}}{C}-NH_2 + 2\,H_2O + NH_4Cl$$

Ammonia (aq.)   **Amide**

$$R-\overset{\overset{\displaystyle O}{\|}}{C}-Cl + \text{⟨benzene ring⟩}-NH_2 \longrightarrow R-\overset{\overset{\displaystyle O}{\|}}{C}-NH-\text{⟨benzene ring⟩} + HCl$$

Aniline   **Anilide**

$$R-\overset{\overset{\displaystyle O}{\|}}{C}-Cl + CH_3-\text{⟨benzene ring⟩}-NH_2 \longrightarrow R-\overset{\overset{\displaystyle O}{\|}}{C}-NH-\text{⟨benzene ring⟩}-CH_3 + HCl$$

***p*-Toluidine**   ***p*-Toluidide**

Procedures for the preparation of these derivatives are given in Appendix Two at the end of the book.

Procedure **50**F _____

PHENOLS

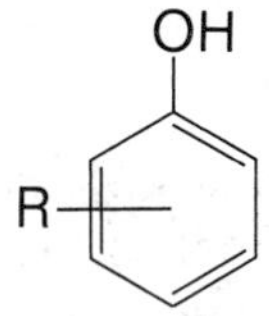

Like carboxylic acids, phenols are acidic compounds. However, except for the nitrosubstituted phenols (discussed in the section covering solubilities) they are not as acidic as the carboxylic acids. The pK_a of a typical phenol is 10, whereas that of a carboxylic acid is usually near 5. Hence, phenols are generally not soluble in the weakly basic sodium bicarbonate solution, but do dissolve in sodium hydroxide solution, which is more strongly basic.

SOLUBILITY CHARACTERISTICS	CLASSIFICATION TESTS
HCl NaHCO$_3$ NaOH H$_2$SO$_4$ ETHER (−) (−) (+) (+) (+) WATER: Most are insoluble, although phenol itself and the nitrophenols are soluble	Colored Phenolate Anion Ferric Chloride Bromine/Water

CLASSIFICATION TESTS

SODIUM HYDROXIDE SOLUTION

Procedure. With phenols, which have a high degree of conjugation possible in their conjugate base (phenolate ion), the anion is often colored. To observe the color, it is only necessary to dissolve a small amount of the phenol in 10% aqueous sodium hydroxide solution. Some phenols do not give a color. Others have an insoluble anion and give a precipitate. The more acidic phenols, like the nitrophenols, have a greater tendency toward colored anions.

FERRIC CHLORIDE

Procedure 1 (Water Soluble Phenols). Add several drops of a 2.5% aqueous solution of ferric chloride to 1 ml of a dilute aqueous solution (about 1 to 3% by weight). Most phenols produce an intense red, blue, purple, or green color. Some colors are transient, and it may be necessary to observe the solution carefully just as the solutions are mixed. The formation of a color is usually immediate, but the color may not be permanent over a period of time. Some phenols do not give a positive result in this test, so a negative test must not be taken as significant without other adequate evidence.

Test Compound. Try this test on phenol.

Procedure 2 (Water Insoluble Phenols). Many phenols do not give a positive result when procedure 1 is used. Often this procedure will give a positive result with these phenols. Dissolve or suspend 20 mg of a solid phenol or 1 drop of a liquid phenol in 1 ml of chloroform. Add one drop of pyridine and 3 to 5 drops of a 1% (weight/volume) solution of ferric chloride in chloroform.

The colors observed in this test result from the formation of a complex of the phenols with Fe(III) ion. Carbonyl compounds which have a high enol content also give a positive result in this test.

BROMINE/WATER

Procedure. Prepare a 1% aqueous solution of the unknown, and then add a saturated solution of bromine in water to it, drop by drop with shaking, until the bromine color is no longer discharged. A positive test is indicated by the precipitation of a substitution product at the same time that the bromine color of the reagent is discharged.

Test Compound. Try this test on phenol.

Aromatic compounds with ring activating substituents give a positive test with bromine in water. The reaction is an aromatic substitution reaction which introduces bromine atoms into the aromatic ring at the positions ortho and para to the hydroxyl group. All available positions are substituted. The precipitate is the brominated phenol, which is generally insoluble due to its large molecular weight.

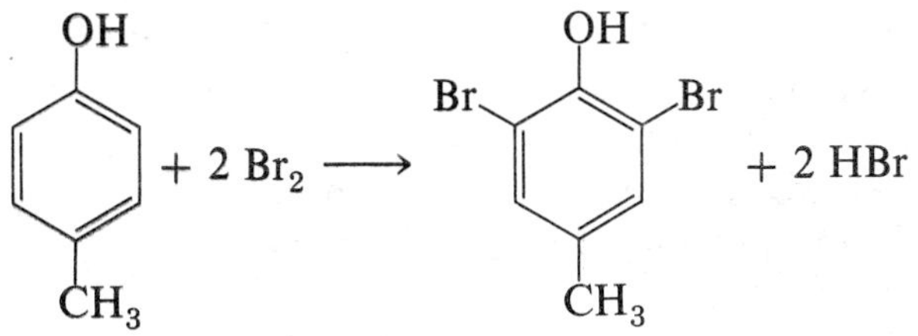

Other compounds which give a positive result with this test include aromatic compounds with activating substituents other than hydroxyl. These compounds include anilines and alkoxy ethers.

SPECTROSCOPY

INFRARED

O—H   Stretch is observed near 3600 cm^{-1} (2.8 μ)
C—O   Stretch is observed near 1200 cm^{-1} (8.3 μ)
The typical aromatic ring absorptions between 1400 and 1650 cm^{-1} (6 to 7 μ) are also found
Aromatic CH is observed near 3100 cm^{-1} (3.2 μ)

NUCLEAR MAGNETIC RESONANCE

Aromatic protons are observed near 7 δ. The hydroxyl proton has a resonance position which is concentration dependent.

DERIVATIVES

Phenols form the same derivatives as do alcohols (Procedure 50H). They form urethanes by reaction with isocyanates, but while phenylurethanes are used for alcohols, the α-naphthylurethanes are more useful for phenols. Like alcohols, phenols also yield 3,5-dinitrobenzoates.

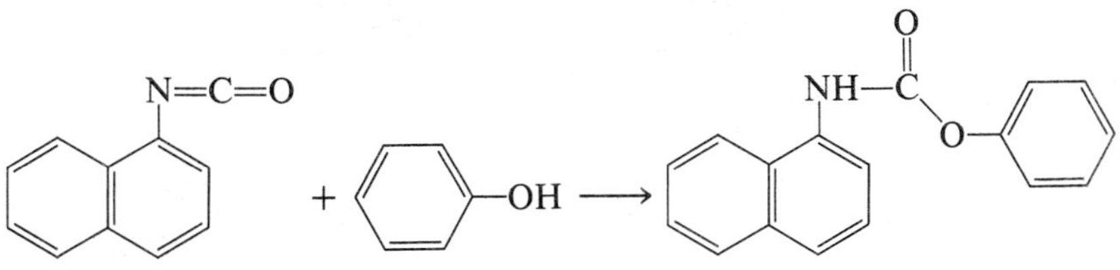

α-Naphthyl Isocyanate   An α-Naphthylurethane

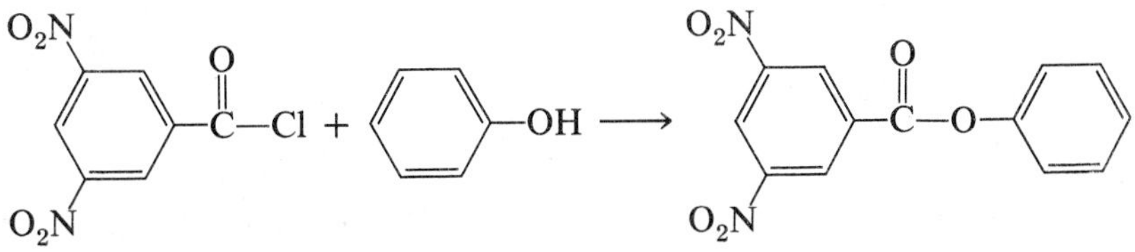

3,5-Dinitrobenzoyl Chloride   A 3,5-Dinitrobenzoate

The bromine/water reagent yields solid bromo derivatives of phenols in a number of cases. These solid derivatives can be used to characterize an unknown phenol.

Procedures for the preparation of these derivatives are given in Appendix Two at the end of the book.

Procedure 50G

AMINES

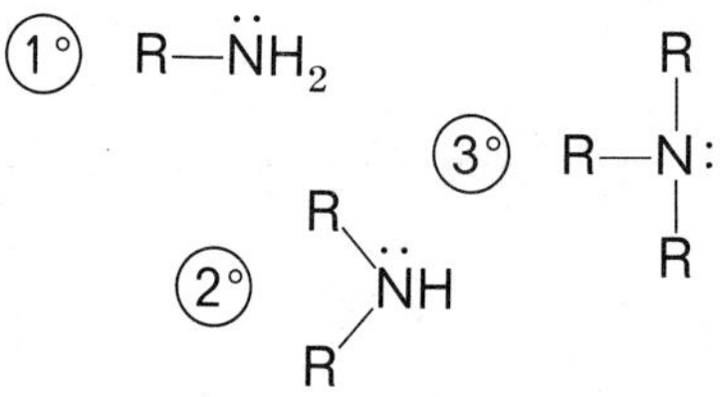

Amines are detected best by their solubility behavior and their basicity. They are the only basic compounds that will be issued for this experiment. Hence, once the compound has been identified as an amine, the main problem which remains will be to decide whether it is primary, secondary, or tertiary. This can usually be decided using the Hinsberg and nitrous acid tests.

SOLUBILITY CHARACTERISTICS	CLASSIFICATION TESTS
HCl NaHCO$_3$ NaOH H$_2$SO$_4$ ETHER (+) (−) (−) (+) (+) WATER: <C$_6$ (+) >C$_6$ (−)	pH of an Aqueous Solution Hinsberg Test Nitrous Acid Test Acetyl Chloride

CLASSIFICATION TESTS

HINSBERG TEST

Procedure.   Place 0.1 ml of a liquid amine or 0.1 g of a solid amine, 0.2 g of p-toluenesulfonyl chloride, and 5 ml of 10% sodium hydroxide solution in a small test tube. Stopper the test tube tightly, and shake it intermittently for a period of from 3 to 5 minutes. Remove the stopper and warm the test tube, with shaking, on a steam bath for 1 minute. Cool the solution and test a drop of it with pH paper to see if it is still basic; if it is not, add more sodium hydroxide. If a precipitate has formed, dilute the basic mixture with

5 ml of water and shake it well. If the precipitate is insoluble, a disubstituted sulfonamide is probably present, which indicates that the unknown was a 2° amine. (CAUTION: The precipitate may also be unreacted *p*-toluenesulfonyl chloride.) If no precipitate remains after diluting the mixture, or if none formed initially, carefully add 5% hydrochloric acid until the solution is just acidic to litmus. If a precipitate forms at this point, it should be the monosubstituted sulfonamide, indicating that the original compound was a 1°amine. If no reaction occurred during the test, the original compound was probably a 3° amine.

If the above procedure gives confusing results(!), the procedure may be repeated using 0.2 ml of benzenesulfonyl chloride instead of the *p*-toluenesulfonyl chloride. However, this reagent is likely to lead to the production of oils rather than solids.

Test Compounds. Try this test on aniline. *N*-methylaniline, and *N,N*-dimethyl-aniline. All three tests should be run simultaneously to allow for easy comparison of the results.

This test is based on the production of mono and disubstituted sulfonamides from primary and secondary amines, respectively. Mono-substituted sulfonamides are soluble in base, while disubstituted sulfonamides are not soluble since they have no acidic hydrogens which may be removed to form a soluble salt. Tertiary amines are unreactive under these conditions since a tertiary amine has no amino hydrogens to be replaced.

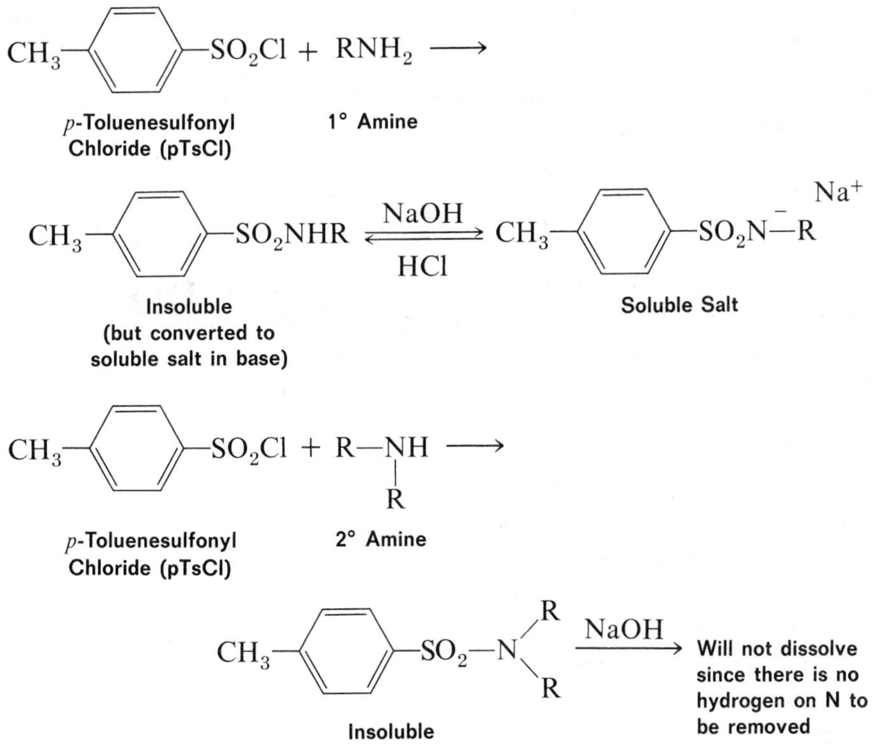

Some sulfonamides of primary amines form insoluble sodium salts. This may lead to the mistaken assumption that the amine is secondary.

HINSBERG TEST SUMMARY

Primary Amine + pTsCl $\xrightarrow{\text{NaOH}}$ Solution $\xrightarrow{\text{HCl}}$ Precipitate

Secondary Amine + pTsCl $\xrightarrow{\text{NaOH}}$ Precipitate

Tertiary Amine + pTsCl $\xrightarrow{\text{NaOH}}$ No Reaction $\xrightarrow{\text{HCl}}$ Clear Solution

---

NITROUS ACID TEST

Procedure. Dissolve 0.1 g of an amine in 2 ml of water to which 8 drops of concentrated sulfuric acid have been added. Use a large test tube. Cool the solution to 5° or less in an ice bath. Also cool 2 ml of 10% aqueous sodium nitrite in another test tube. In a third test tube, prepare a solution of 0.1 g β-naphthol in 2 ml of 10% aqueous sodium hydroxide, and place it in an ice bath to cool. Add the cold sodium nitrite solution, drop by drop with shaking, to the cooled solution of the amine. Look for the evolution of bubbles of nitrogen gas. Be careful not to confuse the evolution of the **colorless** nitrogen gas with an evolution of **brown** nitrogen oxide gas. Substantial evolution of gas at 5° or below indicates a primary aliphatic amine (RNH_2). The formation of a yellow oil or solid usually indicates a secondary amine (R_2NH). Tertiary amines either do not react, or they behave similarly to secondary amines.

If little or no gas evolution occurs at 5°, take **one half** of the solution and warm it gently to about room temperature. The evolution of nitrogen gas bubbles at this elevated temperature indicates that the original compound was a primary **aromatic** amine ($ArNH_2$). Take the remaining solution and, drop by drop, add the solution of β-naphthol in base. If a red dye precipitates, the unknown has been conclusively shown to be a primary aromatic amine ($ArNH_2$).

Test Compounds. Try this test with aniline, *N*-methylaniline, and butylamine.

Before performing this test, it should definitely be proven that the unknown is an amine by some other method. Many other compounds react with nitrous acid (phenols, ketones, thiols, amides), and a positive result with one of these could lead to an incorrect interpretation of the test.

The test is best used to distinguish **primary** aromatic and aliphatic amines from secondary and tertiary amines. It will also differentiate between aromatic and aliphatic primary amines. It cannot distinguish between secondary and tertiary amines. Primary aliphatic amines lose nitrogen gas at low temperatures under the conditions of this test. Aromatic amines yield a more stable diazonium salt, and do not lose nitrogen until the temperature is elevated. In addition, aromatic diazonium salts produce a red azo dye when β-naphthol is added. Secondary and tertiary amines produce yellow nitroso compounds, which may be soluble, may be oils, or may even be solids. Many nitroso compounds have been shown to be carcinogenic. Avoid contact and dispose of all such solutions immediately.

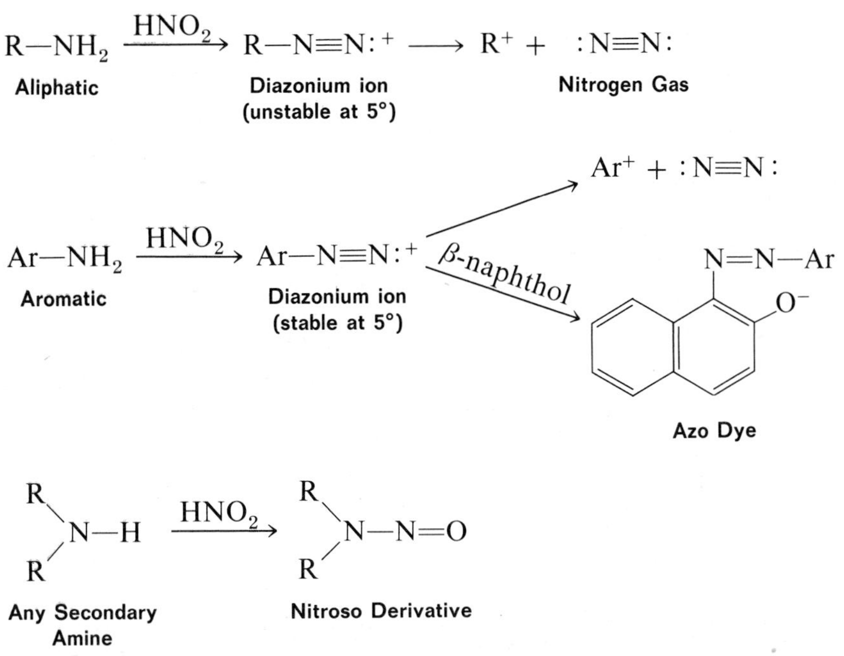

R—NH$_2$ $\xrightarrow{\text{HNO}_2}$ R—N≡N:$^+$ ⟶ R$^+$ + :N≡N:

Aliphatic   **Diazonium ion**   **Nitrogen Gas**

(unstable at 5°)

Ar—NH$_2$ $\xrightarrow{\text{HNO}_2}$ Ar—N≡N:$^+$ $\xrightarrow{\text{β-naphthol}}$ Ar$^+$ + :N≡N:

Aromatic   **Diazonium ion**

(stable at 5°)

Azo Dye

$\underset{R}{\overset{R}{\diagdown}}$N—H $\xrightarrow{\text{HNO}_2}$ $\underset{R}{\overset{R}{\diagdown}}$N—N=O

Any Secondary   **Nitroso Derivative**
Amine

pH OF AN AQUEOUS SOLUTION

Procedure. If the compound is soluble in water, simply prepare an aqueous solution and check the pH by using pH paper. If the compound is an amine, it will be basic, and the solution will have a high pH. Compounds which are insoluble in water may be dissolved in ethanol/water or dioxane/water.

ACETYL CHLORIDE

Procedure. Amines will give a positive acetyl chloride test (liberation of heat). This test is described for alcohols in Experiment 50H. When the test mixture is diluted with water, primary and secondary amines often give a solid acetamide derivative; tertiary amines do not.

SPECTROSCOPY

INFRARED

N—H Stretch   Both aliphatic and aromatic **primary** amines show two absorptions (doublet due to symmetric and asymmetric stretches) in the region 3500 to 3300 cm^{-1} (2.86 to 3.03 μ)

     Secondary amines show a single absorption in this region

     Tertiary amines have no N—H bonds

N—H Bend   **Primary** amines have a strong absorption 1640 to 1560 cm^{-1} (6.10 to 6.41 μ)

Secondary amines have an absorption 1580 to 1490 cm^{-1} (6.33 to 6.71 μ)

Aromatic amines will show bands typical for the aromatic ring in the region 1400 to 1650 cm^{-1} (6 to 7 μ). Aromatic CH is observed near 3100 cm^{-1} (3.2 μ)

NUCLEAR MAGNETIC RESONANCE

The resonance position of amino hydrogens is extremely variable. The resonance may also be very broad (quadrupole broadening). Aromatic amines will give resonances near 7 δ due to the aromatic ring hydrogens.

DERIVATIVES

The easiest derivatives of amines to prepare are the acetamides and the benzamides. These derivatives work well for both primary and secondary amines, but not for tertiary amines.

$$CH_3\text{—}\overset{\overset{\displaystyle O}{\|}}{C}\text{—}Cl + RNH_2 \longrightarrow CH_3\text{—}\overset{\overset{\displaystyle O}{\|}}{C}\text{—}NH\text{—}R + HCl$$

Acetyl Chloride   **An Acetamide**

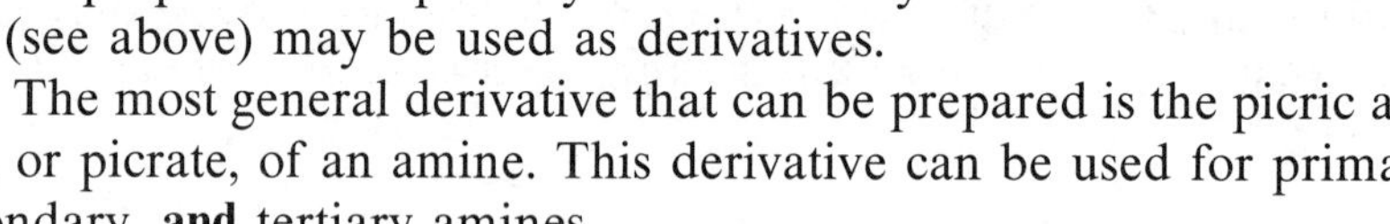

$$\text{Ph}\text{—}\overset{\overset{\displaystyle O}{\|}}{C}\text{—}Cl + RNH_2 \longrightarrow \text{Ph}\text{—}\overset{\overset{\displaystyle O}{\|}}{C}\text{—}NH\text{—}R + HCl$$

Benzoyl Chloride   **A Benzamide**

In some cases the solid *p*-toluenesulfonamides and benzenesulfonamides prepared from primary and secondary amines in the Hinsburg test (see above) may be used as derivatives.

The most general derivative that can be prepared is the picric acid salt, or picrate, of an amine. This derivative can be used for primary, secondary, **and** tertiary amines.

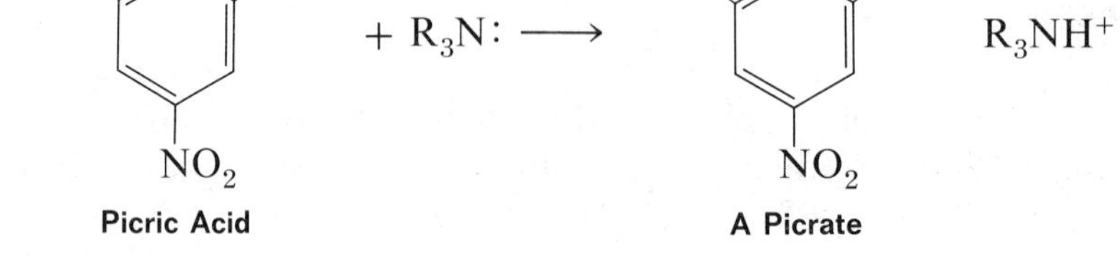

$$\text{Picric Acid} + R_3N\text{:} \longrightarrow \text{A Picrate} \quad R_3NH^+$$

Picric Acid   **A Picrate**

For tertiary amines, the methiodide salt is often useful.

$$CH_3I + R_3N\text{:} \longrightarrow CH_3\text{—}NR_3^+\ I^-$$

A Methiodide

Procedures for preparing derivatives from amines can be found in Appendix Two at the end of the book.

Procedure 50H

ALCOHOLS

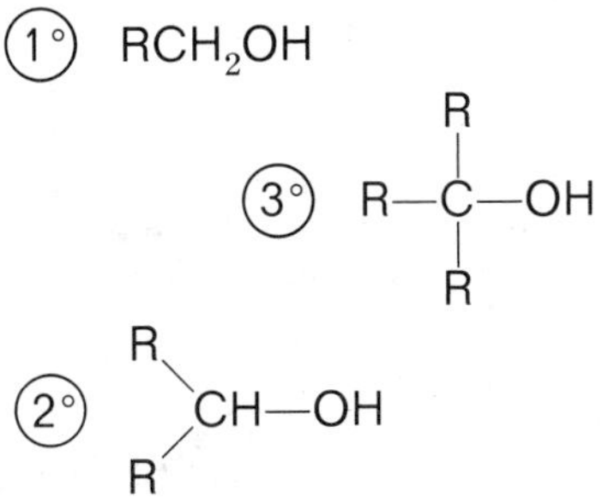

(1°) RCH₂OH

(3°) $R-\underset{\underset{R}{|}}{\overset{\overset{R}{|}}{C}}-OH$

(2°) $\underset{R}{\overset{R}{\diagdown}}CH-OH$

Alcohols are neutral compounds. The only other classes of neutral compounds used in this experiment are those of the aldehydes and ketones, and of the esters. Alcohols and esters usually do not give a positive 2,4-dinitrophenylhydrazine test, whereas aldehydes and ketones do. Esters do not react with acetyl chloride or with Lucas reagent, as do alcohols, and they are easily distinguished from alcohols on this basis. Primary and secondary alcohols are easily oxidized, whereas esters and tertiary alcohols are not. A combination of the Lucas test and the chromic acid test will differentiate between primary, secondary, and tertiary alcohols.

SOLUBILITY CHARACTERISTICS	CLASSIFICATION TESTS
HCl NaHCO₃ NaOH H₂SO₄ ETHER (−) (−) (−) (+) (+) WATER: <C₆ (+) >C₆ (−)	Acetyl Chloride Lucas Test Chromic Acid Test

CLASSIFICATION TESTS

ACETYL CHLORIDE

Procedure. Cautiously add about 10 to 15 drops of acetyl chloride, drop by drop, to about 0.5 ml of the alcohol contained in a small test tube. Evolution of heat and hydrogen chloride gas indicates a positive reaction. Addition of water will sometimes precipitate the acetate.

Acid chlorides react with alcohols to form esters. Acetyl chloride forms acetate esters.

$$CH_3-\overset{\overset{O}{\|}}{C}-Cl + ROH \longrightarrow CH_3-\overset{\overset{O}{\|}}{C}-O-R + HCl$$

Usually the reaction is exothermic and the heat evolved is easily detected. Phenols react with acid chlorides in a manner analogous to that of alcohols. Hence, phenols should be eliminated as possibilities before this test is attempted. Amines also react with acetyl chloride to evolve heat (see Procedure 50G).

LUCAS TEST

Procedure. Place 2 ml of Lucas reagent in a small test tube and add 3 to 4 drops of the alcohol. Stopper the test tube and shake it vigorously. Tertiary (3°), benzylic, and allylic alcohols will give an immediate cloudiness in the solution as the insoluble alkyl halide separates from the aqueous solution. After a short time, the immiscible alkyl halide will form a separate layer. Secondary (2°) alcohols will produce a cloudiness after 2 to 5 minutes. Primary (1°) alcohols will dissolve in the reagent to give a clear solution. Some secondary alcohols may have to be heated slightly to encourage reaction with the reagent.

This test will only work for alcohols which are soluble in the reagent. This often means that alcohols with more than 6 carbon atoms may not be tested.

Test Compounds. Try this test with 1-butanol (*n*-butyl alcohol), 2-butanol (*sec*-butyl alcohol), and 2-methyl-2-propanol (*t*-butyl alcohol).

Reagent. Cool 10 ml of concentrated hydrochloric acid in a beaker using an ice bath. While still cooling, and with stirring, dissolve 16 g of anhydrous zinc chloride in the acid.

This test depends on the appearance of an alkyl chloride as an insoluble second layer when an alcohol is treated with a mixture of hydrochloric acid and zinc chloride (Lucas reagent).

$$R\text{—}OH + HCl \xrightarrow{\text{ZnCl}_2} R\text{—}Cl + H_2O$$

Primary alcohols do not react at room temperature and, therefore, the alcohol is simply seen to dissolve. Secondary alcohols will react slowly, while tertiary, benzylic, and allylic alcohols react instantly. These relative reactivities are explained on the same basis as for the silver nitrate reaction which is discussed in Procedure 50B. Primary carbonium ions are quite unstable and do not form under the conditions of this test. Hence, no test is observed for primary alcohols.

$$
\underset{\underset{R}{|}}{\overset{\overset{R}{|}}{R\text{—}C}}\text{—}OH + ZnCl_2 \longrightarrow \underset{\underset{R}{|}\ \underset{H}{}}{\overset{\overset{R}{|}}{R\text{—}C}}\overset{\delta^+}{\text{—}}\overset{\delta^-}{O}\text{---}ZnCl_2 \longrightarrow \left[\underset{\underset{R}{|}}{\overset{\overset{R}{|}}{R\text{—}C^+}} \right] \xrightarrow{Cl^-} \underset{\underset{R}{|}}{\overset{\overset{R}{|}}{R\text{—}C}}\text{—}Cl
$$

CHROMIC ACID TEST

Procedure. Dissolve one drop of a liquid or about 10 mg of a solid alcohol in 1 ml of **reagent grade** acetone. Add one drop of the chromic acid reagent and note the result that occurs within 2 seconds. A positive test for the appearance of a primary or a secondary alcohol is the appearance of a blue-green color. Tertiary alcohols will not give the test within 2 seconds, and the solution will remain orange. To make sure that the acetone solvent is pure and does not give a positive test, add one drop of chromic acid to 1 ml of acetone which does not have an unknown dissolved in it. The orange color of the reagent should persist for **at least** 3 seconds. If it does not, a new bottle of acetone should be used.

Test Compounds.   Try this test with 1-butanol (*n*-butyl alcohol), 2-butanol (*sec*-butyl alcohol), and 2-methyl-2-propanol (*t*-butyl alcohol).

Reagent.   Dissolve 1 g of chromic oxide in 1 ml of concentrated sulfuric acid. Carefully add the mixture to 3 ml of water.

This test is based on the reduction of chromium (VI), which is orange, to chromium (III), which is green, when an alcohol is oxidized by the reagent. A change in color of the reagent from orange to green represents a positive test. Primary and secondary alcohols are oxidized by the reagent to carboxylic acids and ketones respectively.

$$2\ CrO_3 + 2\ H_2O \xrightarrow{H^+} 2\ H_2CrO_4 \xrightarrow{H^+} H_2Cr_2O_7 + H_2O$$

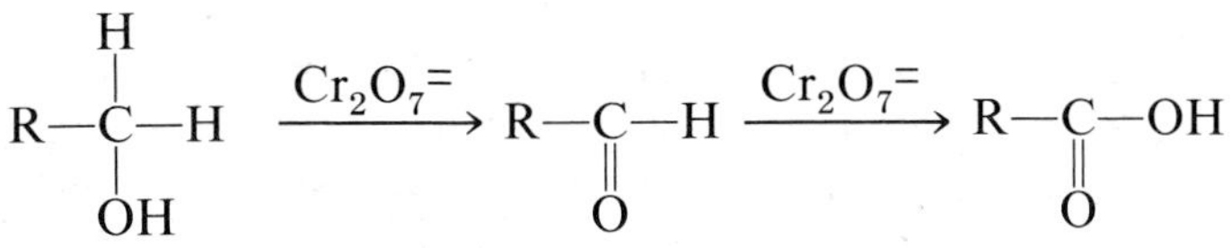

Primary Alcohol

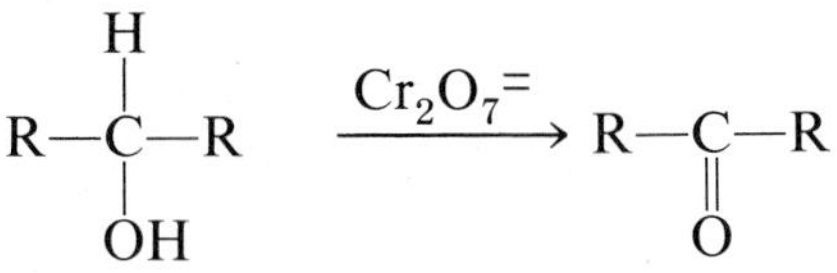

Secondary Alcohol

Although primary alcohols are first oxidized to aldehydes, the aldehydes are further oxidized to carboxylic acids. The ability of chromic oxide to oxidize aldehydes, but not ketones, is taken advantage of in a test using chromic oxide to distinguish between aldehydes and ketones (Procedure 50D). Secondary alcohols are oxidized to ketones, but no further. Tertiary alcohols are not oxidized at all by the reagent. Hence, this test can be used to distinguish primary and secondary alcohols from tertiary alcohols. Unlike the Lucas test, this test can be used with all alcohols regardless of their molecular weight and solubility.

SPECTROSCOPY

INFRARED

O—H Stretch   A medium to strong, and usually broad, absorption in the region 3600 to 3200 cm^{-1} (2.8 to 3.1 μ)
In dilute solutions or with little hydrogen bonding, it is a sharp absorption near 3600 cm^{-1} (2.8 μ). In more concentrated solutions, or with considerable hydrogen bonding, it is a broad absorption near 3400 cm^{-1} (2.9 μ). Sometimes both bands appear.

C—O Stretch   Strong absorption in the region 1050 to 1200 cm^{-1} (9.5 to 8.3 μ)

Primary alcohols absorb nearer 1050 cm^{-1} (9.5 μ), tertiary alcohols and phenols nearer 1200 cm^{-1} (8.3 μ). Secondary alcohols absorb in between this range.

NUCLEAR MAGNETIC RESONANCE

The hydroxyl resonance is extremely concentration dependent, but it is usually found between 1 to 5 δ. Under normal conditions, the hydroxyl proton does not couple with protons on adjacent carbon atoms.

DERIVATIVES

The most commonly used derivatives for alcohols are the 3,5-dinitrobenzoate esters and the phenylurethanes. Occasionally, the α-naphthylurethanes (Procedure 50F) are also prepared, but these latter derivatives are more commonly used for phenols.

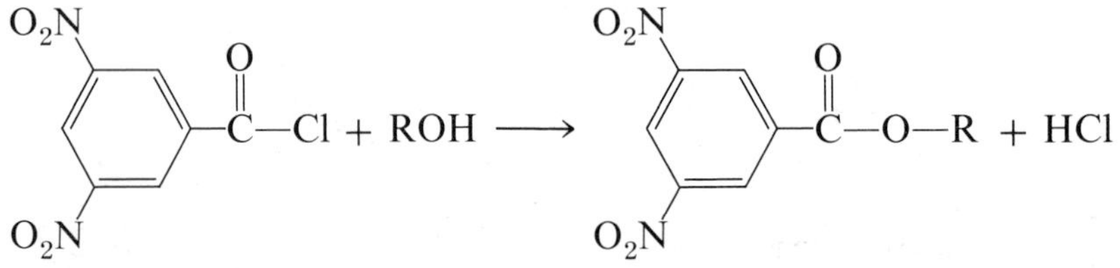

3,5-Dinitrobenzoyl Chloride + ROH $\longrightarrow$ **A 3,5-Dinitrobenzoate** + HCl

Phenyl Isocyanate + ROH $\longrightarrow$ **A Phenylurethane**

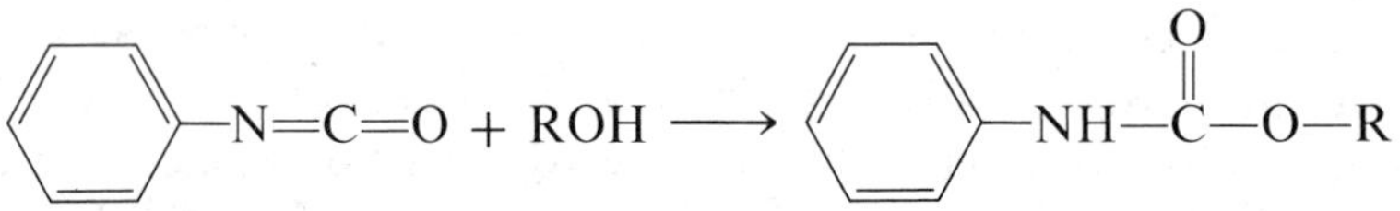

Procedures for the preparation of these derivatives are given in Appendix Two at the end of the book.

Procedure 50I

ESTERS

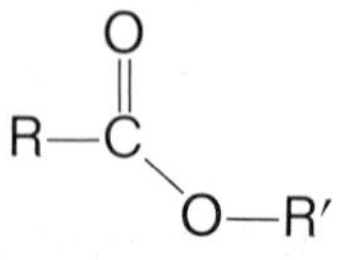

$$R-\overset{\overset{\displaystyle O}{\|}}{C}\diagdown_{O-R'}$$

Esters are formally considered as "derivatives" of the corresponding carboxylic acid. They are frequently synthesized from the carboxylic acid and the appropriate alcohol.

$$R-COOH + R'-OH \overset{H^+}{\rightleftharpoons} R-COOR' + H_2O$$

Thus, esters are sometimes referred to as though they were composed of an acid part and an alcohol part.

Although esters, like aldehydes and ketones, are neutral compounds which possess a carbonyl group, they do not usually give a 2,4-dinitrophenylhydrazine test. The two most common tests used to identify esters are the basic hydrolysis and ferric hydroxamate tests. The **saponification equivalent** is also used. However, it usually requires a difficult and time consuming procedure and will not be discussed here. Procedures for determining the saponification equivalent can be found in the references at the beginning of this experiment.

SOLUBILITY CHARACTERISTICS	CLASSIFICATION TESTS
HCl NaHCO$_3$ NaOH H$_2$SO$_4$ ETHER (−) (−) (−) (+) (+) WATER: $<$C$_4$ (+) $>$C$_5$ (−)	Ferric Hydroxamate Test Basic Hydrolysis

CLASSIFICATION TESTS

FERRIC HYDROXAMATE TEST

Procedure. Before starting, it must be determined if the compound to be tested already has enough enolic character in acid solution to give a positive ferric chloride test. Dissolve one drop of a liquid unknown or a few crystals of a solid unknown in 1 ml of 95% ethanol and add 1 ml of 1 N hydrochloric acid. Add a drop or two of 5% ferric chloride solution. If a definite color, other than yellow, is observed, the ferric hydroxamate test (described below) cannot be used.

If the compound did not show enolic character, proceed as follows. Dissolve 2 or 3 drops of a liquid ester, or about 40 mg of a solid ester, in a mixture of 1 ml of 0.5 N hydroxylamine hydrochloride (dissolved in 95% ethanol) and 0.2 ml of 6 N sodium hydroxide. Heat the mixture to boiling for a few minutes. Cool the solution and then add 2 ml of 1 N hydrochloric acid. If the solution becomes cloudy, add 2 ml of 95% ethanol to clarify it. Add a drop of 5% ferric chloride solution and note whether a color is produced. If the color fades, continue to add ferric chloride until it persists. A positive test should give a deep burgundy or magenta color.

On heating with hydroxylamine, esters are converted to the corresponding hydroxamic acids.

$$\underset{\text{}}{R-\overset{\overset{\textstyle O}{\|}}{C}-O-R'} + \underset{\text{Hydroxylamine}}{H_2N-OH} \longrightarrow \underset{\text{A Hydroxamic Acid}}{R-\overset{\overset{\textstyle O}{\|}}{C}-NH-OH} + R'-OH$$

The hydroxamic acids form strong, colored complexes with ferric ion.

$$R-\overset{\overset{\textstyle O}{\|}}{C}-NH-OH + FeCl_3 \longrightarrow \left(R-C\overset{O}{\underset{NH-O}{=\!\!\!<}}\right)_3 \!\!\!-\!Fe + 3\,HCl$$

BASIC HYDROLYSIS

Procedure.   Place 1 g of the ester in a small flask with 10 ml of 25% aqueous sodium hydroxide. Add a boiling stone and attach the reflux condenser. Reflux the mixture for about 30 minutes. Stop the heating and observe the solution to determine if the oily ester layer has disappeared, or if the odor of the ester (usually pleasant) has disappeared. Low boiling esters (below 110°) will usually dissolve within 30 minutes if the alcohol part has a low molecular weight. If the ester has not dissolved, reheat the mixture to reflux for 1 to 2 hours. After that time, the oily ester layer should have disappeared along with the characteristic odor. Esters boiling up to 200° should hydrolyze during this time. Compounds remaining after this extended period of heating are either unreactive esters or are **not** esters. For esters derived from solid acids, the acid part can, if desired, be recovered after hydrolysis by careful neutralization of the solution with hydrochloric acid. Either it will precipitate or else it can easily be recovered by extraction of the mixture with ether. The melting point of the parent acid can provide valuable information to the identification process.

This procedure converts the ester to its separate acid and alcohol parts. The ester dissolves because the alcohol part (if small) is usually soluble in the aqueous medium as is the sodium salt of the acid. Acidification produces the parent acid.

$$
\begin{array}{ccccc}
\overset{\displaystyle O}{\overset{\|}{R-C-O-R'}} & \xrightarrow{\text{NaOH}} & \overset{\displaystyle O}{\overset{\|}{R-C-O^-}} \ Na^+ + R'OH & \xrightarrow{\text{HCl}} & \overset{\displaystyle O}{\overset{\|}{R-C-O-H}} + R'OH \\
\text{Ester} & & \begin{array}{c}\text{Salt of}\\\text{Acid Part}\end{array} \quad \begin{array}{c}\text{Alcohol}\\\text{Part}\end{array} & &
\end{array}
$$

All derivatives of carboxylic acids will be converted to the parent acid upon basic hydrolysis. Thus, amides, which are not covered in this experiment, would also dissolve in this test, liberating the free amine and the sodium salt of the carboxylic acid.

SPECTROSCOPY

INFRARED

The ester-carbonyl group (C=O) peak is usually a strong absorption, as is that of the carbonyl-oxygen link (C—O) to the alcohol part.

C=O   Stretch at approximately 1735 cm^{-1} (5.75 μ) is normal†

C—O   Stretch usually gives two or more absorptions, one stronger than the others, in the region 1280 to 1050 cm^{-1} (7.8 to 9.5 μ)

NUCLEAR MAGNETIC RESONANCE

Hydrogens which are alpha to an ester carbonyl group have resonance in the region 2 to 3 δ. Hydrogens which are alpha to the alcohol oxygen of an ester have resonance in the region 3 to 5 δ.

---

†Conjugation with the carbonyl group moves the carbonyl absorption to lower frequencies (higher wavelength). Conjugation with the alcohol oxygen raises the carbonyl absorption to higher frequencies (lower wavelength). Ring strain (lactones) moves the carbonyl absorption to higher frequencies (shorter wavelengths).

DERIVATIVES

Esters present a double problem when trying to prepare derivatives. To completely characterize an ester, one needs to prepare derivatives of **both** the acid part and the alcohol part.

ACID PART

The most commonly used derivative of the acid part is the *N*-benzylamide derivative.

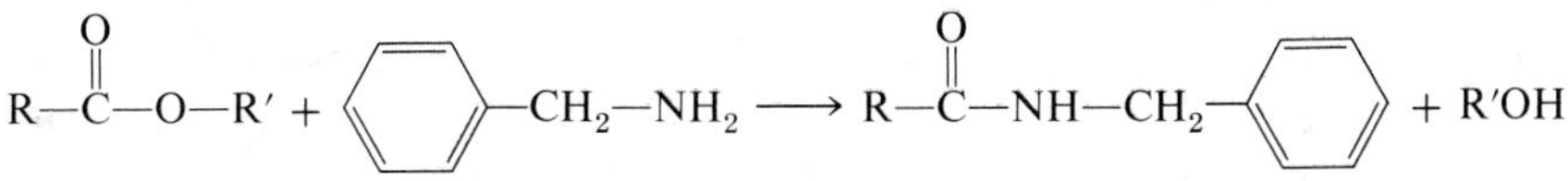

$$R-\overset{\overset{\displaystyle O}{\|}}{C}-O-R' + \text{(benzene)}-CH_2-NH_2 \longrightarrow R-\overset{\overset{\displaystyle O}{\|}}{C}-NH-CH_2-\text{(benzene)} + R'OH$$

An *N*-Benzylamide

The reaction does not proceed well unless R' is methyl or ethyl. For alcohol portions which are larger, the ester must be transesterified to a methyl or an ethyl ester before preparation of the derivative.

$$R-\overset{\overset{\displaystyle O}{\|}}{C}-OR' + CH_3OH \longrightarrow R-\overset{\overset{\displaystyle O}{\|}}{C}-O-CH_3 + R'OH$$

Hydrazine also reacts well with methyl and ethyl esters to give acid hydrazides.

$$R-\overset{\overset{\displaystyle O}{\|}}{C}-OR' + NH_2NH_2 \longrightarrow R-\overset{\overset{\displaystyle O}{\|}}{C}-NHNH_2 + R'OH$$

An Acid Hydrazide

The saponification equivalent (mentioned above) is also sometimes used. This value gives the molecular weight of the ester divided by the number of its ester groups.

ALCOHOL PART

The best derivative of the alcohol part of an ester is the 3,5-dinitro-benzoate ester, which is prepared by an acyl interchange reaction.

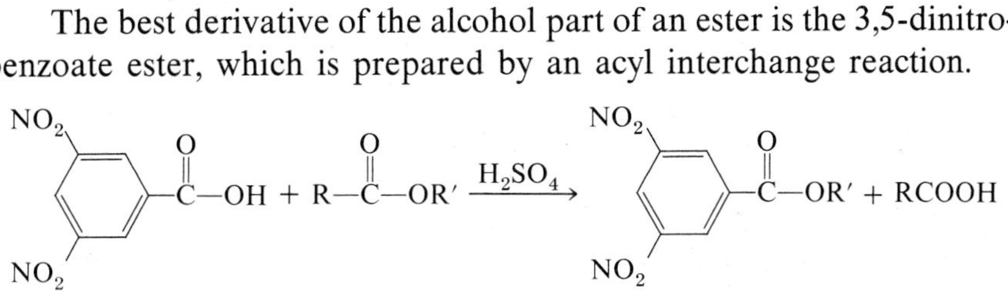

$$\text{(NO}_2\text{ benzene)}-\overset{\overset{\displaystyle O}{\|}}{C}-OH + R-\overset{\overset{\displaystyle O}{\|}}{C}-OR' \xrightarrow{H_2SO_4} \text{(NO}_2\text{ benzene)}-\overset{\overset{\displaystyle O}{\|}}{C}-OR' + RCOOH$$

A 3,5-Dinitrobenzoate
Ester

Most esters are composed of very simple acid and alkyl portions. For this reason, spectroscopy is usually a better method of identification than is the preparation of derivatives. Not only is it necessary to prepare two derivatives with an ester, but all esters with the same acid portion, or all those with the same alcohol portion, give identical derivatives of those portions.

Carbohydrates

The class of compounds known as the carbohydrates includes polyhydroxyaldehydes and polyhydroxyketones, or substances which, when hydrolyzed, give these compounds as products. The carbohydrates include sugars, starches, cellulose, and other substances found in roots, stems, and leaves of all plants, as well as glycogen, chitin, and other similar substances found in animals.

The **monosaccharides** are the carbohydrates which contain one polyhydroxyaldehyde or polyhydroxyketone unit, while the **oligosaccharides** contain several monosaccharide units linked together. **Polysaccharides** contain many monosaccharide units linked together into long chains.

MONOSACCHARIDES

Most of our knowledge of the structures of the carbohydrates comes from the pioneering work of Emil Fischer, who won the Nobel Prize in 1902 for his research. Fischer's classification of the monosaccharides according to their stereochemistry is based on the structure of the simplest monosaccharide, glyceraldehyde, whose naturally-occurring form rotates the plane of polarized light to the right. Since there is one chiral (asymmetric) carbon atom in glyceraldehyde, there are two possible stereoisomers, shown as I and II below.

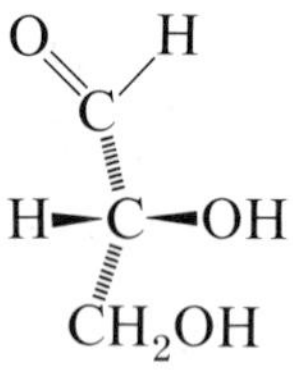

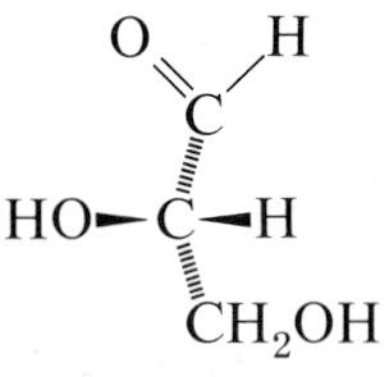

Fischer invented a method of representing the three-dimensional structure of a carbohydrate by means of a two-dimensional projection formula. Using the Fischer projection formula, structures I and II would be written:

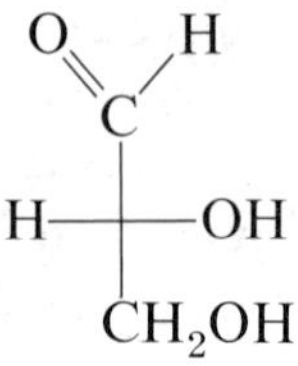

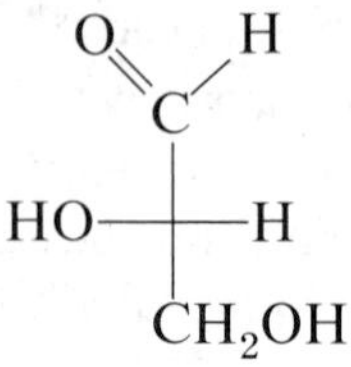

In this essay, projection formulas will be used to represent the structures of the more complicated monosaccharides.

Because structure I has the hydroxyl group pointed to the right, Fischer **arbitrarily** chose that structure as representing (+)-glyceraldehyde. He labeled this configuration D, and the configuration corresponding to structure II he labeled L. In 1951, the Dutch X-ray crystallographer J. M. Bijvoet showed that the correct absolute configuration of (+)-glyceraldehyde was indeed the D-configuration, thereby confirming Fischer's guess. Therefore, structure I corresponds to D-(+)-glyceraldehyde, and structure II corresponds to L-(−)-glyceraldehyde.

D-(+)-Glyceraldehyde is a three-carbon monosaccharide, so it is a **triose.** It is also an aldehyde, so it is an **aldose,** or an **aldotriose.** There is also a triose which is a ketone, and is called a **ketose** or a **ketotriose.** This ketotriose is dihydroxyacetone, which does not possess any chiral carbon atoms.

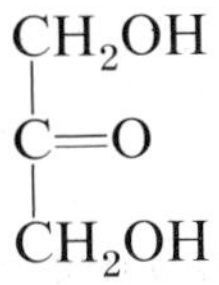

$$\begin{array}{c} CH_2OH \\ | \\ C{=}O \\ | \\ CH_2OH \end{array}$$

Dihydroxyacetone

A useful method for lengthening the carbon chain of a carbohydrate is the Fischer-Kiliani method, which involves the addition of cyanide to a carbonyl group. If such a procedure is attempted with D-(+)-glyceraldehyde, we obtain:

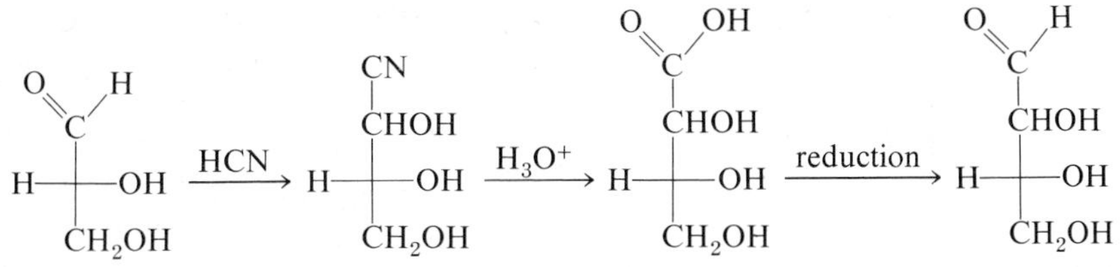

Notice that the configuration around carbon-2 of glyceraldehyde is not changed in this procedure. The reaction provides a mixture of two diastereomeric monosaccharides which are aldotetroses. Since they were prepared from an aldotriose belonging to the D-stereochemical series, both aldotetroses also belong to the D-series. In the Fischer projection, the second hydroxyl group from the bottom is pointed to the right. Each of the D-aldotetroses has an enantiomer which belongs to the L-series.

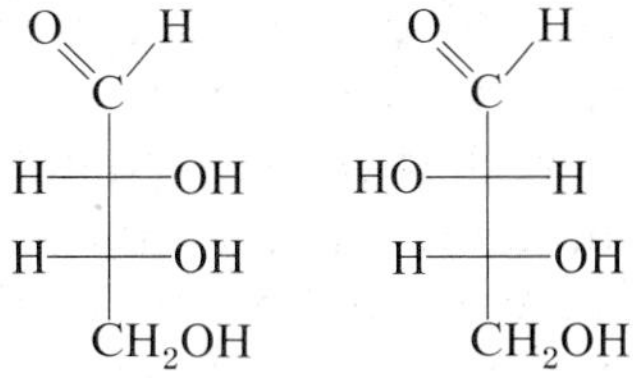

D-(−)-Erythrose D-(−)-Threose

A Fischer-Kiliani procedure, using the two aldotetroses as substrates, would yield four D-aldopentoses. Of course, each D-aldopentose has an enantiomer which is an *L*-aldopentose. Because the second hydroxyl group from the bottom in the Fischer projection formula is the one which derives from *D*-(+)-glyceraldehyde and because it points to the right, the aldopentoses derived from the *D*-aldotetroses belong to the *D*-series. Of these aldopentoses, *D*-(−)-ribose and *D*-(−)-arabinose are the most commonly-occurring in nature. *D*-(−)-Ribose forms the sugar backbone of the nucleic acids called RNA (see the Essay preceding Experiment 6 for a partial structure of a nucleic acid).

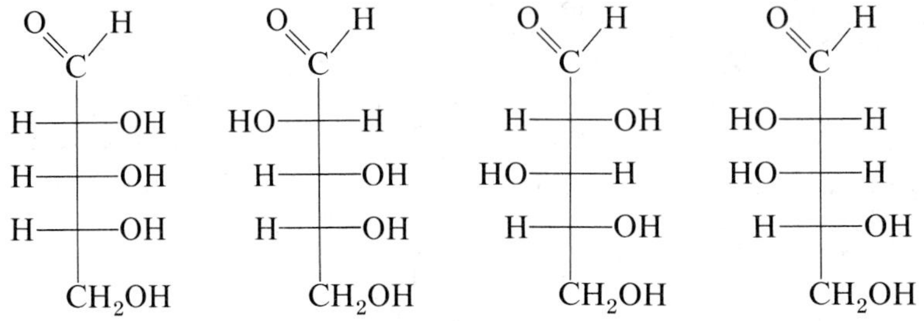

D-(−)-Ribose   *D*-(−)-Arabinose   *D*-(+)-Xylose   *D*-(−)-Lyxose

Of the eight possible *D*-aldohexoses, three are important. By tracing the stereochemical course of the Fischer-Kiliani synthesis, it can be demonstrated that carbon-5 of the aldohexoses is the same as carbon-2 of *D*-(+)-glyceraldehyde and that it has the same configuration. So, all of the aldohexoses shown above legitimately belong to the *D*-family.

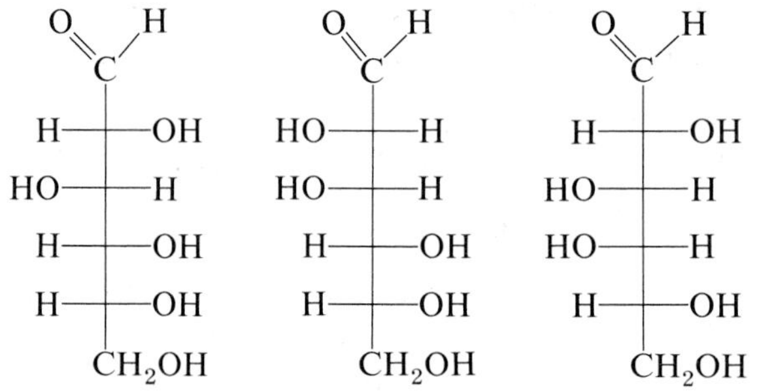

D-(+)-Glucose   *D*-(+)-Mannose   *D*-(+)-Galactose

Each of the aldohexoses is a diastereomer of all of the others. Each possesses an enantiomer, a sugar which is an exact mirror-image and which belongs to the *L*-series. Of the *D*-aldohexoses, only *D*-(+)-glucose, *D*-(+)-mannose, and *D*-(+)-galactose are commonly found in nature. Glucose is found in nearly all living systems, mannose is found to some extent in plants, and galactose is found in milk. Galactose also appears to play an important role in the structure of bacterial cell walls.

During the time that much of the work by Emil Fischer concerning the structures of the monosaccharides was progressing; certain facts were being accumulated which appeared to be inconsistent with the structures of the monosaccharides as free aldehydes or ketones. First, the monosaccharides fail to undergo certain reactions typical of aldehydes or ketones. Second, the monosaccharides exist in two isomeric forms which interconvert. Third, the monosaccharides can form two isomeric methyl acetals.

The apparent inconsistencies can be explained by observing that one of the hydroxyl groups of the monosaccharide is capable of reacting with the carbonyl group of the same molecule to form a cyclic hemi-acetal. A new chiral (asymmetric) center is generated in the process, and it can assume either of two possible configurations. Using D-$(+)$-glucose as an example, we may illustrate this cyclization process as follows.

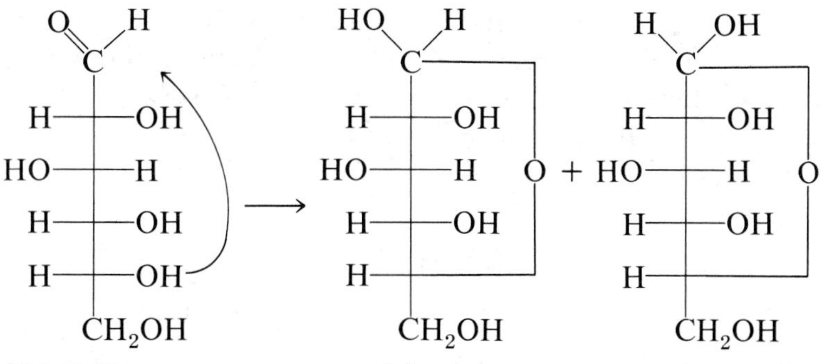

These structures differ in configuration at the hemiacetal position. They are known as **anomers** of each other. The anomers are designated α or β, depending upon the direction which the anomeric hydroxyl group points. A better way of depicting the stereochemistry of the hemiacetal forms of the monosaccharides is by means of conformational structures. The anomers are capable of interconverting through the intermediacy of the open-chain form.

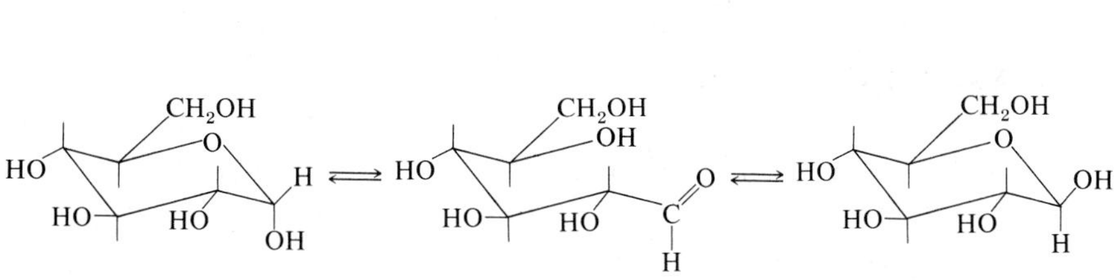

If a pure sample of the α-anomer of glucose is placed in aqueous solution, the magnitude of the specific rotation will decrease until a specific rotation of $+52.5°$ is reached. If a pure sample of the β-anomer of glucose is placed in aqueous solution, the magnitude of the specific rotation will increase until a specific rotation of $+52.5°$ is reached. This process, where either anomer converts to the other by means of the open-chain form to arrive at an equilibrium mixture of both anomers, is called **mutarotation.**

The small concentration of the open-chain form of monosaccharides present in solution at any given moment is sufficient to permit monosaccharides to reduce metal ions such as Ag^+ and Cu^{++} ions. This reducing property is observed by means of the Tollen's and Fehling's test or Benedict's test. Monosaccharides are **reducing sugars,** since they are capable of reducing the metal ions contained in these reagents.

The most important ketohexoses are D-$(-)$-fructose and L-$(-)$-sorbose. Fructose occurs in fruit and honey, as well as occurring as one of the units of the disaccharide sucrose, or table sugar. Sorbose is derived from the naturally-occurring sugar alcohol, D-$(-)$-sorbitol, and is used in the manufacture of vitamin C.

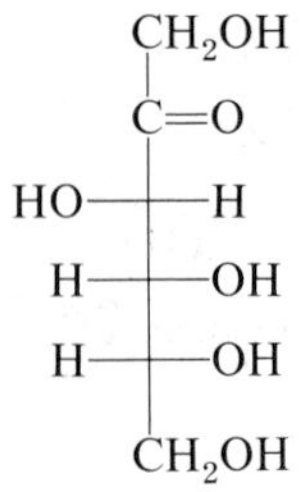

D-(−)-Fructose 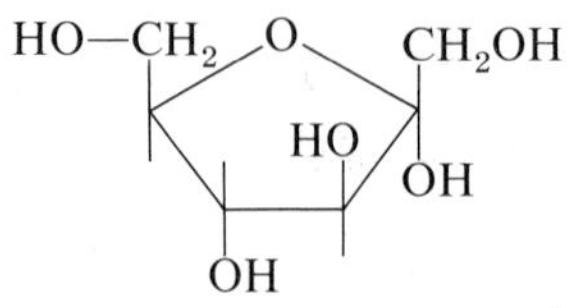   α-Anomer 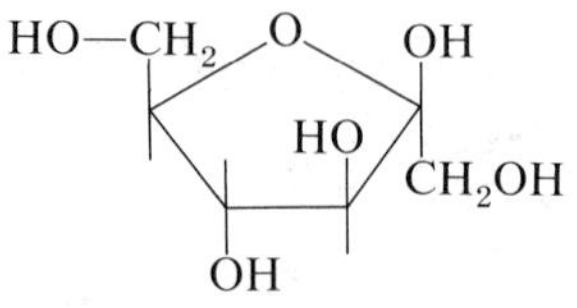   β-Anomer

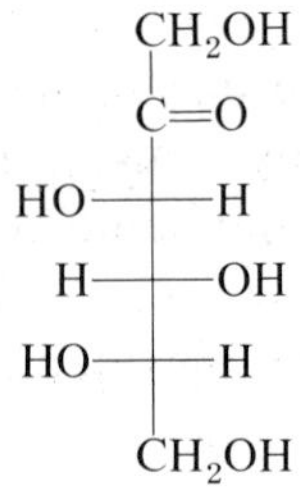

L-(−)-Sorbose 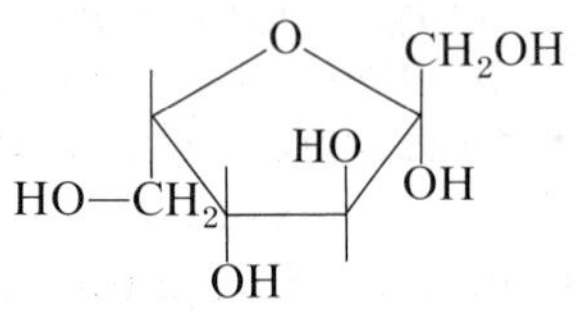   α-Anomer 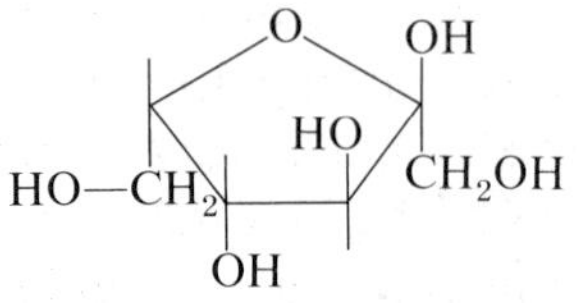   β-Anomer

OLIGOSACCHARIDES

A hydroxyl group of one molecule of a monosaccharide can react with the hemiacetal position of a second molecule of a monosaccharide to yield a molecule which contains the two monosaccharide units linked together by means of an **acetal** structure. The resulting product is a disaccharide. The process is illustrated using the α-anomeric form of D-$(+)$-glucose to form the disaccharide **maltose.** Because the two glucose units were the α-anomeric forms, the linkage between the two rings is called an **α-glycosidic linkage.** Maltose is obtained from the hydrolysis of starch. It is a reducing sugar because it still possesses a hemiacetal position capable of existing in equilibrium with the open-chain aldehyde form that can reduce Tollen's, Fehling's, or Benedict's solutions.

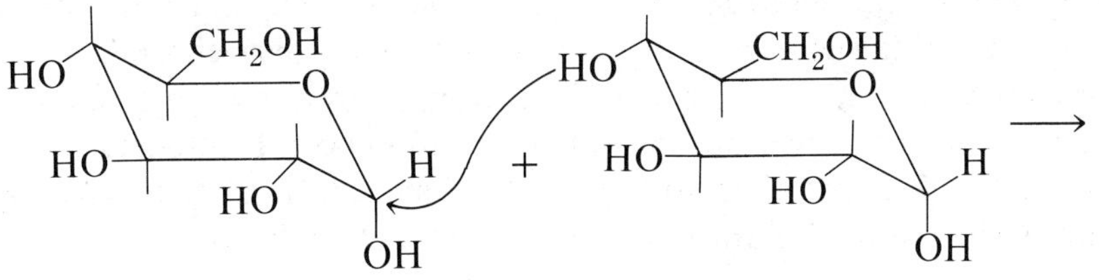

α-D-(+)-Glucose   α-D-(+)-Glucose

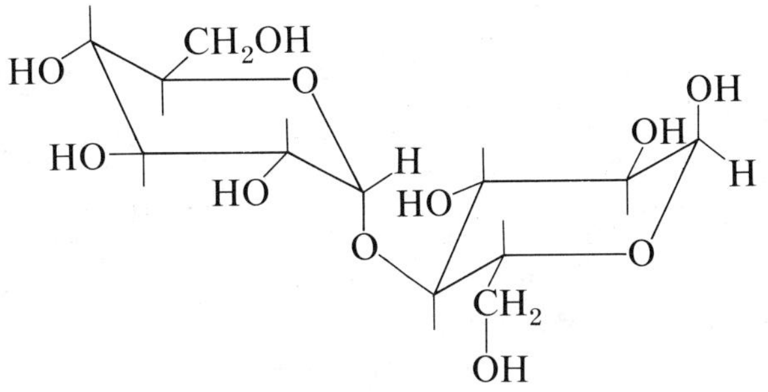

(+)-Maltose

Joining two molecules of the β-anomeric forms of D-(+)-glucose together gives the disaccharide (+)-**cellobiose,** which is obtained from the hydrolysis of cellulose. The linkage between the two rings is an example of a **β-glycosidic linkage.** Cellobiose is also a reducing sugar, since it possesses a free hemiacetal position.

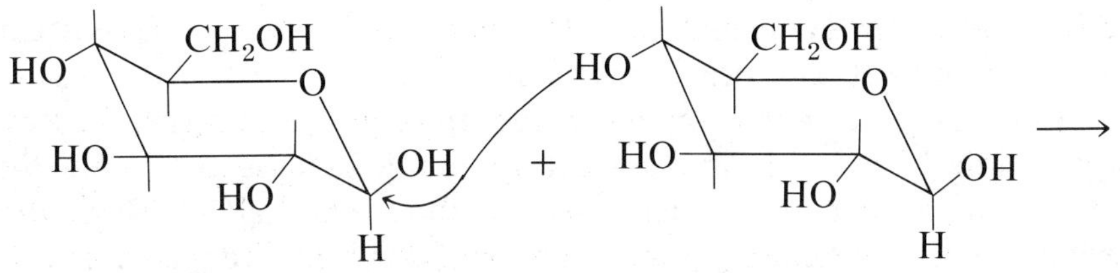

β-D-(+)-Glucose   β-D-(+)-Glucose

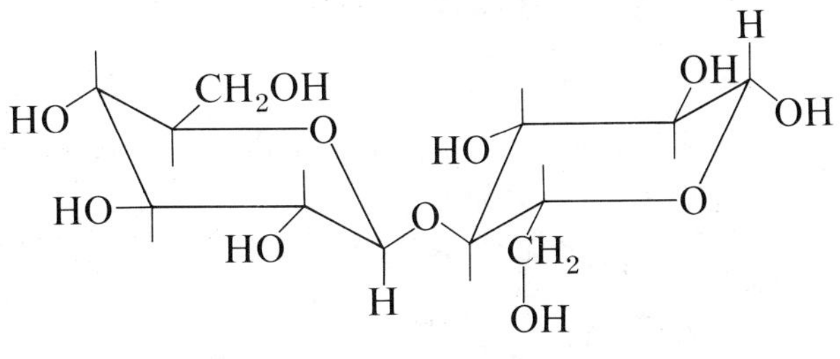

(+)-Cellobiose

Other common disaccharides are **lactose** and **sucrose,** whose structures are shown below. Lactose is formed from the β-anomer of galactose and the α-anomer of glucose. Since it is the principal carbohydrate of milk, it is known as **milk sugar.** Lactose is a reducing sugar, since it possesses a free hemiacetal position. Sucrose is formed from the α-anomer of glucose and the β-anomer of fructose. Sucrose is produced in large quantities from sugar cane and sugar beets, and it is known as **table sugar.** Sucrose is a **non-reducing** disaccharide, since it does not possess a free hemiacetal position. In forming the glycosidic linkage between the two monosaccharide units, both hemiacetal positions react with each other and both are converted to acetals. Acetals do not exist in equilibrium with open-chain aldehyde forms.

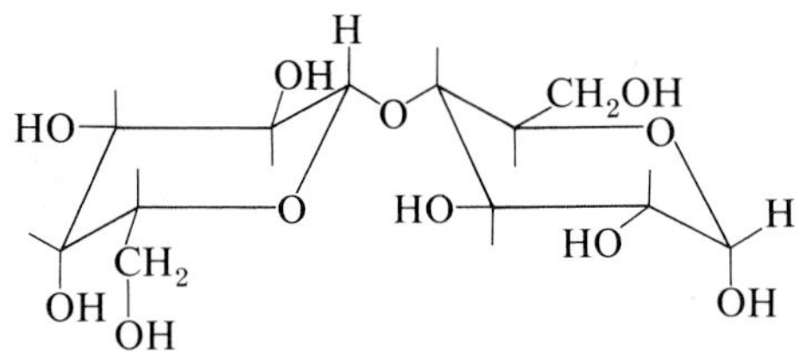

(+)-Lactose

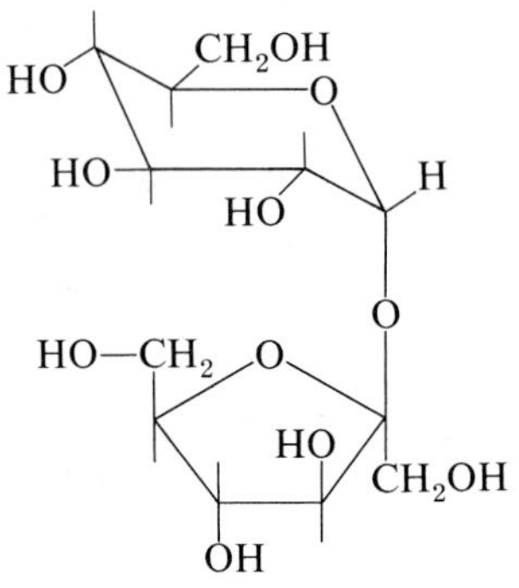

(+)-Sucrose

POLYSACCHARIDES

Naturally-occurring polysaccharides frequently are large polymers of monosaccharide units containing anywhere from 200 to several hundred thousand of these monosaccharide units in each molecule. The number of monosaccharide units per molecule may vary among samples derived from the same material. The polysaccharides are important structural and energy-storage materials.

Cellulose is a polysaccharide made up of glucose units linked by β-glycosidic linkages. It is the most abundant organic compound of natural origin on earth. It accounts for about one-half of all of the carbon atoms contained in the plant kingdom. In the plant world, cellulose is primarily found as a component of the cell wall, functioning chiefly as a structural material.

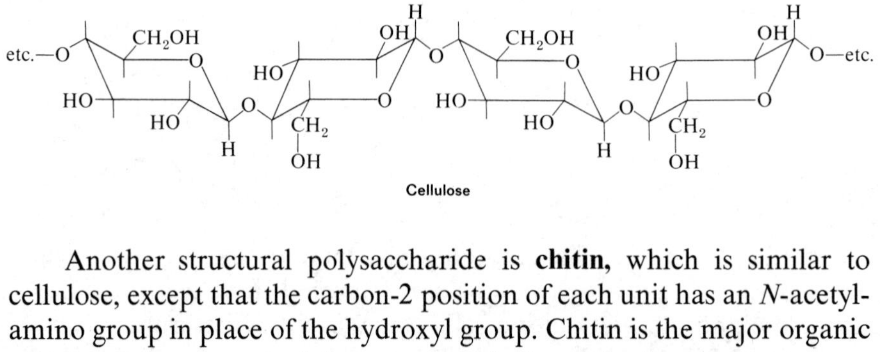

Cellulose

Another structural polysaccharide is **chitin,** which is similar to cellulose, except that the carbon-2 position of each unit has an *N*-acetyl-amino group in place of the hydroxyl group. Chitin is the major organic structural component of the external shells of the invertebrates, e.g., crabs and lobsters.

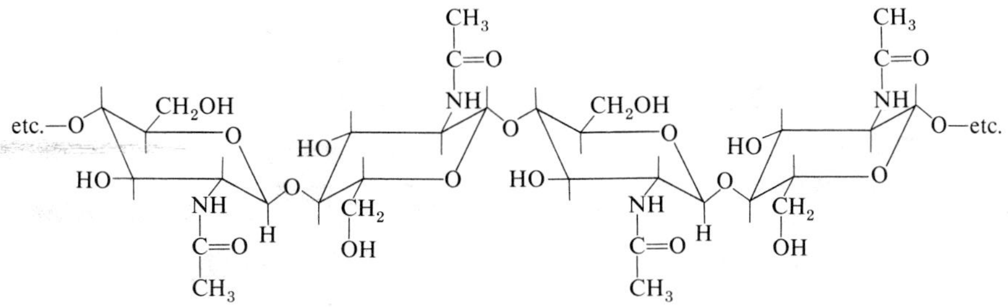

Chitin

Starches are polysaccharides in which the monosaccharide units are linked by α-glycosidic linkages. They serve as the major source of food for some animals, and they are found in potatoes, rice, wheat, and cereal grains. Actually, starch is not a single molecule, but a mixture of two structurally distinct polysaccharides. One component is **amylose,** which is a linear polymer of the α-anomers of D-($+$)-glucose. The other component is **amylopectin,** which is a branched molecule with a small number of α-glycosidic linkages connecting carbon-1 of one glucose unit to carbon-6 of another unit, as well as the usual linkages between carbon-1 of one unit and carbon-4 of another unit. For the sake of simplicity, neither the stereochemistry of these starch components nor all the hydroxyls are shown in the structures below. Only the rings are shown.

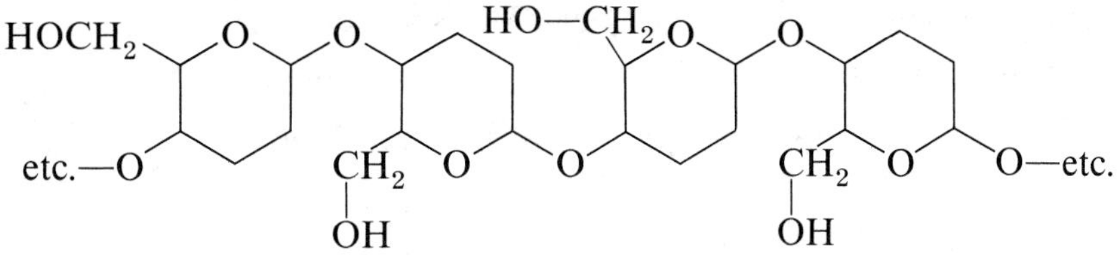

Amylose

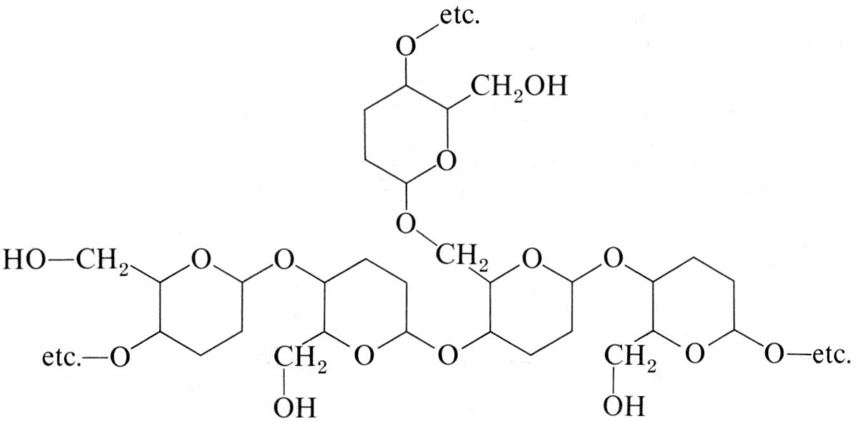

Amylopectin

The starches are molecular storage reservoirs. When the proper functioning of the plant so requires, the starches are enzymatically degraded to simpler sugars, which are then metabolized further to supply the energy and carbon atoms required for synthesis. Starch is present to varying degrees in the diet of all animals.

Glycogen plays the same role in animals that starch plays in plants; it is the carbon atom and energy storage material. Glycogen is most commonly found in the liver and muscle of higher animals. Structurally, glycogen is similar to amylopectin, except that it is more highly branched. Branch points on glycogen occur every 8 or 10 residues along the central core, whereas in amylopectin, branch points occur every 25 to 30 residues.

REFERENCES

Lehninger, A. L. *A Short Course in Biochemistry*. New York: Worth Publishers, 1973. Chapter 5.
Sharon, N. "The Bacterial Cell Wall." *Scientific American, 220* (May, 1969), 92.
White, A., Handler, P., and Smith, E. L. *Principles of Biochemistry* (5th edition). New York: McGraw-Hill Book Co., 1973. Chapters 2 and 3.

Experiment **51**
CARBOHYDRATES

In this experiment, you will perform tests that distinguish between various carbohydrates. The carbohydrates which are included in this study and the classes they represent are as follows:

Aldopentoses:	xylose and arabinose
Aldohexoses:	glucose and galactose
Ketohexoses:	fructose
Disaccharides:	lactose and sucrose
Polysaccharides:	starch and glycogen

The structures of these carbohydrates are given or are discussed in the essay which precedes this experiment.

The tests are divided into the following groups:

A. Tests based on the production of furfural or a furfural derivative: Molisch test, Bial's test, and Seliwanoff's test

B. Tests based on the reducing property of a carbohydrate (sugar): Benedict's test, and Barfoed's test

C. Osazone formation

D. Iodine test for starch

E. Hydrolysis of sucrose

F. Mucic acid test for galactose and lactose

G. Tests on unknowns

SPECIAL INSTRUCTIONS

Read the essay which precedes this experiment. All of the procedures in this experiment involve simple test tube reactions. Most of the tests are rather short. However, Seliwanoff's test, osazone formation, and the mucic acid test take a somewhat longer time to complete. You will need a minimum of 10 test tubes numbered in order. Clean them carefully each time they are used. The 1% solutions of carbohydrates and the reagents needed for the tests have been prepared in advance by the laboratory instructor or assistant. Be sure to shake the starch solution before using it.

A. TESTS BASED ON PRODUCTION OF FURFURAL OR A FURFURAL DERIVATIVE

Under acidic conditions, aldopentoses and ketopentoses **rapidly** undergo dehydration to give furfural by equation 1. Ketohexoses **rapidly** yield 5-hydroxymethylfurfural by equation 2. Disaccharides and polysaccharides may first be hydrolyzed in an acid medium to produce monosaccharides, which may then react to give furfural or 5-hydroxymethylfurfural.

(1)

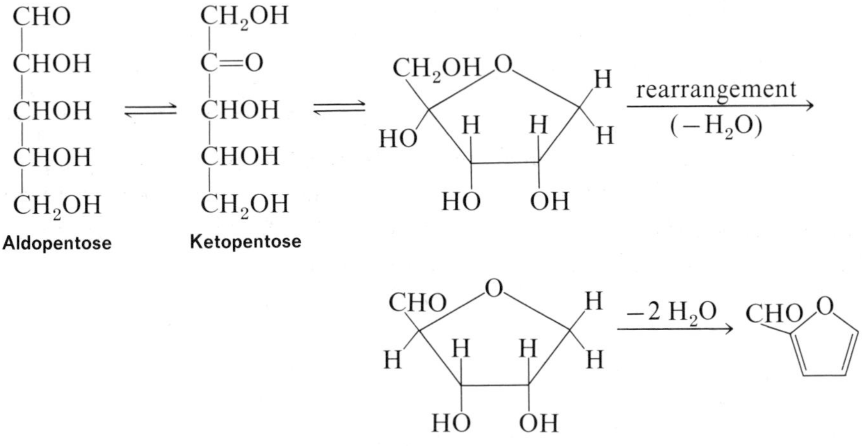

Aldopentose   Ketopentose

Furfural

(2)

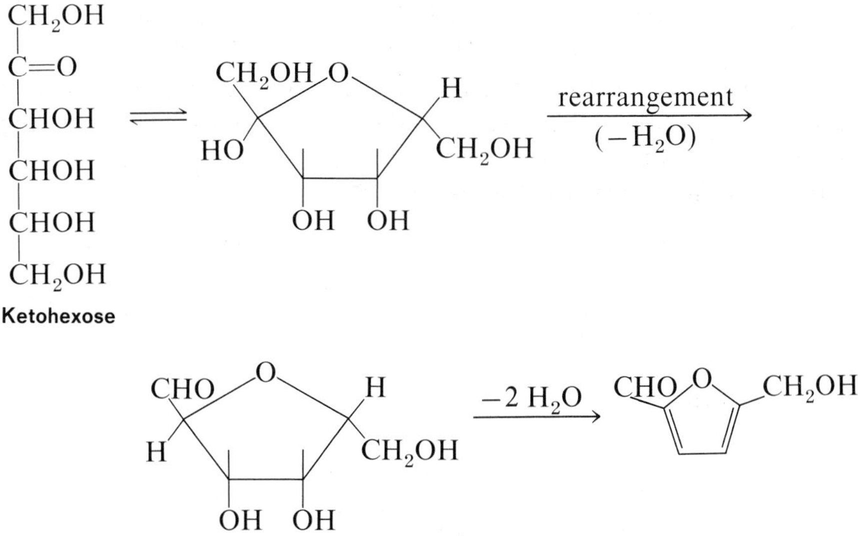

Ketohexose

5-Hydroxymethylfurfural

Aldohexoses are **slowly** dehydrated to 5-hydroxymethylfurfural. One possible mechanism is shown in equation 3. The mechanism is different from that given in equations 1 and 2 in that dehydration occurs at an early step, and the rearrangement step is absent.

(3)

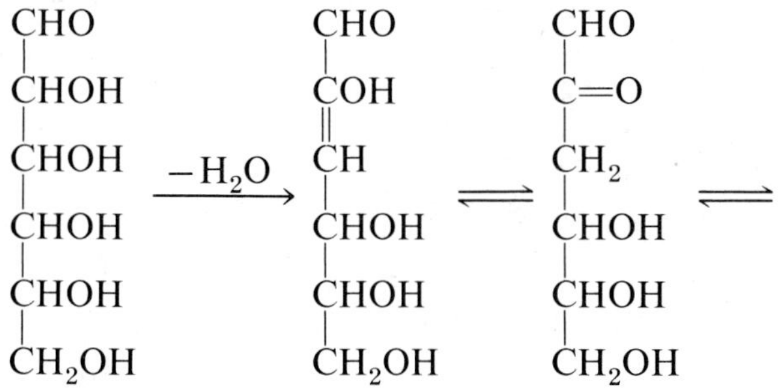

$$
\begin{array}{c}
\text{CHO} \\
\text{CHOH} \\
\text{CHOH} \\
\text{CHOH} \\
\text{CHOH} \\
\text{CH}_2\text{OH}
\end{array}
\quad \xrightarrow{-\text{H}_2\text{O}} \quad
\begin{array}{c}
\text{CHO} \\
\text{COH} \\
\text{CH} \\
\text{CHOH} \\
\text{CHOH} \\
\text{CH}_2\text{OH}
\end{array}
\quad \rightleftharpoons \quad
\begin{array}{c}
\text{CHO} \\
\text{C}=\text{O} \\
\text{CH}_2 \\
\text{CHOH} \\
\text{CHOH} \\
\text{CH}_2\text{OH}
\end{array}
\quad \rightleftharpoons
$$

Aldohexose

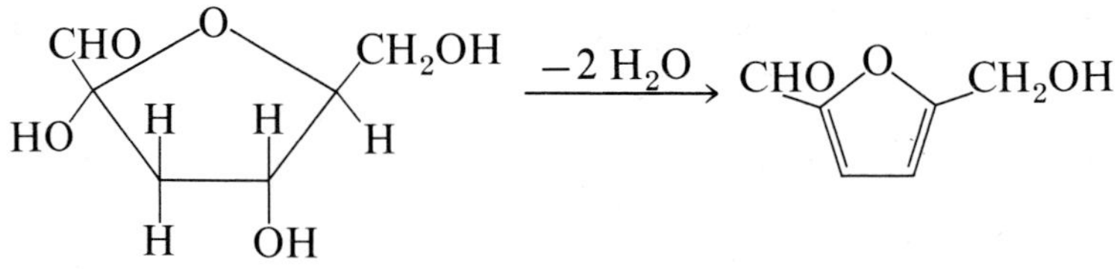

$$\xrightarrow{-2\,\text{H}_2\text{O}}$$

5-Hydroxymethylfurfural

Once furfural or 5-hydroxymethylfurfural is produced by equations 1, 2, or 3, either will then react with a phenol to produce a colored condensation product. α-Naphthol is used in the Molisch test, orcinol in Bial's test, and resorcinol in Seliwanoff's test.

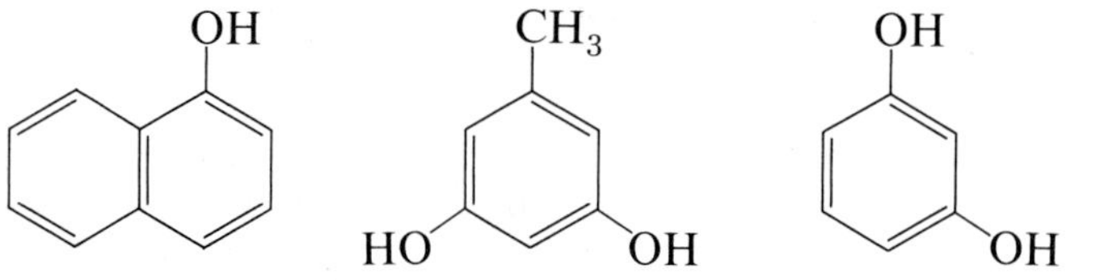

| α-Naphthol (Molisch Test) | Orcinol (Bial's Test) | Resorcinol (Seliwanoff's Test) |

The colors and the rates of formation of these colors are used to differentiate between the carbohydrates. The various color tests are discussed in sections 1, 2, and 3. A typical colored product formed from furfural and α-naphthol (Molisch test) is the following (equation 4):

(4)

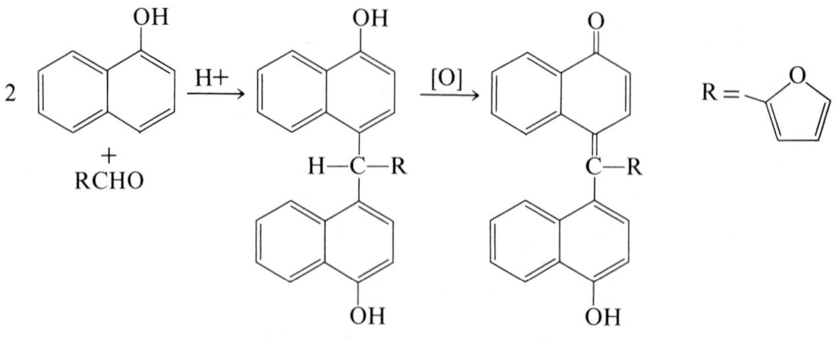

Purple

1. MOLISCH TEST FOR CARBOHYDRATES

This test is a **general** test for carbohydrates. Most carbohydrates are dehydrated with concentrated sulfuric acid to form furfural or 5-hydroxyfurfural. These furfurals react with the α-naphthol present in the test reagent to give a purple colored product. Compounds other than carbohydrates may react with the reagent to give a positive test. However, a negative test usually indicates an absence of a carbohydrate.

Procedure for Molisch Test. Place 4 ml of each of the following 1% carbohydrate solutions in nine separate test tubes: xylose, arabinose, glucose, galactose, fructose, lactose, sucrose, starch (shake it), and glycogen. Also add 4 ml of distilled water to another tube to serve as a control.

Add 2 drops of the Molisch reagent[1] to each test tube and thoroughly mix the contents of the tube. Tilt each test tube slightly, and cautiously add 5 ml of concentrated sulfuric acid down the sides of the tubes. An acid layer forms at the bottom of the tubes. Note and record the color at the interface between the two layers in each tube. A purple color constitutes a positive test.

2. BIAL'S TEST FOR PENTOSES

This test is used to differentiate pentose sugars from hexose sugars. Pentose sugars yield furfural upon dehydration in acidic solution. Furfural reacts with orcinol and ferric chloride to give a blue-green condensation product. Hexose sugars give 5-hydroxymethylfurfural which reacts with the reagent to yield colors such as green, brown, and reddish brown.

Procedure for Bial's Test. Place 2 ml of each of the following 1% carbohydrate solutions in separate test tubes: xylose, arabinose, glucose, galactose, fructose, lactose, sucrose, starch (shake it), and glycogen. Also add 2 ml of distilled water to another tube to serve as a control.

Add 3 ml of Bial's reagent[2] to each test tube. Carefully heat each tube over a Bunsen burner flame until the mixture just begins to boil. Note and record the color which is produced in each test tube. If the color is not distinct, add 5 ml of water and 1 ml of 1-pentanol to the test tube. After shaking them, again observe and record the color. The colored condensation product will be concentrated in the 1-pentanol layer.

---

[1]Dissolve 5 g of α-naphthol in 100 ml of 95% ethanol.
[2]Dissolve 3 g of orcinol in 1 liter of concentrated hydrochloric acid and add 3 ml of 10% aqueous ferric chloride.

3.   SELIWANOFF'S TEST FOR KETOHEXOSE

This test depends on the relative rates of dehydration of carbohydrates. A ketohexose reacts rapidly by equation 2 to give 5-hydroxy-
methylfurfural, whereas an aldohexose reacts more slowly by equation 3
to give the same product. Once 5-hydroxymethylfurfural is produced, it
reacts with resorcinol to give a dark red condensation product. If the
reaction is followed for some time, sucrose will hydrolyze to give
fructose which will eventually react to produce a dark red color.

Procedure for Seliwanoff's Test. Prepare a boiling water bath for
use in this experiment. Place 1 ml of each of the following 1% carbohydrate solutions in separate test tubes: xylose, arabinose, glucose,
galactose, fructose, lactose, sucrose, starch (shake it), and glycogen.
Add 1 ml of distilled water to another tube to serve as a control.

Add 4 ml of Seliwanoff's reagent[3] to each test tube. Place all 10
tubes in a beaker of boiling water for **60 seconds.** Remove them and
note the results in the notebook.

For the remainder of Seliwanoff's test, it is convenient to place a
group of 3 or 4 tubes in the boiling water bath, and to complete the
observations before proceeding to the next group of tubes. Place 3 or 4
tubes in the boiling water bath. Observe the color in each of the tubes at
one minute intervals for 5 minutes beyond the original one minute.
Record the results at each one minute interval. Leave the tubes in the
boiling water bath during the entire 5 minute period. After the first
group has been observed, remove that set of test tubes, and place the
next group of 3 or 4 tubes in the bath. Again, follow the color changes
as before. Finally, place the last group of tubes in the bath and follow
the color changes over the 5 minute period.

B.   TESTS BASED ON THE REDUCING
PROPERTY OF A CARBOHYDRATE (SUGAR)

Monosaccharides, and those disaccharides which possess a potential
aldehyde group, will reduce reagents such as Benedict's solution to
produce a red precipitate of cuprous oxide.

$$RCHO + 2\,Cu^{++} + 4\,OH^- \longrightarrow RCOOH + \ Cu_2O + 2\,H_2O$$
$$\text{(Red Precipitate)}$$

Glucose, for example, is a typical aldohexose which shows reducing
properties. The two diastereomeric α- and β-D-glucoses are in equilibrium with each other in aqueous solution. α-D-Glucose opens at the

---

[3]Dissolve 0.5 g of resorcinol in 1 liter of dilute hydrochloric acid (1 volume of concentrated
hydrochloric acid and 2 volumes of distilled water).

anomeric carbon atom (hemiacetal) to produce the free aldehyde. This aldehyde rapidly closes to give β-*D*-glucose, and a new hemiacetal is produced. It is the presence of this free aldehyde which makes glucose a reducing carbohydrate (sugar). It reacts with Benedict's reagent to produce a red precipitate, which is the basis of the test. Carbohydrates which possess the hemiacetal functional group exhibit reducing properties.

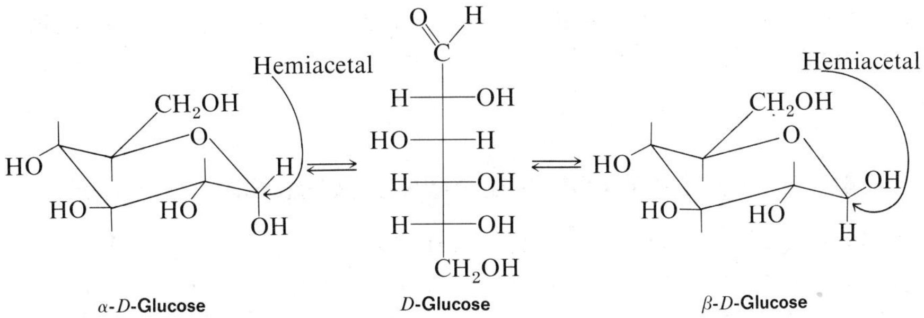

α-*D*-Glucose   *D*-Glucose   β-*D*-Glucose

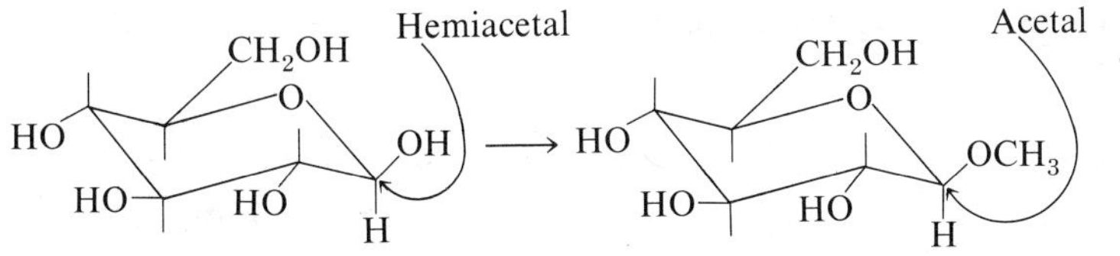

If the hemiacetal is converted to an acetal by methylation, the carbohydrate (sugar) will no longer reduce Benedict's reagent.

With disaccharides, two situations may arise. If the anomeric carbon atoms are bonded to each other (head to head) to give an acetal, then the sugar will not reduce Benedict's reagent. If, however, the sugar molecules are joined head to tail, then one end will still be able to equilibrate through the free aldehyde form (hemiacetal). Examples of a reducing and a non-reducing disaccharide are given below.

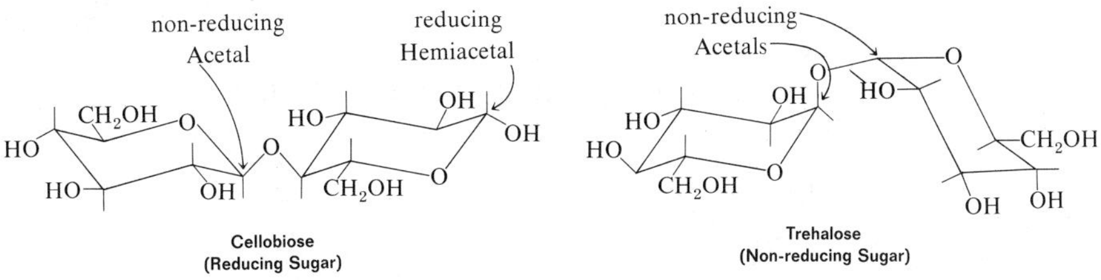

Cellobiose
(Reducing Sugar)

Trehalose
(Non-reducing Sugar)

1. BENEDICT'S TEST FOR REDUCING SUGARS

Benedict's test is performed under mildly basic conditions. The reagent reacts with all reducing sugars to produce a red precipitate of cuprous oxide, as shown above. It also reacts with water-soluble aldehydes which are not sugars. Ketoses, such as fructose, also react with Benedict's reagent. Benedict's test is considered one of the classical tests for determining the presence of an aldehyde functional group.

Procedure for Benedict's Test. Prepare a boiling water bath for use in this experiment. Place 1 ml of each of the following 1% carbohydrate solutions in separate test tubes: xylose, arabinose, glucose, galactose, fructose, lactose, sucrose, starch (shake it), and glycogen. Add 1 ml of distilled water to another tube to serve as a control.

Add 5 ml of Benedict's reagent[4] to each test tube. Place the test tubes in a boiling water bath for 2 to 3 minutes. Remove the tubes and note the results in a notebook. A red, brown, or yellow precipitate indicates a positive test for a reducing sugar. Ignore a change in color of the solution. A precipitate must form in order to constitute a positive test.

2. BARFOED'S TEST FOR REDUCING MONOSACCHARIDES

This test distinguishes between reducing monosaccharides and reducing disaccharides on the basis of a difference in the rate of reaction. The reagent consists of cupric ions, as does Benedict's reagent. However, in this test, Barfoed's reagent reacts with reducing monosaccharides, to produce cuprous oxide, at a faster rate than with reducing disaccharides.

$$RCHO + 2\,Cu^{++} + 2\,H_2O \longrightarrow RCOOH + \underset{\text{(Red precipitate)}}{Cu_2O} + 4\,H^+$$

Procedure for Barfoed's Test. Place 1 ml of each of the following 1% carbohydrate solutions in separate test tubes: xylose, arabinose, glucose, galactose, fructose, lactose, sucrose, starch (shake it), and glycogen. Add 1 ml of distilled water to another tube to serve as a control.

Add 5 ml of Barfoed's reagent[5] to each test tube. Place the tubes in a boiling water bath for 10 minutes. Remove the tubes and note the results in a notebook.

C. OSAZONE FORMATION

Carbohydrates react with phenylhydrazine to form crystalline derivatives, called **osazones.**

$$
\begin{array}{ccccc}
CHO & & HC{=}NNHPh & & HC{=}NNHPh \\
CHOH & & CHOH & & C{=}NNHPh \\
CHOH & \xrightarrow{PhNHNH_2} & CHOH & \xrightarrow{2PhNHNH_2} & CHOH \\
CHOH & & CHOH & & CHOH \quad + NH_3 + PhNH_2 \\
CHOH & & CHOH & & CHOH \\
CH_2OH & & CH_2OH & & CH_2OH
\end{array}
$$

---

[4] Dissolve 173 g of hydrated sodium citrate and 100 g of anhydrous sodium carbonate in 800 ml of distilled water, with heating. Filter the solution. Add to it a solution of 17.3 g of cupric sulfate ($CuSO_4 \cdot 5\,H_2O$) dissolved in 100 ml of distilled water. Dilute the combined solutions to 1 liter.

[5] Dissolve 66.6 g of cupric acetate in 1 liter of distilled water. Filter the solution, if necessary, and add 9 ml of glacial acetic acid.

An osazone may be isolated as a derivative, and its melting point may be determined. However, some of the monosaccharides give **identical** osazones (glucose, fructose, and mannose). Also, the melting points of different osazones are often similar in magnitude. This limits the usefulness of an isolation of the osazone derivative.

A good experimental use for osazone formation is to observe the rate of formation of an osazone. The rates of reaction vary greatly even though the **same** osazone may be produced from different sugars. As an example, fructose forms a precipitate in about 2 minutes, while glucose forms a precipitate about 5 minutes later. The osazone is the same in each case. The crystal structure of the osazone is often distinctive, as well. Arabinose, for example, produces a fine precipitate, while glucose produces a coarse precipitate.

Procedure for Osazone Formation. A boiling water bath is needed for this experiment. Place 0.1 g of each of the following carbohydrates in separate test tubes: xylose, arabinose, glucose, galactose, fructose, lactose, sucrose, starch (shake it), and glycogen. Add 1 ml of distilled water to each tube. Then add 5 ml of phenylhydrazine reagent[6] to each tube. Place the tubes in a boiling water bath simultaneously. Watch for a formation of a precipitate, or in some cases, cloudiness. Note the time at which the precipitate begins to form. After 30 minutes, cool the tubes and record the crystalline form of the precipitates. Reducing disaccharides will not precipitate until the tubes are cooled. Non-reducing disaccharides will hydrolyze first, and then the osazone(s) will precipitate.

D. IODINE TEST FOR STARCH

Starch forms a typical blue color with iodine. This color is due to the absorption of iodine into the open spaces of the amylose molecules (helices) present in starch. Amylopectins, which are the other types of molecules present in starch, form a red to purple color with iodine.

Procedure for the Iodine Test. Place 2 ml of each of the following 1% carbohydrate solutions in three separate test tubes: glucose, starch (shake it), and glycogen. Add 2 ml of distilled water to another tube to serve as a control.

Add 1 drop of iodine solution[7] to each test tube and observe the results. Add a few drops of sodium thiosulfate[7] to the solutions and note the results.

---

[6]Dissolve 50 g of phenylhydrazine hydrochloride and 75 g of sodium acetate trihydrate in 500 ml of distilled water. The reagent is somewhat unstable.

[7] *The iodine solution is prepared as follows:* Dissolve 2 g of potassium iodide in 50 ml of distilled water. Add 1 g of iodine and shake the solution until the iodine dissolves. Dilute the solution to 100 ml.

The sodium thiosulfate solution is prepared as follows: Dissolve 2.5 g of sodium thiosulfate in 100 ml of water.

E. HYDROLYSIS OF SUCROSE

Sucrose can be hydrolyzed in acid solution to its component parts, fructose and glucose. The component parts may then be tested with Benedict's reagent.

Procedure for the Hydrolysis of Sucrose. Place 5 ml of a 1% solution of sucrose in a test tube. Add 2 drops of concentrated hydrochloric acid and heat the tube in a boiling water bath for 10 minutes. Cool the tube and neutralize the contents with 10% sodium hydroxide solution until the mixture is just basic to litmus (about 20 drops are needed). Test the mixture with Benedict's reagent (Part B). Note the results and compare them to the results obtained on sucrose.

F. MUCIC ACID TEST FOR GALACTOSE AND LACTOSE

Procedures are given in Experiment 52 for the oxidation of galactose and lactose to mucic acid. This test confirms the presence of galactose or a galactose moiety in a carbohydrate (sugar).

G. TESTS ON UNKNOWNS

Obtain an unknown solid carbohydrate from the laboratory instructor or assistant. The unknown will be one of the following carbohydrates: xylose, arabinose, glucose, galactose, fructose, lactose, sucrose, starch, or glycogen. Carefully dissolve part of the unknown in distilled water to prepare 1% solution (0.25 g carbohydrate in 25 ml water). Save the remainder for the tests requiring solid material. Apply whatever tests are necessary to identify the unknown.

At the instructor's option, the optical rotation may be obtained as part of the experiment. Experimental details are given in Experiment 53 and Technique 15. Optical rotation data and decomposition points for carbohydrates and osazones are given in the standard reference works on identification of organic compounds (Experiment 50).

QUESTIONS

1. Find the structures for the following carbohydrates (sugars) in a reference work or textbook, and decide if they are reducing or non-reducing carbohydrates (sugars): sorbose, mannose, ribose, maltose, raffinose, and cellulose.

2. Mannose gives the same osazone as glucose. Explain.

3. Predict the results of the following tests with the carbohydrates listed in question 1: Molisch, Bial's, Seliwanoff's (after 1 minute and 6 minutes), Barfoed's, and mucic acid tests.

4. Give a mechanism for the hydrolysis of the acetal linkage in sucrose.

5. The rearrangement that occurs in equations 1 and 2 may be considered a type of pinacol rearrangement. Give a mechanism for that step.

6. Give a mechanism for the acid catalyzed condensation of furfural with 2 moles of α-naphthol shown in equation 4.

ESSAY

Chemistry of Milk

Milk is a food of exceptional interest. Not only is milk an excellent food for the very young, but man has also adopted milk, specifically cow's milk, as a food substance for persons of all ages. Many specialized milk products like cheese, yogurt, butter, and ice cream are staples of our diet.

Milk is probably the most nutritionally complete food which can be found in nature. This property is, of course, required of milk, as it is the only food consumed by young mammals in the nutritionally important weeks following birth. Whole milk contains vitamins (principally thiamine, riboflavin, pantothenic acid, and vitamins A, D, and K), minerals (calcium, potassium, sodium, phosphorous, and trace metals), proteins (which include all of the essential amino acids), carbohydrates (chiefly lactose), and lipids (fats). The only important elements in which milk is seriously deficient are iron and vitamin C. Infants are usually born with a storage supply of iron large enough to meet their needs for several weeks. Vitamin C is easily obtained through an orange juice supplement. The average composition of the milk of several mammals is summarized in the table which follows.

AVERAGE PERCENT COMPOSITION OF THE MILK FROM VARIOUS MAMMALS

	COW	MAN	GOAT	SHEEP	HORSE
Water	87.1	87.4	87.0	82.6	90.6
Protein	3.4	1.4	3.3	5.5	2.0
Fats	3.9	4.0	4.2	6.5	1.1
Carbohydrates	4.9	7.0	4.8	4.5	5.9
Minerals	0.7	0.2	0.7	0.9	0.4

Fats. Whole milk is an oil-water type emulsion, containing about 4% fat dispersed as very small (5 to 10 microns in diameter) globules. The globules are so small that a drop of milk contains about a million of them. Because the fat is so finely dispersed, the fat in milk is digested more easily than the fat from any other source. The fat emulsion is stabilized to some extent by complex phospholipids and proteins which are adsorbed on the surfaces of the globules. On standing, the fat globules, which are lighter than water, coalesce and eventually rise to the surface of the milk, forming a layer of **cream.** Since vitamins A and D are fat-soluble vitamins, they are carried to the surface with the cream. Commercially, the cream is often removed by centrifugation and skimming, and either diluted to form coffee cream ("half and half"), sold as **whipping cream,** converted to **butter,** or converted to **ice cream.** The milk which remains is called **skim milk.** Skim milk, other than lacking the fats and vitamins A and D, has approximately the same composition as whole milk. If milk is **homogenized,** its fatty content will not separate. Milk is homogenized by forcing it through a small hole. This breaks up the fat globules and reduces their size to about 1 to 2 microns in diameter.

The structure of fats and oils is discussed in the essay which precedes Experiment 14. The fats present in milk are primarily triglycerides. For the saturated fatty acids, the following percentages have been reported:

C_2 (3%) C_8 (2.7%) C_{14} (25.3%) $>C_{18}$ (~5%)
C_4 (1.4%) C_{10} (3.7%) C_{16} (9.2%)
C_6 (1.5%) C_{12} (12.1%) C_{18} (1.3%)

Thus, about two thirds of all the fatty acids in milk are saturated, and about one third are unsaturated. Milk is unusual in that about 12% of the fatty acids are **short** chain fatty acids (C_2—C_{10}) like butyric, caproic, and caprylic acids.

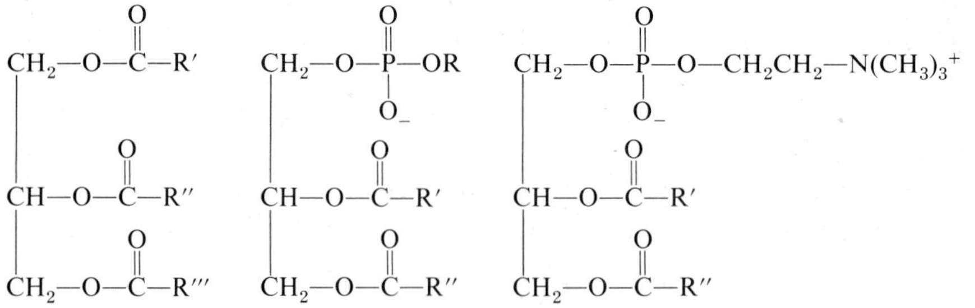

A Triglyceride   A Phospholipid   A Lecithin

Additional lipids (fats and oils) found in milk include small amounts of cholesterol (see the essay preceding Experiment 8), phospholipids, and lecithins (phospholipids conjugated with choline). The latter two types of lipids are illustrated in the figures. As mentioned above, the phospholipids help to stabilize the whole milk emulsion, the phosphate groups helping to achieve partial water solubility for the fat globules. All of the fat can be removed from milk by extraction with petroleum ether or a similar organic solvent.

Proteins. Proteins are described in the essay which precedes Experiment 54. They may be divided broadly into two general categories, fibrous and globular. Globular proteins are those which tend to fold back on themselves into compact units that approach nearly spheroidal shapes. These types of proteins do not form intermolecular interactions between protein units (H-bonds, etc.) as do fibrous proteins, and are more easily solubilized as colloidal suspensions. The proteins of milk are globular, and of three types: the **caseins,** the **lactalbumins,** and the **lactoglobulins.**

Casein is a phosphoprotein, meaning that phosphate groups are attached to some of the amino acid side chains. These are attached mainly to the hydroxyl groups of the serine and threonine moieties. Actually, casein is a mixture of at least three similar proteins, principally α, β, and κ caseins. These three proteins differ primarily in terms of their molecular weights and the amount of phosphorous which they contain (number of phosphate groups).

α Casein	MW	27,300	~9 Phosphate Groups/Molecule
β Casein	MW	24,100	~4 to 5 Phosphate Groups/Molecule
κ Casein	MW	~8,000	~1.5 Phosphate Groups/Molecule

Casein exists in milk in the form of its calcium salt, **calcium caseinate.** The structure of calcium caseinate is quite complex. It is composed of the previously mentioned α, β, and κ caseins, which form a **micelle,** or solubilized unit. Neither α nor β casein is soluble in milk, and neither is soluble either singly or in combination. However, if κ casein is added to either one, or to a combination of the two, the result is a casein complex which is soluble due to the formation of the micelle.

A proposed structure for the casein micelle is illustrated below. The κ casein is thought to stabilize the micelle. Both α and β casein are precipitated by calcium ions since they are phosphoproteins. Recall that $Ca_3(PO_4)_2$ is fairly insoluble.

$$\boxed{\text{PROTEIN}}-\text{O}-\overset{\overset{\text{O}}{\|}}{\underset{\underset{\text{O}_-}{|}}{\text{P}}}-\text{O}^- + \text{Ca}^{++} \longrightarrow \boxed{\text{PROTEIN}}-\text{O}-\overset{\overset{\text{O}}{\|}}{\underset{\underset{\text{O}_-}{|}}{\text{P}}}-\text{O}^-\text{Ca}^{++}\downarrow$$

Insoluble

The κ casein protein, however, has fewer phosphate groups, and, in addition, a high content of carbohydrate bound to it. It is also thought to have all of its serine and threonine residues (which have hydroxyl

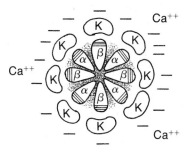

A casein micelle (Average diameter: 1200 Å)

groups), as well as its bound carbohydrates, on only one side of its outer surfaces. This portion of its outer surface is easily solubilized in water due to the presence of these polar groups. The other portion of its surface binds well to the water insoluble α and β caseins, and solubilizes them by forming a protective colloid or micelle around them. Since the entire outer surface of the micelle can be solubilized in water, the unit is solubilized **as a whole,** thus bringing the α and β caseins, as well as κ casein, into solution.

Calcium caseinate has its isoelectric (neutrality) point at pH 4.6. Therefore, it is insoluble in solutions of pH less than 4.6. The pH of milk is about 6.6 and, therefore, casein has a negative charge at this pH, and is solubilized as a salt. If acid is added to milk, the negative charges on the outer surface of the micelle are neutralized (the phosphate groups are protonated) and the neutral protein precipitates.

$$Ca^{++}\text{-Caseinate} + 2\,HCl \longrightarrow Casein \downarrow + CaCl_2$$

The calcium ions remain in solution. When milk sours, lactic acid is produced by bacterial action (see below), and the consequent lowering of the pH causes the same **clotting** reaction. The isolation of casein from milk is described in Experiment 52.

The casein in milk may also be clotted by the action of an enzyme called **rennin.** Rennin is found in the fourth stomach of young calves. However, both the nature of the clot formed and the mechanism of clotting differ when rennin is used. The clot formed using rennin, calcium paracaseinate, contains calcium.

$$Ca^{++}\text{-Caseinate} \xrightarrow{\text{Rennin}} Ca^{++}\text{-Paracaseinate} + A\ Small\ Peptide$$

Rennin is a hydrolytic enzyme (peptidase) and acts specifically to cleave peptide bonds between phenylalanine and methionine residues. It attacks the κ casein, breaking the peptide chain so as to release a small segment of it. This destroys the water solubilizing surface of the κ casein, which protects the inner α and β caseins, and causes the entire micelle to precipitate as calcium paracaseinate. Milk can be decalcified by treatment with oxalate ion which forms an insoluble calcium salt. If the calcium ions are removed from milk, a clot will not be formed when it is treated with rennin.

The clot or **curd** formed by the action of rennin is sold commercially as **cottage cheese.** The liquid remaining is called the **whey.** The curd may also be used to produce various types of **cheese.** It is washed, pressed to remove any excess whey, and chopped. After this treatment, it is melted, hardened, and ground. The ground curd is then salted, pressed into molds, and set aside to age.

Albumins are globular proteins that are soluble in water and in dilute salt solutions. They are, however, denatured and coagulated by heat. The second most abundant protein types in milk are the **lactalbumins.** Once the caseins have been removed, and the solution has been made acidic, the lactalbumins can be isolated by heating the mixture to precipitate them. The typical albumin has a molecular weight of about 41,000.

A third type of protein present in milk is the **lactoglobulins.** They are present in smaller amounts than the albumins, and generally denature and precipitate under the same conditions as the albumins. The lactoglobulins carry the immunological properties of milk. They protect the young mammal until its own immune systems have developed.

Carbohydrates. When the fats and the proteins have been removed from milk, the carbohydrates remain, as they are soluble in aqueous solution. Carbohydrates are discussed in the essay which precedes Experiment 51. The major carbohydrate in milk is lactose.

Lactose
D-Galactose + D-Glucose

A Glycolipid

Lactose, a dissacharride, is the **only** carbohydrate which is synthesized by mammals. When hydrolyzed, it yields one molecule of D-glucose and one of D-galactose. It is synthesized in the mammary glands. In this process, one molecule of glucose is converted to galactose and joined to another of glucose. The galactose is apparently needed by the developing infant to build developing brain and nervous tissue. Brain cells contain **glycolipids** as a part of their structure. A glycolipid is a triglyceride where one of the fatty acid groups has been replaced by a sugar, in this case galactose. Galactose is more stable (to metabolic oxidation) than is glucose and affords a better material for forming structural units in cells.

Although almost all infants have the ability to digest lactose, some adults lose this ability on reaching maturity, since milk is no longer an important part of their diet. To digest lactose, an enzyme called **lactase** is necessary. Lactase is secreted by the cells of the small intestine, and it cleaves lactose into its two component sugars, which are easily digested. Persons lacking the lactase enzyme do not digest lactose properly. As it is poorly absorbed by the small intestine, it remains in the digestive tract where its osmotic potential causes an influx of water. This results in cramps and diarrhea for the affected individual. Persons with a lactase deficiency cannot tolerate more than one glass of milk per day. The deficiency is most common among blacks, but it is also quite common among older whites.

Lactose can be removed from whey by adding ethanol. It is insoluble in ethanol, and when the ethanol is mixed with the aqueous solution, the lactose is forced to crystallize. The isolation of lactose from milk is described in Experiment 52.

When milk is allowed to stand at room temperature, it sours. There are many bacteria present in milk, particularly the **lactobacilli.** These bacteria act on the lactose in milk to produce the sour tasting **lactic acid.** These microorganisms actually **hydrolyze** lactose, and produce lactic acid only from the galactose portion of the lactose. Since the production of the lactic acid also lowers the pH of the milk, the milk clots when it sours.

$$C_{12}H_{22}O_{11} + H_2O \longrightarrow C_6H_{12}O_6 + C_6H_{12}O_6$$

     Lactose   **Galactose**   **Glucose**

$$C_6H_{12}O_6 \xrightarrow{\text{lactobacilli}} CH_3-\underset{\underset{\displaystyle OH}{|}}{CH}-COOH$$

     Galactose   **Lactic Acid**

Many "cultured" milk products are manufactured by allowing milk to sour before it is processed. For instance, milk or cream is usually allowed to sour somewhat by lactic acid bacteria before it is churned to make butter. The fluid left after the milk is churned is sour, and is called **buttermilk.** Other cultured milk products include sour cream, yogurt, and certain types of cheese.

REFERENCES

Fox, B. A., and Cameron, A. G. *Food Science—A Chemical Approach.* New York: Crane, Russak, and Company, 1973. Chapter 6, "Oils, Fats, and Colloids."

Kleiner, I. S., and Orten, J. M. *Biochemistry* (7th edition). St. Louis: C. V. Mosby, 1966. Chapter 7, "Milk."

McKenzie, H. A., editor. *Milk Proteins.* 2 volumes. New York: Academic Press, 1970.

Experiment 52

ISOLATION OF CASEIN AND LACTOSE FROM MILK

Isolation Of A Protein
And A Sugar

In this experiment you will isolate several of the chemical substances found in milk. You will first isolate a phosphorus containing protein, casein (Procedure 52A). The chemistry of casein may be studied in Experiment 54. The remaining milk mixture will then be used as a source of a sugar, α-lactose (Procedure 52B). Following the isolation of the milk sugar, several chemical tests may be performed on this material. Fats, which are present in whole milk, are not isolated in this experiment, since non-fat (skim) milk is used.

The procedure which you will follow is as follows: First, the casein is precipitated by warming the non-fat milk and adding dilute acetic acid. It is important that the heating not be excessive or the acid too strong, since these conditions will also hydrolyze the lactose into its components, glucose and galactose. After the casein has been removed, the excess acetic acid is neutralized with calcium carbonate, and the solution is heated to its boiling point to precipitate the initially soluble protein, albumin. The albumin is removed by filtration, and the filtrate is concentrated. Alcohol is added to the concentrate, and the solution is decolorized. Following vacuum filtration of the solution through Filter Aid, α-lactose crystallizes upon cooling. The crystallization of sugars is quite sensitive to the presence of trace colloidal impurities, and the final filtrate must be clear in order to obtain crystalline lactose. The impurities are removed from the solution by treatment with decolorizing carbon and filtration through Filter Aid.

Lactose is an example of a disaccharide. It is made up of two sugar molecules (moieties), galactose and glucose. In the structures shown, the galactose moiety is on the left and glucose is on the right.

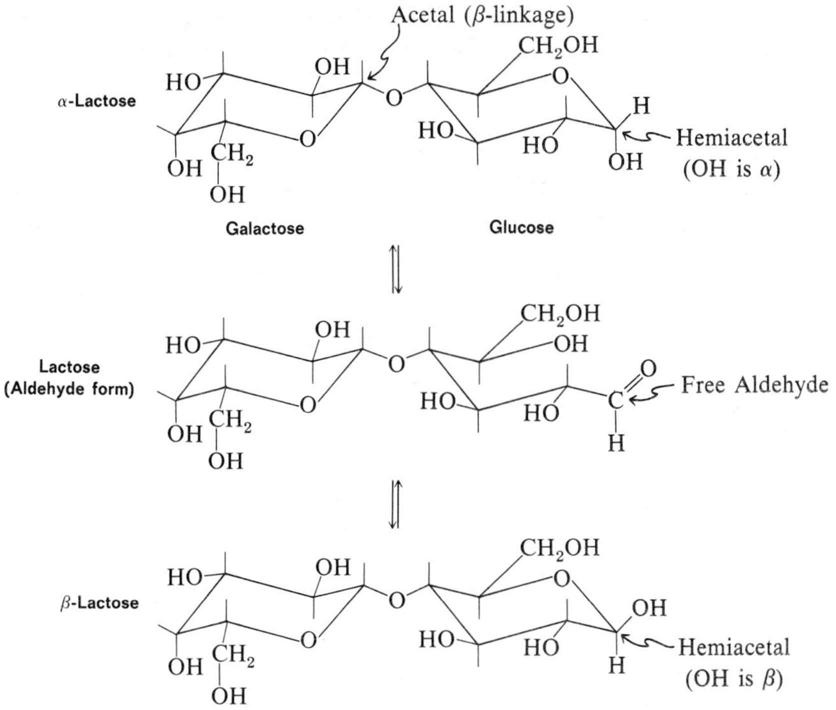

Galactose is bonded through an acetal linkage to glucose. One should notice that the glucose moiety may exist in one of two isomeric hemiacetal structures, α- and β-lactose. The glucose moiety may also exist in a free aldehyde form. This aldehyde form (open form) is an intermediate in the equilibration (interconversion) of α- and β-lactose. Very little of this free aldehyde form exists in the equilibrium mixture. The isomeric α- and β-lactoses are diastereomers since they differ in the configuration at one carbon atom, called the anomeric carbon atom.

α-Lactose is easily obtained by crystallization from a water-ethanol mixture at room temperature. On the other hand, β-lactose must be obtained by a more difficult process, which involves a crystallization from a concentrated solution of lactose at temperatures above 93.5°. In the present experiment, α-lactose will be isolated by the simpler experimental procedure indicated above.

α-Lactose undergoes a number of interesting reactions. First, α-lactose interconverts, **via** the free aldehyde form, to a large extent to the β-isomer in aqueous solution. This results in a change in the rotation of polarized light from +92.6° to +52.3° with increasing time. The process, which results in a change in optical rotation with time, is called mutarotation. Mutarotation of lactose may be studied in Experiment 53.

A second reaction of lactose is the oxidation of the free aldehyde form by use of Benedict's reagent. Lactose is referred to as a reducing sugar because it reduces Benedict's reagent (cupric ion to cuprous ion) and produces a red precipitate (Cu_2O). In the process, the aldehyde group is oxidized to a carboxyl group. The reaction which takes place in Benedict's test is:

$$R—CHO + 2\ Cu^{++} + 4\ OH^- \longrightarrow RCOOH + Cu_2O + 2\ H_2O$$

A third reaction of lactose is the oxidation of the galactose moiety by the mucic acid test. In this test, the acetal linkage between galactose and glucose moieties is cleaved by the acidic medium to give free galactose and glucose. Galactose is oxidized with nitric acid to the dicarboxylic acid, galactaric acid (mucic acid). Mucic acid is an insoluble, high melting solid, which precipitates from the reaction mixture. On the other hand, glucose is oxidized to a diacid (glucaric acid) which is more soluble in the oxidizing medium and does not precipitate.

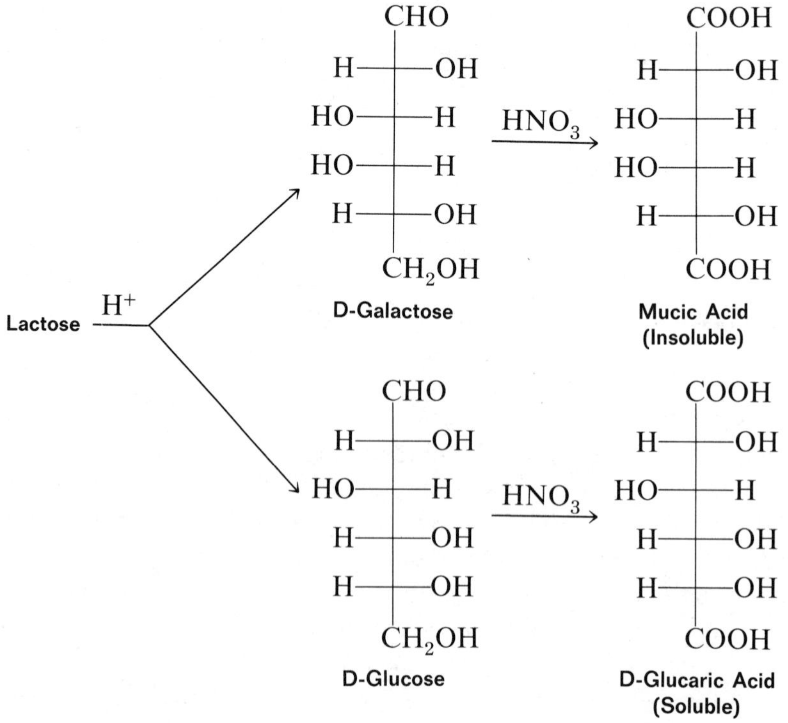

SPECIAL INSTRUCTIONS

It is necessary to have read Techniques 1, 2, and 3 before starting this experiment. Read the essays which precede this experiment and Experiment 51. Procedures 52A and 52B should both be performed during one laboratory period. The lactose solution must be allowed to stand until the following laboratory period.

Procedure **52**A

ISOLATION OF CASEIN FROM MILK

Place 200 ml of non-fat (skim) milk[1] in a 600 ml beaker. Warm the milk to about 40° and add dropwise a dilute acetic acid solution (1 volume of glacial acetic acid to 10 volumes of water) with a pipet. Stir the mixture continuously with a glass rod during the addition process. Continue to add dilute acetic acid until the casein no longer separates. An excess of acid should be avoided because it may hydrolyze some of the lactose to glucose and galactose. Stir the casein until it forms into a large amorphous mass. With the stirring rod, remove the casein and place it in another beaker.[2] Immediately add 5 g of **powdered** calcium carbonate to the **first** beaker containing the liquid from which the casein was removed. Stir this mixture for a few minutes and save it for use in Procedure 52B. Use it as soon as possible during the same laboratory period. This mixture contains the lactose.

Vacuum filter the casein mass for about 15 minutes to remove as much liquid as possible. Press the casein with a spatula during the filtering operation. Place the product between several layers of paper towels to help dry the casein. Transfer the product at least three or four times to fresh paper towels, until the casein is fairly dry. Allow the casein to dry completely in air for one or two days and weigh it. Submit the casein in a labeled vial to the instructor, or save it for Experiment 54. The density of milk is 1.03 g/ml. Using this value, calculate the weight percentage yield of casein isolated from milk.

---

[1] The milk should not be allowed to stand too long before being used in this experiment. Lactose may be slowly converted to lactic acid even if it is stored in a refrigerator.

[2] Even though the crude casein is of adequate purity for chromatographic analysis in Procedure 54A, a small amount of fat may be isolated from casein at this stage. Extract the casein with two successive 15 ml portions of ether. Break up any lumps with a spatula. Stir the casein in the ether with a stirring rod for about 10 minutes. Following each extraction, decant the ether extracts into another container and remove the solvent by distillation in a hood. About 0.1 g of fat is obtained.

ISOLATION OF LACTOSE FROM MILK

Heat the mixture, which was saved from Procedure 52A, to a gentle boil for about 10 minutes. This heating procedure results in a nearly complete precipitation of the albumins. Filter the hot mixture by vacuum to remove the precipitated albumins and the remaining calcium carbonate. With a Bunsen burner concentrate the filtrate in a 600 ml beaker to about 30 ml. Use several applicator sticks to help control the bumping which is caused by further precipitation of the albumins. The solution may also tend to foam out of the beaker if the mixture boils too vigorously. The foaming action may be controlled by blowing gently on the surface of the lactose solution.

Add 175 ml of 95% ethanol **(no flames!)** and 1 or 2 g of decolorizing carbon to the hot solution. After it has been mixed well, filter the warm solution by means of a gentle vacuum through a layer of wet Filter Aid (Technique 2, Section 2.4). The filtrate should be clear.[3] Lactose is unlikely to crystallize unless the solution is clear. However, one may be deceived by cloudiness caused by rapid crystallization of lactose after the vacuum filtration. If the cloudiness increases relatively rapidly on standing, further filtration should be avoided since it will remove the product.

Transfer the solution to an Erlenmeyer flask, stopper it, and allow it to stand overnight or until the next laboratory period. In some cases, several days are required for complete crystallization. Lactose crystallizes on the wall and bottom of the flask. Dislodge the crystals and vacuum filter them. Wash the product with a few milliliters of cold 25% aqueous ethanol. Lactose crystallizes with one water molecule of hydration, $C_{12}H_{22}O_{11} \cdot H_2O$. Weigh the product after it is thoroughly dry. The density of milk is 1.03 g/ml. Using this value, calculate the weight percentage yield of lactose isolated.

At the instructor's option, perform the following tests on the isolated lactose. Submit the remaining sample to the instructor or save it for the mutarotation studies (Experiment 53).

BENEDICT'S TEST (OPTIONAL)

This test is described in Experiment 51. Perform the test on a 1% solution of the isolated lactose. At the same time perform the test on 1% solutions of glucose and galactose.

---

[3]Add 1 g of Filter Aid to the solution and refilter it by gravity if the solution is not clear or has a dark appearance caused by traces of decolorizing carbon. The solution should be heated before it is filtered to keep the lactose in solution.

MUCIC ACID TEST (OPTIONAL)

Place 0.2 g of the isolated lactose, 0.1 g of glucose (dextrose), and 0.1 g of galactose in three separate test tubes. Add 2 ml of water to each tube and dissolve the solids with heating, if necessary. Add 2 ml of concentrated nitric acid to each of the tubes. Heat the tubes in a boiling water bath (hot plate) for 1 hour in a hood (nitrogen oxide gases are evolved). Remove the tubes and allow them to cool slowly after the heating period. Scratch the test tubes with clean stirring rods to induce crystallization. After the test tubes are cooled to room temperature, place them in an ice bath. A fine precipitate of mucic acid should begin to form in the galactose and lactose tubes about one-half hour after the tubes are removed from the water bath. Allow the test tubes to stand until the next laboratory period to complete the crystallization process. Confirm the insolubility of the solid formed by adding about 2 ml of water, then shaking the resulting mixture. If the solid remains, it is mucic acid.

QUESTIONS

1. A student decided to determine the optical rotation of mucic acid. What should be expected as a value? Why?

2. Give a mechanism for the acid-catalyzed hydrolysis of the acetal bond in lactose.

3. β-Lactose is present to a larger extent in an aqueous solution when at equilibrium. Why is this to be expected?

4. Very little of the free aldehyde form is present in an equilibrium mixture of lactose. However, a positive test is obtained with Benedict's test. Explain.

5. Outline a separation scheme for the isolation of casein and lactose from milk. Use a flow chart similar in format to that shown in the **Advance Preparation and Laboratory Records** section at the beginning of the book.

Experiment 53
MUTAROTATION OF LACTOSE

Polarimetry

In this experiment the mutarotation of lactose will be studied by polarimetry. α-Lactose, a disaccharide made up of galactose and glucose, can be isolated from milk (Experiment 52). As is seen in the structures drawn in Experiment 52, the glucose moiety may exist in one of two isomeric hemiacetal structures, α- and β-lactose. These isomers are diastereomers since they differ in configuration at one carbon atom. The glucose moiety may also exist in a free aldehyde form. This aldehyde form (open form in the equation below) is an intermediate in the equilibration of α- and β-lactose. Very little of this free aldehyde form exists in the equilibrium mixture.

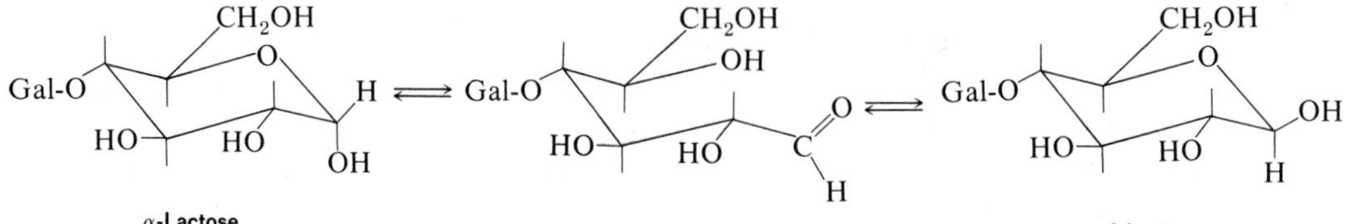

α-Lactose has a specific rotation at 20° C of $+92.6°$. However, when it is placed in water, the optical rotation **decreases** until it reaches an equilibrium value of $+52.3°$. β-Lactose has a specific rotation of $+34°$. The optical rotation of β-lactose **increases** in water until it reaches the same equilibrium value obtained for α-lactose. At the equilibrium point, both the α- and β-isomers are present. However, since the equilibrium rotation is closer in value to the initial rotation of β-lactose, the mixture must contain a larger amount of this isomer. The process, which results in a change in optical rotation over time to approach an equilibrium value, is called **mutarotation.**

SPECIAL INSTRUCTIONS

Read Technique 15, especially Sections 15.2, 15.3, and 15.4, before starting this experiment. The procedure for the preparation of the cells and for the operation of the instrument are those appropriate for the Zeiss polarimeter. Your instructor will provide instructions for the use of another type of polarimeter if a Zeiss instrument is not available. A 2 dm cell is used for this experiment. If a cell of a different path length is used, adjust the concentrations appropriately. Approximately 1 hour may be required to complete the mutarotation study.

PROCEDURE

Turn on the polarimeter in order to warm up the sodium lamp. After about 10 minutes, adjust the instrument so that the scale reads about

$+9°$. This scale reading provides an adjustment of the instrument to the approximate range of rotation which will be observed at the initial reading. Set the timer to zero. Clean and dry a 2 dm cell (Technique 15, Figure 15–6). Weigh 1.25 g of α-lactose (Experiment 52) and transfer it completely to a **dry** 25 ml volumetric flask.

The operations which are described in the next paragraph should be studied carefully **before** starting this part of the experiment. It is essential to complete carefully the described operations in **2 minutes** or less. The reason for speedy operation is that the α-lactose immediately begins to mutarotate when it comes in contact with water. The initial rotations obtained are necessary in order to obtain a precise value of the rotation at zero time. Practice with the necessary equipment in a location near the polarimeter before performing the actual operations. Study the scale on the polarimeter so that it may be read rapidly.

Add about one-half of the volume of distilled water to the volumetric flask containing the α-lactose and swirl it to dissolve the solid. When about one-half of the solid is dissolved (a rough estimate), start the timer. As soon as the solid is dissolved (about **20** to **25** seconds), carefully fill the flask to the mark with distilled water. Use an eye dropper to complete the addition of water. Stopper the flask and shake it about 5 times to mix the contents. Using a funnel, fill the polarimeter cell with the lactose solution. Screw the end piece on the cell and tilt it to transfer any remaining bubbles to the enlarged ring. Place the cell in the polarimeter, close the cover, and adjust the analyzer until the split field is of uniform density (Technique 15, Figure 15–7). Record the time and rotation in the notebook.

Obtain the optical rotation at 1 minute intervals for 8 additional minutes (10 total minutes from the time of initial mixing) and record these values, along with the times at which they were determined. After the 10 minute period, obtain readings at 2 minute intervals for the next 20 minutes. Record the optical rotations and times.

Remove the cell from the polarimeter, and add 2 drops of concentrated ammonium hydroxide to the lactose solution. The ammonia rapidly catalyzes the mutarotation of lactose to its equilibrium value. If the ammonia is not added, the equilibrium value will not be obtained until after about 22 hours. Shake the tube and replace it in the polarimeter. Follow the decrease in rotation until it no longer changes with time. This final value, which is the equilibrium optical rotation, should remain constant for about 5 minutes. Place a thermometer in the polarimeter and determine the temperature in the cell compartment.

Plot the data on a piece of graph paper ruled in millimeters, with the optical rotation plotted on the vertical axis and time plotted (up to 30 minutes) on the horizontal axis. Draw the best possible curved line through the points and extrapolate the line back to $t = 0$. Remember that some scattering of points may occur about the line, especially at the longer time values. The extrapolated value at $t = 0$ corresponds to the optical rotation of α-lactose at the time of initial mixing.

Using the equation in Technique 15, Section 15.2, calculate the specific rotation, $[\alpha]_D^T$, of α-lactose at t = 0. Likewise, calculate the specific rotation of the equilibrium mixture of α- and β-lactose.

Calculate the percentage of each of the diastereomers at equilibrium using the experimentally determined specific rotation values for α-lactose and the equilibrium mixture, and the literature value for the specific rotation of β-lactose ($+34°$). Assume a linear relationship between the specific rotations and the concentrations of the species.

QUESTIONS

1. Explain why β-lactose predominates in the equilibrium mixture of α- and β-lactose.

2. The following data have been obtained for D-glucose at 20°:

α-D-glucose $+112.2°$
β-D-glucose $+18.7°$
equilibrium mixture $+52.7°$

Using these values, calculate the percentage composition of the α- and β-isomers at equilibrium. Inspect the structures of α- and β-D-glucose in Experiment 51 and rationalize the values which were obtained in the calculations.

ESSAY

Proteins and Amino Acids

The amino acids are the simple building blocks from which proteins are synthesized. Proteins are very important molecules in the structure and in the chemical functioning of living systems. Proteins may play a structural role, as in the proteins of hair and connective tissue. Other proteins provide a chemical function, as in the hormones or the oxygen transport proteins such as hemoglobin. Many proteins act as biological catalysts assisting in the process of chemical transformation. These are called enzymes.

The amino acids, or more properly the α-amino acids, possess the general structure

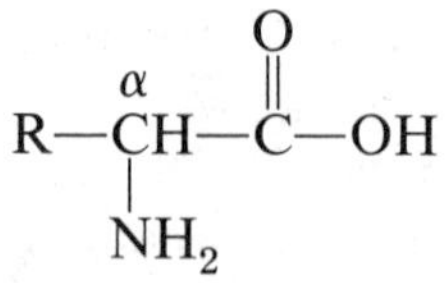

$$\underset{\underset{NH_2}{|}}{R-\overset{\alpha}{C}H}-\overset{\overset{O}{\|}}{C}-OH$$

The R group may range from H, as in glycine, to long carbon chains containing additional functional groups, as in lysine or glutamic acid. For the details of the structures of these compounds, see the table given in Experiment 54.

Because the α-carbon, which bears the amino group, is asymmetrically substituted in all of the amino acids except glycine, these amino acids are chiral molecules. They may exist in either of two enantiomeric forms, and each enantiomeric form is capable of rotating the plane of polarized light. Fischer projections of the enantiomeric forms, together with their stereochemical designations, are given below.

$$\begin{array}{cc}
\underset{\displaystyle \underset{R}{\overset{\displaystyle H\!-\!\!\!\!-\!\!\!\!-\!\!NH_2}{|}}}{\overset{\displaystyle \underset{O}{\diagdown}\underset{}{\overset{OH}{\diagup}}}{C}} &
\underset{\displaystyle \underset{R}{\overset{\displaystyle H_2N\!-\!\!\!\!-\!\!\!\!-\!\!H}{|}}}{\overset{\displaystyle \underset{O}{\diagdown}\underset{}{\overset{OH}{\diagup}}}{C}}\\[3ex]
\text{D-amino acid} & \text{L-amino acid}
\end{array}$$

In virtually all instances, except in the cell walls of bacteria, the amino acids found in living organisms belong to the L-series of absolute configurations. This high degree of stereospecificity has important consequences, as will be seen when proteins are discussed.

Because the amino acids possess both acidic and basic functional groups in close proximity to each other, it is reasonable to expect these groups to exchange a proton with each other. In aqueous solution, amino acids exist to a large extent in a dipolar ionic form, or as a **zwitterion** (from the German **zwitter,** meaning "hybrid"). Consequently, amino acids share many of the characteristics of ionic compounds. They tend to be high-melting materials, and they are quite soluble in water. Since they possess charges, the magnitude of each of the charges may be influenced by the pH of the medium in which the amino acids are located.

The reaction of a carboxylic acid with an amine to form an amide is a well-known process.

$$R\overset{\displaystyle O}{\overset{\|}{-C}}-OH + R'-NH_2 \longrightarrow R\overset{\displaystyle O}{\overset{\|}{-C}}-NH-R' + H_2O$$

In the case of the amino acids, both of the functional groups are contained within the same molecule. As a result, it is possible for two amino acid molecules to react together in a head-to-tail fashion to produce an amide, which still possesses a free amino group at one end and a free carboxyl group at the other end. Such an amide is called a **peptide,** and the amide linkage is called a peptide bond.

$$H_2N-\underset{R}{\overset{}{CH}}-\overset{\displaystyle O}{\overset{\|}{C}}-OH + H_2N-\underset{R'}{\overset{}{CH}}-\overset{\displaystyle O}{\overset{\|}{C}}-OH \longrightarrow H_2N-\underset{R}{\overset{}{CH}}-\overset{\displaystyle O}{\overset{\|}{C}}-NH-\underset{R'}{\overset{}{CH}}-\overset{\displaystyle O}{\overset{\|}{C}}-OH$$

peptide bond

A Dipeptide

The particular example of a peptide formed from two amino acid units is known as a **dipeptide.** A peptide formed from three amino acid units is known as a **tripeptide,** and so on up to **polypeptides.** For all their apparent complexity, it must be remembered that the peptides are merely rather complicated looking amides. Their chemical reactions and their methods of formation are those of the typical amide.

Although there are many substances of biological importance which are peptides made up of several amino acids, the most important peptide compounds are formed when 50 or more amino acids are linked together by means of peptide bonds. The polypeptides which form from such a large number of amino acids are known as **proteins.** In the proteins, the chain of amino acids linked by means of peptide bonds is capable of being folded into a specific three-dimensional shape. It is this ability to form a specific three-dimensional structure which distinguishes the proteins from the polypeptides.

The sequence in which the amino acids are linked together to form the protein is called the **primary structure.** In each protein chain, there is one end which bears a free amino group (the N-terminal end) and the other end which bears a free carboxyl group (the C-terminal end). The determination of the primary structure of a protein is an extremely difficult procedure, since the possible combinations of 20 amino acids is virtually infinite. The primary structures are known for only a relatively few proteins.

If various segments of the protein chain are allowed to lie next to each other, then the electronegative oxygen of the carboxyl group may find itself in close proximity to the N—H bond of the amide functional group. Hydrogen bonding, linking the N—H group to the oxygen, is possible, and such hydrogen bonding holds the segments of the peptide chain in a relationship which is known as the secondary structure of the protein. The most important type of secondary structure is the **α-helix,** first postulated by Pauling and Corey in 1951. In the α-helix, the peptide chain forms a right-handed helix which contains, on the average, 3.7 amino acid residues per turn. The helix must be right-handed because the amino acids from which it is formed each possess the L absolute configuration. A right-handed helix allows the bulky substituent groups to lie toward the outside of the helical structure. Pauling and Corey also postulated another secondary structure, the β-structure, or the **pleated sheet.** The α-helix occurs in most globular proteins. Globular proteins are proteins with a compact and somewhat spherical shape, and they are generally fairly soluble in water. The pleated sheet structure occurs in fibrous proteins. Fibrous proteins are proteins with an elongated, thread-like shape, and they tend to be insoluble in water. The protein of silk contains pleated sheet portions. Both the α-helix and the pleated sheet are illustrated in the figure below.

The polypeptide chain, with its segments of α-helical structure, may be further folded into complex three-dimensional shapes. These shapes are the tertiary structures of the proteins. The tertiary structure is

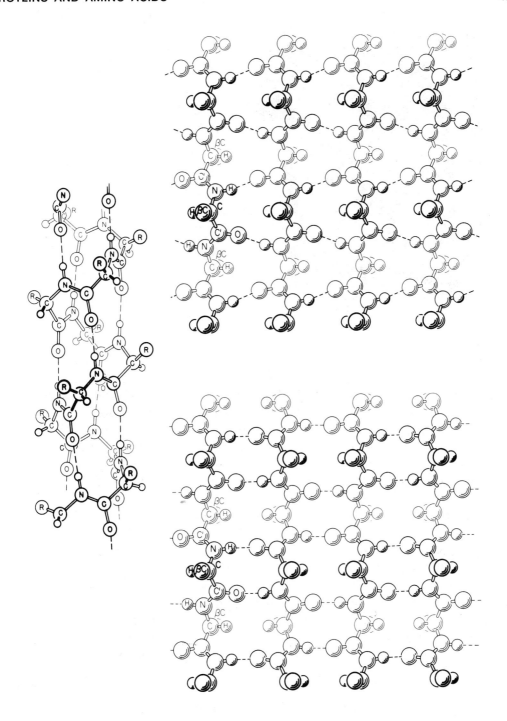

α-Helix and Pleated sheet structures

maintained by a variety of linkages, including covalent linkages between the sulfur atoms of cysteine amino acid residues at various positions in the chain, ionic bonds between charged groups, polar attractions between functional groups, hydrogen bonding, and van der Waals

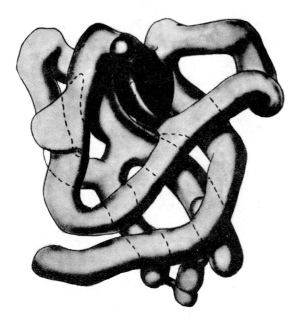

Tertiary structure of a protein

attractions between nonpolar groups. Many of these types of linkages are rather weak, and extremes of temperature, acidity, basicity, or ionic strength can destroy them. When some of these linkages are broken, the chain of the protein is allowed to change its three-dimensional shape. When this happens, the protein loses many of its original properties and is **denatured.** The heat associated with cooking causes the natural proteins of foods to become denatured. When they are thoroughly denatured, we say that the food is "cooked," and we serve it. Occasionally, proteins are made up of combinations of subunits, each of which possesses its own particular primary, secondary, and tertiary structure. The manner in which these subunits fit together is known as the **quaternary structure** of a protein. An example of a tertiary structure of a protein is shown in the figure above.

At this point, it must be mentioned that the chemical reactions of the proteins are precisely those of the simple amides. In the presence of aqueous acid or base, the proteins can be hydrolyzed into smaller peptides, which, in turn, can be hydrolyzed into the constituent amino acids.

Many of the proteins are capable of catalyzing certain chemical processes which occur in the cell. These proteins are called **enzymes.** In the tertiary structure, the functional groups of certain specific amino acids are held in close proximity to one another. This region is capable of binding a particular substrate molecule and catalyzing a chemical change. The region at which this catalytic activity occurs is called the **active site.** Frequently, the enzymes will require the presence of some other small molecule in order to carry out the catalytic function. Such a necessary molecule is a **cofactor.** If the cofactor is an organic compound, it is called a **coenzyme.** When a coenzyme cannot be synthesized in the living system, but is required in the diet, it is called a **vitamin.** The enzymes, being proteins, are composed of large numbers of amino acids, each of which possesses the **L** absolute configuration. In addition, the complex tertiary structure of the enzyme will permit only those molecules with certain specific shapes to act as substrates. These two facts, taken together, make the enzymes the ultimate stereospecific reagents. This extreme stereospecificity explains why naturally occurring substances usually occur as one particular stereoisomer, instead of several possible stereoisomers.

Besides the major roles which proteins play in living systems, it is important to mention several less apparent, but nevertheless necessary, uses of proteins by living systems. Many of the chemical processes in living organisms are regulated by chemical mediators called **hormones.** While many hormones are related to the steroids (see the essay preceding Experiment 8), other hormones are either proteins or peptides. One hormone, whose primary structure is known, is ACTH (adrenocorticotropic hormone), which stimulates the adrenal cortex to secrete steroid hormones. ACTH contains 39 amino acids arranged in a single chain. In the body's immune system, the antibodies which form to counteract the effect of a foreign substance (antigen) in the blood stream are also proteins. Specifically, they are variations of **gamma globulin,** a protein designed to precipitate the antigen from the blood stream, thus rendering it harmless.

REFERENCES

Dickerson, R. E., and Geis, I. *The Structure and Action of Proteins.* New York: Harper and Row, 1969.

Dickerson, R. E. "The Structure and History of an Ancient Protein." *Scientific American, 226* (April, 1972), 58.

Doty, P. "Proteins." *Scientific American, 197* (September, 1957), 173.

Guillemin R., and Burgus, R. "The Hormones of the Hypothalamus." *Scientific American, 227* (November, 1972), 24.

Kendrew, J. C. "The Three-dimensional Structure of a Protein Molecule." *Scientific American, 205* (December, 1961), 96.

Neurath, H. "Protein-digesting Enzymes." *Scientific American, 211* (December, 1964), 68.

Pauling, L., Corey, R. B., and Hayward, R. "The Structure of Protein Molecules." *Scientific American, 191* (July, 1954), 51.

Perutz, M. F. "The Hemoglobin Molecule." *Scientific American, 211* (November, 1964), 64.

Phillips, D. C. "The Three-dimensional Structure of an Enzyme Molecule." *Scientific American, 215* (November, 1966), 78.

Stein, W. H., and Moore, S. "The Chemical Structure of Proteins." *Scientific American, 204* (February, 1961), 81.

Thompson, E. O. P. "The Insulin Molecule." *Scientific American, 192* (May, 1955), 36.

White, A., Handler, P., and Smith, E. L. *Principles of Biochemistry* (5th edition). New York: McGraw-Hill Book Co., 1973. Chapters 5, 6, 7, and 8.

PAPER CHROMATOGRAPHY AND TESTS ON AMINO ACIDS AND PROTEINS

Paper Chromatography

In this experiment, you will determine the composition of proteins by paper chromatographic analysis of the hydrolysates (Procedure 54A). In addition, you will perform chemical tests on proteins and amino acids (Procedure 54B). It is necessary to read the essay which precedes this experiment before starting the laboratory work.

Proteins may be hydrolyzed in acidic or basic solution, or with enzymes. During the hydrolysis, the peptide bonds break to give shorter length polymers (polypeptides) which in turn are further degraded to amino acids. Total hydrolysis of protein can be achieved in 20% hydrochloric acid solution at 100° for 12 to 48 hours. However, adequate hydrolysis for purposes of this experiment can be achieved in refluxing acid in less than one hour.

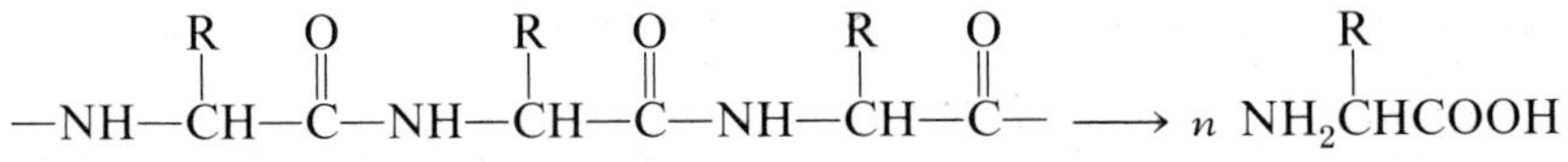

$$\underset{\text{Protein}}{-\text{NH}-\underset{\underset{R}{|}}{\text{CH}}-\underset{\underset{O}{\|}}{\text{C}}-\text{NH}-\underset{\underset{R}{|}}{\text{CH}}-\underset{\underset{O}{\|}}{\text{C}}-\text{NH}-\underset{\underset{R}{|}}{\text{CH}}-\underset{\underset{O}{\|}}{\text{C}}-} \longrightarrow n\ \underset{\text{Amino Acids}}{\text{NH}_2\underset{\underset{R}{|}}{\text{CH}}\text{COOH}}$$

These hydrolysates may be analyzed for their amino acid contents by chromatographic methods, such as paper chromatography. The most common amino acids are listed in the table which follows. They are listed in the order of increasing R_f values. The amino acid contents for the proteins casein, gelatin, silk, and hair are also listed. The major amino acid constituents in each of the proteins are italicized. One should notice that there are large differences in the amino acid content of the various proteins. There are variations in amino acid content between samples and, therefore, the values given in the table are to be considered **approximate.**

The individual amino acids on a chromatogram are made visible with ninhydrin. Ninhydrin reacts with amino acids to produce characteristic deep blue colors. However, a few amino acids produce a different color; for example, proline produces a pale yellow color with

AMINO ACID	FORMULA	R_f VALUE	APPROXIMATE PERCENTAGE COMPOSITION OF PROTEINS			
			Casein	Gelatin	Silk	Hair
Cystine	$\begin{array}{c}NH_2\\ S-CH_2CHCOOH\\ S-CH_2CHCOOH\\ NH_2\end{array}$	0.16	0.4	0.1		*18.0*
Aspartic Acid	HOOCCH$_2$CHCOOH | NH$_2$	.32	*6.8*	*6.7*		3.9
Glutamic Acid	HOOCCH$_2$CH$_2$CHCOOH | NH$_2$	.40	*22.4*	*11.5*		*13.1*
Glycine	H$_2$NCH$_2$COOH	.42	2.6	*25.5*	*42.3*	4.1
Serine	HOCH$_2$CHCOOH | NH$_2$	.43	*7.4*	0.4	*12.6*	10.6
Threonine	CH$_3$CH—CHCOOH | | OH NH$_2$	.51	4.7	1.9	1.5	*8.5*
Lysine (cation)	$\overset{+}{H_3N}$CH$_2$CH$_2$CH$_2$CH$_2$CHCOOH | NH$_2$	.53	*7.9*	4.1	0.4	1.9
Alanine	CH$_3$CHCOOH | NH$_2$	.59	2.9	*8.7*	*24.5*	2.8
Arginine (cation)	H$_2\overset{+}{N}$=CNHCH$_2$CH$_2$CH$_2$CHCOOH | | NH$_2$ NH$_2$	.60	3.9	*8.0*	1.1	*8.9*
Tyrosine	HO—⟨ ⟩—CH$_2$CHCOOH | NH$_2$	.62	*6.1*	0.4	*10.6*	2.2
Valine	CH$_3$CH—CHCOOH | | CH$_3$ NH$_2$	.75	*6.9*	2.5	3.2	5.5
Methionine	CH$_3$SCH$_2$CH$_2$CHCOOH | NH$_2$	.77	3.3	1.0		0.7
Leucine	CH$_3$CHCH$_2$CHCOOH | | CH$_3$ NH$_2$	.79	*8.8*	4.6	0.8	*11.2*
Phenylalanine	⟨ ⟩—CH$_2$CHCOOH | NH$_2$	.82	4.8	2.2		2.4
Proline	(pyrrolidine ring) N—COOH | H	.85	*10.9*	*18.0*	1.5	4.3

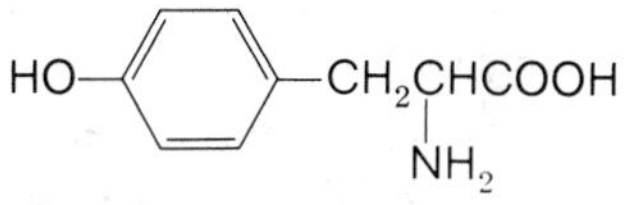

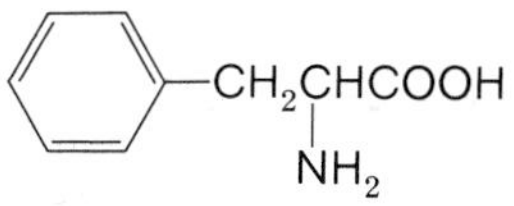

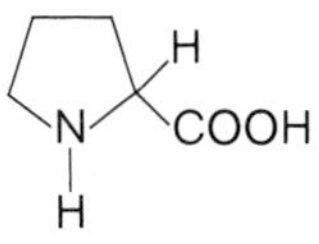

ninhydrin. The reactions which are involved in the production of the color are as follows:

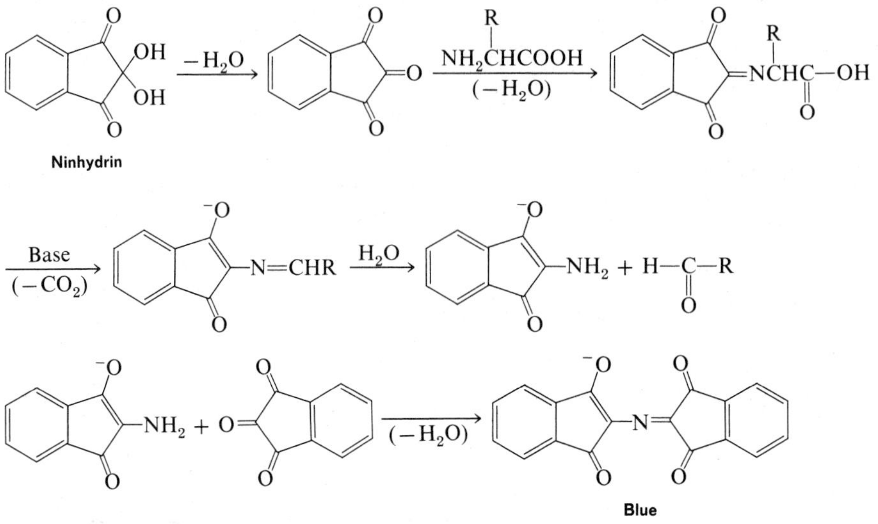

Ninhydrin

Blue

SPECIAL INSTRUCTIONS

It is necessary to have read the essay which precedes this experiment, and Technique 11, "Thin Layer Chromatography." Paper chromatography is similar to thin layer chromatography. The casein isolated in Procedure 52A may be used in this experiment. The hydrolysates of casein, gelatin, silk, and hair may best be provided by the laboratory instructor or assistant. The procedure for preparing them is given in Procedure 54A. The chromatographic development takes about 4 hours to complete. During this time, one should perform the tests given in Procedure 54B. The laboratory instructor or assistant may need to remove the chromatogram at the end of the 4 hour development period.

Procedure 54A

PAPER CHROMATOGRAPHY
OF AMINO ACIDS

The purpose of this experiment is to identify individual, unknown amino acids and to identify the constituent amino acids in the hydrolysates of some common proteins. This identification is accomplished by comparison of R_f values of known amino acids to those of the unknown. If the hydrolysates are provided by the laboratory instructor, proceed directly to the paper chromatography part of the experiment.

PROCEDURES

Hydrolysis of a Protein. Assemble a small scale reflux apparatus using a 100 ml round bottom flask. Place 20 ml of 19% hydrochloric acid (equal volumes of concentrated hydrochloric acid and distilled water), 0.5 g of the protein (casein, gelatin, hair, or silk), and a boiling stone in the flask. For casein or gelatin, carefully heat the mixture under reflux for 35 minutes using a heating mantle or a very small flame. For hair or silk, heat the mixture under reflux for 50 minutes. After the reflux period is completed, the hydrolysate must be decolorized. Add about 0.5 g of decolorizing carbon to the hot hydrolysate and swirl the mixture. Gravity filter the hydrolysate into a 50 ml Erlenmeyer flask. The filtrate should be colorless or pale yellow. Check the hydrolysate with the biuret test to see if the reaction is complete (Procedure 54B, part VI-E). The test should be negative. If hydrolysis is not complete (a positive biuret test), add 3 to 5 ml of 19% hydrochloric acid and heat the mixture for an additional 15 minutes. Again check the hydrolysate with the biuret test to see if it is negative. After the hydrolysis is complete, cool the hydrolysate, stopper it, and save the solution for the analysis by paper chromatography.

Separation of Amino Acids by Paper Chromatography. Obtain a 24 × 15.5 cm rectangle of Whatman #1 filter paper. Handle this paper by the top edge only. If the paper is handled carelessly, fingerprints may appear as colored spots on the chromatogram. Place the rectangle on a clean paper towel or notebook paper with the "x" in the upper right hand corner (see the figure). This "x" was placed on the rectangle when the paper was cut. Since the thickness of the paper is somewhat uneven, this marking procedure assures that the direction of the solvent flow on each piece of chromatography paper is the same for all students. The R_f values are more reproducible when the chromatograms are developed in the same direction.

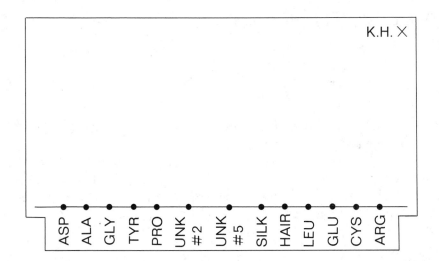

Cut a one centimeter square from the two lower corners. About 1.5 cm from the bottom of the paper, draw a **pencil** line (not a pen line) across the paper. With a pencil, mark thirteen evenly spaced dots 1.5 cm from each other along this line, with the first dot 3 cm from the left hand edge. Nine standard amino acids, 2 hydrolysates, and 2 unknown amino acids will be placed on the dots. One may use the sixth and seventh dots for the unknowns, the eighth and ninth dots for the hydrolysates, and the rest of the dots for the individual standard amino acids. One possible sequence is shown in the figure. The standard amino acids which will be used are 0.1 M solutions of aspartic acid (ASP), alanine (ALA), glycine (GLY), tyrosine (TYR), proline (PRO), leucine (LEU), glutamic acid (GLU), cystine (CYS), and arginine (ARG). Each of these 0.1 M solutions has been acidified with 10 drops of 19% hydrochloric acid for each 10 ml of solution. Record the sequence in the laboratory notebook. Place your name or initials in the upper right hand corner of the chromatogram.

Prepare capillary micropipets for applying the solutions (Technique 11, Section 11.3) or use the micropipets inserted in the standard amino acids, unknowns, or hydrolysate samples. It may be wise to practice the spotting technique on a small piece of Whatman #1 paper before attempting to spot the actual chromatogram. The correct method of spotting is described in Technique 11, Section 11.3. It is important that the spots be made as small as possible and that the paper not be overloaded. If either of these errors is made, the spots will tail and overlap one another after development. The applied spot should be 1 to 2 mm in diameter. Apply the appropriate samples to the dots with the micropipets, taking extreme care to avoid contamination of the samples. After applying the samples to each dot, allow the spots to dry completely and then make a second application at the same point.

The hydrolysates should be spotted a third time. Allow the spots to dry completely before the chromatogram is developed.

CAUTION: Phenol will cause burns if it comes in contact with skin. Immediately wash the area with copious quantities of soap and water. Clean up all spills.

Use a 32 oz wide-mouth screw cap jar for a developing chamber. Insert a pipet **below** the protective layer of ligroin and remove 20 ml of an 80% aqueous phenol solution with a bulb.[1] Carefully transfer the phenol solution to the developing chamber so that it is not splattered on the sides of the jar. Check to see that the depth of the phenol solution does not exceed 1.5 cm. If it does, carefully pipet enough of the solution so that the depth is less than 1.5 cm. Curl the paper into a large cylinder so that the line of spots are on the bottom and are on the inner surface. Overlap the top about 0.5 cm and make sure that the bottom edge is even. Hold the chromatogram together with a paper clip or a staple. Insert the cylinder into the jar so that the bottom end of the paper is immersed in the solvent. Tighten the lid securely and allow the development to proceed for 4 hours without disturbing the jar.

Remove the chromatogram (no fingers), mark the solvent front with a pencil, and allow the solvent to evaporate in air.[2] In the following laboratory period, spray the paper uniformly with ninhydrin spray and place the paper in a 110° oven. Colored spots will begin to appear within 5 minutes. Remove the chromatogram and outline all of the spots with a pencil. With a millimeter ruler, measure the distance each spot has traveled from the point of origin to the **front (top)** of the spot. Also measure the distance that the solvent traveled. Calculate the R_f value for each of the spots (Technique 11, Section 11.8). In some cases, the separation may not be complete enough to make an accurate calculation of the R_f value. Record the calculated values.

From the R_f values and colors of the standard amino acids, identify the two unknown amino acids and the principal constituents of the two protein hydrolysates. The table of protein composition may be useful in helping to make the analysis of the hydrolysates. Remember that amino acids with similar R_f values may not separate well enough to be seen on the chromatogram. One may have to place the amino acids in groups. Record your findings. Submit the chromatogram with your laboratory report.

---

[1] The 80% aqueous phenol is prepared by mixing 80 g of phenol per 20 ml of distilled water. Heat the mixture until it dissolves completely. Add a protective layer of ligroin so that air is excluded from the aqueous phenol. If the ligroin is not added, the solution should be used as soon as possible.

[2] This portion of the procedure may have to be done by the laboratory instructor or assistant if there is not enough time.

TESTS ON PROTEINS AND AMINO ACIDS

I. Amphoteric Properties of Proteins

Amino acids can function as acids or bases by forming salts with strong acids and bases. Compounds which have such behavior are said to be **amphoteric.**

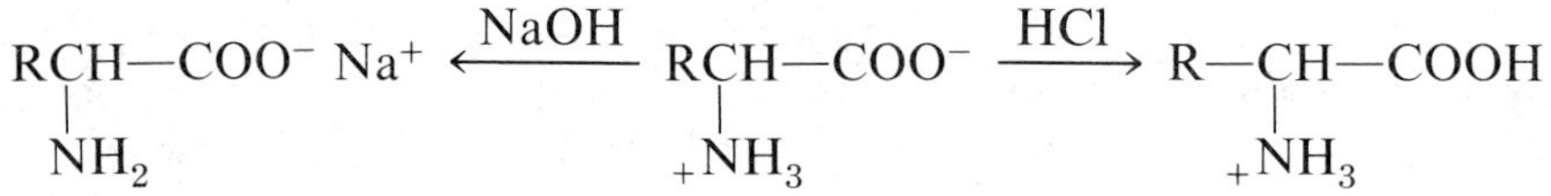

$$RCH—COO^-\ Na^+ \xleftarrow{\ NaOH\ } RCH—COO^- \xrightarrow{\ HCl\ } R—CH—COOH$$
$$\ \ \ |\qquad\qquad\qquad\ \ |\qquad\qquad\qquad\qquad |$$
$$\ \ NH_2\qquad\qquad\qquad _+NH_3\qquad\qquad\qquad _+NH_3$$

**Dipolar Structure
of Amino Acid**

Proteins can also have amphoteric behavior and may dissolve to some extent in an acidic or basic solution. This results from the fact that some amino acid components of the protein have **free** amino groups (lysine and arginine) and **free** carboxyl groups (aspartic and glutamic acids). These groups may react with acids or bases to form soluble salts of a protein.

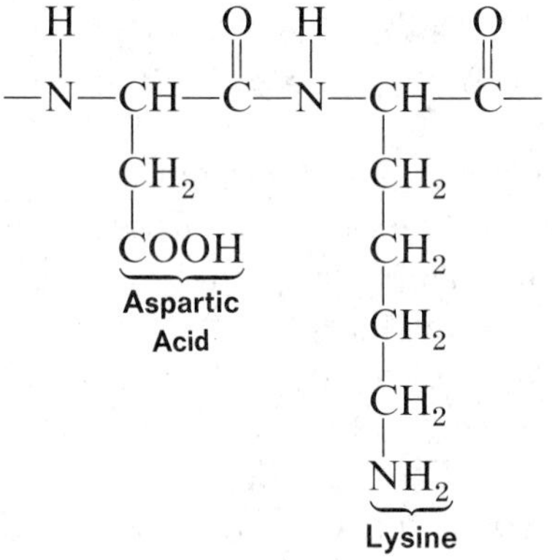

PROCEDURES

A. Place 0.1 g of casein in a test tube and add to it 5 ml of water and 2 ml of 10% sodium hydroxide solution. Stopper the test tube, shake it vigorously and observe the result. Save 2 ml of this mixture for use in part VI.

B. Add concentrated hydrochloric acid dropwise, with shaking, to the remaining solution from part A and observe the result. Continue until 4 ml of the acid have been added, stopper the test tube, shake it vigorously and observe the result.

468

II. Coagulation of Proteins

The α-helix structure of a protein is maintained by hydrogen bonds between the amino acid in one part of the protein with another amino acid in another part. The stable nature of such a protein can be disturbed by physical and chemical means by breaking the hydrogen bonds. The α-helix may then unfold and precipitate. Such a protein is said to be **denatured.** Proteins can be precipitated (coagulated) with heat, strong acids, or with alcohol.

PROCEDURES

A. Place about 2 ml of egg albumin solution[3] in a test tube and boil it gently for a few minutes. Observe what happens to the heated solution.

B. Place about 2 ml of an egg albumin solution in a test tube and add 7 ml of 95% ethanol. Observe what happens to the solution.

III. Precipitation by Heavy Metal Ions

Heavy metal ions, such as silver, lead, and mercury, precipitate proteins by combination of a metal cation with the free carboxylate groups of the protein. The antiseptic action of silver nitrate and mercuric chloride depends upon the precipitation of the proteins present in bacteria.

$$\text{protein}-\underset{\underset{O}{\|}}{C}O^- + Ag^+ \longrightarrow \text{protein}-\underset{\underset{O}{\|}}{C}O^-\,Ag^+$$

Silver Precipitate

PROCEDURES

A. Place about 2 ml of egg albumin solution in a test tube and add 2% silver nitrate dropwise. Observe what happens to the solution.

B. Repeat the above with 5% mercuric chloride solution.

---

[3]This solution is prepared by adding 2 g of dried albumin to 100 ml of distilled water and allowing the mixture to stand overnight. Gravity filter the resulting mixture.

IV. Xanthoproteic Test

Some amino acids incorporated into a protein have aromatic rings. These rings undergo nitration to give yellow compounds. In basic solution the color intensifies.

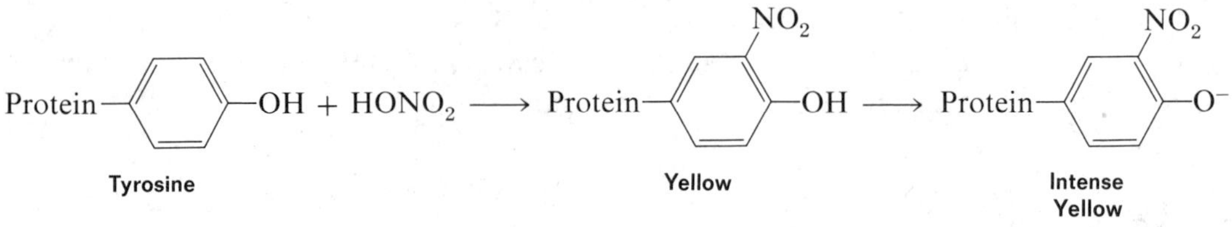

Tyrosine   Yellow   Intense Yellow

PROCEDURE

Place 2 ml of egg albumin solution in a test tube and add 10 drops of concentrated nitric acid. Gently warm the mixture and observe any change in color. Cool the mixture and add 10% sodium hydroxide solution dropwise until the solution is basic. Note any change in color.

V. Sulfur Test

Proteins which have sulfur-containing amino acids, such as cystine (disulfide bond), are cleaved in basic solution to give an inorganic sulfide. In the presence of lead acetate, a black precipitate of lead sulfide is produced.

$$\text{Sulfur-containing protein} \xrightarrow{\text{NaOH}} \text{S}^= \xrightarrow{\text{Pb}^{++}} \text{PbS}$$

Black Precipitate

PROCEDURE

Add 2 ml of egg albumin solution, 5 ml of 10% sodium hydroxide solution, and 2 drops of 5% lead acetate solution to a small Erlenmeyer flask. Carefully boil (froth) the mixture, with mixing, for a few minutes and observe the result.

VI. Biuret Test

When urea is heated above its melting point, it is converted to biuret with liberation of ammonia,

$$2\ NH_2-\overset{\overset{\displaystyle O}{\|}}{C}-NH_2 \xrightarrow{\text{heat}} NH_2-\overset{\overset{\displaystyle O}{\|}}{C}-NH-\overset{\overset{\displaystyle O}{\|}}{C}-NH_2 + NH_3$$

Urea   Biuret

Biuret forms a pink or violet colored complex with cupric ion in basic solution. When the pink-violet color is obtained, the test is said to be positive. A positive biuret test is also obtained with all compounds which contain two or more peptide bonds (tripeptide or larger peptide). Amino acids (except serine and threonine), dipeptides, and urea do not give violet colors with cupric ion and are said to give negative tests. Amino acids, in a similar fashion to ammonia, produce a blue solution in the presence of cupric ion. The test is quite useful in following the hydrolysis of proteins in order to determine when the hydrolysis reaction is completed.

PROCEDURES

A. Place 1 g of urea in a dry test tube. Gently heat the test tube until the urea melts. Cautiously note the odor of the gas which evolves and test it with a piece of moist red litmus paper held over the mouth of the test tube. Gently continue to heat the mixture until the material appears to be solidifiying. Dissolve the white residue in 4 ml of warm water and filter the mixture. Add 4 ml of 10% sodium hydroxide solution to the filtrate, with mixing. Add 10 drops of a 2% copper sulfate solution. Shake the mixture and observe the color.

B. Place 2 ml of egg albumin solution in a test tube and add 2 ml of 10% sodium hydroxide. Add 5 drops of 2% copper sulfate solution and shake the mixture. Observe the color of the solution.

C. Add 2 ml of water and 2 drops of 2% copper sulfate solution to the 2 ml of casein solution from part IA. Shake the mixture and observe the color.

D. Place 0.1 g of glycine in a test tube and add 3 ml of 10% sodium hydroxide solution to dissolve the solid. Add 10 drops of 2% copper sulfate solution, shake the mixture and observe the color.

E. Place 5 drops of hydrolysate from Procedure 54A and 10 drops of 10% sodium hydroxide solution in a test tube. Check the solution with litmus paper to make sure that it is definitely alkaline. Add more 10% sodium hydroxide solution if necessary. Add 4 or 5 drops of 2% copper sulfate solution. A blue color indicates that the hydrolysis is complete. A pink or violet color indicates that the hydrolysis is not complete.

VII. Nitrous Acid Test

Amino acids react with nitrous acid to give an α-hydroxy acid and **nitrogen gas.**

$$R\!-\!\underset{\underset{NH_2}{|}}{CH}\!-\!COOH + HONO \longrightarrow \underset{\underset{OH}{|}}{RCH}\!-\!COOH + H_2O + N_2$$

Since most amino groups in a protein are involved in peptide bonds, they are unavailable for this reaction. Proteins, however, do contain some free amino groups, mostly from the amino acid lysine. These free amino groups react with nitrous acid to produce nitrogen gas. However, they do not liberate as much nitrogen as the amino acids themselves.

PROCEDURES

Conduct each of the tests at approximately the same time.

A. Place 0.1 g of glycine in a test tube and add 5 ml of 10% hydrochloric acid. Carefully add 1 ml of 5% sodium nitrite solution to the test tube. Shake it well and observe the rate of gas evolution.

B. Place 4 or 5 grains of gelatin in a test tube and add 3 drops of 19% hydrochloric acid. Add 3 drops of 5% sodium nitrite solution to the mixture. Shake it well and observe the rate of evolution of gas.

C. Place about 10 drops of hydrolysate from Procedure 54A in a test tube and add about 20 drops of 5% sodium nitrite solution. Shake the mixture well and observe the rate of gas evolution.

QUESTIONS

1. Explain why silver nitrate and mercuric chloride are good germicides.

2. Why is egg white or milk used as an antidote to treat patients who have swallowed heavy metal salts, such as silver nitrate or mercuric chloride?

3. Why is alcohol used to disinfect areas of the skin prior to surgery?

4. When nitric acid is spilled on the skin, a yellow stain is produced. Explain.

5. The xanthoproteic test gives a positive test when tyrosine is present in a protein. What other amino acid moieties may be detected with this test?

ESSAY

Amino Acid Biosynthesis

The α-amino acids, whose general structure is:

$$R-\underset{\underset{NH_2}{|}}{CH}-\overset{\overset{O}{\|}}{C}-OH \quad \text{or} \quad R-\underset{\underset{NH_3^+}{|}}{CH}-\overset{\overset{O}{\|}}{C}-O^-$$

are essential substances for growth because they are the basic substances from which proteins are synthesized in living organisms. An interesting question for scientists is how these substances are themselves synthesized in plants and animals. Obviously, one must postulate the incorporation of an atom of nitrogen into some organic molecule, and it is at this stage that the process begins. Reduction of nitrogen gas to ammonia can be accomplished only by microorganisms. Plants cannot reduce N_2 by themselves, but certain plants, the **legumes** (soybeans are an example), are able to convert nitrogen to ammonia through the action of symbiotic bacteria which reside in the root nodules. The reduction of nitrogen is a six-electron reduction process, and two enzymes are known to catalyze the reduction. One of these enzymes uses molybdenum and iron as cofactors, and the other uses iron.

Both plants and microorganisms are capable of reducing nitrate ion to ammonia, and it is this reduction which provides most of the inorganic nitrogen which plants utilize in amino acid synthesis. The enzymes, called **nitrate reductases,** which catalyze this reduction utilize a molybdenum ion as part of the electron transport system. The enzyme, acting with a biological reducing agent, reduces nitrate ion to nitrite ion.

$$2\,e^- + 2\,H^+ + NO_3^- \longrightarrow NO_2^- + H_2O$$

The next step is carried out with **nitrite reductases** as catalysts. These enzymes are less well understood than the nitrate reductases. In any event, through the action of these enzymes, nitrite ion is reduced to ammonia. An iron ion appears to be involved in the electron-transport system in this reaction.

$$6\,e^- + NO_2^- + 7\,H^+ \longrightarrow NH_3 + 2\,H_2O$$

Ammonia, of whatever origin, can be combined into organic structures by reacting it, by a process known as **reductive amination,** with α-ketoglutaric acid and a reducing agent to produce the amino acid, glutamic acid. The experiment to be performed in this section is a non-enzymatic example of a reductive amination, and this example will be discussed in Experiment 55. The details of the reaction as found in the cell are outlined below:

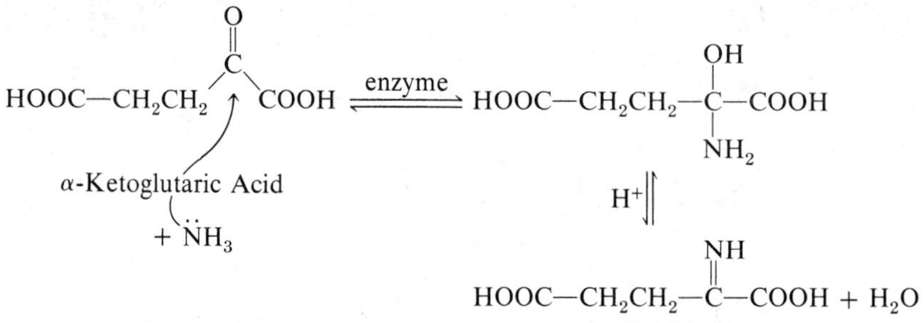

α-Ketoglutaric Acid
+ $\ddot{N}H_3$

Imine (Schiff base)

This intermediate structure, a member of the class of compounds called **imines** or **Schiff bases,** is reduced by the biological reducing agent nicotinamide adenine dinucleotide phosphate, NADPH, which acts as a hydride ion transfer agent.

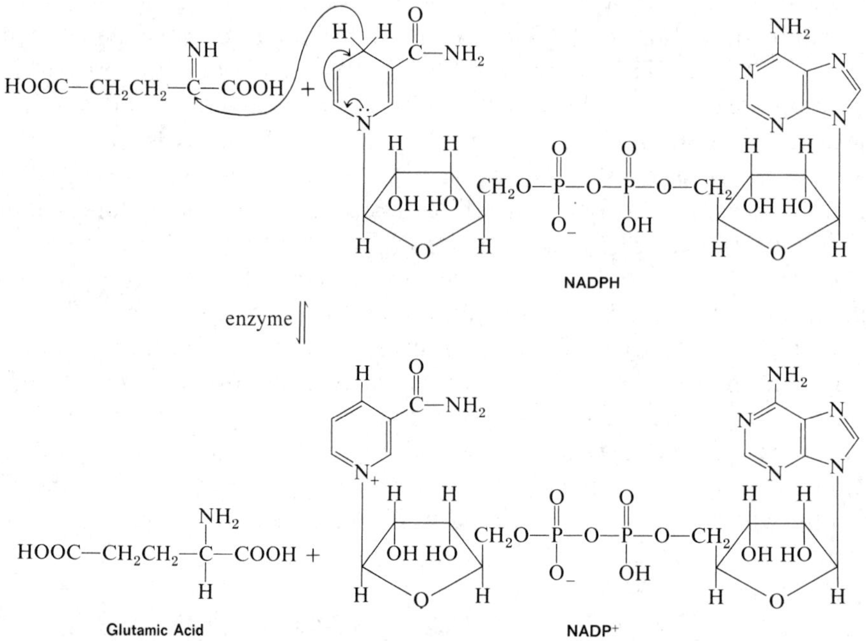

NADPH

enzyme

Glutamic Acid   NADP+

In plants, glutamic acid is the only amino acid which can be formed by direct incorporation of ammonia from the soil. The glutamic acid thus formed serves as the source of nitrogen for the synthesis of all of the other amino acids, **via** the reaction known as **transamination,** where the amino group is transferred from glutamic acid to another organic structure, usually an α-keto acid. The general reaction for this process is:

Glutamic Acid   α-Keto Acid   α-Ketoglutaric Acid   New Amino Acid

The reaction is catalyzed by a family of enzymes known as **aminotransferases.**

The mechanism of the reaction invariably involves the participation of **pyridoxal phosphate** as a coenzyme. This substance is derived from Vitamin B_6, and it alternates with **pyridoxamine phosphate** in each reaction cycle. The aldehyde group of pyridoxal phosphate reacts with the amino group of glutamic acid to form an imine. Within the imine,

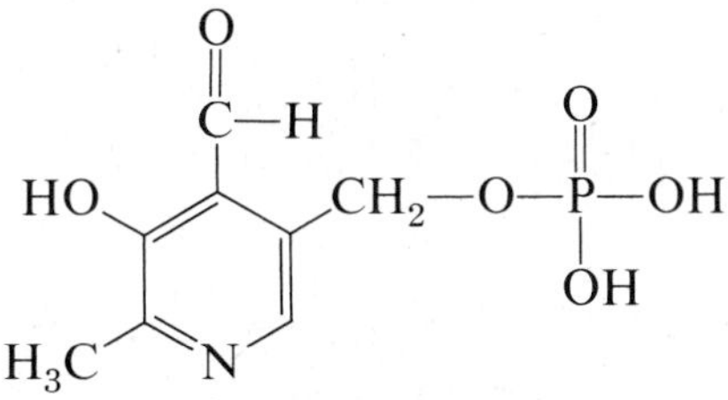

Pyridoxal Phosphate 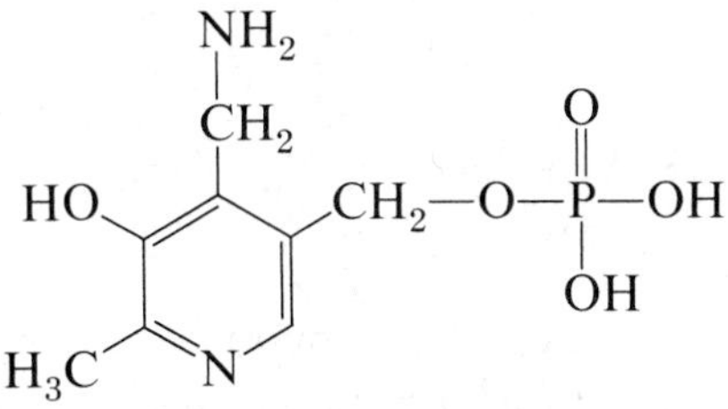   Pyridoxamine Phosphate

hydrogen atoms exchange place by the process of **tautomerism,** and then the imine is hydrolyzed to give α-ketoglutaric acid plus pyridoxamine phosphate. The enzyme is complexed with the pyridoxal-pyridoxamine coenzyme at all times.

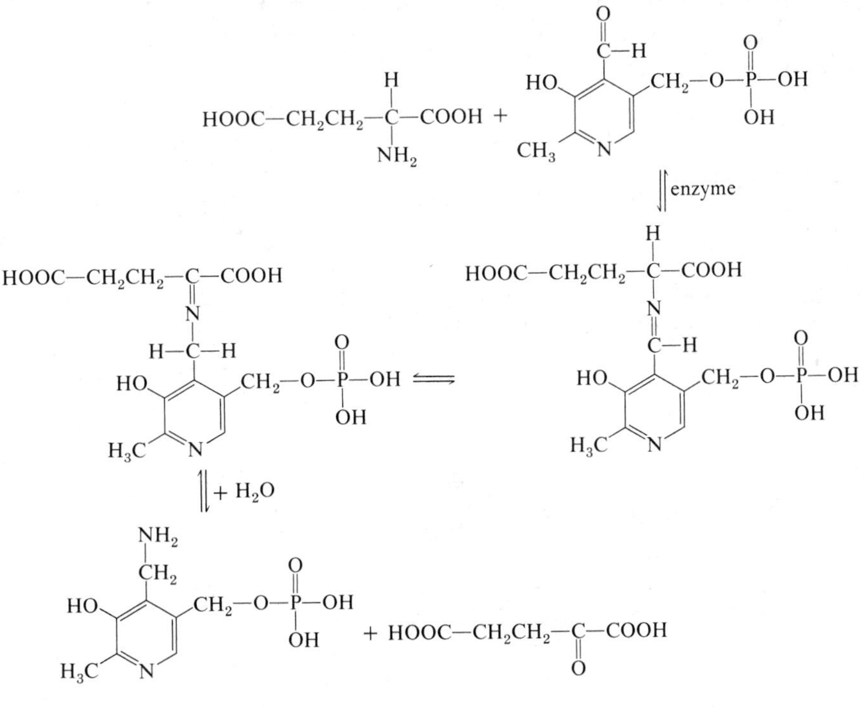

The pyridoxamine phosphate now uses its amino group to react with the carbonyl group of an α-keto acid to form another imine. Within this imine, hydrogen atoms are again transferred by tautomerism, and then the second imine hydrolyzes to yield pyridoxal phosphate back again plus the new amino acid.

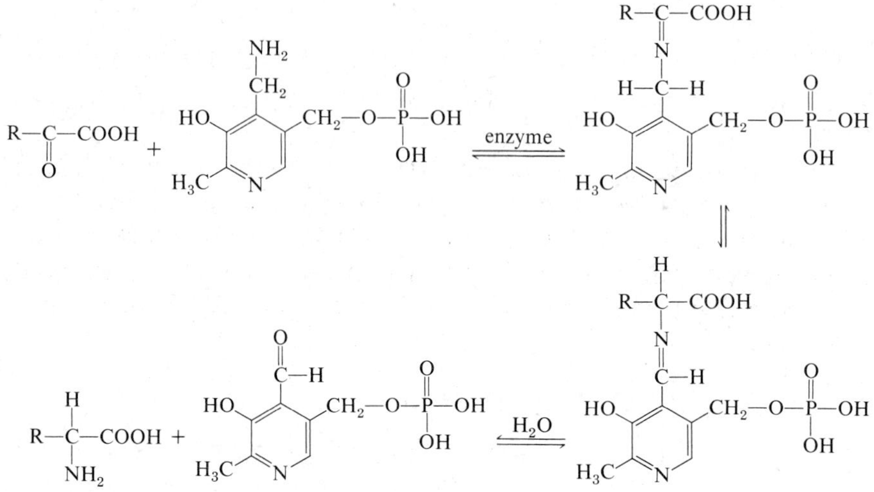

Animal and plant organisms are capable of converting glutamic acid into those amino acids which they require for growth by means of transamination reactions. Of course, not all amino acids required by animals for growth can be synthesized by that animal; some cannot be synthesized so they must be included in the animal's diet. These are the so-called "essential" amino acids. In spite of the animal's capability of converting glutamic acid into some other amino acids, it must be remembered that only plants and microorganisms have the capability of directly incorporating inorganic nitrogen into the amino acid structure.

REFERENCES

Lehninger, A. L. *Biochemistry*. New York: Worth Publishers, 1970. Chapter 24.
Mahler, H. R., and Cordes, E. H. *Biological Chemistry* (2nd edition). New York: Harper and Row, 1971. Chapter 17.
White, A., Handler, P., and Smith, E. L. *Principles of Biochemistry* (5th edition). New York: McGraw-Hill Book Co., 1973. Chapter 22.

SYNTHESIS OF (±)-α-PHENYLETHYLAMINE

Reductive Amination

The synthesis of racemic α-phenylethylamine is performed using a reductive amination reaction, similar to the one by which glutamic acid is formed from ammonia and α-ketoglutaric acid in the cell. In this case the reaction is not enzyme catalyzed and does not use NADPH as the reducing agent, but in all other respects serves as a useful model for the biochemical reaction. Reductive aminations may be carried out in a variety of ways, which may be schematicized as follows:

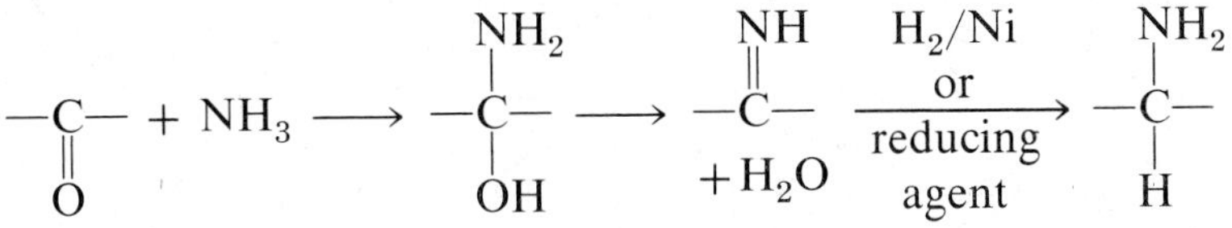

$$\underset{O}{\overset{}{-C-}} + NH_3 \longrightarrow \underset{OH}{\overset{NH_2}{-C-}} \longrightarrow \underset{+H_2O}{\overset{NH}{-C-}} \xrightarrow[\text{reducing agent}]{H_2/Ni \atop \text{or}} \underset{H}{\overset{NH_2}{-C-}}$$

It can be seen that this reaction has the essential features of the biochemical reaction described in the previous essay. Ammonia acts as a nucleophile and attacks the carbonyl group to form an α-amino alcohol, which being unstable dehydrates to form an imine. The imine is then reduced to form the amine product. The reaction conditions which are to be used in this experiment do not involve the use of hydrogen gas and a nickel catalyst, but rather use formic acid as the reducing agent. The reaction directed under these conditions is called the **Leuckart reaction,** and it can be summarized as follows:

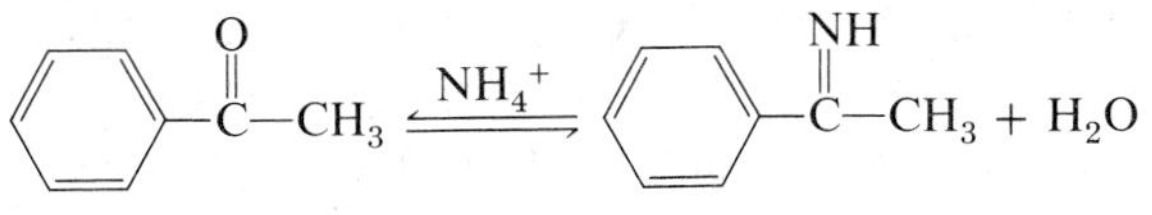

Acetophenone

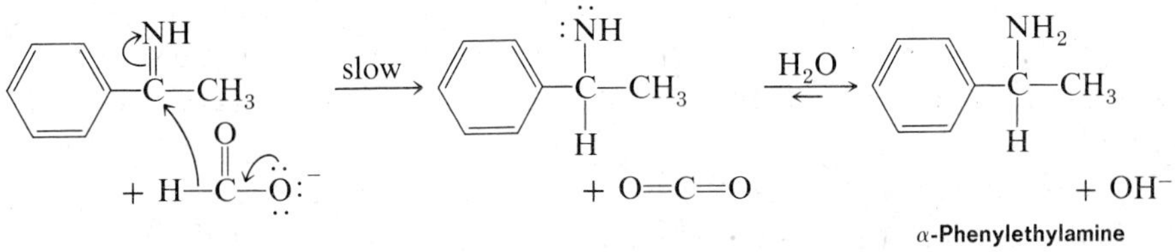

α-Phenylethylamine

The formic acid, or formate ion, is acting as a reducing agent because its hydrogen atom is being transferred to the imine as a **hydride ion,** or as a hydrogen with a pair of electrons. Hydrides are powerful reducing agents.

Actually, because of the presence of the formate ion, the free α-phenylethylamine is not formed directly, but rather the hydride transfer process takes place with the formation of α-phenylethylform-amide. The formamide must then be hydrolyzed in a subsequent step. The balanced equations for the reaction as it is run in this experiment are:

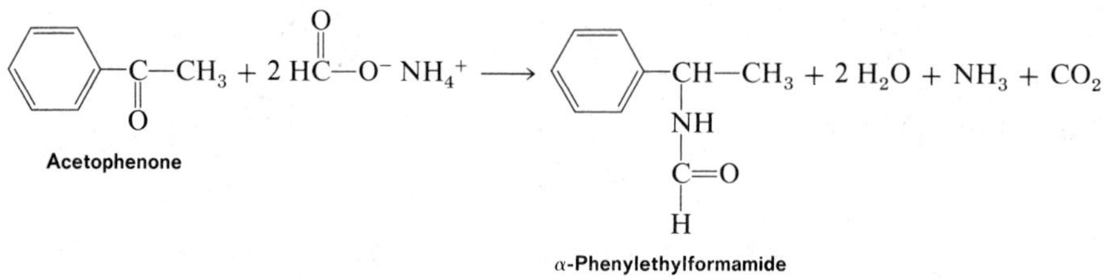

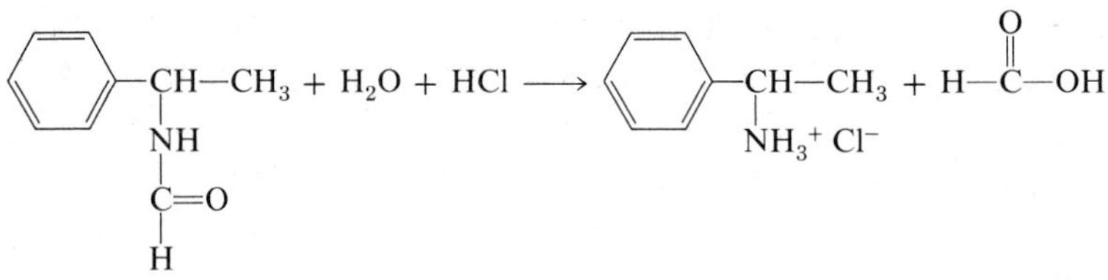

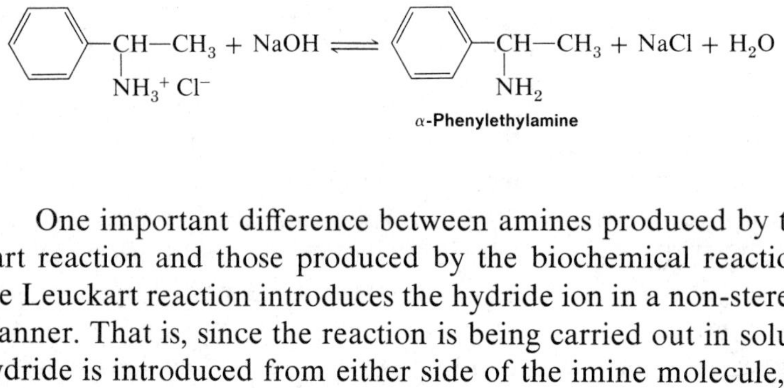

$$\text{C}_6\text{H}_5\text{—C(=O)—CH}_3 + 2\,\text{HC(=O)—O}^-\,\text{NH}_4^+ \longrightarrow \text{C}_6\text{H}_5\text{—CH(—NH—CHO)—CH}_3 + 2\,\text{H}_2\text{O} + \text{NH}_3 + \text{CO}_2$$

Acetophenone   α-**Phenylethylformamide**

$$\text{C}_6\text{H}_5\text{—CH(—NH—CHO)—CH}_3 + \text{H}_2\text{O} + \text{HCl} \longrightarrow \text{C}_6\text{H}_5\text{—CH(—NH}_3^+\,\text{Cl}^-)\text{—CH}_3 + \text{H—C(=O)—OH}$$

$$\text{C}_6\text{H}_5\text{—CH(—NH}_3^+\,\text{Cl}^-)\text{—CH}_3 + \text{NaOH} \rightleftharpoons \text{C}_6\text{H}_5\text{—CH(—NH}_2)\text{—CH}_3 + \text{NaCl} + \text{H}_2\text{O}$$

α-**Phenylethylamine**

One important difference between amines produced by the Leuckart reaction and those produced by the biochemical reaction is that the Leuckart reaction introduces the hydride ion in a non-stereospecific manner. That is, since the reaction is being carried out in solution, the hydride is introduced from either side of the imine molecule, with the result that the chiral carbon generated is formed in each of its enantiomeric forms. The product of the reaction, then, is racemic and must be resolved. In the cell, the reductive amination reaction always produces optically active amines. Throughout a study of biochemical reactions, one always notices that they are invariably highly stereospecific. Enzymes are highly selective in the choice of substrates, complexing with only one isomer of an enantiomeric pair. Similarly, in those reactions involving a chiral center, enzyme-catalyzed reactions usually produce only one form of an enantiomeric pair. The highly complex tertiary structure of the protein serves as a sort of mold into which one enantiomer fits, while the other does not. As the diagram below attempts to demonstrate, the enzyme holds the imine substrate in such a position that the transfer of hydride can occur from only one side. Back side approach by hydride is hindered by the complicated shape of the protein molecule.

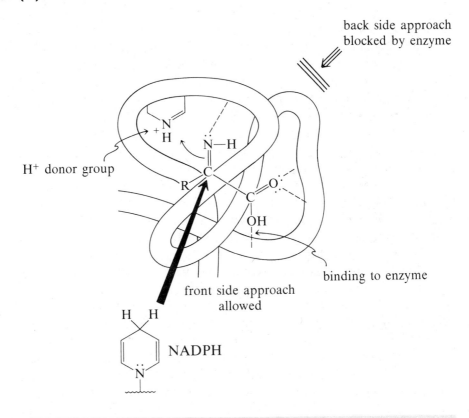

SPECIAL INSTRUCTIONS

Before beginning this experiment, you should read the essay which precedes it. Also read Techniques 6 and 8. For maximum yield, two 1-1/2 hour reflux periods are required. If time is at a premium, the second reflux period may be either shortened or eliminated, although the yields are decreased somewhat as a result.

PROCEDURE

Assemble a 250 ml round bottom flask, fitted with a distillation head and a condenser as for a simple distillation (Technique 6, Section 6.4 and Figure 6–6), except that the thermometer should be placed in the small neck of the flask in such a position that it extends into the reaction mixture. Place 50 g of acetophenone, 85 g of ammonium formate, and one or two boiling stones into the flask. Heat the reaction mixture in the hood, using a heating mantle as a heat source. During this period of heating, the mixture will melt into two phases, and the distillation will begin. When the temperature of the reaction mixture reaches 150 to 155°, the mixture will become homogeneous, and the reaction will take place. Continue gentle heating until the temperature of the reaction mixture reaches 185°. Do not heat the mixture beyond this temperature. About 1 1/2 hours of heating are usually required to reach this stage.

Some solid ammonium carbonate may form in the condenser during the reaction; do not allow the condenser to become plugged with this solid. Stop the heating and transfer the distillate to a separatory funnel. Remove the acetophenone layer and return it to the reaction flask. If time permits, resume the heating, maintaining a temperature of 185° (no higher) for an additional 1 1/2 hours.

Cool the reaction mixture and extract it with 50 ml of water. Separate the crude α-phenylethylformamide and return it to the original reaction flask. Extract the aqueous layer with two 25 ml portions of benzene. Save the benzene extracts and discard the aqueous layer. Combine the benzene extracts with the α-phenylethylformamide and add 50 ml of concentrated hydrochloric acid. Heat the mixture (no flames!) until all of the benzene has distilled. Reflux the solution for an additional 40 minutes.

Cool the reaction mixture to room temperature. Some crystals may form at this stage; if so, add the **minimum** amount of water required to dissolve them. Extract the reaction mixture with three 25 ml portions of benzene (be careful not to form emulsions). Transfer the aqueous layer to a 500 ml three neck, round bottom flask which has been equipped for steam distillation using the live steam method (Technique 8, Section 8.3A and Figure 8–2). Add 85 ml of a 50% solution of sodium hydroxide and begin the steam distillation. Be sure that the ground glass joints are well greased to prevent freezing of the joints caused by the basic solution. Continue the steam distillation until the distillate, which is initially basic, is no longer basic when tested with pH paper. About 300 ml of distillate will be obtained.

Extract the distillate, which contains the free amine, with three 50 ml portions of benzene. Transfer the benzene extracts to an Erlenmeyer flask, add some sodium hydroxide pellets, and stopper the flask in order to dry the solution. The free amine absorbs carbon dioxide from the air, and care should be used to minimize exposure of the amine to air. Distill the dry solution using a simple distillation (Technique 6, Section 6.4), and collect the amine in the temperature range of 180 to 190°C. Calculate the percentage yield and either submit the sample in a labeled vial to the instructor or save it for Experiment 56. Store the amine in a tightly stoppered flask to minimize exposure to atmospheric carbon dioxide.

QUESTIONS

1. Outline a reaction sequence, using sodium borohydride and ammonium hydroxide, which would lead to the same product as found in this experiment.

2. Write a mechanism for the hydrolysis of α-phenylethylformamide to yield α-phenylethylamine.

3. Why is the solution made basic before the steam distillation of the reaction mixture?

RESOLUTION OF
(±)-α-PHENYLETHYLAMINE

Resolution
Polarimetry

After racemic α-phenylethylamine has been synthesized (Experiment 55), the next step is to obtain one of the enantiomers in optically pure form. This is accomplished through a **resolution** process, or a separation of the enantiomers. In this experiment, you will isolate only one of the enantiomers, the levorotatory one, since its isolation is easier than that of the dextrorotatory enantiomer. The resolving agent to be used is (+)-tartaric acid, which forms diastereomeric salts with racemic α-phenylethylamine. The important reactions for this experiment are:

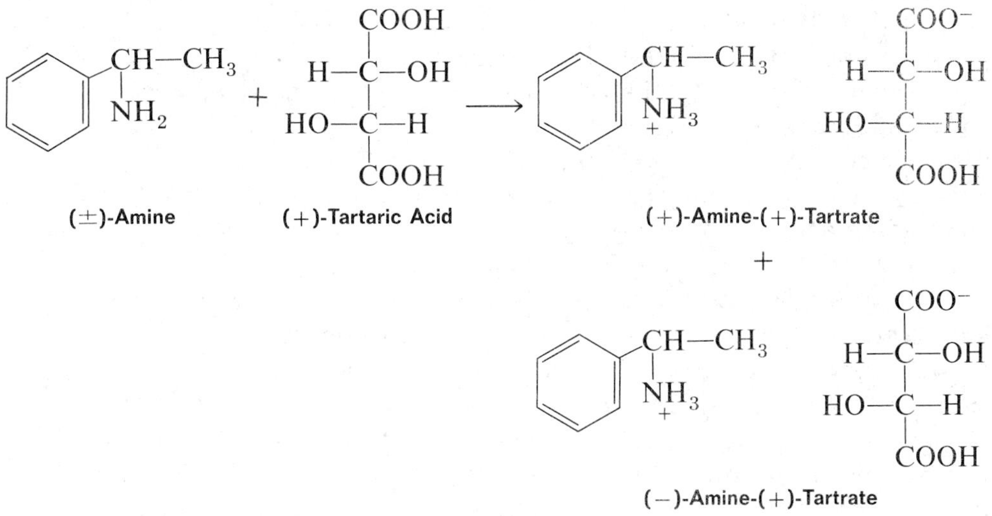

(±)-Amine   (+)-Tartaric Acid   (+)-Amine-(+)-Tartrate

+

(−)-Amine-(+)-Tartrate

Optically pure (+)-tartaric acid is quite abundant in nature. It is readily obtained as a by-product in the wine-making process. Crystals of the (−)-amine-(+)-tartrate diastereomeric salt have a lower solubility than the (+)-amine-(+)-tartrate salt, and they separate from the solution as crystals. The crystals are removed by filtration and purified. The salt is then treated with dilute base to regenerate the free (−)-amine. In principle, the mother liquor, which contains mostly the (+)-amine-(+)-tartrate salt, can also be purified to eventually yield the other diastereomeric salt. Hydrolysis of this salt would yield the (+)-amine.

SPECIAL INSTRUCTIONS

Before beginning this experiment, it is important to read Technique 15. Technique 6 should also be read. The racemic α-phenylethylamine used to perform this reaction may be either that material prepared in Experiment 55 or that supplied by the instructor. Once the amine-tartrate salt is precipitated, the instructor may choose to have students work in pairs to complete the experiment.

PROCEDURE

Place 31.25 g of L-(+)-tartaric acid and 450 ml of methanol in a one-liter Erlenmeyer flask. Heat this mixture on a steam bath until the solution is nearly boiling. Slowly add 25 g of racemic α-phenylethyl-amine (Experiment 55) to this hot solution. **NOTE:** Caution should be exercised at this step, because the mixture is very likely to froth and boil over. Next, stopper the flask and allow it to stand overnight. The crystals which form should be prismatic. If needles are obtained, they should be redissolved and recrystallized by seeding the mixture with a prismatic crystal, if one is available. Alternatively, the mixture may be heated until **most** of the solid has dissolved. The needle crystals dissolve easily and usually a small amount of the prismatic crystals will remain to seed the solution. The solution is allowed to cool slowly, forming primatic crystals. The needles do not have a sufficiently high optical purity to give a complete resolution of enantiomers, and they must be redissolved and crystallized again. Filter the crystals using a Büchner funnel and wash them with a few portions of cold methanol.

At this point, two students should combine their yields of amine-tartrate adduct in order that approximately 25 g of crystals will be available for the next step. An alternate approach would be to reduce the volume of mother liquor to 250 ml by evaporation (in the hood; no flames), and to allow an additional amount (about 3.5 g) of adduct to crystallize. Partially dissolve the salt in 100 ml of water, add 15 ml of 50% sodium hydroxide solution, and extract this mixture with three 30 ml portions of diethyl ether. Dry the ether layer over anhydrous magnesium sulfate for about 10 minutes. Assemble a simple distillation apparatus using a 100 ml distillation flask (Technique 6, Figure 6–6). Remove the ether by simple distillation, using a heating mantle or an oil bath as a heat source (no flames!). Then, allow the temperature to rise and distill the amine over a temperature range of 180 to 190°C. NOTE: Frothing can sometimes be a problem at this stage. Use a 25 ml Erlenmeyer flask, which has previously been weighed, with its stopper, to within ±0.003 g, to collect the amine. Stopper the flask tightly as soon as the product is collected. Weigh the flask to determine the weight of amine collected.

Since this reaction does not generally yield enough pure amine to fill a polarimeter cell, it will have to be diluted with methanol. Using a pipet, add 10 ml of absolute methanol to the stoppered flask and shake the solution in order to mix the contents thoroughly. At this point the total volume of the solution is very nearly equal to 10 ml plus the volume of the amine, or its weight divided by its density (0.9395 g/ml). The approximation of additivity of volumes introduces a negligible error in this procedure. This total volume of the solution is needed in order to calculate the new concentration (in g/ml) of the solution. Transfer the solution to a 2 dm polarimeter tube and determine its observed rotation (Technique 15, Sections 15.3 and 15.4). The published value is $[\alpha]_D^{22} = -40.3°$. Be sure to keep the pure amine tightly stoppered. Report the value of the specific rotation and the optical purity to the instructor. Calculate the percentage of **each** of the enantiomers in the sample.

REFERENCES

Ault, A. "Resolution of D,L-α-Phenylethylamine." *Journal of Chemical Education, 42* (1965), 269.
Jacobus, J., and Raban, M. "An NMR Determination of Optical Purity." *Journal of Chemical Education, 46* (1969), 351.

QUESTIONS

1. Using a reference text, find examples of reagents used in performing chemical resolutions of acidic, basic, and neutral racemic compounds.

2. Propose methods of resolving each of the following racemic compounds:

a. 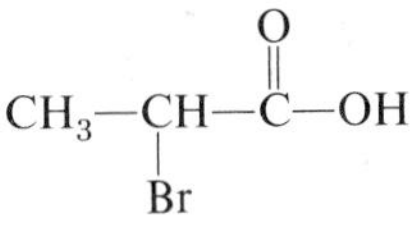
$$CH_3-\underset{\underset{Br}{|}}{CH}-\overset{\overset{O}{||}}{C}-OH$$

b.

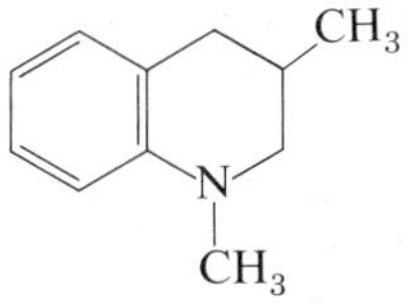

c.

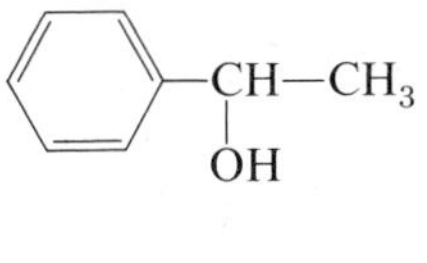

PART TWO

THE TECHNIQUES

Technique **1**

SOLVENTS AND METHODS
OF HEATING
REACTION MIXTURES

1.1 SAFETY PRECAUTIONS WITH SOLVENTS

One should always bear in mind that organic solvents are all toxic to at least some degree and that many of them are flammable. Great care must be exercised when using these materials. One should also be thoroughly familiar with the section entitled "Laboratory Safety," located in the introductory portion of this book.

The most commonly used organic solvents are listed in Table 1–1, along with their boiling points. Solvents marked in bold face type will burn. Ether, pentane, and hexane are especially dangerous since, in combination with the correct amount of air, they will explode.

The terms **petroleum ether** and **ligroin** are often confusing. Petroleum ether is a mixture of hydrocarbons, with isomers of the formulas C_5H_{12} and C_6H_{14} predominating; it is not an ether at all, since there are no oxygen-bearing compounds present in the mixture. Care should be exercised when instructions call for either **ether** or **petroleum ether** in order that the two do not become accidentally confused. Confusion is particularly easy when one is selecting the container of solvent from the side shelf.

Ligroin, or high-boiling petroleum ether, is similar in composition

TABLE 1–1. COMMON ORGANIC SOLVENTS*

HYDROCARBONS		ALCOHOLS	
pentane	36°	**methanol**	65°
hexane	69°	**ethanol**	78°
benzene	80°		
toluene	111°	ETHERS	
		ether (diethyl)	35°
HYDROCARBON MIXTURES		**dioxane**	101°
petroleum ether	30–60°		
ligroin	60–90°	OTHERS	
		acetic acid	118°
CHLOROCARBONS		acetic anhydride	140°
		pyridine	115°
methylene chloride	40°	**acetone**	56°
chloroform	61°	**ethyl acetate**	77°
carbon tetrachloride	77°		

*Bold face type indicates flammability.

to petroleum ether, except that the alkanes in ligroin generally include higher boiling alkane isomers than petroleum ether. Depending upon the supplier, ligroin may have different boiling ranges. While some brands of ligroin have boiling points ranging from about 60°C to about 90°C, other brands have boiling points ranging from about 60°C to about 75°C.

1.2 FLAMES

The simplest technique for heating mixtures is by means of the Bunsen burner. However, because of the high danger of fires, the use of the Bunsen burner should be severely limited to those cases where the danger of fire is low or where no reasonable alternative source of heat is available. Generally speaking, a flame should be used only to heat aqueous or very high boiling solutions. Even under these circumstances, great care should be taken to ensure that persons in the vicinity are not using flammable solvents.

When heating a flask with a Bunsen burner, the use of wire gauze can produce more even heating over a broader area. The wire gauze, when placed under the object being heated, spreads the flame to prevent the flask from being heated in one small area only.

1.3 STEAM CONES

The steam cone or steam bath is a safe source of heat for use with most reaction mixtures. The steam cone is depicted in Figure 1-1. This

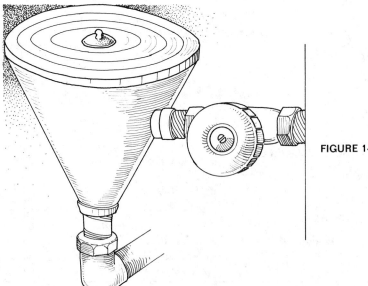

FIGURE 1-1. Steam cone

device has the disadvantage that water vapor may be introduced through condensation of the steam into the mixture being heated. A slow flow of steam may minimize this difficulty.

Because water condenses in the steam line when it is not in use, it is necessary to purge the line of water before the steam will begin to flow. This purging should be accomplished before the flask to be heated is placed on the steam cone. The steam flow should be started with a high rate in order to purge the line; then the flow should be reduced to the desired rate. Once the steam cone is heated, a slow steam flow will maintain the temperature of the mixture being heated unless that mixture contains a high-boiling solvent. There is no advantage to having a Vesuvius on your desk! An excessive steam flow may cause problems with condensation, both in the flask being heated and elsewhere in the laboratory room. This condensation problem may often be avoided by selecting the correct place to locate the flask on top of the steam cone.

The top of the steam cone, as shown in Figure 1-1, consists of a number of flat concentric rings. The amount of heat delivered to the flask being heated may be controlled by selecting the correct sizes of these rings. Heating is most efficient when the largest opening which will still support the flask is used. Heating large flasks on a steam bath while using the smallest opening provides slow heating and wastes laboratory time.

1.4   OIL BATHS

In some laboratories oil baths may be available. These can be used instead of a flame when one is carrying out a distillation or when one is heating a reaction mixture which requires a temperature above 100°C. An oil bath with a variable transformer is shown in Figure 1-2. Since

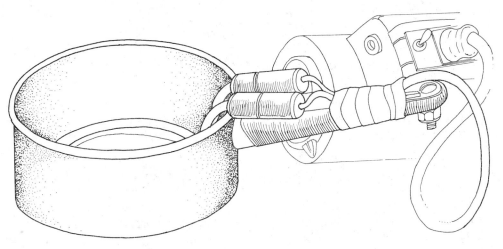

FIGURE 1-2. Oil bath

the oil is heated electrically by means of an immersion coil, the danger of fires is minimized. The heating coil must be plugged into a variable transformer, or Variac, which is a source of variable AC voltage. Control of the immersion coil voltage controls the degree of heating. Because oil baths have a high heat capacity and heat slowly, it is advisable to preheat the oil bath partially before the actual time at which it is to be used.

An oil bath using ordinary mineral oil cannot be used above 200 to 220°C. Above this temperature the oil bath may "flash" or suddenly burst into flame. A hot oil fire is not extinguished easily. If the oil starts smoking it may be near its flash temperature; discontinue heating. Old oil, which is dark colored, is more likely to flash than is new oil. Also one should remember that hot oil causes bad burns. Water should be kept away from a hot oil bath, since water in the oil will cause it to splatter. Never use an oil bath when it is obvious that there is water in the oil. If there should be water present, replace the oil before using the heating bath. One should also remember that an oil bath has only a finite lifetime. While the new oil is clear and colorless, after extended use, oxidation causes it to become dark brown and somewhat gummy.

In addition to ordinary mineral oil, there is a variety of other types of oils that may be used in an oil bath. Silicone oil does not begin to decompose at as low a temperature as mineral oil. When silicone oil is heated high enough to decompose, however, its vapors are far more hazardous than those of mineral oil. The polyethylene glycols may be used in oil baths. They are water-soluble, which makes clean-up after using an oil bath much easier than with mineral oil. One may select any one of a variety of polymer sizes with polyethylene glycol, depending upon the temperature range which is required. The larger molecular weight polymers are often solid at room temperature. Wax may also be used for higher temperatures, but of course this material also becomes solid at room temperature. Some workers prefer to use a material that solidifies when not in use since it minimizes both storage and spillage problems. Vegetable shortening is occasionally used in heating baths.

1.5 HEATING MANTLES

A useful source of heat for situations that require temperatures above 100°C is the heating mantle, which is illustrated in Figure 1–3. The heating mantle consists of a blanket of spun fiberglass with electric heating coils embedded within the blanket. This blanket fits snugly around the flask, providing an even source of heat. The temperature of a heating mantle is controlled by the use of a Variac. Some heating mantles are designed to fit only specific sizes of flasks; other types fit a range of flask sizes. The heating mantle is safe, because it does not produce flames. It is rapid, since it does not have the high thermal inertia of the oil bath. It is convenient because it is not subject to

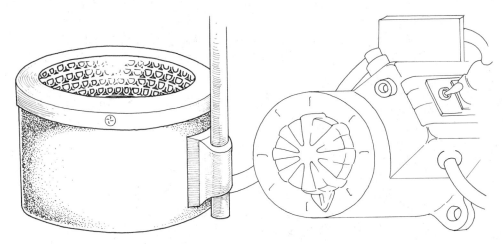

FIGURE 1–3. Heating mantle

contamination as is the oil bath, and it does not produce the problem of oil clinging to the outside of the flask. One must be careful, however, when using heating mantles, since there is the danger that the heating mantle may burn out if it is used to heat an empty flask. One should avoid spilling chemicals into the heating mantle. Finally, one additional disadvantage to the heating mantle is its great initial cost.

1.6 HOT PLATES

Occasionally, one might use a hot plate for heating small quantities of solvents when temperatures around 100°C are required. Care must be taken with flammable solvents to ensure against fires caused by solvent vapors "flashing" when they come into contact with the hot plate surface. One should never evaporate large quantities of a solvent using this method; there is too large a fire hazard involved.

Some hot plates also have built-in magnetic stirring motors. The use of these is described in Section 1.8.

1.7 HEATING UNDER REFLUX

Often it is desired to heat a mixture for a long time and to be able to leave it untended. The reflux apparatus allows such heating. It also prevents the loss of solvent due to evaporation. A condenser is attached to the boiling flask, and cooling water is circulated to condense escaping vapors. A reflux apparatus is shown in Figure 1–4. One should always use a boiling stone or a magnetic stirrer (see Section 1.8) to prevent "bumping" of the boiling solution. The direction of the water flow should be such that the condenser will fill with cooling water; the water should enter the bottom of the condenser and exit from the

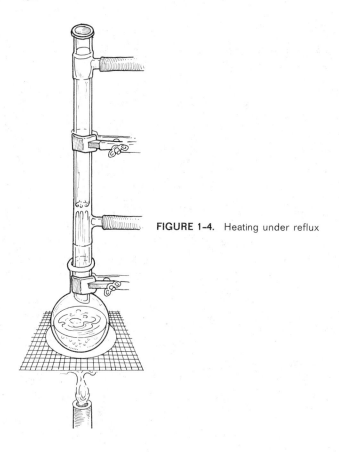

FIGURE 1-4. Heating under reflux

top. The flow rate of the water should be sufficient to withstand changes in pressure which may occur in the water lines, but it should not be any higher than is absolutely necessary. An excessive flow rate greatly increases the chance of a flood, and high water pressure may force the hose from the condenser. When a flame is used as a source of heat it is convenient to use a wire gauze beneath the flask to provide an even distribution of heat from the flame. It is essential that the cooling water be flowing before the heating has begun! If the water is to remain flowing overnight, it is advisable to fasten the rubber tubing securely with wire to the condenser.

The liquid which is being heated under reflux will travel only a partial distance up the condenser tube before condensing when the heating rate is correctly adjusted. Below this point, solvent will be seen running back into the flask; above it, the condenser will appear dry. The boundary between these two zones will be clearly demarcated, and a "reflux ring" or a ring of liquid will appear there. In heating under reflux, the rate of heating should be adjusted so that the reflux ring is no higher than 1/3 to 1/2 the distance to the top of the condenser.

It is possible to heat small amounts of a solvent under reflux in an Erlenmeyer flask. With gentle heating the evaporated solvent will condense in the relatively cold neck of the flask and return to the solution. This technique, which is illustrated in Figure 1-5, requires constant attention. The flask must be swirled frequently and removed from the heating stage for a short period if the boiling action becomes too vigorous. When heating, the reflux ring should not rise above the base of the flask's neck.

In Figure 1–5 still another technique for heating small amounts of solvent under reflux is illustrated. A **cold-finger condenser** is inserted into a test tube or a small flask. As the vapors rise they contact the cold surface of the condenser. The vapors are condensed, and the resulting liquid drips back to the bottom of the container. Some commercial cold-finger condensers are designed to rest on top of the test tube or flask. If pressure builds up in the container, it can escape by lifting the condenser slightly. With cold-finger condensers which fit through cork stoppers, such as the one in the illustration, it is necessary to provide a slot in the stopper to prevent pressure from building within the container. Without the slotted stopper, one would be heating a closed system, thus creating a potential "bomb"!

1.8 BOILING STONES

A boiling stone, also known as a boiling chip or Boileezer, is a small lump of porous material which, when it is heated in a solvent, produces a steady stream of fine air bubbles. This stream of bubbles, and the turbulence which accompanies it, breaks up large bubbles of gases in the liquid. It also reduces the tendency of the liquid to become

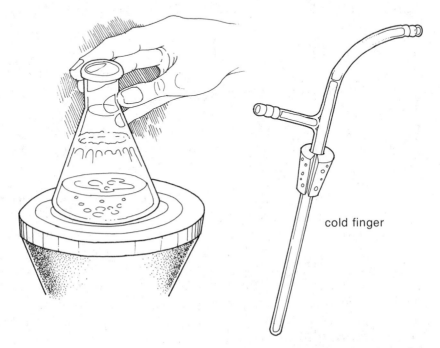

cold finger

FIGURE 1–5. Tended reflux of small quantities

superheated and promotes the smooth boiling of the liquid. If the liquid becomes superheated, very large bubbles may erupt rather violently from the solution; this is called **bumping.** The boiling stone, by its action, decreases the chance that bumping will occur.

Boiling stones are generally made from pieces of pumice, carborundum, or marble. Wooden applicator sticks are also used.

Because boiling stones act to promote the smooth boiling of liquids, one should always make certain that the boiling stone has been placed in the liquid **before** the heating is begun. If one waits until the liquid is hot, it may have become superheated, at which time a boiling stone would cause all of the liquid to try to boil at once. The liquid, as a result, would erupt entirely out of the flask, or at least froth violently. As soon as boiling ceases, the liquid is drawn into the pores of the boiling stone. When this happens, the boiling stone is no longer capable of producing a fine stream of bubbles. It is spent. A new boiling stone must be added each time the boiling stops.

Magnetic stirrers provide much the same action as boiling stones, since they produce a great deal of turbulence in the solution. The turbulence breaks up the large bubbles which form in hot solutions. A magnetic stirring system consists of a magnet which is rotated by an electric motor. The rate at which this magnet rotates may be adjusted by means of a potentiometer. One places a small bar magnet, which is coated with some nonreactive material such as Teflon or glass, into the flask. The magnet within the flask rotates in response to the rotating magnetic field caused by the motor-driven magnet. The result is that the inner bar magnet stirs the solution as it rotates. As previously mentioned, some hot plates which are available incorporate a magnetic stirring motor so that heating and stirring operations can be performed simultaneously.

1.9 EVAPORATION TO DRYNESS

Frequently one may desire to evaporate a solution to dryness or to concentrate a solution by removing the solvent either completely or to some desired extent. This may be accomplished by evaporating the solvent from an open Erlenmeyer flask. Such an evaporation must be conducted in a hood, since many solvent vapors are toxic or flammable. A boiling stone must be used. A gentle stream of air directed toward the surface of the liquid will remove the vapors which are in equilibrium with the solution, and accelerate the evaporation process. An eyedropper tube or capillary pipet, connected by a short piece of rubber tubing to the compressed air line, will act as a convenient air nozzle. A tube or an inverted funnel connected to an aspirator may also be used. In this case the vapors are removed by suction. Both methods are illustrated in Figure 1–6. It is better to use an Erlenmeyer flask than a beaker for this procedure, since deposits of solid will usually

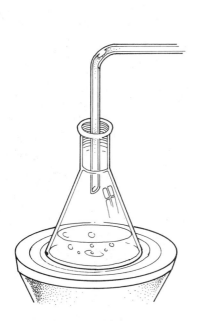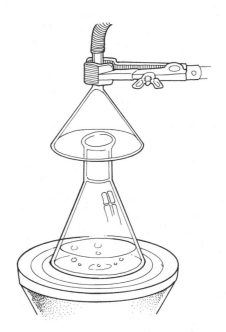

FIGURE 1-6. Rapid evaporation of a solvent

build up on the sides of the beaker where the solvent evaporates. The refluxing action in an Erlenmeyer flask will not allow this build-up to occur.

It is also possible to remove low boiling solvents by reduced pressure. In this method, the solution is placed in a filter flask, along with a wooden applicator stick. The flask is stoppered, and the side arm is connected to an aspirator (using a trap), as is described in Technique 2, Section 2.3. Under reduced pressure, the solvent begins to boil. The wooden stick serves the same function as a boiling stone. By this method, the solvent can be evaporated from the solution without the use of heat. Often, this technique, which is illustrated in Figure 1–7, is used when heating the solution might decompose thermally sensitive substances. The method has the disadvantage that, when low boiling solvents are used, evaporation of the solvent will cool the flask below the freezing point of water. When this happens, a layer of frost will form on the outside of the flask. Since frost is insulating, it must be removed to keep the evaporation process going at a reasonable rate. The solution must take heat from the surrounding air to evaporate. The frost prevents this necessary heat transfer. Removal of the frost is achieved best by one of two methods. Either the flask is placed in a bath of warm water (with constant swirling) or it is heated on the steam bath (again with swirling). Either method promotes efficient heat transfer.

To remove large amounts of a solvent, a distillation should be employed (see Technique 6). ONE SHOULD NEVER EVAPORATE ETHER SOLUTIONS TO DRYNESS, except on a steam bath or by the use of the reduced pressure method. The tendency of ether to form

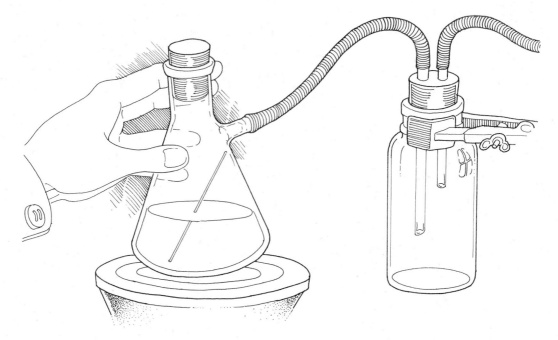

FIGURE 1-7. Reduced pressure–solvent evaporation

explosive peroxides is a serious potential hazard. If peroxides should be present, the large and rapid temperature increase in the flask once the ether evaporates could bring about the detonation of any residual peroxides. The temperature of the steam bath is not sufficiently high to cause such a detonation.

1.10 COLD BATHS

At times, one may require a medium in which a flask or some other piece of apparatus may be cooled rapidly to a temperature below room temperature. A **cold bath** is used for this purpose. The most common cold bath is an **ice bath.** An ice bath serves as a highly convenient source of 0°C temperatures. The ice bath should actually be called an ice-water bath, since it requires water to work well. An ice bath made up of nothing but ice is not very effective, because the large pieces of ice do not make good contact with the outside walls of the vessel immersed in the bath. Some liquid water must be added to the ice in order to create a much more efficient cooling medium. There must be enough water to ensure that the flask being cooled is totally surrounded by water, but there must not be so much water that the amount of ice present is no longer sufficient to maintain a temperature of 0°C. If too much water is added to the ice bath, the buoyancy of a flask resting in the ice bath may cause it to tip over. There should be enough ice in the bath to permit the flask to rest firmly.

When temperatures lower than 0°C are desired, one may add some solid sodium chloride to the ice-water mixture. The ionic salt lowers the freezing point of the ice, so that temperatures in the range of 0° to $-10°C$ may be attained. The lowest temperatures are attained with ice-water mixtures that contain relatively little water.

Very low temperatures may be obtained through the use of solid carbon dioxide, or Dry Ice, whose temperature is $-78.5°C$. Again, the large chunks of Dry Ice do not provide uniform contact with a flask being cooled. Liquid is required, along with the Dry Ice, to provide uniform contact with the flask and efficient cooling. The liquid used most often with Dry Ice is isopropyl alcohol, although acetone or ethanol may also be used. One should be cautious when handling Dry Ice or cooling baths made from it, since Dry Ice is capable of inflicting very severe cases of frostbite.

Extremely low temperatures may be obtained with liquid nitrogen ($-195.8°C$). Its application in organic chemistry is not as common as that of Dry Ice.

Technique 2
FILTRATION

Filtration is a technique used for two major purposes. The first of these purposes is to remove solid impurities from a liquid or a solution. The second is to collect a solid product from the solution from which it was precipitated or crystallized. Two somewhat different kinds of filtration are in general use: gravity filtration and vacuum (or suction) filtration.

2.1 GRAVITY FILTRATION

The most familiar filtration technique is probably filtration of a solution through a paper filter held in a funnel, allowing gravity to draw the liquid through the paper. In general, it is best to use a short stem or a wide stem funnel. In these types of funnels there is less likelihood that the stem of the funnel will become clogged due to the accumulation of solid material in the stem. This clogging is a particular problem if a hot solution saturated with a dissolved solid is being filtered. If the hot saturated solution comes in contact with a relatively cold funnel (or a cold flask, for that matter), the solution will become cooled. The rapidly cooled solution will be supersaturated, and crystallization will begin. The crystals will form in the filter and either fail to pass through the filter paper or clog the stem of the funnel.

There are four other measures which may be taken to prevent clogging of the filter. The first measure is to keep the solution to be filtered at or near its boiling point at all times. The second measure is to preheat the funnel by pouring hot solvent through it prior to the actual filtration. This prevents the cold glass from causing instantaneous crystallization. The third way is to keep the **filtrate** (filtered solution) in the receiver hot enough to continue boiling **slightly** (e.g., by setting it on a steam bath). The refluxing solvent heats the receiving flask and the funnel stem and washes them clean of solids. This boiling of the filtrate also serves to keep warm the liquid in the funnel. Finally, it is useful to accelerate the filtration process by using **fluted filter paper,** as described below. A gravity filtration is shown in Figure 2–1.

A. Filter Cones

The simplest way to prepare filter paper for gravity filtration is to prepare a filter cone, as outlined in Figure 2–2. The filter cone is

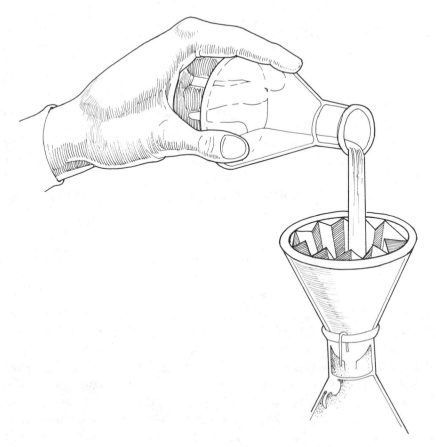

FIGURE 2–1. Gravity filtration

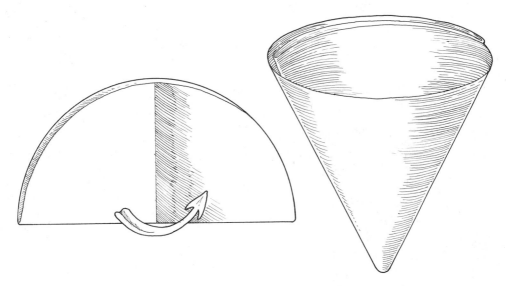

FIGURE 2–2. Folding a filter cone

particularly useful when the solid material being filtered from a mixture is to be collected and used in a later procedure. The filter cone, because of its smooth sides, can be scraped free of collected solids easily. Because of the many folds, fluted filter paper, described in the next section, cannot be scraped easily.

With filtrations using a simple filter cone, solvent may form seals between the filter and the funnel and between the funnel and the lip of the flask. When a seal forms, the filtration stops, because there is no possibility of the displaced air escaping. In order to avoid the formation of a solvent seal, a small piece of paper or a paper clip or other bent wire is inserted between the funnel and the lip of the flask to permit the escape of displaced air. Alternatively, one may support the funnel by a ring clamp fixed **above** the flask, rather than by replacing it in the neck of the flask.

B. Fluted Filters

The technique for folding a fluted filter paper is illustrated in Figure 2–3. The fluted filter increases the speed of filtration in two ways. First, it increases the surface area of the filter paper through which the solvent seeps; second, it allows air to enter the flask along its sides to permit rapid pressure equalization. If pressure builds up in the flask due to the presence of hot vapors, filtering action will slow down. This problem is especially pronounced with filter cones. The fluted filter tends to reduce this problem considerably, but it may be a good idea to use a piece of paper, paper clip, or wire between the funnel and the lip of the flask as an added precaution against solvent seals. Fluted filters are used when the desired material is expected to remain in

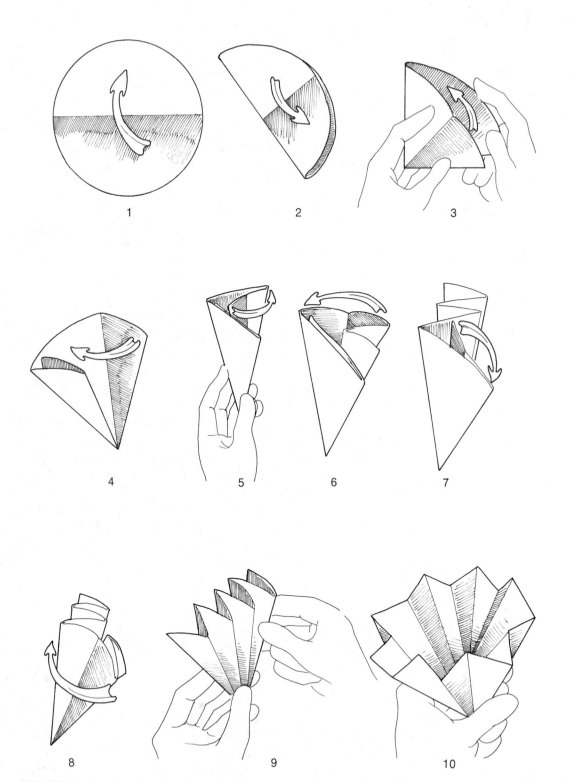

FIGURE 2–3.   Folding a fluted filter paper, or Origami at work in the organic lab

solution. These filters are used to remove undesired solid materials, such as dirt particles, activated charcoal, or undissolved impure crystals.

2.2 FILTER PAPER

Many kinds and grades of filter paper are available. In general, paper is available in fine, medium, and coarse porosities. Fine porosity paper will catch very fine solid particles but will generally give very slow rates of filtration. Coarse paper increases the rate of filtration but may not catch all of the particles. The correct paper must be chosen for a given application. In making such a choice, one should be aware of various properties which filter paper possesses. **Porosity** is a measure of the particle size which the paper will permit to pass through. Highly porous paper does not remove small particles from the solution; paper with low porosity will remove very small particles. **Retentivity** is a property which is the opposite of porosity. Paper with low retentivity will not remove small particles from the filtrate. The **speed** of filter paper is a measure of the time which a liquid requires to drain through the filter. Fast paper allows the liquid to drain quickly; slow paper requires a much longer time to complete the filtration. Since all of these properties are somewhat related, fast filter paper usually has a low retentivity and high porosity, while slow filter paper usually has high retentivity and low porosity.

Table 2–1 compares some commonly available qualitative filter paper types and ranks them according to porosity, retentivity, and speed. Eaton-Dikeman (E&D), Schleicher and Schuell (S&S), and Whatman are the most common brands of filter paper. The numbers in the table refer to the grades of paper as used by each company.

2.3 VACUUM FILTRATION

Vacuum, or suction, filtration is more rapid than gravity filtration, but without specially prepared filter media it will not catch fine particles

TABLE 2–1. SOME COMMONLY AVAILABLE QUALITATIVE FILTER PAPER TYPES AND THEIR APPROXIMATE RELATIVE SPEEDS AND RETENTIVITIES

				TYPE	
SPEED	E&D	S&S	Whatman
Very Slow	#610	#576	#5
Slow	#613	#602	#3
Medium	#615	#597	#2
Fast	#617	#595	#1
Very Fast	–	#604	#4

Porosity: Fine → Coarse
Retentivity: High → Low
Speed: Slow → Fast

without clogging the paper pores. In this technique a receiver flask with a side arm, a **filter flask,** is used. The side arm is connected by means of a **heavy-walled** rubber tubing to a source of vacuum. Thin-walled tubing will collapse under vacuum, due to atmospheric pressure on its outside walls, and will seal the vacuum source from the flask. A **Büchner funnel** (see Figure 2–4) is sealed to the filter flask by the use of a rubber stopper or a rubber gasket (Neoprene adapter) cone. The flat bottom of the Büchner funnel is covered with an unfolded piece of circular filter paper which is held in place by suction. In order to prevent the unfiltered mixture to pass around the edges of the filter paper and contaminate the filtrate in the flask below, it is advisable to moisten the paper with a small amount of solvent before beginning the actual filtration. The moistened filter paper adheres more strongly to the bottom of the Büchner funnel. Since the filter flask is attached to a source of vacuum, a solution poured into the Büchner funnel is literally "sucked" through the filter paper at a rapid rate. To prevent the escape of solid materials from the Büchner funnel, it is important to be certain that the filter paper fits the Büchner funnel exactly. The paper must be neither too big nor too small. It must cover all of the holes in the bottom of the funnel, but it must not extend up the sides.

There are actually two types of funnel which may be used for vacuum filtrations. The Büchner funnel, which has already been discussed, is used for the filtration of a fairly large amount of crystals from a solution. The Hirsch funnel, which is also shown in Figure 2–4, operates on the same principle as the Büchner funnel, except that it is smaller and its sides are sloped, rather than vertical. The Hirsch funnel is used for the isolation of smaller quantities of solid materials from a solution. Again, the filter paper must cover all of the holes in the bottom of the Hirsch funnel, but it must not extend up the sides.

2.4   FILTER AID

It was mentioned that specially prepared filter beds are needed to separate fine particles when using the vacuum filtration technique. Often, fine particles either pass right through a paper filter or they clog it up so completely that the filtering action stops. This is avoided by the use of a substance called Filter Aid or Celite. This material is also called **diatomaceous earth,** because of its source. It is a finely-divided inert material derived from the microscopic shells of dead diatoms (a type of phytoplankton that grows in the sea). Filter Aid will not clog the fiber pores of filter paper. It is slurried, or mixed with a solvent to form a rather thin paste, and filtered through a Büchner funnel (with filter paper in place) until a layer of diatoms about 3 mm thick is formed on top of the filter paper. The solvent in which the diatoms were slurried is poured from the filter flask and, if necessary, the filter flask is cleaned before the actual filtration is begun. Finely divided particles may now

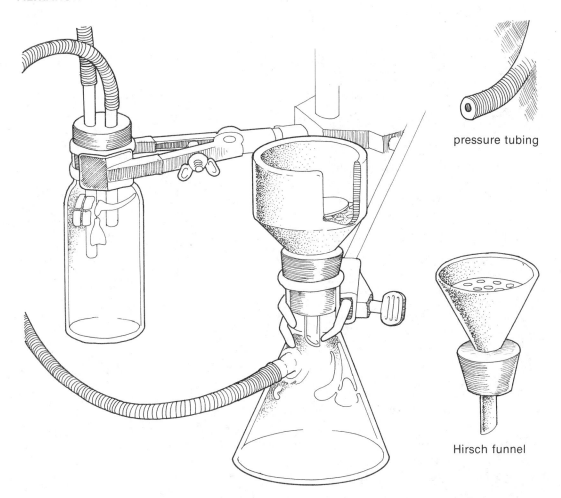

pressure tubing

Hirsch funnel

FIGURE 2–4. Vacuum filtration

be suction filtered through this layer and will be caught in the Filter Aid. Of course, this technique is used to remove impurities and not to collect a product. The filtrate (filtered solution) is the desired material in this procedure. If the material caught in the filter was the desired material, one would have to try to separate the product from all of those diatoms! Filtrations with Filter Aid must not be used when the desired substance is likely to precipitate or crystallize from solution.

2.5 THE ASPIRATOR

The most common source of vacuum (approx. 10 to 20 mm Hg) in the laboratory is the water aspirator or "water pump," which is illustrated in Figure 2–5. This device passes water rapidly past a small hole to which a side arm is attached. The "Bernoulli effect" gives rise to a reduced pressure along the side of the rapidly moving water stream and creates a partial vacuum in the side arm.

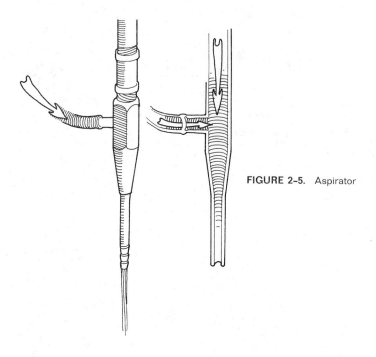

FIGURE 2–5. Aspirator

A water aspirator can never lower the pressure beyond the vapor pressure of the water used to create the vacuum. Hence, there is a lower limit to the pressure (on cold days) of about 9 or 10 mm Hg. In the summer a water aspirator will not provide as high a vacuum as in the winter, due to this water temperature effect.

A trap must be used with an aspirator. A trap is illustrated in Figure 2–4. If the water pressure in the laboratory line drops suddenly, the pressure in the filter flask may suddenly become less than that in the water aspirator. This would cause water to be drawn from the aspirator stream into the filter flask and would contaminate the filtrate. The trap stops this reverse flow. A similar flow will occur if the water flow at the aspirator is stopped before disconnecting the tubing connected to the aspirator side arm. ALWAYS DISCONNECT THE TUBING BEFORE STOPPING THE ASPIRATOR. If a "back-up" begins to occur, disconnect the tubing as rapidly as possible before the trap fills with water. Some workers like to fit a stopcock into the stopper on top of the trap. A three-hole stopper is required for this purpose. With a stopcock in the trap, the system may be vented before shutting off the aspirator. If the system is vented before the water is shut off, water cannot back up into the trap.

Aspirators do not work well if too many people use the water line at the same time, since the water pressure is lowered. Also, the sinks at the ends of the lab benches or the lines which carry away the water flow may have a limited capacity to drain the resultant water flow from a large number of aspirators. Care must be maintained that floods do not occur.

2.6 CRUDE FILTRATIONS

In many instances one may desire to perform a very rapid filtration to remove dirt or impurities of fairly large particle size from a solution. This is accomplished most easily by laying a loose mat of glass wool in the bottom of an ordinary funnel and pouring the solution through the mat. It may be helpful to decant, or pour off, the clear liquid gently before performing this crude filtration on the solid residue at the bottom of the flask.

Small amounts of solution may be filtered in a somewhat similar way. A plug of glass wool is packed loosely into an eye dropper pipet, and the solution is dropped into the pipet by means of a second pipet.

Technique 3

CRYSTALLIZATION: THE PURIFICATION OF SOLIDS

Organic compounds which are solid at room temperature are usually purified by crystallization. The general technique involves dissolving the material to be crystallized in a **hot** solvent (or solvent mixture) and cooling the solution slowly. The dissolved material has a decreased solubility at lower temperatures and will precipitate from the solution on cooling. This phenomenon is called **crystallization** if the crystal growth is relatively slow and selective, or **precipitation** if the process is rapid and nonselective. The former process is an equilibrium process and results in very pure material. A small seed crystal is formed initially, and it then grows layer by layer in a reversible manner. In a sense, the crystal "selects" the correct molecules from the solution. In the latter process the crystal lattice is formed so rapidly that impurities are trapped within the lattice. Therefore, in any attempt at purification, too rapid a process should be avoided. Too slow a process should also be avoided. The time scale for crystal formation should cover tens of minutes or hours, rather than seconds or days. The two principal mistakes which can be made are 1) cooling the solution too rapidly, and 2) suddenly adding an "incompatible" solvent to the solution. Both of these mistakes will be discussed in this chapter.

3.1 SOLUBILITY

The first problem in performing a crystallization is selecting a solvent in which the material to be crystallized exhibits the desired

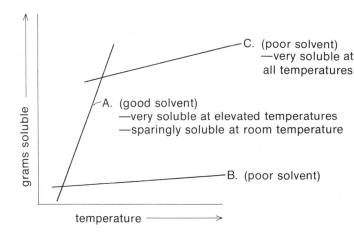

FIGURE 3-1. Graph of solubility versus temperature

solubility behavior. Ideally, the material should be sparingly soluble at room temperature and yet quite soluble at the boiling point of the solvent selected. The solubility curve should be quite steep, as can be seen in line **A** of Figure 3-1. A curve with a low slope (line **B**, Figure 3-1) would not cause significant crystallization when the temperature of the solution was lowered. A solvent in which the material was very soluble at all temperatures (line **C**, Figure 3-1) would not be a suitable crystallization solvent. The basic problem in performing a crystallization is to select a solvent (or mixed solvent) which exhibits a steep solubility **versus** temperature curve for the material to be crystallized. That is, a solvent that gives the behavior shown in line **A** is an ideal crystallization solvent.

The solubility of organic compounds is a function of the polarities of both the solvent and the solute (dissolved material). There is a general rule which states "like dissolves like." If the solute is very polar, a very polar solvent will be required to dissolve it; if it is nonpolar, a nonpolar solvent will be required. Usually compounds which have functional groups capable of forming hydrogen bonds (e.g., —OH, —NH, —COOH, —CONH) will be more soluble in hydroxylic solvents such as water or methanol than in hydrocarbon solvents such as benzene or hexane. However, if the functional group is not a major part of the molecules, this solubility behavior may be reversed. For instance, dodecyl alcohol, $CH_3(CH_2)_{10}CH_2OH$, is almost insoluble in water; its twelve carbon chain causes it to behave more like a hydrocarbon than an alcohol. The list found in Table 3-1 gives an approximate order for decreasing polarity of organic functional groups.

The stability of the crystal lattice also affects solubility. Often, with other things being equal, the higher the melting point (more stable crystal) the less soluble a compound will be. For instance, *p*-nitrobenzoic acid (mp 242°C) is, by a factor of ten, less soluble than are the *ortho* (mp 147°C) and *meta* (mp 141°C) isomers in a fixed amount of ethanol.

TABLE 3-1. DECREASING ORDER OF POLARITY OF SOLVENTS

DECREASING POLARITY (APPROXIMATE)		
	H_2O	water
	RCOOH	organic acids (acetic acid)
	$RCONH_2$	amides (N,N-dimethylformamide)
	ROH	alcohols (methanol, ethanol)
	RNH_2	amines (triethylamine, pyridine)
	RCOR	aldehydes, ketones (acetone)
	RCOOR	esters (ethyl acetate)
	RX	halides ($CHCl_3 > CH_2Cl_2 > CCl_4$)
	ROR	ethers (diethyl ether)
	ArH	aromatics (benzene, toluene)
	RH	alkanes (hexane, petroleum ether)

3.2 THEORY OF CRYSTALLIZATION

A successful crystallization depends upon a large difference in the solubility of a material in a hot solvent and its solubility in the same solvent when it is cold. Naturally, when the impurities in a substance are equally soluble in both the hot and cold solvent, an effective purification is not easily achieved through crystallization. However, a material may be purified by crystallization when both the desired substance and the impurity have similar solubilities, but only in cases where the impurity represents a fairly small fraction of the total solid. The desired substance will crystallize on cooling but the impurities will not. An example may serve to illustrate this latter statement. Consider a case where the solubilities of substance **A** and its impurity **B** are both 1 g/100 ml of solvent at 20°C and 10 g/100 ml of solvent at 100°C. In an impure sample of **A,** the composition is given to be 9 g of **A** and 2 g of **B** for this particular example. At 20°C this total amount of material will not dissolve. However, if the solvent is heated to 100°C, all 11 g will dissolve, since the solvent has the capacity to dissolve 10 g of **A** and 10 g of **B** at this temperature. If the solution is cooled now to 20°C, only 1 g of each solute can remain dissolved, so 8 g of **A** and 1 g of **B** will crystallize, leaving 2 g of material in the solution. This crystallization process is illustrated in Figure 3–2. The solution which remains after a

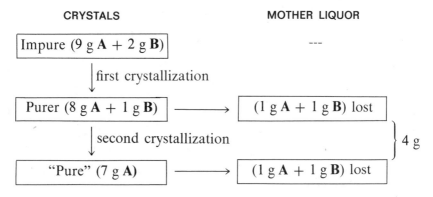

FIGURE 3-2. Purification of a mixture by crystallization

crystallization is called the **mother liquor.** If the process is now repeated by treating the crystals with 100 ml of fresh solvent, 7 g of **A** will crystallize again, leaving 1 g of **A** and 1 g of **B** in the mother liquor. As a result of these operations, 7 g of pure **A** are obtained, but with the loss of 4 g of material. Again, this second crystallization step is illustrated in Figure 3–2. This result illustrates an important aspect of crystallization—it is wasteful. Nothing can be done to prevent this waste; some **A** must be lost along with the impurity **B** for the method to be successful. Of course, if the impurity were **more** soluble than **A** in the solvent, the losses would be reduced. Losses could also be reduced if the impurity were present in **much smaller** amounts than the desired material.

It should be noticed that for the above case the method operated successfully because **A** was present in substantially larger quantity than its impurity **B.** If there had been a 50/50 mixture of **A** and **B** initially, no separation would have been achieved. In general, a crystallization procedure is successful only if there is a **small** amount of impurity. As the amount of impurity increases, the loss of material must also increase. Two substances with nearly equal solubility behavior, present in equal amounts, cannot be separated. If the solubility behavior of two components present in equal amounts is different, however, a separation or purification frequently may be achieved.

3.3 SELECTION OF SOLVENT

A few points should be made regarding the question of selection of a solvent for a crystallization procedure. Obviously, a solvent which dissolves little of the material to be crystallized when it is cold, but dissolves a great deal of the material when it is hot, is a good solvent for this procedure. But the realities of the matter are slightly more complicated than the above statement would suggest.

Frequently, one selects a solvent for crystallization by experimenting with a variety of solvents using a very small amount of the material to be crystallized. Such experiments are conducted on a small test tube scale before the entire quantity of material is committed with a particular solvent. Such trial-and-error methods are common when one is attempting to purify a solid material which has not been extensively studied.

With compounds which are well-known, such as the compounds which are either isolated or prepared in this text, the correct crystallization solvent is already known through the experiments of earlier workers. In such cases, the chemical literature may be consulted to determine which solvent should be used. Such sources as handbooks or tables may also provide this information. Quite often, the correct crystallization solvent is indicated in the experimental procedures in this text.

One note of caution should be mentioned regarding the choice of crystallization solvent. Care should be taken not to choose a solvent whose boiling point is higher than the melting point of the substance to be crystallized. If the boiling point of the solvent is too high, the solid may melt in the solvent rather than dissolve. In such a case, the solid may "oil out." "Oiling out" occurs when the solid substance melts to form a liquid which is not soluble in the solvent. Upon cooling, the liquid refuses to crystallize, but rather it becomes a supercooled liquid or oil. Such an oil may be solidified if the temperature is lowered sufficiently, but it will not crystallize. A solidified oil becomes an amorphous solid or a hardened mass—a situation which does not result in the purification of the substance. Oils are very difficult to deal with in the laboratory. One must attempt to redissolve them and hope that they will reprecipitate as crystals upon careful cooling.

One additional criterion which may be considered in selecting the correct crystallization solvent is the **volatility** of that solvent. A solvent with a low boiling point may be removed from the crystals through evaporation without much difficulty. A solvent with a high boiling point will be difficult to remove from the crystals without heating them under vacuum.

Table 3–2 lists common crystallization solvents. The solvents which are used most commonly are found earlier in the Table.

3.4 TECHNIQUE AND METHOD

Dissolving the Solid. To minimize losses of material to the mother liquor it is desirable to **saturate** the boiling solvent with solute. This solution, when cooled, will return the maximum possible amount of solute as crystals. To achieve this high return, the solvent is brought to its boiling point, and the solute is dissolved in the MINIMUM

TABLE 3–2. COMMON SOLVENTS FOR CRYSTALLIZATIONS

	Boils	Freezes	Soluble in H_2O	Flammability
Water	100°	0°	+	−
Methanol	65°	*	+	+
95% Ethanol	78°	*	+	+
Ligroin	60–90°	*	−	+
Benzene	80°	5°	−	+
Chloroform	61°	*	−	−
Acetic Acid	118°	17°	+	+
Dioxane	101°	11°	+	+
Acetone	56°	*	+	+
Diethyl Ether	35°	*	Slightly	+ +
Petroleum Ether	30–60°	*	−	+ +
Methylene Chloride	41°	*	−	−
Carbon Tetrachloride	77°	*	−	−

*Lower than 0°C (ice temperature)

AMOUNT(!) OF BOILING SOLVENT. For this procedure it is advisable to maintain a container of boiling solvent. From this container, a small portion (a few milliliters) of the solvent is added to the flask containing the solid to be crystallized, and this mixture is heated until it resumes boiling. If the solid has totally dissolved in this small portion of solvent, too much has been used and the solvent must be partially evaporated. If the solid has not dissolved in this portion of boiling solvent, then another small portion of boiling solvent is added to the flask. The mixture is heated again until it resumes boiling. If the solid has dissolved, no more solvent is added. But if the solid has not dissolved, another portion of boiling solvent is added, as before, and the process is repeated until the solid dissolves. It is important to stress that the portions of solvent added each time are small so that only the **minimum** amount of solvent necessary to dissolve the solid is added. It is also important to stress that the procedure requires the addition of solvent to the solid. One must never add portions of solid to a fixed quantity of boiling solvent. By this latter method it may be impossible to tell when saturation has been achieved.

Occasionally, one encounters an impure solid which contains small particles of insoluble impurities, pieces of dust, or paper fibers which will not dissolve in the hot crystallizing solvent. A common error is to add too much of the hot solvent in an attempt to dissolve these small particles, without realizing that they are not soluble. In such cases, one must be careful not to add too much solvent. It is probably better to add too little solvent and not dissolve all of the desired solid than to add too much solvent and lower the yield of solid which is returned as crystals. Once the solid has dissolved, decolorizing charcoal is added **if it is required** (see Section 3.5).

Filtration. The hot solution is filtered if any insoluble matter remains or if charcoal has been used. A gravity filtration through a fluted filter paper is preferred for this step (see Technique 2, Section 2.1). Use a stemless funnel and preheat it with boiling solvent prior to using it. Figure 3–3 illustrates some ways in which a funnel might be preheated. The funnel may be preheated by pouring hot solvent through it as it rests within a beaker. After this step, the funnel is fitted with a fluted filter as rapidly as possible and installed at the top of the Erlenmeyer flask to be used for the actual filtration. Alternatively, or in addition, the funnel and filter paper can be preheated by placing the assembly in the neck of the Erlenmeyer flask and by resting the entire apparatus on top of a steam bath or hot plate. The material to be filtered is brought to its boiling point and is poured into the filter. The solvent which first passes through the filter into the flask will begin to boil. The hot vapors will rise up around the funnel and heat it. In the former method, the solvent which has been used to preheat the funnel is discarded. These preheating operations prevent the formation of crystals in the filter and also prevent the clogging of the filter which such crystallization would cause.

During the filtration, the **hot** solution is poured through the filter in portions. It is necessary to keep the solutions in both flasks at their

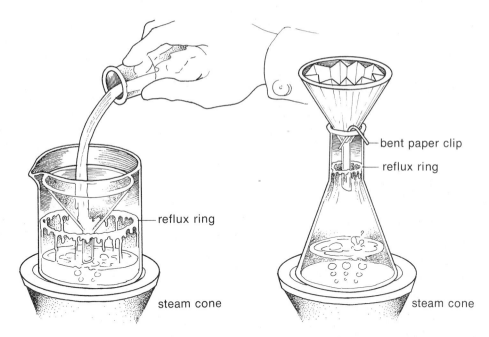

FIGURE 3-3. Methods of preheating a funnel

boiling temperatures to prevent premature crystallization. The refluxing action of the filtrate keeps the funnel warm and reduces the chance that the filter will clog with crystals which may have formed during the filtration. It is advisable to place a small piece of wire between the funnel and the mouth of the flask to relieve any build-up of pressure caused by the boiling solvent. The filtration procedure is illustrated in Figure 3-4. It should be stressed that if the solution is clear and colorless, if all of the solid material has dissolved, or if decolorizing charcoal was not used, this gravity filtration is not necessary.

If the crystals begin to crystallize in the filter during gravity filtration, a minimum amount of boiling solvent is added to redissolve the crystals and to allow the solution to pass through the funnel. After the filtration, the filtrate is boiled until the extra amount of solvent required to redissolve the crystals caught in the filter has evaporated. Once this extra amount of solvent has evaporated, the solution is set aside to cool until crystals are formed.

Crystallization. It is always a good idea to use an Erlenmeyer flask, not a beaker, for crystallization. The large open top of a beaker makes it an excellent dust catcher. The narrow opening of the Erlenmeyer flask reduces the danger of contamination by dust and allows the flask to be stoppered if it is to be set aside for a long period. Mixtures set aside for long periods must, in fact, be stoppered to prevent the evaporation of solvents. If all of the solvent evaporates, no purification is achieved, and the crystals originally formed will become coated with the dried contents of the mother liquor.

If the cooled solution will not crystallize, several techniques may be used to induce crystallization. First, one should try vigorous scratch-

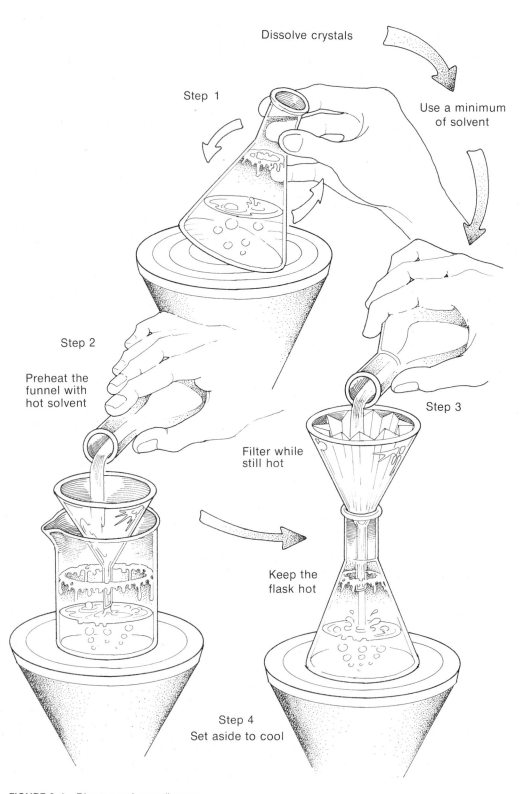

FIGURE 3–4. Filtration and crystallization

ing of the inside surface of the flask with a glass rod that **has not been** fire-polished. The motion of the rod should be vertical (in and out of the solution) and should be vigorous enough to produce an audible scratching noise. This scratching will often induce crystallization. The effect is not well understood although there are two proposed explanations. The high-frequency vibrations may have something to do with initiating crystallization; or, perhaps, it is far more likely that small amounts of solution dry by evaporation on the side of the flask, and the solute is pushed into the solution. These small amounts of material provide "seed crystals" or nuclei on which crystallization may begin.

A second technique which may be used to induce crystallization is to cool the solution in an ice bath. This decreases the solubility of the solute. A third technique is useful when small amounts of the original material to be crystallized are saved. The saved material may be used to "seed" the cooled solution. A small crystal dropped into the cooled flask often will start the crystallization process—this is called **seeding.**

Isolation of Crystals. The crystals are collected by vacuum filtration through a Büchner or Hirsch funnel (see Technique 2, Section 2.3 and Figure 2-4). The crystals should be washed with a small amount of **cold** solvent to remove any mother liquor adhering to their surface. Hot or warm solvent will dissolve some of the crystals. The crystals should then be left for a short time in the funnel where the passage of air will dry them free of solvent. It is often wise to cover the Büchner (or Hirsch) funnel with an oversize filter paper or towel during this air-drying process. This prevents the accumulation of dust in the crystals.

3.5 DECOLORIZATION

Small amounts of highly colored impurities may make the original crystallization solution appear colored; this color often may be removed by **decolorization** utilizing activated charcoal (called Norit). As soon as the solute is dissolved in the minimum amount of boiling solvent, a small amount of Norit is added to the boiling mixture. The Norit adsorbs the impurities. A reasonable amount of Norit would be that which could be held on the end of a small nickel spatula, or about 0.25 g (1/8 tsp). If too much Norit is used it will adsorb product as well as impurities. A small amount of Norit should be used, and its use should be repeated if necessary. Caution should be exercised that the solution does not froth or erupt when the finely-divided charcoal is added. The boiling mixture is filtered by gravity using a fluted filter (see Technique 2, Section 2.1 and Technique 3, Section 3.4), and the crystallization is carried forward as described in Section 3.4. The Norit is very finely divided and is removed most effectively by gravity filtration through fluted filter paper. If a Büchner funnel is used to suction filter the mixture, a layer of Celite should be used to trap the fine particles of charcoal (see Technique 2, Section 2.4). Decolorizing charcoal will usually pass right through filter paper when suction is used. Filtration by suction through Celite is used only when the solute cannot precipitate or if excess solvent has been used. In the latter case, the excess solvent is removed by evaporation after the filtration has been completed. The Norit preferentially adsorbs the colored impurities and removes them from the solution. The technique seems to be most effective with hydroxylic solvents.

3.6 DRYING CRYSTALS

The most common method of drying crystals involves placing them in a watch glass or an evaporating dish and allowing the air to dry them. While the advantage of this method is that heat is not required, thus reducing the danger of decomposition or melting, exposure to atmospheric moisture may cause the hydration of strongly **hygroscopic** materials. A hygroscopic substance is one which absorbs moisture from the air.

Another method of drying crystals is to place the crystals on a watch glass or on a piece of absorbant paper in an oven. While this method is simple, some possible difficulties deserve mention. Crystals which sublime readily should not be dried in an oven because they might pass into the vapor state and disappear. Care should be taken that the temperature of the oven does not exceed the melting point of the crystals. One must remember that the melting point of crystals is lowered by the presence of solvent, and one must allow for this melting point depression when selecting a suitable oven temperature. Some materials decompose upon exposure to heat, and they should not be dried in an oven. Finally, when many different samples are being dried in the same oven, crystals might be lost due to confusion or reaction with another person's sample. It is important to label the crystals when they are placed in the oven.

A third method, which requires neither heat nor exposure to atmospheric moisture, is the use of a vacuum desiccator. In a desiccator, the sample is placed under vacuum in the presence of a drying agent. Two potential problems must be noted, however. The first problem deals with samples which sublime readily. Under vacuum, the likelihood of sublimation is increased. The second problem deals with the vacuum desiccator itself. Since the surface area of glass which is under vacuum is large, there is a certain danger that the desiccator could implode. A vacuum desiccator should never be used unless it has been placed within a protective metal container.

3.7 MIXED SOLVENTS

Often the desired solubility characteristics for a particular compound may not be found in a single solvent. In these cases a mixed solvent may be used. One simply selects a first solvent in which the

TABLE 3-3. COMMON SOLVENT PAIRS FOR CRYSTALLIZATION

Methanol–Water	Ether–Acetone
Ethanol–Water	Ether–Petroleum Ether
Acetic Acid–Water	Benzene–Ligroin
Acetone–Water	Methylene Chloride–Methanol
Ether–Methanol	Dioxane–Water

solute is soluble and a second solvent, miscible with the first, in which the solute is relatively insoluble. The compound is dissolved in a minimum amount of the boiling solvent in which it is soluble. Following this, the second hot solvent is added to the boiling mixture, dropwise, until the mixture barely becomes cloudy. The cloudiness indicates precipitation. At this point, more of the first solvent should be added. Just enough is added to clear up the cloudy solution. At that point the solution is saturated and, on cooling, crystals should separate. Common solvent mixtures are listed in Table 3–3. Benzene-ligroin and methanol-water are the most commonly used solvent mixtures.

Steps in a Crystallization

A. DISSOLVING

1. Find a solvent with a steep solubility vs. temperature characteristic. (Done by trial and error using small amounts of material, or by consulting a handbook.)
2. Heat the desired solvent to its boiling point.
3. Dissolve the solid in a **minimum** of boiling solvent.
4. Add decolorizing charcoal if necessary.
5. Filter the hot solution through a preheated funnel to remove insoluble impurities and/or charcoal. If no decolorizing charcoal has been added or if there are no undissolved particles, this filtration may be omitted.
6. Allow the solution to cool.
7. If crystals do not appear, proceed to part B; if crystals do appear, proceed to part C.

B. INDUCING CRYSTALLIZATION

1. Scratch the flask with a glass rod . . . or . . .
2. Seed the solution . . . or . . .
3. Cool the solution in an ice-water bath.

C. COLLECTING

1. Collect crystals by vacuum filtration using a Büchner (or Hirsch) funnel.
2. Rinse crystals with a small portion of **cold** solvent.
3. Continue suction until crystals are dry.

D. DRYING

1. Air dry the crystals . . . or . . .
2. Place the crystals in a drying oven . . . or . . .
3. Dry the crystals in a vacuum desiccator.

It is important not to add an excess of the second solvent or to cool the solution too rapidly. Either of these errors may cause the solute to "oil out," or separate as a viscous liquid. If this happens, one should reheat the solution and add more of the first solvent.

PROBLEMS

1. Listed below are solubility **versus** temperature data for an organic substance A dissolved in water.

Temperature (°C)	Solubility of A in 100 ml of water
0	1.5 g
20	3.0
40	6.5
60	11.0
80	17.0

a. Construct a graph of the solubility of A **versus** temperature. Use the data given above. Connect the data points with a smooth curve.

b. Suppose 0.5 g of A and 5 ml of water were mixed and heated to 80°C. Would all of the substance A dissolve?

c. This solution, prepared in part b, is cooled. At what temperature will crystals of A appear?

d. Suppose the cooling described in part c were continued to 0°C. How many grams of A would come out of solution? Explain how you obtained your answer.

2. What would be likely to happen if a hot saturated solution were filtered by vacuum filtration using a Büchner funnel?

3. a. Draw a graph of a cooling curve (temperature **vs.** time) for a solution of a solid substance which shows no supercooling effects. Assume that the solvent does not freeze.

b. Repeat the above instructions for a solution of a solid substance which shows some supercooling behavior, but which eventually yields crystals if the solution is cooled sufficiently.

4. A solid substance A is soluble in water to the extent of 1 g/100 ml of water at 25°C and 10 g/100 ml of water at 100°C. One has a sample which contains 10 g of A and an impurity B.

a. Assuming that 0.2 g of an impurity, B, is present along with 10 g of A, describe how you could purify A if B is completely insoluble in water.

b. Assuming that 0.2 g of an impurity, B, is present along with 10 g of A, describe how you could purify A if B had the same solubility behavior as A. Would one crystallization produce absolutely pure A?

c. Assume that 3 g of impurity, B, is present along with 10 g of A. Describe how you could purify A if B had the same solubility behavior as A. Each time use the correct amount of water to just dissolve the solid. Would one crystallization produce absolutely pure A? How many crystallizations would be required to produce pure A? How much A would have been recovered when the crystallizations were completed?

Technique **4**

THE MELTING POINT:
AN INDEX OF PURITY

4.1 INTRODUCTION

The primary index of purity used by an organic chemist for a crystalline compound is its melting point. A small amount of material is heated **slowly** in a special apparatus equipped with a thermometer, a heating coil or heating bath and, usually, a magnifying eyepiece for observing the sample. Two temperatures are noted. The first is the point at which the first drop of liquid forms among the crystals, and the second is the point at which the whole mass of crystals turns to a **clear** liquid. The melting point is then recorded giving this range of melting. For example, one might say that the melting point of a substance is 51 to 54°C. That is, the substance melted over a 3°C range.

The melting point of a **pure** crystalline substance is a physical property of that substance. Since the vapor pressure of a solid is low, unlike that of a liquid, the melting point is usually insensitive to changes of pressure (within reasonable limits). This melting point may be used to identify a given substance.

4.2 MELTING POINT BEHAVIOR

The melting point (or range) indicates purity in two ways. First, the purer the material, the higher its melting point will be. Second, the purer the material, the narrower the melting point range will be. In general, the addition of successive amounts of an impurity to a pure substance will cause its melting point to decrease in proportion to the amount of impurity. This is a result of the fact that the freezing point of a substance is lowered by the addition of a foreign substance. The freezing point is simply the melting point (solid $\longrightarrow$ liquid) being approached from the opposite direction (liquid $\longrightarrow$ solid). In Figure 4–1 the usual melting point behavior of various mixtures of two substances, **A** and **B** is depicted. The two extreme temperatures of the melting range for various mixtures of **A** and **B** are shown. The upper curves indicate the temperature at which all of the sample has melted. The lower curve indicates the temperature at which melting is observed to begin. If one begins with pure **A**, the melting point will decrease as impurity **B** is added. At some particular proportion of the two substances, a minimum temperature will be reached, and the melting point will begin to increase to that of a pure substance, **B**. In general, the melting point depression curves for (**A** + impurity **B**) and (**B** + impurity **A**) will always intersect at some particular composition, and a minimum melt-

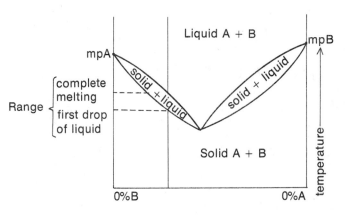

FIGURE 4-1. Melting point-composition curve

ing point will be reached. In the curve in Figure 4–1 the vertical distance between the convex and concave curves represents the **melting range.** One may consider a mixture containing substance A together with a relatively small amount of **B.** In this mixture the melting point would be lowered, and the range of the melting point would be increased. This example illustrates that for mixtures containing small amounts of impurity (<15%), the **melting point range often will indicate purity.** A substance which melts with a narrow range should be pure. However, at the minimum point of the melting point-composition curve, the mixture often will form a **eutectic,** which also melts sharply. Not all binary mixtures form eutectics, and some caution must be exercised in assuming that every binary mixture follows the behavior described above. Some compounds form more than one eutectic. In spite of these variations, both the melting point and its range are useful indications of purity and are obtained experimentally with ease.

4.3   THEORY

Figure 4–2 is a phase diagram describing the behavior of the usual two component mixture **(A + B)** on melting. The behavior on melting

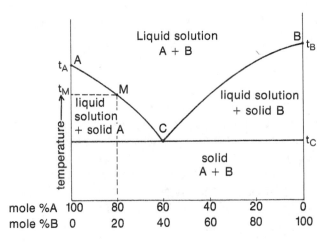

FIGURE 4-2. Phase diagram for melting in a two component system

depends on the relative amounts of **A** and **B** in the mixture. If **A** is a pure substance (no **B**), then **A** melts sharply at its melting point, t_A. This is represented by point **A** on the left hand side of the diagram. When **B** is a pure substance, it melts at t_B; its melting point is represented by point **B** on the right hand side of the diagram. At either point **A** or **B** the pure solid passes cleanly, with a narrow range, from solid to liquid.

In mixtures of **A** and **B** the behavior is quite different. Using Figure 4–2, consider a mixture of 80% **A** and 20% **B** on a mole/mole basis (i.e., mole percentage). The melting point of this mixture is given by t_M at point **M** on the diagram. That is, addition of **B** to **A** has lowered the melting point of **A** from t_A to t_M. It has also expanded the melting range. The temperature of t_M corresponds to the **upper limit** of the melting range.

Lowering of the melting point of **A** by addition of impurity **B** comes about in the following way. **A** has the lower melting point in the phase diagram shown, and on heating it begins to melt first. As **A** begins to melt, solid **B** begins to dissolve in the liquid **A** formed. When solid **B** dissolves in liquid **A**, the melting point is depressed. To understand this, consider the melting point from the opposite direction. When a liquid starts at a high temperature, as it cools it reaches a point where it solidifies or "freezes." The temperature at which the liquid freezes is identical to its melting point. It will be recalled that the freezing point of a liquid can be depressed by the addition of an impurity substance. Since the freezing point and the melting point are identical, a lowering of the freezing point corresponds to a lowering of the melting point. As with the lowering of freezing point, the more impurity substance which is added to a solid, the lower its melting point becomes. There is, however, a limit to how far the melting point may be depressed. One cannot dissolve an infinite amount of impure substance in a liquid. At some point the liquid will become saturated with the impure substance. The solubility of **B** in **A** has an upper limit. In Figure 4–2 the solubility limit of **B** in liquid **A** is reached at point C, the **eutectic point.** The melting point of the mixture cannot be depressed below t_C, the melting temperature of the eutectic.

Now consider what happens when the melting point of a mixture of 80% **A** and 20% **B** is approached. As the temperature is increased, **A** begins to "melt." This is not really a visible phenomenon; it happens prior to the visible appearance of liquid. It is a softening of the compound to the point where it can begin to mix with the impurity. As **A** begins to soften it dissolves **B**. As it dissolves **B** the melting point is lowered. The lowering continues until all of **B** is dissolved, or until the eutectic composition is reached. When the maximum amount of **B** possible has been dissolved, actual melting will begin, and one will observe the first appearance of liquid. The initial temperature of melting will be below t_A. The amount below t_A at which melting begins will be determined by the amount of **B** dissolved in **A**, but will never be below t_C. Once all of **B** has been dissolved, the melting point of the mixture will begin to rise as more **A** begins to melt. As more **A** melts, the semi-solid solution will be diluted by **A,** and its melting point will rise. While all of this is happening, one will observe **both** solid and liquid in the melting point capillary. Once all of **A** has begun to melt, the composition of the mixture **M** will become uniform throughout and will reach 80% **A** and 20% **B**. At this point the mixture will finally melt sharply, giving a clear solution. The maximum melting point will be t_M, since t_A is depressed by the impurity **B** which is present. The lower end of the melting range will always be t_C; however, melting will not always be observed at this temperature. An observable melting at t_C only comes about when a fairly large amount of **B** is present. Otherwise, the amount of liquid formed at t_C will be too small to observe. Therefore, the melting behavior which is **actually** observed will have a smaller range, as shown in Figure 4–1.

4.4   MIXED MELTING POINTS

The melting point may be used as supporting evidence in the identification of a given compound if an authentic sample of that compound is available for comparison. If the object of the experiment is to confirm that **A** and **B** are identical substances, because they each individually exhibit similar or identical melting points, they are pulverized finely and mixed in equal quantities. Then the melting point of the mixture is determined. If there is a melting point depression or if the range of melting is expanded by a large amount, one may conclude that one substance has acted as an impurity toward the other and they are not the same compound. If there is no depression of the melting point for the mixture (the melting point is identical with that of either pure **A** or pure **B**), then **A** and **B** are almost certainly the same compound.

4.5   PACKING THE MELTING POINT TUBE

Melting points usually are determined by heating the sample in a piece of thin-walled capillary tubing (1 mm × 100 mm) which has been sealed at one end. To pack the tube, the open end is pressed gently into a pulverized sample of the crystalline material. Crystals will stick in the open end of the tube. The amount of solid which is pressed into the tube should correspond to a column no more than 1 to 2 mm high. In order to transfer the crystals to the closed end of the tube, the capillary is dropped, closed end first, down a 2/3 meter length of glass tubing, which is held upright on the desk top. When the capillary tube hits the desk top the crystals will pack down into the bottom of the tube. This procedure is repeated if necessary. Tapping the capillary on the desk top with the fingers is not recommended, since it is easy to drive the small tubing into a finger if the tubing should break.

4.6 DETERMINING THE MELTING POINT

There are two principal types of melting point apparatus available: the Thiele tube and a commercially available, electrically heated apparatus. The Thiele tube, which is shown in Figure 4–3, is the simpler device, and it is widely available. It is a glass tube designed to contain heating oil and a thermometer to which a capillary tube containing the sample is attached. The shape of the Thiele tube allows the formation of convection currents in the oil when it is heated. These currents maintain a fairly uniform temperature distribution throughout the oil in the tube. The side arm of the tube is designed to generate these convection currents and thus transfer the heat from the flame evenly and rapidly throughout the heating oil. The sample, which is in a capillary tube attached to the thermometer, is held by means of a rubber band or a small slice of rubber tubing. It is important that this rubber band be above the level of the oil (allowing for expansion of the oil on heating) so that the oil does not soften the rubber and allow the capillary tubing to fall into the oil.

The Thiele tube is usually heated by a microburner. When heating, the rate of temperature increase should be regulated. Usually one holds the burner by its base and, using a gentle flame, moves the burner slowly back and forth along the bottom of the arm of the Thiele tube.

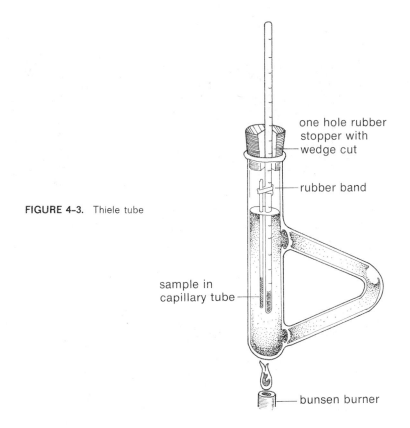

FIGURE 4–3. Thiele tube

one hole rubber stopper with wedge cut

rubber band

sample in capillary tube

bunsen burner

If the heating rate is too fast, the burner is removed for a few seconds before resuming the heating process. The rate of heating should be **slow** near the melting point (about 1°C per minute) to ensure that the rate of temperature increase is not faster than the ability of the heat to be transferred to the sample being observed. At the melting point it is necessary that the mercury in the thermometer and the sample in the capillary tube be at temperature equilibrium.

The need for careful heating near the melting point also applies when using the electrical apparatus. Such an apparatus is operated by moving the switch to the ON position, adjusting the potentiometric control for the desired rate of heating, and observing the sample through the magnifying eyepiece. The temperature is read from a thermometer. Two examples of such an apparatus are illustrated in Figure 4–4. There are many types and styles of melting point apparatus.

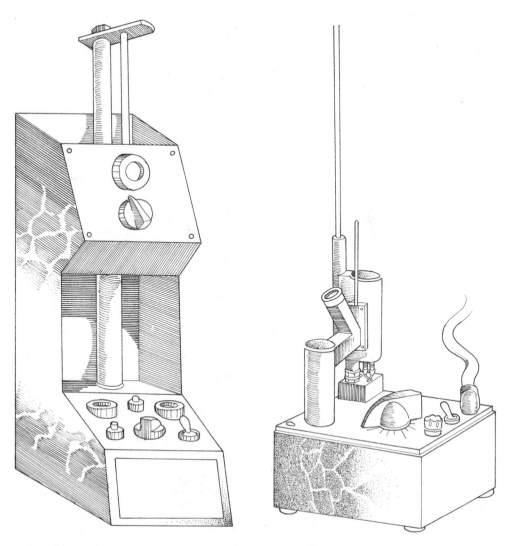

FIGURE 4–4. Melting point apparatus

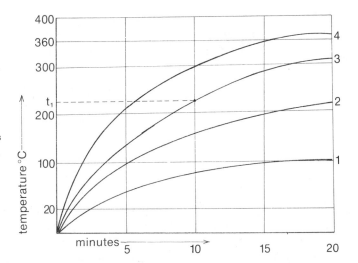

FIGURE 4-5. Heating rate curves

Your instructor will demonstrate and explain the type used in your laboratory.

Most electrical apparatuses do not heat or increase the temperature of the substance in a linear fashion. Although in the early heating stages the rate of heating is linear, it usually decreases in rate and leads to a constant temperature at some upper temperature limit. The upper limit temperature will be determined by the setting of the heating control. Thus, a family of heating curves is usually obtained for various control settings, as is shown in Figure 4–5. Four hypothetical curves (**1** through **4**) are shown which might correspond to different control settings. For a compound melting at temperature t_1, the setting corresponding to curve **3** would be ideal. In the early part of the curve the temperature is increasing too rapidly to allow the determination of an accurate melting point but, after the change in slope, the temperature increase will have slowed to a more useable rate.

If the melting point of the substance is unknown, one can often save time by preparing two samples for melting point determination. With one sample, one rapidly obtains a crude melting point value. Then one repeats the experiment more carefully using the second sample. For the second determination, one already has some approximate idea of what the melting temperature should be.

4.7 DECOMPOSITION, DISCOLORATION, SOFTENING, AND SHRINKAGE

Many solid substances undergo some degree of unusual behavior prior to melting. At times it may be difficult to distinguish these other types of behavior from actual melting. One should learn, through experience, how to recognize melting and how to distinguish it from decomposition, discoloration, and particularly softening and shrinkage.

Some compounds decompose upon melting. This decomposition is usually evidenced by discoloration of the sample. Frequently this decomposition point may serve as a reliable physical property to be used in lieu of an actual melting point. Such decomposition points are listed in tables of physical properties. An example of a decomposition point is given for **thiamine hydrochloride,** whose melting point would be listed as **248° d,** indicating that this substance melts with decomposition at 248°C.

Some substances may begin to decompose at a temperature **below** their melting point. Thermally unstable substances may undergo elimination reactions or anhydride formation reactions during the heating process. The decomposition products which are thus formed represent impurities in the original sample, so the melting point of the substance is artificially lowered due to such decomposition.

It is normal for compounds to soften or shrink immediately prior to melting. Such behavior does not represent decomposition, but rather it represents a change in the crystal structure. Some substances tend to "sweat" or release solvent of crystallization prior to melting. Such changes are not indicators of the beginning of melting. The actual melting begins when the first drop of liquid becomes visible, and the melting range continues until the temperature where all of the solid has been converted to the liquid state. With experience, one soon learns to distinguish between softening or "sweating" and actual melting.

Some substances have such a high vapor pressure that they sublime at or before the melting piont. In such cases, the melting point determination must be conducted in sealed capillary tubes.

4.8 THERMOMETER DEVIATIONS

When the melting point determination has been completed, one expects to obtain a result which exactly duplicates the result recorded

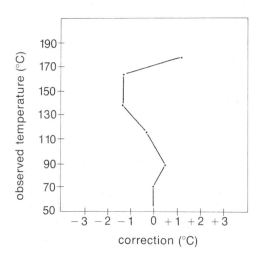

FIGURE 4–6. Thermometer calibration curve

TABLE 4-1. MELTING POINT STANDARDS

COMPOUND	MELTING POINT (°C)
Ice (solid-liquid water)	0
Acetanilide	115
Benzamide	128
Urea	132
Succinic Acid	189
3,5-Dinitrobenzoic Acid	205

in the handbook or in the original chemical literature. It is not infrequent, however, that one observes a discrepancy of one or two degrees. Such a discrepancy does not necessarily indicate that the experiment was incorrectly performed; rather it may indicate that the thermometer used for the determination is a source of systematic error. Most thermometers used in the laboratory do not measure the temperature with perfect accuracy; rather, their readings are likely to be somewhat in error.

In order to obtain accurate melting point values, it is necessary that one calibrate the thermometer that is going to be used for the melting point determination. This calibration is accomplished by determining the melting point of a variety of standard substances with the thermometer. A plot is drawn of the observed temperature **versus** the correction required to obtain the published melting point of each standard substance, as in the example shown in Figure 4-6. The correction obtained in this way is applied to each melting point which is determined with that particular thermometer. A list of suitable standard substances for the calibration of thermometers is given in Table 4-1.

Technique **5**

EXTRACTION
THE SEPARATORY FUNNEL
DRYING AGENTS

5.1 DISTRIBUTION COEFFICIENT

When a solution (solute **A** in solvent **1**) is shaken with a second solvent (solvent **2**) with which it is immiscible, the solute will distribute itself between the two liquid phases. When the two phases have separated again into two distinct solvent layers, an equilibrium situation will have been achieved such that the ratio of the concentrations of the solute in each layer will define a constant. This constant, called the **distribution coefficient** (or partition coefficient), **K,** is then defined by

$$K = \frac{C_2}{C_1}$$

where C_1 and C_2 are the concentrations, at equilibrium, in grams/liter, of the solute **A** in solvent **1** and in solvent **2,** respectively. This relationship is independent of the total concentration and the actual amounts of the two solvents mixed. The distribution coefficient has a constant value for each solute considered and is dependent on the nature of the solvents used in each case. The solute distributes itself between the two solvents so that its chemical activity (effective concentration) is the same in each phase.

5.2 EXTRACTION

Transfer of a solute from one solvent to another is called **extraction.** The solute is extracted from one solvent into the other by means of the distribution process described in Section 5.1 (see Figure 5–1). Extraction is used for many purposes in organic chemistry. Many **natural products** (chemicals which occur in nature) occur in animal and plant tissues having high water content. Extraction of these tissues with a water-immiscible solvent is useful for isolation of the natural products. Often ether is used for this purpose. Sometimes, alternative water-immiscible solvents such as hexane, petroleum ether, ligroin, benzene, chloroform, methylene chloride, and carbon tetrachloride are used. For instance, caffein, a natural product, may be extracted from an aqueous tea solution by shaking it successively with several portions of chloroform. On the other hand, water may be used to extract impurities from an organic reaction mixture.

Given the distribution coefficient as described in Section 5.1, it should be pointed out that not all of the solute will be transferred to solvent **2** in a single extraction unless **K** is very large. Usually several extractions are required to remove all the solute from solvent **1.**

To extract a solute from a solution, it is always better to use several small portions of the second solvent than to use a single extraction with a large portion. As an illustration, suppose a particular extraction will proceed with a distribution coefficient of 10. The system consists of 5.0 g of organic compound dissolved in 100 ml of water (solvent **1**). In this illustration, the effectiveness of three 50 ml extractions with ether (solvent **2**) will be compared with one 150 ml extraction with ether. In the first 50 ml extraction, the amount extracted into the ether layer is given by the following calculation. The amount of compound remaining in the aqueous phase is given by **x.**

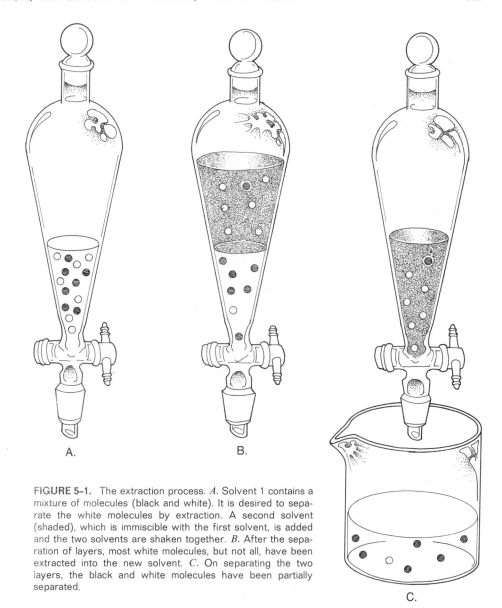

FIGURE 5–1. The extraction process. A. Solvent 1 contains a mixture of molecules (black and white). It is desired to separate the white molecules by extraction. A second solvent (shaded), which is immiscible with the first solvent, is added and the two solvents are shaken together. B. After the separation of layers, most white molecules, but not all, have been extracted into the new solvent. C. On separating the two layers, the black and white molecules have been partially separated.

A.

B.

C.

$$K = 10 = \frac{C_2}{C_1} = \frac{\left(\dfrac{5.0 - x}{50}\ \dfrac{g}{ml\ ether}\right)}{\left(\dfrac{x}{100}\ \dfrac{g}{ml\ H_2O}\right)}; \quad 10 = \frac{(5.0 - x)(100)}{50x}$$

$$500x = 500 - 100x$$
$$600x = 500$$
$$x = 0.83 \text{ g remaining in the aqueous phase}$$
$$5.0 - x = 4.17 \text{ g in the ether layer}$$

As a check on the calculation, it is possible to resubstitute the value of 0.83 g for x in the original equation and demonstrate that the concentration in the ether layer divided by the concentration in the water layer equals the distribution coefficient.

$$\frac{\left(\dfrac{5.0 - x}{50} \dfrac{g}{ml\ ether}\right)}{\left(\dfrac{x}{100} \dfrac{g}{ml\ H_2O}\right)} = \frac{\dfrac{4.17}{50}}{\dfrac{0.83}{100}} = \frac{0.083\ g/ml}{0.0083\ g/ml} = 10 = K$$

A second extraction with another 50 ml portion of fresh ether performed on the aqueous phase, which now contains 0.83 g of the solute, will extract an amount of solute given by the calculation:

$$K = 10 = \frac{\left(\dfrac{0.83 - x}{50} \dfrac{g}{ml\ ether}\right)}{\left(\dfrac{x}{100} \dfrac{g}{ml\ H_2O}\right)};\ 10 = \frac{(0.83 - x)(100)}{50x}$$

$$500x = 83 - 100x$$
$$600x = 83$$
$$x = 0.14\ g\ \text{remaining in the water layer}$$
$$0.83 - x = 0.69\ g\ \text{in the ether layer}$$

By a similar calculation, it can be shown that a third extraction with another fresh 50 ml portion of ether will remove 0.12 g of solute into the ether layer, leaving 0.02 g of solute remaining in the water layer. The total amount extracted into the combined ether layers, 4.17 + 0.69 + 0.12, is equal to 4.98 g of solute.

If an extraction were performed using an equivalent amount of ether (150 ml) in **one** extraction, the amount extracted would be given by:

$$K = 10 = \frac{\left(\dfrac{5.0 - x}{150} \dfrac{g}{ml\ ether}\right)}{\left(\dfrac{x}{100} \dfrac{g}{ml\ H_2O}\right)};\ 10 = \frac{(5.0 - x)(100)}{150x}$$

$$1500x = 500 - 100x$$
$$1600x = 500$$
$$x = 0.31\ g\ \text{remaining in the water layer}$$
$$5.0 - x = 4.69\ g\ \text{in the ether layer}$$

It can be seen that the three extractions which used smaller amounts of ether succeeded in extracting 0.29 g **more** solute from the aqueous phase than one large extraction could remove. This differential represents 5.8% of the total material. If the material of interest is present in only small quantities or is very expensive, such a difference in efficiency of extraction becomes quite important.

5.3 PURIFICATION AND SEPARATION METHODS

In nearly all of the synthetic experiments undertaken in the first part of this textbook, a series of operations involving extractions are utilized after the actual reaction has been concluded. These extractions form an important part of the purification procedure. Using them, the desired product is separated from unreacted starting materials or from undesired side products in the reaction mixture. These extractions may be grouped into three categories, depending upon the nature of the impurity which they are designed to remove.

The first of these categories involves extraction of an organic material with **water.** Water extractions are designed to remove such highly polar materials as inorganic salts, strong acids or bases, and such **low molecular weight,** polar substances as alcohols, carboxylic acids, and amines. Many organic compounds containing fewer than five carbons are water soluble. Water extractions are also used immediately following extractions with either acid or base to ensure that all traces of acid or base have been removed.

The second category concerns extraction of an organic material with a dilute **acid,** usually five or ten percent hydrochloric acid. Such acid extractions are intended to remove basic impurities, especially such basic impurities as organic amines. The bases are converted to their corresponding cationic salts by the acid used in the extraction. If an amine is one of the reactants, or if pyridine or another amine is a solvent, such an extraction might be used to remove any excess amine which is present at the end of a reaction.

$$RNH_2 + HCl \longrightarrow RNH_3^+ \, Cl^-$$
$$\text{(water-soluble salt)}$$

The cationic salts are usually soluble in the aqueous solution, and they are thus extracted from the organic material. A water extraction may be used immediately following the acid extraction to ensure that all traces of the acid have been removed from the organic material.

The third category is extraction of an organic material with a dilute base, usually five percent sodium bicarbonate, although extractions with dilute sodium hydroxide may also be used. Such basic extractions are intended to convert acidic impurities, such as organic acids, into their corresponding anionic salts. In the preparation of an ester, a sodium bicarbonate extraction might be used to remove any excess carboxylic acid which might be present.

$$RCOOH + NaHCO_3 \longrightarrow RCOO^-Na^+ + H_2O + CO_2$$
$$pK_a \sim 5 \qquad\qquad \text{(water soluble salt)}$$

The anionic salts, being highly polar, would be expected to be soluble in the aqueous phase. As a result, these acidic impurities are extracted from the organic material into the basic solution. Again, a water extraction may be used after the basic extraction to ensure that all of the base has been removed from the organic material.

Occasionally, phenols may be present in a reaction mixture as impurities, and removing them by means of extraction may be desired. Because phenols, although they are acidic, are about 10^5 times less acidic than carboxylic acids, basic extractions may be used to separate phenols from carboxylic acids by means of a careful selection of the base. If sodium bicarbonate is used as a base, carboxylic acids will be extracted into the aqueous base, but phenols will not. Phenols are not sufficiently acidic to be deprotonated to any substantial degree by the rather weak base, bicarbonate. Extraction with sodium hydroxide, on the other hand, will extract **both** carboxylic acids and phenols into the aqueous basic solution, since hydroxide ion is a sufficiently strong base to deprotonate phenols.

$$R-\!\!\!\bigcirc\!\!\!-OH + NaOH \longrightarrow R-\!\!\!\bigcirc\!\!\!-O^- Na^+ + H_2O$$

pK$_a \sim 10$ (water-soluble salt)

It would be a useful exercise for the student to examine the experimental instructions for some of the preparative experiments in Part One of this textbook. While examining these procedures, an attempt should be made to identify which impurities are being removed at each extraction step.

Mixtures of acidic, basic, and neutral compounds are easily separated by extraction techniques. One such example is shown in Figure 5–2.

Materials which have been extracted may be regenerated by neutralizing the extraction reagent. If an acidic material has been extracted with aqueous base, the material can be regenerated by acidifying the extract until the solution becomes acidic to blue litmus. The material will separate from the acidified solution. A basic material can be recovered from an acidic extract by adding base to the extract. These substances can then be removed from the neutralized aqueous solution by extraction with an organic solvent such as ether. Evaporation of the ether extract yields the isolated compound.

5.4 THE SEPARATORY FUNNEL

The separatory funnel is the piece of apparatus used in the extraction procedure. This apparatus is illustrated in Figure 5–3. There is an art to using a separatory funnel correctly, and it is best learned by observing a person, such as the instructor, who is thoroughly familiar with its use. To fill the separatory funnel one usually supports it in an iron ring attached to a ring stand. Since it is easy to break a separatory funnel by "clanking" it against the metal ring, it is recommended that

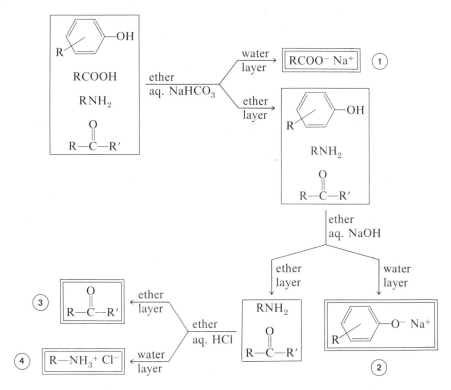

FIGURE 5-2. Separation of a four component mixture using extractions

three short lengths (about 3 cm each) of rubber tubing be cut and split open along their length. When slipped over the inside of the ring these pieces of tubing will stay in place and cushion the funnel. Next the stopcock is closed. This simple maneuver can easily be forgotten—with disastrous consequences for the experiment! An ordinary funnel is placed in the top opening, and both the solution and the extraction solvent are poured into the funnel. The separatory funnel is swirled gently by holding it by its upper neck, and then it is stoppered. The separatory funnel is picked up with **two** hands and held as shown in Figure 5-4. It is essential to hold the stopper in place firmly because the two immiscible solvents build up pressure when they mix, and this pressure may force the stopper out of the separatory funnel. The pressure results from the two partial vapor pressures of the solvents adding together on mixing; the vapors of **both** solvents are now in equilibrium with the solution. The pressure problem becomes especially great with sodium bicarbonate extractions, where the acidic impurities react with the sodium bicarbonate to produce carbon dioxide gas. As this gas is liberated, it causes increased pressure within the separatory funnel. To release this build-up of pressure, the funnel is vented by holding it upside-down (hold the stopper securely) and **slowly** opening the stop-cock. Usually the rush of vapors out of the opening can be heard. Shaking and frequent venting should be continued until the "whoosh" is no longer audible. At this point the mixture is equilibrated, and further

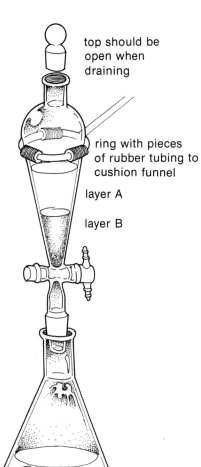

top should be
open when
draining

ring with pieces
of rubber tubing to
cushion funnel

layer A

layer B

FIGURE 5-3. The separatory funnel

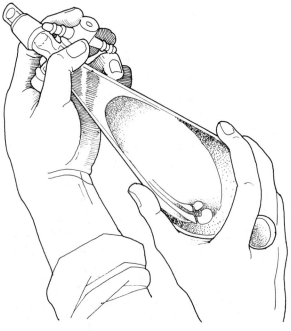

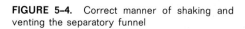

FIGURE 5-4. Correct manner of shaking and
venting the separatory funnel

shaking will lead to no improvement. The funnel is then placed in the ring, and the top stopper is immediately removed. The two immiscible solvents will separate into two layers after a short time, and they may be separated from one another by draining the lower layer through the stopcock. If the stopper is left in the separatory funnel, the resulting partial vacuum will prevent drainage from the funnel. Approximately three-fourths of the lower layer is allowed to drain before the stopcock is closed. A few minutes are allowed to pass so that any of the lower phase adhering to the inner glass surfaces of the separatory funnel is allowed to drain down. The stopcock is again opened carefully, and the remainder of the lower layer is allowed to drain until the interface between the upper and lower phases just begins to enter the bore of the stopcock. At this moment, the stopcock is closed. The remaining upper layer is removed by pouring it from the top opening of the separatory funnel. To minimize contamination of the two layers, the lower layer should always be removed from the **bottom** of the separatory funnel and the upper layer from the **top** of the funnel.

At this point some caution should be exercised in the identification of the aqueous and organic layers. Common sense usually will suffice if relative densities of the two solvents are considered. For example, in the extraction of an aqueous solution, the extraction solvent may be either "lighter," e.g., ether or benzene, or "heavier," e.g., chloroform, than water. Ether or benzene layers will float on top of water, while a chloroform layer will sink below the water. However, if a solvent dissolves a large amount of material, its density will be increased, and it may not have the relative density originally predicted. If there is any doubt, a few drops of each layer should be tested by adding them to a small test tube containing a little water. The water-miscible layer can be identified easily in this way. Table 5-1 lists the densities of common extraction solvents.

5.5 EMULSIONS

An **emulsion** is a colloidal suspension of one liquid in another liquid. Minute droplets of an organic solvent often are held in suspen-

TABLE 5-1. DENSITIES OF COMMON EXTRACTION SOLVENTS

SOLVENT	DENSITY
Ligroin	0.67–0.69 g/ml
Diethyl ether	0.713
Benzene	0.879
Water	1.000
Saturated NaCl	1.198
Methylene chloride	1.335
Chloroform	1.498
Sulfuric acid (conc.)	1.84

sion in an aqueous solution when the two are mixed vigorously, and form an emulsion. This is especially true if any gummy or viscous material was present in the solution. Emulsions are encountered often when performing extractions. Emulsions often may require a long time to separate into two layers and are a nuisance to the organic chemist. Fortunately, several "tricks" may be used to break up emulsions. If one of the solvents is water, addition of a saturated aqueous sodium chloride solution will help destroy the emulsion. This makes the aqueous and organic layers, which are usually somewhat mutually soluble to form the emulsion in the first place, less compatible, thereby forcing separation. Addition of a very small amount of a water soluble detergent may also help. One may recall that this method has been used in the past to combat oil spills. The detergent helps to solubilize the tightly bound oil droplets. Often gravity filtration (see Technique 2, Section 2.1) will help destroy an emulsion by removing gum particles. In many cases, once the gum is removed, the emulsion will break up rapidly. Getting the liquid to swirl in the separatory funnel will help sometimes. **Gentle** stirring also may be tried. If all else fails, this is definitely a case where patience and/or the correct vituperations are beneficial in one way or another.

If a solution is known, through prior experience, to have a tendency to form an emulsion, the mixing should be gentle and the shaking non-vigorous. Extractions should be performed with gentle swirling, rather than shaking, or several gentle inversions of the separatory funnel may also be used. The separatory funnel must not be shaken vigorously in these cases. It is important to stress that observing these precautions will require a longer time to carry out an extraction than it would if the danger of forming an emulsion were lower.

5.6 DRYING AGENTS

After an organic solvent has been shaken with an aqueous solution, it will be "wet," that is, it will have dissolved some water even though its misciblity with water is not great. The amount of water dissolved varies from solvent to solvent, with ether representing a case where a fairly large amount of water dissolves. Ether will hold 1.5% of its weight of water. To remove water from the organic layer a **drying agent** is used. A drying agent is an anhydrous inorganic salt which acquires waters of hydration when exposed to moist air or a wet solution. Anhydrous sodium sulfate crystals are added to the wet solution, which is usually allowed to stand for at least 15 minutes. Enough sodium sulfate is added to make a 2 or 3 mm layer in the bottom of the flask, depending on the volume of the solution. After a period of standing, the crystals are removed by filtration or decantation (see Technique 2, Sections 2.1 and 2.6), and the solution then is relatively free of water. At times the drying operation will have to be repeated more than once to obtain a relatively dry solution. In this case, the liquid is decanted into another **dry** flask, and more drying agent is added.

TABLE 5-2. SOLID DRYING AGENTS THAT CAN BE REMOVED BY FILTRATION

	ACIDITY	HYDRATED	CAPACITY[1]	COMPLETENESS[2]	RATE[3]	USE
Magnesium sulfate	neutral	$MgSO_4 \cdot 7\,H_2O$	high	medium	rapid	general
Sodium sulfate	neutral	$Na_2SO_4 \cdot 7\,H_2O$ $Na_2SO_4 \cdot 10\,H_2O$	high	low	medium	general
Calcium chloride	neutral	$CaCl_2 \cdot 2\,H_2O$ $CaCl_2 \cdot 6\,H_2O$	low	high	rapid	hydrocarbons halides
Calcium sulfate (Drierite)	neutral	$CaSO_4 \cdot 1/2\,H_2O$ $CaSO_4 \cdot 2\,H_2O$	low	high	rapid	general
Potassium carbonate	basic	$K_2CO_3 \cdot 1\,1/2\,H_2O$ $K_2CO_3 \cdot 2\,H_2O$	medium	medium	medium	amines, esters bases, ketones
Potassium hydroxide	basic	—	—	—	rapid	amines only
Molecular sieves (3Å or 4Å)	neutral	—	high	extremely high	—	general

[1] Amount of water removed per given weight of drying agent.
[2] Refers to the amount of H_2O still in solution at equilibrium with drying agent.
[3] Refers to rate of action (drying).

Several simple observations allow one to determine if a solution is "dry." If a solution is wet, the drying agent usually will clump together and stick to the flask. In extreme cases the drying agent may even be seen to dissolve in the aqueous phase which has formed at the bottom of the flask. If the solution is dry, the drying agent will shift or move freely on the bottom of the flask. A wet solution usually will appear cloudy; a dry one will be clear.

Other drying agents frequently used are magnesium sulfate, calcium chloride, calcium sulfate (Drierite), and potassium carbonate. The **anhydrous** salts must be used. These have varying properties and applications. For instance, not all will absorb the same amount of water for a given weight, nor will they dry the solution to the same extent. **Capacity** is a term used to refer to the amount of water a drying agent will absorb per unit weight. Sodium and magnesium sulfates will absorb a large amount of water (high capacity), but the magnesium compound will dry a solution more completely. **Completeness** refers to a compound's effectiveness in removing **all** of the water from a solution when equilibrium has been reached. Magnesium ion has the disadvantage that it sometimes causes rearrangements of compounds such as epoxides; it is a strong Lewis acid. Calcium chloride is a good drying agent, but cannot be used with most compounds containing oxygen or nitrogen since it forms complexes. Calcium chloride does absorb methanol and ethanol, in addition to water, so it is useful for the removal of these materials as well as being a drying agent. Potassium carbonate is a base and is used for drying basic solutions. Calcium sulfate dries completely but has a low total capacity.

Sodium sulfate is the best all around drying agent. It is mild and effective, but it will not free a solution completely of water. It must also be used at room temperature to be effective; it cannot be used with boiling solvents. Table 5–2 compares the various drying agents.

At room temperature, ether dissolves 1.5% by weight of water, and water dissolves 7.5% of ether. Ether, however, will dissolve a much smaller amount of water from a saturated aqueous sodium chloride solution. Hence, the bulk of water in ether, or ether in water, can be removed by shaking it with a saturated aqueous sodium chloride solution. Any salt will act similarly, but not many are as cheap as sodium chloride or nearly as soluble in water. A solution of high ionic strength is usually not compatible with an organic solvent and forces separation of it from the aqueous layer.

PROBLEMS

1. Suppose solute **A** has a distribution coefficient of 1.0 between water and diethyl ether. Demonstrate that if 100 ml of a solution of 5 g of **A** in water were extracted with two 25 ml portions of ether, a smaller amount of **A** would remain in the water than if the solution were extracted with one 50 ml portion of ether.

2. In a preparative experiment from Part One of the textbook, outline the purification method described, and explain the purpose of each extraction step.

Technique **6**

BOILING POINTS, SIMPLE DISTILLATION, AND VACUUM DISTILLATION

Distillation is the process of vaporizing a substance, condensing the vapor, and collecting the condensate in another container. This technique is useful for **separating** a mixture when the components have different boiling points. It is the principal method of **purifying** a liquid. Four basic distillation methods are available to the chemist: simple distillation, vacuum distillation (distillation at reduced pressure), fractional distillation, and steam distillation. Technique 6 discusses simple distillation and vacuum distillation. Fractional distillation is discussed in Technique 7, and steam distillation is discussed in Technique 8.

6.1 BOILING POINTS

As a liquid is heated, the vapor pressure of the liquid increases to the point where it just equals the applied pressure (usually atmospheric pressure). At this point the liquid will be observed to boil. The normal boiling point is measured at 760 mm Hg (1 atmosphere). At a lower applied pressure, the vapor pressure needed for boiling is also lowered, and the liquid will now boil at a lower temperature. The relationship between the applied pressure and the temperature of boiling for a liquid is determined by its vapor-pressure-temperature behavior. Figure 6-1 is an idealization of the typical vapor pressure-temperature behavior of a liquid.

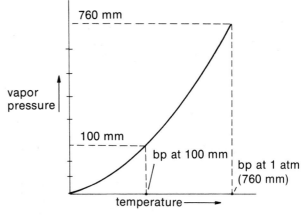

FIGURE 6-1. The vapor pressure-temperature curve for a typical liquid

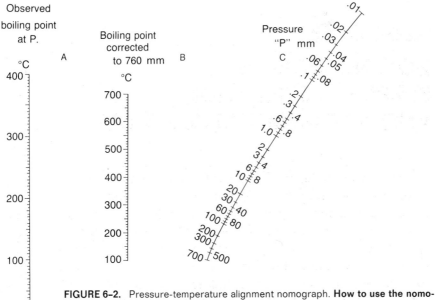

FIGURE 6–2. Pressure-temperature alignment nomograph. **How to use the nomograph:** Assume a reported boiling point of 100° at 1 mm. To determine the boiling point at 18 mm, connect 100° (column A) to 1 mm (Column C) with a transparent plastic ruler and observe where this line intersects column B (about 280°). This value would correspond to the normal boiling point. Next, connect 280° (column B) with 18 mm (column C) and observe where this intersects column A (151°). The approximate boiling point will be 151° at 18 mm.

Since the boiling point is quite sensitive to pressure, it is important to record the barometric pressure if the distillation is being conducted at an elevation significantly above or below sea level. If it is being conducted at a reduced pressure, such as that obtained with a vacuum pump or an aspirator, the pressure should be recorded.

As a rule of thumb, the boiling point of many liquids will drop about 0.5° for a 10 mm decrease in pressure in the vicinity of 760 mm Hg. At lower pressures, a 10° drop in boiling point is observed for each halving of the pressure. For example, if the observed boiling point for a liquid is 150°C at 10 mm pressure, then the boiling point would be about 140°C at 5 mm Hg.

A more accurate estimate of the change in boiling point with a change of pressure can be obtained by the use of a **nomograph.** In Figure 6–2, a nomograph is given, and a method is described for using it to obtain boiling points at various pressures when the boiling point is known at some other pressure.

6.2   DETERMINATION OF A BOILING POINT: METHODS

Two experimental methods of determining boiling points are easily available. When large quantities of material are available, one can simply record the boiling point (or boiling range) on a thermometer as the substance distills during a simple distillation (see Section 6.4). With smaller amounts of material, one can carry out a micro determination of the boiling point by use of the apparatus illustrated in Figure 6–3.

To carry out the determination, a piece of 5 mm glass tubing sealed at one end is attached to a thermometer with a rubber band or a thin slice of rubber tubing. The liquid whose boiling point is to be determined is introduced by pipet or eye dropper into this piece of tubing, and a short piece of melting point capillary tubing (sealed at one end) is dropped in with the open end down. The whole unit is then placed in a Thiele tube (see Technique 4, Figure 4–3). The rubber band should be placed well above the level of the oil in the Thiele tube. If it is not, the band may soften in the hot oil. When positioning the band, one should bear in mind that the oil will expand when it is heated. Next, the Thiele tube is heated as if one were determining a melting point until a **rapid** and continuous stream of bubbles emerges from the inverted capillary tube. At this point the heating is stopped. Soon the stream of bubbles will slow down and stop. When they stop, the liquid will enter the capillary tube. The moment when the liquid enters the capillary corresponds to the boiling point of the liquid, and the temperature is recorded.

The explanation of this method is a simple one. During the initial heating, the air trapped in the capillary tube expands and leaves the tube, giving rise to a stream of bubbles. Once the heating is stopped, the only vapor left in the capillary comes from the heated liquid which seals its open end. There is always vapor in equilibrium with a heated liquid. If the temperature of the liquid is above its boiling point, the pressure of the trapped vapor will either exceed or be equal to the atmospheric pressure. As the liquid cools, its vapor pressure will decrease. When the

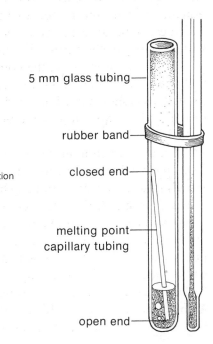

FIGURE 6–3. Micro boiling point determination

5 mm glass tubing

rubber band

closed end

melting point capillary tubing

open end

vapor pressure drops just below atmospheric pressure, the liquid will be forced into the capillary tube.

Two problems are common to this method. The first arises when the liquid is heated so strongly that it evaporates or is boiled away. The second arises when the liquid is not heated above its boiling point. If the heating is stopped at any point below the boiling point of the liquid, the liquid will enter the tube **immediately.** It will enter the tube because the trapped vapor will have a pressure less than that of the atmosphere.

The temperature observed on the thermometer during distillation or micro boiling point determination may often be **lower** than the actual boiling point. This is especially true when measuring the temperatures higher than 150°C. Manufacturers design most thermometers to read correctly only when they are immersed in the medium to be measured so as to at least cover all of the mercury thread. Since this situation rarely occurs when using a thermometer, a **stem correction** should be added to the observed temperature. This correction may be fairly large when high temperatures are being measured. A method of stem correction is described in Section 6.8.

6.3 SIMPLE DISTILLATION

When a pure liquid is distilled, vapor rises from the distilling flask and comes in contact with a thermometer. The vapor then passes through a condenser which reliquefies the vapor and passes it into the receiving flask. The temperature observed during the distillation of a **pure substance** will remain constant throughout the distillation as long as both vapor **and** liquid are present in the system (see Figure 6-5, Part A). When a **liquid mixture** is distilled, often the temperature will not remain constant, but rather it will increase throughout the distillation. This occurs because the composition of the vapor which is distilling varies continuously during the distillation (see Figure 6-5, Part B).

For a liquid mixture, the composition of the vapor in equilibrium with the heated solution is different than the composition of the solution itself. This is shown in Figure 6-4, which is a phase diagram illustrative of the typical vapor-liquid relationships for a two component system (**A + B**).

On this diagram, horizontal lines represent constant temperatures. The upper curve represents vapor composition, the lower curve represents liquid composition. For any horizontal line (constant temperature), as that shown at **t,** the intersections of the line with the curves give the compositions of the liquid and the vapor which are in equilibrium with one another at that temperature. In the diagram, at temperature **t,** the intersection of the curve at **X** indicates that liquid of composition, **W,** will be in equilibrium with vapor of composition **Z,** which corresponds to the intersection at **Y.** Composition is given as mole percent of **A** and **B** in the mixture. Pure **A,** which boils at temperature t_A, is represented at the left. Pure **B,** which boils at temperature t_B, is represented on the

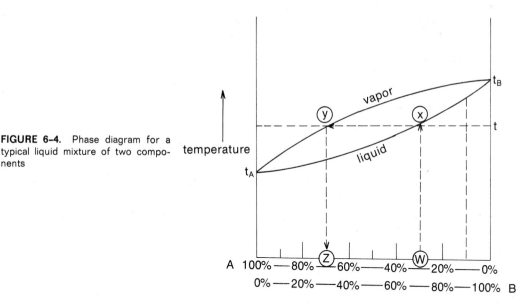

FIGURE 6-4. Phase diagram for a typical liquid mixture of two components

right. For either pure **A** or pure **B**, the vapor and liquid curves meet at the boiling point. Thus, either pure **A** or pure **B** will distill at a constant temperature (t_A or t_B). Both the vapor and the liquid must have the same composition in either of these cases. This is not the case for mixtures of **A** and **B**.

A mixture of **A** and **B** of composition **W** will have the following behavior when heated. The temperature of the liquid mixture will increase until the boiling point of the mixture is reached. This corresponds to following line **WX** from **W** to **X,** the boiling point of the mixture **(t).** At temperature **t** the liquid begins to vaporize, which corresponds to line **XY.** The vapor will have the composition corresponding to **Z.** In other words, the first vapor obtained in distilling a mixture of **A** and **B** does not consist of pure **A.** It is richer in **A** than the original mixture, but still contains a significant amount of the higher boiling component **B, even from the very beginning of the distillation.** The result of this is that it is never possible to separate a mixture completely by use of a simple distillation. However, in two cases, it is possible to obtain an acceptable separation into relatively pure components. In the first case, if the boiling points of **A** and **B** differ by a large amount ($>100°$), and if the distillation is carried out carefully, it will be possible to obtain a fair separation of **A** and **B.** In the second case, if **A** contains a fairly small amount of **B** ($<10\%$), a reasonable separation of **A** from **B** can be achieved. When boiling point differences are not large, and when high purity components are desired, it is necessary to do a **fractional distillation.** Fractional distillation is described in Technique 7, where the behavior during a simple distillation is also discussed in more detail. Note only that as vapor distills from the mixture of composition **W** (Figure 6-4), it is richer in **A** than in the solution. Thus, the composition of the material left behind in the distillation becomes richer in **B** (moves to the right from **W** toward pure **B** in the graph). A mixture which is 90% **B** (dotted line to right of Figure 6–4) has a higher boiling point than that at **W**. Hence, the temperature of the liquid in the distilling flask will increase during the distillation, and the composition of the distillate will change (as is shown in Figure 6–5, part B).

When two components which have a large boiling point difference are distilled, one will observe that the temperature remains constant while the first component distills. If the temperature remains constant, a relatively pure substance is being distilled. After the first substance distills, the temperature of the vapors will rise, and then the second component will distill, again at a constant temperature. This is shown in Figure 6–5, part C. A typical application of this type of distillation might be an instance of a reaction mixture containing the desired component **A** (bp 140°C) contaminated with a small amount of undesired component **B** (bp 250°) and mixed with a solvent such as diethyl ether (bp 36°). The ether is easily removed at low temperature. Pure **A** is removed at a higher temperature and collected in a separate receiver. Component **B** could then be distilled, but it usually is left as a residue and not distilled. This separation would not be difficult and would represent a case where simple distillation might be used to advantage.

6.4 SIMPLE DISTILLATION: METHODS

For a simple distillation, the apparatus shown in Figure 6–6 is used. The distilling flask, condenser, and vacuum adapter should be clamped. It is best if the receiving flask is supported by means of wooden blocks or with a wire gauze supported by an iron ring which is attached to a ring stand. Either of these two methods will facilitate removal or change of the receiving flask during the distillation. If an oil bath or heating mantle is used for heating, it should likewise be supported by wooden blocks adjusted to such a height that the heat source can be removed

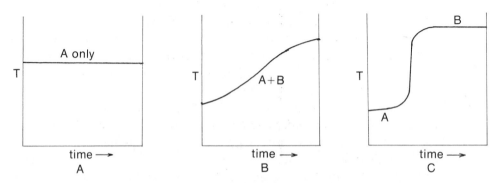

FIGURE 6–5. Three types of temperature behavior during a simple distillation: *A*, a relatively pure component is being distilled; *B*, a mixture of two components of similar boiling point is being distilled; *C*, a mixture of two components with widely differing boiling points is being distilled. Good separations are achieved in A and C.

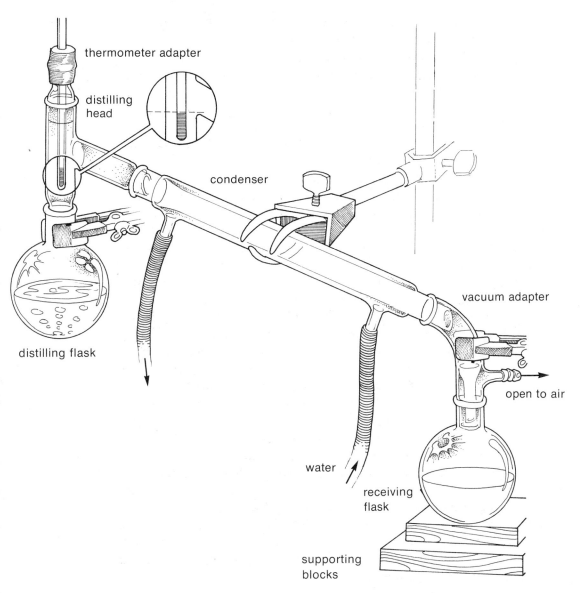

FIGURE 6-6. Apparatus for a simple distillation

from the distillation flask, by removal of the blocks, to stop the distillation.

Each standard taper ground glass joint should be lightly greased with stopcock grease to ensure that the joints do not "freeze" together. An excess of grease should not be used. When properly seated, greased, and clamped, the joint should appear to be nearly transparent, with no air bubbles visible. The tightness of the joints should be checked frequently during the distillation. If the joints are not tightly sealed, material will be lost.

The thermometer is inserted through the rubber adapter so that the bulb will extend to a depth just **below** the sidearm of the distilling head (see inset in Figure 6–6). The thermometer bulb must be **in the vapor stream,** not above it, to give a correct reading.

When attaching the water lines to the condenser, the water should pass into the lower end of the condenser and exit from the upper end. The condenser will not remain full if the water flows from the top down.

Before the liquid to be distilled is heated, one or two boiling stones should be added. These boiling stones prevent superheating of the liquid and reduce the tendency of the liquid to "bump." (See Technique 1, Section 1.8.)

The distilling flask must not be filled to more than two-thirds of its capacity because the surface area of boiling liquid should be kept as large as possible. On the other hand, the use of too large a distilling flask should also be avoided. With too large a flask, the "hold up," i.e., the amount of material which cannot distill due to the fact that some vapor must fill the empty flask, will be excessive.

Based on information provided in Technique 1, the appropriate heating source for a distillation can be chosen. In general, an oil bath or a heating mantle will be used. When the boiling point of the liquid to be distilled is reached, a ring of condensate (or **reflux ring**) will move up through the apparatus and come in contact with the thermometer bulb. At this point, the thermometer will record a rapid rise in temperature, and soon condensate will begin to pass from the condenser into the receiver. The heating is then adjusted to provide the proper rate of **take off,** which is the rate at which distillate exits from the condenser. One drop per second is considered a proper rate of take off for most applications. At a greater rate, equilibrium is not established within the distillation apparatus, and the separation may be poor. A slower rate of take off is also unsatisfactory since the temperature recorded on the thermometer is not maintained by a constant vapor stream, thus leading to the observation of a boiling temperature which is too low. Generally, the material is distilled, and fractions (portions collected in separate flasks) are collected over narrow ranges of temperature. If a good separation has been achieved, the temperature of the vapors will drop between fractions and then rise dramatically when the next component begins to distill.

One should never distill to dryness. A dry residue may explode on overheating, or the flask may melt or crack. With no liquid in the flask the heat applied is not carried away, and the flask becomes **very** hot.

6.5 VACUUM DISTILLATION

Vacuum distillation (distillation at reduced pressures) is used when compounds have high boiling points (above 200°) and/or decompose at the high temperatures required for atmospheric pressure distillation. As seen in Figure 6–2, the boiling point is reduced substantially by lowering the pressure. For example, a liquid with a boiling point of 200° at

760 mm Hg would boil at about 90° at 20 mm Hg. Thus, it may be advantageous to use a vacuum when distilling such a material. However, counterbalancing this advantage is the fact that often a separation will not be as good under reduced pressure as it was at atmospheric pressure.

6.6 VACUUM DISTILLATION: METHODS

An apparatus, such as that shown in Figure 6–7, is assembled for reduced pressure distillation. It is important that all of the rubber tubing (except for water connections to the condenser) be of the heavy wall type, so that the tubing will not collapse under the vacuum. If the rubber tubing shows evidence of cracks when it is bent or extended by pulling from both ends, it should be discarded and replaced with new tubing.

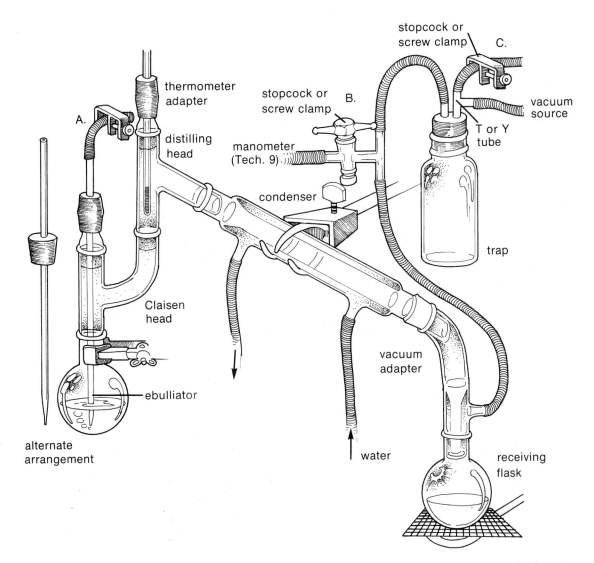

FIGURE 6–7. Apparatus for a vacuum distillation

Similarly, rubber stoppers should be inspected to make sure that they fit well, are crack-free, and are pliable. The connections between glass and rubber tubing must be secure and air-tight. It is advisable to use the shortest possible lengths of rubber tubing in the system.

All ground glass equipment must be free of cracks. Careful examination of the distilling flask to make sure that it does not have small cracks is recommended. If there is doubt about the condition of the apparatus, the laboratory supervisor should be consulted. Any cracked equipment or, for that matter, **thin wall** glassware of any type (large Erlenmeyer flasks, etc.), may implode because of the large pressure differences between the outside and the inside of the glassware. SAFETY GLASSES MUST BE WORN AT ALL TIMES. The ground glass joints should be lubricated as for the simple distillation, and one should determine that all joints fit well.

When distilling, a method is needed to form small bubbles to prevent superheating of the liquid, or to prevent "bumping." Conventional boiling stones will not work under a vacuum. Although wooden applicator sticks and microporous boiling stones have been used to prevent bumping, **ebulliators** (bubblers) are generally employed in most vacuum systems. Most standard ground glass kits contain an ebulliator. If one is not available, an ebulliator can be prepared easily by heating a section of glass tubing and drawing it out to a length of about 3 cm. The glass is then scored in the center of this drawn-out section and is broken. The ebulliator can be inserted into a thermometer adapter, if available, or, alternatively, into a rubber stopper, as is shown in Figure 6–7. The ebulliator tube is used in conjunction with a **Claisen head.** The Claisen head prevents carry-over if bumping should occur. The ebulliator is adjusted so that it comes close to the bottom of the flask. A short section of heavy wall rubber tubing, with an open screw clamp, is attached to the ebulliator at **A,** as shown in Figure 6–7. This clamp is used to regulate the amount of air admitted to the system and the rate of production of bubbles.

Normally, water should be used as a coolant in the condenser. Some very high-boiling materials may crack the condenser because of the large differences in the temperature of the vapor and cold water. In those cases, an "air condenser" may be used. An air condenser is a normal condenser, but no water is allowed to pass through it. The condenser is cooled solely by the air in the room. Some high-boiling substances may solidify in the condenser. In these cases, the condenser should not be water-cooled. The use of an air condenser or the use of steam, rather than water, circulating in the condenser should be considered.

When more than one fraction is expected from a vacuum distillation, it is considered good practice to have several pre-weighed receiver flasks, including the original, available before the distillation is begun. Such preparation permits the rapid changing of receiving flasks during the distillation. The pre-weighing procedure permits easy calculation of the weight of distillate in each fraction without the need for transferring the distillate to yet another flask.

With the apparatus shown in Figure 6–7, the vacuum must be stopped in order to change receiving flasks when a new substance (fraction) begins to distill. Although this situation can be easily tolerated in distillations where only two substances are present in the initial mixture, it is rather inconvenient when more substances are present. In Figure 6–8, two pieces of apparatus are shown where fraction collecting under vacuum is greatly simplified. All one has to do is rotate the device to collect the various fractions.

If the vacuum is interrupted while fractions are to be changed, the heat must be removed from the apparatus and the clamp controlling the air flow into the ebulliator must be opened. If heating is continued while air has been admitted to the distilling flask, oxidation of the material being distilled may occur. If air is admitted into the apparatus at some other point while the screw clamp controlling the ebulliator remains closed, liquid may back up into the ebulliator tube.

The trap assembly shown in Figure 6–7, or an alternative assembly shown in Technique 9 (Figure 9–6), is quite necessary to prevent water from backing up into the receiving flask and manometer when there is a change in pressure in the distillation system. A suitable manometer (pressure measuring device) should be included in the system at least part of the time during the distillation in order to measure the pressure at which the distillation is being conducted. A boiling point is of little value unless the pressure is known! In Figure 6–7 the stopcock at point **B** may be closed when the manometer is not in use or not attached

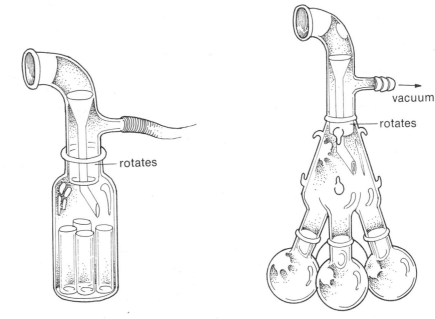

FIGURE 6–8. Rotating receiving flasks

to the system. Several types of manometers are discussed in Technique 9. The screw clamp at point **C** can be opened slowly to allow air to pass into the apparatus to bring the system back to atmospheric pressure. The manometer may break if the screw clamp is opened too rapidly.

The most convenient source of vacuum for a reduced pressure distillation is the aspirator. The aspirator, or other vacuum source, is attached to the trap. The aspirator theoretically can pull a vacuum equal to the vapor pressure of the water flowing through it. The vapor pressure of the flowing water depends on its temperature (24 mm at 25°C; 18 mm at 20°C; 9 mm at 10°C). However, in the typical laboratory, the pressures attained are higher than expected due to the reduced water pressure when many students are using their aspirators simultaneously. Good laboratory practice requires that only a few students on a given bench use the aspirator at any given time. In addition, the pressure should be checked early in the assembly procedure to make sure that the aspirator is working properly. Movement to another location in the laboratory may be necessary. Generally, a 10 to 30 mm pressure is considered adequate. Lower pressures can be obtained by using a mechanical vacuum pump.

6.7 STEPWISE DIRECTIONS FOR VACUUM DISTILLATION

The following procedures may be followed in applying the technique of vacuum distillation.

Evacuation of the Apparatus

1. Assemble the apparatus as shown in Figure 6–7 and as described in Section 6.6. REMEMBER TO USE SAFETY GLASSES! Weigh each of the empty receiving flasks to be used in the collection of the various fractions to be obtained during the distillation.

2. Concentrate the material to be distilled in an Erlenmeyer flask or beaker by removing all volatile solvents, such as ether, on a steam bath in the hood. Use boiling stones. Transfer the concentrate to a distilling flask, and complete the transfer with a small amount of solvent. Again concentrate until no additional volatile solvent can be removed (boiling will cease). The flask should be no more than one-half full after concentration. Attach the flask to the apparatus, and secure it with a clamp. Make sure all joints are tight.

3. Open the stopcock at **B** (Figure 6–7).

4. Turn on the aspirator to the maximum extent.

5. Tighten the screw clamp at **A** until the tubing is **nearly** closed.

6. **Slowly** tighten the screw clamp at **C** until it is closed completely. Watch the bubbling action of the ebulliator to see that it is not too

vigorous or too slow. Adjust **A** until a fine steady stream of bubbles is formed with **C** closed.

7. Record the pressure obtained after waiting a few minutes to allow any residual solvent to be removed. Readjust **A** if necessary. If the pressure is not satisfactory, check all connections to see if they are tight. **Do not proceed until a good vacuum is obtained.**

Beginning Distillation

8. Raise the heat source (see Technique 1) into position with wooden blocks, or other means, and begin to heat.

9. Increase the temperature of the heat source. Eventually a reflux ring will contact the thermometer bulb and distillation will begin. Record the temperature range as well as the pressure range during the distillation. The distillate should be collected at the rate of about 1 drop per second. The Claisen head and distilling head may have to be wrapped with glass wool or aluminum foil (shiny side in) for insulation during the distillation, if it is slow. The boiling point should be relatively constant as long as the pressure is constant. A rapid increase in pressure may be due to increased usage of the aspirators in the laboratory or due to rapid decomposition of the material being distilled. In the latter case, a dense white fog will be produced in the distilling flask. If this happens, reduce the temperature of the heat source, or remove it, and **stand back** until the system cools. Investigate the cause.

Changing Receiving Flasks

10. To change receiving flasks during distillation when a new component begins to distill (higher boiling point at the same pressure), the clamp at **C** must be opened slowly, and the heat source must be lowered immediately. (Watch the ebulliator for **excessive** back-up. It may be necessary to open clamp **A**). The wooden blocks under the receiver are removed, or the clamp is released, and the flask is replaced with a clean, pre-weighed receiver.

11. Reclose the clamp at **C** and allow several minutes for the system to reestablish the reduced pressure. Bubbling will commence after the liquid is drawn back out of the ebulliator. This liquid may have been forced into the ebulliator when the vacuum was interrupted.

12. Raise the heating source back into position under the distilling flask, and continue with the distillation. When the temperature falls at the thermometer, this usually indicates that the distillation is complete. However, if a significant amount of liquid remains, the bubbling may have stopped, the pressure may have risen, the heating source may not be hot enough, or perhaps insulation of the distillation head is required. Adjust accordingly.

Shutdown

13. At the end of the distillation, remove the heat source, and **slowly** open the clamps at **A** and **C. Then** turn off the water at the aspirator. Remove the receiving flask, and clean all glassware as soon as possible after disassembly to prevent the ground glass joints from sticking.

6.8 THERMOMETER STEM CORRECTIONS

The temperature observed on the thermometer during distillation or micro boiling point determination may be **lower** than the actual boiling point due to a stem error in the thermometer. When it is desired to make a stem correction for a thermometer, the formula given below may be used. It is based upon the fact that the portion of the mercury thread in the stem is cooler than the portion immersed in the vapor. The mercury in the cool stem will not have expanded to the same extent as that in the warmed section of the thermometer, and therefore the reading must be corrected. The following equation is used:

$$(0.000154)(T - t_1)(T - t_2) = \text{Correction to be added to the observed temperature, } \mathbf{T}$$

1. The factor 0.000154 is a constant which corresponds to a co-efficient of expansion for mercury in the thermometer.

2. $(T - t_1)$ corresponds to the length of the mercury thread not immersed in vapor. It is convenient to use the temperature scale for this measurement, where **T** is the observed temperature and t_1 is the **approximate** place where the heated part of the stem ends and the cooler part begins. In simple distillation cases, t_1 would correspond approximately to the lower part of the thermometer adapter, which is inserted into the distilling head. See Figure 6–6.

3. $(T - t_2)$ corresponds to the difference between the temperature of the mercury in the vapor, **T,** and the temperature of the mercury in the air outside the distillation apparatus, t_2. **T** is the observed temperature, while t_2 is measured by hanging another thermometer so that the bulb is close to the stem of the main thermometer. Figure 6–9 illustrates the method of correcting a thermometer and indicates where the various temperatures are determined.

Using the formula given above, it can be shown that high boiling substances are the most likely to require a stem correction, while low boiling substances need not be corrected. The calculations given below illustrate this point.

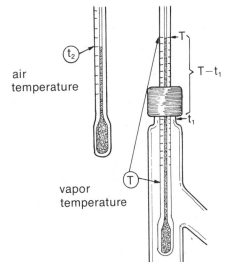

air
temperature

vapor
temperature

$T - t_1$

t_1

T

t_2

FIGURE 6-9. Measurement for a thermometer stem correction

EXAMPLE 1:

$T = 200°$

$t_1 = 0°$

$t_2 = 35°$

$(0.000154)(200)(165) = 5.1°$ stem correction

$200° + 5° = 205°$ corrected boiling point

EXAMPLE 2:

$T = 100°$

$t_1 = 0°$

$t_2 = 35°$

$(0.000154)(100)(65) = 1.0°$ stem correction (negligible)

PROBLEMS

1. Using the pressure-temperature alignment chart in Figure 6–2, determine the following:

a. What is the normal boiling point (at 760 mm) for a compound which boils at 150° at 10 mm Hg pressure?
b. Where would the compound in part a boil if the pressure were 40 mm?
c. A compound was distilled at atmospheric pressure and had a boiling range of 290 to 300°. Since extensive decomposition occurred, it was decided that a vacuum distillation was required. What would be the approximate boiling range observed at 15 mm Hg?

2. Calculate the corrected boiling point for nitrobenzene by using the method given in Section 6.8. The apparatus used is shown in Figure 6–6. The observed boiling point was 205°. The thermometer was inserted in the thermometer adapter so that the 0° graduation was visible at the bottom of the adapter. The air temperature just above the distilling head was measured on a second thermometer and was 35°.

3. In Section 6.4 it was stated that the distilling flask should not be too large because of extensive "hold up" (material that cannot distill due to the fact that some

vapor must fill the empty flask). Using the ideal gas equation, prove that the statement is correct by performing the following calculations:

a. A 15 g sample of acetone (M.W. = 58) was placed in a one-liter distilling flask. The acetone was distilled at 1 atm at the boiling point of 56° until the flask appeared to be empty. Assuming that the flask contained only acetone vapor at that point, calculate how many grams of acetone remained in the "empty" flask.

b. Repeat the above calculation in part **a,** using a distilling flask with a volume of 50 ml.

c. In each of the above cases, calculate the percentage loss due to "hold up" with the 15 g sample of acetone. Should you always use the smallest possible distilling flask?

Technique **7**

FRACTIONAL DISTILLATION, AZEOTROPES

The technique of **simple distillation** was described in Technique 6. Simple distillation works well for most routine separations and purifications of organic compounds. However, when the boiling point differences of components to be separated are not large, the technique of **fractional distillation** must be employed in order to obtain a good separation.

7.1 DIFFERENCES BETWEEN SIMPLE AND FRACTIONAL DISTILLATION

When an ideal solution of two liquids, such as benzene (bp 80°) and toluene (bp 110°) is distilled by **simple distillation,** using the apparatus shown in Technique 6, Figure 6–6, the first vapor produced from this solution will be richer in the lower boiling component (benzene). However, when the vapor is condensed and analyzed, it is unlikely that the distillate will be pure benzene, because the boiling point difference of benzene and toluene is only 30°. Likewise, the remaining liquid in the distilling flask will contain a larger amount of the higher boiling component, but will be far from being pure toluene.

In principle, one could distill a solution of 50% benzene and 50% toluene by simple distillation and collect the distillate in **fractions** (portions collected in separate flasks). The first fraction would contain the largest amount of benzene and the least amount of toluene, and would have the lowest boiling point range. The second fraction would contain less benzene and more toluene than the previous one, and

TABLE 7-1. SIMPLE DISTILLATION OF A MIXTURE OF BENZENE AND TOLUENE

FRACTION	BOILING RANGE	PERCENT COMPOSITION	
		Benzene	Toluene
1	80–85	90	10
2	85–90	72	28
3	90–95	55	45
4	95–100	45	55
5	100–105	27	73
6	105–110	10	90

would have a higher boiling point range. The trend of decreasing benzene and increasing toluene would continue until the last fraction was removed. This last fraction would have the smallest amount of benzene, have the largest amount of toluene, and have the highest boiling point range. The results of such a hypothetical distillation are given in Table 7-1.

As such a distillation proceeded, a plot of boiling points **vs.** volume of condensate (distillate) collected might appear as in Figure 7-1. Clearly, the separation obtained by this method would be poor. The continuously increasing temperature observed in Figure 7-1 indicates that the composition of the vapor itself was also continuously changing. At no time did a pure substance distill.

However, each of the fractions indicated in Table 7-1 could be redistilled. On redistillation, each of these fractions would yield vapor and a resulting condensate which would contain **more** benzene than was initially present. The residue left in the distilling flask would contain **more** toluene than initially present. The distillates and residues of similar composition (similar boiling ranges) could be combined and

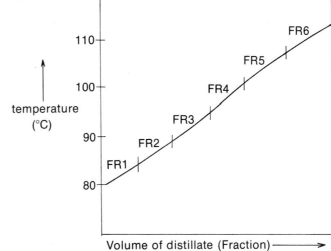

FIGURE 7-1. Temperature-distillate plot during simple distillation of a benzene-toluene mixture

redistilled again. Eventually, one should obtain distillate which would be essentially pure benzene, and residue which would be essentially pure toluene.

Obviously the above procedure would be **very tedious,** and fortunately need not be done in usual laboratory practice. The technique of **fractional distillation** accomplishes the same result. One simply has to employ a **column** inserted between the distilling flask and the distilling head as shown in Figure 7–2. The column is filled with a suitable **packing,** such as a stainless steel sponge. This packing allows a mixture of benzene-toluene to be subjected continuously to many vaporization-condensation cycles as the material moves up the column. **As each cycle occurs within the column, the composition of the vapor is progressively enriched in the lower boiling component (benzene).** Finally, nearly pure benzene (bp 80°) emerges from the top of the column, condenses, and

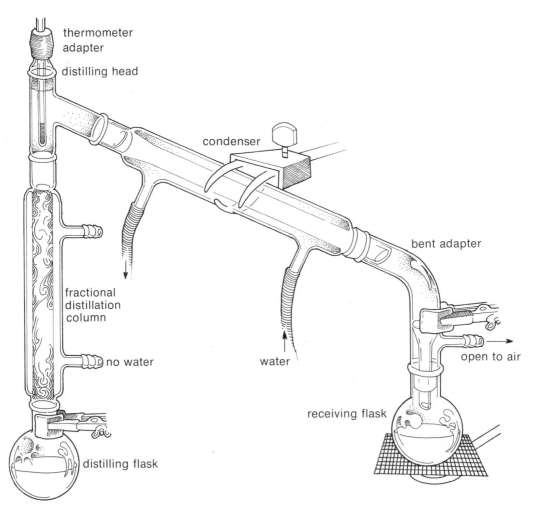

FIGURE 7–2. Fractional distillation apparatus

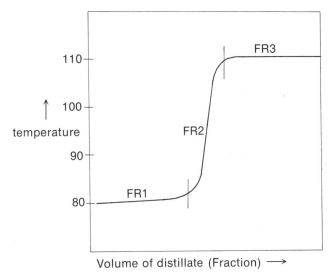

FIGURE 7–3. Temperature-distillate plot during fractional distillation of a benzene-toluene mixture

temperature →

Volume of distillate (Fraction) ⟶

passes into the receiving flask as the first fraction. The distillation process must be carried out slowly to insure that numerous vaporization-condensation cycles occur. When nearly all of the benzene is removed, the temperature begins to rise and a small amount of a second fraction is collected which contains some benzene and toluene. When the temperature reaches 110°, the boiling point of pure toluene, the vapor is condensed and collected in another receiving flask as the third fraction.

A plot of the boiling point **vs.** volume of condensate (distillate) would appear as in Figure 7–3. Clearly the separation would be much better than that obtained by a simple distillation (Figure 7–1).

7.2   VAPOR-LIQUID COMPOSITION DIAGRAMS

A vapor-liquid composition phase diagram such as the one shown in Figure 7–4 can be used to explain the operation of a fractionating column with an **ideal solution** of two liquids, **A** and **B.** An ideal solution is one in which the two liquids are chemically similar but do not interact with each other and are totally miscible in all proportions. Such solutions are said to obey Rauolt's law. This law will be explained in detail in Section 7.3. In the example given, pure **A** boils at 50° while pure **B** boils at 90° at atmospheric pressure.

The phase diagram relates the composition of the boiling liquid (lower curve) to that of its vapor (upper curve) as a function of temperature. Any horizontal line drawn across the diagram (a constant temperature line) intersects the diagram in two places. These intersections relate the vapor composition to the composition of the boiling liquid which produces that vapor. By convention, composition is expressed in either

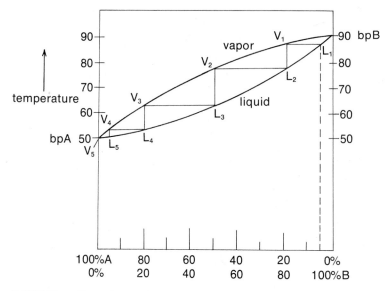

FIGURE 7–4. Phase diagram for a fractional distillation of an ideal two component system

mole fraction or in **mole per cent.** The mole fraction and mole per cent are defined as follows:

$$\text{mole } \textbf{fraction A} = N_A = \frac{\text{moles } \textbf{A}}{\text{moles } \textbf{A} + \text{moles } \textbf{B}}$$

$$\text{mole } \textbf{fraction B} = N_B = \frac{\text{moles } \textbf{B}}{\text{moles } \textbf{A} + \text{moles } \textbf{B}}$$

$$N_A + N_B = 1$$
$$\text{mole } \textbf{percent A} = N_A \times 100$$
$$\text{mole } \textbf{percent B} = N_B \times 100$$

The horizontal and vertical lines shown in Figure 7–4 represent the processes occurring during a fractional distillation. The **horizontal lines** (L_1V_1, L_2V_2, etc.) each represent the **vaporization** steps of a given vaporization-condensation cycle and indicate the composition of the vapor in **equilibrium** with liquid at a given temperature. For example, at 63° a liquid with a composition of 50% **A** (L_3 on diagram) would yield vapor of composition 80% **A** (V_3 on diagram) at equilibrium. The vapor is richer in the lower boiling component, **A**, than was the original liquid.

The **vertical lines** (V_1L_2, V_2L_3, etc.) each represent the **condensation** step of a given vaporization-condensation cycle. The composition does not change as the temperature drops upon condensation. For example, the vapor at V_3 condenses to give a liquid (L_4) of composition 80% **A** with a drop in temperature from 63° to 53°.

Now consider the phase diagram in Figure 7–4 for a solution initially 5% in component **A**. The solution of composition 5% **A** (95% **B**) is heated (following the dotted line) until it is observed to boil at 87° (L_1). The vapor has a composition (V_1) of 20% **A** (80% **B**). The vapor is richer in **A** than was the original liquid, but is by no means pure **A**. In a simple distillation apparatus this liquid would have been condensed and passed into the receiving flask in a highly unpurified state. However, with a fractionating column in place, the vapor is condensed in the column to give liquid (L_2) of composition 20% **A** (80% **B**). This liquid (L_2) is revaporized (bp 78°) to give vapor of composition V_2 (50% **A**) which is condensed to give liquid L_3. Liquid L_3 (50% **A**) is revaporized (bp 63°) to give vapor of composition V_3 (80% **A**) which is condensed to give liquid L_4. Liquid L_4 (80% **A**) is revaporized (bp 53°) to give vapor of composition V_4 (95% **A**) which is condensed to give liquid L_5. Finally, liquid L_5 (95% **A**) is once again revaporized (bp 51°) to give vapor (V_5) which is essentially pure **A**. The nearly pure liquid emerges from the column, is condensed and passes into the receiving flask as a pure liquid **A**. In the meantime component **B** has been concentrated in the distilling flask and is now removed as a pure liquid by distillation.

The above vaporization-condensation process in a fractional distillation column is shown in Figure 7–5. The compositions of the liquids, their boiling points and the composition of the vapor are shown.

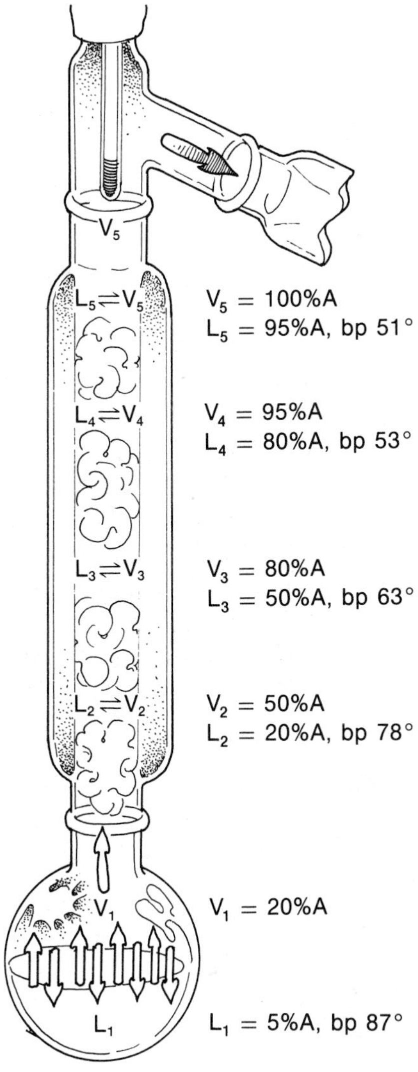

FIGURE 7–5. Vaporization-condensation process in a fractionation column

V_5

$L_5 \rightleftharpoons V_5$ $V_5 = 100\%A$
$L_5 = 95\%A$, bp 51°

$L_4 \rightleftharpoons V_4$ $V_4 = 95\%A$
$L_4 = 80\%A$, bp 53°

$L_3 \rightleftharpoons V_3$ $V_3 = 80\%A$
$L_3 = 50\%A$, bp 63°

$L_2 \rightleftharpoons V_2$ $V_2 = 50\%A$
$L_2 = 20\%A$, bp 78°

V_1 $V_1 = 20\%A$

L_1 $L_1 = 5\%A$, bp 87°

7.3   RAOULT'S LAW

Two liquids **(A and B)** which are mutually soluble (miscible) and which do not interact with each other will form an **ideal solution** and follow Raoult's law. The law states that the partial vapor pressure of component **A** in the solution (P_A) is equal to the vapor pressure of pure **A** ($P_A°$) times its mole fraction (N_A) in the solution (equation 1). Likewise, an expression can be written for component **B** (equation 2). The mole fractions, N_A and N_B, were defined in Section 7.2.

$$\text{Partial vapor pressure of } \mathbf{A} \text{ in solution} = P_A = (P_A°)(N_A) \quad (1)$$
$$\text{Partial vapor pressure of } \mathbf{B} \text{ in solution} = P_B = (P_B°)(N_B) \quad (2)$$

$P_A°$ = vapor pressure of pure **A,** independent of **B**
$P_B°$ = vapor pressure of pure **B,** independent of **A**

The partial vapor pressures are **added** to give a total pressure above the solution (equation 3). When the applied pressure equals the sum of the partial vapor pressures (P_{total}), the solution will be observed to boil.

$$P_{total} = P_A + P_B = P_A°N_A + P_B°N_B \quad (3)$$

As an example of an application of Raoult's law, consider the following problems:

a. What is the partial vapor pressure of **A** in a solution of $N_A = 0.5$, where the vapor pressure of pure **A** at 100° is 1020 mm? Answer: $P_A = P_A°N_A = (1020)(0.5) = 510$ mm.

b. What is the partial vapor pressure of **B** in a solution of $N_B = 0.5$, where the vapor pressure of pure **B** at 100° is 500 mm? Answer: $P_B = P_B°N_B = (500)(0.5) = 250$ mm.

c. Would the solution boil at 100° if the applied pressure were 760 mm? Answer: Yes, $P_{total} = P_A + P_B = 510 + 250 = 760$ mm.

d. What is the composition of the **vapor** at the boiling point (100°)? Answer: The fraction of each component is **proportional** to the partial vapor pressure of that component compared to the total pressure. Therefore, N_A(vapor) $= 510/760 = 0.67$, N_B(vapor) $= 250/760 = 0.33$.

One should note in Part d that the **vapor is richer ($N_A = 0.67$) in the lower boiling (higher vapor pressure) component (A) than was the liquid prior to vaporization ($N_A = 0.5$).** This proves mathematically what was described in Section 7.2.

The consequences of Raoult's law are shown schematically in Figure 7–6. In Figure 7–6A the boiling points are identical (vapor pressures are the same), and no separation is attained regardless of how the distillation is conducted. In Figure 7–6B a fractional distillation is required, while in Figure 7–6C a simple distillation will provide an adequate separation.

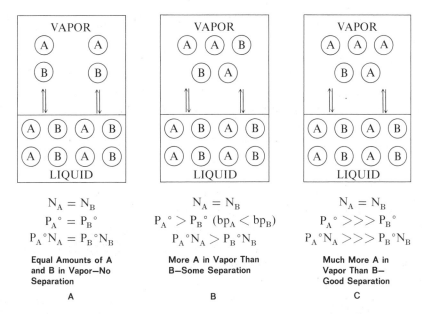

$$N_A = N_B$$
$$P_A^\circ = P_B^\circ$$
$$P_A^\circ N_A = P_B^\circ N_B$$

Equal Amounts of A and B in Vapor—No Separation

A

$$N_A = N_B$$
$$P_A^\circ > P_B^\circ \; (bp_A < bp_B)$$
$$P_A^\circ N_A > P_B^\circ N_B$$

More A in Vapor Than B—Some Separation

B

$$N_A = N_B$$
$$P_A^\circ >>> P_B^\circ$$
$$P_A^\circ N_A >>> P_B^\circ N_B$$

Much More A in Vapor Than B— Good Separation

C

FIGURE 7-6. Consequences of Raoult's Law. *A*, Boiling points (vapor pressures) are identical—**No Separation**; *B*, Boiling point of A somewhat less than B—**Requires Fractional Distillation**; *C*, Boiling point of A much less than B—**Simple Distillation Will Suffice.**

In the extreme case of a negligible vapor pressure for **B** the vapor will be pure **A.** For example, the distillation of an aqueous salt solution will behave in this manner:

$$P_{total} = P_{H_2O}^\circ N_{H_2O} + P_{salt}^\circ N_{salt}$$
$$P_{salt}^\circ = 0$$
$$P_{total} = P_{H_2O}^\circ N_{H_2O}$$

A solution whose mole fraction of water is 0.7 will not boil at 100° because $P_{total} = (760)(0.7) = 532$ mm and is less than atmospheric pressure. Eventually, the solution can be heated to the boiling point. At 110° the solution will boil because $P_{total} = (1085)(0.7) = 760$ mm. The vapor will be pure water, and its observed boiling point will be 100°. (The vapor pressure of pure water at 110° is 1085 mm.)

7.4 COLUMN EFFICIENCY

A measure of column efficiency is given in terms of **theoretical plates.** A column would have one theoretical plate if the first distillate (condensed vapor) had the composition located at L_2 (20% **A**) starting with a liquid with composition L_1 (5% **A**) as shown in Figure 7–4. This would correspond to a simple distillation or one vaporization-condensation cycle. A column would have two theoretical plates if the distillate (vapor) had the composition L_3 (50% **A**) starting with a liquid with composition L_1 (5% **A**). The two theoretical plate column has essentially carried out "two simple distillations" corresponding to lines $L_1V_1L_2$ and $L_2V_2L_3$ in Figure 7–4. Thus, **five theoretical plates** would be necessary to separate nearly pure **A** from **B** (lines $L_1V_1L_2$, $L_2V_2L_3$, $L_3V_3L_4$, $L_4V_4L_5$, and L_5V_5 in the example shown in Figure 7–4). In effect a five plate column would correspond to "five simple distillations." In practice, as shown in Figure 7–5, the first theoretical plate corresponds to the initial vaporization from the distilling flask, while the remaining four plates would be obtained from the condensation-vaporization cycles in the column.

It should be noted that most columns do not allow distillation to proceed in **discrete steps** as indicated in Figure 7–4. Instead, the process is a **continuous** one which allows vapors to be continuously in contact with liquid of changing composition as it passes through the column. In principle, almost any material can be used to pack the column as long as it can be wetted by the liquid.

The relationship between the number of **theoretical plates** needed to separate an ideal two component mixture and the difference between the boiling points of the components is given in Table 7–2. The values have been calculated assuming an average boiling point of 150° for each mixture and for a column operating at equilibrium. Since columns are seldom operated at equilibrium, more theoretical plates than listed may be necessary for a complete separation. For example, three or four plates rather than the listed two plates may be necessary to separate a mixture with a boiling point difference of 72°. The table still is useful in giving an approximate number of plates needed to separate two component mixtures. For example, one can separate compound A (bp 100°) from compound B (bp 200°) by simple distillation (approximately one plate is required). A fractional distillation with a five plate column would be required to separate a mixture of A (bp 130°) from B (bp 166°). The **efficiency** (number of theoretical plates) of the column must

**TABLE 7–2. THEORETICAL PLATES REQUIRED
TO SEPARATE MIXTURES WHEN GIVEN THE
BOILING POINT DIFFERENCES OF THE COMPONENTS**

BOILING POINT DIFFERENCE	THEORETICAL PLATES REQUIRED FOR SEPARATION
108	1
72	2
54	3
43	4
36	5
20	10
10	20
7	30
4	50
2	100

TABLE 7-3. TYPES OF FRACTIONATING COLUMNS AND PACKINGS

TYPE (20 cm LONG × 1 cm DIAMETER)	THEORETICAL PLATES	BOILING POINT DIFFERENCES (°C)	HOLD-UP (ml)
Vigreux	2.5	60	1
Packed (Glass Helices)	5	36	4
Packed (Heli-Pak)	13	17	7

increase as the boiling point differences between the components decrease in order to provide an adequate separation.

7.5 TYPES OF FRACTIONATING COLUMNS AND PACKINGS

A number of different types of fractionating columns have been developed over the years to carry out separations. Only a few of the most common types of columns will be discussed in this text.

In Table 7-3 a comparison is made of three different types of columns and packings for columns. A Vigreux column, shown in Figure 7-7A, has indentations which are inclined downward at angles of 45° and which appear in pairs on opposite sides of the column. A column of this type can be prepared by heating a small spot on a piece of glass tubing until it melts and then pushing the glass inward with the tip of a pencil with a long point. The projections into the column give increased possibilities for condensation to occur and for the vapor to equilibrate with the liquid. With this type of column one can distill material rapidly, but the efficiency is not very high. For example, a 20 cm column would have only 2.5 plates and would only separate materials with a 60° boiling point difference. A longer column, such as a 40 cm column, would give 5 plates and would allow a separation of compounds with a 36° boiling point difference. Vigreux columns are often used in small scale distillations because of their small **hold-up** (the amount of liquid retained by the column). The 20 cm column mentioned above would only retain 1 ml of liquid at the end of the distillation.

The most common type of column is made by packing a tube with some material such as glass (beads, short sections of tubing) or metal (stainless steel sponge, copper scouring pad, or Heli-Pak). The glass has the advantage that it is unreactive with organic compounds while the metal may react with such compounds as alkyl halides.

Two commercial packings (helices and Heli-Pak) listed in Table 7-3 are relatively expensive. Glass beads or short sections of glass tubing (Raschig rings) are often used as an alternative packing. These packings are held in place with tight fitting sections of glass tubing or a small amount of stainless steel sponge (Figure 7-7B). An inexpensive metal

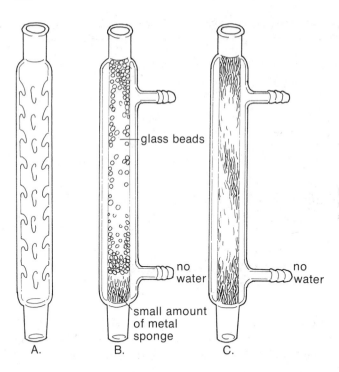

glass beads

no
water

no
water

small amount
of metal
sponge

A.   B.   C.

FIGURE 7–7. Types of fractionating columns: *A*. Vigreux; *B*. glass beads; *C*. metal
sponge

packed column is prepared by inserting part of a stainless steel sponge
or copper scouring pad into one of the condensers available in a typical
organic kit and tamping it gently with a pencil (Figure 7–7C).

It is important to note that many fractionating columns must be
insulated so that temperature equilibrium is established at all times.
When this is necessary a simple way to prepare insulation is to sandwich
fine glass wool in between aluminum foil with the shiny side out. The
edges of the foil are crimped so that the glass wool is not visible. The
column is then wrapped with this insulation. A piece of cloth may also
be used as insulation.

It is also important to distill the mixture as slowly as possible. Much
of the liquid should be allowed to return through the column in order
to establish good vapor-liquid equilibrium. The values given in Table
7-3 depend on establishing good equilibrium. The distillation must be
conducted fast enough to maintain a constant **take-off** (rate at which
material collects in the receiver) so that the temperature at the thermometer bulb remains constant. In other words, a distillation conducted
too slowly will also be unsatisfactory.

7.6 FRACTIONAL DISTILLATION: METHODS

For a fractional distillation, the apparatus shown in Figure 7–2
and the column shown in Figure 7–7C are used. The distilling flask,
condenser, and vacuum adapter should be clamped so that the **column**
is perpendicular to the desk top. No cooling water is used in the

fractionating column. The receiving flask will need to be supporte
an iron ring with wire gauze attached to a ring stand. If an oil b
or heating mantle is used wooden blocks should be placed under tl.
heat source so that it can be easily removed.

The methods given in Technique 6, Sections 6.4 for simple distillation are also used in fractional distillation. In addition to those methods,
care must often be taken to **insulate** the column with glass wool/
aluminum foil, or a cloth towel if the material to be distilled has a
high boiling point, and to distill as **slowly** as possible, as indicated in
Section 7.5.

In order to obtain the best possible separation, the temperature
of the material in the distilling flask should be raised slowly so that
the liquid/vapor can move up the column and be equilibrated. If the
contents of the flask are heated too quickly, the column will become
filled with liquid (flooding). Flooding will decrease the efficiency of the
separation. If this happens, the heat source must be lowered in order
for the liquid to return to the distilling flask.

The temperature at the thermometer bulb should remain constant
as the pure low boiling component is removed (Figure 7–3). When most
of this component is distilled, the distillation rate will decrease. At this
point the distilling flask is heated to a higher temperature, and an
intermediate fraction is collected until the temperature of the vapor
at the thermometer bulb stabilizes at the higher value. One collects the
higher boiling component in another container. Ideally, one should
obtain results as shown in Figure 7–3.

7.7 NONIDEAL SOLUTIONS: AZEOTROPES

Many mixtures of compounds do not exhibit ideal behavior because of intermolecular attractions or repulsions. Because of this non-
ideal behavior, Raoult's law is not followed. There are two types of
vapor-liquid composition diagrams which result from this non-ideal
behavior; the minimum boiling point and the maximum boiling point
diagrams. The minimum or maximum points in such diagrams correspond to a constant boiling mixture called an **azeotrope.** An azeotrope
has a fixed composition, which cannot be altered by normal distillation
(simple or fractional) procedures, and a fixed boiling point. Hence, an
azeotrope acts as if it were a **pure** compound.

A. Minimum Boiling Point Diagrams

The most common two-component mixture which gives a minimum boiling point azeotrope is the ethanol-water system shown in
Figure 7–8. A minimum boiling point azeotrope results from a slight
incompatability of the substances which leads to higher than expected
combined vapor pressures from the solution. The higher combined
vapor pressures result in a lower boiling point for the mixture than

that of either of the two components. One notes that the azeotrope V_3 in Figure 7–8 has a composition of about 96% ethanol-4% water, and a boiling point of 78.1°. This boiling point is not much lower than the boiling point of pure ethanol (78.3°). However, this small difference means that one can obtain only 96% ethanol-4% water in a simple or fractional distillation procedure. Even with the **best** fractionating column one cannot obtain 100% ethanol! The remaining 4% of water can be removed by adding benzene and removing a different azeotrope, the benzene-water-ethanol azeotrope (bp 65°). Once the water is removed, the excess benzene is removed as an ethanol-benzene azeotrope (bp 68°). The resulting material will be free of water and is called "absolute" ethanol.

The behavior on fractional distillation of an ethanol-water mixture of composition X (Figure 7–8) can be described as follows. The mixture is heated (line XL_1) until it is observed to boil (L_1). The vapor (V_1) will be richer in the lower boiling component, ethanol. The condensate (L_2) is vaporized to give V_2. The process continues, following the lines to the right, until the azeotrope (V_3) is obtained. The distillate is not pure ethanol but contains 96% ethanol and 4% water. The contents of the distilling flask become progressively richer in the higher boiling component, water, as the distillation proceeds. When all of the alcohol is removed as the azeotrope, pure water distills at 100°. A 96% ethanol solution acts as if it were a **pure** liquid and boils at 78.1° with no change in the composition of the vapor.

Some common minimum boiling azeotropes are given in Table 7–4. There are numerous other azeotropes that are formed in two and three component systems. In fact, such azeotropes are quite common. One notes that water forms azeotropes with many substances. Thus, water must be carefully removed with **drying agents** before compounds are

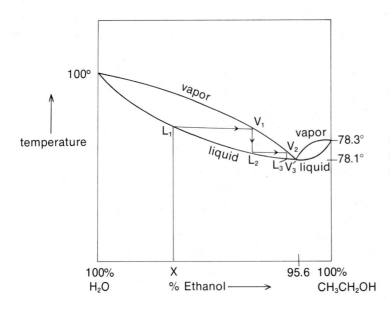

FIGURE 7–8. The ethanol-water minimum boiling point phase diagram

TABLE 7-4. COMMON MINIMUM BOILING AZEOTROPES

AZEOTROPE	COMPOSITION (WEIGHT PERCENTAGE)	BOILING POINT
Ethanol-Water	95.6% C_2H_5OH, 4.4% H_2O	78.17°
Benzene-Water	91.1% C_6H_6, 8.9% H_2O	69.4°
Benzene-Water-Ethanol	74.1% C_6H_6, 7.4% H_2O, 18.5% C_2H_5OH	64.9°
Methanol-Carbon Tetrachloride	20.6% CH_3OH, 79.4% CCl_4	55.7°
Ethanol-Benzene	32.4% C_2H_5OH, 67.6% C_6H_6	67.8°
Methanol-Toluene	72.4% CH_3OH, 27.6% $C_6H_5CH_3$	63.7°
Methanol-Benzene	39.5% CH_3OH, 60.5% C_6H_6	58.3°
Cyclohexane-Ethanol	69.5% C_6H_{12}, 30.5% C_2H_5OH	64.9°
2-Propanol-Water	87.8% $(CH_3)_2 CHOH$, 12.2% H_2O	80.4°
Butyl Acetate-Water	72.9% $CH_3COOC_4H_9$, 27.1% H_2O	90.7°
Phenol-Water	9.2% C_6H_5OH, 90.8% H_2O	99.5°

distilled. Extensive listings of azeotropic data are available in references such as the **Handbook of Chemistry and Physics.**

B. Maximum Boiling Point Diagrams

A two component maximum boiling point phase diagram is shown in Figure 7-9. A maximum boiling point azeotrope results from a slight attraction between the component molecules which leads to lower combined vapor pressures in the solution. The lower combined vapor pressures result in a higher boiling point than would be observed for either of the two components. Since the azeotrope has a higher boiling point than any of the components, it will be concentrated in the distilling flask, as the distillate (pure **B**) is removed. The distillation of a solution of composition **X** would follow to the right along the lines in

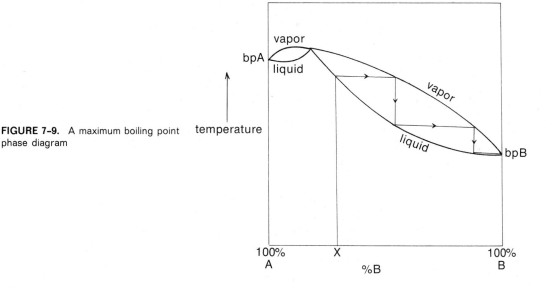

FIGURE 7-9. A maximum boiling point phase diagram

TABLE 7-5. MAXIMUM BOILING AZEOTROPES

AZEOTROPE	COMPOSITION (WEIGHT PERCENTAGE)	BOILING POINT
Acetone-Chloroform	20.0% CH_3COCH_3, 80.0% $CHCl_3$	64.7°
Chloroform-Methyl Ethyl Ketone	17.0% $CHCl_3$, 83.0% $CH_3COCH_2CH_3$	79.9°
Hydrochloric Acid	20.2% HCl, 79.8% H_2O	108.6°
Acetic Acid-Dioxane	77.0% CH_3COOH, 23.0% $C_4H_8O_2$	119.5°
Benzaldehyde-Phenol	49.0% C_6H_5CHO, 51.0% C_6H_5OH	185.6°

Figure 7-9. Once the composition of the material remaining in the distilling flask has reached that of the azeotrope, the temperature will rise and the azeotrope will begin to distill. The azeotrope will continue to distill until the material in the distillation flask has been exhausted.

Some maximum boiling azeotropes are listed in Table 7-5. They are not nearly as common as minimum boiling azeotropes.

PROBLEMS

1. The following are approximate vapor pressures for benzene and toluene at various temperatures:

TEMP.		mm Hg	TEMP.		mm Hg
Benzene	30°C	120	Toluene	30°C	37
	40	180		40	60
	50	270		50	95
	60	390		60	140
	70	550		70	200
	80	760		80	290
	90	1010		90	405
	100	1340		100	560
				111	760

a. What is the mole fraction of each component if 39 g of benzene (C_6H_6) is dissolved in 46 g of toluene (C_7H_8)?
b. Assuming this mixture is ideal, i.e., it follows Raoult's law, what is the partial vapor pressure of benzene in this mixture at 50°?
c. Estimate to the nearest degree the temperature at which the vapor pressure of the solution will be equal to one atmosphere (b.p. of the solution).
d. At the boiling point of this solution, calculate the composition of the vapor (mole fraction of each component) that is in equilibrium with the solution.
e. Calculate the composition of the vapor in weight percent that is in equilibrium with the solution.

2. Make an estimate of the number of theoretical plates that are necessary to separate a mixture which has a mole fraction of **B** equal to 0.70 (70% **B**) in Figure 7-4.

3. Two moles of sodium chloride are dissolved in 8 moles of water. Assume that the solution follows Raoult's law, and that the vapor pressure of sodium chloride is negligible. The boiling point of water is 100°. The distillation is carried out at 1 atm (760 mm Hg).

a. Calculate the vapor pressure of the solution when the temperature reaches 100°.

b. What temperature would be observed during the entire course of the distillation?

c. What would be the composition of the distillate?

d. If a thermometer were immersed below the surface of the liquid of the boiling flask, what would be observed?

4. Explain why the boiling point of a two component mixture rises slowly throughout a simple distillation when the boiling point differences are not large.

5. Given the boiling points of several known mixtures of **A** and **B** (mole fractions are known) and the vapor pressures of **A** and **B** in the pure state ($P_A°$ and $P_B°$) at these same temperatures, how would you construct a boiling point-composition phase diagram for **A** and **B?**

6. Describe the behavior on distillation of a 98% ethanol solution through an efficient column. Refer to Figure 7-8.

7. Construct an approximate boiling point-composition diagram for a benzene-methanol system. The mixture shows azeotropic behavior. Include on the graph the boiling points of pure benzene and pure methanol and the boiling point of the azeotrope. Qualitatively describe the behavior for a mixture which is initially rich in benzene and a mixture which is initially rich in methanol.

8. Construct an approximate boiling point-composition diagram for an acetone-chloroform system (maximum boiling mixture). Include on the graph the boiling points of each of the pure components and the boiling point of the azeotrope. Qualitatively describe the behavior upon distillation of a mixture which is rich in acetone (90%) and of a mixture which is rich in chloroform (90%).

9. Two compounds have a difference in boiling point of 20°C. How many theoretical plates may be required to separate these substances? How long must a **Vigreux column** be in order to separate this mixture? How many milliliters of liquid will be held up in the column?

Technique **8**

STEAM DISTILLATION

Simple and fractional distillations were described in Techniques 6 and 7, respectively. These techniques are applied to completely soluble (miscible) mixtures only. When liquids are not mutually soluble (immiscible), they can also be distilled. A mixture of immiscible liquids will boil at a lower temperature than the boiling points of any of the separate components as pure compounds. When steam is used to provide one of the immiscible phases, the process is called **steam distillation.** The advantage of this technique is that the desired material distills at a temperature below 100°. Thus, if unstable or very high boiling substances are to be removed from a mixture, decomposition is avoided. Since all gases mix, the two substances can mix together in the vapor and co-distill. Once the distillate is cooled, the desired component separates from the water since it is not miscible. Steam distillation is used widely in isolating liquids and solids from natural sources. It is also used in removing a reaction product from a tarry reaction mixture.

8.1 DIFFERENCES BETWEEN DISTILLATION OF MISCIBLE AND IMMISCIBLE MIXTURES

$$\text{Miscible Liquids} \qquad P_{total} = P_A{}^{\circ}N_A + P_B{}^{\circ}N_B \qquad (1)$$

Two liquids (A and B) which are mutually soluble (miscible), and which do not interact with each other, will form an ideal solution and follow Raoult's law as shown in equation (1). One notes that the vapor pressures of pure liquids ($P_A{}^{\circ}$ and $P_B{}^{\circ}$) are not added directly to give the total pressure (P_{total}), but are reduced by their respective mole fractions (N_A and N_B). The total pressure above a miscible or homogeneous solution will be dependent upon $P_A{}^{\circ}$ and $P_B{}^{\circ}$ as well as N_A and N_B. Thus, the composition of the vapor will also depend upon **both** the vapor pressures and mole fractions of each component.

$$\text{Immiscible Liquids} \qquad P_{total} = P_A{}^{\circ} + P_B{}^{\circ} \qquad (2)$$

On the other hand, when two mutually insoluble (immiscible) liquids are "mixed" to give a heterogeneous mixture, each will exert its own vapor pressure, independently of the other, as shown in equation (2). One notes that the mole fraction term does not appear in this equation because the compounds are not miscible. One simply adds the vapor pressures of the pure liquids ($P_A{}^{\circ}$ and $P_B{}^{\circ}$) at a given temperature to obtain the total pressure above the mixture. When the total pressure equals 760 mm, the mixture will boil. Unlike the case with miscible liquids, the composition of the vapor is determined **only** by the vapor pressures of the two substances co-distilling. Equation (3) defines the composition of the vapor from an immiscible mixture. Calculations involving this equation are given in Section 8.2.

$$\frac{\text{Moles } \mathbf{A}}{\text{Moles } \mathbf{B}} = \frac{P_A{}^{\circ}}{P_B{}^{\circ}} \qquad (3)$$

A mixture of two immiscible liquids will boil at a lower temperature than the boiling points of either of the components. The explanation for this behavior is similar to that given for minimum boiling azeotropes (Technique 7, Section 7.7). One may regard the behavior of immiscible liquids as being due to an extreme incompatability between the two liquids leading to higher combined vapor pressures than would be predicted by Raoult's law. The higher combined vapor pressures result in a lower boiling point for the mixture than for either single component. Thus, one may think of steam distillation as a special type of azeotropic distillation where the substance is completely insoluble in water.

The differences in the behavior of miscible and immiscible liquids, where $P_A{}^{\circ}$ equals $P_B{}^{\circ}$, are shown in Figure 8–1. One notes that with

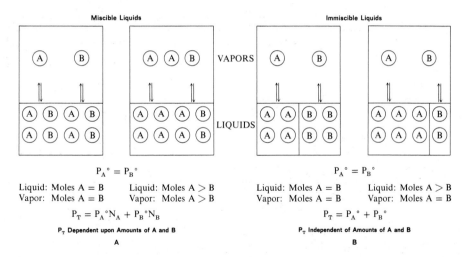

FIGURE 8-1. Total Pressure Behavior for Miscible and Immiscible Liquids. *A*, Ideal miscible liquids follow Raoult's Law: P_T is dependent upon the mole fractions and vapor pressures of A and B; *B*, Immiscible liquids do not follow Raoult's Law: P_T is dependent only upon the vapor pressures of A and B.

miscible liquids the composition of the vapor depends on the relative amounts of A and B present (Figure 8–1A). Thus, the composition of the vapor must change during the course of a distillation. In contrast, the composition of the vapor with immiscible liquids is independent of the amount of A and B present (Figure 8–1B). Hence, the vapor composition must remain **constant** during the distillation of such liquids as predicted by equation (3). Immiscible liquids act as if they were being distilled simultaneously from separate compartments, as shown in Figure 8–1B, even though in practice they are "mixed" together when conducting a steam distillation. Since all gases mix, they do give rise to a homogeneous vapor and co-distill.

8.2 IMMISCIBLE MIXTURES: CALCULATIONS

In the previous section it was stated that the composition of a distillate is constant during the course of a steam distillation. This would also require that the boiling point of the mixture be constant. The boiling point of the mixtures will be below the boiling points of water (100°) and the pure substance. Some representative boiling points and compositions of steam distillates are given in Table 8–1. Note that the higher the boiling point of a pure substance, the more closely the temperature of the steam distillate approaches, but does not exceed, 100°. A substance may be co-distilled with water at a temperature below 100°. This avoids the decomposition that might otherwise result at high temperatures with a simple distillation technique.

For immiscible liquids, the molar proportions of two components in a distillate are equal to the ratio of their vapor pressures in the boiling

TABLE 8-1. BOILING POINTS AND COMPOSITIONS OF STEAM DISTILLATES

MIXTURE	BOILING POINT OF PURE SUBSTANCE (°C)	BOILING POINT OF MIXTURE (°C)	COMPOSITION (% WATER)
Benzene-water	80.1	69.4	8.9%
Toluene-water	110.6	85.0	20.2%
Hexane-water	69.0	61.6	5.6%
Heptane-water	98.4	79.2	12.9%
Octane-water	125.7	89.6	25.5%
Nonane-water	150.8	95.0	39.8%
1-Octanol-water	195.0	99.4	90.0%

mixture, as given in equation (3). When the general equation (3) is rewritten for an immiscible mixture involving water, equation (4) results. Equation (4) can be modified by substituting the relationship, moles = (weight/molecular weight), to give equation (5).

$$\frac{\text{Moles Substance}}{\text{Moles water}} = \frac{P^{\circ}_{\text{substance}}}{P^{\circ}_{\text{water}}} \tag{4}$$

$$\frac{\text{Wt Substance}}{\text{Wt Water}} = \frac{(P^{\circ}_{\text{substance}})(\text{Molecular weight}_{\text{substance}})}{(P^{\circ}_{\text{water}})(\text{Molecular weight}_{\text{water}})} \tag{5}$$

For a sample calculation, consider the steam distillation of 1-octanol. From Table 8–1, one observes that the mixture boils at 99.4°. From a handbook, one obtains the vapor pressure of pure water at 99.4° (744 mm). Since P_{total} must equal 760 mm, the vapor pressure of 1-octanol at 99.4° must be equal to 16 mm. Equation (5) is now solved as follows:

$$\frac{\text{Wt 1-octanol}}{\text{Wt water}} = \frac{(16)(130)}{(744)(18)} = 0.155 \text{ grams/gram water}$$

Thus, 0.155 g of 1-octanol will co-distill with each gram of water. It will require 100 g of water to remove 15.5 g of 1-octanol from a distilling flask. The distillate will have the calculated composition of 87% water and 13% 1-octanol, which is very close to the experimental value given in Table 8–1.

8.3 STEAM DISTILLATION: METHODS

Two methods for steam distillation are in general use in the laboratory. The first method uses live steam, from a steam line, passed into the flask containing the compound. In the second method, steam is generated **in situ** by heating a flask containing the compound and water.

A. Live Steam Method

This method is the most widely used, especially with high molecular weight (low vapor pressure) substances. It may even be used with volatile solids. The apparatus is assembled as shown in Figure 8-2. A piece of 6 mm glass tubing is bent to fit the flask and serves as a steam inlet tube. The ground glass joints must be lubricated, and the apparatus must be fastened securely to a ring stand. During the course of the distillation, the joint between the condenser and the distilling head sometimes separates due to vibration. If this happens, vapor will be lost, unless the joint is immediately resealed. The distilling flask should never be filled more than one-half full. The flask should be positioned about 15 cm above the desk top. This distillation method does not usually require the use of an external heating source since the mixture is maintained at the boiling point with steam. Sometimes the distilling flask may begin to fill with water during the distillation. If this happens, it may be necessary to employ a heating source such as a Bunsen burner to heat the liquid to its boiling point. The heat from the burner should be dispersed by means of a wire gauze. The entering steam helps to prevent bumping of the mixture. The Claisen head helps to reduce the possibility of material being transferred to the receiving flask through excessive bumping.

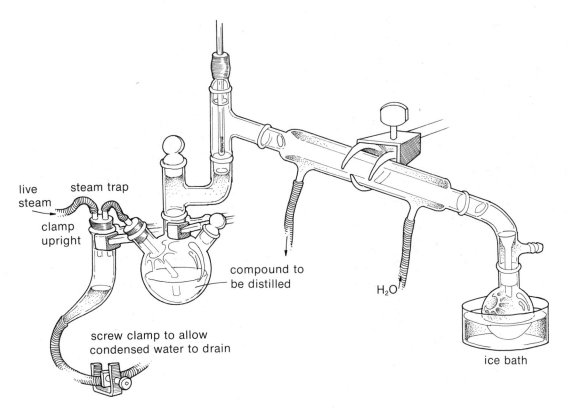

FIGURE 8-2. Steam distillation using live steam

As shown in the figure, a steam/water trap must be placed in the steam line. When the steam valve is first turned on, a considerable amount of water (condensate) exits from the valve. To prevent this condensate from entering the distilling flask and filling it with water, the screw clamp on the trap is **opened fully** to allow water to drain before reaching the flask. Once the water is drained, steam will begin to enter the distilling flask. The screw clamp on the trap may then be closed. Occasionally, the clamp should be opened to allow condensate to be removed during the course of the distillation. An ice bath can be used to increase the efficiency of the condensation of the distillate in the receiving flask. One should adjust the rate of flow of the steam so that vapor passes over into the condenser as quickly as possible but continues to be condensed. The flow of cooling water through the condenser should be faster than in other types of distillations to help cool the vapor. The vacuum adapter should be cool to the touch during the distillation. The condensate will be cloudy while the steam volatile material is distilling. The substances which co-distill will separate on cooling to give this cloudiness. Once the distillate is clear, the distillation is nearly complete. It is considered good practice to remove at least 10 ml more distillate after this point. When the distillation is to be stopped, the screw clamp

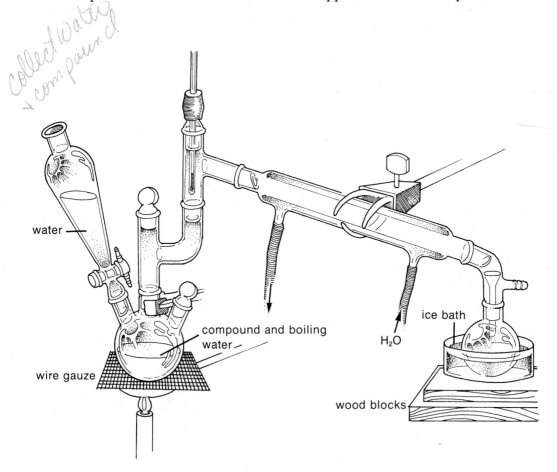

FIGURE 8–3. Direct steam distillation

on the steam trap must be opened completely and must be removed from the distilling flask. If this is will back up into the tube and steam trap.

B. Direct Method

This method is experimentally more convenient. Steam is produced in the distilling flask (in situ) by heating water to the boiling point in the presence of the compound to be distilled (Figure 8–3). As the steam is removed with the compound, water is added dropwise from the separatory funnel. This method works well for volatile liquids or for small amounts of material. It is mainly used for mixtures which do not have solids present. Solids may cause excessive bumping. In addition, foaming may be somewhat more troublesome. The live steam method often eliminates these two problems.

PROBLEMS

1. Calculate the weight of benzene co-distilled with each gram of water and the percentage composition of the vapor produced during a steam distillation. The boiling point of the mixture is 69.4°. The vapor pressure of water at 69.4° is 227.7 mm. Compare the result with the data in Table 8–1.

2. Calculate the approximate boiling point of a mixture of bromobenzene and water at atmospheric pressure. A table of vapor pressures of water and bromobenzene are given at various temperatures.

TEMPERATURE (°C)	VAPOR PRESSURES (mm)	
	Water	Bromobenzene
93	588	110
94	611	114
95	634	118
96	657	122
97	682	127
98	707	131
99	733	136

3. Calculate the weight of nitrobenzene which co-distills (bp 99°) with each gram of water during a steam distillation. You may need the data given in the previous problem.

4. A mixture of p-nitrophenol and o-nitrophenol can be separated by steam distillation. The o-nitrophenol is steam volatile while the para isomer is not volatile. Explain. Base your answer upon the ability of the isomers to form hydrogen bonds internally.

Technique **9**

MANOMETERS;
MEASURING PRESSURE

The atmosphere is a mixture of gases, primarily nitrogen (78.08%), oxygen (20.95%), argon (0.9%), and carbon dioxide (0.03%). There is also water vapor, the amount depending on weather conditions. In various localities, pollutants are present as well. This layer of gases is more dense near the surface of the earth than at higher elevations, due mainly to the influence of gravitational attraction. These gaseous constituents of air exert pressure on all objects within the earth's atmosphere. The pressure arises from the bombardment of objects by the atoms and molecules in the air.

9.1 THE BAROMETER

Atmospheric pressure is most often measured with a device known as a **barometer.** The barometer was invented by Torricelli in the 17th century. It is constructed by filling a glass tube, sealed at one end, with mercury and then inverting the tube in another container of mercury such that the open end is below the surface of the mercury in that container. When this is done, the level of the mercury in the sealed tube will fall due to the attraction of gravity exerted on the column of mercury. Recall that weight = mgh = (mass)(gravitational force constant)(height). The tube of mercury will not empty completely. At some point the weight of the descending column of mercury (its downward force) will be equaled by the force that the atmosphere exerts on the surface of the mercury in the container. This force is transmitted through the liquid and pushes upward on the descending column. Since the voided space at the top of the column contains no air, it is a vacuum, and the pressure is zero. The only downward force comes from the weight of the column of mercury. Hence, as shown in Figure 9–1, the atmospheric pressure supports the column of mercury, and the height of the column supported is directly proportional to the pressure exerted by the atmosphere.

The height of the column of mercury in a barometer will vary both with the elevation (it becomes less at higher elevation where air is less dense) and the prevailing weather conditions. Under normal conditions and at sea level, the pressure exerted by the atmospheric gases will support a column of mercury 760 mm high. Hence, 1 atmosphere is defined to correspond to 760 mm of mercury or 760 torr (1 mm Hg = 1 torr), a unit named in honor of Torricelli.

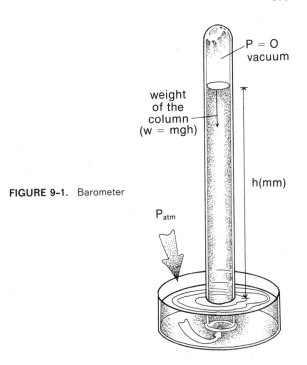

FIGURE 9-1. Barometer

P = O vacuum

weight
of the
column
(w = mgh)

h(mm)

P_{atm}

9.2 THE OPEN-END MANOMETER

To measure the pressure of a gas sample, one uses a **manometer.** There are two basic types of manometers, the **open-end** and the **closed-end** manometer. The closed-end manometer is described in Section 9.3.

An open-end manometer is basically a U-tube filled with mercury. One end is opened to the atmosphere (hence, open-end), and the other end is connected to a flask or bulb filled with a gas at a pressure higher or lower than that exerted by the atmosphere. In Figure 9–2, atmospheric pressure supports the column on the right-hand side of the U-tube, and the gas pressure in the bulb supports the column on the left-hand side. The column on the left side is higher than that on right by exactly the pressure difference between the sample of gas and the atmosphere: Δh (mm) = ($P_{gas} - P_{atm}$). The atmospheric pressure is obtained by reading a barometer. The gas pressure, P_{gas}, is then calculated as follows: $P_{gas} = P_{atm} + \Delta h$. Thus, the gas pressure in the bulb is **greater** than that of the atmosphere by the amount Δh. On the other hand, if the pressure in the bulb were **less** than the pressure of the atmosphere, then the right-hand column would be higher than the left. In this latter case, the gas pressure in the bulb is calculated as follows: $P_{gas} = P_{atm} - \Delta h$.

A simple type of open-end manometer is shown in Figure 9–3. It is simply a glass U-tube mounted on a board to which a meter stick is affixed for the measurement of Δh. Using such a device, one can measure the pressure of a gas sample which has pressure either greater than or less than atmospheric pressure by adding or subtracting, respectively, Δh from the known barometric pressure.

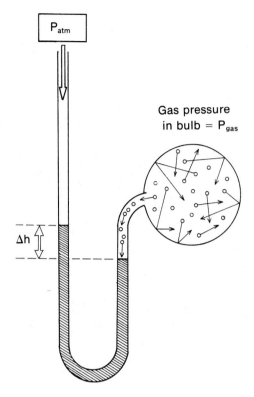

FIGURE 9-2. The open-end manometer

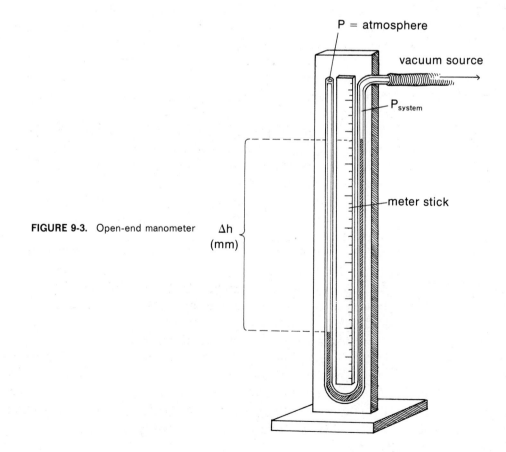

FIGURE 9-3. Open-end manometer

9.3 THE CLOSED-END MANOMETER

The other principal type of manometer is called the **closed-end manometer.** It can only be used to measure pressures which are **less** than atmospheric pressure. Two basic types are shown in Figure 9–4. The manometer shown in Figure 9–4A consists of a U-tube which is closed at one end and is mounted on a wooden support. This type of manometer is more difficult to construct than an open-end manometer because it must be filled under vacuum. One may construct the manometer from 9 mm capillary tubing, and fill it as shown in Figure 9–5. The U-tube is evacuated with a good vacuum pump, and then the mercury is introduced by tilting the mercury reservoir. When the vacuum is interrupted by admitting air, the mercury is forced by atmospheric pressure to the end of the evacuated U-tube. The manometer is then ready for use. The constriction shown in Figure 9–5 helps to protect the manometer against breakage when the pressure is released. When using an aspirator or any other vacuum source, a manometer may be connected into the system. As the pressure is lowered, the mercury will rise in the right-hand tube and will drop in the left-hand tube until Δh corresponds to the pressure of the system (see Figure 9–4A). A short piece of a metric ruler or a piece

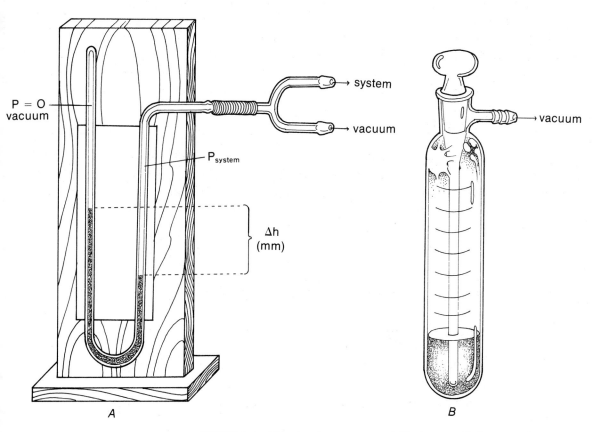

FIGURE 9–4. Closed-end manometers: *A*, U-tube type; *B*, Commercial type

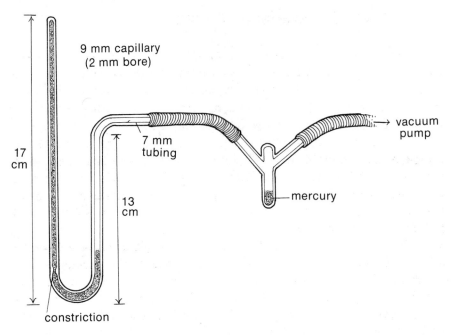

FIGURE 9-5. Constructing and filling the U-tube manometer

of graph paper ruled in millimeter squares is mounted on the support board to allow Δh to be read. No addition or subtraction is necessary since the reference pressure (created by the initial evacuation when filling) is zero. Hence, the height difference (Δh) gives the pressure in the system directly.

It should be noticed that a barometer must be at least 760 mm tall and that an open end manometer is often one meter tall. A closed-end manometer is often less than 20 cm long and is thus very convenient for laboratory use. These manometers are conveniently used with an aspirator which only rarely creates pressures lower than 10 to 20 mm. Since neither the open-end nor the closed-end manometer can be read to better than ±1 mm, they cannot be used for pressures lower than this, i.e., high vacuum. Other types of manometers must be used with high vacuum. The manometer commonly used with high vacuum systems is called a McLeod gauge. It will not be discussed in this text.

9.4 CONNECTING AND USING A MANOMETER

The most common use of a closed-end manometer is to monitor pressure during a reduced pressure distillation, as discussed in Technique 6, Sections 6.5, 6.6, and 6.7. The manometer is placed in a vacuum distillation system as shown in Figure 9–6. Generally an aspirator is used as the source of vacuum. Both the manometer and the distillation apparatus should be protected from possible back-ups in the water line by a trap. An alternate trap arrangement is shown in Technique 6,

Figure 6–7. The trap assembly shown in Figure 9–6 is more easily constructed. It may be noticed in the figure that the trap has a device for opening the system to the atmosphere. This is especially important when using a manometer since one must always make pressure changes in a slow fashion. If this is not done there is danger of spraying mercury throughout the system, breaking the manometer, or spurting mercury into the room. In the closed-end manometer, if the system is suddenly opened, the mercury will rush to the closed end of the U-tube with such speed and force that the end will often be broken out of the manometer. Air should be admitted **slowly** by opening the valve (Figure 9–6) on the trap cautiously. Similarly, when starting the vacuum, the valve should be closed slowly so as not to suck mercury suddenly out of the gauge. Rapid opening of an open-end manometer system will inevitably result in a fountain of mercury spraying into the room through the open end.

When conducting a reduced pressure distillation, if the pressure is lower than desired, it is possible to adjust it by means of a **bleed** valve. One removes the screw clamp on the valve shown in Figure 9–6 and

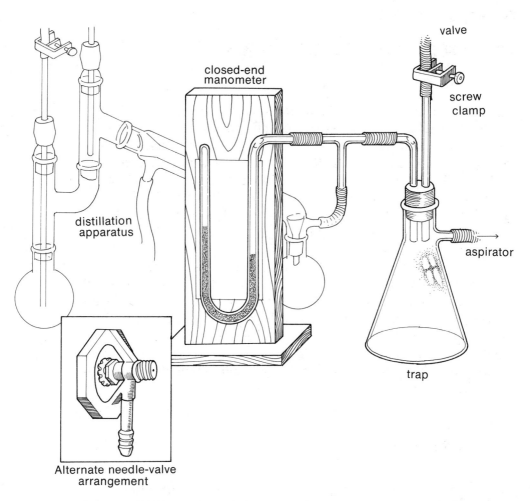

FIGURE 9–6. Method of connecting the manometer to a vacuum distillation system

connects the base of a Tirrill style Bunsen burner to the water trap as shown in the alternate arrangement. The needle valve in the burner can be used to adjust precisely the amount of air bleeding into the system and, hence, the pressure. When it is opened fully, atmospheric pressure is admitted; conversely, when it is fully closed, the maximum vacuum is achieved.

The boiling point of a liquid is a function of the applied pressure. The boiling point is often reported at atmospheric pressure or some other pressure which cannot be exactly reproduced by use of the aspirator. By the use of the nomograph chart in Technique 6 (Figure 6–2), it is possible to determine the boiling point of a liquid at any given pressure if it is known at least at one other pressure.

Technique **10**

COLUMN CHROMATOGRAPHY

The most modern and sophisticated methods of separating mixtures which are available to the organic chemist all involve some sort of **chromatography.** Chromatography is defined as the separation of a mixture of two or more different compounds (in some cases, ions) by distribution between two phases, one of which is stationary and one of which is moving. Various types of chromatography are possible, depending on the nature of the two phases involved; **solid-liquid** (column, thin-layer, and paper), **liquid-liquid,** and **gas-liquid** (vapor phase) chromatographic methods are commonly available.

All chromatography works on much the same principle as solvent extraction (Technique 5). Basically, the methods depend on differential solubilities (or absorptivities) of the substances to be separated with respect to the two phases between which they are to be partitioned. In this technique section, column chromatography—a solid-liquid method —will be discussed. Thin layer chromatography will be discussed in Technique 11. Gas chromatography, a gas-liquid method, will be discussed in Technique 12.

10.1 ADSORBENTS

Column chromatography is a technique based on both adsorptivity and solubility. It is a solid-liquid phase partitioning technique. The solid may be almost any material that will not dissolve in the associated liquid phase, but those solids most commonly used are silica gel ($SiO_2 \cdot xH_2O$), also called silicic acid, and alumina ($Al_2O_3 \cdot xH_2O$). These compounds are used in their powdered or finely ground (usually 200 to 400 mesh) forms.

Most alumina used for chromatography is prepared from the impure ore bauxite ($Al_2O_3 \cdot xH_2O + Fe_2O_3$). The bauxite is dissolved in hot sodium hydroxide and filtered to remove the insoluble iron oxides; the alumina in the ore forms the soluble amphoteric hydroxide $Al(OH)_4^-$. The hydroxide is precipitated by CO_2 (which reduces the pH) as $Al(OH)_3$. When heated, the $Al(OH)_3$ loses water to form pure alumina, Al_2O_3.

$$\text{bauxite (crude)} \xrightarrow{\text{hot NaOH}} Al(OH)_4^-(aq) + Fe_2O_3 \text{ (insoluble)}$$

$$Al(OH)_4^-(aq) + CO_2 \longrightarrow Al(OH)_3 + HCO_3^-$$

$$2\, Al(OH)_3 \xrightarrow{\Delta} Al_2O_3 \text{ (s)} + 3\, H_2O$$

Alumina prepared in this way is called **basic alumina** because it still contains some hydroxides. Basic alumina cannot be used for chromatography of compounds that are base-sensitive. Therefore, it is washed with acid to neutralize the base, giving **acid-washed alumina.** This material is unsatisfactory unless it has been washed with enough water to remove **all** the acid, thus giving the best chromatographic material, called **neutral alumina.** If a compound is acid-sensitive, either basic or neutral alumina must be used. One should be careful to ascertain what type of alumina is being used for chromatography. Silica gel is not available in any form other than that suitable for chromatography.

10.2 INTERACTIONS

If powdered or finely-ground alumina (or silica gel) is added to a solution containing an organic compound, some of the organic compound will **adsorb** onto or stick to the fine particles of alumina. Many kinds of intermolecular forces cause organic molecules to bind to alumina. These forces vary in strength according to their type. Non-polar compounds bind to the alumina utilizing only Van der Waals forces. These are weak forces, and non-polar molecules do not bind strongly unless they have extremely high molecular weights. The more important interactions are those utilized by polar organic compounds. These forces are either of the dipole-dipole type or they involve some sort of direct interaction (coordination, hydrogen bonding, or salt formation). These types of interactions are illustrated in Figure 10–1 where, for convenience, only a portion of the alumina structure is shown.

Similar types of interactions occur on silica gel. The strengths of such interactions vary in the approximate order:

salt formation > coordination > hydrogen bonding >
dipole-dipole > Van der Waals

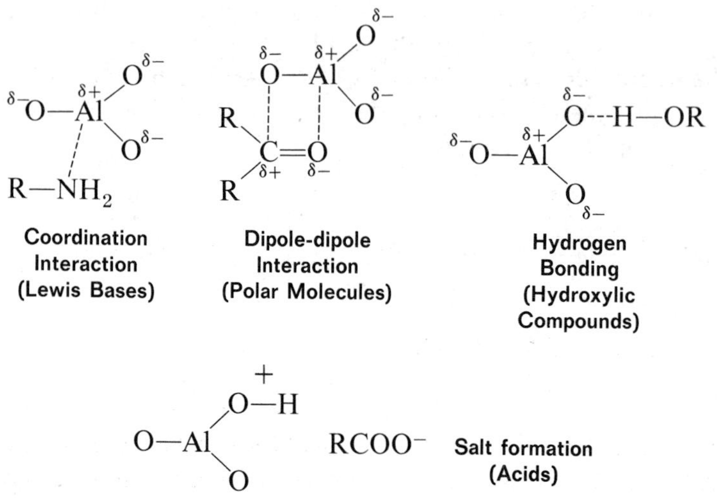

FIGURE 10–1. Possible interactions of organic compounds with alumina

Of course, the strength of interaction varies for any given compound. For instance, a strongly basic amine would bind more strongly than a weakly basic one (by coordination). In fact, strong bases and strong acids often interact so strongly that they **dissolve** alumina to some extent. One can use the following rule of thumb: THE MORE POLAR THE FUNCTIONAL GROUP, THE MORE STRONGLY IT WILL BIND TO ALUMINA (OR SILICA GEL).

A similar rule holds for solubility. Polar solvents dissolve polar compounds more effectively than non-polar solvents; non-polar compounds are dissolved best by non-polar solvents. Thus, the extent to which any given solvent has the ability to wash an adsorbed compound from alumina depends almost directly on the solvent's relative polarity. For example, while a ketone adsorbed on alumina might not be removed by hexane, it might be removed completely by chloroform. For any adsorbed material a kind of **distribution** equilibrium can be envisioned between the adsorbent material and the solvent. This is illustrated in Figure 10–2.

The distribution equilibrium is a situation of **dynamic** equilibrium, with molecules constantly **adsorbing** from and **desorbing** into the solution. The average number of molecules remaining adsorbed on the particle at equilibrium will depend both on the particular molecule (RX) involved and the dissolving power of the solvent with which the adsorbent must compete.

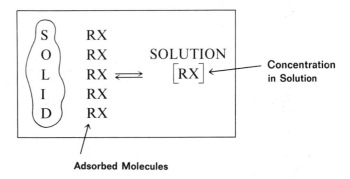

Adsorbed Molecules

FIGURE 10–2. Dynamic adsorption equilibrium

10.3 THE PRINCIPLE OF THE COLUMN CHROMATOGRAPHIC SEPARATION METHOD

The dynamic equilibrium mentioned above, and the fact that the different compounds adsorb on alumina (or silica gel) to different extents, can be used in a versatile and ingenious method to **separate** mixtures of organic compounds. In this method, the mixture of compounds to be separated is introduced onto the top of a cylindrical glass column (Figure 10–3) **packed** or filled with fine alumina particles (stationary solid phase). The adsorbent is then continuously washed by a flow of solvent (moving phase) passing through the column.

Initially the components of the mixture adsorb onto the alumina particles at the top of the column. The continuous flow of solvent through the column **elutes** or washes the solutes off the alumina and sweeps them down the column. The solutes (or materials to be separated) are called **elutants** and the solvents used are called **eluents.** As the solutes pass down the column to fresh alumina, new equilibria are established between the adsorbent, the solutes, and the solvent. The constant equilibration means that different compounds will move down the column at differing rates depending on their relative affinity for the adsorbent on one hand and for the solvent on the other. Since the number of alumina particles is large, since they are closely packed, and since fresh solvent is being added continuously, the number of equilibrations between adsorbent and solvent which the solutes experience is enormous.

As the components of the mixture are separated they begin to form moving bands (or zones), each band containing a single (hopefully!) component. If the column is long enough and the various other parameters (column diameter, adsorbent, solvent, and rate of flow) are correctly chosen, the bands will separate from one another leaving gaps of pure solvent in between them. As each band (solvent and solute) passes out the bottom of the column, it can be collected completely before the

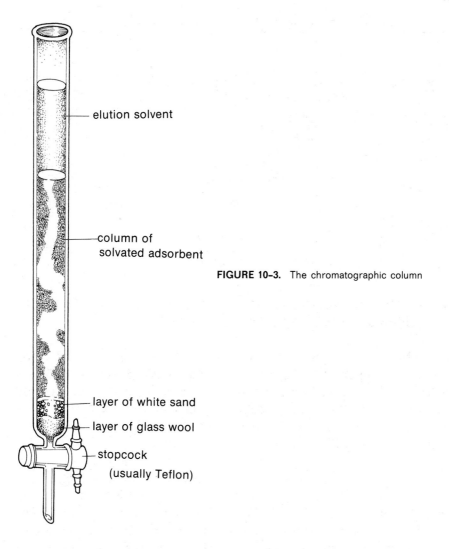

elution solvent

column of
solvated adsorbent

FIGURE 10–3. The chromatographic column

layer of white sand

layer of glass wool

stopcock
(usually Teflon)

next band arrives. If the parameters mentioned above are poorly chosen, the various bands will either overlap or coincide, in which case either a poor separation or no separation at all will result. The chromatographic separation process is illustrated in Figure 10–4.

10.4 PARAMETERS AFFECTING SEPARATION

Chromatography is truly a sophisticated method of separating mixtures. Its versatility results from the many factors which can be varied. These include:

1. the adsorbent chosen
2. the polarity of the solvent(s) chosen
3. the size of the column (both length and diameter) relative to the amount of material to be chromatographed
4. the rate of elution (or flow).

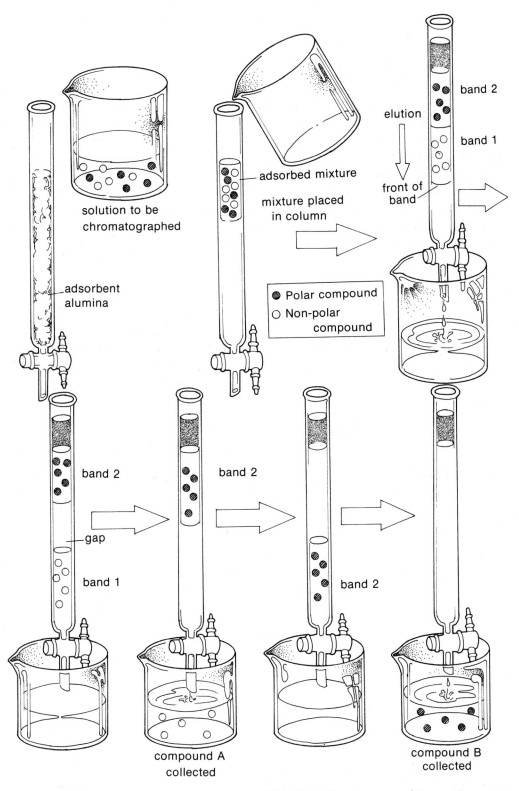

solution to be
chromatographed

adsorbent
alumina

adsorbed mixture

mixture placed
in column

● Polar compound
○ Non-polar
    compound

elution

front of
band

band 2

band 1

band 2

gap

band 1

band 2

compound A
collected

band 2

compound B
collected

FIGURE 10–4. The chromatographic separation process

By careful choice of conditions almost any mixture can be separated. Recently the technique has even been used to separate optical isomers. An optically active solid phase adsorbent was used to separate the enantiomers.

Two fundamental choices to be made by anyone attempting a chromatographic separation are the choice of the adsorbent and the choice of the solvent system. In general, non-polar compounds will pass through the column at a faster rate than polar compounds since they have a smaller affinity for the adsorbent. If the adsorbent chosen binds all of the solute molecules (both polar and non-polar) strongly, they will not move down the column. On the other hand, if too polar a solvent is chosen, all the solutes (polar and non-polar) may simply be washed through the column without any separation being achieved. The adsorbent and the solvent should be chosen so that neither is excessively favored in the equilibrium competition for solute molecules.[1]

A. Adsorbents

In Table 10–1 various kinds of adsorbents (solid phases) used in column chromatography are listed. The choice of adsorbent will often depend on the types of compounds to be separated. Cellulose, starch, and sugars are used for polyfunctional plant and animal materials (natural products) very sensitive to acid-base interactions. Magnesium silicate is often used for the separation of acetylated sugars, steroids, and essential oils. Silica gel and Florisil are relatively mild toward most compounds and are widely used for a variety of functional groups

---

[1] Often the chemist uses thin layer chromatography (tlc), which is described in Technique 11, to arrive at the best choices of solvents and adsorbents for optimum separation. The tlc experimentation can be done quickly and with extremely small amounts (microgram quantities) of the mixture to be separated. This results in a great savings of time and materials. Technique 11 will have to be read to acquire the details of this use of tlc.

**TABLE 10–1. SOLID ADSORBENTS FOR
COLUMN CHROMATOGRAPHY**

Paper	
Cellulose	
Starch	
Sugars	
Magnesium Silicate	**INCREASING STRENGTH OF**
Calcium Sulfate	**BINDING INTERACTIONS**
Silicic Acid	**TOWARD POLAR COMPOUNDS**
Silica Gel	
Florisil	
Magnesium Oxide	
Aluminum Oxide (Alumina)*	
Activated Charcoal (Norit)	

*Basic, Acid-Washed, Neutral

—hydrocarbons, alcohols, ketones, esters, acids, azo compounds, amines. Alumina is the most widely used adsorbent and is obtained in the three forms mentioned in Section 10.1: acidic, basic, and neutral. The pH of acidic or acid-washed alumina is approximately 4. This adsorbent is particularly useful for separation of acidic materials such as carboxylic acids and amino acids. Basic alumina has a pH of 10 and is useful in separating amines. Neutral alumina can be used to separate a variety of non-acidic and non-basic materials.

The approximate strength of the various adsorbents listed in Table 10–1 is also given. It should be noted especially that the order is approximate and may vary. For instance, the strength or separating abilities of alumina and silica gel depend to a large extent on the amount of water present. Water binds very tightly to either adsorbent taking up sites on the particles that could otherwise be used for equilibration. If one adds water to the adsorbent it is said to have been **deactivated.** Anhydrous alumina or silica gel are said to be highly **activated.** High activity is usually avoided in these adsorbents since use of the more active forms may lead to certain types of destruction or decomposition of the compounds to be separated. Use of the highly active forms of either alumina or silica gel, or of the acidic or basic forms of alumina, can often lead to molecular rearrangement in certain types of solute compounds.

B. Solvents

In Table 10–2 some common chromatographic solvents are listed along with their relative ability to dissolve polar compounds. Sometimes a single solvent can be found which will separate all of the components of a mixture. Sometimes a mixture of solvents can be found that will achieve separation. More often, one must start elution with a non-polar

TABLE 10–2. SOLVENTS (ELUENTS) FOR CHROMATOGRAPHY

Petroleum ether	
Cyclohexane	
Carbon tetrachloride	
Benzene	
Methylene chloride	**INCREASING POLARITY**
Chloroform	**AND "SOLVENT POWER"**
Diethyl ether	**TOWARD POLAR FUNCTIONAL**
Ethyl acetate	**GROUPS**
Pyridine	
Acetone	
Ethanol	
Methanol	
Water	
Acetic acid	↓

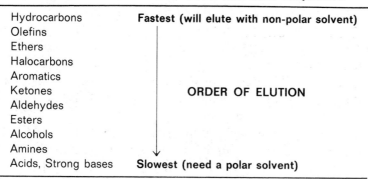

TABLE 10-3. COMPOUND TYPE ELUTION SEQUENCE

Hydrocarbons	**Fastest (will elute with non-polar solvent)**
Olefins	
Ethers	
Halocarbons	
Aromatics	
Ketones	**ORDER OF ELUTION**
Aldehydes	
Esters	
Alcohols	
Amines	
Acids, Strong bases	**Slowest (need a polar solvent)**

solvent to remove relatively non-polar compounds from the column and then gradually increase the solvent polarity to force compounds of greater polarity to come down the column or elute. The approximate order in which various classes of compounds elute when utilizing this procedure is given in Table 10-3. In general, non-polar compounds travel through the column fastest (elute first) and polar compounds travel more slowly (elute last). However, molecular weight is also a factor in determining the order of elution. A non-polar compound of high molecular weight will travel more slowly than a non-polar compound of low molecular weight and may even be passed by some polar compounds.

When trying to change the polarity of the solvent during the course of a chromatographic separation, some precautions must be taken. Rapid changes from one solvent to another are to be avoided (especially when using silica gel or alumina). Usually, small percentages of a new solvent are mixed slowly into the one in use until the percentage reaches the desired level. If this is not done the column packing will often "crack." This occurs because heat is liberated when alumina or silica gel is mixed with a solvent. The solvent solvates the adsorbent and the formation of a weak bond generates heat.

$$\text{Solvent} + \text{Alumina} \longrightarrow (\text{Alumina} \cdot \text{Solvent}) + \text{Heat}$$

Often enough heat is generated locally to evaporate the solvent. The formation of vapor creates bubbles which force a separation of the column packing called "cracking." A "cracked column" does not give a good separation since it has discontinuities in the **packing** (column of adsorbent). In fact, the way in which a column is packed or filled with adsorbent is quite important and is discussed in the next section.

That the solvent itself has a tendency to adsorb on the alumina is an important factor in the movement of compounds down the column. The solvent can displace the adsorbed compound if it is more polar than the compound and, hence, can move it down the column. Thus, a more polar solvent not only dissolves more compound, but is effective in removing the compound from the alumina since it displaces the compound from its site of adsorption.

Certain solvents should be avoided when using alumina or silica gel, especially the acidic, basic, and high activity forms. For instance, with any of these adsorbents, acetone will dimerize via an aldol condensation to give diacetone alcohol. Mixtures of esters will transesterify (exchange their alcoholic portions) when ethyl acetate or alcohols are used as eluents. And finally, the more active solvents (pyridine, methanol, water, and acetic acid) will dissolve and elute some of the adsorbent itself. Generally, try to avoid going to more polar solvents than diethyl ether or chloroform in the eluent series (Table 10–2).

C. Column Size And Adsorbent Quantity

The correct column size and the correct amount of adsorbent must also be selected to achieve a good separation for a given amount of sample. As a rule of thumb, the amount of adsorbent should be from 25 to 30 times as large, by weight, as the amount of material to be separated by chromatography. Additionally, the column used should have a height to diameter ratio of about 8:1. Some typical relationships of this sort are given in Table 10–4.

It should be mentioned as a cautionary note that the size and length of the column, as well as the amount of adsorbent required, will also depend on the **difficulty** of the separation. Compounds which do not separate easily may require larger columns and more adsorbent than is specified in Table 10–4. For easily separated compounds, reduced column size and less adsorbent may suffice.

D. Flow Rate

The rate at which solvent flows through the column also plays a role in determining how good a separation will be. In general, the longer the mixture to be separated remains on the column the more extensive the equilibration between stationary and moving phases will be. As a result, rather similar compounds will eventually separate if they remain on the

TABLE 10–4. SIZE OF COLUMN AND AMOUNT OF ADSORBENT FOR TYPICAL SAMPLE SIZES

AMOUNT OF SAMPLE	AMOUNT OF ADSORBENT	COLUMN DIAMETER	COLUMN HEIGHT
0.01 g	0.3 g	3.5 mm	30 mm
0.10	3.0	7.5	60
1.00	30.0	16.0	130
10.00	300.0	35.0	280

column long enough. The time a material remains on the column depends upon the flow rate of the solvent. However, if the flow rate is too slow, the rate of diffusion of the substances in the mixture, when they are in the solvent, may become greater than the rate at which the substances move down the column. In this case, the bands will grow wider and more diffuse, and the separation will become poorer.

10.5 PACKING THE COLUMN: TYPICAL PROBLEMS

Perhaps the most critical operation in column chromatography is the **packing** (filling with adsorbent) of the column. The column of alumina (or other solid adsorbent), called the column packing, must be evenly packed and free of irregularities, air bubbles, and gaps. As a compound travels down the column it will do so in an advancing zone or **band.** It is important that the leading edge or **front** of this band be horizontal or perpendicular to the long axis of the column. If two bands are close together and do not have horizontal band fronts it will be impossible to collect each band to the exclusion of the other. The leading edge of the second band will begin to elute before the first band has finished eluting. This can be seen from Figure 10–5. Two major factors cause this problem. First, if the top surface edge of the adsorbent packing is not level, non-horizontal bands will result. Second, non-horizontal bands may also result if the column is not held in an exactly vertical position in both planes (front-to-back and side-to-side). When

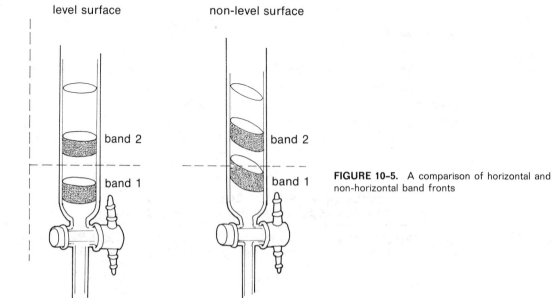

level surface   non-level surface

band 2   band 2

band 1   band 1

FIGURE 10–5. A comparison of horizontal and non-horizontal band fronts

Horizontal bands
good separation

Non-horizontal bands
bad separation

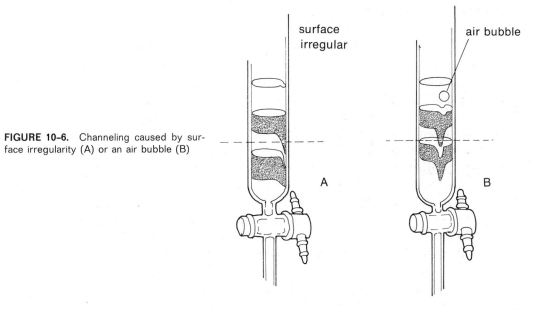

FIGURE 10–6. Channeling caused by surface irregularity (A) or an air bubble (B)

preparing a column for use, both of these factors must be watched carefully.

Another phenomenon called **streaming** or **channeling** occurs when part of the band front advances ahead of the major part of the band. This can occur if there are any irregularities in the adsorbent surface or if there are any irregularities or air bubbles in the packing. Often large channels or air gaps form in the column packing, and a part of the advancing front gets ahead of the rest by use of the channel. Two examples of channeling are illustrated in Figure 10–6.

10.6 PACKING THE COLUMN: METHODS

The following methods are used to avoid the problems resulting from uneven packing and column irregularities. These procedures should be followed carefully when preparing a chromatographic column. If one fails to pay close attention to the preparation of the column, it may well affect the quality of the separation (or non-separation) that is achieved.

Preparation of a column can be divided into two distinct stages. In the first stage a support base on which the packing will rest is prepared. This must be done so that the packing, a finely divided material, will not wash out of the bottom of the column. In the second stage, the column of adsorbent is deposited on top of the supporting base. There are several alternative methods for the second process, and these are detailed below.

A. *Preparing The Support Base*

First, the chromatographic column is clamped upright in a vertical position. The column (Figure 10–3) is a piece of cylindrical glass tubing with a stopcock attached at one end. The stopcock usually has a Teflon plug since stopcock grease (used on glass plugs) will dissolve in many of the organic solvents used as eluents. The admixture of stopcock grease in the eluent will contaminate the elutants.

Instead of a stopcock, a piece of flexible tubing is often attached to the bottom of the column, and a screw clamp is used to stop or regulate the flow (Figure 10–7). When this alternate arrangement is used, care must be taken that the tubing used will not be dissolved by the chromatographic solvents used to achieve the separation. Rubber, for instance, will dissolve in chloroform, benzene, or tetrahydrofuran (THF). Tygon tubing will dissolve (actually, the plasticizer is removed) in a wide variety of solvents including benzene, methylene chloride, chloroform, ether, ethyl acetate, and THF. Polyethylene tubing is the best choice since it is inert to most solvents.

Next, the column is partially filled with a quantity of solvent, usually a non-polar solvent like hexane, and a support for the finely divided adsorbent is prepared in the following way. A loose plug of glass wool is tamped down into the bottom of the column with a long glass rod until all entrapped air is forced out as bubbles. Care should be taken not to totally plug the column by tamping the glass wool too hard. A small layer of clean, white sand is formed on top of the glass wool by pouring sand into the column. The column is tapped to level the surface of the sand. Any sand adhering to the side of the column is washed

FIGURE 10–7. Tubing with screw clamp to regulate solvent flow on a chromatography column

down with a small quantity of solvent. The sand forms a base which will support the column of adsorbent and prevent it from washing through the stopcock. Then the column is best packed in one of two ways: the "slurry method" or the "dry pack method."

B. Depositing The Adsorbent:

The Slurry Method

In this method the adsorbent is packed into the column as a **slurry.** A slurry is a mixture of a solvent and an undissolved solid. The slurry is prepared in a separate container by adding the solid adsorbent, a little at a time, to a quantity of the solvent. This order of addition (adsorbent to solvent) should be followed strictly since the adsorbent will solvate and liberate heat. If the solvent is added to the adsorbent, it may boil away almost as fast as it is added due to heat evolved, especially if ether or another low boiling solvent is used. When this happens the final mixture will be uneven and lumpy. Enough adsorbent is added to the solvent, with swirling, to form a thick, but flowing, slurry. The slurry should be swirled until it is homogeneous and relatively free of entrapped air bubbles.

When the slurry has been prepared, the column is filled about half full with solvent, and the stopcock is opened to allow solvent to drain slowly into a large beaker. Alternately, the slurry is mixed by swirling and is poured in portions into the top of the draining column (a wide-neck funnel may be useful here). The column is tapped constantly and **gently** on the side, during the pouring operation, with a pencil fitted with a rubber stopper. A short piece of large diameter pressure tubing may also be used for tapping. The tapping promotes even settling and mixing and gives an evenly packed column free of air bubbles. Tapping is continued until all the material has settled to give a well defined level at the top of the column. Solvent from the collecting beaker may be re-added to the slurry if it becomes too thick to be poured into the column at one time. In fact, the collected solvent should be recycled through the column several times to assure that settling is complete and that the column is firmly packed. The downward flow of solvent has a tendency to compact the adsorbent.

The Dry Pack Method

In this method, the column is filled with solvent and allowed to drain **slowly.** The dry adsorbent is added, a little at a time, from a beaker while the column is tapped constantly as described above. When the column has the desired length no more adsorbent is added. This method also gives an evenly packed column. For the same reasons as described above solvent should be recycled through this column several times before each use.

In another dry pack method, the entire column is packed dry, i.e., without any solvent. Then the solvent is allowed to percolate down the column at a slow rate until the column is entirely moistened. **This method is not recommended for use with silica gel or alumina** since it leads to uneven packing, air bubbles, and cracking with these adsorbents, especially if a solvent which has a highly exothermic process of solvation is used.

10.7 APPLYING THE SAMPLE TO THE COLUMN

The solvent (or solvent mixture) used to pack the column is normally the least polar elution solvent one intends to use during the chromatographic process. The compounds to be chromatographed will, of course, not be infinitely soluble in this solvent. If they were, they would probably have a greater affinity for the solvent than for the adsorbent and pass right on through the column without equilibrating. Since it would require a large amount of the initial chromatographic solvent to dissolve the compound in this case, it would be difficult to get the mixture to form a narrow band on top of the column. Ideally a narrow band is desired to give an optimum separation of components. To achieve this, the compound may be applied to the top of the column, undiluted if it is a liquid, or in a very small amount of a highly polar solvent if it is a solid.

In adding the sample to the column one adheres to the following procedure. The solvent level is lowered to the top of the adsorbent column by draining. The liquid (diluted or neat) or dissolved solid is added (usually with a small pipet) to form a small layer on top of the adsorbent. Care is taken not to disturb the surface. This is done best by touching the pipet to the inside of the column and slowly draining it so as to allow the sample to spread into a thin film which slowly descends to cover the entire adsorbent surface. The pipet is drained close to the surface of the adsorbent. When all the sample has been added, this small layer of liquid is drained into the column until the top surface of the column **just begins** to dry. A small layer of the chromatographic solvent is then added carefully with a pipet, again taking care not to disturb the surface. This small layer of solvent is drained into the column until the column just dries. Another small layer of fresh solvent is added, if necessary, and the process is repeated until it is clear that the sample is strongly adsorbed on the top of the column. If the sample is colored and the fresh layer of solvent acquires this color, the sample has not been properly adsorbed. Once the sample has been properly applied, the level surface of the adsorbent may be protected by carefully filling the top of the column with solvent and sprinkling clean, white sand into the column so as to form a small protective layer on top of the adsorbent.

Better separations are often achieved if the sample is allowed to stand a short time on the column before elution is started. This allows a true equilibrium to be established. It should be noted, however, that in columns which stand for a long time, the adsorbent can often compact or even swell, and the flow rate can become annoyingly slow. Diffusion of the sample to widen the bands also becomes a problem if a column is allowed to stand over an extended period.

10.8 ELUTION TECHNIQUES

Solvents used for analytical and preparative chromatography should be pure reagents. Commercial grade solvents often contain small amounts of residue which remain when the solvent is evaporated. For normal work and for relatively easy separations where only small amounts of solvent are used, the residue usually presents very few problems. When using commercial grade solvents, it may be necessary to redistill them before use. This is especially true for hydrocarbon solvents, which tend to have more residue than other solvent types. It should be noted that most of the experiments in this laboratory manual have been designed to avoid this particular problem.

One usually begins elution of the products with a non-polar solvent like hexane or petroleum ether. The polarity of the elution solvent can be increased gradually by addition of successively greater percentages of either ether or benzene (for instance 1%, 2%, 5%, 10%, 15%, 25%, 50%, 100%) or some other solvent of greater solvent power (polarity) than hexane. The transition from one solvent to another should not be too rapid in most solvent changes. If the two solvents to be changed differ greatly in their heats of solvation in binding to the adsorbent, enough heat can be generated to crack the column. Ether is especially bad in this respect as it has both a low boiling point and a relatively high heat of solvation. Most organic compounds can be separated on silica gel or alumina using hexane-ether or hexane-benzene combinations for elution, and following these by pure chloroform. Solvents of greater polarity are usually avoided for the various reasons mentioned above.

The rate of flow of solvent through the column should not be too rapid or the solutes will not have time to equilibrate with the adsorbent as they pass down the column. If the rate of flow is too slow or stopped over a period of time, diffusion can become a problem—the solute band will diffuse or spread out in all directions. In either of these cases, a poor separation will result. As a general (and very approximate) rule, most columns are run with flow rates ranging from five to fifty drops of effluent per minute. A steady flow of solvent is usually avoided. To avoid diffusion of the bands, do not stop the column or set it aside overnight.

10.9 RESERVOIRS

When large quantities of solvent are used in a chromatographic separation, it is often convenient to use a solvent reservoir to forestall the necessity of continuously adding small portions of fresh solvent. The simplest type of reservoir, employed in many columns, is created by fusing the top of the column to a round bottom flask (Figure 10–8A). If the column has a standard taper joint at its top, a reservoir may be created by joining a standard taper separatory funnel to the column (Figure 10–8B). In this arrangement the stopcock is left open, and no stopper is placed in the top of the separatory funnel. A third common arrangement is shown in Figure 10–8C. A separatory funnel is filled with solvent; its stopper is wetted with solvent and put **firmly** in place. The funnel is inserted into the empty filling space at the top of the chromatographic column, and the stopcock is opened. Solvent will flow out of the funnel filling this space until the solvent level is well above the outlet at the end of the funnel's stem. As solvent leaves the capped separatory funnel, a partial vacuum is drawn in the funnel, and solvent cannot drain unless air is admitted. Air cannot be admitted until the solvent level drops below the opening in the funnel's stem. At that point air can enter the funnel, and a quantity of solvent will be discharged so that the solvent level in the column will rise to once again cover the

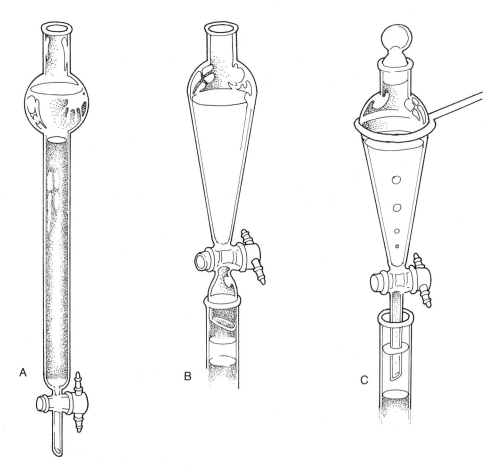

FIGURE 10–8. Various types of solvent reservoir arrangements for chromatographic columns

outlet. Air cannot enter the funnel again until the solvent level in the column drops. The column is thus self-filling. For this arrangement, the funnel should have a **long** stem, with a correspondingly long length of the column unfilled to accommodate the stem and solvent discharges. Generally, the funnel will discharge a constant, but fairly large, solvent volume each time the column is filled. The available volume at the top of the column must be large enough that the column will not overflow each time it is filled.

10.10 MONITORING THE COLUMN

It is a happy instance when the compounds to be separated are colored. The separation can then be followed visually and the various bands collected separately as they elute from the column. However, for the majority of organic compounds, this lucky circumstance does not occur, and other methods must be used to determine the positions of the various bands. The most common method of following a separation of uncolored compounds is to collect **fractions** of constant volume in pre-weighed flasks, to evaporate the solvent from each fraction, and to reweigh the flask plus any residue. A plot of fraction number versus the weight of the residues after evaporation of solvent gives a plot like that shown in Figure 10–9. Clearly, fractions 2 through 7 (Peak 1), may be combined as a single compound, as can fractions 8 through 11 (Peak 2) and 12 through 15 (Peak 3). The size of the fractions collected (10 ml, 100 ml, or 500 ml) will depend on the size of the column and the ease of separation.

Another common method of monitoring the column is to mix an inorganic phosphor into the adsorbent used to pack the column. When the column is illuminated with an ultraviolet light, the adsorbent treated in this way will fluoresce. However, many solutes have the ability to

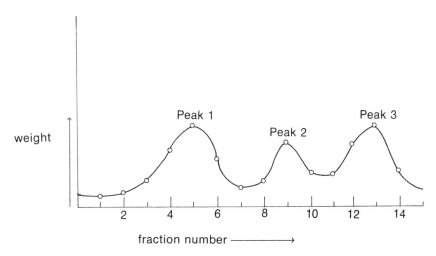

FIGURE 10–9. A typical elution graph

quench the fluorescence of the indicator phosphor. In those areas where solutes are present the adsorbent will not fluoresce, and a dark band will be seen. Thus, in this type of column, the separation may also be followed visually.

Thin layer chromatography is often used to monitor a column. This method is described in Technique 11 (Section 11.9). Several sophisticated instrumental and spectroscopic methods can also be used to monitor a chromatographic separation. These, however, will not be discussed here.

10.11 TAILING

Often when a single solvent is used for elution, an elution curve (weight vs. fraction) similar to that shown as a solid line in Figure 10–10 will be observed. An ideal elution curve is shown using dashed lines. In the non-ideal curve, the compound is said to be **tailing.** Tailing can interfere with the beginning of a curve or peak of a second component and lead to a poor separation. One way to avoid this is to constantly increase the polarity of solvent while eluting. In this way, at the tail of the peak where the solvent polarity is increasing, the compound will move slightly faster than at the front and allow the tail to squeeze forward, forming a more nearly ideal band.

10.12 RECOVERING THE SEPARATED COMPOUNDS

To recover each of the separated compounds of a chromatographic separation when they are solids, the various correct fractions are combined, evaporated, and recrystallized. If the compounds are liquids, the correct fractions are combined, evaporated, and distilled. The combination of chromatography-crystallization or chromatography-distillation usually yields very pure compounds.

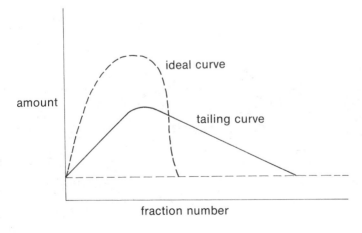

FIGURE 10–10. An ideal elution curve and one which "tails"

Technique **11**

THIN LAYER CHROMATOGRAPHY

Thin layer chromatography (tlc) is a very important technique for the rapid separation and qualitative analysis of small amounts of material. The technique is closely related to column chromatography. In fact, tlc can most simply be considered as column chromatography **in reverse,** with the solvent ascending the adsorbent rather than descending. Because of this close relationship to column chromatography, and because the principles governing the two techniques are similar, one should have read Technique 10 (Column Chromatography) prior to reading this one.

11.1 PRINCIPLES OF THE METHOD

Like column chromatography, thin layer chromatography is a solid-liquid partitioning technique. However, rather than allowing the moving liquid phase to percolate down the adsorbent, it is allowed to **ascend** a thin layer of adsorbent coated onto a backing support. The most typical backing support is a glass plate, but other materials are also used. A thin layer of the adsorbent is spread onto the plate and allowed to dry. A coated and dried plate of glass is called a **thin layer plate** or a **thin layer slide.** (The latter terminology comes about because microscope slides are often used to prepare small thin layer plates.) When a thin layer plate is placed upright in a container which contains a shallow layer of solvent, the solvent will ascend the layer of adsorbent on the plate by capillary action.

In performing thin layer chromatography, the sample is applied to the plate before allowing the solvent to ascend the adsorbent layer. The sample is usually applied as a small spot near the base of the plate, and this technique is often referred to as "spotting." The plate is spotted by repeated applications of a sample solution from a small capillary pipet. When the filled pipet touches the plate, capillary action delivers its contents to the plate, and a small spot is formed.

As the solvent ascends the plate, the sample is partitioned between the moving liquid phase and the stationary solid phase. During this process, one is said to be **developing** or **running** the thin layer plate. In development, the various components in the applied mixture are separated. The separation process occurs due to the many equilibrations which the solutes experience between the moving and the stationary phases. The nature of these equilibrations was thoroughly discussed in Technique 10, Sections 10.2 and 10.3. As with column chromatography,

599

the least polar substances will advance faster than the most polar substances. A separation results from the differences in the rates at which the individual components of the mixture advance upwards on the plate. When there are many substances present in a mixture, each will have its own characteristic solubility and adsorptivity properties depending on the functional groups present in its structure. In general, the stationary phase is strongly polar and will strongly bind polar substances. The moving liquid phase is usually less polar than the adsorbent and will most easily dissolve those substances which are less polar or even non-polar. Thus, while the most polar substances will travel slowly upwards, or not at all, if the solvent is sufficiently non-polar, the non-polar substances will travel more rapidly.

When the thin layer plate has been developed, it is removed from the developing tank and allowed to dry until it is free of solvent. If the mixture which was originally spotted on the plate was separated, there will be a vertical series of spots on the plate. Each spot will correspond to a separate component or compound from the original mixture. If the components of the mixture are colored substances the various spots will be clearly visible after development. More often, however, the "spots" will be invisible because they correspond to colorless substances. In this latter case, the spots can be perceived visually only if a **visualization method** is used. Often the spots can be seen when the thin layer plate is held under an ultraviolet light, and the use of a uv lamp is a common visualization method. Also common is the use of iodine vapor. The plates are placed in a chamber containing iodine crystals and left to stand for a short time. The iodine will react with the various compounds adsorbed on the plate to give colored complexes which are clearly visible. Because iodine often changes the compounds by reaction, the mixture components cannot be recovered from the plate when this method is used. Other methods of visualization are discussed in Section 11.6.

11.2 PREPARATION OF THIN LAYER SLIDES AND PLATES

The two adsorbent materials most often used for tlc are alumina G (aluminum oxide) and silica gel G (silicic acid). The G designation stands for gypsum (calcium sulfate). Calcined gypsum ($CaSO_4 \cdot 1/2 \ H_2O$) is better known as plaster of Paris. When exposed to water or moisture, gypsum sets into a rigid mass ($CaSO_4 \cdot 2 \ H_2O$) which binds the adsorbent together and to the glass plates used as a backing support. In the adsorbents used for tlc, about 10 to 13% by weight of gypsum is added as a binder. The adsorbent materials are otherwise similar to those used in column chromatography. It should be noted, however, that the adsorbents used in column chromatography have a larger particle size. The material for thin layer work is a fine powder.

The small particle size, along with the added gypsum, makes it impossible to use silica gel G or alumina G for column work. In a column these adsorbents generally set so rigidly that the flow of solvent through the column is virtually stopped.

A. Microscope Slide TLC Plates

For qualitative work, such as attempting to identify the number of components in a mixture or trying to establish that two compounds are identical, small tlc plates made from microscope slides are especially convenient. Coated microscope slides are easily made by dipping the slides into a container holding a slurry of the adsorbent material. Although it is possible to use a great many solvents to prepare the slurry for making tlc slides, chloroform is probably the most convenient solvent. It has the two advantages of low boiling point (61°) and inability to cause the adsorbent to set or form lumps. The low boiling point means that it is not necessary to dry the coated slides in the oven. Its inability to cause the gypsum binder to set means that slurries made with it are stable for several days. It has the disadvantage that the layer of adsorbent formed is rather fragile and must be treated carefully. For this reason, some persons prefer to add a small amount of methanol to the chloroform to enable the gypsum to set more firmly. The methanol solvates the calcium sulfate in much the same way as water. More durable plates can be made by dipping plates into a slurry prepared from water. These plates must be oven dried prior to use. Also, a slurry prepared from water must be used soon after its preparation. If it is not, it will begin to set and to form lumps. Thus, an aqueous slurry must be prepared **just prior** to use. It cannot be used after it has stood for any length of time. For microscope slides a slurry of silica gel G in chloroform is not only convenient but also adequate for most purposes.

Preparing the Slurry. The slurry is most conveniently prepared in a 4 oz wide-mouth screw cap jar. About 3 ml of chloroform is required for each gram of silica gel G. In order to form a smooth slurry without lumps, the silica gel should be added to the solvent while either stirring or swirling the mixture. Adding solvent to the adsorbent usually results in the formation of lumps in the mixture. When the addition is complete, the cap should be placed on the jar, tightened, and the jar shaken vigorously to assure thorough mixing. The slurry may be stored, tightly capped in the screw cap jar, until it is to be used.

Preparing the Slides. If new microscope slides are available, they may be used without any special treatment. However, it is more economical to reuse or recycle used microscope slides. The slides should be washed with soap and water, rinsed with water, and then rinsed with 50% aqueous methanol. The plates should be allowed to dry thoroughly on paper towels. They should be handled by the edge because fingerprints on the plate surface will make it difficult for the adsorbent to bind to the glass.

Coating the Slides.   The slides are coated with adsorbent by dipping them into the container of slurry. Two slides may be coated simultaneously by sandwiching them together prior to dipping them in the slurry. The slurry should be shaken vigorously **just prior** to dipping the slides. Since the slurry settles on standing, it should be mixed in this way before each set of slides is dipped. The depth of the slurry in the jar should be about three inches, and the plates should be dipped into the slurry until only about one-quarter inch at the top remains uncoated. The dipping operation should be performed smoothly. The plates may be held at the top (see Figure 11–1) where they will not be coated. They are dipped into the slurry and withdrawn with a slow and steady motion. About 2 seconds will be required for the dipping operation. Some practice may be required to get the correct timing. After dipping, the cap should be replaced on the jar, and the plates should be held for a minute until the majority of the solvent has evaporated. The plates may then be separated and placed on paper towels to complete the drying.

The plates should have an even coating. There should be no streaks and no thin spots where glass shows through the adsorbent. Neither should the plates have a thick and lumpy coating. Two conditions can cause thin and streaked plates. First, the slurry may not have been thoroughly mixed prior to the dipping operation. In this case, the adsorbent might have settled to the bottom of the jar, and the thin slurry at the top would not have coated the slides properly. Second, the slurry simply may not have been thick enough. In this case, more silica gel G should be added to the slurry until the proper consistency is obtained. If the slurry is too thick, the coating on the plates will be thick, uneven, and lumpy. To correct this, the slurry should be diluted with enough solvent to obtain the proper consistency.

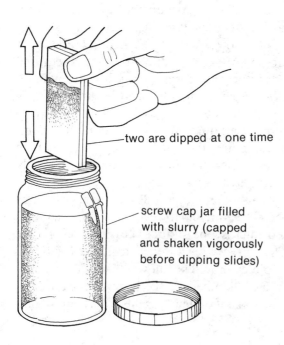

two are dipped at one time

screw cap jar filled
with slurry (capped
and shaken vigorously
before dipping slides)

FIGURE 11–1.   Dipping slides to coat them

Plates with an unsatisfactory coating may be wiped clean with a paper towel and redipped. Care must be taken to handle the plates only from the top or by the sides.

B. Larger Thin Layer Plates

For separations involving large amounts of material, or for separations which are more difficult, it may be necessary to use larger thin layer plates. Plates with dimensions up to 20 to 25 cm square are common. With larger plates it is desirable to have a more durable coating, and a water slurry of the adsorbent should be used to prepare them. If silica gel is used, the slurry should have a ratio of about 1 g of silica gel G to each 2 ml of water. The glass plate used to prepare the thin layer plate should be washed, dried, and placed on a sheet of newspaper. Along two edges of the plate are placed two strips of masking tape. More than one layer of masking tape is used if a thicker coating is desired on the plate. A slurry is prepared, shaken well, and poured along one of the untaped edges of the plate. A heavy piece of glass rod, long enough to span across the taped edges, is used to level and spread the slurry over the plate. While resting on the layers of tape on opposite edges of the plate, the rod is pushed along the plate from the end where the slurry was poured to the opposite end. This is illustrated in Figure 11–2. After spreading, the masking tape strips are removed, and the plates are dried in a 110° oven for about one hour. Plates up to about 20 to 25 cm square in size are easily prepared by this method. Larger plates present more difficulties. Many laboratories have a commercially manufactured spreading machine which makes the entire operation simpler.

C. Pre-Prepared Plates

Many manufacturers supply glass plates which are precoated with a durable layer of silica gel or alumina. More conveniently, plates are

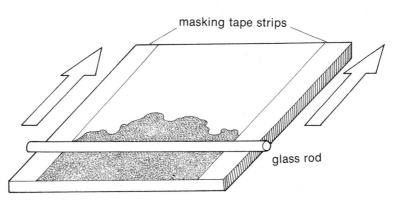

FIGURE 11–2. Preparing a large plate

also available which have either a flexible plastic backing or an aluminum backing. The latter two types of plate are becoming increasingly common. They are somewhat expensive, but they are made quite uniformly and, being flexible, have the advantage that they do not flake easily. They may also be cut with a pair of scissors into whatever size is required.

11.3 SAMPLE APPLICATION: SPOTTING THE PLATES

Preparing a Micropipet. To apply the sample which is to be separated to the thin layer plate, one needs a micropipet. A micropipet is easily made from a short length of thin walled capillary tubing similar to that used for melting point determinations. The capillary tubing is heated at its midpoint with a microburner and is rotated until it is soft. When the tubing is soft the heated portion of the tubing is drawn out until a constricted portion of tubing about 4 to 5 cm long is formed. After cooling, the constricted portion of tubing is scored at its center with a file or a scorer and broken. The two halves yield two capillary micropipets. The construction of the pipets is illustrated in Figure 11-3.

Spotting the Plate. To apply the sample to the plate, about 1 mg of a solid test substance, or one drop of a liquid test substance, is placed in a small container like a watch glass or test tube. The sample is then dissolved in a few drops of a volatile solvent. Chloroform, acetone, or methylene chloride are usually suitable solvents. If a solution is to be tested, it may often be used directly. The small capillary pipet, prepared as above, is filled by dipping the pulled end into the solution to be examined. Capillary action will fill the pipet. It is emptied by touching it

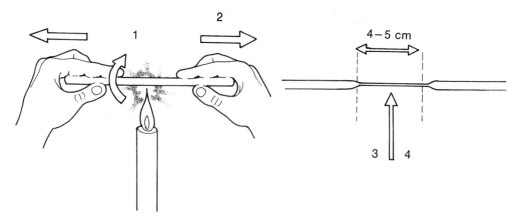

① Rotate in flame until soft. ③ Score lightly in center of pulled section.

② Remove from flame and pull. ④ Break in half to give two pipets.

FIGURE 11-3. Construction of two capillary micropipets

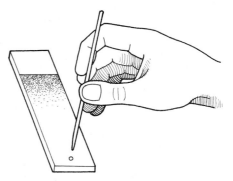

FIGURE 11-4. Spotting the plate with a drawn capillary pipet

lightly to the thin layer plate at a point about 1 cm from the bottom. See Figure 11–4. The spot must be high enough that it does not dissolve in the developing solvent. It is important to touch the plate very lightly and not to gouge a hole in the adsorbent. When the pipet touches the plate, solution is transferred to the plate in the form of a small spot. The pipet should be touched to the plate **very briefly** and then removed. If the pipet is held to the plate, its entire contents will be delivered to the plate. Only a small amount of material is required. It is often helpful to blow gently on the plate as the sample is applied. This helps to keep the spot small by evaporating the solvent before it can spread out on the plate. The smaller the spot formed, the better the separation obtainable. If needed, additional material may be applied to the plate by repeating the spotting procedure. It is best to repeat the procedure with several small amounts, rather than to apply one large amount. The solvent should be allowed to evaporate between applications. If the spot is not small (about 2 mm in diameter), a new plate should be prepared. The capillary pipet may be used several times if it is rinsed between uses. It is repeatedly dipped into a small portion of solvent to rinse it and touched to a paper towel to empty it.

As many as three different spots may be applied to a microscope slide tlc plate. Each spot should be about 1 cm from the bottom of the plate, and they should be evenly spaced with one spot in the center of the plate. Due to diffusion, spots will often increase in diameter as the plate is developed. To avoid merging of spots containing different materials, or the confusion of samples, a maximum of only three spots is recommended. Larger plates, of course, can accommodate many more samples.

11.4   DEVELOPING (RUNNING) TLC PLATES

Preparing a Development Chamber.   A convenient developing chamber for microscope slide tlc plates can be made from a 4 oz wide mouth screw cap jar. The inside of the jar should be lined with a piece of filter paper cut so that it does not quite extend completely around the inside of the jar. A small vertical opening (2 to 3 cm) should be left for observation of the development process. Before development, the filter paper liner inside the jar should be thoroughly moistened with the development solvent. The solvent-saturated liner helps to keep the chamber saturated with solvent vapors, thereby speeding up the development process. Once the liner is saturated, the level of solvent in the bottom of the jar is adjusted to a depth of about 5 mm, and the jar is capped and set aside until it is to be used. A correctly prepared development chamber (with the slide in place) is shown in Figure 11–5.

Developing the TLC Slide. Once the spot has been applied to the thin layer plate and the solvent has been selected (see Section 11.5), the plate is placed in the chamber for development. The plate must be placed in the chamber carefully so that none of the coated portion touches the filter paper liner. In addition, the solvent level in the bottom of the chamber must not be above the spot which was applied to the plate, or the spotted material will dissolve in the pool of solvent rather than be subjected to chromatography. Once the plate has been correctly placed, one replaces the cap on the developing chamber and waits for the solvent to advance up the plate by capillary action. This will generally occur quite rapidly, and one should watch carefully. As the solvent rises, the plate will become quite visibly moist. When the solvent has advanced to within 5 mm from the end of the coated surface, the plate should be removed, and the position of the solvent front should be marked **immediately** by scoring the plate along the solvent line with a pencil. The solvent front must not be allowed to travel beyond the end of the coated surface. The plate should be removed before this happens. The solvent will not actually advance beyond the end of the plate, but spots which are allowed to stand on a completely moistened plate on which the solvent is not in motion will expand in size by diffusion. Once the plate has dried, any visible spots should be outlined on the plate

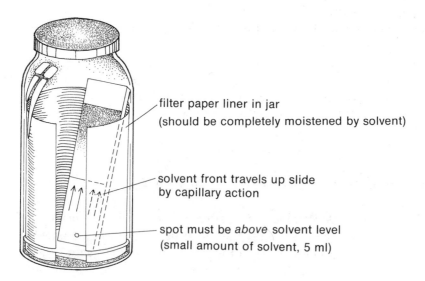

filter paper liner in jar
(should be completely moistened by solvent)

solvent front travels up slide
by capillary action

spot must be *above* solvent level
(small amount of solvent, 5 ml)

FIGURE 11–5. A development chamber with a thin layer plate undergoing development

with a pencil. If no apparent spots are visible, a visualization method (Section 11.6) may be needed.

11.5 CHOOSING A SOLVENT FOR DEVELOPMENT

The choice of a development solvent will depend on the materials to be separated. It may be necessary to actually try several solvents before a satisfactory separation is achieved. Since microscope slides can be prepared and developed fairly rapidly, an empirical choice is usually not hard to make. A solvent which causes all the spotted material to move with the solvent front is too polar. One which will not cause any of the material in the spot to move is not polar enough. As a guide to the relative polarity of solvents, Table 10–2 in Technique 10 should be consulted.

Chloroform and benzene are solvents of intermediate polarity and are good choices for a wide variety of functional groups to be separated. For hydrocarbon materials, hexane, petroleum ether (ligroin), or benzene are good first choices. Hexane or petroleum ether with varying proportions of benzene or ether give solvent mixtures of moderate polarity which are useful for many common functional groups. Polar materials may require the use of ethyl acetate, acetone, or methanol.

A rapid way to determine a good solvent is to apply several sample spots to a single plate. The spots should be placed a minimum of about 1 cm apart. A capillary pipet is filled with a solvent and gently touched to one of the spots. The solvent will expand outward in a circle. The solvent front should be marked with a pencil. A different solvent is applied to each spot. As the solvents expand outward, the spots will expand as concentric rings. From the appearance of the rings, one can make an approximate judgment as to the suitability of the solvent. Several types of behavior often encountered in this method of testing are illustrated in Figure 11–6.

11.6 VISUALIZATION METHODS

If the compounds separated by tlc are colored, it is a fortunate result, because the separation process can be followed visually. More

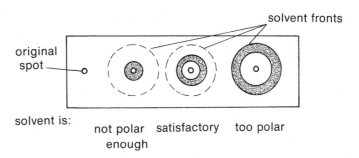

FIGURE 11–6. The concentric ring method of testing solvents

often than not, however, the compounds will be colorless. In this case, the separated materials must be made visible by the use of some reagent or some method that makes the separated compounds visible. Reagents which give rise to colored spots are called **visualization reagents.** Methods of viewing which make the spots apparent are called **visualization methods.**

The most commonly used visualization reagent is iodine. Iodine reacts with many organic materials to form complexes which are either brown or yellow. In this method of visualization, the developed and dried tlc plate is placed in a 4 oz wide mouth screw cap jar along with a few crystals of iodine. The jar is capped and gently warmed on a steam bath at low heat. The jar fills with iodine vapors, and the spots begin to appear. When the spots are sufficiently intense, the plate is removed from the jar, and the spots are outlined with a pencil. The spots are not permanent. Their appearance results from the formation of complexes which the iodine makes with the organic substances. As the iodine sublimes off the plate, the spots fade. Hence, they should be marked immediately. Nearly all compounds except saturated hydrocarbons and alkyl halides will form complexes with iodine. The intensities of the spots are not accurate indications of the amount of material present, except in the crudest way.

The second most common method of visualization is the use of an ultraviolet lamp. Under a uv light, compounds will often appear as bright spots on the plate. This often gives a clue to the structure of the compounds, because certain types of compounds will shine very brightly under a uv light due to the fact that they fluoresce.

Another method which gives good results is accomplished by adding a fluorescent indicator to the adsorbent used to coat the plates. A mixture of zinc and cadmium sulfides is often used. When treated in this way, the entire plate will fluoresce when held under a uv light. However, dark spots will appear on the plate where the separated compounds are seen to quench this fluorescence.

In addition to the above methods, several chemical methods are available which either destroy or permanently alter the separated compounds through reaction. Many of these methods are specific only for a given type of functional group.

Alkyl halides can be visualized by spraying a dilute solution of silver nitrate on the plates. Silver halides are formed. These halides decompose on exposure to light, giving rise to dark spots (free silver) on the tlc plate.

Most organic functional groups can be made visible by charring them with sulfuric acid. Concentrated sulfuric acid is sprayed on the plate, and it is then heated in an oven at 110° to complete the charring. Permanent spots are created in this manner.

Colored compounds may be prepared from colorless compounds by making derivatives before spotting them on the plate. An example of this is the preparation of 2,4-dinitrophenylhydrazones from aldehydes and ketones to prepare yellow and orange compounds. One may also spray the 2,4-dinitrophenylhydrazine reagent on the plate after the ketones or aldehydes have separated. Red and yellow spots will form at the location of compounds. Other examples of this method are the use of ferric chloride for visualizing phenols or bromocresol green for detecting carboxylic acids. Chromium trioxide, potassium dichromate, and potassium permanganate can be used for visualizing compounds that are easily oxidized. *p*-Dimethylaminobenzaldehyde will easily detect the presence of amines. Ninhydrin will react with amino acids to visualize them. A number of other methods and reagents are available from various supply outlets that will be specific for certain types of functional groups. These will only visualize the class of compounds of interest.

11.7 PREPARATIVE PLATES

If larger plates, like those described in Section 11.2B, are used, materials may be separated and the separated components individually recovered from the plates. Plates used in this way are said to be **preparative plates.** For preparative plates, a thicker layer of adsorbent is generally used. Rather than being applied as a spot, or as a series of spots, the mixture to be separated is applied as a line of material about 1 cm from the bottom of the plate. As the plate is developed, the separated materials form into bands. After development, the separated bands are observed, usually by uv light, and the zones are outlined in pencil. If a destructive method of visualization is used, the majority of the plate is covered with paper to protect it, and the reagent is applied only to the extreme edge of the plate.

Once the zones have been identified, the adsorbent in those bands is scraped from the plate and is extracted with solvent to remove the adsorbed material. Filtration removes the adsorbent, and evaporation of the solvent gives the recovered component from the mixture.

11.8 THE R_f VALUE

Under an established set of tlc conditions which include:

1. the solvent system used
2. the adsorbent used
3. the thickness of the adsorbent layer, and
4. the relative amount of material spotted,

a given compound will always travel a fixed distance relative to the distance the solvent front travels. This ratio of the distance the compound travels to the distance the solvent front travels is called the R_f

value. R_f stands for "ratio to front," and it is expressed as a decimal fraction:

$$R_f = \frac{\text{distance traveled by substance}}{\text{distance traveled by solvent front}}$$

When the conditions of measurement are completely specified, the R_f value is constant for any given compound, and it corresponds to a physical property of that compound.

The R_f value can be used to identify an unknown compound but, like any other identification based on a single piece of data, it is best to confirm such an identification with some additional data. Many compounds could have the same R_f value, just as many different compounds have the same melting point.

In measuring an R_f value, it is not always possible to duplicate exactly the conditions of measurement which another worker has used. Therefore, R_f values tend to be of more use to a single worker in his own laboratory than they are to workers in different laboratories. The only exception to this occurs when both workers use tlc plates from the same source, i.e., commercial plates, or know the **exact** details of how the plates were prepared. Nevertheless, the R_f value can be a useful guide. Where exact values cannot be relied on, the relative values can provide another worker with useful information as to what to expect. When using published R_f values, it is always a good idea to check them by comparing them with standard substances whose identity and R_f values are known.

To calculate the R_f value for a given compound, one measures the distance that it has traveled from the point where it was originally spotted. For spots which are not too large, one measures to the center of the migrated spot. Where spots are large, the measurement should be repeated on a new plate, using less material. In cases where spots show tailing, the measurement is made to the "center of gravity" of the spot. This first distance measurement is then divided by the distance that the solvent front has traveled from the same original spot. A sample calculation of the R_f values of two compounds is illustrated in Figure 11–7.

11.9 APPLICATIONS OF THIN LAYER CHROMATOGRAPHY TO ORGANIC CHEMISTRY

Thin layer chromatography has several important uses in organic chemistry. It can be used to

1. establish that two compounds are identical
2. determine the number of components in a mixture
3. determine the appropriate solvent for a column chromatographic separation

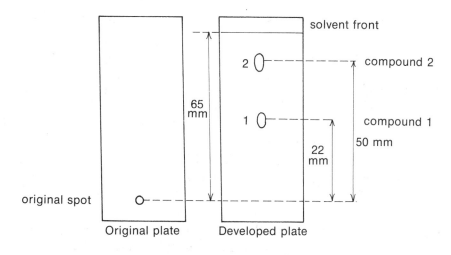

$$R_f \text{ (compound 1)} = \frac{22}{65} = 0.34 \qquad R_f \text{ (compound 2)} = \frac{50}{65} = 0.77$$

FIGURE 11–7. A sample calculation of R_f values

4. monitor a column chromatographic separation
5. check the effectiveness of a separation achieved on a column, by crystallization, or by extraction
6. monitor the progress of a reaction.

In all of these applications, thin layer chromatography has the advantage that only small amounts of material are utilized. Material is not wasted. With many of the visualization methods, less than a tenth of a microgram (10^{-7} g) of material can be detected. On the other hand, samples as large as a milligram may also be used. With preparative plates which are large (about 9 inches on a side) and have a relatively thick ($>500\,\mu$m) coating of adsorbent, it is often possible to separate from 0.2 to 0.5 of a gram of material at one time. The major disadvantage of tlc is that volatile materials may not be used, since they will evaporate from the plates.

Thin layer chromatography can establish that two compounds suspected to be identical are in fact identical. One simply spots both compounds side by side on a single plate and develops the plate. If both compounds travel the same distance on the plate (have the same R_f value), they are probably identical. If the spot positions are not the same, the compounds are definitely not identical. It is important to spot both compounds **on the same plate.** This is especially important when using hand-dipped microscope slides, since they vary widely from plate to plate, no two plates having exactly the same thickness of adsorbent. If commercial plates are used, this precaution is not necessary, but nevertheless a good idea.

Thin layer chromatography can be used to establish if a compound is a single substance or if it is a mixture. A single substance will give a single spot no matter what solvent is used to develop the plate. On the other side of the coin, by trying various solvents on a mixture, the number of components in a mixture can be established. A word of caution should be given. When dealing with compounds which have very similar properties, e.g., isomers, it may be difficult to find a solvent that will separate the mixture. Inability to achieve a separation is not an absolute proof that a compound is a single pure substance. Many compounds can be separated only by **multiple developments** of the tlc slide, using a fairly non-polar solvent. In this method, the plate is removed after the first development and allowed to dry. After drying, it is placed in the chamber again and developed once more. This effectively doubles the length of the slide. At times, several developments may be necessary.

When given a mixture, tlc can be used to choose the best solvent to separate it when using column chromatography. Various solvents can be tried on a plate coated with the same adsorbent as used in the column. The solvent that gives the best resolution of the components will probably work well on the column. These small scale experiments are done quickly, use very little material, and save time which would be wasted by doing a more time-consuming separation of the entire mixture on the column. Similarly, tlc plates can be used to **monitor** a column. A hypothetical situation is shown in Figure 11–8. A solvent was found that would separate the mixture into four components (A-D). A column was run using this solvent and eleven fractions of 15 ml each were collected. Thin layer analysis of the various fractions showed that fractions 1 through 3 contained component A, fractions 4 through 7 component B, fractions 8 through 9 component C, and fractions 10 through 11 component D. A small amount of cross contamination was observed in fractions 3,4,7 and 9.

In another example of the use of tlc, a worker found a product obtained from a reaction to be a mixture. It gave two spots, A and B, on

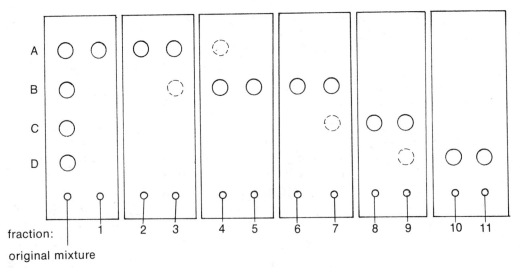

FIGURE 11–8. Monitoring a column

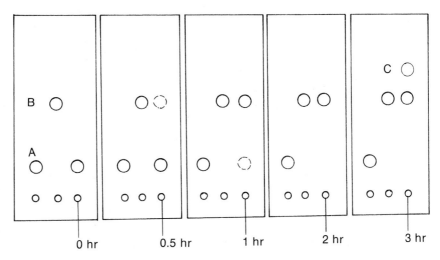

FIGURE 11-9. Monitoring a reaction

a tlc slide. After crystallizing, the product, the crystals were found to be pure A by tlc analysis, whereas the mother liquor contained a mixture of both A and B. The crystallization was judged to have given a satisfactory purification of A.

Finally, it is often possible to monitor the progress of a reaction by the use of tlc. At various points during a reaction, samples of the reaction mixture are taken and subjected to tlc analysis. An example is given in Figure 11-9. In this case, the desired reaction was the conversion of A to B. At the beginning of the reaction (0 hr), a tlc slide was prepared that was spotted with pure A, pure B, and the reaction mixture. Similar slides were prepared at 0.5 hr, 1 hr, 2 hrs, and 3 hrs after the start of the reaction. The slides showed that the reaction was complete in 2 hours. When the reaction was run longer than 2 hours, a new compound, side product C, began to appear. Thus, the optimum reaction time was judged to be 2 hours.

11.10 PAPER CHROMATOGRAPHY

Paper chromatography is often considered to be a technique related to thin layer chromatography. The experimental techniques are somewhat similar to those of tlc, but the principles are more closely related to those of extraction. Paper chromatography is actually a liquid-liquid partitioning technique, rather than a solid-liquid technique. For paper chromatography, a spot is placed near the bottom of a piece of high grade filter paper (Whatman #1 is often used). Then the paper is placed in a developing chamber. The development solvent ascends the paper by capillary action and moves the components of the spotted mixture upward at differing rates. Although paper consists mainly of pure cellulose, the cellulose itself does not function as the stationary phase. Rather, the cellulose absorbs water from the atmosphere, especially from an atmosphere saturated with water vapor. Cellulose can absorb up to about 22% of water. It is this water adsorbed on the cellulose that functions as the stationary phase. To assure that the cellulose is kept saturated with water, many of the development solvents used in paper chromatography will contain water as a component. As the solvent ascends the paper, the compounds are partitioned between the stationary water phase and the moving solvent. Since the water phase is stationary, the components in a mixture which are most water soluble, or those which have the greatest hydrogen bonding capability, are the ones which are held back and move slowest. Paper chromatography is applicable mostly to highly polar compounds, or to those which are polyfunctional. The most common use of paper chromatography is for sugars, amino acids, and natural pigments. Since filter paper is manufactured with good uniformity, R_f values can often be relied upon in paper chromatographic work. However, the R_f values are customarily measured from the leading edge (top) of the spot, rather than from its center as is customary in tlc.

Technique 12

GAS CHROMATOGRAPHY

Gas chromatography is similar in principle to column chromatography, but it differs in three respects. First, the partitioning processes for the compounds to be separated are carried out between a **moving gas phase** and a **stationary liquid phase.** (Recall that in column chromatography the moving phase is a liquid and the stationary phase is a solid adsorbent.) Second, the solubility of any given compound in the gas phase is a function of its vapor pressure only. And third, the temperature of the system can be controlled since the column is contained in an insulated oven. These differences will be discussed in detail in the sections which follow.

Gas chromatography (gc) is also known as vapor phase chromatography (vpc) and as gas-liquid partition chromatography (glpc). All three names, as well as their indicated abbreviations, are often found in the literature of organic chemistry. When referring to the technique, the last term, glpc, is the most strictly correct, and is preferred by most authors.

12.1 THE GAS CHROMATOGRAPH

The apparatus used to carry out a gas liquid chromatographic separation is generally called a **gas chromatograph.** A typical student model gas chromatograph, the Carle model 8000, is illustrated (with a partial cutaway view) in Figure 12–1. A schematic block diagram of a basic gas chromatograph is shown in Figure 12–2. The basic elements of the apparatus are easily seen. In short, the sample is injected into the

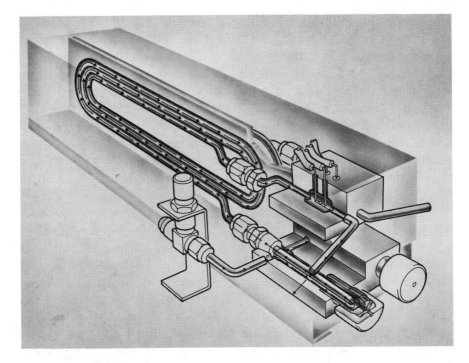

FIGURE 12–1. Gas chromatograph

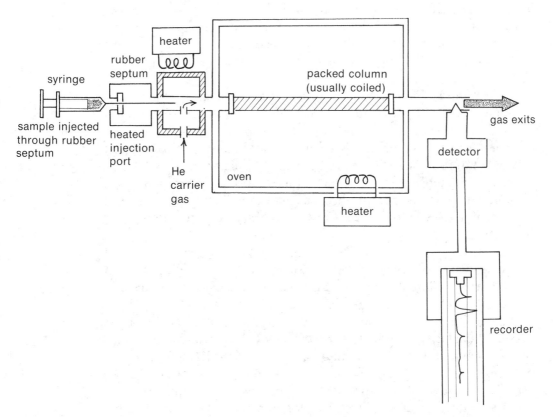

FIGURE 12–2. Schematic diagram of a gas chromatograph

chromatograph, and it is immediately vaporized and introduced into a moving stream of gas, called the carrier gas. The vaporized sample is then swept into a column filled with particles coated with a liquid adsorbent. The column is contained in a temperature controlled oven. As the sample passes through the column, it is subjected to many gas-liquid partitioning processes and is separated. As each component emerges from the column, its presence is detected by an electrical detector that generates a signal which is recorded on a strip chart recorder. Each element of the gas chromatographic process will be subsequently examined in detail in the following sections.

12.2 THE COLUMN

The heart of the gas chromatograph is the packed column. This column is usually made of copper or stainless steel tubing, but sometimes glass is used. Various diameters of tubing are used, but 1/8 in (3 mm) and 1/4 in (6 mm) tubing are probably most common. To construct a column, one selects the length of tubing desired (1 m, 2 m, 3 m, 10 m, etc.), cuts it to the desired length, and attaches the proper fittings on each of the two ends to connect it to the apparatus.

Next, the tubing (column) is packed with the **stationary phase.** The material chosen for the stationary phase is usually a liquid, a wax, or a low-melting solid. This material should be relatively non-volatile, i.e., have a low vapor pressure and a high boiling point. Liquids commonly used are high boiling hydrocarbons, silicone oils, waxes, and polymeric esters, ethers, and amides. Some typical substances are listed in Table 12–1.

The liquid phase is usually coated onto a **support material.** A commonly used support material is crushed firebrick. There are many methods of coating the high-boiling liquid phase onto the support particles. The easiest method is to dissolve the liquid (or low-melting wax or solid) in a volatile solvent like methylene chloride (bp 40°). The firebrick (or other support) is added to this solution which is then slowly evaporated (rotary evaporator) in such a way as to leave each particle of support material evenly coated. Commonly used support materials are listed in Table 12–2.

Finally, the liquid-phase-coated support material is packed into the tubing as evenly as possible. Then, the tubing is bent or coiled in a way that allows it to fit into the oven of the gas chromatograph and have its two ends connected to the gas entrance and exit ports.

Selection of a liquid phase usually revolves about two factors. First, most of them have an upper temperature limit above which they cannot be used. Above the specified limit of temperature, the liquid phase itself will begin to "bleed" off the column. Second, the materials to be separated must be taken into consideration. For polar samples, it is best to use a rather polar liquid phase; for non-polar samples, a rather non-polar liquid phase is indicated.

TABLE 12-1. TYPICAL LIQUID PHASES

TYPE	COMPOSITION	MAXIMUM TEMPERATURE	POLAR	TYPICAL USE
APIEZONS (L,M,N,etc.) — Hydrocarbon greases (varying MW)	Hydrocarbon Mixtures	250–300°	no	hydrocarbons
CARBOWAXES (400–6000) — Polyethylene glycols (varying chain length)	Polyether $HO-(CH_2CH_2-O)_n-CH_2CH_2OH$	up to 250°	yes	alcohols ethers
DC-200 — Silicone oil ($R=CH_3$)	$R_3Si-O-\left[\begin{array}{c} R \\ \mid \\ Si-O \\ \mid \\ R \end{array}\right]_n-SiR_3$	225°	med.	aldehydes ketones halocarbons
DEGS — Diethylene glycol succinate	Polyester $\left(CH_2CH_2-O-\overset{\displaystyle O}{\underset{\displaystyle \parallel}{C}}-(CH_2)_2-\overset{\displaystyle O}{\underset{\displaystyle \parallel}{C}}-O\right)_n$	225°	yes	esters acids
SE-30 — Methyl Silicone Rubber	Like oil, but cross-linked	350°	no	steroids
NUJOL — Mineral Oil	Hydrocarbon mixture	200°	no	pesticides

TABLE 12-2. TYPICAL SOLID SUPPORTS

Crushed Firebrick	Chromosorb T
Nylon Beads	(Teflon Beads)
Glass Beads	Chromosorb P
Silica	(Pink diatomaceous earth,
Alumina	highly absorptive, pH 6–7)
Charcoal	Chromosorb W
Molecular Sieves	(White diatomaceous earth,
	medium absorptivity, pH 8–10)
	Chromosorb G
	(similar to the above,
	low absorptivity, pH 8.5)

12.3 PRINCIPLES OF THE SEPARATION PROCESS

After selecting, packing, and installing a column, the **carrier gas** (usually helium, argon, or nitrogen) is allowed to flow through the column supporting the liquid phase. The mixture of compounds to be separated is introduced into the carrier gas stream, where its components are equilibrated between the moving gas phase and the stationary liquid phase. The latter is held stationary by virtue of its being adsorbed onto the surfaces of the support material.

At room temperature, most organic compounds are not very soluble in a gas like helium or nitrogen, and they would never make it through the column unless they were caused to develop a significant vapor pressure, usually by heating. The solubility of a compound in the gas phase is determined almost solely by its vapor pressure. As may be

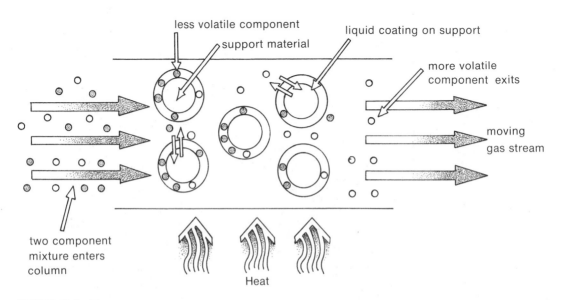

less volatile component
support material
liquid coating on support
more volatile component exits
moving gas stream
two component mixture enters column
Heat

FIGURE 12-3. The separation process

recalled from general chemistry, all gases mix with one another, and, therefore, if the mixture to be separated is vaporized as it is introduced into the carrier gas stream, it will be swept along with the carrier gas and introduced into the column. To accomplish this, the sample is introduced into the gas chromatograph by means of a hypodermic-like syringe. The sample is injected as a liquid or a solution through a rubber septum into a heated chamber, called the **injection port,** where it is vaporized and mixed with the carrier gas. As this mixture reaches the column it will begin to equilibrate between the liquid and gas phases. This process, however, cannot last very long unless the column is maintained at a temperature high enough to allow the components of the mixture to eventually revaporize back into the moving gas stream from the liquid phase in which they have become dissolved. For this reason, the column is placed in an insulated oven, the temperature of which may be adjusted. At the correct temperature, and if the correct liquid phase has been selected, the compounds in the injected mixture will travel through the column at different rates and be separated.

12.4 THE FACTORS AFFECTING THE SEPARATION

There are a number of factors that will determine the rate at which a given compound will travel through a gas chromatograph. First of all, compounds of lower boiling point will in general travel through the gas chromatograph at rates faster than those of higher boiling point. Obviously, this is a result of the fact that the column is heated and that low boiling compounds will always have higher vapor pressures than those of higher boiling point. If the column is heated to a temperature which is too high, the entire mixture to be separated will be flushed through the column at the same rate as the carrier gas, and no equilibrations will take place with the liquid phase. At too low a temperature, the mixture will dissolve in the liquid phase and never revaporize. Thus, it will be retained on the column.

Second, the rate of flow of the carrier gas can be important. The carrier gas must not pass over the liquid phase so rapidly that molecules "dissolved" in the gas phase cannot equilibrate with the liquid-coated particles. On the other hand, the rate of flow must be great enough to prevent hold up.

Third, the particular liquid phase chosen to prepare a column is quite important. The molecular weights, functional groups, and polarities of the component molecules in the mixture to be separated must be taken into account when selecting a liquid phase. One generally uses a different type of material for hydrocarbons, for instance, than for esters. The useful temperature limit of the liquid phase selected must also be considered.

Fourth, the length of the column is important. Compounds which are very similar to one another will, in general, require longer columns than those which are very dissimilar. Many kinds of isomeric mixtures fit into the "difficult" category. The components of isomeric mixtures are so similar in structure that they travel through the column at very similar rates. A longer column, therefore, is required to capitalize on any differences that do exist.

12.5 ADVANTAGES OF THE METHOD

All of the above factors must be adjusted by the chemist for any mixture he wishes to separate. A good amount of preliminary investigation is required before a mixture can successfully be separated into its components by gas chromatography (or at least a good amount of prior experience). Nevertheless, the advantages of the technique are many.

First, many mixtures can be separated by this technique when no other method will suffice. Second, unlike other methods, as little as 10 to 20 μl ($1\ \mu$l $=\ 10^{-6}$ l) of a mixture can be separated using this technique. It can be used on a very small scale indeed! Third, when coupled with an electronic recording device (see below) a **quantitative** estimate of the amount of each component present in the separated mixture can be obtained.

The range of compounds that may be usefully separated by gas chromatography extends from gases such as oxygen (bp $-183°$) and nitrogen (bp $-196°$) to organic compounds with boiling points in excess of 400°. The only requirement for the compounds to be separated is that they have an appreciable vapor pressure at a temperature at which they can be separated, and that they be thermally stable at this temperature.

12.6 MONITORING THE COLUMN
(THE DETECTOR)

To follow the progress of the separation of the mixture injected into the gas chromatograph, it is necessary to use an electrical device called a **detector.** The most commonly used detector is the **thermal conductivity detector.** This is simply a hot wire placed in the gas stream at the exit of the column. The wire is heated by a constant electrical voltage. When a steady stream of carrier gas passes over this wire, the rate at which it loses heat and its electrical resistance will each have a constant value. When the composition of the vapor stream changes, the rate of heat flow from the wire, and, hence, its resistance will change. Helium, which has a higher thermal conductivity than most common organic substances, is commonly used as a carrier gas. Thus, when a substance elutes in the vapor stream, the wire will generally heat up, and its resistance will decrease.

A typical thermal conductivity detector operates by difference. Two detectors are used, one exposed to the actual effluent gas, and the other exposed to a reference flow of carrier gas only. To achieve this situation, a portion of the carrier gas stream is diverted **before** it enters the injection port. The diverted gas is routed through a reference column, into which no sample has been admitted. The detectors mounted in the sample and reference columns are arranged so as to form the arms of a Wheatstone bridge circuit as is shown in Figure 12–5. As long as the carrier gas alone flows over both detectors, the circuit will be in balance. However, when a sample elutes from the sample column, the bridge circuit will become unbalanced, creating an electrical signal. This signal can be amplified and used to activate a strip chart recorder. The recorder is an instrument which plots, by means of a moving pen, the unbalanced bridge current versus time on a continuously moving roll of chart paper. This record of detector response (current) versus time is called a **chromatogram.** A typical gas chromatogram is illustrated in Figure 12–4. Deflections of the pen are called peaks.

At the time the sample is injected, a peak called the **air peak** usually appears. This is due to the fact that some air (CO_2, H_2O, N_2, and O_2) is introduced along with the sample. The air travels through the column almost as rapidly as the carrier gas, and as it passes the detector it causes a small pen response, thereby giving a peak. At later times (t_1, t_2, t_3), the components of the mixture also give rise to peaks on the chromatogram as they pass out of the column and past the detector.

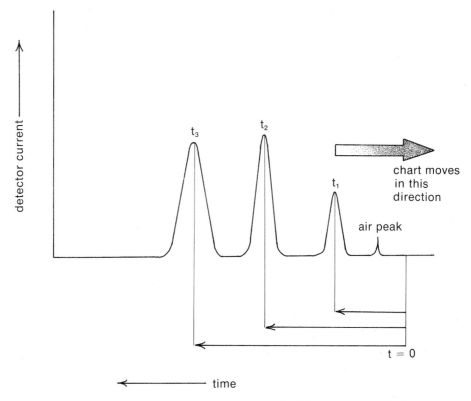

FIGURE 12-4. A typical chromatogram

12.7   RETENTION TIME

The period of time, following injection, required for a compound to pass through the column, is called the **retention time** of that compound. For a given set of constant conditions (flow rate of carrier gas, column temperature, column length, liquid phase selected, injection port temperature, carrier gas used), the retention time of any compound will always be constant (much like the R_f value in thin layer chromatography). The retention time is measured from the time of injection to the time of maximum pen deflection (detector current) for the component being observed. This value, when obtained under controlled conditions, may be used to identify a compound by a direct comparison of it to the values for known compounds determined under the same conditions. To facilitate the measurement of retention times, most strip chart recorders are adjusted to move the paper at a rate which corresponds to time divisions calibrated on the chart paper. The retention times (t_1, t_2, t_3) are indicated in Figure 12–5 for the three peaks illustrated.

12.8   QUALITATIVE ANALYSIS

Unfortunately, the gas chromatograph gives **no** information whatever as to the identities of the substances which it has separated. The little information that it does provide is given by the retention time. This quantity is hard to reproduce from day to day, however, and complete duplications of separations performed last month may be difficult to obtain this month. It is usually necessary to **calibrate** the column each time it is used. That is, one must run pure samples of all known and suspected components of a mixture individually, just prior to chroma-

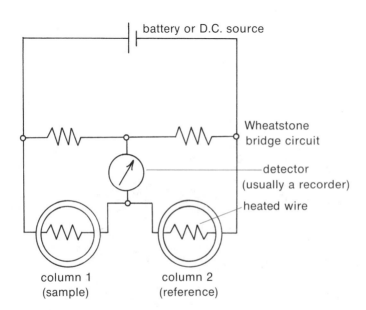

battery or D.C. source

Wheatstone bridge circuit

detector (usually a recorder)

heated wire

column 1 (sample)

column 2 (reference)

FIGURE 12–5. A typical thermal conductivity detector

tographing the mixture, so as to obtain the retention times of each of the known compounds. Alternatively, each of the suspected components may be added, one by one, to the unknown mixture while looking to see which of the peaks will have its intensity increased relative to the unmodified mixture. Finally, the components may be individually **collected** as they exit from the gas chromatograph. Each component may then be identified by other means, such as by the use of infrared or nuclear magnetic resonance spectroscopy.

12.9 COLLECTION OF THE SAMPLE

Samples passed through the gas chromatograph may be collected or recovered by means of a cooled trap connected to the outlet of the column. The simplest form of trap is just a U-shaped tube dipped into a bath of coolant (ice-water, liquid nitrogen, or Dry Ice-acetone).

For instance, if the coolant is liquid nitrogen (bp $-196°$) and the carrier gas is helium (bp $-269°$), compounds boiling above the temperature of liquid nitrogen will generally be condensed or trapped, while the carrier gas will continue to flow. However, if the exiting compound is frozen in the form of a fine mist, it may be carried through the trap even though frozen. When this happens, special trapping arrangements are required. To collect each component of the mixture, it is necessary to change the trap each time a new peak begins to appear on the chromatogram. In fact, since even a gas chromatograph with large diameter columns can rarely handle more than about 1/2 ml of a mixture at a time, a chemist or student having 50 ml of a mixture would have to make 50 to 100 separate injections, changing the trap for each component as it comes off the column. Fortunately, there are machines which can be attached to a gas chromatograph which will perform this task automatically.

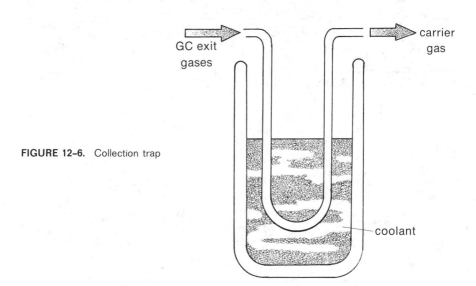

FIGURE 12-6. Collection trap

12.10 QUANTITATIVE ANALYSIS

The area under a gas chromatographic peak is proportional to the amount (moles) of compound eluted. Hence, the molar percentage composition of a mixture can be approximated by comparing relative peak areas. This method of analysis does make the assumption that the detector has equal sensitivity to all of the compounds eluted, and that it gives a linear response with regard to amount. Nevertheless, it gives reasonably accurate results.

The simplest method of measuring the area of a peak is by geometrical approximation, or triangulation. In this method, one multiplies the height (h) of the peak above the base line of the chromatogram by the width of the peak at half of its height ($w_{1/2}$). This is illustrated in Figure 12–7. The base line is approximated by drawing a line between the two "side arms" of the peak. This method works well only if the peak is symmetrical. If the peak has tailed or is unsymmetrical, it is best to cut out the peaks with a scissors and weigh them (the paper pieces) on an **analytical** balance. Since the weight per area of a piece of good chart paper is reasonably constant from place to place, the ratio of the areas is the same as the ratio of the weights. To obtain a percentage composition for the mixture, one first adds all the peak areas (weights). Then, to calculate the percentage of any component in the mixture, one divides its individual area by the total area and multiplies the result by 100. A sample calculation is illustrated in Figure 12–8. If peaks **overlap** (see Figure 12–9A) or are not well **resolved,** either the gas chromatographic conditions must be readjusted to achieve better resolution of the peaks (Figure 12–9B), or an **estimate** of the peak shape must be made.

Poor resolution is often caused by using too much sample, too high a column temperature, too short a column, a liquid phase that does not discriminate well between the two components, too large a diameter column, or in short, almost any wrongly adjusted parameter. The best resolution is always obtained with smaller diameter columns; however, it is always necessary to use less sample with these. For analytical work, it is not uncommon to use Golay or capillary columns of diameter 0.1 to 0.2 mm. With these, no solid support is used, the liquid is coated directly on the inner walls of the tubing. Such columns are often quite long, with lengths of 200 to 300 feet being quite common.

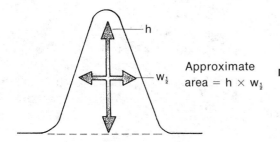

Approximate area = h × $w_{\frac{1}{2}}$

FIGURE 12–7. Triangulation of a peak

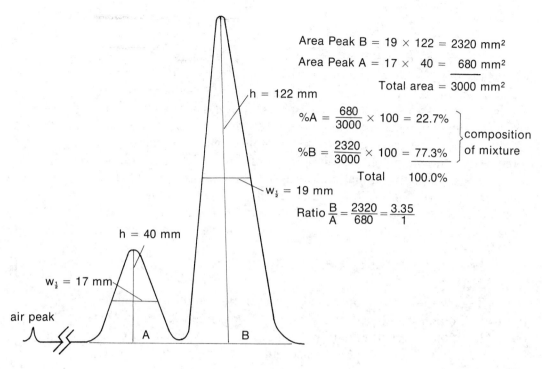

Area Peak B = 19 × 122 = 2320 mm²
Area Peak A = 17 × 40 = 680 mm²
Total area = 3000 mm²

$\%A = \dfrac{680}{3000} \times 100 = 22.7\%$

$\%B = \dfrac{2320}{3000} \times 100 = 77.3\%$

composition of mixture

Total   100.0%

$\text{Ratio} \dfrac{B}{A} = \dfrac{2320}{680} = \dfrac{3.35}{1}$

h = 122 mm

$w_{\frac{1}{2}}$ = 19 mm

h = 40 mm

$w_{\frac{1}{2}}$ = 17 mm

air peak

A   B

FIGURE 12–8. A sample percent composition calculation

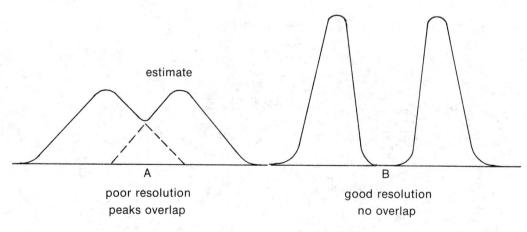

estimate

A

poor resolution
peaks overlap

B

good resolution
no overlap

FIGURE 12–9.

Technique **13**

SUBLIMATION

In Technique 6, the influence of temperature on the change in vapor pressure of a liquid was discussed (see Figure 6–1). It was shown that the vapor pressure of a liquid increases with temperature. At atmospheric pressure, the vapor pressure of a liquid will be equal to 760 mm at its boiling point. The vapor pressure of a solid also varies with temperature. Because of this behavior, some solids can readily pass directly into the vapor phase, without going through a liquid phase. This process is called **sublimation.** Since the vapor can be resolidified, the overall vaporization-solidification cycle can be used as a purification method. The purification can be successful only if the impurities have significantly lower vapor pressures than the material being sublimed.

13.1 VAPOR PRESSURE BEHAVIOR OF SOLIDS AND LIQUIDS

In Figure 13–1, vapor pressure curves for solid and liquid phases for two different substances are shown. Along lines **AB** and **DF,** the sublimation curves, the solid and vapor are at equilibrium. To the left of these lines the solid phase exists, and to the right of these lines

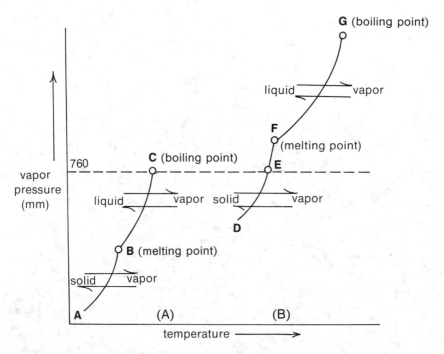

FIGURE 13–1. Vapor pressure curves for solids and liquids. *A*, Substance shows normal solid to liquid to gas transitions at 760 mm pressure; *B*, substance shows a solid to gas transition at 760 mm pressure.

the vapor phase is present. Along lines **BC** and **FG,** the liquid and vapor are at equilibrium. To the left of these lines, the liquid phase exists and to the right, the vapor is present. The two substances vary greatly in their physical properties as seen in Figure 13-1A and 13-1B.

In the first case (Figure 13-1A), the substance shows normal change of state behavior on heating, going from solid to liquid to gas. The dashed line, which represents an atmospheric pressure of 760 mm, is drawn **above** the melting point, **B** in Figure 13-1A. Thus, the applied pressure is **greater** than the vapor pressure of the solid-liquid phase at the melting point. Starting at **A,** as the temperature of the solid is raised, the vapor pressure increases along **AB** until the solid is observed to melt at **B.** At **B,** the vapor pressures of **both** the solid and liquid are identical. As the temperature continues to rise, the vapor pressure will increase along **BC** until the liquid is observed to boil at **C.** The description given is for the "normal" behavior expected for a solid substance. All three states (solid, liquid and gas) are sequentially observed during the change in temperature.

In the second case (Figure 13-1B), the substance develops enough vapor pressure to vaporize completely at a temperature below its melting point. The substance shows a solid to gas transition only. The dashed line is now drawn **below** the melting point, **F,** of this substance. Thus, the applied pressure is **less** than the vapor pressure of the solid-liquid phase at the melting point. Starting at **D,** the vapor pressure of the solid rises as the temperature increases along line **DF.** However, the vapor pressure of the solid reaches atmospheric pressure (point **E**) **before** the melting point (point **F**) is attained. Therefore, sublimation occurs at **E.** No melting behavior will be observed at atmospheric pressure for this substance. In order to observe a melting point and to have the behavior along line **FG,** a sealed pressure apparatus would have to be used.

The sublimation behavior described above is a relatively rare occurrence for substances at atmospheric pressure. Several compounds exhibiting this behavior, carbon dioxide, perfluorocyclohexane, and hexachloroethane, are listed in Table 13-1. Notice that

TABLE 13-1. VAPOR PRESSURES OF SOLIDS AT THEIR MELTING POINTS

COMPOUND	VAPOR PRESSURE OF SOLID AT MP (mm)	MELTING POINT (°C)
Carbon Dioxide	3876 (5.1 atm)	−57
Perfluorocyclohexane	950	59
Hexachloroethane	780	186
Camphor	370	179
Iodine	90	114
Naphthalene	7	80
Benzoic Acid	6	122
p-Nitrobenzaldehyde	0.009	106

these compounds have vapor pressures **above** 760 mm at their melting point. In other words, their vapor pressures reach 760 mm below their melting points and they sublime rather than melt. If an attempt to determine the melting point of hexachloroethane at atmospheric pressure is made, the observation will be vapor pouring from the end of the melting point tube! With a sealed capillary tube, the observed melting point of 186° is obtained.

13.2 SUBLIMATION BEHAVIOR OF SOLIDS

Sublimation is usually a property of relatively non-polar substances which also have a highly symmetrical structure. Symmetrical compounds have relatively high melting points and high vapor pressures. The ease with which a substance can escape from the solid is determined by the strength of the intermolecular forces between the molecules. Symmetrical molecular structures will have a relatively uniform distribution of electron density and a small dipole moment. A smaller dipole moment means a higher vapor pressure because of lower electrostatic forces in the crystal.

Solids will sublime if they possess appreciable vapor pressures at their melting points. A number of compounds are listed in Table 13–1 together with the vapor pressures at their melting points. The first three entries in the table were discussed in Section 13.1. At atmospheric pressure they would sublime rather than melt, as shown in Figure 13–1B.

The next four entries in Table 13–1, camphor, iodine, naphthalene, and benzoic acid exhibit typical change of state behavior (solid, liquid, and gas) at atmospheric pressure as shown in Figure 13–1A. However, these compounds will sublime readily under reduced pressure. Vacuum sublimation is discussed in Section 13.3.

It should be stated that, compared to many other organic compounds, camphor, iodine, and naphthalene all have relatively high vapor pressures at relatively low temperatures. For example, they each have a vapor pressure of 1 mm at 42°, 39°, and 53°, respectively. While this vapor pressure does not seem very large, it is high enough to lead, after a period of time, to **evaporation** of the solid from an open container. Moth balls (naphthalene and 1,4-dichlorobenzene) show this behavior. When iodine is allowed to stand in a closed container over a period of time, one observes movement of crystals from one part of the container to another.

While chemists often refer to any solid-vapor transition as sublimation, the process described above for camphor, iodine and naphthalene is really an **evaporation** of a solid. Strictly speaking, a sublimation point is like a melting point or boiling point. It is defined as the point when the vapor pressure of the solid **equals** the applied pressure. Many liquids readily evaporate at temperatures far below their boiling points. It is, however, much less common for solids to evaporate. Solids which readily sublime (evaporate) must be stored in tightly stoppered containers. When the melting point of these solids is being determined, some of the solid may sublime and collect towards the open end of the melting point tube while the rest of the sample melts. To solve the sublimation problem, one seals the capillary tube or makes a rapid determination of the melting point. One may make use of the sublimation behavior to purify camphor. For example, at atmospheric pressure camphor may be readily sublimed, just below its melting point, at 175°. At 175°, the vapor pressure of camphor is 320 mm. The vapor will solidify on a cool surface.

13.3 VACUUM SUBLIMATION

Many organic compounds sublime readily under reduced pressure. When the vapor pressure of the solid equals the applied pressure, sublimation occurs, and the behavior will be identical to that shown in Figure 13–1B. The solid phase will pass directly into the vapor phase. From the data given in Table 13–1, one would expect camphor, naphthalene and benzoic acid to sublime at or below the applied pressures of 370, 7, and 6 mm, respectively. In principle, one could sublime p-nitrobenzaldehyde (last entry in Table 13–1), but it would not be practical because of the low applied pressure required.

13.4 SUBLIMATION METHODS

The process of sublimation can be used to purify solids. The solid is warmed until its vapor pressure becomes high enough for it to vaporize and to condense as a solid on a cooled surface placed close above. Several types of apparatus are illustrated in Figure 13–2. The upper surface provided for collection may be cooled by a continuous flow of water using a "cold finger" condenser (Figure 13–2A,B), by an ice/water (or Dry Ice/acetone) mixture (Figure 13–2D), or by a flow of air directed from a nozzle (Figure 13–2C). The water hoses must be securely attached to the inlet and outlet of the cold finger condenser. Otherwise, the connections may leak and allow water to pass into the sublimation apparatus. Many solids will not develop sufficient vapor pressure at 760 mm, but can be sublimed at reduced pressure. Thus, most sublimation equipment has provision for connection to an aspirator or vacuum pump. Reduction of the pressure is also advantageous in preventing thermal decomposition of substances that require high temperatures to sublime at ordinary pressures.

An inexpensive sublimation apparatus, shown in Figure 13–2B, can be prepared by shortening a 200 × 25 mm side arm test tube to 160 mm. A 150 × 20 mm test tube is wrapped with a 27 × 2.5 cm piece of rubber dam material and is inserted into the side arm test tube until it is secure. The test tube should be positioned about 2 cm from the bottom. Two sections of 6 mm glass tubing are inserted into a number 2 rubber stopper so that one piece extends to the bottom of the inner test tube. The rubber stopper is placed securely into the test tube. The inner test tube serves as the cold finger condenser. The outer side arm test tube serves as the container for the solid to be sublimed.

In practice, the term sublimation is loosely applied and is often used to describe the process where the solid actually melts prior to vaporization. This is not really a sublimation. It should be kept in mind while performing a sublimation that it is important to keep the temperature below the melting point of the solid.

After sublimation, the material that has collected on the cooled surface is recovered by removal of the central tube (cold finger) from the apparatus. Some care must be used when removing this tube in order to avoid dislodging the crystals which have collected. The deposit of crystals is scraped from the cold finger with a spatula. If reduced pressure has been used, the release of the pressure must also be done carefully to prevent a blast of air from dislodging the crystals.

13.5 ADVANTAGES OF SUBLIMATION

The advantage of this method is that no solvent is used and, therefore, that it need not be removed later. This method also removes

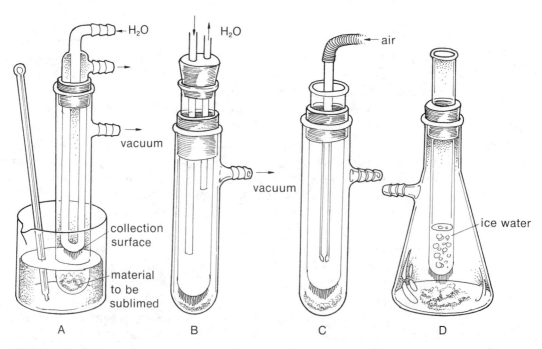

FIGURE 13–2. Sublimation apparatus

occluded material, e.g., molecules of solvation, from the sublimed substance. For instance, caffein (sublimes 178°, melts 236°) absorbs water gradually from the atmosphere to form a hydrate. During sublimation this water is lost and anhydrous caffein is obtained. If too much solvent is present in a sample to be sublimed, however, it will condense on the cooled surface rather than be lost and thus interfere with the sublimation.

Sublimation is a faster method of purification than crystallization, but it is not as selective. Vapor pressures are often similar when dealing with solids that sublime, consequently little separation can be achieved. For this reason, solids are far more commonly purified by crystallization. Sublimation is most effective in removing a volatile substance from non-volatile compounds, particularly salts and other inorganic materials. Sublimation is also an effective method for removing highly volatile bicyclic or other rather symmetrical molecules from less volatile reaction products. Examples of volatile bicyclic compounds are borneol, isoborneol and camphor.

PROBLEMS

1. Why is solid carbon dioxide called "Dry Ice"? How does its behavior differ from solid water?

2. Under what conditions can one have **liquid** carbon dioxide?

3. A solid substance has a vapor pressure of 800 mm at its melting point (80°). Describe the behavior of this solid as the temperature is raised at atmospheric pressure (760 mm).

4. A solid substance has a vapor pressure of 100 mm at the melting point (100°). Describe the behavior of this solid as the temperature is raised at atmospheric pressure (760 mm),

5. A substance has a vapor pressure of 50 mm at the melting point (100°). Describe how one would experimentally carry out the sublimation of this substance.

TECHNIQUES FOR HANDLING SODIUM METAL

> EXTREME CARE AND CAUTION SHOULD ALWAYS BE EXERCISED WHEN WORKING WITH SODIUM METAL

Elemental sodium is a metal which is low melting (98°C) and which has a low density (0.96 g/cc). It is metallic silver in color when pure. It is extremely soft, and it can be cut with ease by a spatula or a dull knife. It is very malleable; that is, small pieces can be "squashed" easily by pressing against them with the flat side of a spatula and by using only thumb pressure.

14.1 REACTIVITY OF SODIUM

Sodium is very reactive. It is one of the most electropositive elements and will react explosively with elements such as fluorine, chlorine, or pure oxygen. Sodium easily gives up its lone valence electron to more electronegative elements or other suitable acceptors to form the Na^+ ion. In particular, sodium reacts vigorously, often violently, with organic compounds having acidic hydrogens. Hydrogen is formed in these reactions.

$$Na\cdot + H^+ \longrightarrow [H\cdot] + Na^+$$
$$2[H\cdot] \longrightarrow H_2$$

With moderately to very acidic compounds this reaction is often sufficiently exothermic to ignite the hydrogen as it forms. Even water is sufficiently acidic to fall into this category:

$$2\,Na + 2\,H_2O \longrightarrow 2\,NaOH + \underbrace{H_2 + Heat}_{\text{Ignites}}$$

14.2 CAUTION WHEN HANDLING SODIUM

SODIUM REACTS VIOLENTLY WITH WATER, and care should always be taken when working with sodium metal to avoid contact with water or any wet apparatus. It should be noted especially that SODIUM SHOULD NEVER BE DISPOSED OF IN THE SINK. It is certain that organic solvents will have been poured into the sink in the organic laboratory. If these solvents are flammable, throwing sodium into the sink, where it reacts with water to produce and ignite hydrogen, might result in a colossal fire in the sink.

$$2\,Na + 2\,ROH \longrightarrow 2\,NaOR + H_2$$

Sodium reacts less violently with alcohols than with water—they are less acidic. As the hydrocarbon chain length of the alcohols increases, the rate of reaction decreases. Sodium may be added to ethanol, for instance, if some care is taken. The ethanol will become hot enough to boil, but it will not ignite if sodium is added slowly and if the solution is cooled. The rate of reaction with 1-butanol is quite slow; therefore, excess sodium is disposed of most prudently by placing it in a beaker with enough 1-butanol to cover it. The sodium should be left until **all** of the metal has dissolved and no bubbling action is seen. Then either water or methanol is added to the 1-butanol solution, and the material is disposed of normally. Other functional groups which react with sodium include: acids, amines, terminal acetylenes, enols, phenols, and mercaptans.

When working with sodium metal, one should avoid all contact of the sodium with the skin. Sodium reacts with moisture on the surface of the skin to produce sodium hydroxide which is, of course, highly caustic.

14.3 STORING SODIUM

Freshly cut sodium reacts with the moisture and oxygen in the air to form an oxide layer, consisting of Na_2O and $NaOH$, on the surface of the metal. A sample of freshly cut silvery sodium metal quickly tarnishes to grey when exposed to moist air. Eventually a thick white crust will form on all the surfaces. To prevent this crust formation, sodium metal usually is stored in a shallow, wide-mouth jar under an inert, high-boiling solvent. Xylene or ligroin, or sometimes mineral oil, are used most often for this purpose. Even so, some oxidation takes place, and sodium metal usually must be "pared" of its oxide coating before use.

14.4 CUTTING AND WEIGHING SODIUM

Since sodium tarnishes and oxidizes when exposed to air, it is usually manipulated under xylene or ligroin. To weigh sodium, a beaker containing xylene or ligroin is placed on a balance and tared (its weight recorded). Then, pieces of sodium are added directly to this beaker, resting on the balance, until the proper weight of sodium has been obtained. The following procedure is followed. The sodium is transferred from the storage container to a crystallizing dish or a paper towel, and it is cut as quickly as possible with a spatula or knife in such a way as to remove the oxide coating. The freshly cut pieces of sodium are transferred quickly to the tared beaker of xylene or ligroin to forestall oxidation. When enough sodium has been trimmed, as judged by the increase in weight of the beaker, the excess is returned to the storage container. Small pieces of sodium may be destroyed by treating them with 1-butanol in the manner described in Section 14.2. The crystallizing dish may be filled with 1-butanol to decompose the small amounts of sodium metal that remain. If a paper towel was used, it should be placed in a beaker of 1-butanol. Additional 1-butanol should be poured over the paper towel; after this treatment, the towel may be removed and discarded as usual.

The sodium weighed in the beaker is usually cut into small pieces in the beaker under the covering liquid. A knife or a spatula is used. If mineral oil is used, the pieces of sodium must be rinsed before use; a good solvent for this rinsing is petroleum ether. The small pieces may be pulled from the liquid using forceps or tweezers (never with the fingers), plunged into another beaker containing petroleum ether, and quickly transferred to the reaction mixture.

In general, for use in organic reactions, the sodium should be cut into pieces about the size of a large pea, or under less common circumstances the size of a BB. Pieces which correspond to a cube anywhere from 2 to 4 mm on a side are commonly used.

14.5 OTHER ALKALI METALS

Lithium metal (mp 186°C, density 0.53 g/cc) should be handled in a manner similar to that given for sodium. It is **slightly** less reactive than sodium. Potassium metal (mp 62.3°C, density 0.86 g/cc) is extremely reactive and must be handled with extra caution.

Lithium is sold often as a wire of known weight-per-length. It is stored in mineral oil. To weigh a given amount of lithium is easy when it is obtained in this form; one only need cut off a measured length of the wire.

14.6 THE SODIUM PRESS

Many labs have a **sodium press** which extrudes the metal into a wire about 1/8 inch in diameter. This sodium wire is then handled easily in the same manner mentioned for lithium. The weight per cm is usually constant from one sodium wire to the next coming from the same press, and measuring a length of wire may again be substituted for a weight determination.

Technique **15**

POLARIMETRY

15.1 THE NATURE OF POLARIZED LIGHT

Light possesses a dual nature, since it exhibits the properties of waves as well as those of particles. The wave nature of light may be demonstrated by two experiments, polarization and interference. Of the two, polarization is the more interesting to organic chemists because one can take advantage of polarization experiments to learn something about the structure of an unknown molecule.

Ordinary white light consists of wave motion in which the waves have a variety of wavelengths and are vibrating in all possible planes perpendicular to the direction of propagation. Light can be made to be **monochromatic** (of one wavelength or color) by the use of filters or special light sources. Frequently, a sodium lamp (sodium D line = 5893 Å) is used. Although the light from this lamp consists of waves of only one wavelength, the individual light waves are still vibrating in all possible planes perpendicular to the direction of the beam. If we imagine that the beam of light is aimed directly at the viewer, the ordinary light can be represented by showing the edges of the planes oriented randomly around the path of the beam, as in the left part of Figure 15-1.

A Nicol prism, which consists of a specially prepared crystal of Iceland spar (or calcite), has the property of serving as a screen with which to restrict the passage of the light waves. Waves that are vibrating in one plane are transmitted, while those in a perpendicular plane are rejected (either refracted in another direction or absorbed). The light which passes through the prism is called **plane-polarized light,** and it consists of waves which vibrate only in one plane. Again, considering a beam of plane-polarized light aimed directly at the viewer, it can be represented by showing the edges of the plane oriented in one particular direction, as in the right part of Figure 15-1.

Iceland spar has the property of **double refraction,** that is, an entering beam of ordinary light is split, or doubly refracted, into two separate emerging beams of light. The two emerging beams (labeled **A** and **B** in Figure 15-2) both have only a single plane of vibration, and

FIGURE 15-1. Ordinary versus plane-polarized light

the plane of vibration in beam **A** is perpendicular to that of beam **B**. In other words, the crystal has separated the incident beam of ordinary light into two beams of plane-polarized light, with the plane of polarization of beam **A** being perpendicular to that of beam **B**. Figure 15–2 illustrates the phenomenon of double refraction.

In order to generate a single beam of plane-polarized light, one can take advantage of the double refracting property of Iceland spar. A Nicol prism, invented by the Scottish physicist, William Nicol, consists of two crystals of Iceland spar cut to specified angles and cemented by Canada balsam. This prism transmits one of the two beams of plane-polarized light while reflecting the other at a sharp angle so that it does not interfere with the transmitted beam. Plane-polarized light may also be generated by a Polaroid filter, a device invented by the American E. H. Land. Polaroid filters consist of certain types of crystals, imbedded in a transparent plastic, capable of producing plane-polarized light.

After passing through a first Nicol prism, plane-polarized light can also pass through a second Nicol prism, but only if the second prism has its axis oriented so that it is **parallel** to the incident light's plane of polarization. Plane-polarized light will be **absorbed** by a Nicol prism which is oriented so that its axis is **perpendicular** to the incident light's plane of polarization. These situations can be illustrated by using the picket-fence analogy, as shown in Figure 15–3. Plane-polarized light can pass through a fence whose slats are oriented in the proper direction, but is blocked out by a fence whose slats are oriented in the perpendicular direction.

An optically active substance is one which interacts with polarized light to rotate the plane of polarization through some angle α. Figure 15–4 illustrates this phenomenon.

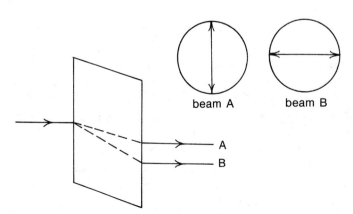

beam A   beam B   **FIGURE 15-2.** Double refraction

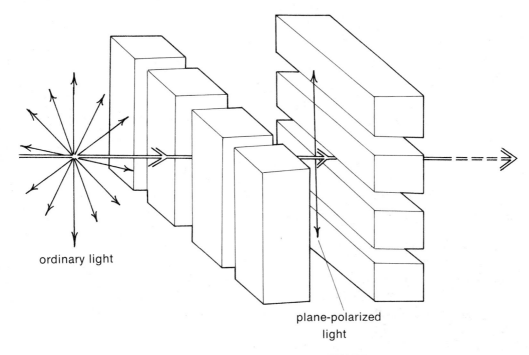

FIGURE 15-3. The picket fence analogy

ordinary light

plane-polarized
light

15.2 THE POLARIMETER

An instrument called a **polarimeter** is used to measure the extent to which a substance interacts with polarized light. A schematic diagram of a polarimeter is shown in Figure 15-5. The light from the source lamp is polarized by passing it through a fixed Nicol prism, called a polarizer. This light passes through the sample where it may or may not interact to have its plane of polarization rotated in one direction or the other. A second, rotatable Nicol prism, called the analyzer, is adjusted to allow the maximum amount of light to pass through. The number of degrees and the direction of rotation required for this adjustment are measured to give the **observed rotation, α.**

In order to permit comparisons with data determined by other persons under different conditions, a standardized means of presenting

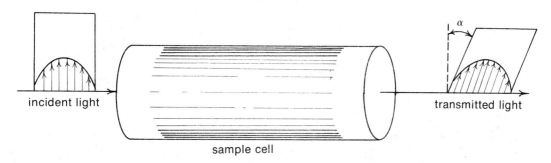

incident light

transmitted light

sample cell

FIGURE 15-4. Optical activity

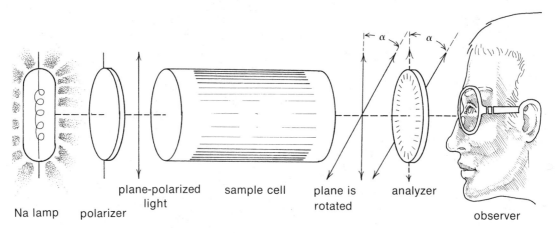

FIGURE 15-5. A schematic diagram of a polarimeter

optical rotation data is required. The most common means of presenting such data is by recording the specific rotation, $[\alpha]_\lambda^t$, which has been corrected for differences in concentration, cell path length, temperature, solvent, and wavelength of the light source. The equation defining the specific rotation of a compound in solution is:

$$[\alpha]_\lambda^t = \frac{\alpha}{cl}$$

α = observed rotation on degrees
c = concentration in grams/milliliter of solution
l = length of sample tube in decimeters
λ = wavelength of light (usually indicated as "D" for the sodium D line)
t = temperature in degrees Centigrade

For pure liquids, the density, d, of the liquid in grams per milliliter replaces c in the above formula. Occasionally one wishes to compare compounds of different molecular weights, so a **molecular rotation,** based on moles rather than grams, is more convenient than a **specific rotation.** The molecular rotation, M_λ^t, is obtained from the specific rotation $[\alpha]_\lambda^t$ by:

$$M_\lambda^t = \frac{[\alpha]_\lambda^t \times \text{Molecular Weight}}{100}$$

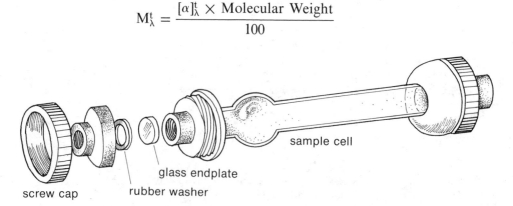

FIGURE 15-6. Polarimeter cell assembly

Usually measurements are performed at 25°C with the sodium D line as a light source, so specific rotations are reported as $[\alpha]_D^{25°}$.

15.3 THE SAMPLE CELLS

It is important that the solution whose optical rotation is to be determined contain no suspended particles of dust or dirt which might disperse the incident polarized light. For this reason, it is necessary to clean the sample cell carefully and to make certain that there are no air bubbles trapped in the path of the light. The sample cells contain an enlarged ring near one end into which the air bubbles may be trapped. The sample cell, shown in Figure 15-6, is tilted upwards and tapped until the air bubbles move into the enlarged ring. In reassembling the cell, it is important not to get fingerprints on the glass endplate. The cell assembly is shown in Figure 15-6.

Most commonly, the sample is prepared by dissolving 0.1 to 0.5 grams of the substance to be studied in 25 milliliters of solvent, usually water, ethanol, or chloroform. Of course, if the specific rotation of the substance is very high or very low, it may be necessary to make the solution respectively lower or higher in concentration, but usually this is determined after an initial attempt at a concentration range such as that suggested above.

15.4 OPERATION OF THE POLARIMETER

The procedures for the preparation of the cells and for the operation of the instrument are those appropriate for the Zeiss Polarimeter with the circular scale, but other models of polarimeter are operated in a quite similar manner. Before beginning the measurements, it is necessary to throw the power switch to the ON position and wait for five to ten minutes until the sodium lamp is properly warmed up.

The instrument should be checked initially by making a zero reading with a sample cell filled only with solvent. If the zero reading does not correspond with the zero degree calibration mark, then the difference in readings must be used to correct all subsequent readings. The reading is determined by laying the sample tube into the cradle, enlarged end up (while making sure that there are no air bubbles in the light path), closing the cover, and turning the knob until the proper angle of the analyzer is reached. Most instruments, including the Zeiss Polarimeter, are of the double-field type, where the eye sees a split field whose sections must be matched in light intensity. The value of the angle through which the plane of polarized light has been rotated (if any) is read directly from the scale which may be seen through the eyepiece directly below the split-field image. Figure 15-7 illustrates how this split field might appear.

 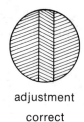 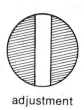

FIGURE 15-7. Split field image in the polarimeter

adjustment   adjustment   adjustment
incorrect   correct   incorrect

The cell containing the solution of the sample is then placed in the polarimeter, and the observed angle of rotation is obtained in the same fashion. Be sure to record not only the numerical value of the angle of rotation in degrees, but also the direction of rotation. Rotations clockwise are due to **dextrorotatory** substances and are indicated by the sign, "+." Rotations counterclockwise are due to **levorotatory** substances and are indicated by the sign, "−." In making a determination, it is best to take several readings, including readings where the actual value was approached from both sides. In other words, where the actual reading might be +75°, first approach this reading upward from a reading near zero, then on the next measurement approach this reading downward from an angle greater than +75°. Duplication of readings and approach of the observed value from both sides serve to reduce errors. The readings are then averaged to obtain the observed rotation, α. This rotation is then corrected by the appropriate factors, according to the formulas given in Section 15.2, to provide the specific rotation. The specific rotation is always reported as a function of temperature, indicating the wavelength by "D" if a sodium lamp is used, and by reporting the concentration and solvent used. For example, $[\alpha]_D^{20} = +43.8°$ (c = 7.5 g/100 ml in absolute ethanol).

15.5 OPTICAL PURITY

When one prepares a sample of an enantiomer by a resolution method, the sample is not always 100.0% pure enantiomer. It frequently is contaminated by residual amounts of the opposite stereoisomer. In order to determine the amount of the desired enantiomer in the sample, one calculates the **optical purity,** or the **excess** of one enantiomer in a mixture expressed as a percentage of the total. In a racemic (±) mixture, there is no excess enantiomer and the optical purity is zero; in a completely resolved material, the excess enantiomer is equal in weight to the total material, and the optical purity is 100%. Although the following is not the most precise equation for determining the optical purity, it is an equation which should prove useful in most simple applications.

$$\text{Optical purity} = \frac{\text{observed specific rotation}}{\text{specific rotation of pure substance}} \times 100$$

A compound which is X% optically pure contains X% of one enantiomer and $(100 - X)$% of a **racemic mixture.**

Given the optical purity, the relative percentages of the enantiomers may be calculated easily. Assuming that the predominant form in the impure, optically active mixture was the $(+)$ enantiomer, the percentage of the $(+)$ enantiomer would be $\left[X + \left(\dfrac{100 - X}{2}\right)\right]$%, and the percentage of the $(-)$ enantiomer would be $\left(\dfrac{100 - X}{2}\right)$%. The relative percentages of $(+)$ and $(-)$ forms in a partially resolved mixture of enantiomers may be calculated in the manner illustrated below. Consider a partially resolved mixture of camphor enantiomers. The specific rotation for pure $(+)$-camphor is $+43.8°$ in absolute ethanol, but the mixture shows a specific rotation of $+26.3°$.

$$\text{Optical purity} = \frac{+26.3°}{+43.8°} \times 100 = 60\% \text{ optically pure}$$

$$\% \ (+) \text{ enantiomer} = 60 + \left(\frac{100 - 60}{2}\right) = 80\%$$

$$\% \ (-) \text{ enantiomer} = \left(\frac{100 - 60}{2}\right) = 20\%$$

PROBLEMS

1. Calculate the specific rotation of a substance dissolved in a solvent (0.4 g/ml) which has an observed rotation of $-10°$ as observed using a 0.5 dm cell.

2. Calculate the observed rotation for a solution of a substance (2.0 g/ml) which is 80% optically pure. A 2 dm cell is used. The specific rotation for the optically pure substance is $+20°$.

3. Calculate the optical purity of a partially racemized product if the calculated specific rotation is $-8°$, while the pure enantiomer has a specific rotation of $-10°$. Calculate the percentage of each of the enantiomers in the partially racemized product.

Technique 16

REFRACTOMETRY

The **refractive index** is a useful physical property of liquids. Often a liquid can be identified by measuring its refractive index. Alternatively, a comparison of the experimentally measured refractive index to that reported in the literature for an ultrapure sample provides a measure of the purity of the sample being examined. The closer the measured sample's value approaches the literature value, the purer the sample will be.

16.1 THE REFRACTIVE INDEX

The refractive index derives from the fact that light travels at a different velocity in condensed phases (liquids, solids) than it does in air. The refractive index (n) is defined as the ratio of the velocity of light in air to the velocity of light in the medium being measured.

$$n = \frac{V_{air}}{V_{liquid}} = \frac{Sin\,\theta}{Sin\,\phi}$$

The ratio of the velocities is not difficult to measure experimentally. It corresponds to $Sin\,\theta/Sin\,\phi$, where θ is the angle of incidence for a beam of light striking the surface of the medium, and ϕ is the angle of refraction of the beam of light **within** the medium. This is illustrated in Figure 16–1.

The refractive index for a given medium is dependent on two variable factors. First, it is **temperature** dependent. The density of the medium changes with temperature and, hence, the speed of light in the medium also changes. Second, the refractive index is **wavelength** dependent. Beams of light with different wavelengths will be refracted to different extents in the same medium and will give different refractive indices for that medium. It is usual to report refractive indices measured at 20°C, using a sodium discharge lamp as a source of illumination. The sodium lamp gives off yellow light of 589 nm wavelength, the so-called "sodium D line." Under these conditions, the refractive index is reported in the following form:

$$n_D^{20} = 1.4892$$

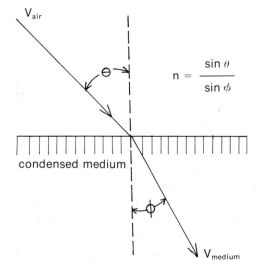

$$n = \frac{sin\,\theta}{sin\,\phi}$$

FIGURE 16–1. The refractive index

The superscript indicates the temperature, and the subscript indicates that the sodium D line was used for the measurement. If another wavelength is used for the determination, the D is replaced by the appropriate value, usually in nanometers ($1 \text{ nm} = 10^{-9} \text{ m}$).

It should be noticed that the hypothetical value reported has four decimal places. It is easy to determine the refractive index to within several parts in 10,000. Therefore, n_D is a very accurate physical constant for a given substance and can be used for identification. However, it is sensitive to even small amounts of impurity in the substance measured. Unless the substance is purified **extensively,** it will not usually be possible to reproduce the last two decimal places given in a handbook or other literature source.

16.2 THE ABBE REFRACTOMETER

The instrument used to measure the refractive index is called a **refractometer.** Although many styles of refractometer are available, by far the most common instrument is the Abbé refractometer. This style of refractometer has the following advantages:

1. White light may be used for illumination, but the instrument is

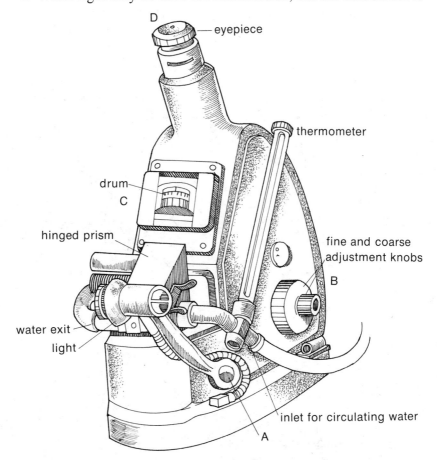

FIGURE 16–2. Abbé refractometer (Bausch and Lomb Abbé 3L)

compensated so that the index of refraction obtained is actually that for the sodium D line.

2. The prisms can be temperature controlled.
3. Only a small sample is required (a few drops of a liquid).

A common type of Abbé refractometer is shown in Figure 16–2.

Although the optical arrangement of the refractometer is very complex, a simplified diagram of the internal workings is given in Figure 16–3. The letters **A, B, C,** and **D** label corresponding parts in both Figures 16–2 and 16–3. A complete description of refractometer optics is too difficult to attempt here, but Figure 16–3 gives a simplified diagram of the essential operating principles.

The sample to be measured is introduced between the two prisms. If it is a "free-flowing" liquid, it may be introduced into a channel along the side of the prisms, injecting it with a blunt pipet or eye dropper. If it is a viscous sample, the prisms must be opened (they are hinged) by lifting the upper one, and a few drops of liquid are applied to the lower prism with a wooden applicator or an eyedropper. If an eyedropper is used, care must be taken not to touch the prisms, since they scratch easily. On closing the prisms, the liquid should spread evenly to make a thin film.

Next, one turns on the light and looks into the eyepiece, **D.** The

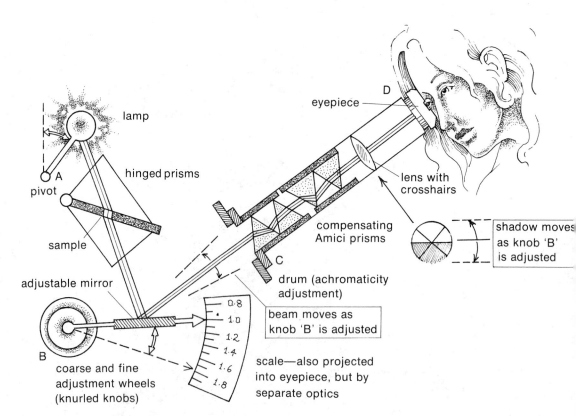

FIGURE 16–3. Simplified diagram of a refractometer

FIGURE 16–4. *A,* Refractometer is incorrectly adjusted; *B,* correct adjustment of refractometer

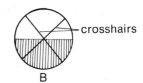

A   B

hinged lamp and the coarse adjustment knob at **B** are adjusted to give the most uniform illumination to the visible field seen in the eyepiece (no dark areas). The light rotates at pivot **A.**

Once a uniform field is found, one rotates the coarse and fine adjustment knobs at **B** until the dividing line between the light and dark halves of the visual field coincides with the center of the cross hairs (Figure 16–4). If the cross hairs are not in "sharp" focus, it will be necessary to adjust the eyepiece to focus them. If the horizontal line dividing the light and dark areas appears as a colored band, as in Figure 16–5, the refractometer is showing **chromatic aberration** (color dispersion). This can be adjusted with the knob labelled **C.** This knurled knob rotates a series of prisms, called Amici prisms, that color-compensate the refractometer and cancel out dispersion. The knob should be adjusted to give a sharp, uncolored division between the light and dark segments. When one has the correct adjustments made (as in Figure 16–4B), the refractive index is read. In the instrument described here, a small button on the left side of the housing is pressed, and the scale becomes visible in the eyepiece. In other refractometers the scale is visible at all times, frequently through a separate eyepiece.

Occasionally the refractometer will be so far out of adjustment that it may be difficult to measure the refractive index of an unknown. When this happens, it is wise to place a pure sample of known refractive index in the instrument, set the scale to the correct value of refractive index, and adjust the controls for the sharpest line possible. Once this has been done, it will be easier to measure an unknown sample. Typical organic liquids have refractive index values between 1.3400 and 1.5600.

There are many styles of refractometer, but most will have adjustments similar to those described here.

16.3   CLEANING THE REFRACTOMETER

When using the refractometer, it should always be remembered that if the prisms are scratched the instrument will be ruined. DO NOT

FIGURE 16–5. Refractometer showing chromatic aberration (color dispersion). The dispersion is incorrectly adjusted

TOUCH THE PRISMS WITH ANY HARD OBJECT. This admonition includes eyedroppers and glass rods.

When measurements are completed, the prisms should be cleaned with ethanol or petroleum ether. **Soft** tissues are moistened with the solvent, and the prisms are wiped **gently.** When the solvent has evaporated from the prism surfaces, they should be locked together. The refractometer should be left with the prisms closed to avoid collection of dust in the space between them. The instrument should also be turned off when it is no longer being used.

16.4 TEMPERATURE CORRECTIONS

If the refractive index is not determined in a room whose temperature is 20°C, or if 20° cooling water is not used to circulate through the instrument, a temperature correction must be used. Although the magnitude of the temperature correction to be used may vary from one class of compounds to another, a value of 0.00045 per degree Celsius serves as a useful approximation for most substances. The index of refraction of a substance **decreases** with **increasing** temperature. Therefore, one adds the correction to the observed n_D value for temperatures higher than 20°C and subtracts it for temperatures lower than 20°C. As an example, the reported n_D^{20} value for nitrobenzene is 1.5529. At 25°C, one would observe a value of 1.5506. The temperature correction would be made as follows:

$$n_D^{20} = 1.5506 + 5(0.00045) = 1.5529$$

Technique 17

PREPARATION OF SAMPLES FOR SPECTROSCOPY

Modern organic chemistry requires the use of sophisticated scientific instruments. Most important among these instruments are the two spectroscopic instruments: the infrared and nuclear magnetic resonance spectrometers. The instruments are indispensable to the modern organic chemist in proving the structures of unknown substances, in verifying that reaction products are indeed what had been predicted, and in characterizing organic compounds. The theory underlying the operation of these instruments may be found in most standard lecture textbooks in organic chemistry. Additional information, including correlation charts, to assist in the interpretation of spectra may be found in this textbook in Appendix Three (Infrared Spectroscopy) and in Appendix Four (Nuclear Magnetic Resonance). This technique chapter will concentrate on the preparation of samples for these spectroscopic methods.

PART A. INFRARED

17.1 SAMPLE PREPARATION: COMMON METHODS

In order to determine the infrared spectrum of a compound, one must place it in a sample holder or cell of some kind. In infrared spectroscopy this immediately poses a problem. Glass, quartz, and plastics absorb strongly throughout the infrared region of the spectrum (any compound with covalent bonds usually absorbs) and cannot be used to construct sample cells. Ionic substances must be employed in cell construction. Metal halides (sodium chloride, potassium bromide, silver chloride) are commonly used for this purpose. Single crystals of sodium chloride are cut and polished to give plates which are transparent throughout the infrared region. These plates are then used to fabricate sample cells. The only disadvantages are that the plates cleave easily (break) when too much pressure is applied and that sodium chloride is water soluble. This means that samples must be **dry** before a spectrum can be obtained. Since water absorbs in the infrared, water should be removed from the samples. Silver chloride is water insoluble, but **expensive,** and can be used for aqueous solutions.

17.2 LIQUID SAMPLES

Salt Plates

The simplest method of preparing the sample, if it is a liquid, is to place a thin layer of the liquid between two sodium chloride plates which have been flat-ground and polished. A drop of the liquid is placed on the surface of one plate, and the second plate then placed on top. The pressure of this second plate causes the liquid to spread out and form a thin capillary film between the two plates. The plates are then mounted in a holder designed to allow them to be placed in the sample beam of the spectrophotometer. This holder is illustrated in Figure 17–1. All parts of the holder (plates and mounting holder) are usually stored in a desiccator or oven near the infrared spectrophotometer. The large polished NaCl plates are expensive because they are cut from a large single crystal of NaCl, and break easily (specifically they can be cleaved). They are also water soluble. Handle these plates carefully; **they should only be touched on their edges.** The moisture from the

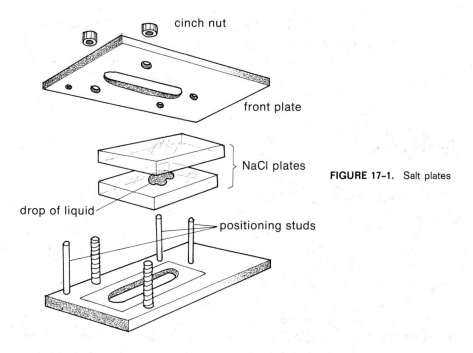

cinch nut

front plate

NaCl plates

FIGURE 17–1. Salt plates

drop of liquid

positioning studs

fingers will mar and occlude the polished surfaces. Also, a spectrum of water or any solution containing water can obviously not be taken, as water will damage the plates. **It must be certain that the sample is dry or free from water.**

When assembling the holder **do not tighten the cinch nuts excessively** because too great a pressure will cleave or split the NaCl plates. Just tighten the nuts firmly but do not use any force at all to turn them. Spin them with the fingers until they stop, then turn them just another fraction of a full turn, and they will be tight enough.

Once the spectrum is determined, the NaCl plates should be washed with a volatile dry solvent such as chloroform or carbon tetrachloride. A soft tissue, moistened with solvent, may be used to wipe them off, and this will be especially useful if one has run a Nujol mull (see Section 17.4).

To some extent the thickness of the film obtained between the two plates is a function of two factors: a. the amount of liquid placed on the first plate (one drop, two drops, etc.), and b. the pressure used to hold the plates together. If more than one or two drops of liquid have been used, it will probably be too much, and the resulting spectrum will show such strong absorptions everywhere that they are off the scale of the chart paper. Only enough liquid is needed to wet both surfaces. The capillary layer of liquid can be made thinner by increasing the pressure on the two plates, but one must be cautious of exerting too much pressure because the plates will break.

Occasionally a sample will have too low a viscosity or be too volatile to give a practicable capillary film; the liquid will either form too thin a film or evaporate too quickly. Should this happen, the spectrum will have to be determined using solution cells. These are described in Section 17.3.

17.3 SOLID SAMPLES

KBr Pellets

The easiest method of preparing a solid sample is to make a KBr pellet. Effectively, KBr does not absorb in the infrared, and it can be used to make a solid "solution" of an unknown sample whose spectrum is desired. A good grade of KBr must be used, and it must be dry. Potassium bromide acquires waters of hydration on standing, and it is, therefore, stored in an oven. One weighs out about 1 to 2 mg of the solid sample and mixes it with about 100 mg of KBr. The first few times this mixture is prepared it should be weighed out on an analytical balance. After some experience, these quantities can be quite closely estimated by eye. The mixture is then ground to a fine powder in an agate or glass mortar and pestle. Be sure the sample gets uniformly mixed into the KBr. The powder must be finely ground or it will scatter the infrared radiation excessively. The total grinding time will be 3 to 5 minutes.

A **portion** of the finely ground powder (usually not more than half) is placed into a die which compresses it into a translucent pellet. The simplest die consists of two stainless steel bolts and a large nut, as is illustrated in Figure 17–2. The bolts have their ends ground flat.[1] To use this die, screw one of the bolts into the nut, but not all the way; leave 1 or 2 turns. Carefully add the powder into the open end of the partly assembled die and tap it lightly on the desk top to give an even layer of filling. Then, carefully screw the second bolt into the open nut until it is just firm. The die is then transferred to a holder which is bolted to the table and keeps the head of one bolt from turning. The final tightening

_____

[1]A Minipress of this type is available from Wilks Scientific Company.

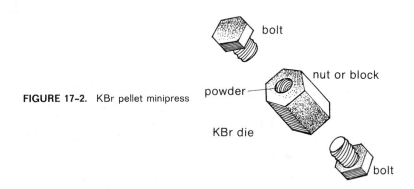

FIGURE 17–2. KBr pellet minipress

of the die to compress the KBr mixture is done using a torque wrench. Continue to turn the torque wrench until a loud click is heard (the ratchet mechanism makes softer clicks) or until the appropriate torque value (20 ft-lbs) is reached. If the bolt is tightened beyond this point, the head may be twisted off one of the bolts or the block (nut) may be split. Leave the die under pressure for 30 to 60 seconds and then reverse the ratchet on the torque wrench or pull the torque wrench in the opposite direction to open the assembly. Remove the two bolts, but leave the compressed KBr disc in the center of the block. The block rests in a holder which allows the sample to sit in the sample beam of the spectrophotometer.

The KBr pellet should be either clear or translucent. If it is not, the sample will scatter the beam of infrared radiation to such an extent that a spectrum cannot be obtained. If the pellet is found to be cloudy, one of several things may have been wrong:

a. The KBr mixture may not have been ground finely enough, and the particle size may be too big.
b. The sample may not have been thoroughly dried.
c. The ratio of sample to KBr may have been too high, i.e., too much sample was used.
d. The pellet may be too thick, i.e., too much of the mixture was put into the die.
e. The KBr may have been "wet" or have acquired moisture from the air.
f. The die may not have been tightened enough (this is improbable when a torque wrench is used).
g. The sample may have a low melting point. Low melting solids are not only difficult to dry, but also melt under pressure.

Often a cloudy pellet cannot be avoided. When this happens it is often possible to compensate (at least partially) for the scattering by placing a wire screen in the reference beam, thereby balancing the lowered transmittance of the pellet.

When the determination of the spectrum has been completed, punch the pellet out of the block and wash both the block and the bolts thoroughly with water. Then rinse them with acetone to dry them and return them to the oven for storage.

Solution Spectra

Method A. For substances which are soluble in carbon tetrachloride, a quick and easy method of determining the spectra of solids is available. The solid is dissolved in carbon tetrachloride (weight ratio = 1:10). One or two drops of the solution is placed between sodium chloride plates in precisely the same manner as that used for pure liquids (Section 17.2). The spectrum is determined in the way described for pure liquids using salt plates (Section 17.2). Since the spectrum contains the absorptions of the solute superimposed on the absorptions of carbon tetrachloride, it is important to remember that any absorption which appears to the right of about 900 cm^{-1} (11.1 μ) may be due to the stretching of the C—Cl bond of the solvent. Information which is contained to the right of 900 cm^{-1} is not usable in this method. Chloroform solutions may not be studied using this method, since the solvent has too many interfering absorptions.

Method B. The spectra of solids or those liquids which are too volatile or of too low a viscosity to be run between salt plates may be determined in a type of permanent sample cell called a **solution cell.** The solution cell, which is shown in Figure 17–3, is constructed from two salt plates which are mounted with a Teflon spacer between them to control the thickness of the sample. The top sodium chloride plate has two holes drilled in it so that the sample may be introduced into the cavity between the two plates. These holes are extended through the face plate by two tubular extensions which are designed to hold two Teflon plugs which seal the internal chamber and prevent evaporation. These tubular extensions are tapered so that a syringe body (Luer lock without a needle) will fit snugly into them from the outside. The cells are thus filled from a syringe, usually by holding them upright and filling from the bottom entrance port.

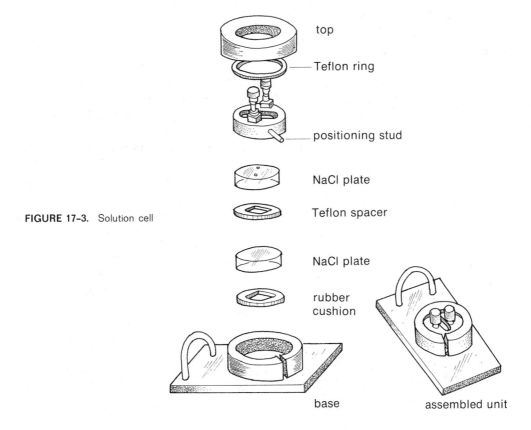

FIGURE 17–3. Solution cell

top

Teflon ring

positioning stud

NaCl plate

Teflon spacer

NaCl plate

rubber cushion

base

assembled unit

These cells are expensive (over $125.00 a pair), and, hence, your instructor may wish to limit their availability for use by large classes. It is strongly suggested that the instructor's permission to use the solution cells be secured before beginning to handle them. The cells are purchased in matched pairs. The dissolved sample is placed in one cell (the sample cell), and the pure solvent is placed in the other cell (the reference cell). The spectrum of the solvent is thus subtracted from the spectrum of the solution (not always completely), thus providing a spectrum of the solute. For this solvent compensation to be done as exactly as possible, it is essential that the same cell be used as a reference and that the other cell be used as a sample cell without ever interchanging them. When the spectrum is determined, it is important to clean the cells by flushing them with clean solvent.

Common solvents used to determine infrared spectra are carbon tetrachloride, chloroform, and carbon disulfide. The spectra of these substances are shown in Figures 17–4 through 17–6. For solution work, a 5 to 10 percent solution usually gives a good spectrum.

> Before using the solution cells, the instructor's permission and his instruction in how to fill and clean the cells must be obtained.

17.4 OTHER METHODS

Nujol Mulls

If an adequate KBr pellet cannot be obtained, the spectrum of a solid may be determined as a Nujol mull. In this method, about 5 mg of

INFRARED SPECTRA OF SOLVENTS COMMONLY USED FOR SAMPLE PREPARATION

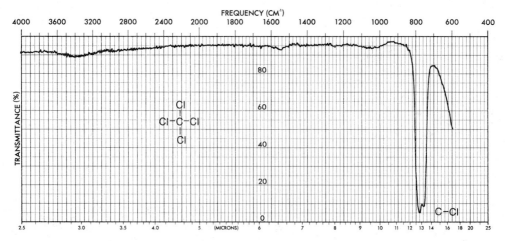

FIGURE 17–4. Carbon tetrachloride

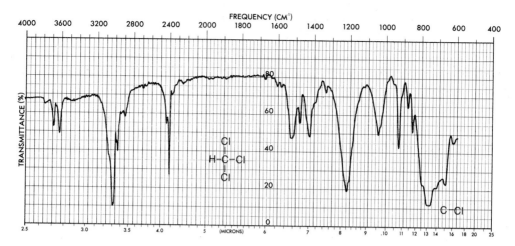

FIGURE 17-5. Chloroform

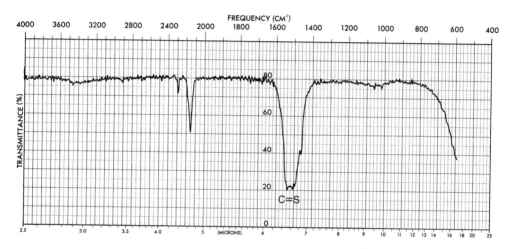

FIGURE 17-6. Carbon disulfide

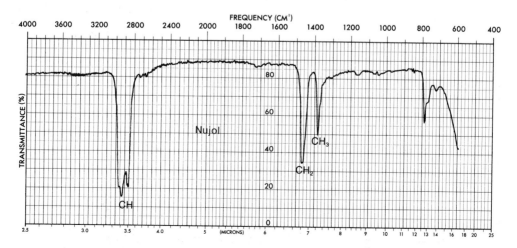

FIGURE 17-7. Nujol (mineral oil)

the solid sample are ground finely in a glass or agate mortar and pestle. One or two drops of Nujol mineral oil, white) is then added and the mixture is ground to a very fine dispersion about the consistency of Milk of Magnesia. This "mull" is then placed between two salt plates in the same manner as with liquid samples (see Section 17.2).

Nujol is a mixture of high molecular weight hydrocarbons. Hence, it has absorptions in the C—H stretch and CH_2 and CH_3 bending areas of the spectrum (see Figure 17–7). Clearly, if Nujol is used, no information can be obtained in these portions of the spectrum. When interpreting the spectrum, these Nujol peaks must be ignored. It is important to label the spectrum immediately after it was determined, noting that it was determined as a Nujol mull. Otherwise it might be forgotten that the C—H peaks belong to Nujol and not the solute.

17.5 RECORDING THE SPECTRUM

The instructor will describe the method of operating the infrared spectrophotometer, since the controls vary, depending upon the manufacturer and model of the instrument. In all cases, it is important that the sample, the solvent, and the type of cell or method used, and any other pertinent information be written on the spectrum immediately after determination. This information may be quite important, and it is easily forgotten if not recorded.

17.6 CALIBRATION

In order to know the frequency, or wavelength, of each absorption peak precisely, it is necessary to calibrate the frequency scale of the spectrum. This calibration is accomplished by rerecording a partial spectrum of a standard substance, which is always polystyrene, over the spectrum of the compound being studied. The complete spectrum of polystyrene is shown in Figure 17–8. A thin film of polystyrene is mounted in a card designed to fit into the sample cell holder on the spectrophotometer. The card is inserted into the sample cell holder, and the tips (not the entire spectrum) of the important peaks are recorded over the sample spectrum. The most important of these peaks is at 1603 cm^{-1} (6.238 μ), while other useful peaks are at 2850 cm^{-1} (3.509 μ) and 906 cm^{-1} (11.035 μ).

While some modern instruments record spectra so precisely that calibration is usually not necessary, it is always good practice to calibrate a spectrum whenever possible. The time spent in calibration is insignificant, and it minimizes frequency assignment errors caused by a poor fit of the paper to the chart recorder bed.

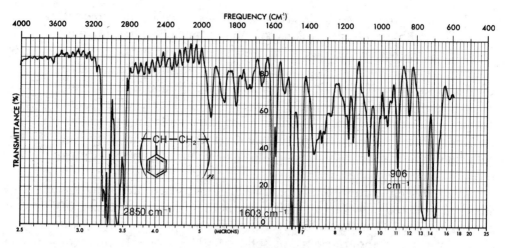

FIGURE 17–8. Infrared spectrum of polystyrene (thin film)

PART B. NUCLEAR MAGNETIC RESONANCE

17.7 PREPARING AN NMR SAMPLE

The nmr sample tubes used in most 60 MHz instruments are approximately $3/8'' \times 6''$ in overall dimension and are fabricated of uniformly thin glass tubing. The sample tube is spun around its cylindrical axis while being suspended in the magnet gap by a holder. The lower tip of the tube is positioned between the magnet pole pieces and the oscillator and detector coils by use of a depth gauge. To be sure that the sample will be aligned correctly, the tube should be filled to a minimum depth of about one and one-half inches from the bottom. This usually requires about a 0.5 to 1.0 ml quantity of the sample solution (sample dissolved in a suitable solvent). Figure 17–9 depicts an nmr sample tube showing the correct sample level.

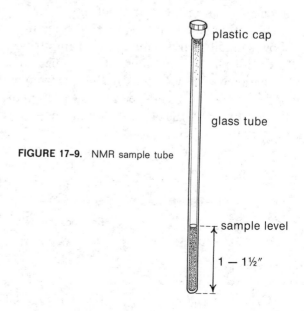

FIGURE 17–9. NMR sample tube

Liquid compounds **which are not viscous** may be determined "neat," i.e., without solvent. Viscous compounds must be determined in solution. It is generally best to determine **all** samples in solution. Solutions which are 10 to 30 percent sample (weight/weight) generally give satisfactory spectra. Typically, a 20 percent concentration is used. Usually the sample can be prepared "by eye" if a sample of a liquid compound is being prepared. Visually divide the sample tube into fifths and fill it with a disposable pipet or eye dropper to the one-fifth level with liquid sample, and then bring the total volume to the five-fifths level by addition of the solvent.

A solid sample, however, is best weighed out when preparing the solution. A deceivingly large amount (by volume), or so it seems, of solid is required to make a 20 percent w/w solution of a solid. One should weigh out about 150 mg of solid per 0.5 ml of solvent used.

To prepare the solution, one must of course choose the appropriate solvent *first!* There are two requirements for the solvent: 1) the sample should dissolve in it and 2) the solvent should have no nmr absorption peaks of its own, i.e., no protons. Carbon tetrachloride (CCl_4) is the most widely used solvent for this purpose. Most compounds will dissolve in CCl_4. It has the ability to dissolve a wide variety of functional groups and has **no protons.** If the sample will not dissolve in CCl_4, the more polar solvent deuterochloroform, or chloroform-d ($CDCl_3$), can often be used to advantage. Deuterium (^2_1H) does not absorb in the proton region and is thus "invisible" or not seen in the nmr spectrum. An even more polar solvent is deuterium oxide (D_2O). This solvent is often used for very polar compounds. Since the deuterated solvents are **expensive,** one usually determines the spectrum in carbon tetrachloride if possible. It should also be kept in mind that the chemical shift values commonly reported for various types of protons refer to the chemical shift as determined in either CCl_4 or $CDCl_3$. If another solvent is used, the chemical shift values can, and often do, change markedly owing to the different solvent environment. When $CDCl_3$ is used, there is often a low intensity peak in the nmr spectrum at 7.27δ owing to an unavoidably small amount of $CHCl_3$ impurity. Spectra determined in D_2O often show a small peak because of OH impurity. If the sample compound has acidic hydrogens, these may **exchange** with D_2O, leading to the appearance of an OH peak in the spectrum and the **loss** of the absorption owing to the exchanged hydrogen. In many cases this will also alter the splitting patterns of a compound since J_{HD} vicinal is often negligible ($J_{HD} \approx 0$ to 1.5 Hz).

$$R-\overset{\overset{O}{\|}}{C}\diagdown_{O-H} + D_2O \rightleftharpoons R-\overset{\overset{O}{\|}}{C}\diagdown_{O-D} + D-OH$$

~12.0δ   becomes invisible   OH peak appears

$$\underset{CH_3\diagup \diagdown CH_3}{\overset{\overset{O}{\|}}{C}} + D_2O \rightleftharpoons \underset{D-CH_2\diagup \diagdown CH_3}{\overset{\overset{O}{\|}}{C}} + D-OH$$

$$CH_3CH_2OH + D_2O \rightleftharpoons CH_3CH_2OD + D-OH$$

When the above solvents fail, other special solvents may be used. Acetone, acetonitrile, dimethylsulfoxide, pyridine, benzene, and dimethylformamide can be used if one is not interested in the region(s) of the nmr spectrum where they give rise to absorption. The deuterated, but **expensive,** analogs of these compounds are also used in special instances (e.g., acetone-d_6, dimethylsulfoxide-d_6, dimethylformamide-d_7, and benzene-d_6). If the sample is not sensitive to acid, trifluoroacetic acid (which has no protons with $\delta < 12$) can be used. Once again, one has to be aware that these solvents often lead to different chemical shift values than those determined in CCl_4 or $CDCl_3$. Variations of as much as 0.5 to 1.0 ppm have been observed. In fact, it is sometimes possible to separate peaks which are overlapped in CCl_4 or $CDCl_3$ solutions by switching to pyridine, benzene, acetone, or dimethylsulfoxide as solvents.

> Before using any deuterated solvent, check the solubility of the compound in its undeuterated analog. Deuterated compounds are expensive and should not be wasted.

17.8 RECORDING THE SPECTRUM

In most instances the instructor or some qualified laboratory assistant will actually record the nmr spectrum of a student sample. If students should be permitted to operate the nmr spectrometer, the instructor will provide instruction in its operation. Since the controls of nmr spectrometers vary, depending upon the make or model of the instrument, no attempt to describe these controls will be made here. **Do not operate the nmr spectrometer unless proper instruction has been obtained.**

17.9 REFERENCE SUBSTANCES

To provide the internal reference standard, TMS must be added to the sample solution. TMS, or tetramethylsilane, is $(CH_3)_4Si$. The concentration of TMS should be about 1 to 3 percent. Some people prefer to add one or two drops of TMS to the sample just prior to determining the spectrum. (Not much TMS is needed since it has 12 equivalent protons!) For this purpose a small pipet or a syringe is used. It is far easier to make (in bulk) standard solvents already containing 1 to 3 percent of TMS dissolved in them. Since TMS is highly volatile (bp 26.5°) such solutions must be stored, tightly stoppered, in a refrigerator. TMS itself is best stored in a refrigerator as well.

TMS will not dissolve in D_2O. For spectra determined in D_2O a

different internal standard, sodium 2,2-dimethyl-2-silapentane-5-sulfo-
nate, must be used. This standard is water soluble and gives a resonance
peak at essentially the same δ value as TMS.

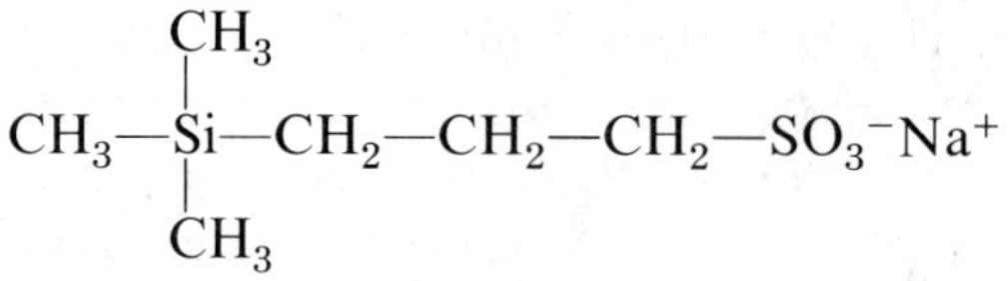

$$
\begin{array}{c}
\quad\;\; CH_3 \\
\quad\;\; | \\
CH_3-Si-CH_2-CH_2-CH_2-SO_3^-Na^+ \\
\quad\;\; | \\
\quad\;\; CH_3
\end{array}
$$

Sodium 2,2-Dimethyl-2-silapentane-5-sulfonate (DSS)

Appendix 1

TABLES OF UNKNOWNS AND DERIVATIVES

ALDEHYDES

COMPOUND	BP	MP	SEMI-CARBAZONE*	2,4-DINITROPHENYL-HYDRAZONE*
Ethanal (Acetaldehyde)	21	—	162	168
Propanal (Propionaldehyde)	48	—	89	148
Propenal (Acrolein)	52	—	171	165
2-Methylpropanal (Isobutyraldehyde)	64	—	125	187
Butanal (Butyraldehyde)	75	—	95	123
3-Methylbutanal (Isovaleraldehyde)	92	—	107	123
Pentanal (Valeraldehyde)	102	—	—	106
2-Butenal (Crotonaldehyde)	104	—	199	190
2-Ethylbutanal (Diethylacetaldehyde)	117	—	99	95
Hexanal (Caproaldehyde)	130	—	106	104
Heptanal (Heptaldehyde)	153	—	109	108
2-Furaldehyde (Furfural)	162	—	202	212
2-Ethylhexanal	163	—	254	114
Octanal (Caprylaldehyde)	171	—	101	106
Benzaldehyde	179	—	222	237
Phenylethanal (Phenylacetaldehyde)	195	33	153	121
2-Hydroxybenzaldehyde (Salicylaldehyde)	197	—	231	248
4-Methylbenzaldehyde (*p*-Tolualdehyde)	204	—	234	234
3,7-Dimethyl-6-octenal (Citronellal)	207	—	82	77
2-Chlorobenzaldehyde	213	11	229	213
4-Methoxybenzaldehyde (*p*-Anisaldehyde)	248	2.5	210	253
trans-Cinnamaldehyde	250 d.	—	215	255
3,4-Methylenedioxybenzaldehyde (Piperonal)	263	37	230	266 d.
2-Methoxybenzaldehyde (*o*-Anisaldehyde)	245	38	215 d.	254
4-Chlorobenzaldehyde	214	48	230	254
3-Nitrobenzaldehyde	—	58	246	293
4-Dimethylaminobenzaldehyde	—	74	222	325
Vanillin	285 d.	82	230	271
4-Nitrobenzaldehyde	—	106	221	320 d.
4-Hydroxybenzaldehyde	—	116	224	280 d.
(±)-Glyceraldehyde	—	142	160 d.	167

*See "Procedures for Preparing Derivatives," Appendix Two.

KETONES

COMPOUND	BP	MP	SEMI-CARBAZONE*	2,4-DINITROPHENYL HYDRAZONE*
2-Propanone (Acetone)	56	—	187	126
2-Butanone (Methyl ethyl ketone)	80	—	146	117
3-Methyl-2-butanone (Isopropyl methyl ketone)	94	—	112	120
2-Pentanone (Methyl propyl ketone)	101	—	112	143
3-Pentanone (Diethyl ketone)	102	—	138	156
Pinacolone	106	—	157	125
4-Methyl-2-pentanone (Isobutyl methyl ketone)	117	—	132	95
2,4-Dimethyl-3-pentanone (Diisopropyl ketone)	124	—	160	95
2-Hexanone (Methyl butyl ketone)	128	—	125	106
4-Methyl-3-penten-2-one (Mesityl oxide)	130	—	164	205
Cyclopentanone	131	—	210	146
2,3-Pentanedione	134	—	122 (mono) 209 (di)	209
2,4-Pentanedione (Acetylacetone)	139	—	—	122 (mono) 209 (di)
4-Heptanone (Dipropyl ketone)	144	—	132	75
2-Heptanone (Methyl amyl ketone)	151	—	123	89
Cyclohexanone	156	—	166	162
2,6-Dimethyl-4-heptanone (Diisobutyl ketone)	168	—	122	92
2-Octanone	173	—	122	58
Cycloheptanone	181	—	163	148
2,5-Hexanedione (Acetonylacetone)	191	−9	185 (mono) 224 (di)	257 (di)
Acetophenone (Methyl phenyl ketone)	202	20	198	238
Phenyl-2-propanone (Phenylacetone)	216	27	198	156
Propiophenone (Ethyl phenyl ketone)	218	21	182	191
4-Methylacetophenone	226	—	205	258
2-Undecanone	231	12	122	63
4-Chloroacetophenone	232	12	204	236
4-Phenyl-2-butanone (Benzylacetone)	235	—	142	127
4-Chloropropiophenone	—	36	176	223
4-Phenyl-3-buten-2-one	—	37	187	227
4-Methoxyacetophenone	258	38	198	228
Benzophenone	305	48	167	238
4-Bromoacetophenone	225	51	208	230
2-Acetonaphthone	—	54	235	262
Desoxybenzoin	320	60	148	204
3-Nitroacetophenone	202	80	257	228
9-Fluorenone	345	83	234	283
Benzoin	344	136	206	245
4-Hydroxypropiophenone	—	148	—	229
(±)-Camphor	205	179	237	177

*See "Procedures for Preparing Derivatives," Appendix Two.

CARBOXYLIC ACIDS

COMPOUND	BP	MP	p-TOLUIDIDE*	ANILIDE*	AMIDE*
Formic acid	101	8	53	47	43
Acetic acid	118	17	148	114	82
Propenoic acid (Acrylic acid)	139	13	141	104	85
Propanoic acid (Propionic acid)	141	—	124	103	81
2-Methylpropanoic acid (Isobutyric acid)	154	—	104	105	128
Butanoic acid (Butyric acid)	162	—	72	95	115
2-Methylpropenoic acid (Methacrylic acid)	163	16	—	87	102
Pyruvic acid	165 d.	14	109	104	124
3-Methylbutanoic acid (Isovaleric acid)	176	—	109	109	135
Pentanoic acid (Valeric acid)	186	—	70	63	106
2-Methylpentanoic acid	186	—	80	95	79
2-Chloropropanoic acid	186	—	124	92	80
Dichloroacetic acid	194	6	153	118	98
Hexanoic acid (Caproic acid)	205	—	75	95	101
2-Bromopropanoic acid	205	24	125	99	123
Octanoic acid (Caprylic acid)	237	16	70	57	107
Nonanoic acid	254	12	84	57	99
Decanoic acid (Capric acid)	268	32	78	70	108
4-Oxopentanoic acid (Levulinic acid)	246	33	108	102	108 d.
Dodecanoic acid (Lauric acid)	299	43	87	78	100
3-Phenylpropanoic acid (Hydrocinnamic acid)	279	48	135	98	105
Bromoacetic acid	208	50	—	131	91
Tetradecanoic acid (Myristic acid)	—	54	93	84	103
Trichloroacetic acid	198	57	113	97	141
Hexadecanoic acid (Palmitic acid)	—	62	98	90	106
Chloroacetic acid	189	63	162	137	121
Octadecanoic acid (Stearic acid)	—	69	102	95	109
trans-2-Butenoic acid (Crotonic acid)	—	72	132	118	158
Phenylacetic acid	—	77	136	118	156
2-Methoxybenzoic acid (o-Anisic acid)	200	101	—	131	129
2-Methylbenzoic acid (o-Toluic acid)	—	104	144	125	142
Nonanedioic acid (Azelaic acid)	—	106	201 (di)	107 (mono) 186 (di)	93 (mono) 175 (di)
3-Methylbenzoic acid (m-Toluic acid)	263 s.	110	118	126	94
(±)-Phenylhydroxyacetic acid (Mandelic acid)	—	118	172	151	133
Benzoic acid	249	122	158	163	130
2-Benzoylbenzoic acid	—	127	—	195	165
Maleic acid	—	130	142 (di)	198 (mono) 187 (di)	172 (mono) 260 (di)
Decanedioic acid (Sebacic acid)	—	133	201 (di)	122 (mono) 200 (di)	170 (mono) 210 (di)
Cinnamic acid	300	133	168	153	147
2-Chlorobenzoic acid	—	140	131	118	139
3-Nitrobenzoic acid	—	140	162	155	143
2-Aminobenzoic acid (Anthranilic acid)	—	146	151	131	109

*See "Procedures for Preparing Derivatives," Appendix Two.

CARBOXYLIC ACIDS (Continued)

COMPOUND	BP	MP	p-TOLUIDIDE*	ANILIDE*	AMIDE*
Diphenylacetic acid	—	148	172	180	167
2-Bromobenzoic acid	—	150	—	141	155
Benzilic acid	—	150	190	175	154
Hexanedioic acid	—	152	239	151 (mono)	125 (mono)
(Adipic acid)				241 (di)	220 (di)
Citric acid	—	153	189 (tri)	199 (tri)	210 (tri)
4-Chlorophenoxyacetic acid	—	158	—	125	133
2-Hydroxybenzoic acid	—	158	156	136	142
(Salicyclic acid)					
Trimethylacetic acid	—	164	—	127	178
(Pivalic acid)					
5-Bromo-2-hydroxybenzoic acid	—	165	—	222	232
(5-Bromosalicylic acid)					
Methylenesuccinic acid	—	166 d.	—	152 (mono)	191 (di)
(Itaconic acid)					
(+)-Tartaric acid	—	169	—	180 (mono)	171 (mono)
				264 (di)	196 (di)
4-Chloro-3-nitrobenzoic acid	—	180	—	131	156
4-Methylbenzoic acid	—	180	160	145	160
(p-Toluic acid)					
4-Methoxybenzoic acid	280	184	186	169	167
(p-Anisic acid)					
Butanedioic acid	235 d.	188	180 (mono)	143 (mono)	157 (mono)
(Succinic acid)			255 (di)	230 (di)	260 (di)
3-Hydroxybenzoic acid	—	201	163	157	170
3,5-Dinitrobenzoic acid	—	202	—	234	183
Phthalic acid	—	210 d.	150 (mono)	169 (mono)	144 (mono)
			201 (di)	253 (di)	220 (di)
4-Hydroxybenzoic acid	—	214	204	197	162
Pyridine-3-carboxylic acid	—	236	150	132	128
(Nicotinic acid)					
4-Nitrobenzoic acid	—	240	204	211	201
4-Chlorobenzoic acid	—	242	—	194	179
Fumaric acid	—	300	—	233 (mono)	270 (mono)
				314 (di)	266 (di)

*See "Procedures for Preparing Derivatives," Appendix Two.

PHENOLS

COMPOUND	BP	MP	α-NAPHTHYL-URETHANE*	BROMO DERIVATIVE*			
				Mono	Di	Tri	Tetra
2-Chlorophenol	176	7	120	48	76	—	—
3-Methylphenol (*m*-Cresol)	203	12	128	—	—	84	—
2-Methylphenol (*o*-Cresol)	191	32	142	—	56	—	—
2-Methoxyphenol (Guaiacol)	204	32	118	—	—	116	—
4-Methylphenol (*p*-Cresol)	202	34	146	—	49	—	198
Phenol	181	42	133	—	—	95	—
4-Chlorophenol	217	43	166	33	90	—	—
2,4-Dichlorophenol	210	45	—	68	—	—	—
4-Ethylphenol	219	45	128	—	—	—	—
2-Nitrophenol	216	45	113	—	117	—	—
2-Isopropyl-5-methylphenol (Thymol)	234	51	160	55	—	—	—
3,4-Dimethylphenol	225	64	141	—	—	171	—
4-Bromophenol	238	64	169	—	—	95	—
3,5-Dimethylphenol	220	68	109	—	—	166	—
2,5-Dimethylphenol	212	75	173	—	—	178	—
1-Naphthol (α-Naphthol)	278	96	152	—	105	—	—
2-Hydroxyphenol (Catechol)	245	104	175	—	—	—	192
3-Hydroxyphenol (Resorcinol)	281	109	275	—	—	112	—
4-Nitrophenol	—	112	150	—	142	—	—
2-Naphthol (β-Naphthol)	286	121	157	84	—	—	—
1,2,3-Trihydroxybenzene (Pyrogallol)	309	133	—	—	158	—	—
4-Phenylphenol	305	164	—	—	—	—	—

*See "Procedures for Preparing Derivatives," Appendix Two.

PRIMARY AMINES

COMPOUND	BP	MP	BENZAMIDE*	PICRATE*	ACETAMIDE*
t-Butylamine	46	—	134	198	101
Propylamine	48	—	84	135	—
Allylamine	56	—	—	140	—
sec-Butylamine	63	—	76	139	—
Isobutylamine	69	—	57	150	—
Butylamine	78	—	42	151	—
Cyclohexylamine	135	—	149	—	104
Furfurylamine	145	—	—	150	—
Benzylamine	184	—	105	194	60
Aniline	184	—	163	198	114
2-Methylaniline (*o*-Toluidine)	200	—	144	213	110
3-Methylaniline (*m*-Toluidine)	203	—	125	200	65
2-Chloroaniline	208	—	99	134	87
2,6-Dimethylaniline	216	11	168	180	177
2-Methoxyaniline (*o*-Anisidine)	225	6	60	200	85
3-Chloroaniline	230	—	120	177	74
2-Ethoxyaniline (*o*-Phenetidine)	231	—	104	—	79
4-Chloro-2-methylaniline	241	29	142	—	140
4-Ethoxyaniline (*p*-Phenetidine)	250	2	173	69	137
4-Methylaniline (*p*-Toluidine)	200	43	158	182	147
2-Ethylaniline	210	47	147	194	111
2,5-Dichloroaniline	251	50	120	86	132
4-Methoxyaniline (*p*-Anisidine)	—	58	154	170	130
4-Bromoaniline	245	64	204	180	168
2,4,5-Trimethylaniline	—	64	167	—	162
4-Chloroaniline	—	70	192	178	179
2-Nitroaniline	—	72	110	73	92
Ethyl *p*-aminobenzoate	—	89	148	—	110
o-Phenylenediamine	258	102	301 (di)	208	185 (di)
2-Methyl-5-nitroaniline	—	106	186	—	151
2-Chloro-4-nitroaniline	—	108	161	—	139
3-Nitroaniline	—	114	157	143	155
4-Chloro-2-nitroaniline	—	118	—	—	104
2,4,6-Tribromoaniline	300	120	200	—	232 (mono) 127 (di)
2-Methyl-4-nitroaniline	—	130	—	—	202
2-Methoxy-4-nitroaniline	—	138	149	—	153
p-Phenylenediamine	267	140	128 (mono) 300 (di)	—	162 (mono) 304 (di)
4-Nitroaniline	—	148	199	100	215
4-Aminoacetanilide	—	162	—	—	304
2,4-Dinitroaniline	—	180	202	—	120

*See "Procedures for Preparing Derivatives," Appendix Two.

SECONDARY AMINES

COMPOUND	BP	MP	BENZAMIDE*	PICRATE*	ACETAMIDE*
Diethylamine	56	—	42	155	—
Diisopropylamine	84	—	—	140	—
Pyrrolidine	88	—	oil	112	—
Piperidine	106	—	48	152	—
Dipropylamine	110	—	—	75	—
Morpholine	129	—	75	146	—
Diisobutylamine	139	—	—	121	86
N-Methylcyclohexylamine	148	—	85	170	—
Dibutylamine	159	—	—	59	—
Benzylmethylamine	184	—	—	117	—
N-Methylaniline	196	—	63	145	102
N-Ethylaniline	205	—	60	132	54
N-Ethyl-m-toluidine	221	—	72	—	—
Dicyclohexylamine	256	—	153	173	103
N-Benzylaniline	298	37	107	48	58
Indole	254	52	68	—	157
Diphenylamine	302	52	180	182	101
N-Phenyl-1-naphthylamine	335	62	152	—	115

*See "Procedures for Preparing Derivatives," Appendix Two.

TERTIARY AMINES

COMPOUNDS		BP	MP	PICRATE*	METHIODIDE*
Triethylamine		89	—	173	280
Pyridine		115	—	167	117
2-Methylpyridine	(α-Picoline)	129	—	169	230
3-Methylpyridine	(β-Picoline)	144	—	150	92
Tripropylamine		157	—	116	207
N,N-Dimethylbenzylamine		183	—	93	179
N,N-Dimethylaniline		193	—	163	228 d.
Tributylamine		216	—	105	186
N,N-Diethylaniline		217	—	142	102
Quinoline		237	—	203	133

*See "Procedures for Preparing Derivatives," Appendix Two.

ALCOHOLS

COMPOUND	BP	MP	3,5-DINITRO-BENZOATE*	PHENYL-URETHANE*
Methanol	65	—	108	47
Ethanol	78	—	93	52
2-Propanol (Isopropyl alcohol)	82	—	123	88
2-Methyl-2-propanol	83	26	142	136
(*t*-Butyl alcohol)				
2-Propen-1-ol (Allyl alcohol)	97	—	49	70
1-Propanol	97	—	74	57
2-Butanol (*sec*-Butyl alcohol)	99	—	76	65
2-Methyl-2-butanol	102	−8.5	116	42
(*t*-Pentyl alcohol)				
2-Methyl-3-butyn-2-ol	104	—	112	—
2-Methyl-1-propanol	108	—	87	86
(Isobutyl alcohol)				
2-Propyn-1-ol	114	—	—	—
(Propargyl alcohol)				
3-Pentanol	115	—	101	48
1-Butanol	118	—	64	61
2-Pentanol	119	—	62	—
3-Methyl-3-pentanol	123	—	96	43
2-Methoxyethanol	124	—	—	(113)[a]
2-Chloroethanol	129	—	95	51
2-Methyl-1-butanol	130	—	70	31
(Isoamyl alcohol)				
4-Methyl-2-pentanol	132	—	65	143
1-Pentanol	138	—	46	46
Cyclopentanol	140	—	115	132
2-Ethyl-1-butanol	146	—	51	—
2,2,2-Trichloroethanol	151	—	142	87
1-Hexanol	157	—	58	42
Cyclohexanol	160	—	113	82
(2-Furyl)-methanol	170	—	80	45
(Furfuryl alcohol)				
1-Heptanol	176	—	47	60
2-Octanol	179	—	32	114
1-Octanol	195	—	61	74
3,7-Dimethyl-1,6-octadien-3-ol	196	—	—	66
(Linalool)				
Benzyl alcohol	204	—	113	77
1-Phenylethanol	204	20	92	95
2-Phenylethanol	219	—	108	78
1-Decanol	231	7	57	59
3-Phenylpropanol	236	—	45	92
1-Dodecanol (Lauryl alcohol)	—	24	60	74
3-Phenyl-2-propen-1-ol	250	34	121	90
(Cinnamyl alcohol)				
1-Tetradecanol	—	39	67	74
(Myristyl alcohol)				
(−)-Menthol	212	41	158	111
1-Hexadecanol (Cetyl alcohol)	—	49	66	73
1-Octadecanol (Stearyl alcohol)	—	59	77	79
Diphenylmethanol (Benzhydrol)	288	68	141	139
Benzoin	—	133	—	165
Cholesterol	—	147	—	168
(+)-Borneol	—	208	154	138

*See "Procedures for Preparing Derivatives," Appendix Two.

[a] α-naphthylurethane

ESTERS

COMPOUND	BP	MP	COMPOUND	BP	MP
Methyl formate	34	—	Pentyl acetate	142	—
Ethyl formate	54	—	(n-Amyl acetate)		
Vinyl acetate	72	—	3-Methylbutyl acetate	142	—
Ethyl acetate	77	—	(Isoamyl acetate)		
Methyl propanoate	77	—	Ethyl chloroacetate	143	—
(Methyl propionate)			Ethyl lactate	154	—
Methyl acrylate	80	—	Ethyl hexanoate	168	—
2-Propyl acetate	85	—	(Ethyl caproate)		
(Isopropyl acetate)			Methyl acetoacetate	170	—
Ethyl chloroformate	93	—	Dimethyl malonate	180	—
Methyl 2-methylpropanoate	93	—	Ethyl acetoacetate	181	—
(Methyl isobutyrate)			Diethyl oxalate	185	—
2-Propenyl acetate	94	—	Methyl benzoate	199	—
(Isopropenyl acetate)			Ethyl octanoate	207	—
2-(2-Methylpropyl) acetate	98	—	(Ethyl caprylate)		
(t-Butyl acetate)			Ethyl cyanoacetate	210	—
Ethyl acrylate	99	—	Ethyl benzoate	212	—
Ethyl propanoate	99	—	Diethyl succinate	217	—
(Ethyl propionate)			Methyl phenylacetate	218	—
Methyl methacrylate	100	—	Diethyl fumarate	219	—
Methyl trimethylacetate	101	—	Methyl salicylate	222	—
(Methyl pivalate)			Diethyl maleate	225	—
Propyl acetate	102	—	Ethyl phenylacetate	229	—
Methyl butanoate	102	—	Ethyl salicylate	234	—
(Methyl butyrate)			Dimethyl suberate	268	—
2-Butyl acetate	111	—	Ethyl cinnamate	271	—
(sec-Butyl acetate)			Diethyl phthalate	298	—
Methyl 3-methylbutanoate	117	—	Dibutyl phthalate	340	—
(Methyl isovalerate)			Methyl cinnamate	—	36
Ethyl butanoate	120	—	Phenyl salicylate	—	42
(Ethyl butyrate)			Methyl p-chlorobenzoate	—	44
Butyl acetate	127	—	Ethyl p-nitrobenzoate	—	56
Methyl pentanoate	128	—	Phenyl benzoate	314	69
(Methyl valerate)			Methyl m-nitrobenzoate	—	78
Methyl chloroacetate	130	—	Methyl p-bromobenzoate	—	81
Ethyl 3-methylbutanoate	132	—	Ethyl p-aminobenzoate	—	90
(Ethyl isovalerate)			Methyl p-nitrobenzoate	—	94

Appendix 2

PROCEDURES FOR PREPARING DERIVATIVES

ALDEHYDES AND KETONES

Semicarbazones. Place 0.5 ml of a 2 M stock solution of semicarbazide hydrochloride (or 0.5 ml of a solution prepared by dissolving 1.11 g of semicarbazide hydrochloride (MW 111.5) in 5 ml of water) in a small test tube. Add an estimated 1 millimole of the unknown compound to the test tube. If the unknown does not dissolve in the solution, or if a cloudy solution results, add enough methanol to dissolve the solid and to produce a clear solution. Using a disposable capillary pipet, add 10 drops of pyridine and heat the mixture gently on a steam bath for about 5 minutes, at which time the product should have begun to crystallize. Collect the product by vacuum filtration. The product may be recrystallized from ethanol if necessary.

Semicarbazones (Alternate Method). Dissolve 1 g of semicarbazide hydrochloride and 1.5 g of sodium acetate in 5 ml of water. Then dissolve 1.0 g of the unknown in 10 ml of ethanol. Mix the two solutions together in a 50 ml Erlenmeyer flask and heat the mixture to boiling for about 5 minutes. After heating, place the reaction flask in a beaker of ice and scratch the sides of the flask with a glass rod to induce crystallization of the derivative. Collect the derivative by vacuum filtration, and recrystallize it from ethanol.

2,4-Dinitrophenylhydrazones. Place 10 ml of a solution of 2,4-dinitrophenylhydrazine (prepared as described for the classification test in Procedure 50D) in a test tube, and add an estimated 1 millimole of the unknown compound. If the unknown is a solid, it should be dissolved in the minimum amount of 95% ethanol or dioxane before it is added. If crystallization does not occur immediately, gently warm the solution for a minute on a steam bath and then set it aside to crystallize. Collect the product by vacuum filtration.

CARBOXYLIC ACIDS

Using the smallest flask in the organic kit and a reflux condenser, heat a mixture of 0.5 g of the acid and 2 ml of thionyl chloride on a steam bath for about 30 minutes. Allow the mixture to cool and use it for one of the following three procedures:

Amides. Working in a hood, pour the reaction mixture into a beaker containing 10 ml of ice cold, concentrated ammonium hydroxide and stir it vigorously. When the reaction is complete, collect the product by vacuum filtration and recrystallize it from water or from water-ethanol using the mixed solvent method (Technique 3, Section 3.7).

Anilides. Dissolve 1 g of aniline in 25 ml of benzene and carefully add it to the reaction mixture. Warm the mixture for an additional 5 minutes on a steam bath. Then transfer the benzene solution to a separatory funnel and wash it sequentially with 5 ml of water, 5 ml of 5% hydrochloric acid, 5 ml of 5% sodium hydroxide, and a second 5 ml portion of water. Dry the benzene layer over a small amount of anhydrous sodium sulfate. Decant the benzene layer away from the drying agent into a small beaker and evaporate the benzene on a steam bath in the hood. Recrystallize the product from water or from ethanol-water using the mixed solvent methods (Technique 3, Section 3.7).

p-Toluidides. Use the same procedure as that described for the anilide, but substitute p-toluidine for aniline.

PHENOLS

α-Naphthylurethanes. Follow the procedure which is given below for the preparation of phenylurethanes from alcohols, substituting α-naphthylisocyanate for phenylisocyanate.

Bromo Derivatives. First, if a stock brominating solution is not available, prepare one by dissolving 0.75 g of potassium bromide in 5 ml of water and adding 0.5 g of bromine. Dissolve 0.1 g of the phenol in 1 ml of methanol or dioxane and then add 1 ml of water. Add 1 ml of the brominating mixture to the phenol solution and swirl the mixture vigorously. Then, continue adding the brominating solution, dropwise with swirling, until the color of the bromine reagent persists. Finally, add 3 to 5 ml of water and shake the mixture vigorously. Collect the precipitated product by vacuum filtration and wash it well with water. Recrystallize the derivative from methanol-water using the mixed solvent method (Technique 3, Section 3.7).

AMINES

Acetamides. Place an estimated 1 millimole of the amine and 0.5 ml of acetic anhydride in a small Erlenmeyer flask. Heat the mixture for about 5 minutes and then add 5 ml of water and stir the solution vigorously to precipitate the product and hydrolyze the excess acetic anhydride. If the product does not crystallize, it may be necessary to scratch the walls of the flask with a glass rod. Collect the crystals by vacuum filtration and wash them with several portions of cold 5% hydrochloric acid. Recrystallize the derivative from methanol-water using the mixed solvent method (Technique 3, Section 3.7).

Aromatic amines, or those which are not very basic, may require pyridine (2 ml) as a solvent and catalyst for the reaction. If pyridine is used, a longer period of heating will be required (up to 1 hour), and the reaction should be carried out in an apparatus equipped with a reflux condenser. After reflux, the reaction mixture must be extracted with 5 to 10 ml of 5% sulfuric acid to remove the pyridine.

Benzamides. Using a test tube, suspend an estimated 1 millimole of the amine in 1 ml of 10% sodium hydroxide solution and add 0.5 g of benzoyl chloride. Stopper the test tube with a cork and shake the mixture vigorously for a period of about 10 minutes. After shaking, add enough dilute hydrochloric acid to bring the pH of the solution to pH 7 or 8. Collect the precipitate by vacuum filtration, wash it thoroughly with cold water, and recrystallize it from ethanol-water using the mixed solvent method (Technique 3, Section 3.7).

Benzamides (Alternate Method). Dissolve 0.5 g of the amine in a solution of 2.5 ml of pyridine and 5 ml of benzene. Add 0.5 ml of benzoyl chloride to the solution and heat the mixture under reflux for about 30 minutes. Pour the cooled reaction mixture into 50 ml of water and stir the mixture vigorously to hydrolyze the excess benzoyl chloride. Separate the benzene layer and wash it first with 3 ml of water and then with 3 ml of 5% sodium carbonate. Dry the benzene over anhydrous sodium sulfate, decant the benzene into a small beaker, and remove the benzene by evaporation on a steam bath in the hood. Recrystallize the benzamide from ethanol or from ethanol-water using the mixed solvent method (Technique 3, Section 3.7).

Picrates. Dissolve 0.2 g of the unknown in about 5 ml of ethanol and add 5 ml of a saturated solution of picric acid in ethanol. Heat the solution to boiling and then allow it to cool slowly. Collect the product by vacuum filtration and rinse it with a small amount of cold ethanol.

Methiodides. Mix equal volume quantities of the amine and methyl iodide in a large test tube (about 0.5 ml is convenient) and allow the mixture to stand for several minutes. Then, heat the mixture under reflux on a steam bath for about 5 minutes. The methiodide should crystallize on cooling. If it does not, induce crystallization by scratching the walls of the tube with a glass rod. Collect the product by vacuum filtration and recrystallize it from ethanol or ethyl acetate.

ALCOHOLS

3,5-Dinitrobenzoates

Liquid Alcohols.   Dissolve 0.5 g of 3,5-dinitrobenzoyl chloride in 0.5 ml of the alcohol and heat the mixture for about 5 minutes. Allow the mixture to cool and add 3 ml of a 5% sodium carbonate solution and 2 ml of water. Stir the mixture vigorously and crush any solid which forms. Collect the product by vacuum filtration and wash it with cold water. Recrystallize the derivative from ethanol-water using the mixed solvent method (Technique 3, Section 3.7).

Solid Alcohols.   Dissolve 0.5 g of the alcohol in 3 ml of dry pyridine and add 0.5 g of 3,5-dinitrobenzoyl chloride. Heat the mixture under reflux for 15 minutes. Pour the cooled reaction mixture into a cold mixture of 5 ml of 5% sodium carbonate and 5 ml of water. Keep the solution cooled in an ice bath until the product crystallizes and stir it vigorously during the entire period. Collect the product by vacuum filtration, wash it with cold water, and recrystallize it from ethanol-water using the mixed solvent method (Technique 3, Section 3.7).

Phenylurethanes.   Place 0.5 g of the **anhydrous** alcohol in a dry test tube and add 0.5 ml of phenylisocyanate (α-naphthyl isocyanate for a phenol). If the compound is a phenol, add several drops of pyridine to catalyze the reaction. If the reaction is not spontaneous, heat the mixture on a steam bath for 5 to 10 minutes. Cool the test tube in a beaker of ice and scratch the tube with a glass rod to induce crystallization. Decant the liquid from the solid product or, if necessary, collect the product by vacuum filtration. Dissolve the product in 5 to 6 ml of hot ligroin or hexane and filter the mixture by gravity (preheat funnel) to remove any unwanted and insoluble diphenylurea which may be present. Cool the filtrate to induce crystallization of the urethane. Collect the product by vacuum filtration.

ESTERS

The preparations of the derivatives of esters usually require rather complicated procedures. It is recommended that esters be characterized by spectroscopic methods whenever possible. If the preparation of a derivative is required, consult one of the more comprehensive textbooks. Several of these are listed in the first section of Experiment 50.

Appendix **3**

INFRARED SPECTROSCOPY

Almost any compound having covalent bonds, whether organic or inorganic, will be found to absorb various frequencies of electromagnetic radiation in the infrared region of the spectrum. The infrared region of the electromagnetic spectrum lies at wavelengths longer than those associated with visible light, which includes wavelengths

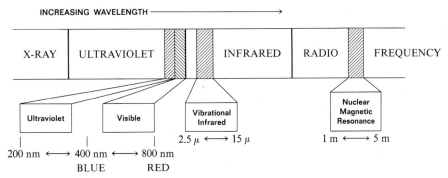

FIGURE IR-1. A portion of the electromagnetic spectrum showing the relationship of the vibrational infrared to other types of radiation

from approximately 400 nm to 800 nm (1 nm = 10^{-9} m), but lies at wavelengths shorter than those associated with radio waves, which have wavelengths longer than 1 cm. For chemical purposes, we will be interested in the **vibrational** portion of the infrared region. This portion is defined as that including radiations with wavelengths (λ) between 2.5 μ and 15 μ (1 μ = 1 micron = 1 μm = 10^{-6} m). Although the more technically correct unit for wavelength in the infrared region of the spectrum is **micrometer,** we shall follow common practice and use **micron** as the unit. The relationship of the infrared region to others included in the electromagnetic spectrum is illustrated in Figure IR-1.

As with other types of energy absorption, molecules are excited to a higher energy state when they absorb infrared radiation. The absorption of infrared radiation is, like other absorption processes, a quantized process. Only selected frequencies (energies) of infrared radiation will be absorbed by a molecule. The absorption of infrared radiation corresponds to energy changes on the order of from 2 to 10 kcal/mole. Radiation in this energy range corresponds to the range encompassing the stretching and bending vibrational frequencies of the bonds in most covalent molecules. In the absorption process, those frequencies of infrared radiation which match the natural vibrational frequencies of the molecule in question will be absorbed, and the energy absorbed will serve to increase the **amplitude** of the vibrational motions of the bonds in the molecule.

Many chemists refer to the radiation in the vibrational infrared region of the electromagnetic spectrum in terms of a unit called **wavenumbers** ($\bar{\nu}$). Wavenumbers are expressed as cm^{-1} (reciprocal centimeters), and are easily computed by taking the reciprocal of the wavelength (λ) expressed in centimeters. This unit has the advantage, for those performing calculations, that it is directly proportional to energy. Thus, in terms of wavenumbers, the vibrational infrared extends from about 4000 to 650 cm^{-1} (or wavenumbers).

Interconversions may be made between wavelengths (μ) and wavenumbers or between wavenumbers and wavelengths by using the following relationships:

$$ cm^{-1} = \frac{1}{(\mu)} \times 10,000 \text{ and } \mu = \frac{1}{(cm^{-1})} \times 10,000 $$

IR.1 USES OF THE INFRARED SPECTRUM

Since every different type of bond has a different natural frequency of vibration, and since the same type of bond in two different compounds is in a slightly different environment, no two molecules of different structure will have exactly the same infrared absorption pattern or **infrared spectrum.** Although some of the frequencies absorbed in the two cases might be the same, in no case of two different molecules will their infrared spectra (the patterns of absorption) be identical. Thus, the infrared spectrum can be used for molecules much as a fingerprint can be used for humans. By comparing the infrared spectra of two substances thought to be identical, one can establish whether or not they are in fact identical. If their infrared spectra coincide peak for peak (absorption for absorption), in most cases the two substances will be identical.

A second and more important use of the infrared spectrum is that it gives structural information about a molecule. The absorptions of each type of bond (N—H, C—H, O—H, C—X, C=O, C—O, C—C, C=C, C≡C, C≡N, etc.) are regularly found only in certain small portions of the vibrational infrared region. A small range of absorption can be defined for each type of bond. Outside this range, absorptions will normally be due to some other type of bond. Thus, for instance, any absorption in the range 3000 ± 150 cm^{-1} (around 3.33 μ) will almost always be due to the presence of a CH bond in the molecule; an absorption in the range of 1700 ± 100 cm^{-1} (around 6.1 μ) will normally be due to the presence of a C=O bond (carbonyl group) in the molecule. The same type of range applies to each type of bond. The way these are spread out over the vibrational infrared is illustrated schematically below. It is a good idea to try to fix this general scheme in one's mind for future convenience.

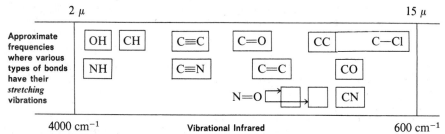

FIGURE IR-2. The approximate regions where various common types of bonds absorb (bending and twisting and other types of bond vibrations have been omitted for clarity)

IR.2 THE MODES OF VIBRATION AND BENDING

The simplest types, or **modes**, of vibrational motion in a molecule which are **infrared active**, that is, give rise to absorptions, are the stretching and bending modes.

However, other more complex types of stretching and bending are also active. In order to introduce several words of terminology, the normal modes of vibration for a methylene group are illustrated below.

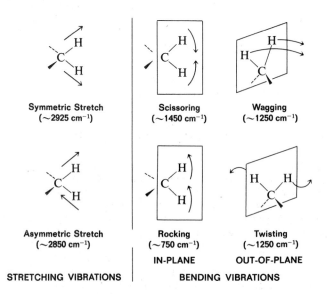

In any group of three or more atoms, at least two of which are identical, there will be **two** modes of stretching and/or bending: the symmetric mode and the asymmetric mode. Examples of such groupings are $-CH_3$, $-CH_2-$, $-NO_2$, $-NH_2$, and anhydrides, $(CO)_2O$. In the case of the anhydride, owing to asymmetric and symmetric modes of stretch, this functional group gives **two** absorptions in the $C=O$ region. A similar phenomenon is seen for the amino group, where a primary amine usually has **two** absorptions in the NH stretch region, while a secondary amine (R_2NH) has only one absorption peak. Amides show similar bands. There are two strong $N=O$ stretch peaks for a nitro group caused by asymmetric and symmetric stretching modes.

IR.3 WHAT TO LOOK FOR WHEN EXAMINING INFRARED SPECTRA

The instrument that determines the absorption spectrum for a compound is called an **infrared spectrophotometer.** The spectrophotometer determines the relative strengths and positions of all the absorptions in the infrared region and plots this on a piece of calibrated chart paper. This plot of absorption intensity versus wavenumber or wavelength is referred to as the **infrared spectrum** of the compound. A typical infrared spectrum, that of methyl isopropyl ketone, is shown in Figure IR-3.

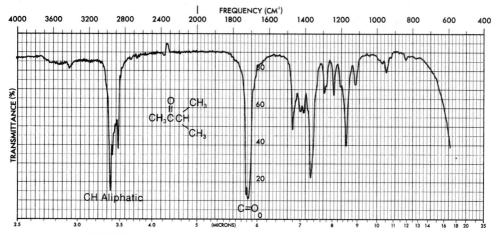

FIGURE IR-3. Infrared spectrum of methyl isopropyl ketone (neat liquid, salt plates)

The strong absorption in the middle of the spectrum corresponds to $C=O$, the carbonyl group. Note that the $C=O$ peak is quite intense. In addition to the characteristic position of absorption, the **shape** and **intensity** of this peak are also unique to the $C=O$ bond. This is true for almost every type of absorption peak; both shape and intensity characteristics can be described, and these characteristics often allow one to distinguish the peak where the situation might be confusing. For instance, to some extent both $C=O$ and $C=C$ bonds absorb in the same region of the infrared spectrum,

$$C=O \ 1850 - 1630 \ cm^{-1} \ (5.5 - 6.1 \ \mu)$$
$$C=C \ 1680 - 1620 \ cm^{-1} \ (5.95 - 6.2 \ \mu)$$

however, the $C=O$ bond is a strong absorber whereas the $C=C$ bond generally absorbs only weakly. Hence, a trained observer would not normally interpret a strong peak at $1670 \ cm^{-1}$ to be a carbon-carbon double bond, nor would he interpret a weak absorption at this frequency to be due to a carbonyl group.

The shape of a peak will often give a clue to its identity as well. Thus, while the NH and OH regions of the infrared overlap,

$$OH \ 3650 - 3200 \ cm^{-1} \ (2.75 - 3.12 \ \mu)$$
$$NH \ 3500 - 3300 \ cm^{-1} \ (2.85 - 3.00 \ \mu)$$

NH usually gives a **sharp** absorption (absorbs a very narrow range of frequencies) while OH, when it is in the NH region, usually gives a **broad** absorption peak. Also, primary amines give **two** absorptions in this region, whereas alcohols give only one.

Therefore, while studying the sample spectra in the pages that follow, one should also take notice of shapes and intensities. They are as important as the frequency at which an absorption occurs, and the eye must be trained to recognize these features. Often, when reading the literature of organic chemistry, one will find absorptions referred to as strong (s), medium (m), weak (w), broad and/or sharp. The author is trying to convey some idea of what the peak looks like without actually drawing the spectrum.

IR.4 CORRELATION CHARTS AND TABLES

To extract structural information from infrared spectra, one must be familiar with the frequencies or wavelengths at which various functional groups absorb. To facilitate this, tables exist, called infrared **correlation tables,** which have tabulated as much information as is known about where the various functional groups absorb. The books listed at the end of this appendix present extensive lists of correlation tables. Sometimes the absorption information is given in the form of a chart, called a **correlation chart.** A simplified correlation table is given in Table IR–1.

TABLE IR–1. A SIMPLIFIED CORRELATION TABLE

TYPE OF VIBRATION			FREQUENCY (cm^{-1})	WAVELENGTH (μ)	INTENSITY
C—H	Alkanes	(stretch)	3000–2850	3.33–3.51	s
	—CH$_3$	(bend)	1450 and 1375	6.90 and 7.27	m
	—CH$_2$—	(bend)	1465	6.83	m
	Alkenes	(stretch)	3100–3000	3.23–3.33	m
		(bend)	1700–1000	5.88–10.0	s
	Aromatics	(stretch)	3150–3050	3.17–3.28	s
		(out-of-plane bend)	1000–700	10.0–14.3	s
	Alkyne	(stretch)	ca. 3300	ca. 3.03	s
	Aldehyde		2900–2800	3.45–3.57	w
			2800–2700	3.57–3.70	w
C—C	Alkane	not interpretatively useful			
C=C	Alkene		1680–1600	5.95–6.25	m-w
	Aromatic		1600–1400	6.25–7.14	m-w
C≡C	Alkyne		2250–2100	4.44–4.76	m-w
C=O	Aldehyde		1740–1720	5.75–5.81	s
	Ketone		1725–1705	5.80–5.87	s
	Carboxylic acid		1725–1700	5.80–5.88	s
	Ester		1750–1730	5.71–5.78	s
	Amide		1700–1640	5.88–6.10	s
	Anhydride		ca. 1810	ca. 5.52	s
			ca. 1760	ca. 5.68	s
C—O	Alcohols, Ethers, Esters, Carboxylic acids		1300–1000	7.69–10.0	s
O—H	Alcohols, Phenols				
	Free		3650–3600	2.74–2.78	m
	H-Bonded		3400–3200	2.94–3.12	m
	Carboxylic acids		3300–2500	3.03–4.00	m
N—H	Primary and secondary amines		ca. 3500	ca. 2.86	m
C≡N	Nitriles		2260–2240	4.42–4.46	m
N=O	Nitro (R—NO$_2$)		1600–1500	6.25–6.67	s
			1400–1300	7.14–7.69	s
C—X	Fluoride		1400–1000	7.14–10.0	s
	Chloride		800–600	12.5–16.7	s
	Bromide, Iodide		<600	>16.7	s

Although the mass of data in Table IR–1 seems as though it may be difficult to assimilate, it is easy if a modest start is made, and then a gradual increase is made in familiarity with the data. An ability to interpret the finer details of an infrared spectrum will follow. This is most easily done by first establishing the broad visual patterns of Figure IR–2 quite firmly in mind. Then, as a second step, a "typical absorption value" can be memorized for each of the functional groups in this pattern. This value will be a single number that can be used as a pivot value for the memory. For instance, start with a simple aliphatic ketone as a model for all typical carbonyl compounds. The typical aliphatic ketone has a carbonyl absorption of about 1715 ± 10 cm^{-1}. Without worrying about the variation, memorize 1715 cm^{-1} as the base value for carbonyl absorption. Then, more slowly, familiarize yourself with the extent of the carbonyl range and the visual pattern of the way in which the different kinds of carbonyl groups are arranged throughout this region. See, for instance, Figure IR–14 which gives typical values for the various types of carbonyl compounds. Also learn how factors as ring size (when the functional group is contained in a ring) and conjugation affect the base values (i.e., which direction the values are shifted). Learn the trends—always keeping the memorized base value (1715 cm^{-1}) in mind. As a beginning, it might prove useful to memorize the following base values for this approach. Notice that there are only eight of them.

TABLE IR–2. BASE VALUES FOR ABSORPTIONS OF BONDS

OH	3600 cm^{-1}	2.78 μ	C$\equiv$C	2150 cm^{-1}	4.65 μ
NH	3500	2.86	C=O	1715	5.83
CH	3000	3.33	C=C	1650	6.06
C$\equiv$N	2250	4.44	C—O	1100	9.09

IR.5 HOW TO APPROACH THE ANALYSIS OF A SPECTRUM (OR, WHAT YOU CAN TELL AT A GLANCE)

In trying to analyze the spectrum of an unknown you should concentrate your first efforts on trying to determine the presence (or absence) of a few major functional groups. C=O, O—H, N—H, C—O, C=C, C$\equiv$C, C$\equiv$N and NO$_2$ are the most conspicuous peaks and give immediate structural information if they are present. Do not try to make a detailed analysis of the CH absorptions near 3000 cm^{-1} (3.33 μ); almost all compounds have these absorptions. Do not worry about subtleties of the exact type of environment in which the functional group is found. Below is a major check list of the important gross features.

1. Is a carbonyl group present?
   The C=O group gives rise to a strong absorption in the region 1820 to 1660 cm^{-1} (5.5 to 6.1 μ). The peak is often the strongest in the spectrum and of medium width. You can't miss it.
2. If C=O is present, check the following types (if absent, go to 3).

ACIDS	Is OH also present?
	–**broad** absorption near 3300 to 2500 cm^{-1} (3.0 to 4.0 μ) (usually overlaps C—H)
AMIDES	is NH also present?
	–medium absorption near 3500 cm^{-1} (2.85 μ), sometimes a double peak, equivalent halves
ESTERS	is C—O also present?
	–medium intensity absorptions near 1300 to 1000 cm^{-1} (7.7 to 10 μ)
ANHYDRIDES	have **two** C=O absorptions near 1810 and 1760 cm^{-1} (5.5 and 5.7 μ)
ALDEHYDES	is aldehyde CH present?
	–two weak absorptions near 2850 and 2750 cm^{-1} (3.50 and 3.65 μ) on the right hand side of CH absorptions
KETONES	The above 5 choices have been eliminated

3. If C=O is absent

ALCOHOLS or   Check for OH
PHENOLS   –**broad** absorption near 3600 to 3300 cm^{-1} (2.8 to 3.0 μ)
               –confirm this by finding C—O near 1300 to 1000 cm^{-1} (7.7 to 10 μ)
AMINES   Check for NH
               –medium absorption(s) near 3500 cm^{-1} (2.85 μ)
ETHERS   Check for C—O (and absence of OH) near 1300 to 1000 cm^{-1}
               (7.7 to 10 μ)

4. Double Bonds and/or Aromatic Rings

               –C=C is a **weak** absorption near 1650 cm^{-1} (6.1 μ)
               –medium to strong absorptions in the region 1650 to 1450 cm^{-1}
               (6 to 7 μ) often imply an aromatic ring
               –confirm the above by consulting the CH region
               aromatic and vinyl CH occurs to the left of 3000 cm^{-1} (3.33 μ)
               (aliphatic CH occurs to the right of this value.)

5. Triple Bonds

               –C≡N is a medium, sharp absorption near 2250 cm^{-1} (4.5 μ)
               –C≡C is a weak but sharp absorption near 2150 cm^{-1} (4.65 μ)
               Check also for acetylenic CH near 3300 cm^{-1} (3.0 μ)

6. Nitro Groups

               –**two** strong absorptions 1600 to 1500 cm^{-1} (6.25 to 6.67 μ) and
               1390 to 1300 cm^{-1} (7.2 to 7.7 μ)

7. Hydrocarbons

               –none of the above are found
               –major absorptions are in CH region near 3000 cm^{-1} (3.33 μ)
               –very simple spectrum, only other absorptions near 1450 cm^{-1}
               (6.90 μ) and 1375 cm^{-1} (7.27 μ)

The beginning student should resist the idea of trying to assign or interpret **every** peak in the spectrum. You simply will not be able to do this. Concentrate first on learning these **major** peaks and recognizing their presence or absence. This is best done by carefully studying the illustrative spectra in the section that follows.

NOTE. In describing the shifts of absorption peaks or their relative positions, we have used the terms "to the left" and "to the right." This was done to facilitate economy of use when using **both** microns and reciprocal centimeters. The meaning is clear since all spectra are conventionally presented left to right from 4000 cm^{-1} to 600 cm^{-1} or from 2.5 μ to 16 μ. "To the right" avoids saying each time "to lower frequency (cm^{-1}) or to longer wavelength (μ)" which is confusing since cm^{-1} and μ have an inverse relationship; as one goes up, the other goes down.

IR.6 A SURVEY OF THE IMPORTANT FUNCTIONAL GROUPS WITH EXAMPLES

Alkanes

Spectrum is usually simple with few peaks.
C—H   Stretch occurs around 3000 cm^{-1} (3.33 μ):
      a. in alkanes (except strained ring compounds) absorption always occurs to the right of 3000 cm^{-1} (3.33 μ);
      b. if a compound has vinylic, aromatic, acetylenic or cyclopropyl hydrogens, the CH absorption is to the left of 3000 cm^{-1} (3.33 μ).
CH$_2$ Methylene groups have a characteristic absorption of approximately 1450 cm^{-1} (6.90 μ).
CH$_3$ Methyl groups have a characteristic absorption at approximately 1375 cm^{-1} (7.27 μ).
C—C   Stretch—not interpretatively useful—has many peaks.
The spectrum of decane is shown in Figure IR-4.

Alkenes

=C—H Stretch occurs to the left of 3000 cm^{-1} (3.33 μ).
=C—H Out of plane (oop) bending 1000 to 650 cm^{-1} (10 to 15 μ).
C=C   Stretch 1600 to 1675 cm^{-1} (6.25 to 5.95 μ), often weak.
      Conjugation moves C=C stretch to the right.
      Symmetrically substituted bonds, e.g., 2,3-dimethyl-2-butene, do not absorb
      in the infrared (no dipole change).
      Also, highly substituted double bonds are often vanishingly weak in absorp-
      tion.

The spectrum of styrene is shown in Figure IR–5. The spectrum of cyclohexene is shown
in Experiment 26.

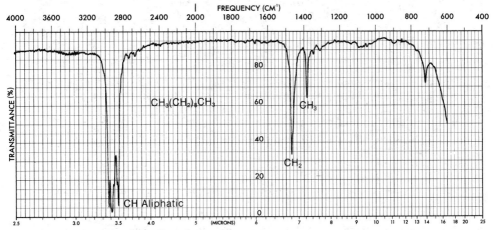

FIGURE IR–4. Infrared spectrum of decane (neat liquid, salt plates)

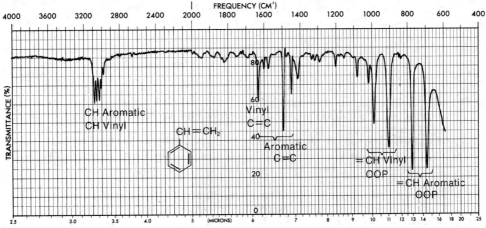

FIGURE IR–5. Infrared spectrum of styrene (neat liquid, salt plates)

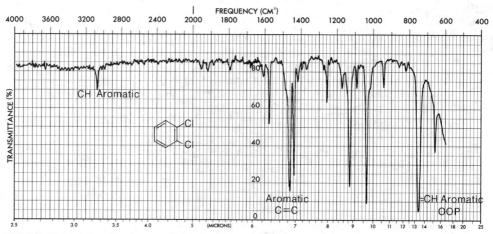

FIGURE IR-6. Infrared spectrum of *o*-dichlorobenzene (neat liquid, salt plates)

Aromatic Rings

=C—H Stretch is always to the left of 3000 cm^{-1} (3.33 μ).

CH Out of plane (oop) bending 900 to 690 cm^{-1} (11.0 to 14.5 μ).

The CH out-of-plane absorptions often allow one to determine the type of ring substitution by their numbers, intensities, and positions. The following correlation chart (Figure IR-7) indicates the positions of these bands.

The patterns are generally reliable, but are most reliable for rings with alkyl substituents and least reliable for polar substituents.

Ring Absorptions (C=C). There are often four sharp absorptions that occur in pairs at 1600 cm^{-1} (6.25 μ) and 1450 cm^{-1} (6.90 μ) and are characteristic of an aromatic ring. See, for example, the spectra of anisole (Figure IR-10), benzonitrile (Figure IR-12) and methyl benzoate (Figure IR-16).

There are many **weak** absorptions around 2000 to 1600 cm^{-1} (5 to 6 μ). The relative shapes and numbers of these peaks can be used to tell whether a ring is mono-, di-, tri-, tetra-, penta-, or hexa-substituted. Positional isomers can also be distinguished. We will not consider these bands in detail here (see References).

The spectra of styrene and *o*-dichlorobenzene are shown in Figures IR-5 and IR-6.

Alkynes

≡C—H Stretch is usually near 3300 cm^{-1} (3.0 μ).

C≡C Stretch is near 2150 cm^{-1} (4.65 μ).

Conjugation moves C≡C stretch to the right.

Disubstituted or symmetrically substituted triple bonds give either no absorption or weak absorption.

The spectrum of propargyl alcohol is shown in Figure IR-8.

Alcohols and Phenols

OH Stretch is a sharp peak 3650 to 3600 cm^{-1} (2.74 to 2.78 μ) if no hydrogen-bonding takes place. (This is usually only observed in dilute solutions.)

If there is hydrogen-bonding (usual in neat or concentrated solutions) the absorption is **broad** and occurs more to the right at 3500 to 3200 cm^{-1} (2.85 to 3.12 μ), sometimes overlapping with C—H stretch absorptions.

C—O Stretch is usually in the range 1300 to 1000 cm^{-1} (7.7 to 10 μ).

Phenols are similar to alcohols. The 2-naphthol shown in Figure IR-9 has some molecules H-bonded and some free. The spectrum of cyclohexanol is given in Experiment 26. This alcohol, which was determined neat, would also have a free OH spike to the left of its H-bonded band if it had been determined in dilute solution. The solution spectra of borneol and isoborneol are shown in Experiment 20.

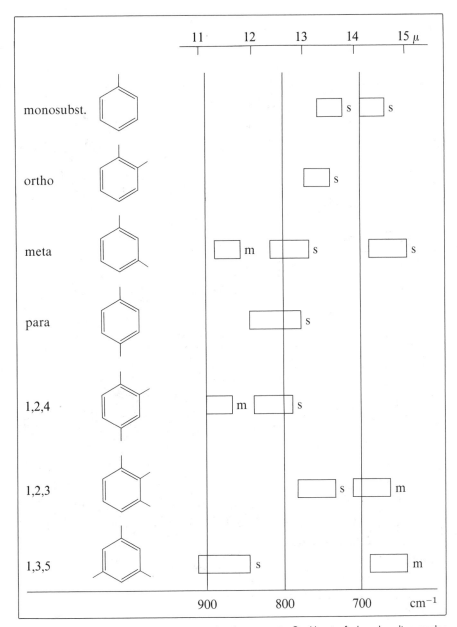

FIGURE IR-7. Correlation chart for the aromatic C—H out-of-plane bending modes

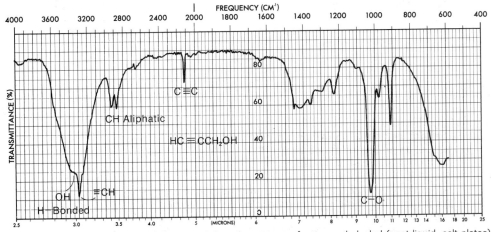

FIGURE IR-8. Infrared spectrum of propargyl alcohol (neat liquid, salt plates)

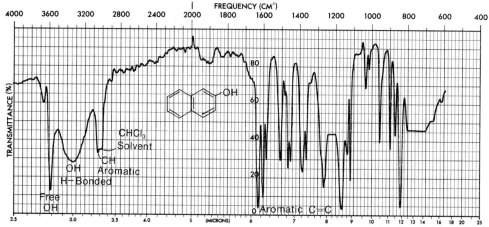

FIGURE IR-9. Infrared spectrum of 2-naphthol showing both free and hydrogen-bonded OH (CHCl₃ solution)

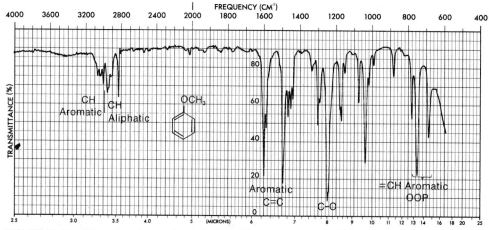

FIGURE IR-10. Infrared spectrum of anisole (neat liquid, salt plates)

Ethers

C—O The most prominent band is that due to C—O stretch, 1300 to 1000 cm⁻¹ (7.7 to 10.0 μ). Absence of C=O and O—H bands is required to be sure C—O stretch is not due to an alcohol or ester. Phenyl and vinyl ethers are found in the left hand portion of the range, aliphatic ethers to the right. (Conjugation with the oxygen moves the absorption to the left.)

The spectrum of anisole is shown in Figure IR–10.

Amines

N—H Stretch occurs in the range 3500 to 3300 cm⁻¹ (2.86 to 3.03 μ).
Primary amines have **two** bands typically 30 cm⁻¹ (0.03 μ) apart.
Secondary amines have one band, often vanishingly weak.
Tertiary amines have no NH stretch.

C—N Stretch is weak and occurs in the range 1350 to 1000 cm⁻¹ (7.4 to 10 μ).

N—H Scissoring mode occurs in the range 1640 to 1560 (6.1 to 6.4 μ) (broad).
An out-of-plane bending absorption can sometimes be observed at about 800 cm⁻¹ (12.5 μ).

The spectrum of *n*-butylamine is shown in Figure IR–11. In addition, the spectrum of aniline is shown in Experiment 30.

Nitro Compounds

N=O   Stretch is usually two strong bands 1600 to 1500 cm^{-1} (6.25 to 6.67 μ) and 1390
         to 1300 cm^{-1} (7.2 to 7.7 μ).
The spectrum of nitrobenzene is shown in Experiment 29.

Nitriles

C≡N   Stretch is a sharp absorption near 2250 cm^{-1} (4.5 μ).
         Conjugation with double bonds or aromatic rings moves the absorption to the
         right.
The spectrum of benzonitrile is shown in Figure IR–12.

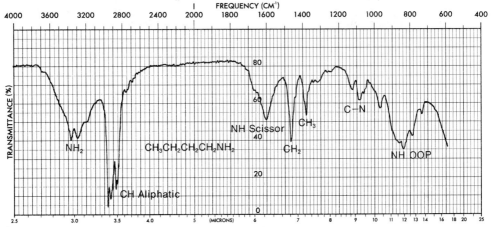

FIGURE IR–11.   Infrared spectrum of n-butylamine (neat liquid, salt plates)

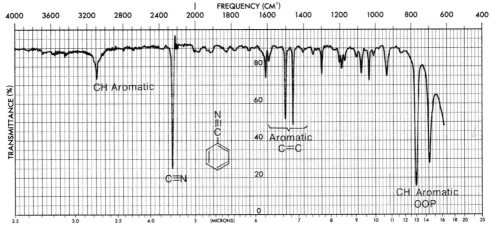

FIGURE IR–12.   Infrared spectrum of benzonitrile (neat liquid, salt plates)

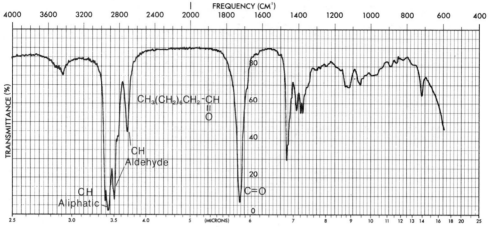

FIGURE IR-13. Infrared spectrum of nonanal (neat liquid, salt plates)

Carbonyl Compounds

The carbonyl group is one of the most strongly absorbing groups in the infrared region of the spectrum. This is mainly due to its large dipole moment. It absorbs in a variety of compounds (aldehydes, ketones, acids, esters, amides, anhydrides, etc.) in the range of 1850 to 1650 cm^{-1} (5.41 to 6.06 μ). In Figure IR–14, the normal values for the various types of carbonyl groups are compared. In the sections which follow, each type will be examined separately.

5.52	5.68	5.76	5.80	5.83	5.85	5.92	μ
1810	1760	1735	1725	1715	1710	1690	cm^{-1}
ANHYDRIDE (Band 1)		ESTERS		KETONES		AMIDES	
	ANHYDRIDE (Band 2)		ALDEHYDES		CARBOXYLIC ACIDS		

FIGURE IR-14. Normal values (± 10 cm^{-1}) for various types of carbonyl groups

Aldehydes

C=O   Stretch at approximately 1725 cm^{-1} (5.80 μ) is normal.
       Aldehydes **seldom** absorb to the left of this value.
       Conjugation moves the absorption to the right.

C—H   Stretch, aldehyde hydrogen (—CHO), consists of **weak** bands at about 2750 cm^{-1} (3.65 μ) and 2850 cm^{-1} (3.50 μ). Note that the CH stretch in alkyl chains does not usually extend this far to the right.

The spectrum of nonanal is shown in Figure IR–13. In addition, the spectrum of benzaldehyde is shown in Experiment 37.

Ketones

C=O   Stretch at approximately 1715 cm^{-1} (5.83 μ) is normal.
       Conjugation moves the absorption to the right.
       Ring strain moves the absorption to the left in cyclic ketones.

The spectra of methyl isopropyl ketone and mesityl oxide are shown in Figures IR-3 and IR-15. The spectrum of camphor is shown in Experiment 20.

5.83	5.83	5.90	5.97	6.01	6.10 μ
1715	1715	1695	1675	1665	1640 cm^{-1}

Normal α,β-Unsaturated   Enolic β-diketones

CONJUGATION   ⟶

5.51	5.62	5.73	5.83	5.83	5.86 μ
1815	1780	1745	1715	1715	1705 cm^{-1}

Normal

⟵ RING STRAIN

Acids

O—H   Stretch, usually **very broad** (strongly H-bonded) 3300 to 2500 cm^{-1} (3.0 to 4.0 μ), often interferes with C—H absorptions.

C=O   Stretch, broad, 1730 to 1700 cm^{-1} (5.8 to 5.9 μ).
Conjugation moves the absorption to the right.

C—O   Stretch, in range of 1320 to 1210 cm^{-1} (7.6 to 8.3 μ), strong.

The spectrum of benzoic acid is shown in Experiment 28B.

Esters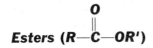

C=O   Stretch occurs at about 1735 cm^{-1} (5.76 μ) in normal esters.
       a. conjugation in the R part moves the absorption to the right;
       b. conjugation with the O in the R′ part moves the absorption to the left;
       c. ring strain (lactones) moves the absorption to the left.

C—O   Stretch, 2 bands or more, one stronger than the others, is in the range of 1300 to 1000 cm^{-1} (7.69 to 10.0 μ).

The spectrum of methyl benzoate is shown in Figure IR–16. In addition, the spectra of isoamyl acetate and methyl salicylate are shown in Experiments 11 and 12.

Amides

C=O   Stretch is at approximately 1670 to 1640 cm^{-1} (6.0 to 6.1 μ).
Conjugation and ring size (lactams) have the usual effects.

N—H   Stretch (if monosubstituted or unsubstituted) 3500 to 3100 cm^{-1} (2.85 to 3.25 μ).
Unsubstituted amides have two bands (—NH$_2$) in this region.

N—H   Bending around 1640 to 1550 cm^{-1} (6.10 to 6.45 μ).

The spectrum of benzamide is shown in Figure IR–17.

Anhydrides

C=O   Stretch always has **two** bands 1830 to 1800 cm^{-1} (5.46 to 5.56 μ) and 1775 to 1740 cm^{-1} (5.63 to 5.75 μ).
Unsaturation moves the absorptions to the right.
Ring strain (cyclic anhydrides) moves the absorptions to the left.

C—O   Stretch is 1300 to 900 cm^{-1} (7.7 to 11 μ).

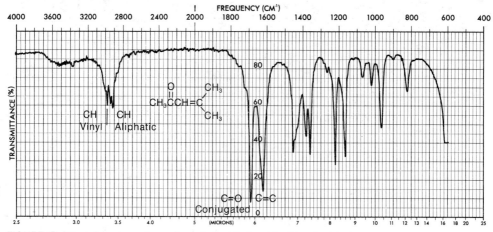

FIGURE IR-15. Infrared spectrum of mesityl oxide (neat liquid, salt plates)

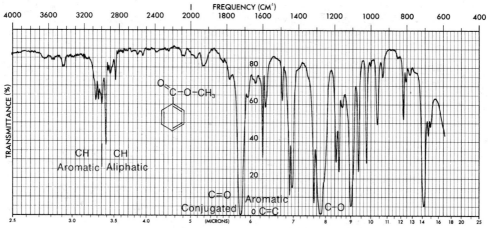

FIGURE IR-16. Infrared spectrum of methyl benzoate (neat liquid, salt plates)

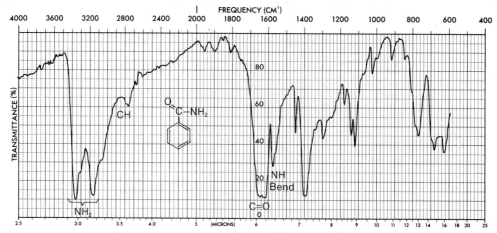

FIGURE IR-17. Infrared spectrum of benzamide (solid phase, KBr)

Halides

It is often quite difficult to determine either the presence or the absence of a halide in a compound by means of infrared spectroscopy. The absorption bands cannot be relied upon, especially if the spectrum is being determined with the compound dissolved in CCl_4 or $CHCl_3$ solution.

C—F Stretch, 1350 to 960 cm^{-1} (7.41 to 10.4 μ).
C—Cl Stretch, 850 to 500 cm^{-1} (11.8 to 20.0 μ).
C—Br Stretch, to the right of 667 cm^{-1} (15.0 μ).
C—I Stretch, to the right of 667 cm^{-1} (15.0 μ).

The spectra of chloroform and carbon tetrachloride are shown in Technique 17, Figures 17-4 and 17-5.

REFERENCES

Bellamy, L. J. *The Infrared Spectra of Complex Molecules* 2nd edition. New York: Wiley, 1958.
Dyer, J. R. *Applications of Absorption Spectroscopy of Organic Compounds.* Englewood Cliffs, New Jersey: Prentice-Hall, 1965.
Nakanishi, K. *Infrared Absorption Spectroscopy.* San Francisco: Holden-Day, 1962.
Pasto, D. J., and Johnson, C. R. *Organic Structure Determination.* Englewood Cliffs, New Jersey: Prentice-Hall, 1969.
Silverstein, R. M., Bassler, G. C., and Morrill, T. C. *Spectrophotometric Identification of Organic Compounds* 3rd edition. New York: Wiley, 1974.

Appendix 4
NUCLEAR MAGNETIC RESONANCE SPECTROSCOPY

NMR.1 THE RESONANCE PHENOMENON

Nuclear magnetic resonance, or nmr, spectroscopy is an instrumental technique that allows the number, type, and relative positions of certain atoms in a molecule to be determined. Nmr spectroscopy is applicable only to those atoms which have nuclear magnetic moments because of their nuclear spin properties. Although a number of atoms meet this requirement, hydrogen atoms (^1_1H) are of the greatest interest to the organic chemist. Atoms of the ordinary isotopes of carbon ($^{12}_6C$) and oxygen ($^{16}_8O$) do not possess nuclear magnetic moments, and ordinary nitrogen atoms ($^{14}_7N$), although they do possess magnetic moments, generally fail to exhibit typical nmr behavior for other reasons. The latter is also true of the halogen atoms, except for fluorine ($^{19}_9F$), which does exhibit active nmr behavior. Atoms other than hydrogen will not be discussed here.

$-1/2$ ——

$+1/2$ +

———— $+ h\nu$ ————→

+

——

Magnetic Field Direction

FIGURE NMR-1. The NMR absorption process

When placed in a magnetic field, the nuclei of nmr active atoms can be thought of as tiny bar magnets. In the case of hydrogen, which has two allowed nuclear spin states ($+1/2$ and $-1/2$), either the nuclear magnets of individual atoms can be aligned with the magnetic field (spin $+1/2$), or they can be opposed to it (spin $-1/2$). A slight majority of the nuclei will be aligned with the field, as this spin orientation constitutes a slightly lower energy spin state. If radiofrequency waves of the appropriate energy are supplied, nuclei aligned with the field can absorb this radiation and reverse their direction of spin, or become reoriented so that the nuclear magnet opposes the applied magnetic field.

The frequency of radiation required to induce spin conversion is a direct function of the strength of the applied magnetic field. When a spinning hydrogen nucleus is placed in a magnetic field, the nucleus begins to precess with angular frequency ω, much like a child's toy top. This precessional motion is depicted in Figure NMR-2. The angular frequency of nuclear precession, ω, increases as the strength of the applied magnetic field is increased. The radiation which must be supplied to induce spin conversion in a hydrogen nucleus of spin $+1/2$ must have a frequency which just matches the angular precessional frequency ω. This is called the resonance condition, and the process of spin conversion is said to be a resonance process.

For the average proton (hydrogen atom), if a magnetic field of approximately 14,000 Gauss is applied, radiofrequency radiation of 60 MHz (60,000,000 cps) is required to induce a spin transition. Fortunately, the magnetic field strength which is required to induce the various protons in a molecule to absorb 60 MHz radiation varies from proton to proton within the molecule and is a sensitive function of the immediate **electronic** environment of each proton. The typical proton nuclear magnetic resonance spectrometer supplies a basic radiofrequency radiation of 60 MHz to the sample being measured and **increases** the strength of the applied magnetic field over a range of several parts per million from the basic field strength. As the field increases, various protons come into resonance (absorb 60 MHz energy), and a resonance signal is generated for each of these protons. An nmr spectrum is a plot of the strength of the magnetic field versus the intensity of the absorptions. A typical nmr spectrum is shown in Figure NMR-3.

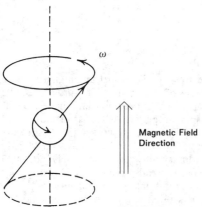

ω

Magnetic Field Direction

FIGURE NMR-2. Precessional motion of a spinning nucleus in an applied magnetic field

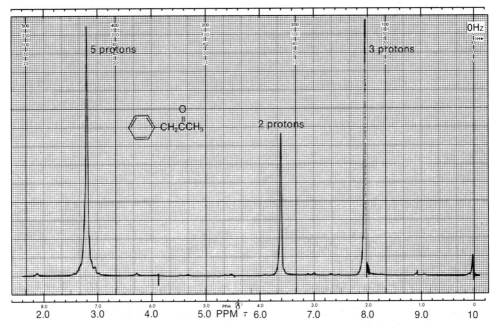

FIGURE NMR-3. The nuclear magnetic resonance spectrum of phenylacetone (the absorption peak at the far right is caused by the added reference substance TMS)

NMR.2 THE CHEMICAL SHIFT

The differences in the applied field strengths at which the various protons in a molecule will absorb 60 MHz radiation are extremely small. The different absorption positions amount to only a few parts per million (ppm) difference in the magnetic field strength. Since it is experimentally difficult to measure the precise field strength at which each proton absorbs to less than one part in a million, a technique has been developed whereby the **difference** between two absorption positions is measured directly. To achieve this measurement, a standard reference substance is utilized and the positions of the absorptions of all other protons are measured relative to those of the reference substance. The reference substance which has been universally accepted is tetramethylsilane, $(CH_3)_4Si$, which is also called TMS. The proton resonances in this molecule appear at a higher field strength than do those in the majority of all other molecules, and all the protons have resonance at the same field strength.

In order to give the position of absorption of a proton a quantitative measurement, a parameter called the **chemical shift** (δ) has been defined. One δ unit corresponds to a 1 ppm change in the magnetic field strength. To determine the chemical shift value for the various protons in a molecule, an nmr spectrum of the molecule is determined with a small quantity of TMS added directly to the sample. That is, both spectra are determined **simultaneously.** The TMS absorption is adjusted to correspond to the $\delta = 0$ position on the recording chart, which is calibrated in δ units, and the δ values of the absorption peaks for all other protons can be read directly from the chart.

Since the nmr spectrometer increases the magnetic field as the pen moves from left to right on the chart, the TMS absorption appears at the extreme right hand edge of the spectrum ($\delta = 0$) or at the **upfield** end of the spectrum. The chart is calibrated in δ units (or ppm), and most other protons absorb at a lower field strength (or **downfield**) from TMS.

Because the frequency at which a proton will precess and, hence the frequency at which it will absorb radiation, is directly proportional to the strength of the applied magnetic field, a second method of measuring an nmr spectrum is possible. One could hold the magnetic field strength constant and vary the frequency of the radiofrequency radiation supplied. Thus, a given proton could be induced to absorb **either** by increasing the field strength, as described earlier, or alternatively, by decreasing the frequency of the radiofrequency oscillator. A 1 ppm decrease in the frequency of the oscillator would have the same effect as a 1 ppm increase in the magnetic field strength. For reasons of instrumental design, it is simpler to vary the strength of the magnetic field than to vary the frequency of the oscillator. Most instruments operate on the former principle. Nevertheless, the recording chart is calibrated not only in δ units, but in Hz as well (1 ppm = 60 Hz when the frequency is 60 MHz), and the chemical shift is customarily defined and computed using units of Hz rather than units of Gauss:

$$\delta = \text{Chemical Shift} = \frac{(\text{Observed Shift from TMS in Hz})}{(60 \text{ MHz})} = \frac{\text{Hz}}{\text{MHz}} = \text{ppm}$$

Although the above equation defines the chemical shift for a spectrometer operating at 14,100 Gauss and 60 MHz, the chemical shift value which is calculated is **independent** of the field strength. For instance, at 23,500 Gauss the oscillator frequency would have to be 100 MHz. Although the observed shifts from TMS (in Hz) would be larger at this field strength, the divisor of the equation would be 100 MHz, instead of 60 MHz, and δ would turn out to be identical under either set of conditions.

NMR.3 CHEMICAL EQUIVALENCE—INTEGRALS

All of the protons in a molecule which are in chemically identical environments will often exhibit the same chemical shift. Thus, all the protons in tetramethylsilane (TMS), or all the protons in benzene, cyclopentane, or acetone have their own respective resonance values all at the same δ value. Each compound gives rise to a single absorption peak in its nmr spectrum. The protons are said to be **chemically equivalent.** On the other hand, molecules which have sets of protons which are chemically distinct from one another may give rise to an absorption peak from each set.

MOLECULES GIVING RISE TO ONE NMR ABSORPTION PEAK—ALL PROTONS CHEMICALLY EQUIVALENT

MOLECULES GIVING RISE TO TWO NMR ABSORPTION PEAKS—TWO DIFFERENT SETS OF CHEMICALLY EQUIVALENT PROTONS

The nmr spectrum given in Figure NMR-3 was that of phenylacetone, a compound having **three** chemically distinct types of protons:

One can immediately see that the nmr spectrum furnishes a valuable type of information on this basis alone. In fact, the nmr spectrum can not only distinguish how many different types of protons a molecule has, but it can also reveal **how many** of each different type are contained within the molecule.

In the nmr spectrum the area under each peak is proportional to the number of hydrogens generating that peak. Hence, in the above compound, the area ratio of the three peaks is 5:2:3, the same as the ratio of the numbers of each type of hydrogen. The

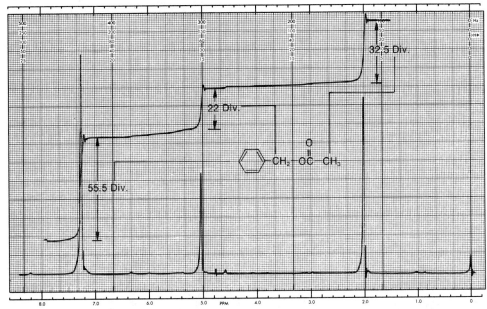

FIGURE NMR–4. Determination of the integral ratios for benzyl acetate

nmr spectrometer has the capacity to electronically "integrate" the area under each peak. It does this by tracing over each peak a vertically rising line which rises in height by an amount which is proportional to the area under the peak. Shown above in Figure NMR–4 is an nmr spectrum of benzyl acetate showing each of the peaks integrated in this way.

It is important to note that the height of the integral line does not give the absolute number of hydrogens. It gives the **relative** numbers of each type of hydrogen. For a given integral to be of any use, there must be a second integral to which it may be referred. The benzyl acetate case gives a good example of this. The first integral rises for 55.5 divisions on the chart paper, the second 22.0 divisions, and the third 32.5 divisions. These numbers are relative and give the **ratios** of the various types of protons. One can find these ratios by dividing each of the larger numbers by the smallest number:

$$\frac{55.5 \text{ div}}{22.0 \text{ div}} = 2.52 \qquad \frac{22.0 \text{ div}}{22.0 \text{ div}} = 1.00 \qquad \frac{32.5 \text{ div}}{22.0 \text{ div}} = 1.48$$

Thus, the number ratio of the protons of each type is 2.52 : 1.00 : 1.48. If we assume that the peak at 5.1 δ is really caused by two hydrogens, and if we assume that the integrals are slightly in error (this can be as much as 10 per cent), then we can arrive at the true ratios by multiplying each figure by two and rounding off: 5 : 2 : 3. Clearly the peak at 7.3 δ, which integrates for 5, arises from the resonance of the aromatic ring protons, while that at 2.0 δ, which integrates for 3, is caused by the methyl protons. The two proton resonance at 5.1 δ arises from the benzyl protons. Notice then that the integrals give the simplest ratios, but not necessarily the true ratios, of the numbers of protons of each type.

NMR.4 CHEMICAL ENVIRONMENT AND CHEMICAL SHIFT

If the resonance frequencies of all protons in a molecule were the same, nmr would be of little use to the organic chemist. However, not only do different types of protons have different chemical shifts, but they also have a value of chemical shift which is characteristic of the type of proton they represent. Every type of proton has only a limited range of δ values over which it gives resonance. Hence, the numerical value of the chemical shift for a proton gives a clue as to the **type of proton** originating the signal, just as the infrared frequency gives a clue as to the type of bond or functional group.

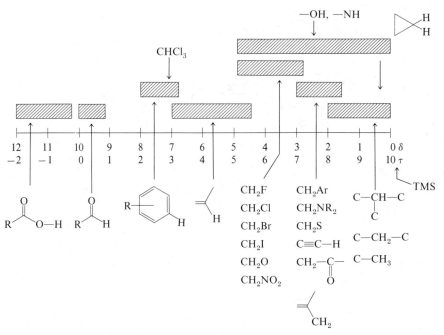

FIGURE NMR-5. A simplified correlation chart for proton chemical shift values.

For instance, notice that the aromatic protons of both phenylacetone (Figure NMR-3) and benzyl acetate (Figure NMR-4) have resonance near 7.3 δ and that both of the methyl groups attached directly to a carbonyl group have a resonance of approximately 2.1 δ. Aromatic protons characteristically have resonance near 7 to 8 δ, while acetyl groups (the methyl protons) have their resonance near 2 δ. These values of chemical shift are diagnostic. Notice also how the resonance of the benzyl (—CH$_2$—) protons comes at a higher value of chemical shift (5.1 δ) in benzyl acetate than in phenylacetone (3.6 δ). Being attached to the electronegative element, oxygen, these protons are more deshielded (see Section NMR.5) than those in phenylacetone. A trained chemist would have readily recognized the probable presence of the oxygen by the value of chemical shift shown by these protons.

It is important for the student to learn the ranges of chemical shifts over which the most common types of protons have resonance. Above, in Figure NMR-5, a correlation chart which contains the most essential and frequently encountered types of protons is presented. For the beginner, it is often difficult to memorize a large body of numbers relating to chemical shifts and proton types. One need actually do this only crudely. It is more important to "get a feel" for the regions and the types of protons than to know a string of factual numbers.

The values of chemical shift given in Figure NMR-5 can be easily understood in terms of two factors: local diamagnetic shielding and anisotropy. These two factors are discussed in Sections NMR.5 and NMR.6 below.

NMR.5 LOCAL DIAMAGNETIC SHIELDING

The easiest trend of chemical shifts to explain is that involving electronegative elements substituted on the same carbon to which the protons of interest are also attached. The chemical shift simply increases as the electronegativity of the attached element increases. This is illustrated in Table NMR-1 for several compounds of the type CH$_3$X.

Multiple substituents have a stronger effect than a single substituent. The influence of the substituent, an electronegative element having little effect on protons which are more than three carbons away, drops off rapidly with distance. These effects are illustrated in Table NMR-2.

TABLE NMR-1. THE DEPENDENCE OF THE CHEMICAL SHIFT OF CH_3X ON THE ELEMENT X

COMPOUND CH_3X		CH_3F	CH_3OH	CH_3Cl	CH_3Br	CH_3I	CH_4	$(CH_3)_4Si$
ELEMENT X		F	O	Cl	Br	I	H	Si
ELECTRONEGATIVITY OF X		4.0	3.5	3.1	2.8	2.5	2.1	1.8
CHEMICAL SHIFT	δ	4.26	3.40	3.05	2.68	2.16	0.23	0
	τ	5.74	6.60	6.95	7.32	7.84	9.77	10

TABLE NMR-2. SUBSTITUTION EFFECTS

	$CHCl_3$	CH_2Cl_2	CH_3Cl	$-C\underline{H}_2Br$	$-C\underline{H}_2-CH_2Br$	$-C\underline{H}_2-CH_2CH_2Br$
δ	7.27	5.30	3.05	3.30	1.69	1.25
τ	2.73	4.70	6.95	6.70	8.31	8.75

Electronegative substituents attached to a carbon atom, because of their electron withdrawing effects, reduce the valence electron density around the protons attached to that carbon. These electrons **shield** the proton from the applied magnetic field. This effect, called **local diamagnetic shielding,** occurs because the applied magnetic field induces the valence electrons to circulate and thus to generate an induced magnetic field which **opposes** the applied field. This is illustrated in Figure NMR-6. Electronegative substituents on carbon reduce the local diamagnetic shielding in the vicinity of the attached protons because they reduce the electron density around those protons. Substituents which have this type of effect are said to **deshield** the proton. The greater the electronegativity of the substituent, the more it deshields the protons, and hence, the greater is the chemical shift of those protons.

NMR.6 ANISOTROPY

Examination of Figure NMR-5 clearly shows that there are a number of types of protons whose chemical shifts are not easily explained by simple considerations of the electronegativity of the attached groups. For instance, consider the protons of benzene or other aromatic systems. Aryl protons generally have a chemical shift which is as large as that for the proton of chloroform! Alkenes, alkynes, and aldehydes also have protons whose resonance values are not in line with the expected magnitude of any electron

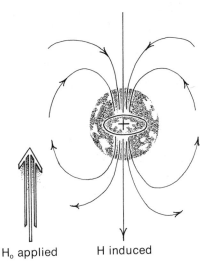

FIGURE NMR-6. Local diamagnetic shielding of a proton owing to its valence electrons

H_o applied   H induced

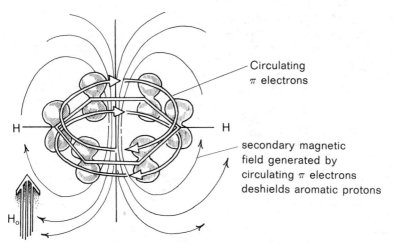

Circulating π electrons

secondary magnetic field generated by circulating π electrons deshields aromatic protons

FIGURE NMR-7. Diamagnetic anisotropy in benzene

withdrawing effects. In each of these cases, the effect is due to the presence of an unsaturated system (π electrons) in the vicinity of the proton in question. Take, for instance, benzene as an example. When placed in a magnetic field, the π electrons in the aromatic ring system are induced to circulate around the ring. This circulation is called a **ring current.** Moving electrons (the ring current) generate a magnetic field much like that generated in a loop of wire through which a current is induced to flow. The magnetic field covers a spatial volume large enough that it influences the shielding of the benzene hydrogens. This is illustrated in Figure NMR-7. The benzene hydrogens are said to be deshielded by the **diamagnetic anisotropy** of the ring. An applied magnetic field is non-uniform (anisotropic) in the vicinity of a benzene molecule because of the labile electrons in the ring which interact with the applied field. Thus, a proton attached to a benzene ring is influenced by **three** magnetic fields: the strong magnetic field applied by the electromagnets of the nmr spectrometer and two weaker fields, one owing to the usual shielding by the valence electrons around the proton and the other owing to the anisotropy generated by the ring system electrons. It is this anisotropic effect that gives the benzene protons a greater chemical shift than is expected. These protons just happen to lie in a **deshielding** region of this anisotropic field. If a proton were placed in the center of the ring rather than on its periphery, it would be found to be shielded since the field lines would have the opposite direction.

All groups in a molecule which have π electrons generate secondary anisotropic fields. In acetylene the magnetic field generated by induced circulation of π electrons has a geometry such that the acetylenic hydrogens are **shielded.** Hence, acetylenic hydrogens come at a higher field than expected. The shielding and deshielding regions due to the various π electron functional groups have characteristic shapes and directions, and they are illustrated in Figure NMR-8. Protons falling within the cones will be shielded, and those falling outside the conical areas will be deshielded. The magnitude of the anisotropic field, of course, diminishes with distance and beyond a certain distance there will be essentially no effect by anisotropy.

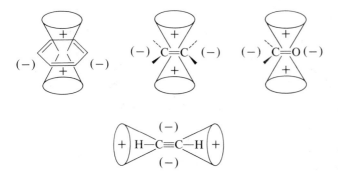

FIGURE NMR-8. Anisotropy caused by the presence of π electrons in some common multiple bond systems

NMR.7 SPIN-SPIN SPLITTING
(N + 1 RULE)

The manner in which the chemical shift and the integral (peak area) can give information about the number and type of hydrogens contained in a molecule has been discussed above. A third type of information that can be obtained from the nmr spectrum is that derived from the spin-spin splitting phenomenon. Even in simple molecules, one finds that each type of proton rarely gives a single resonance peak. For instance, in 1,1,2-trichloroethane there are two chemically distinct types of hydrogen:

$$Cl-C-(CH_2)-Cl$$

On the basis of the information given thus far, one would predict **two** resonance peaks in the nmr spectrum of 1,1,2-trichloroethane with an area ratio (integral ratio) of 2:1. In fact, the nmr spectrum of this compound has **five** peaks. There is a group of three peaks (called a triplet) at 5.77 δ and a group of two peaks (called a doublet) at 3.95 δ. The spectrum is shown in Figure NMR–9. The methine (CH) resonance (5.77 δ) is said to be split into a triplet, and the methylene resonance (3.95 δ) is split into a doublet. The area under the three triplet peaks is **one**, relative to an area of **two** under the two doublet peaks.

This phenomenon is called spin-spin splitting. In an empirical fashion, spin-spin splitting can be explained by the "n + 1 rule." Each type of proton "senses" the number of equivalent protons (n) on the carbon atom(s) next to the one to which it is bonded, and its resonance peak is split into (n + 1) components.

Let's examine the case at hand, 1,1,2-trichloroethane, utilizing the so-called "n + 1 rule." First, the lone methine hydrogen is situated next to a carbon bearing two methylene protons. According to the rule, it has two equivalent neighbors, (n = 2), and is split into n + 1 = 3 peaks (a triplet). The methylene protons are situated next to a carbon bearing only one methine hydrogen. According to rule, they have one neighbor (n = 1) and are split into n + 1 = 2 peaks (a doublet).

Two neighbors give
a triplet (n + 1 = 3)
(area = 1)

One neighbor gives
a doublet (n + 1 = 2)
(area = 2)

Equivalent
protons
behave as
a group

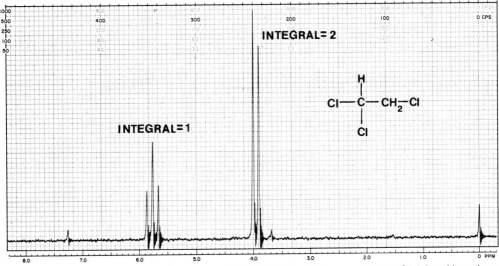

FIGURE NMR–9. The nmr spectrum of 1,1,2-trichloroethane
Courtesy of Varian Associates.

The spectrum of 1,1,2-trichloroethane can be explained easily by the interaction, or coupling, of the spins of protons on adjacent carbon atoms. The position of absorption of proton H_a may be affected by the spins of protons H_b and H_c attached to the neighboring (adjacent) carbon atom. If the spins of these protons are aligned with the applied magnetic field, the small magnetic field generated by their nuclear spin properties will augment the strength of the field experienced by the first mentioned proton H_a. The proton H_a will thus be **deshielded.** If the spins of H_b and H_c are opposed to the applied field, they will decrease the field experienced by proton H_a. It will then be **shielded.** In each of these situations, the absorption position of H_a will be altered. Among the many molecules in the solution, one will find all the various possible spin combinations for H_b and H_c and, hence, the nmr spectrum of the molecular solution will give **three** absorption peaks (a triplet) for H_a, since H_b and H_c have three different possible spin combinations:

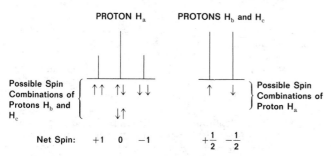

FIGURE NMR–10. An analysis of the spin-spin splitting pattern for 1,1,2-trichloro-ethane

By a similar analysis, it can be seen that protons H_b and H_c should appear as a doublet.

Some common splitting patterns which can be predicted by the n + 1 rule, and which are frequently observed in a number of molecules, are shown in Figure NMR–11. Notice particularly the last entry in the table where **both** methyl groups (6 protons in all) function as a unit and split the methine proton into a septet (6 + 1 = 7).

NMR.8 THE COUPLING CONSTANT

The quantitative amount of spin-spin interaction between two protons can be defined by the coupling constant. The spacings between the component peaks in a simple multiplet is called the coupling constant, J. This distance is measured on the same scale as the chemical shift and is expressed either in cycles per second (cps) or in Hertz (Hz).

⋀	X—CH—CH—Y (X≠Y)	⋀
⋀	—CH₂—CH	⋏⋎
⋏⋎	X—CH₂—CH₂—Y (X≠Y)	⋏⋎
⋀	CH₃—CH	⋰⋰
⋏⋎	CH₃—CH₂—	⋰⋰⋰
⋀	⎰CH₃⧹⎱CH— / CH₃	⣿

FIGURE NMR–11. Some commonly observed splitting patterns

TABLE NMR-3. SOME REPRESENTATIVE COUPLING CONSTANTS AND THEIR APPROXIMATE VALUES (Hz).

	6 to 8		*ortho* 6 to 10		a,a 8 to 14 a,e 0 to 7 e,e 0 to 5
	11 to 18		*meta* 1 to 4		*cis* 6 to 12 *trans* 4 to 8
	6 to 15				*cis* 2 to5 *trans* 1 to 3
	0 to 5		*para* 0 to 2		5 to 7
	4 to 10		8 to 11		
H—C=C—CH	0 to 3				

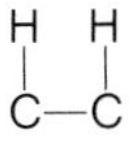

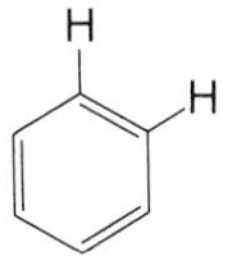

For the interaction of most aliphatic protons in acyclic systems, the magnitudes of the coupling constants are always near 7.5 Hz. See, for instance, the nmr spectrum of 1,1,2-trichloroethane in Figure NMR-9 where the coupling constant is approximately 6 Hz. Different magnitudes of J are found for different types of protons. For instance, the *cis* and *trans* protons substituted on a double bond commonly have values where $J_{trans} \cong 17$ HZ and $J_{cis} \cong 10$ Hz are typical coupling constants. In ordinary compounds, coupling constants may range anywhere from 0–18 Hz. The magnitude of J will often provide structural clues. For instance, one can usually distinguish between a *cis* olefin and a *trans* olefin on the basis of the observed coupling constants for the vinyl protons. The approximate values of some representative coupling constants are given in Table NMR-3.

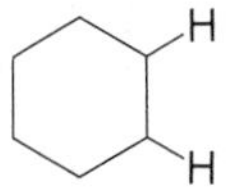

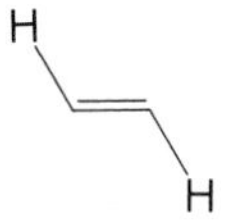

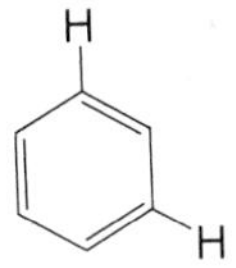

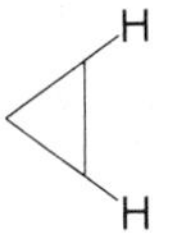

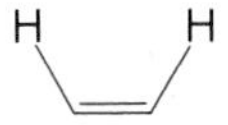

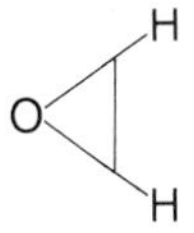

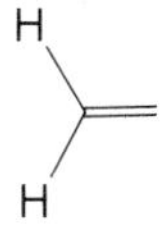

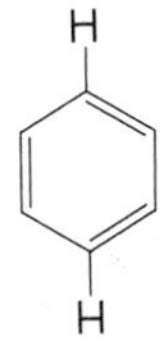

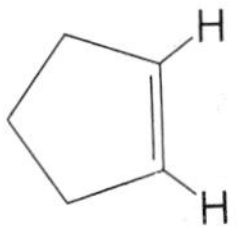

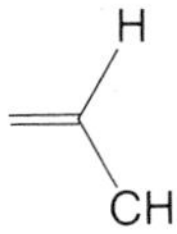

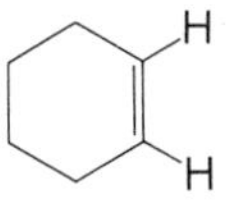

NMR.9 MAGNETIC EQUIVALENCE

In the example of spin-spin splitting dicussed for 1,1,2-trichloroethane, one notices that the two protons H_b and H_c which are attached to the same carbon atom do not split one another. They behave as an integral group. Actually the two protons H_b and H_c **are** coupled to one another; however, for reasons we cannot explain fully here, protons which are attached to the same carbon and both of which have the **same chemical shift** do not show spin-spin splitting. Another way of stating this is to say that protons which are coupled to the same extent to **all** other protons in a molecule do not show spin-spin splitting. Protons which have the same chemical shift and are coupled equivalently to all other protons are said to be **magnetically equivalent** and do not show spin-spin splitting. Thus, in 1,1,2-trichloroethane, protons H_b and H_c have the same value of δ and are both coupled by the same value of J to proton H_a. They are magnetically equivalent.

It is important to note that magnetic equivalence is **not** the same as chemical equivalence. Note the following two compounds.

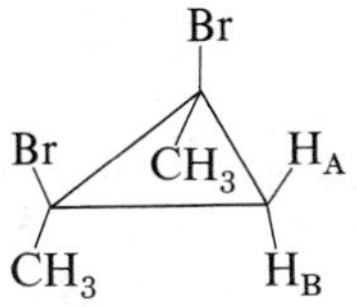

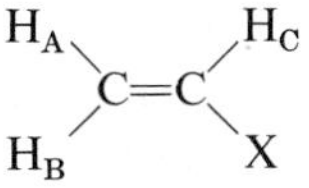

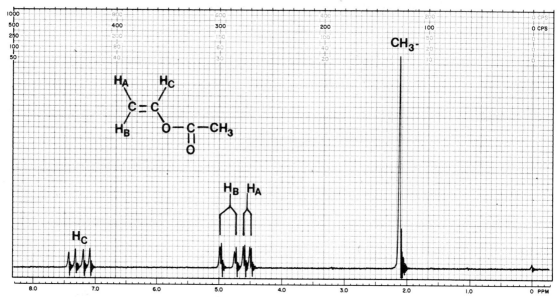

FIGURE NMR-12. The nmr spectrum of vinyl acetate
Courtesy of Varian Associates.

In the cyclopropane compound illustrated on the preceding page, the two geminal hydrogens are chemically equivalent. However, they are not magnetically equivalent. Proton H_A is on the same side of the ring as the two halogens. Proton H_B is on the same side of the ring as the two methyl groups. Protons H_A and H_B will have different chemical shifts, will couple with one another, and will show spin-spin splitting. Two doublets will be observed for H_A and H_B. For cyclopropane rings, $J_{geminal}$ is usually around 5 Hz.

Another situation where protons are chemically equivalent, but not magnetically equivalent, occurs in the vinyl compound. In this example, protons A and B are chemically equivalent, but not magnetically equivalent. Both H_A and H_B have different chemical shifts. In addition, a second distinction can be made between H_A and H_B in this type of compound. H_A and H_B have different coupling constants with H_C. The constant J_{AC} is a *cis* coupling constant, and J_{BC} is a *trans* coupling constant. Whenever two protons have different coupling constants to a third proton they will not be magnetically equivalent. In the vinyl compound, H_A and H_B do not act as a group to split proton H_C. Each protons acts independently of the other. Thus H_B splits H_C with coupling constant J_{BC} into a doublet, and then H_A splits each of the components of the doublet into doublets with coupling constant J_{AC}. In such a case the nmr spectrum must be analyzed graphically, splitting by splitting. An nmr spectrum of a vinyl compound is shown above in Figure NMR-12. The graphical analysis of the vinyl portion of the nmr spectrum is given in Figure NMR-13.

NMR.10   AROMATIC COMPOUNDS

The nmr spectra of protons on aromatic rings are often too complex to explain by simple theories. However, some simple generalizations can be made that are useful in the analysis of the aromatic region of nmr spectrum. First of all, most aromatic protons have resonance near 7.0δ. In monosubstituted rings, where the ring substituent is an alkyl group, all the ring protons often have chemical shifts which are very nearly identical, and the 5 ring protons may appear as if they gave rise to an overly broad singlet (see Figure NMR-14 A). If an electronegative group is attached to the ring, all the ring protons are shifted downfield from where they would appear in benzene. However, often the ortho protons are shifted more than the others as they are more affected by the group. This often gives rise to an absorption pattern like that in Figure NMR-14 B. In a para disubstituted ring with two substituents X and Y which are identical, all the protons in the ring are chemically and magnetically equivalent, and a

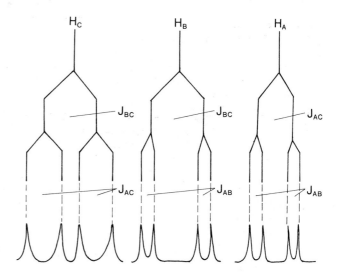

FIGURE NMR-13. An analysis of the splittings in vinyl acetate

singlet is observed. If X is quite different from Y in electronegativity, however, a pattern like that shown in the left hand side of Figure NMR–14 C is often observed, clearly identifying a *p*-disubstituted ring. If X and Y are more nearly similar, a pattern more similar to the one on the right is observed. In monosubstituted rings which have a carbonyl group or a double bond attached directly to the ring, a pattern like that in Figure NMR–14 D is not uncommon. In this case, the ortho protons of the ring are influenced by the anisotropy of the π systems which comprise the CO and CC double bonds and are deshielded by them. In other types of substitution, such as ortho or meta, or polysubstituted ring systems, the patterns may be much more complicated and require an advanced analysis.

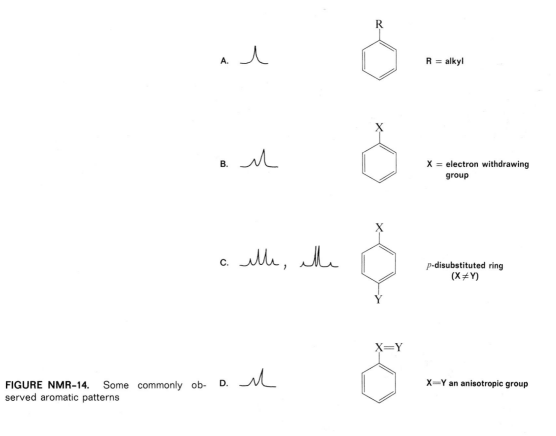

FIGURE NMR-14. Some commonly observed aromatic patterns

NMR.11 PROTONS ATTACHED TO ATOMS
OTHER THAN CARBON

Protons which are attached to atoms other than carbon often have a widely variable range of absorptions. Several of these groups are tabulated in Table NMR–4. In addition, under the usual conditions of determining an nmr spectrum, protons on

TABLE NMR–4. TYPICAL RANGES FOR GROUPS WITH VARIABLE CHEMICAL SHIFT

Acids	RCOOH	10.5–12.0 δ
Phenols	ArOH	4.0–7.0
Alcohols	ROH	0.5–5.0
Amines	RNH$_2$	0.5–5.0
Amides	RCONH$_2$	5.0–8.0
Enols	CH=CH—OH	≥ 15

heteroelements normally do not couple with those on adjacent carbon atoms to give spin-spin splitting. This is due primarily to the fact that such protons often exchange very rapidly with the solvent medium. The variability of the absorption position is explained by the fact that these groups also undergo varying degrees of hydrogen bonding in solutions of different concentrations. The amount of hydrogen bonding that occurs with a proton radically affects the valence electron density around that proton and produces correspondingly large changes in the chemical shift. The absorption peaks for protons that have hydrogen bonding or are undergoing exchange are frequently quite broad relative to other singlets and can often be recognized on that basis. For a different reason, called quadrupole broadening (not explained here), protons attached to nitrogen atoms often exhibit an extremely broad resonance peak, often almost indistinguishable from the baseline.

REFERENCES

Dyer, J. R. *Applications of Absorption Spectroscopy of Organic Compounds.* Englewood Cliffs, New Jersey: Prentice-Hall, 1965.

Jackman L. M., and Sternhell, S. *Applications of Nuclear Magnetic Resonance in Organic Chemistry* 2nd edition. New York: Pergamon Press, 1969.

Pasto D. J., and Johnson, C. R. *Organic Structure Determination.* Englewood Cliffs, New Jersey: Prentice-Hall, 1969.

Paudler, W. W. *Nuclear Magnetic Resonance.* Boston: Allyn and Bacon, 1971.

Silverstein, R. M., Bassler, G. C., and Morrill, T. C. *Spectrometric Identification of Organic Compounds* 3rd edition. New York: Wiley, 1974.

APPENDIX 5
INDEX OF SPECTRA

INDEX

Entries are alphabetized by disregarding prefixes.

CONCENTRATED ACIDS AND BASES

REAGENT	HCl	HNO₃	H₂SO₄	H₃PO₄	HCOOH	CH₃COOH	NH₄OH
Specific Gravity	1.18	1.41	1.84	1.69	1.20	1.06	0.90
% Acid or Base (by weight)	37.3	70.0	96.5	85.0	90.0	99.7	29.0
Molecular Weight	36.47	63.02	98.08	98.00	46.03	60.05	35.05
Molarity of Concentrated Acid or Base	12	16	18	14.7	23.4	17.5	7.4
Normality of Concentrated Acid or Base	12	16	36	44	23.4	17.5	7.4
Volume of Concentrated Reagent Required to Prepare 1 liter of 1M Solution (ml)	83	64	56	69	42	58	134
Volume of Concentrated Reagent Required to Prepare 1 liter of 10% Solution (ml) *	227	101	56	70	93	95	383
Molarity of a 10% Solution *	2.74	1.59	1.02	1.02	2.17	1.67	2.85

*Percent solutions by weight